KU-143-710

Hypersonic and High-Temperature Gas Dynamics

Second Edition

Hypersonic and High-Temperature Gas Dynamics

Second Edition

John D. Anderson, Jr.
National Air and Space Museum
Smithsonian Institution
Washington, DC
and
Professor Emeritus, Aerospace Engineering
University of Maryland
College Park, Maryland

EDUCATION SERIES
Joseph A. Schetz
Series Editor-in-chief
Virginia Polytechnic Institute and State University
Blacksburg, Virginia

Published by the
American Institute of Aeronautics and Astronautics, Inc.
1801 Alexander Bell Drive, Reston, Virginia 20191-4344

American Institute of Aeronautics and Astronautics, Inc., Reston, Virginia

2 3 4 5

Library of Congress Cataloging-in-Publication Data

Anderson, John David.
Hypersonic and high-temperature gas dynamics / John D. Anderson, Jr. -- 2nd ed.
p. cm.
Includes bibliographical references and index.
ISBN-13: 978-1-56347-780-5 (alk. paper)
ISBN-10: 1-56347-780-7 (alk. paper)
1. Aerodynamics, Hypersonic. I. Title
TL571.5.A53 2006
629.132'306--dc22 2006025724

To Sarah-Allen, Katherine, and Elizabeth Anderson, for all their love and understanding

Foreword

It is a great personal pleasure for me to welcome the Second Edition of *Hypersonic and High Temperature Gas Dynamics* by John D. Anderson to the AIAA Education Series. I have known John Anderson for more years than either he or I are comfortable recalling, and I have always found him to be extremely articulate and insightful. The original edition published by McGraw-Hill in 1989 was a very well received, comprehensive, and in-depth treatment of these important topics, and it was reprinted by AIAA in 2000. This new edition has updated the material and expanded the coverage, and we anticipate that it will be equally well received, especially since hypersonics is now enjoying a resurgence of interest. This edition has 18 chapters divided into three main parts and more than 800 pages.

John Anderson is very well-qualified to write this book, first because of his broad and deep expertise in the area. Second, his command of the material is excellent, and he is able to organize and present it in a very clear manner. In addition, John writes in a very readable style, which has made all of his books popular with both students and working professionals. Finally, John Anderson has long played a key role in AIAA publications activities, including books and journal papers, as well as leadership roles, and that makes us particularly pleased to have this book under the AIAA masthead.

The AIAA Education Series aims to cover a very broad range of topics in the general aerospace field, including basic theory, applications and design. A complete list of titles can be found at http://www.aiaa.org. The philosophy of the series is to develop textbooks that can be used in a university setting, instructional materials for continuing education and professional development courses, and also books that can serve as the basis for independent study. Suggestions for new topics or authors are always welcome.

Joseph A. Schetz
Editor-in-Chief
AIAA Education Series

Table of Contents

Preface to the Second Edition

Almost 20 years have passed since the publication of the first edition of this book. During those 20 years, much progress has been made in hypersonic and high-temperature gas dynamics, principally in the extensive use of sophisticated modern computational fluid dynamics and in the design of serious flight hardware. The physical and mathematical fundamentals of hypersonic and high-temperature gas dynamics, however, have by their very nature remained the same.

This book is about the *fundamentals*, which have not changed. Therefore, almost all of the content of the first edition has been carried over the present edition. Indeed, now is a good moment to pause and read the preface of the first edition. Everything said there is appropriate to the second edition. For example, the book remains a self-contained teaching instrument for those students and readers interested in learning hypersonic and high-temperature gas dynamics starting with the basics. This book assumes no prior familiarity with either subject on the part of the reader. If you have never studied hypersonic and/or high-temperature gas dynamics and if you have never worked in the area, *then this book is for you.* On the other hand, if you have worked and/or are working in these areas and you want a cohesive presentation of the fundamentals, a development of important theory and techniques, a discussion of salient results with emphasis on the physical aspects, and a presentation of modern thinking in these areas, *then this book is also for you.*

As with the first edition, this second edition is written in an informal, conversational style. This book *talks to you*, just as if you and I were sitting down together at a table discussing the subject matter. I want you to have fun learning these topics. This is not difficult because the areas of hypersonic and high-temperature gas dynamics are full of interesting and exciting phenomena and applications.

What is new and different about the second edition? A lot! Much new material has been added to accomplish two purposes, namely, to bring the book up to date and to enhance even further the pedagogical goal of helping the reader to learn. For example:

1) A lot of new literature has been published in the discipline over the past 20 years. This new edition draws from the modern literature in order to update the presentations. This is not a book about the state of the art—it is about fundamentals. But the state of the art is used to reinforce the fundamentals.

2) The topic of shock-shock interactions, particularly the important type-IV interaction, has been added as a new extensive section in Chapter 5.

3) Although this book emphasizes the fundamentals, modern hypersonic and high-temperature gas dynamics is moving more and more toward design of viable systems. Therefore, this edition has a design flavor not present in the first edition. At the back of a number of chapters, design examples that illustrate the application of the fundamentals to methods of design are added. A number of these design examples focus on different aspect of hypersonic waverider design. Waveriders were not treated in the first edition. Because they are an interesting configuration for possible future hypersonic vehicles they are extensively treated here.

4) Chapter previews have been added at the beginning of most of the chapters. These are pedagogical tools to provide the reader with insight about what each chapter is about and why the material is so important. They are written in a particularly informal manner—plain speaking—to help turn the reader on to the content. In these previews I am unabashedly admitting to providing some fun for the readers.

5) Road maps have been placed at the beginning of almost every chapter to help guide the reader through the logical flow of the material—another pedagogical tool to enhance the *self-learning nature* of this book.

Special thanks go to Rodger Williams of the AIAA for suggesting, encouraging, and essentially commissioning this second edition, and with whom it is always a pleasure to work, and to the entire AIAA publications team who have always made me feel like one of them. Thanks also to Susan Cunningham who typed the original manuscript of the first edition 20 years ago and who received the call again for the added material for the second edition.

Finally, I want to acknowledge in a very special way the late Rudolph Edse, my mentor and advisor at The Ohio State University, who gave me the true appreciation for the fundamentals, and Dr. John D. Lee and the rest of the faculty of the Department of Aeronautical and Astronautical Engineering at Ohio State during the 1960s who taught me all there was to know about hypersonic flow.

John D. Anderson, Jr.
July 2006

Preface to the First Edition

This book is designed to be a self-contained *teaching instrument* for those students and readers interested in learning hypersonic flow and high-temperature gas dynamics. It assumes no prior familiarity with either subject on the part of the reader. If you have never studied hypersonic and/or high-temperature gas dynamics before, and if you have never worked extensively in the area, *then this book is for you.* On the other hand, if you have worked and/or are working in these areas, and you want a cohesive presentation of the fundamentals, a development of important theory and techniques, a discussion of the salient results with emphasis on the physical aspects, and a presentation of modern thinking in these areas, *then this book is also for you.* In other words, this book is aimed for two roles: 1) as an effective classroom text, which can be used with ease by the instructor and which can be understood with ease by the student; and 2) as a viable, professional working tool on the desk of all engineers, scientists, and managers who have any contact in their jobs with hypersonic and/or high-temperature flow.

The only background assumed on the part of the reader is a basic knowledge of undergraduate fluid dynamics, including a basic introductory course on compressible flow; that is, the reader is assumed to be familiar with material exemplified by two of the author's previous books, namely, *Fundamentals of Aerodynamics* (McGraw-Hill, 1984), and the first half of *Modern Compressible Flow: with Historical Perspective* (McGraw-Hill, 1982). Indeed, throughout the present book, frequent reference is made to basic material presented in these two books. Finally, the present book is pitched at the advanced senior and first-year graduate levels and is designed to be used in the classroom as the main text for courses at these levels in hypersonic flow and high-temperature gas dynamics. Homework problems are given at the ends of most chapters in order to enhance its use as a teaching instrument.

Hypersonic aerodynamics is an important part of the entire flight spectrum, representing the segment at the extreme high velocity of this spectrum. Interest in hypersonic aerodynamics grew in the 1950s and 1960s with the advent of hypersonic atmospheric entry vehicles, especially the manned space program as represented by Mercury, Gemini, and Apollo. Today, many new, exciting vehicle concepts involving hypersonic flight are driving renewed and, in some cases, frenzied interest in hypersonics. Such new concepts are described in Chapter 1. This book is a response to the need to provide a basic education in hypersonic and high-temperature gas dynamics for a new generation of engineers and scientists, as well as to provide a basic discussion of these areas from a modern perspective. Six texts in hypersonic flow were published before 1966; the present book is the first basic classroom text to become available since then. Therefore, the present book is intended to make up for this 20-year hiatus and to provide a *modern* education in hypersonic and high-temperature gas dynamics, while discussing at length the basic fundamentals.

To enhance the reader's understanding and to peak his or her interest, the present book is written in the style of the author's previous ones, namely, it is intentionally written in an informal, conversational style. The author wants the reader to *have fun* while learning these topics. This is not difficult because the areas of hypersonic and high-temperature gas dynamics are full of interesting and exciting phenomena and applications.

The present book is divided into three parts. Part 1 deals with inviscid hypersonic flow, emphasizing purely the fluid-dynamic effects of the Mach number becoming large. High-temperature effects are not included. Part 2 deals with viscous hypersonic flow, emphasizing the purely fluid-dynamic effects of including the transport phenomena of viscosity and thermal conduction at the same time that the Mach number becomes large. High-temperature effects are not included. Finally, Part 3 deals with the influence of high temperatures on both inviscid and viscous flows. In this fashion, the reader is led in an organized fashion through the various physical phenomena that dominate high-speed aerodynamics. To further enhance the organization of the material, the reader is given a "road map" in Fig. 1.24 to help guide his or her thoughts as we progress through our discussions.

When this book was first started, the author's intent was to have a Part 4, which would cover the miscellaneous but important topics of low-density flows, experimental hypersonics, and applied aerodynamics associated with hypersonic vehicle design. During the course of writing this book, it quickly became apparent that including Part 4 would vastly exceed the length constraints allotted to this book. Therefore, the preceding matters are not considered in any detail here. This is not because of a lack of importance of such material, but rather because of an effort to emphasize the basic fundamentals in the present book. Therefore, Parts 1, 2, and 3 are sufficient; they constitute the essence of a necessary fundamental background in hypersonic and high-temperature gas dynamics. The material of the missing Part 4 will have to wait for another time.

The content of this book is influenced in part by the author's experience in teaching such material in courses at the University of Maryland. It is also influenced by the author's three-day short course on the introduction to hypersonic aerodynamics, which he has had the privilege to give at 10 different laboratories, companies, and universities over the past year. These experiences have fine tuned the present material in favor of what the reader wants to know and what he or she is thinking.

Several organizations and people are owed the sincere thanks of the author in aiding the preparation of this book. First, the author is grateful to the National Air and Space Museum of the Smithsonian Institution where he spent an enlightening sabbatical year during 1986–1987 as the Charles Lindbergh Professor in the Aeronautics Department. A substantial portion of this book was written during that sabbatical year at the museum. Secondly, the author is grateful to the University of Maryland for providing the intellectual atmosphere conducive to scholarly projects. Also, many thanks go to the author's graduate students in the hypersonic aerodynamics program at Maryland—thanks for the many enlightening discussions on the nature of hypersonic and high-temperature flows. For the mechanical preparation of this manuscript, the author has used his own word processor named Susan O. Cunningham—a truly "human" human being who has typed the manuscript with the highest professional standards. Finally, once again the author is grateful for the support at home provided by the Anderson family, who allowed him to undertake this project in the first place, and for joining him in the collective sigh of relief upon its completion.

I would like to express my thanks for the many useful comments and suggestions provided by colleagues who reviewed this text during the course of its development, especially to Judson R. Baron, Massachusetts Institute of Technology; Daniel Bershader, Stanford University; John D. Lee, Ohio State University; and Maurice L. Rasmussen, University of Oklahoma.

John D. Anderson, Jr.
October 1987

1
Some Preliminary Thoughts

Almost everyone has their own definition of the term hypersonic. If we were to conduct something like a public opinion poll among those present, and asked everyone to name a Mach number above which the flow of a gas should properly be described as hypersonic there would be a majority of answers round about five or six, but it would be quite possible for someone to advocate, and defend, numbers as small as three, or as high as 12.

P. L. Roe, comment made in a lecture at the Von Kármán Institute. Belgium, January 1970

Chapter Preview

This is the first of many short preview boxes in this book—one for each of the chapters. Their purpose is to tell you in plain language right up front what to expect from each chapter and why the material is important and exciting. They are primarily motivational; they are here for you—to encourage you to actually enjoy reading the chapter.

Part of the success of learning a new subject is simply to get going. The purpose of this chapter is to get going on our journey through hypersonic and high-temperature gas dynamics. It starts with some history. When and how did the first human-made object achieve hypersonic flight? When was the first time that a human flew hypersonically? What is so special about hypersonic flight anyway? Why is it important? Why do aerodynamicists single out hypersonic as a special flight regime separate from the more general regime of supersonic flight? After all, both supersonic and hypersonic flight deal with speeds above Mach 1. So what is so special about hypersonic speeds that we can write a book about the subject? These are more than just interesting questions—they are *important questions* that demand answers. This chapter supplies the answers. The preliminary thoughts supplied here, moreover, set the stage and the tone for the rest of our discussions in this book. A study of hypersonic flow is fun—it can be technically demanding at times as we will see, but it can also be fun. So sit back, relax, and especially enjoy this first chapter. Let it give you a smooth takeoff on our flight through the worlds of hypersonic and high-temperature gas dynamics—a flight that is both intellectually and professionally rewarding.

1.1 Hypersonic Flight—Some Historical Firsts

The day is Thursday, 24 February 1949; the pens on the automatic plotting boards at South Station are busy tracking the altitude and course of a rocket, which just moments before had been launched from a site three miles away on the test range of the White Sands Proving Ground. The rocket is a V-2, one of many brought to the United States from Germany after World War II. By this time, launching V-2s had become almost routine for the crews at White Sands, but on this day neither the launch nor the rocket are "routine." Mounted on top of this V-2 is a slender, needle-like rocket called the WAC Corporal, which serves as a second stage to the V-2. This test firing of the combination V-2/WAC Corporal is the first meaningful attempt to demonstrate the use of a *multi-stage* rocket for achieving high velocities and high altitudes and is part of a larger program labeled "Bumper" by the U.S. Army. All previous rocket launchings of any importance, both in the United States and in Europe, had utilized the single-stage V-2 by itself. Figure 1.1 shows a photograph of the "Bumper" rocket as it lifts off the New Mexico desert on this clear, February day. The pen plotters track the V-2 to an altitude of 100 miles at a velocity of 3500 mph, at which point the WAC Corporal is ignited. The slender upper stage accelerates to a maximum velocity of 5150 mph and reaches an altitude of 244 miles, exceeding by a healthy 130 miles previous record set by a V-2 alone. After reaching this peak, the WAC Corporal noses over and careers back into the atmosphere at over 5000 mph. *In so doing, it becomes the first object of human origin to achieve hypersonic flight*—the first time that any vehicle has flown faster than five times the speed of sound. In spite of the pen plotters charting its course, the WAC Corporal cannot be found in the desert after the test. Indeed, the only remnants to be recovered later are a charred electric switch and part of the tail section, and these are found more than a year later, in April 1950.

The scene shifts to the small village of Smelooka in the Ternov District, Saratov region of Russia. The time is now 1055 hrs (Moscow time) on 12 April 1961. A strange, spherical object has just landed under the canopy of a parachute. The surface of this capsule is charred black, and it contains three small viewing ports covered with heat-resistant glass. Inside this capsule is Flight Major Yuri Gagarin, who just 108 min earlier had been sitting on top of a rocket at the Russian cosmodrome at Baikonur near the Aral Sea. What partly transpired during those 108 min is announced to the world by a broadcast from the Soviet newsagency Tass at 9 : 59 a.m.:

> The world's first spaceship, Vostok (East), with a man on board was launched into orbit from the Soviet Union on April 12, 1961. The pilot space-navigator

Fig. 1.1 V-2/WAC Corporal liftoff on 24 Feb. 1949, the first object of human origin to achieve hypersonic flight (National Air and Space Museum).

> of the satellite-spaceship Vostok is a citizen of the U.S.S.R., Flight Major Yuri Gagarin.
>
> The launching of the multistage space rocket was successful and, after attaining the first escape velocity and the separation of the last stage of the carrier rocket, the spaceship went into free flight on around-the-earth orbit. According to preliminary data, the period of the revolution of the satellite spaceship around the earth is 89.1 min. The minimum distance from the earth at perigee is 175 km (108.7 miles) and the maximum at apogee is 302 km (187.6 miles), and the angle of inclination of the orbit plane to the equator is 65° 4′. The spaceship with the navigator weighs 4725 kg (10,418.6 lb), excluding the weight of the final stage of the carrier rocket.

After this announcement is made, Major Gagarin's orbital craft, called Vostok I, is slowed at 10 : 25 a.m. by the firing of a retrorocket and enters the atmosphere at a speed in excess of 25 times the speed of sound. Thirty minutes later, Major Yuri Gagarin becomes the first man to fly in space, to orbit the Earth, and safely return.

Moreover, on that day, 12 April 1961, Yuri Gagarin becomes the first human being in history to experience *hypersonic flight*. A photograph of the Vostok I capsule is shown in Fig. 1.2.

Later, 1961 becomes a bumper year for manned hypersonic flight. On 5 May, Alan B. Shepard becomes the second man in space by virtue of a suborbital flight over the Atlantic Ocean, reaching an altitude of 115.7 miles, and entering the atmosphere at a speed above Mach 5. Then, on 23 June, U.S. Air Force test pilot Major Robert White flies the X-15 airplane at Mach 5.3, the first X-15 flight to exceed Mach 5. (In so doing, White accomplishes the first "mile-per second" flight in an airplane, reaching a maximum velocity of 3603 mph.) This record is extended by White on 9 November, flying the X-15 at Mach 6.

The preceding events are historical firsts in the annuals of hypersonic flight. They represent certain milestones and examples of the application of hypersonic aerodynamic theory and technology. The purpose of this book is to present and discuss this theory and technology, with the hope that the reader, as a student and professional, will be motivated and prepared to contribute to the hypersonic milestones of the future.

Fig. 1.2 Vostok I, in which Russian Major Yuri Gagarin became the first human to fly at hypersonic speed, during the world's first manned, orbital flight. 12 April 1961 (National Air and Space Museum).

1.2 Hypersonic Flow—Why Is It Important?

The development of aeronautics and spaceflight, from its practical beginnings with the Wright Brothers' first airplane flight on 17 December 1903 and Robert H. Goddard's first liquid-fueled rocket launch on 16 March 1926, has been driven by one primary urge—the urge to always fly faster and higher. Anyone who has traced advancements in aircraft in the 20th century has seen an exponential growth in both speed and altitude, starting with the 35-mph Wright flyer at sea level in 1903, progressing to 400-mph fighters at 30,000 ft in World War II, transitioning to 1200-mph supersonic aircraft at 60,000 ft in the 1960s and 1970s, highlighted by the experimental X-15 hypersonic airplane, which achieved Mach 7 and an altitude of 354,200 ft on 22 August 1963, and finally capped by the space shuttle—the ultimate in manned airplanes with its Mach 25 reentry into the Earth's atmosphere from a 200-mile low Earth orbit. (See [1] for graphs which demonstrate the exponential increase in both aircraft speed and altitude over the past 100 years.) Superimposed on this picture is the advent of high-speed missiles and spacecraft: for example, the development of the Mach 25 intercontinental ballistic missile in the 1950s; the Mach 25 Mercury, Gemini, and Vostok manned orbital spacecraft of the 1960; and of course the historic Mach 36 Apollo spacecraft, which returned men from the moon starting in 1969. The point here is that the extreme high-speed end of the flight spectrum has been explored, penetrated, and utilized since the 1950s. Moreover, flight at this end of the spectrum is called *hypersonic flight*, and the aerodynamic and gas dynamic characteristics of such flight are classified under the label of *hypersonic aerodynamics*—one of the primary subjects of this book.

Hypersonic aerodynamics is *different* than the more conventional and experienced regime of supersonic aerodynamics. These differences will be discussed at length in Sec. 1.3, along with an in-depth definition of just what hypersonic aerodynamics really means. However, we can immediately see that such differences must exist just by comparing the shapes of hypersonic vehicles with those of more commonplace supersonic aircraft. For example, Fig. 1.3 shows a Lockheed F-104, the first fighter aircraft designed for sustained supersonic flight at Mach 2. This aircraft embodies principles for good supersonic aerodynamic design:

Fig. 1.3 Lockheed F-104, a supersonic airplane designed in the early 1950s (National Air and Space Museum).

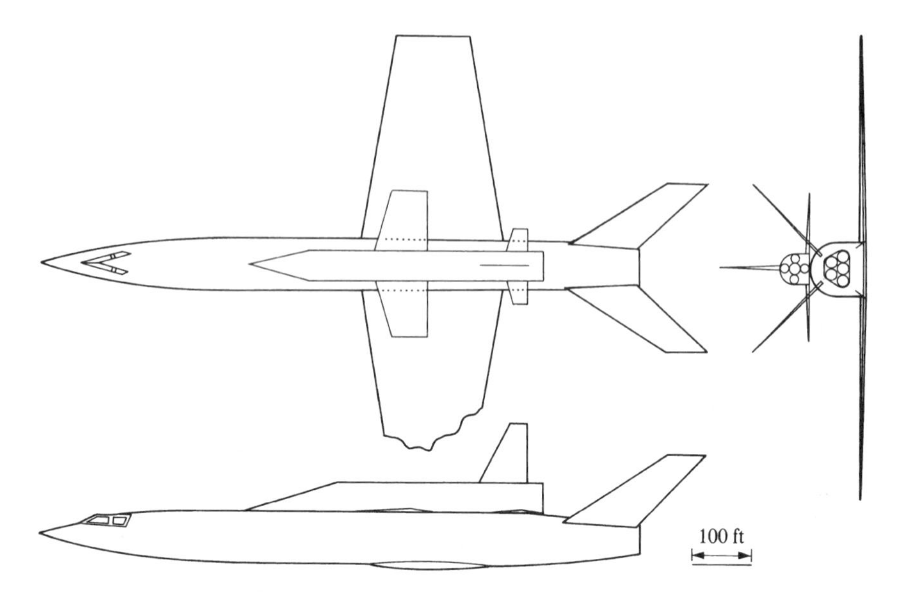

Fig. 1.4 Drake–Carman hypersonic aircraft/orbiter, proposed in 1953 (from [2]).

a sharp, needle-like nose and slender fuselage, very thin wings and tail surfaces (3.36 percent thickness-to-chord ratio) with very sharp leading edges (almost sharp enough to pose a hazard during ground handling), and with a low aspect ratio of 2.45 for the straight wing itself—all designed to minimize wave drag at supersonic speeds. To design a hypersonic airplane for flight at much higher Mach numbers, it is tempting to utilize these same design principles—only more so. Indeed, such was the case for an early hypersonic aircraft concept conceived by Robert Carman and Hubert Drake of the NACA (now NASA) in 1953. One of their hand drawings from an internal NACA memorandum is shown in Fig. 1.4 (see [2] for more details). Here we see an early concept for a hypersonic booster/orbiter combination, where each aircraft has a sharp nose, slender fuselage, and thin, low-aspect-ratio straight wings—the same features that are seen in the F-104—except the aircraft in Fig. 1.4 is designed for Mach 25. However, in 1953 hypersonic aerodynamics was in its infancy. Contrast Fig. 1.4 with another hypersonic airplane designed just seven years later, the X-20 Dynasoar shown in Fig. 1.5. Here we see a completely different-looking aircraft—one embodying new hypersonic principles that were not fully understood in 1953. The X-20 design utilized a sharply swept delta wing with a blunt, rounded leading edge, and a rather thick fuselage with a rounded (rather than sharp) nose. The fuselage was placed on top of the wing, so that the entire undersurface of the vehicle was flat. The X-20 was intended to be an experimental aircraft for rocket-powered flight at Mach 20. Eclipsed by the Mercury, Gemini, and Apollo manned spaceflight program, the X-20 project was cancelled in 1963 without the production of a vehicle. However, the X-20 reflected design features that were uniquely

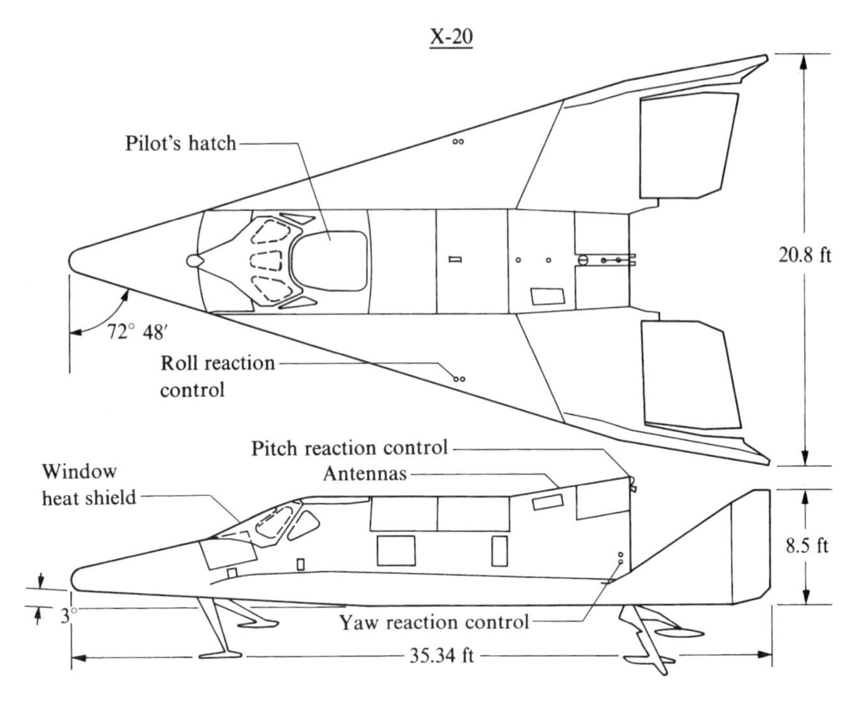

Fig. 1.5 Boeing X-20A Dynasoar orbital hypersonic aircraft, 1963 (from [2]).

hypersonic and that were later contained in the space shuttle. Indeed, the space shuttle is shown in Fig. 1.6 for further comparison with the earlier concepts shown in Figs. 1.3 and 1.4. Clearly, hypersonic vehicles are different configurations from supersonic vehicles, and hence we might conclude (correctly) that hypersonic aerodynamics is different from supersonic aerodynamics. This difference is dramatically reinforced when we examine Fig. 1.7, which shows the Apollo space vehicle, designed to return humans from the moon, and to enter the Earth's atmosphere at the extreme hypersonic speed of Mach 36. Here we see a very blunt body with no wings at all. To be objective, we have to realize that many considerations besides high-speed aerodynamics go into the design of the vehicles shown in Figs. 1.3–1.7; however, to repeat once again, the important point here is that hypersonic vehicles are different than supersonic vehicles, and this is in part because hypersonic aerodynamics is different from supersonic aerodynamics.

Hypersonic flight, both manned and unmanned, has been successfully achieved. However, at the time of this writing, it is by no means commonplace. The era of *practical* hypersonic flight is still ahead of us, and it poses many exciting challenges to the aerodynamicist. Let us briefly examine some new ideas for modern hypersonic vehicles. For example, airbreathing hypersonic vehicles designed for sustained flight in the atmosphere have captured the imagination of aerospace engineers and mission planners alike. One concept is that of an

Fig. 1.6 Space shuttle (National Air and Space Museum).

Fig. 1.7 Artist's conception of the atmosphere entry of the Apollo spacecraft (National Air and Space Museum).

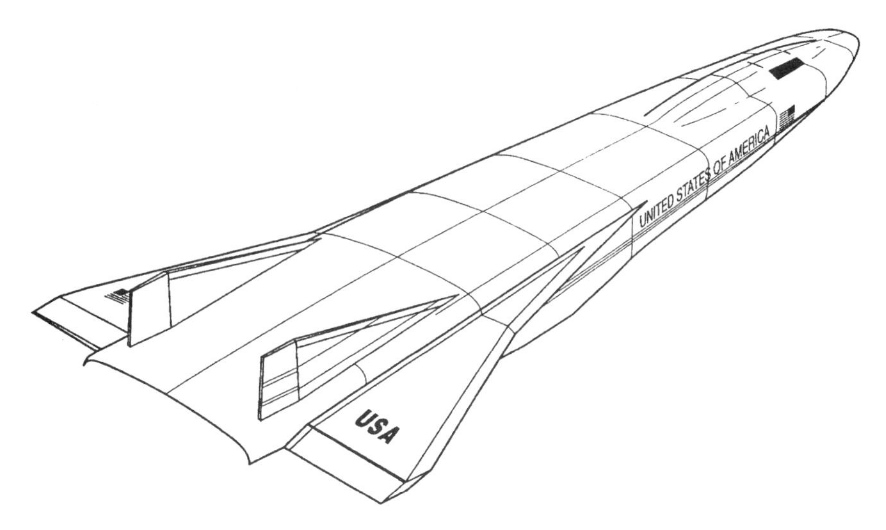

Fig. 1.8 Generic representation of a nominal single-stage-to-orbit vehicle (NASA).

aerospace plane—ideally an aircraft designed to take off horizontally from a runway and then accelerate into orbit around the Earth, for the most part powered by an airbreathing supersonic combustion ramjet engine (scramjet) with most likely a rocket assist for final insertion into orbit. It will subsequently carry out a mission in orbit, or within the outer regions of the atmosphere, and then reenter the atmosphere at Mach 25, finally landing under power on a conventional runway. This idea was first seriously examined by the U.S. Air Force in the early 1960s, and a combination of airbreathing and rocket propulsion was intended to power the vehicle. Work on the early aerospace plane was cancelled in October 1963 mainly because of the design requirements exceeding the state of the art at that time. The idea was resurrected in the middle 1980s by both NASA and the U.S. Department of Defense, as well as by aerospace companies in England and Germany. A generic single-stage-to-orbit aerospace plane is shown in Fig. 1.8, representative of this second generation of aerospace plane technology. In the United States, this effort was labeled the National Aerospace Plane program (NASP). It too was terminated in the mid-1990s after a decade of intensive research and development. The reason was the same—the design requirements exceeded the state of the art. This work lives on, however, in the form of more modest but focused research and technology advancement efforts. A recent success has been the X-43 Hyper-X unmanned research vehicle shown in Fig. 1.9. In November 2004 the X-43 made aeronautical engineering history by achieving sustained flight at Mach 10 powered by a scramjet engine for a period of about 10 s; it was the first sustained atmospheric flight by a scramjet-powered vehicle. Of more importance, the flight data from the X-43 verified the predictions based on wind-tunnel and computational-fluid-dynamic data and proved the basic methodology used for the design of such vehicles. The door to successful sustained hypersonic atmospheric flight has been slightly cracked open.

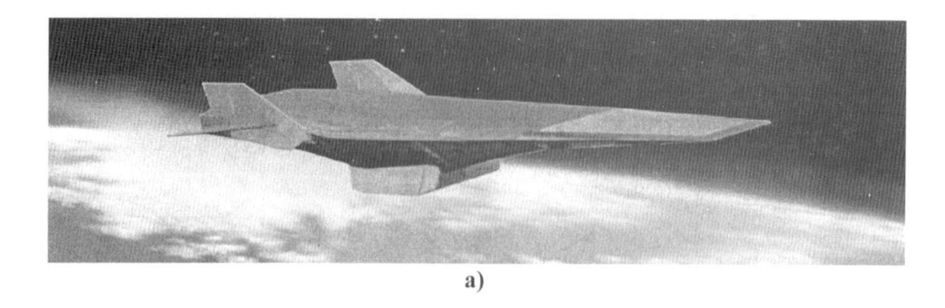

a)

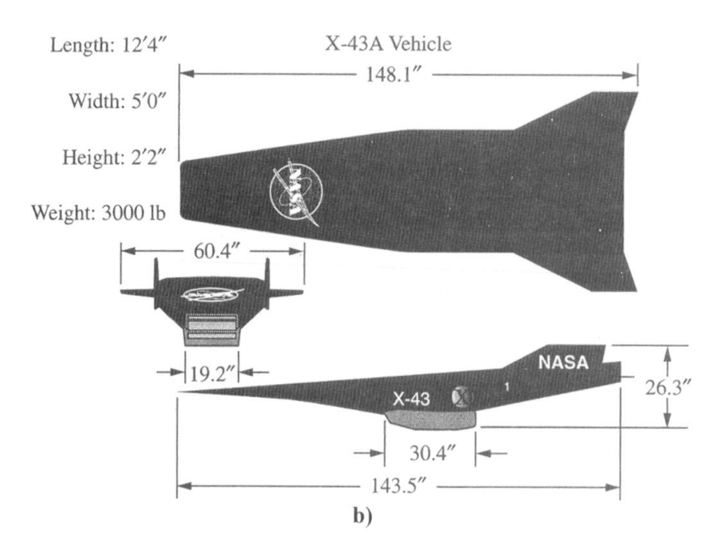

b)

Fig. 1.9 X-43A Hyper-X test vehicle (NASA): a) pictorial view and b) three view.

Other ideas for new hypersonic vehicles include scramjet-powered missiles to fly in the Mach 6–8 range for military use, both tactical and strategic. In the United States, at the time of writing, both the U.S. Air Force and the U.S. Navy have active research and development programs for hypersonic atmospheric flight vehicles. In terms of getting payloads into orbit, the single-stage-to-orbit vehicle represented in Fig. 1.8 has a strong competitor in the form of a two-stage-to-orbit vehicle combination shown generically in Fig. 1.10. Here, the first stage is a hypersonic ramjet/scramjet-powered airbreathing vehicle, and the second stage is a rocket-powered orbiter riding piggyback. Finally, dare we mention the possibility of hypersonic commercial air transportation? At the time of writing, there is no active program to develop even a second-generation supersonic transport to replace the retired Anglo–French Concorde SST. So, is a hypersonic transport only a "pipe dream"? This author thinks not. Figure 1.11 shows an artist's rendering, circa 1985, of a futuristic hypersonic transport. Sometime in the 21st century, but well after the development of a second-generation SST, such an airplane as shown in Fig. 1.11 will be streaking

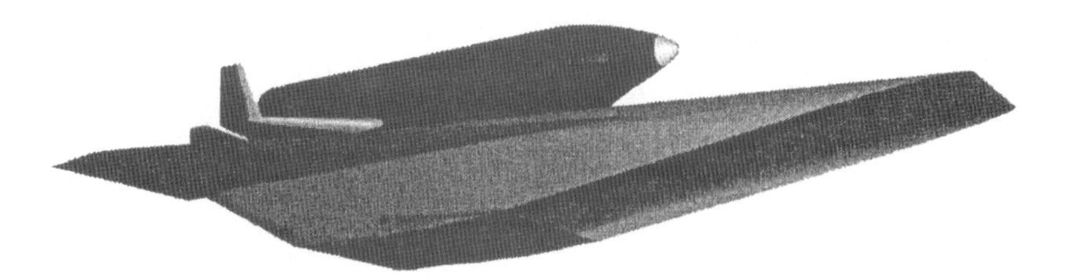

Fig. 1.10 Generic representation of a two-stage-to-orbit vehicle (from [270]).

through the sky, carrying passengers and cargo at Mach numbers of 5 and larger. The technical, economic, and environmental challenges will be enormous, but I cannot believe it will not happen.

It is important to mention an aspect that distinguishes hypersonic atmospheric flight vehicles from conventional subsonic and supersonic airplane design philosophy. For subsonic and supersonic aircraft, the components for providing lift (the wings), propulsion (the engines and nacelles), and volume (the fuselage) are not strongly coupled with each other. They are separate and distinct components, easily identifiable by looking at the airplane; moreover, they can be treated as

Fig. 1.11 Concept for a hypersonic transport (McDonnell-Douglas Aircraft Corporation).

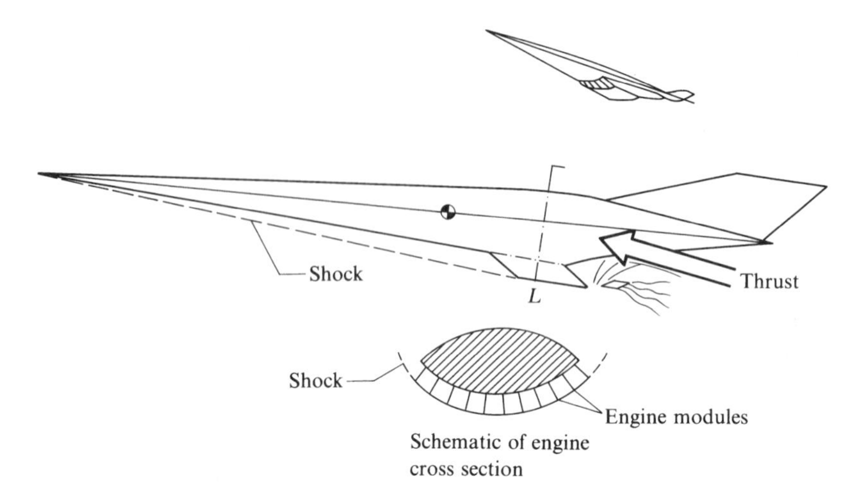

Fig. 1.12 Hypersonic vehicle with integrated scramjet (NASA).

separate aerodynamic bodies with only a moderate interaction when they are combined in the total aircraft. Modern hypersonic aerodynamic design is exactly the opposite. Figure 1.12 is an example of an integrated airframe-propulsion concept for a hypersonic airplane, wherein the entire undersurface of the vehicle is part of the scramjet engine. Initial compression of the air takes place through the bow shock from the nose of the aircraft; further compression and supersonic combustion take place inside a series of modules near the rear of the aircraft, and then expansion of the burned gases is partially realized through nozzles in the engine modules, but mainly over the bottom rear surface of the aircraft, which is sculptured to a nozzle-like shape. Hence, the propulsion mechanism is intimately integrated with the airframe. Moreover, most of the lift is produced by high pressure behind the bow shock wave and exerted on the relatively flat undersurface of the vehicle; the use of large, distinct wings is not necessary for the production of high lift. Finally, the fuel for airbreathing hypersonic airplanes shown in Figs. 1.8–1.12 is liquid H_2, which occupies a large volume. All of these considerations combine in a hypersonic vehicle in such a fashion that the components to generate lift, propulsion, and volume are *not* separate from each other; rather, they are closely integrated in the same overall lifting shape, in direct contrast to conventional subsonic and supersonic vehicle design.

Finally, return to the question asked at the beginning of this section: Hypersonic flow, why is it important? We now have a feeling for the answer. Hypersonic flow is important because of the following:

1) It is physically different from supersonic flow.

2) It is the flow that will dictate many of the new exciting vehicle designs for the 21st century.

Recognizing this importance, the purpose of the present book is to introduce the reader to the basic fundamentals of hypersonic flow, including an emphasis on high-temperature gas dynamics, which, as we will see, is an important aspect of high-speed flows in general. Wherever pertinent, we will also discuss modern experimental and computational-fluid-dynamic applications in hypersonic and high-temperature flow, as well as certain related aspects of hypersonic vehicle design. Such material is an integral part of modern aerodynamics. Moreover, the importance of this material will grow steadily into the 21st century, as we continue to extend the boundaries of practical flight.

1.3 Hypersonic Flow—What Is It?

There is a conventional rule of thumb that defines hypersonic aerodynamics as those flows where the Mach number M is greater than 5. However, this is no more than just a rule of thumb; when a flow is accelerated from $M = 4.99$ to 5.01, there is no "clash of thunder," and the flow does not "instantly turn from green to red." Rather, hypersonic flow is best defined as that regime where certain physical flow phenomena become progressively more important as the Mach number is increased to higher values. In some cases, one or more of these phenomena might become important above Mach 3, whereas in other cases they may not be compelling until Mach 7 or higher. The purpose of this section is to briefly describe these physical phenomena; in some sense this entire section will constitute a "definition" of hypersonic flow. For more details of an introductory nature, see [3].

1.3.1 Thin Shock Layers

Recall from oblique shock theory (for example, see [4] and [5]) that, for a given flow deflection angle, the density increase across the shock wave becomes progressively larger as the Mach number is increased. At higher density, the mass flow behind the shock can more easily "squeeze through" smaller areas. For flow over a hypersonic body, this means that the distance between the body and the shock wave can be small. The flowfield between the shock wave and the body is defined as the *shock layer*, and for hypersonic speeds this shock layer can be quite thin. For example, consider the Mach 36 flow of a calorically perfect gas with a ratio of specific heats, $\gamma = c_p/c_v = 1.4$, over a wedge of 15-deg half-angle. From standard oblique shock theory the shock-wave angle will be only 18 deg, as shown in Fig. 1.13. If high-temperature,

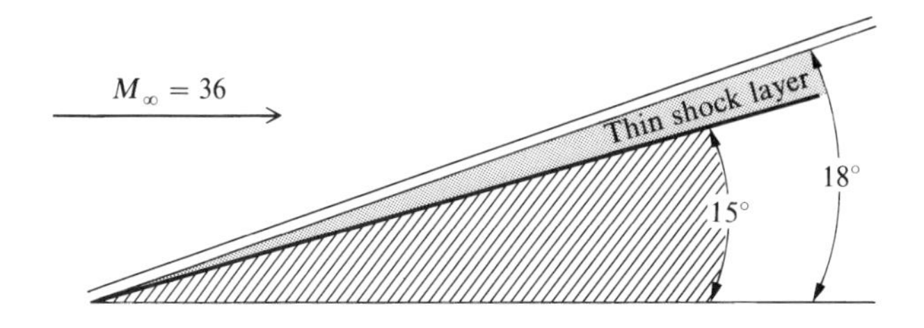

Fig. 1.13 Thin hypersonic shock layer.

chemically reacting effects are included, the shock-wave angle will be even smaller. Clearly, this shock layer is thin. It is a basic characteristic of hypersonic flows that shock waves lie close to the body and that the shock layer is thin. In turn, this can create some physical complications, such as the merging of the shock wave itself with a thick, viscous boundary layer growing from the body surface—a problem that becomes important at low Reynolds numbers. However, at high Reynolds numbers, where the shock layer is essentially inviscid, its thinness can be used to theoretical advantage, leading to a general analytical approach called thin shock-layer theory (to be discussed in Chapter 4). In the extreme, a thin shock layer approaches the fluid-dynamic model postulated by Issac Newton in 1687; such Newtonian theory is simple and straightforward and is frequently used in hypersonic aerodynamics for approximate calculations (to be discussed in Chapter 3).

1.3.2 Entropy Layer

Consider the wedge shown in Fig. 1.13, except now with a blunt nose, as sketched in Fig. 1.14. At hypersonic Mach numbers, the shock layer over the blunt nose is also very thin, with a small shock-detachment distance *d*. In the nose region, the shock wave is highly curved. Recall that the entropy of the flow increases across a shock wave, and the stronger the shock, the larger the entropy increase. A streamline passing through the strong, nearly normal portion of the curved shock near the centerline of the flow will experience a larger entropy increase than a neighboring streamline, which passes through a weaker portion of the shock further away from the centerline. Hence, there are strong entropy gradients generated in the nose region; this entropy layer flows downstream and essentially wets the body for large distances from the nose, as shown in Fig. 1.14. The boundary layer along the surface grows inside this

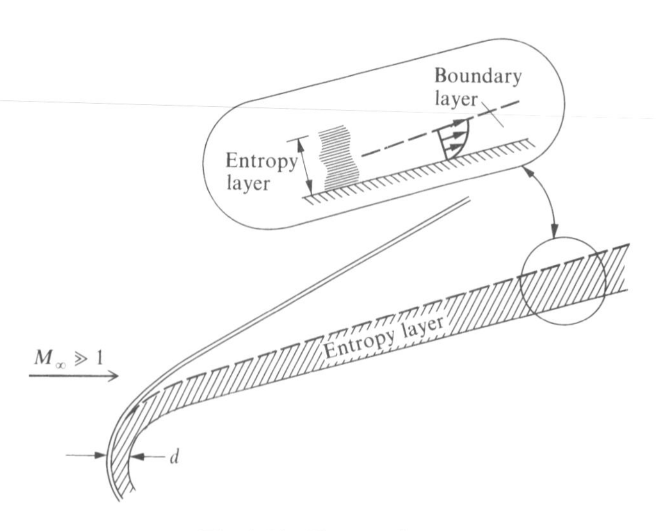

Fig. 1.14 Entropy layer.

entropy layer and is affected by it. Because the entropy layer is also a region of strong vorticity, as related through Crocco's theorem from classical compressible flow (for example, see [4]), this interaction is sometimes called a vorticity interaction. The entropy layer causes analytical problems when we wish to perform a standard boundary-layer calculation on the surface because there is a question as to what the proper conditions should be at the outer edge of the boundary layer.

1.3.3 Viscous Interaction

Consider a boundary layer on a flat plate in a hypersonic flow, as sketched in Fig. 1.15. A high-velocity, hypersonic flow contains a large amount of kinetic energy; when this flow is slowed by viscous effects within the boundary layer, the lost kinetic energy is transformed (in part) into internal energy of the gas—this is called viscous dissipation. In turn, the temperature increases within the boundary layer; a typical temperature profile within the boundary layer is also sketched in Fig. 1.15. The characteristics of hypersonic boundary layers are dominated by such temperature increases. For example, the viscosity coefficient increases with temperature, and this by itself will make the boundary layer thicker. In addition, because the pressure p is constant in the normal direction through a boundary layer, the increase in temperature T results in a decrease in density ρ through the equation of state $\rho = p/RT$, where R is the specific gas constant. To pass the required mass flow through the boundary layer at reduced density, the boundary-layer thickness must be larger. Both of these phenomena combine to make hypersonic boundary layers grow more rapidly than at slower speeds. Indeed, the flat-plate compressible laminar boundary-layer thickness δ grows essentially as

$$\delta \propto \frac{M_\infty^2}{\sqrt{Re_x}}$$

where M_∞ is the freestream Mach number and Re_x is the local Reynolds number. (This relation will be derived in Chapter 6.) Clearly, because δ varies as the square of M_∞, it can become inordinately large at hypersonic speeds.

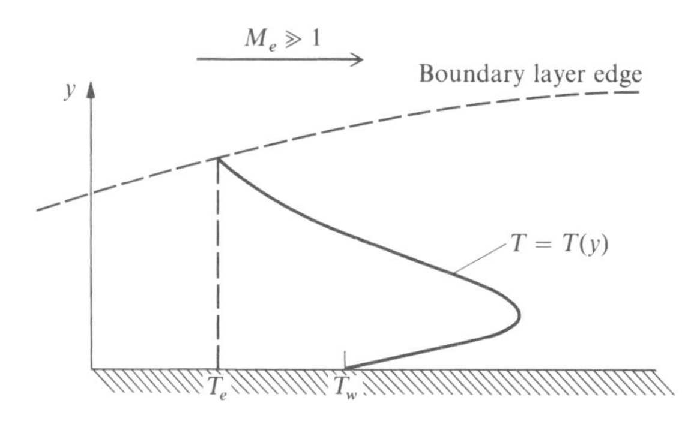

Fig. 1.15 Temperature profile in a hypersonic boundary layer.

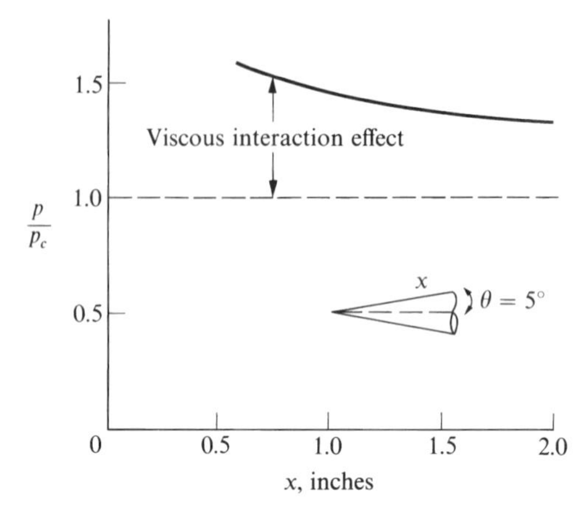

Fig. 1.16 Viscous interaction effect. Induced pressure on a sharp cone at $M_\infty = 11$ and $Re = 1.88 \times 10^5$ per foot.

The thick boundary layer in hypersonic flow can exert a major displacement effect on the inviscid flow outside the boundary layer, causing a given body shape to appear much thicker than it really is. Because of the extreme thickness of the boundary-layer flow, the outer inviscid flow is greatly changed; the changes in the inviscid flow in turn feed back to affect the growth of the boundary layer. This major interaction between the boundary layer and the outer inviscid flow is called *viscous interaction.* Viscous interactions can have important effects on the surface-pressure distribution, hence lift, drag, and stability on hypersonic vehicles. Moreover, skin friction and heat transfer are increased by viscous interaction. For example, Fig. 1.16 illustrates the viscous interaction on a sharp, right-circular cone at zero degree of angle of attack. Here, the pressure distribution on the cone surface p is given as a function of distance from the tip. These are experimental results obtained from [6]. If there were no viscous interaction, the inviscid surface pressure would be constant, equal to p_c (indicated by the horizontal dashed line in Fig. 1.16). However, because of the viscous interaction, the pressure near the nose is considerably greater; the surface-pressure distribution decays further downstream, ultimately approaching the inviscid value far downstream. These and many other aspects of viscous interactions will be discussed in Chapter 7.

The boundary layer on a hypersonic vehicle can become so thick that it essentially merges with the shock wave—a merged shock layer. When this happens, the shock layer must be treated as fully viscous, and the conventional boundary-layer analysis must be completely abandoned. Such matters will be discussed in Chapter 9.

1.3.4 High-Temperature Flows

As discussed earlier, the kinetic energy of a high-speed, hypersonic flow is dissipated by the influence of friction within a boundary layer. The extreme viscous dissipation that occurs within hypersonic boundary layers can create very high temperatures—high enough to excite vibrational energy internally within molecules and to cause dissociation and even ionization within the gas. If the surface of a hypersonic vehicle is protected by an ablative heat shield, the products of ablation are also present in the boundary layer, giving rise to complex hydrocarbon chemical reactions. On both accounts, we see that the surface of a hypersonic vehicle can be wetted by a *chemically reacting boundary layer*.

The boundary layer is not the only region of high-temperature flow over a hypersonic vehicle. Consider the nose region of a blunt body, as sketched in Fig. 1.17. The bow shock wave is normal, or nearly normal, in the nose region, and the gas temperature behind this strong shock wave can be enormous at hypersonic speeds. For example, Fig. 1.18 is a plot of temperature behind a normal shock wave as a function of freestream velocity, for a vehicle flying at a standard altitude of 52 km; this figure is taken from [4]. Two curves are shown: 1) the upper curve, which assumes a calorically perfect nonreacting gas with the ratio of specific heats $\gamma = 1.4$, and which gives an unrealistically high value of temperature; and 2) the lower curve, which assumes an equilibrium chemically reacting gas and which is usually closer to the actual situation. This figure illustrates two important points:

1) By any account, the temperature in the nose region of a hypersonic vehicle can be extremely high, for example, reaching approximately 11,000 K at a Mach number of 36 (Apollo reentry).

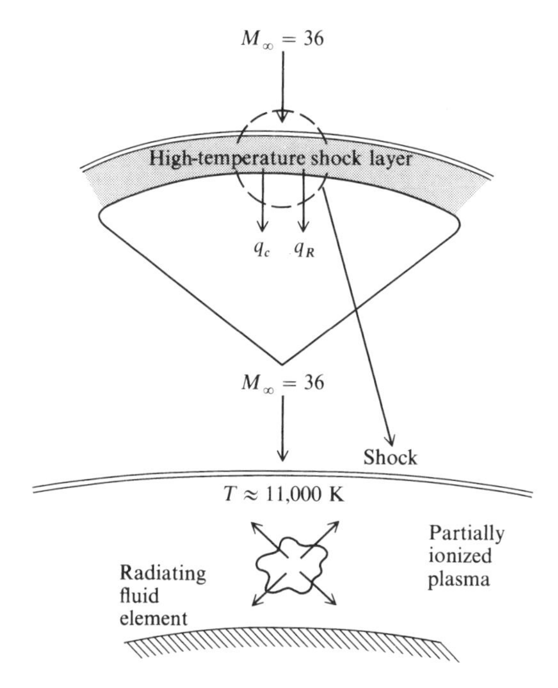

Fig. 1.17 High-temperature shock layer.

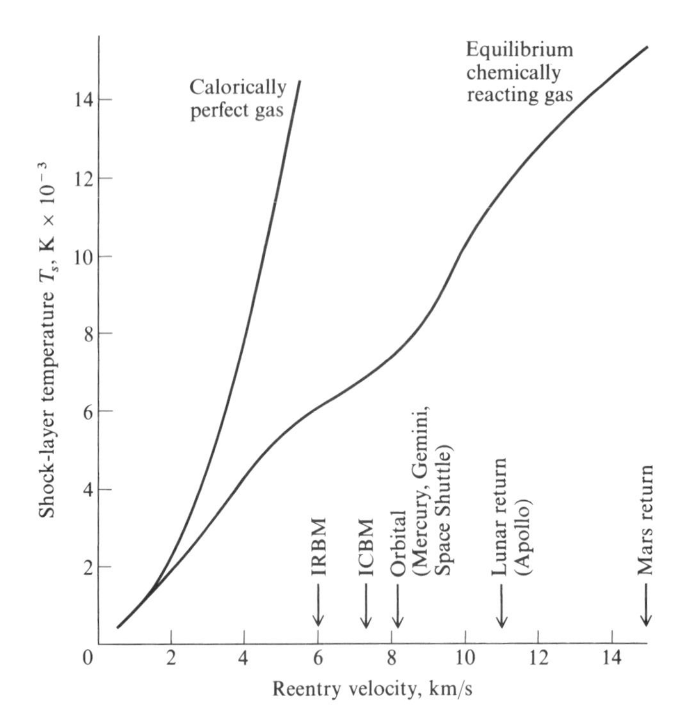

Fig. 1.18 Temperature behind a normal shock wave as a function of freestream velocity at a standard altitude of 52 km (from [4]).

2) The proper inclusion of chemically reacting effects is vital to the calculation of an accurate shock-layer temperature; the assumption that γ is constant and equal to 1.4 is no longer valid.

So we see that, for a hypersonic flow, not only can the boundary layer be chemically reacting, but the entire shock layer can be dominated by *chemically reacting flow.*

For a moment, let us examine the physical nature of a high-temperature gas. In introductory studies of thermodynamics and compressible flow, the gas is assumed to have constant specific heats; hence, the ratio $\gamma = c_p/c_v$ is also constant. This leads to some ideal results for pressure, density, temperature, and Mach-number variations in a flow. However, when the gas temperature is increased to high values, the gas behaves in a "nonideal" fashion, specifically as follows:

1) The vibrational energy of the molecules becomes excited, and this causes the specific heats c_p and c_v to become functions of temperature. In turn, the ratio of specific heats, $\gamma = c_p/c_v$, also becomes a function of temperature. For air, this effect becomes important above a temperature of 800 K.

2) As the gas temperature is further increased, chemical reactions can occur. For an equilibrium chemically reacting gas, c_p and c_v are functions of both temperature and pressure, and hence $\gamma = f(T, p)$. For air at 1 atm pressure, O_2

dissociation ($O_2 \rightarrow 2O$) begins at about 2000 K, and the molecular oxygen is essentially totally dissociated at 4000 K. At this temperature N_2 dissociation ($N_2 \rightarrow 2N$) begins and is essentially totally dissociated at 9000 K. Above a temperature of 9000 K, ions are formed ($N \rightarrow N^+ + e^-$, and $O \rightarrow O^+ + e^-$), and the gas becomes a partially ionized plasma.

All of these phenomena are called *high-temperature effects.* (They are frequently referred to in the aerodynamic literature as real-gas effects, but there are good technical reasons to discourage the use of that label, as we will see later.) If the vibrational excitation and chemical reactions take place very rapidly in comparison to the time it takes for a fluid element to move through the flowfield, we have vibrational and chemical equilibrium flow. If the opposite is true, we have *nonequilibrium flow*, which is considerably more difficult to analyze. All of these effects will be discussed at length in Chapters 10–18.

High-temperature chemically reacting flows can have an influence on lift, drag, and moments on a hypersonic vehicle. For example, such effects have been found to be important for estimating the amount of body-flap deflection necessary to trim the space shuttle during high-speed reentry. However, by far the most dominant aspect of high temperatures in hypersonics is the resultant high heat-transfer rates to the surface. Aerodynamic heating dominates the design of all hypersonic machinery, whether it be a flight vehicle, a ramjet engine to power such a vehicle, or a wind tunnel to test the vehicle. This aerodynamic heating takes the form of heat transfer from the hot boundary layer to the cooler surface—called *convective heating* and denoted by q_c in Fig. 1.17. Moreover, if the shock-layer temperature is high enough the thermal radiation emitted by the gas itself can become important, giving rise to a radiative flux to the surface—called *radiative heating* and denoted by q_R in Fig. 1.17. (In the winter, when you warm yourself beside a roaring fire in the fireplace, the warmth you feel is not hot air blowing out of the fireplace, but rather radiation from the hot bricks or stone of the walls of the fireplace. Imagine how "warm" you would feel standing next to the gas behind a strong shock wave at Mach 36, where the temperature is 11,000 K—about twice the surface temperature of the sun.) For example, for Apollo reentry radiative heat transfer was more than 30% of the total heating. For a space probe entering the atmosphere of Jupiter, the radiative heating will be more than 95% of the total heating.

Another consequence of high-temperature flow over hypersonic vehicles is the "communications blackout" experienced at certain altitudes and velocities during atmospheric entry, where it is impossible to transmit radio waves either to or from the vehicle. This is caused by ionization in the chemically reacting flow, producing free electrons that absorb radio-frequency radiation. Therefore, the accurate prediction of electron density within the flowfield is important.

Clearly, high-temperature effects can be a dominant aspect of hypersonic aerodynamics, and because of this importance Part 3 of this book is devoted entirely to high-temperature gas dynamics. (Part 3 is self-contained and represents a study of high-temperature gas dynamics in general, a field with applications that go far beyond hypersonics, such as combustion, high-energy lasers, plasmas, and laser-matter interaction, to name just a few.)

1.3.5 Low-Density Flow

Consider for a moment the air around you; it is made up of individual molecules, principally oxygen and nitrogen, which are in random motion. Imagine that you isolate one of these molecules, and watch its motion. It will move a certain distance and then collide with one of its neighboring molecules, after which it will move another distance, and collide again with another neighboring molecule, and it will continue this molecular collision process indefinitely. Although the distance between collisions is different for each of the individual collisions, over a period of time there will be some *average* distance the molecule moves between successive collisions. This average distance is defined as the *mean free path*, denoted by λ. At standard sea-level conditions for air, $\lambda = 2.176 \times 10^{-7}$ ft, a very small distance. This implies that, at sea level, when you wave your hand through the air the gas itself "feels" like a continuous medium—a so-called *continuum.* Most aerodynamic problems (more than 99.9% of all applications) are properly addressed by assuming a continuous medium; indeed, all of our preceding discussion has so far assumed that the flow is a continuum.

Imagine now that we are at an altitude of 342,000 ft, where the air density is much lower, and consequently the mean free path is much larger than at sea level; indeed, at 342,000 ft, $\lambda = 1$ ft. Now, when you wave your hand through the air, you are more able to feel individual molecular impacts; the air no longer feels like a continuous substance, but rather like an open region punctuated by individual, widely spaced particles of matter. Under these conditions, the aerodynamic concepts, equations, and results based on the assumption of a continuum begin to break down; when this happens, we have to approach aerodynamics from a different point of view, using concepts from kinetic theory. This regime of aerodynamics is called *low-density flow.*

There are certain hypersonic applications that involve low-density flow, generally involving flight at high altitudes. For example, as noted in [7] the flow in the nose region of the space shuttle cannot be properly treated by purely continuum assumptions for altitudes above 92 km (about 300,000 ft). For any given flight vehicle, as the altitude progressively increases (hence the density decreases and λ increases), the assumption of a continuum flow becomes tenuous. An altitude can be reached where the conventional viscous flow no-slip conditions begin to fail. Specifically, at low densities the flow velocity at the surface, which is normally assumed to be zero because of friction, takes on a finite value. This is called the *velocity-slip* condition. In analogous fashion, the gas temperature at the surface, which is normally taken as equal to the surface temperature of the material, now becomes something different. This is called the *temperature-slip* condition. At the onset of these slip effects, the governing equations of the flow are still assumed to be the familar continuum-flow equations, except with the proper velocity and temperature-slip conditions utilized as boundary conditions. However, as the altitude continues to increase, there comes a point where the continuum-flow equations themselves are no longer valid, and methods from kinetic theory must be used to predict the aerodynamic behavior. Finally, the air density can become low enough that only a few molecules impact the surface per unit time, and after these molecules reflect from the surface they do not interact with the incoming molecules. This is the regime of *free molecule flow*. For the space shuttle, the free molecular regime begins about 150 km (500,000 ft). Therefore, in a simplified sense, we visualize that a hypersonic vehicle moving from a very rarified atmosphere to a denser atmosphere will shift from the *free molecular regime*, where individual molecular impacts on the surface are important, to the *transition regime*, where slip effects are important, and then to the *continuum regime*.

The similarity parameter that governs these different regimes is the *Knudsen number*, defined as $Kn = \lambda/L$, where L is a characteristic dimension of the body. The values of Kn in the different regimes are noted in Fig. 1.19, taken from [7]. Note the region where the continuum Navier–Stokes equations hold is described by $Kn < 0.2$. However, slip effects must be included in these equations when $Kn > 0.03$. The effects of free molecular flow begin around a value or $Kn = 1$ and extend out to the limit of Kn becoming infinite. Hence, the transitional regime is essentially contained within $0.03 < Kn < 1.0$. In a given problem, the Knudsen number is the criterion to examine in order to decide if low-density effects are important and to what extent. For example, if Kn is very small, we have continuum flow; if Kn is very large, we have free molecular flow, and so forth. A hypersonic vehicle entering the atmosphere from space will encounter the full range of these low-density effects, down to an altitude below which the full continuum aerodynamics takes over. Because $Kn = \lambda/L$ is the governing parameter, that altitude below which we have continuum flow is greater or lesser as the characteristic length L is larger or smaller. Hence, large vehicles experience continuum flow to higher altitudes than small vehicles. Moreover, if we let the characteristic length be a running distance x from the nose or leading edge of the vehicle then $Kn = \lambda/x$ becomes infinite when $x = 0$. Hence, for any vehicle at any altitude the flow immediately at the leading edge is governed by low-density effects. For most practical applications in aerodynamics, this leading-edge region is very small and is usually ignored. However, for high-altitude hypersonic vehicles the proper treatment of the leading-edge flow by low-density methods can be important.

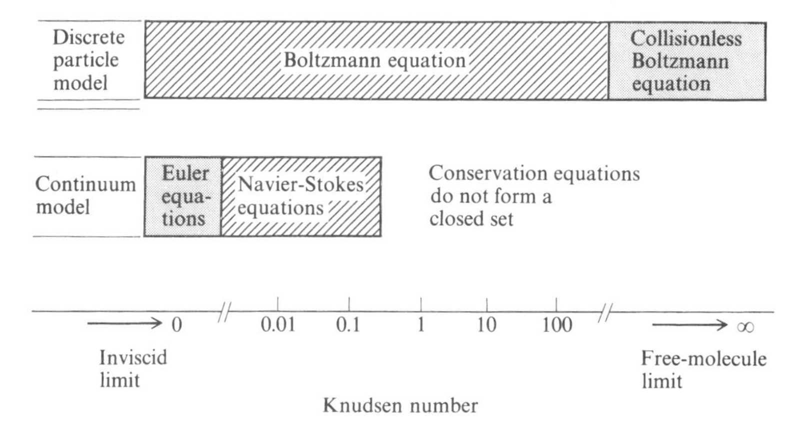

Fig. 1.19 Regimes of applicability of various flow equations for low-density flows (from [7]).

To consider low-density effects as part of the "definition" of hypersonic aerodynamics might be stretching that definition too much. Recall that we are defining hypersonic aerodynamics as that regime where certain physical flow phenomena become progressively more important as the Mach number is increased to high values. Low-density effects are not, per se, high-Mach-number effects. However, low-density effects are included in our discussion because some classes of hypersonic vehicles, as a result of their high Mach number, will fly at or through the outer regions of the atmosphere and hence will experience such effects to a greater or lesser extent.

1.3.6 Recapitulation

To repeat, hypersonic flow is best defined as that regime where all or some of the preceding physical phenomena become important as the Mach number is increased to high values. To help reinforce this definition, Fig. 1.20 summarizes

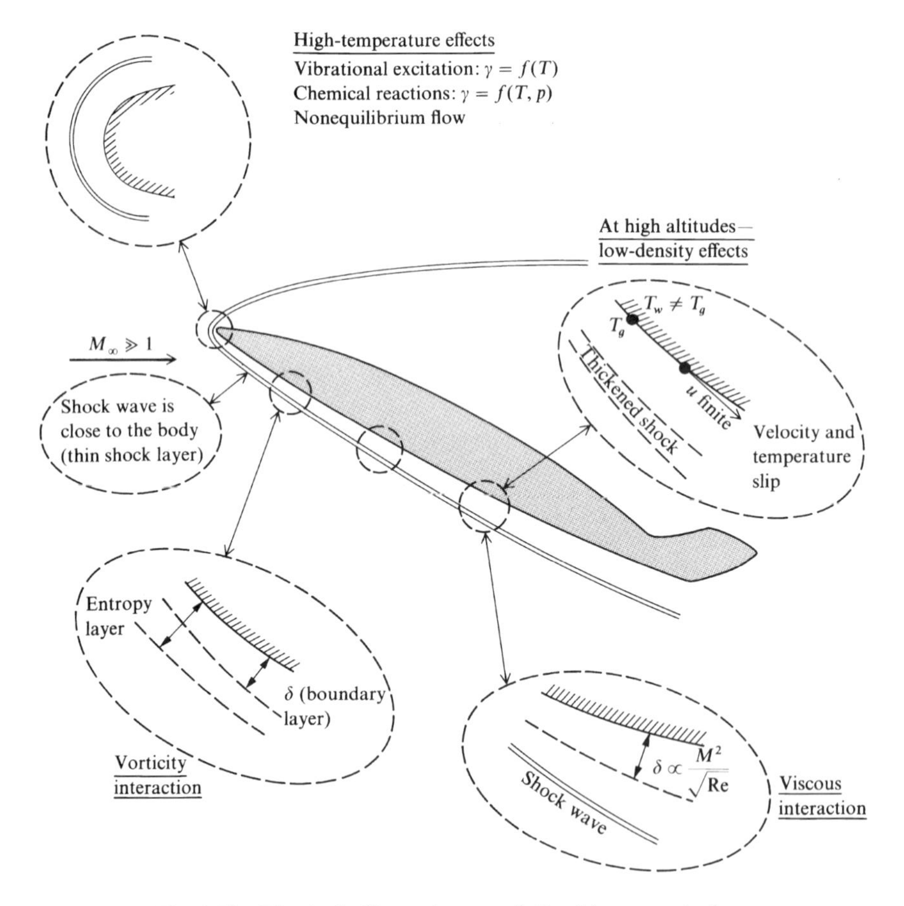

Fig. 1.20 Physical effects characteristic of hypersonic flow.

the important physical phenomena associated with hypersonic flight. Throughout this book, the fundamental aspects and practical consequences of these phenomena will be emphasized.

1.4 Fundamental Sources of Aerodynamic Force and Aerodynamic Heating

Any object immersed in a flowing gas experiences a force as a result of the interaction of the body with the gas—the *aerodynamic force*, illustrated by the vector $\boldsymbol{R}$ in Fig. 1.21. The two components of $\boldsymbol{R}$ perpendicular and parallel to the relative wind far ahead of the body are *lift* L and *drag* D, respectively, as also shown in Fig. 1.21. Similarly, $\boldsymbol{R}$ can be resolved into two components perpendicular and parallel to the chord (or axis) of the body, the *normal force* N and the *axial force* A, respectively, shown in Fig. 1.21. These forces are usually

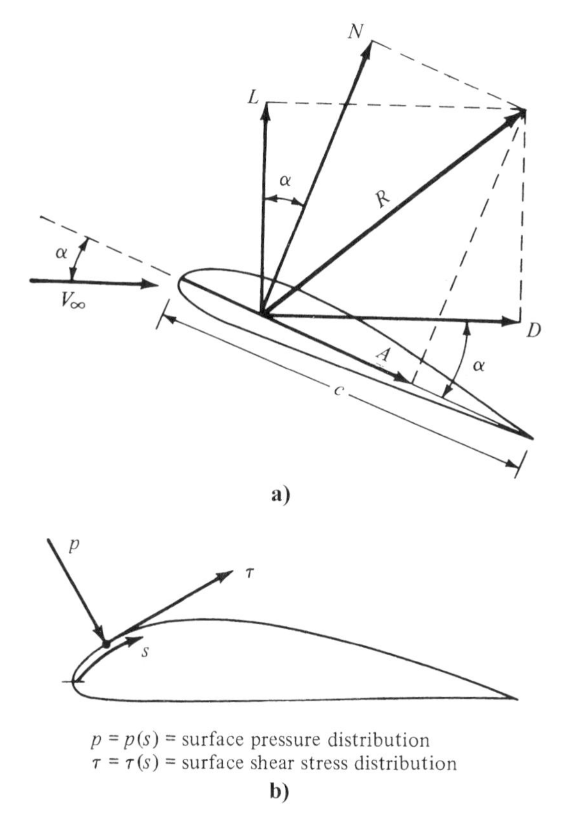

Fig. 1.21 a) Resultant aerodynamic force and the components into which it splits and b) illustration of pressure and shear stress on an aerodynamic surface.

couched in terms of force coefficients. Defining the dynamic pressure q_∞ as

$$q_\infty = \tfrac{1}{2}\rho_\infty V_\infty^2$$

where ρ_∞ and V_∞ are the freestream density and velocity respectively, and choosing some reference area S, we have the following by definition.

Lift coefficient:

$$C_L = \frac{L}{q_\infty S}$$

Drag coefficient:

$$C_D = \frac{D}{q_\infty S}$$

Normal-force coefficient:

$$C_N = \frac{N}{q_\infty S}$$

Axial-force coefficient:

$$C_A = \frac{A}{q_\infty S}$$

In addition, consider any point on the body. The resultant force $\boldsymbol{R}$ can create a moment M about this point, and we define

Moment coefficient:

$$\frac{M}{q_\infty Sc}$$

where c is a characteristic length of the body, such as the chord length shown in Fig. 1.21a.

How does nature exert an aerodynamic force on the body? To help answer this question, pick up this book, and hold it stationary in the palm of your hand. You are exerting a force on the book, equal to its weight. This force is communicated to the book by the skin of your hand being in contact with the bottom cover of the book. This is the only way that the book knows that you are exerting a force on it—by contact of the book's surface with your hand. Similarly, the surface of the body in Fig. 1.21a is in contact with the air—the molecular layer of air immediately adjacent to the body surface. The only way that the body feels the presence of the air is through the pressure distribution and shear-stress distribution exerted on the body surface that is in contact with the air. As illustrated in Fig. 1.21b, the pressure acts locally normal to the body surface, and shear stress, which is caused by friction between the body and the molecular layer of air just adjacent to the body surface, acts locally tangential to the surface. Let s be the distance along the body surface measured from the leading edge (nose) of the body, as shown in Fig. 1.21b. The variation of pressure with distance along the surface, denoted by $p(s)$, is the *pressure distribution*, and the variation of shear stress with distance along the surface, denoted by $\tau_w(s)$, is the *shear-stress distribution*. Clearly, pressure and shear stress are distributed loads acting on the body and when integrated over the complete surface area of the body give rise to the resultant aerodynamic force vector $\boldsymbol{R}$ shown in Fig. 1.21a. The two hands of nature that reach out and grab hold of the body, exerting a force on the body, are the pressure distribution and shear-stress distribution exerted on the body surface. These are the two fundamental sources of aerodynamic force—and that is all there is.

Except in the cases where viscous interaction is dominant and/or flow separation takes place, the surface-pressure distribution is essentially, by nature, an inviscid flow phenomena. Part 1 (Chapters 2–5) of this book deals with hypersonic inviscid flow. Much of the material in Part 1 is focused on the calculation of pressure distributions over the surface of hypersonic aerodynamic shapes, thus leading to the prediction of lift and wave drag (a pressure effect).

Shear stress is a viscous flow phenomena. Part 2 (Chapters 6–8) of this book deals with viscous flow. Some of the material in Part 2 is focused on the calculation of shear-stress distributions over the surface of hypersonic aerodynamic shapes, thus leading to the prediction of skin-friction drag. The surface shear stress is dictated by the velocity gradient at the surface

$$\tau_w = \mu_w \left(\frac{\partial u}{\partial y} \right)_w \tag{1.1}$$

where $(\partial u/\partial y)_w$ is the velocity gradient at the wall and μ_w is the viscosity coefficient at the wall.

Any object immersed in a flowing gas also experiences heat transfer at the body surface. For the majority of practical cases in high-speed aerodynamics, heat is transferred from the gas into the body—*aerodynamic heating*. For low-speed subsonic flows, aerodynamic heating, though present, is small and not generally a player. For supersonic flows, however, aerodynamic heating becomes a consideration. For hypersonic flows, aerodynamic heating is a major player, so much so that it dictates the configuration design of most hypersonic vehicles.

The sources of aerodynamic heating to a body are related to the hot gas in the flowfield around the body. The boundary layer adjacent to the body surface is hot because the high-kinetic-energy hypersonic flow entering the boundary layer is slowed by viscous effects within the boundary layer, dissipating the kinetic energy, and transforming it (in part) into internal energy of the gas. The gas in the inviscid portion of the shock layer between the shock wave and the body is hot because it has come through a strong shock wave. Shock-wave heating and intense viscous dissipation are the causes of the hypersonic flow over a body being hot. Nature, in turn, pumps some of this heat energy into the body in the form of aerodynamic heating. There are two physical sources, or mechanisms, that nature uses to transfer the heat into the body: thermal conduction and radiation. Let us examine each one of these mechanisms in turn.

Thermal conduction takes place when a temperature gradient exists in the gas, and from Fourier's law of heat conduction the heat transfer to the surface

q_w is given by

$$q_w = -k_w \left(\frac{\partial T}{\partial y} \right)_w \tag{1.2}$$

where k_w is the thermal conductivity of the gas at the wall and $(\partial T/\partial y)_w$ is the temperature gradient in the gas at the wall. The minus sign denotes that heat is conducted in the opposite direction of the temperature gradient. For example, reflect again on the boundary-layer temperature profile sketched in Fig. 1.15. The temperature gradient at the wall is clearly seen; at the wall, the gas temperature increases with distance above the wall; hence, $(\partial T/\partial y)_w$ is a positive quantity. Consequently, q_w flows into the wall in the negative y direction. Thermal conduction as quantified by Eq. (1.2) is one of nature's ways to transfer heat to the surface of a body. The hand of nature that does this is not the temperature of the gas at the wall, but rather the *temperature gradient* in the gas at the wall. Part 2 of this book deals in part with the calculation of temperature gradients at the wall and hence the conductive heat transfer at the surface.

Thermal radiation is the other hand that nature uses to heat a body (or to cool a body). If the temperature of the flowfield around a body is hot enough, significant thermal radiation from the gas to the body takes place. How hot is hot enough? For air, if the shock-layer temperature is about 10,000 K or higher radiative heat transfer from the hot shock layer to the cooler body becomes a player in the overall heating of the body. Examples are space vehicles entering the Earth's atmosphere at superorbital velocities, such as the Apollo lunar return capsule, where the shock-layer temperature reached as high as 11,000 K. Almost 30% of the total heat transfer to the Apollo during atmospheric entry was caused by radiative heating. For the Galileo probe entering the Jovian atmosphere at 15 km/s, where the shock-layer temperature exceeded 15,000 K, more than 95% of the total heat transfer was radiative heating. Radiative heating is treated in Chapter 18.

Parenthetically, we note that the solid surface of the body radiates energy away from the surface. This is a form of cooling of the surface—radiative cooling. Solid bodies emit radiation primarily in the infrared part of the spectrum. (When you stand in front of a fireplace on a cold winter day, and enjoy the heat from the fire, most of the heat you are feeling is coming from infrared radiation emitted from the hot bricks or stone of the walls of the fireplace.) Moreover, the radiative energy emitted from the surface of a solid body is a function of the surface temperature, following the T^4 variation of a blackbody. Surface radiation, therefore, can be used effectively as part or all of the thermal protection mechanism for a hypersonic vehicle.

In summary, the two sources of aerodynamic heating are the temperature gradient in the gas at the body surface, adding heat to the body by means of thermal conduction, and the hot temperature that exists globally throughout the shock layer, adding heat to the body by means of thermal radiation from each hot fluid element in the flowfield. Frequently, the former mode of heating is called *convective heating*, although thermal conduction at the surface is the hand that nature uses to transfer the energy to the surface. The latter mode of heating is simply called *radiative heating*.

1.5 Hypersonic Flight Paths: Velocity-Altitude Map

Although this is a book on hypersonic and high-temperature gas dynamics, we must keep in mind that the frequent application of this material is to the design and understanding of hypersonic flight vehicles. In turn, it is helpful to have some knowledge of the flight paths of these vehicles through the atmosphere and the parameters that govern such flight paths. This is the purpose of the present section. In particular, we will examine the flight path of lifting and nonlifting hypersonic vehicles during atmospheric entry from space.

Consider a vehicle flying at a velocity V along a flight path inclined at the angle θ below the local horizontal, as shown in Fig. 1.22. The forces acting on the vehicle are lift L, drag D, and weight W; the thrust is assumed to be zero; hence, we are considering a hypersonic glide vehicle. Summing forces along and perpendicular to the curvilinear flight path, we obtain the following equations of motion from Newton's second law.

Along flight path:

$$W \sin\theta - D = m\frac{\mathrm{d}V}{\mathrm{d}t} \tag{1.3}$$

Perpendicular to flight path:

$$L - W\cos\theta = -m\frac{V^2}{R} \tag{1.4}$$

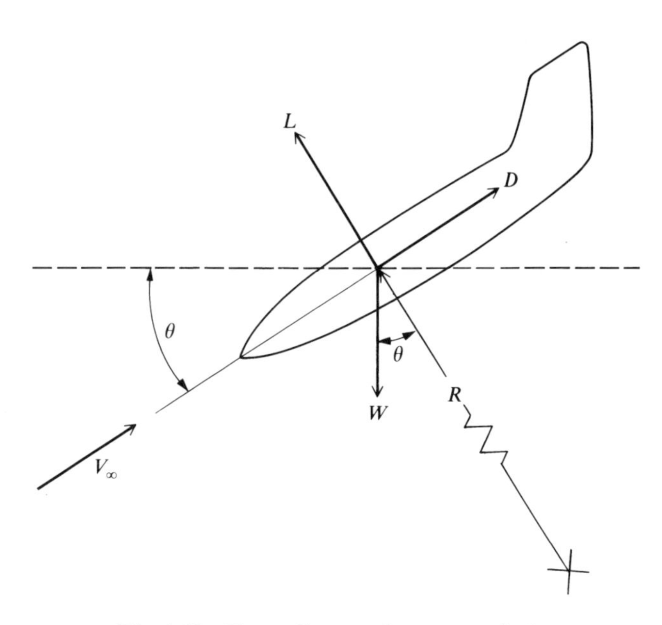

Fig. 1.22 Force diagram for reentry body.

In Eq. (1.4), R is the local radius of curvature of the flight path. For most entry conditions, θ is small; hence, we assume $\sin\,\theta \approx 0$ and $\cos\,\theta \approx 1$. For this case, Eqs. (1.3) and (1.4) become, noting that $m = W/g$

$$-D = \frac{W}{g}\frac{\mathrm{d}V}{\mathrm{d}t} \tag{1.5}$$

$$L - W = -\frac{W}{g}\frac{V^2}{R} \tag{1.6}$$

The drag can be expressed in terms of the drag coefficient C_D as $D = \frac{1}{2}\rho V^2 S C_D$, where ρ is the freestream density and S is a reference area. Hence, Eq. (1.5) becomes

$$-\frac{1}{2}\rho V^2 S C_D = \frac{W}{g}\frac{\mathrm{d}V}{\mathrm{d}t}$$

Rearranging, we obtain

$$-\frac{1}{g}\frac{\mathrm{d}V}{\mathrm{d}t} = \left(\frac{W}{C_D S}\right)^{-1}\frac{\rho V^2}{2} \tag{1.7}$$

In Eq. (1.7), $W/C_D S$ is defined as the ballistic parameter; it clearly influences the flight path of the entry vehicle via the solution of Eq. (1.7). For a purely ballistic reentry (no lift), $W/C_D S$ is the only parameter governing the flight path for a given entry angle.

Returning to Eq. (1.6) and expressing the lift in terms of the lift coefficient C_L as $L = \frac{1}{2}\rho V^2 S C_L$, we obtain

$$\frac{1}{2}\rho V^2 S C_L - W = -\frac{W}{g}\frac{V^2}{R}$$

Rearranging,

$$1 - \frac{1}{g}\frac{V^2}{R} = \left(\frac{W}{C_L S}\right)^{-1}\frac{\rho V^2}{2} \tag{1.8}$$

In Eq. (1.8), $W/C_L\,S$ is the lift parameter; it clearly influences the flight path of a lifting-entry vehicle via the solution of Eq. (1.8).

Equations (1.7) and (1.8) illustrate the importance of $W/C_D S$ and $W/C_L S$ in determining the flight path through the atmosphere of a vehicle returning from space. Such flight paths are frequently plotted on a graph of altitude vs velocity—a *velocity-altitude map*, an example of which is shown in Fig. 1.23. Here, two classes of flight paths are shown: 1) lifting entry, governed mainly by $W/C_L S$, and 2) ballistic entry, governed mainly by $W/C_D S$. The vehicle enters the atmosphere at either satellite velocity (such as from orbit) or at

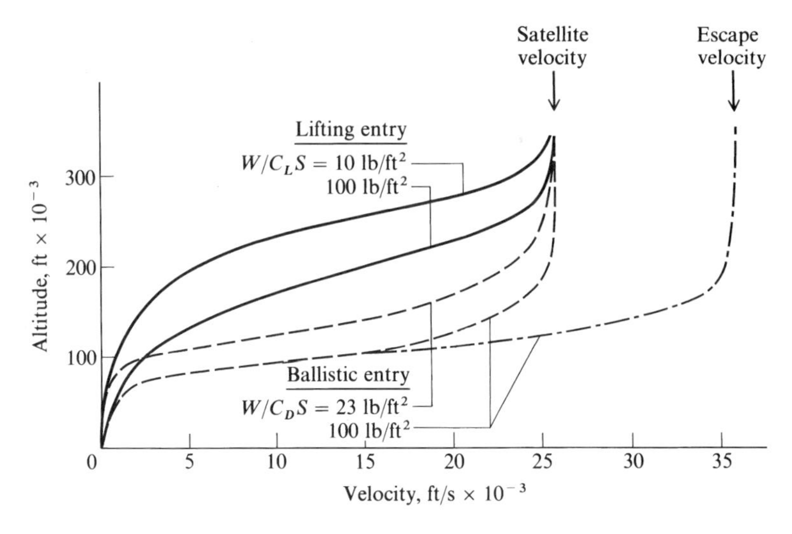

Fig. 1.23 Atmospheric entry flight paths on a velocity-altitude map.

escape velocity (such as a return from a lunar mission). As it flies deeper into the atmosphere, it slows as a result of aerodynamic drag, giving rise to flight paths shown in Fig. 1.23. Note that vehicles with larger values of W/C_LS and/or W/C_DS penetrate deeper into the atmosphere before slowing. The lifting-entry curve for $W/C_LS = 100\ \mathrm{lb/ft^2}$ pertains approximately to the space shuttle; the curve initiated at escape velocity with $W/C_DS = 100\ \mathrm{lb/ft^2}$ pertains approximately to the Apollo entry capsule. Velocity-altitude maps are convenient diagrams to illustrate various aerothermodynamic regimes of supersonic flight, and they will be used as such in some of our subsequent discussion.

1.6 Summary and Outlook

The major purposes of this chapter have been motivation and orientation—motivation as to the importance, interest, and challenge associated with hypersonic aerodynamics, and orientation as to what hypersonics entails. For the remainder of this book, our purpose is to present and discuss the important fundamental aspects of hypersonic and high-temperature gas dynamics and to highlight various practical applications as appropriate. Towards this end, the book is organized into three major parts, as diagramed in Fig. 1.24. These three parts are as follows.

Part 1—Inviscid Flow: Here, the purely fluid-dynamic effect of large Mach number is emphasized, without the added complications of viscous and high-temperature effects. In this part, we examine what happens when the freestream Mach number M_∞ becomes large and how this influences aerodynamic theory at high Mach numbers. Here we calculate pressure distributions on surfaces and obtain the aerodynamic lift and wave drag on hypersonic bodies.

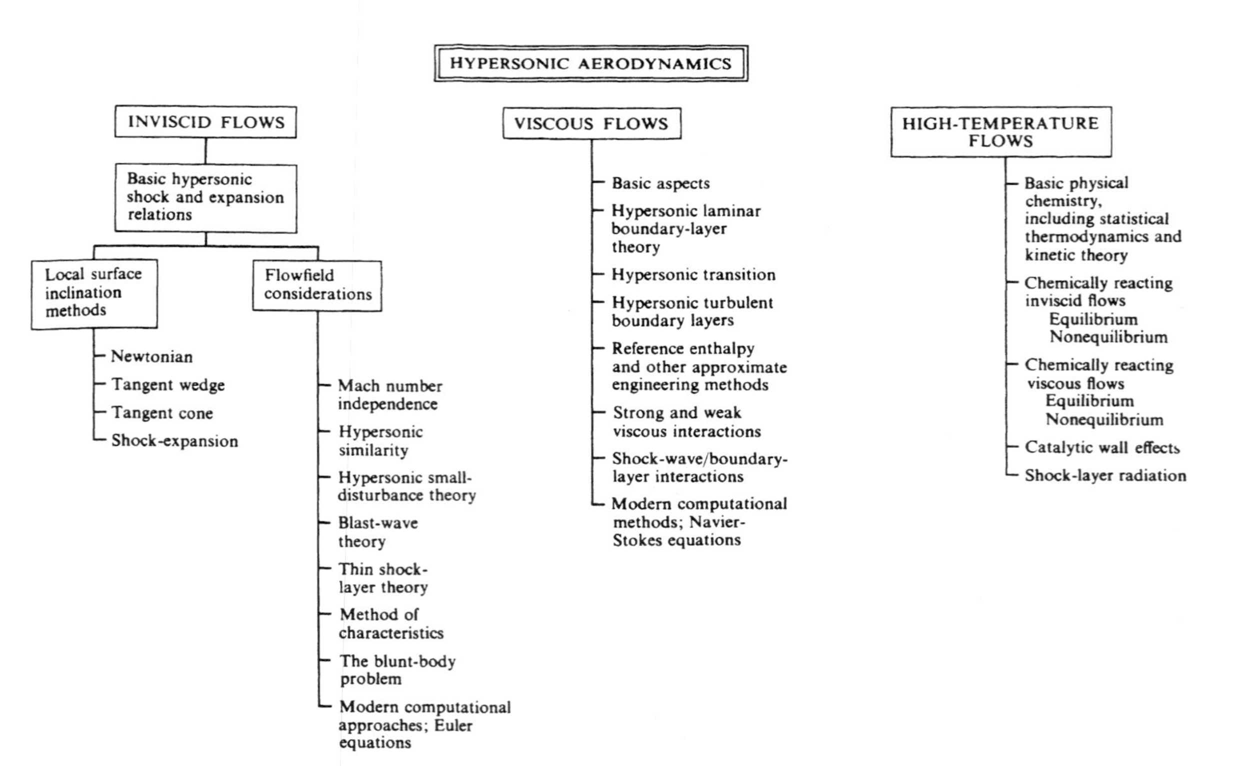

Fig. 1.24 Road map for our study of hypersonic and high-temperature flows.

Part 2—Viscous Flow: Here, the combined effect of high Mach number and finite Reynolds number will be examined. The purely fluid-dynamic effect of hypersonic flow with friction and thermal conduction will be presented; again, high-temperature effects will not be included. Here, we calculate the shear stress at the surface and the resulting skin friction drag on hypersonic bodies. In addition we calculate the temperature gradients in the gas at a surface and the resulting aerodynamic heating to hypersonic bodies as a result of thermal conduction (convective aerodynamic heating).

Part 3—High-Temperature Flow: Here, the important aspects of high-temperature gas dynamics will be presented. Emphasis will be placed on the development of basic physical chemistry principles and how they affect both inviscid and viscous flows. High-temperature flows find application in many fields in addition to hypersonic aerodynamics, such as combustion processes, explosions, plasmas, high-energy lasers, etc. Therefore, Part 3 will be a self-contained presentation of high-temperature gas dynamics in general, along with pertinent applications to hypersonic flow. Here we also calculate the radiative heating of a hypersonic body caused by intense radiation from very high-temperature shock layers.

Figure 1.24 is a block diagram showing each one of the three parts just discussed, along with the major items to be discussed under each part. In essence, this figure is a *road map* for our excursions in hypersonic and high-temperature gas dynamics. Figure 1.24 is important, and we will refer to it often in order to see where we are, where we have been, and where we are going in our presentation.

Problems

1.1 Consider the supersonic and hypersonic flow of air (with constant ratio of specific heats, $\gamma = 1.4$) over a 20-deg half-angle wedge. Let θ denote the wedge half-angle and β the shock-wave angle. Then β-θ is a measure of the shock-layer thickness. Make a plot of β-θ vs the freestream Mach number M_∞ from $M_\infty = 2.0$ to 20.0. Make some comments as to what Mach-number range results in a "thin" shock layer.

1.2 The lifting parameter $W/C_D S$ is given Fig. 1.23 in units of lb/ft^2. Frequently, the analogous parameter $m/C_D S$ is used, when m is the vehicle mass; the units of $m/C_D S$ are usually given in kg/m^2. Derive the appropriate conversion between these two sets of units, that is, what number must $W/C_D S$ expressed in lb/ft^2 be multiplied by to obtain $m/C_D S$ in kg/m^2? (*Comment*: Even at the graduate level, it is useful now and then to go through this type of exercise.)

Part 1
Inviscid Hypersonic Flow

In Part 1 we emphasize the purely fluid-dynamic effects of high Mach number; the complicating effects of transport phenomena (viscosity, thermal conduction, and diffusion) and high-temperature phenomena will be treated in Parts 2 and 3, respectively. In dealing with inviscid, hypersonic flow in Part 1, we are simply examining the question: what happens to the fluid dynamics of an inviscid flow when the Mach number is made very large? We will see that such an examination goes a long way toward the understanding of many practical hypersonic applications.

2
Hypersonic Shock and Expansion-Wave Relations

It is clear that the thorough study of gas-dynamic discontinuities and their structures combines in an essential way the fields of hydrodynamics, physics, and chemistry, and that there is no lack of problems which deserve attention.

Wallace D. Hayes, Princeton University, 1958

Chapter Preview

This chapter is short and sweet. Shock waves and expansion waves are ubiquitous features in supersonic and hypersonic flowfields. The detailed analysis and calculation of shock and expansion wave properties are bread-and-butter topics in the study of compressible flow. This chapter goes a step further; here, the assumption of very high Mach numbers ahead of the waves allows the rather elaborate, but exact, shock- and expansion-wave relations to be reduced to much simpler, albeit approximate, equations that hold for hypersonic flow. You might ask—so what? Indeed, for an accurate calculation of the details of a hypersonic flow you should always use the *exact* shock- and expansion-wave equations. The value of this chapter lies elsewhere—it lies in the world of hypersonic aerodynamic *theory*. This chapter is the front end of the beautiful intellectual triumphs that are the foundation of the hypersonic aerodynamic theories to be developed in subsequent chapters. One such theory is hypersonic similarity, to be explained and developed in Chapter 4. But right away, in the present chapter, the approximate hypersonic shock- and expansion-wave relations expose one of the most important parameters in hypersonic aerodynamic theory—the hypersonic similarity parameter. But we are getting ahead of ourselves. For right now, just sit back and enjoy the clean and neat development of the simplified equations for shock and expansion waves afforded by the assumption of high Mach numbers in the flow ahead of the waves. Then store these hypersonic shock- and expansion-wave relations on your mental bookshelf for future use in subsequent chapters.

2.1 Introduction

Consider an airplane flying at Mach 28 at the outer regions of the Earth's atmosphere, say at an altitude of 120 km (approximately 400,000 ft). Upon descent into the lower regions of the atmosphere, the aircraft can follow one of the lifting trajectories shown on the altitude-velocity map in Fig. 2.1.* Superimposed on this figure are lines of constant Mach number. The purpose of this figure is to emphasize the obvious fact that such hypersonic vehicles encounter exceptionally high Mach numbers. Moreover, the flight path remains hypersonic over most of its extent. Figure 2.1 justifies the study of high-Mach-number flows and underscores the question: what happens in a purely fluid-dynamic sense when the Mach number becomes very large? This question has particular relevance in regard to the basic shock- and expansion-wave relations. In the present chapter, we will obtain and examine the limiting forms of both the conventional shock-wave equations and the Prandtl–Meyer expansion-wave relations when the upstream Mach number increases toward infinity. These limiting forms are interesting in their own right; however, of more importance, they are absolutely necessary for the development of various inviscid hypersonic theories to be discussed in subsequent chapters.

2.2 Basic Hypersonic Shock Relations

Anytime a supersonic flow is turned into itself (such as flowing over a wedge, cone, or compression corner), a shock wave is created. Also, if a sufficiently high backpressure is created downstream of a supersonic flow, a standing shock wave can be established. Such shock waves are extremely thin regions (on the order of 10^{-5} cm in air) across which large changes in density, pressure, velocity, etc. occur. These changes take place in a continuous fashion within the shock wave itself, where viscosity and thermal conduction are important mechanisms. However, because the wave is usually so thin, to the macroscopic observer the changes appear to take place discontinuously. Therefore, in conventional supersonic aerodynamics shock waves are usually treated as mathematical and physical discontinuities. As the Mach number is increased to hypersonic speeds, no dramatic qualitative difference occurs. The same exact shock relations that are obtained in supersonic aerodynamics also hold at hypersonic speeds. However, some interesting approximate and simplified forms of the shock relations are obtained in the limit of high Mach number; these forms are obtained next.

Consider the flow through a straight oblique shock wave, as sketched in Fig. 2.2. Upstream and downstream conditions are denoted by subscripts 1 and 2, respectively. For a calorically perfect gas (constant specific heats, hence $\gamma = c_p/c_v =$ constant), the classical results for changes across the shock are

---

*Velocity-altitude maps are discussed in Sec. 1.5. In that section, the parameters W/C_DS and W/C_LS are introduced. Related parameters are m/C_DS and m/C_LS, where m is the mass of the vehicle. Figure 2.1 is shown in terms of SI units, and the lift parameter is couched in terms of m rather than W. In comparing values of W/C_LS and m/C_LS, for example, note that m/C_LS (in kg/m^2) $= 5 \times W/C_LS$ (in lb/ft^2), that is, a value of $W/C_LS = 1000$ lb/ft^2 is equal to $m/C_LS = 5000$ kg/m^2. The same ratio holds, of course, for W/C_DS and m/C_DS.

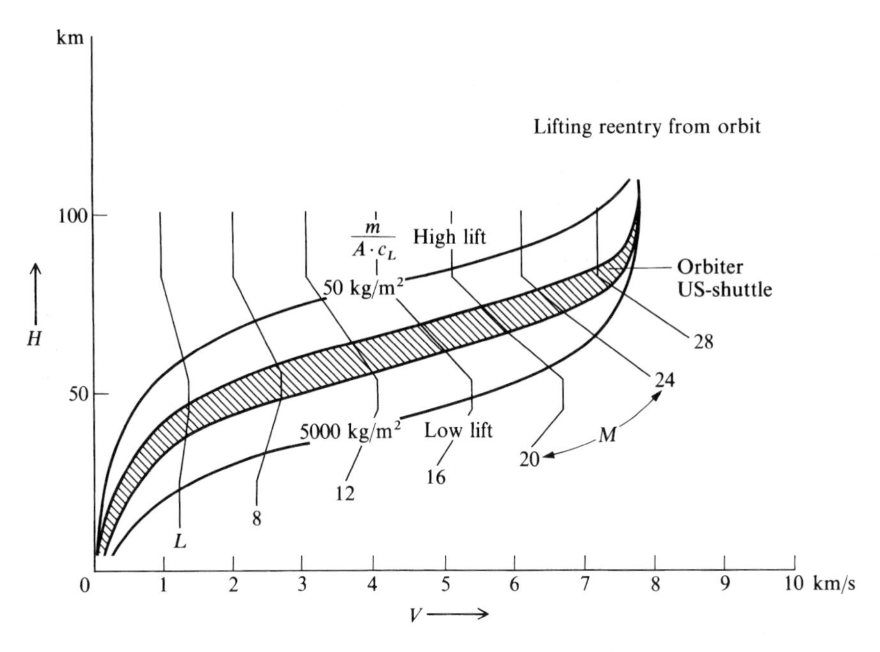

Fig. 2.1 Velocity-altitude map with superimposed lines of constant Mach number.

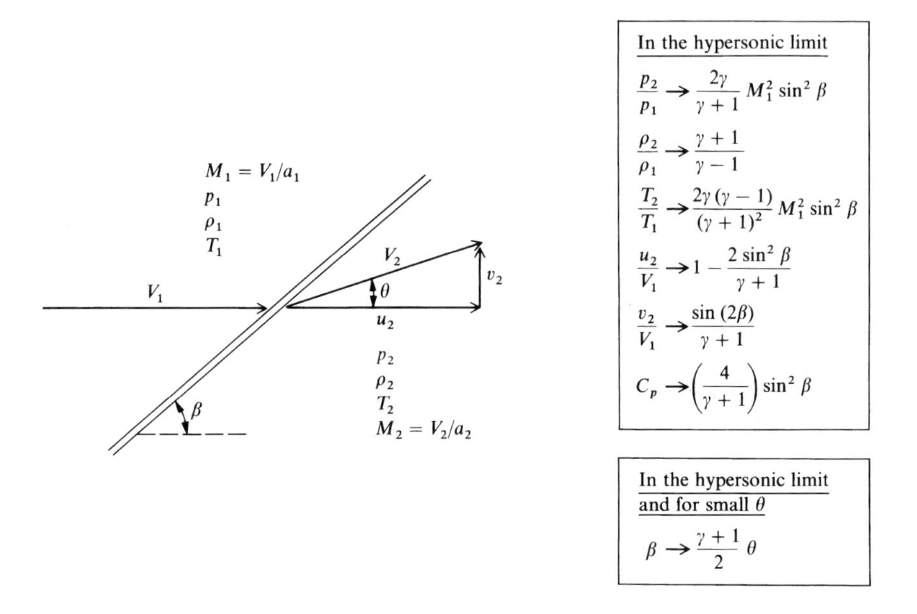

Fig. 2.2 Oblique shock-wave geometry.

given in any standard textbook on compressible flow (for example, see [4] and [5]). To begin, the exact oblique shock relation for pressure ratio across the wave is given by the following.

Exact:

$$\frac{p_2}{p_1} = 1 + \frac{2\gamma}{\gamma + 1}(M_1^2 \sin^2 \beta - 1) \tag{2.1}$$

where β is the wave angle, shown in Fig. 2.2. In the limit as M_1 goes to infinity, the term $M_1^2 \sin^2 \beta \gg 1$, and hence Eq. (2.1) becomes the following.

As $M_1 \to \infty$:

$$\boxed{\frac{p_2}{p_1} = \frac{2\gamma}{\gamma + 1} M_1^2 \sin^2 \beta} \tag{2.2}$$

In a similar vein, the density and temperature ratios are as follows.

Exact:

$$\frac{\rho_2}{\rho_1} = \frac{(\gamma + 1)M_1^2 \sin^2 \beta}{(\gamma - 1)M_1^2 \sin^2 \beta + 2} \tag{2.3}$$

As $M_1 \to \infty$:

$$\boxed{\frac{\rho_2}{\rho_1} = \frac{\gamma + 1}{\gamma - 1}} \tag{2.4}$$

$$\frac{T_2}{T_1} = \frac{(p_2/p_1)}{(\rho_2/\rho_1)} \text{(from the equation of state } p = \rho RT\text{)}$$

As $M_1 \to \infty$:

$$\boxed{\frac{T_2}{T_1} = \frac{2\gamma(\gamma - 1)}{(\gamma + 1)^2} M_1^2 \sin^2 \beta} \tag{2.5}$$

Returning to Fig. 2.2, note that u_2 and v_2 are the components of the flow velocity behind the shock wave *parallel* and *perpendicular* to the upstream flow (not parallel and perpendicular to the shock wave itself, as is frequently done). With this in mind, the following is true.

Exact:

$$\frac{u_2}{V_1} = 1 - \frac{2(M_1^2 \sin^2 \beta - 1)}{(\gamma + 1)M_1^2} \tag{2.6}$$

As $M_1 \to \infty$:

$$\boxed{\frac{u_2}{V_1} = 1 - \frac{2 \sin^2 \beta}{\gamma + 1}} \tag{2.7}$$

Exact:

$$\frac{v_2}{V_1} = \frac{2(M_1^2 \sin^2 \beta - 1)\cos\beta}{(\gamma + 1)M_1^2} \tag{2.8}$$

For large M_1, Eq. (2.8) can be approximated by

$$\frac{v_2}{V_1} = \frac{2(M_1^2 \sin^2 \beta)\cot\beta}{(\gamma + 1)M_1^2} = \frac{2 \sin\beta \cos\beta}{\gamma + 1} \tag{2.9}$$

Because $2 \sin\beta \cos\beta = \sin 2\beta$, then from Eq. (2.9) this follows:

As $M_1 \to \infty$:

$$\boxed{\frac{v_2}{V_1} = \frac{\sin 2\beta}{\gamma + 1}} \tag{2.10}$$

In the preceding, the choice of velocity components parallel and perpendicular to the upstream flow direction rather than to the shock wave is intentional. Equations (2.7) and (2.10) will be used to great advantage in subsequent chapters to demonstrate some physical aspects of the velocity field over slender hypersonic bodies.

Note from Eqs. (2.2) and (2.5) that both p_2/p_1 and T_2/T_1 become infinitely large as $M_1 \to \infty$. In contrast, from Eqs. (2.4), (2.7), and (2.10), ρ_2/ρ_1, u_2/V_1, and v_2/V_1 approach limiting finite values as $M_1 \to \infty$.

In aerodynamics, pressure distributions are usually quoted in terms of the nondimensional pressure coefficient C_p rather than the pressure itself. The pressure coefficient is defined as

$$C_p = \frac{p_2 - p_1}{q_1} \tag{2.11}$$

where p_1 and q_1 are the upstream (freestream) static pressure and dynamic pressure, respectively. (In later chapters we will use the subscript ∞ to denote freestream conditions, such as freestream pressure p_∞ and freestream dynamic pressure q_∞. However, consistent with standard shock-wave nomenclature, we denote the freestream conditions by the subscript 1 in the present section.) By definition, the dynamic pressure is given by

$$q_1 = \tfrac{1}{2}\rho_1 V_1^2$$

This is a definition only—it is used for all flows, from incompressible to hypersonic. (*Note:* For incompressible flow, $q_1 = \frac{1}{2}\rho_1 V_1^2$ is exactly the difference between the total and static pressure of the freestream; for all other aerodynamic speed regimes, $q_1 = \frac{1}{2}\rho_1 V_1^2$ is a definition only, with no exact physical significance.) In high-speed flow theory, it is convenient to express q_1 in terms of Mach number and pressure M_1 and p_1, rather than velocity and density V_1 and ρ_1. This is easily accomplished by recalling that the speed of sound $a_1 = \sqrt{\gamma p_1/\rho_1}$ and that the Mach number $M_1 = V_1/a_1$. Hence,

$$q_1 = \frac{1}{2}\rho_1 V_1^2 = \frac{1}{2}\rho_1 V_1^2 \frac{\gamma p_1}{\gamma p_1} = \frac{\gamma p_1}{2}\frac{V_1^2}{a_1^2}$$

or

$$q_1 = \frac{\gamma}{2} p_1 M_1^2 \tag{2.12}$$

Equation (2.12) is a very convenient expression for dynamic pressure and can be viewed almost as an alternate definition of q_1. We can now write the pressure coefficient as

$$C_p = \frac{p_2 - p_1}{q_1} = \frac{2}{\gamma M_1^2}\left(\frac{p_2}{p_1} - 1\right) \tag{2.13}$$

Combining Eqs. (2.1) and (2.13), we obtain an exact relation for C_p behind an oblique shock wave as follows.

Exact:

$$C_p = \frac{4}{\gamma + 1}\left(\sin^2\beta - \frac{1}{M_1^2}\right) \tag{2.14}$$

In the hypersonic limit, the following is shown.

As $M_1 \to \infty$:

$$\boxed{C_p = \left(\frac{4}{\gamma + 1}\right)\sin^2\beta} \tag{2.15}$$

The relationship between Mach number M_1, shock angle β, and deflection angle θ is expressed by the so-called θ-β-M relation (see [4] and [5]).

Exact:

$$\tan\theta = 2\cot\beta\left[\frac{M_1^2\sin^2\beta - 1}{M_1^2(\gamma + \cos 2\beta) + 2}\right] \tag{2.16}$$

This relation is plotted in Fig. 2.3, which is a standard plot of wave angle vs deflection angle, with Mach number as a parameter. From this figure, note that, in the hypersonic limit, when θ is small, β is also small. Hence, in this limit we can insert the usual small-angle approximations into Eq. (2.16):

$$\sin\beta \approx \beta$$
$$\cos 2\beta \approx 1$$
$$\tan\theta \approx \sin\theta \approx \theta$$

resulting in

$$\theta = \frac{2}{\beta}\left[\frac{M_1^2\beta^2 - 1}{M_1^2(\gamma + 1) + 2}\right] \tag{2.17}$$

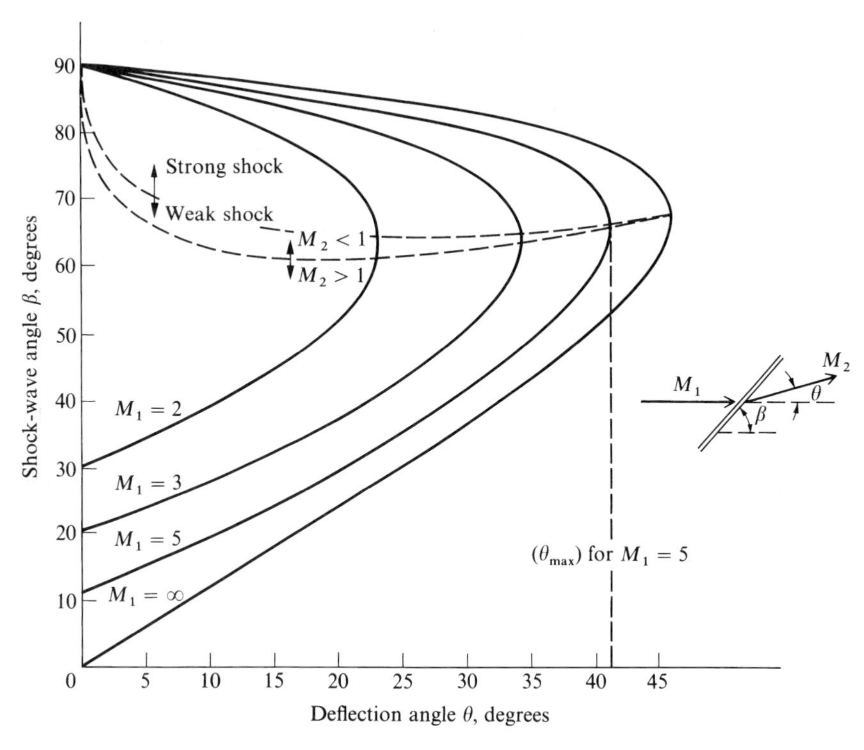

Fig. 2.3 θ-β-M diagram.

Applying the high-Mach-number limit to Eq. (20.17), we have

$$\theta = \frac{2}{\beta}\left[\frac{M_1^2\beta^2}{M_1^2(\gamma+1)}\right] \tag{2.18}$$

In Eq. (2.18) M_1 cancels, and we finally obtain in both the small-angle and hypersonic limits:

$$\left.\begin{array}{l}\text{As } M_1 \to \infty \\ \text{and} \\ \theta \text{ hence } \beta \text{ is small}\end{array}\right\} \quad \boxed{\frac{\beta}{\theta} = \frac{\gamma+1}{2}} \tag{2.19}$$

Note that for $\gamma = 1.4$,

$$\boxed{\beta = 1.2\theta} \tag{2.20}$$

It is interesting to observe that, in the hypersonic limit for a slender wedge, the wave angle is only 20% larger than the wedge angle—a graphic demonstration of a thin shock layer in hypersonic flow. (Check Fig. 1.13, drawn from exact oblique shock results, and note that the 18-deg shock angle is 20% larger than the 15-deg wedge angle at Mach 36—truly an example of the hypersonic limit.)

For your convenience, the limiting hypersonic shock relations obtained in this section are summarized in Fig. 2.2. These limiting relations, which are clearly simpler than the corresponding exact oblique shock relations, will be important for the development of some of our hypersonic aerodynamic techniques in subsequent chapters.

2.3 Hypersonic Shock Relations in Terms of the Hypersonic Similarity Parameter

In the study of hypersonic flow over slender bodies, the product $M_1\theta$ is an important governing parameter, where, as before, M_1 is the freestream Mach number and θ is the flow deflection angle. Indeed, we will demonstrate in Chapter 4 that $M_1\theta$ is a similarity parameter for such flows. Denoting $M_1\theta$ by K, we state

$$\boxed{K \equiv M_1\theta \equiv \text{hypersonic similarity parameter}}$$

In our future discussions, it will be helpful to express the oblique shock relations in terms of K, particularly in the case of pressure ratio p_2/p_1. This is the purpose of the present section.

Return to the exact θ-β-M relation given by Eq. (2.16). As expressed, this is an explicit relation for $\theta = \theta(\beta)$. Obtaining the exact inverse relation, $\beta = \beta(\theta)$, from Eq. (2.16) is tedious. However, in the combined limit of hypersonic flow and small angles, an approximate explicit relation for $\beta = \beta(\theta)$ can be obtained. This will be our first step toward introducing $K = M_1\theta$ into the shock relations. Specifically, for small angles Eq. (2.16) reduces to Eq. (2.17), rewritten here as

$$M_1^2\beta^2 - 1 = \left[\frac{M_1^2(\gamma+1)}{2} + 1\right]\beta\theta \tag{2.21}$$

In Eq. (2.21), assume that M_1 is large and finite; hence, $\frac{1}{2}(\gamma+1)M_1^2 \gg 1$. However, because β is small we *cannot* assume that $M_1^2\beta^2$ is large compared to unity. With this, Eq. (2.21) becomes

$$M_1^2\beta^2 - 1 = \frac{\gamma+1}{2}M_1^2\beta\theta \tag{2.22}$$

Rearranging, we obtain

$$\left(\frac{\beta}{\theta}\right)^2 - \frac{\gamma+1}{2}\left(\frac{\beta}{\theta}\right) - \frac{1}{M_1^2\theta^2} = 0 \tag{2.23}$$

This is a quadratic equation in terms of β/θ; solving by means of the quadratic formula,

$$\frac{\beta}{\theta} = \frac{\gamma+1}{4} + \sqrt{\left(\frac{\gamma+1}{4}\right)^2 + \frac{1}{M_1^2\theta^2}} \tag{2.24}$$

[In Eq. (2.24), the minus sign on the radical has been ruled out; it would produce the nonphysical result of a negative β/θ.] [*Note*: Equation (2.24) is the desired explicit relations for $\beta = \beta(\theta)$, good for the limit of hypersonic Mach numbers and small angles.]

Now return to Eq. (2.1), which is an exact relation for the pressure ratio across an oblique shock wave. Assuming small angles, this becomes

$$\frac{p_2}{p_1} = 1 + \frac{2\gamma}{\gamma+1}(M_1^2\beta^2 - 1) \tag{2.25}$$

If we wish to apply Eq. (2.25) at hypersonic but finite Mach numbers, we repeat again that, although M_1 is large, the product $M_1\beta$ might *not* be large; hence, for this case Eq. (2.25) cannot be reduced further. However, within the framework of these assumptions Eq. (2.24) gives an explicit relation for $\beta = \beta(\theta)$, which can be introduced into Eq. (2.25) to obtain an expression for p_2/p_1 in terms of the deflection angle θ. This is carried out as follows. Combining Eqs. (2.23) and (2.24),

$$\left(\frac{\beta}{\theta}\right)^2 = \frac{\gamma+1}{2}\left[\frac{\gamma+1}{4} + \sqrt{\left(\frac{\gamma+1}{4}\right)^2 + \frac{1}{M_1^2\theta^2}}\right] + \frac{1}{M_1^2\theta^2}$$

or

$$\beta^2 = \left[\frac{(\gamma+1)}{2}\frac{(\gamma+1)}{4} + \frac{\gamma+1}{2}\sqrt{\left(\frac{\gamma+1}{4}\right)^2 + \frac{1}{M_1^2\theta^2}}\right]\theta^2 + \frac{1}{M_1^2} \tag{2.26}$$

Substituting Eq. (2.26) into (2.25), we obtain

$$\frac{p_2}{p_1} = 1 + \frac{\gamma(\gamma+1)}{4}M_1^2\theta^2 + \gamma\sqrt{\left(\frac{\gamma+1}{4}\right)^2 + \frac{1}{M_1^2\theta^2}}M_1^2\theta^2 \tag{2.27}$$

Again denoting $M_1\theta$ by K, Eq. (27) is written as

$$\boxed{\frac{p_2}{p_1} = 1 + \frac{\gamma(\gamma+1)}{4}K^2 + \gamma K^2\sqrt{\left(\frac{\gamma+1}{4}\right)^2 + \frac{1}{K^2}}} \tag{2.28}$$

Equation (2.28) is the desired result; it gives the pressure ratio across an oblique shock wave in terms of the hypersonic similarity parameter, subject to the combined assumptions of high (but finite) Mach number and small angles. Because the pressure field behind a two-dimensional oblique shock is constant, Eq. (2.28) also gives the pressure p_2 on the surface of a wedge of deflection angle θ.

To round out our present discussion associated with the hypersonic similarity parameter, consider the pressure coefficient, defined in Eq. (2.13). Substituting Eq. (2.28) into (2.13), we obtain

$$C_p = \frac{2}{\gamma M_1^2}\left[\frac{\gamma(\gamma+1)}{4}K^2 + \gamma K^2\sqrt{\left(\frac{\gamma+1}{4}\right)^2 + \frac{1}{K^2}}\right]$$

or, multiplying and dividing by θ^2,

$$C_p = \frac{2\theta^2}{\gamma K^2}\left[\frac{\gamma(\gamma+1)}{4}K^2 + \gamma K^2\sqrt{\left(\frac{\gamma+1}{4}\right)^2 + \frac{1}{K^2}}\right]$$

or

$$\boxed{C_p = 2\theta^2\left[\frac{\gamma+1}{4} + \sqrt{\left(\frac{\gamma+1}{4}\right)^2 + \frac{1}{K^2}}\right]} \tag{2.29}$$

Note from Eq. (2.29) that, for hypersonic flow over wedges with small deflection angles,

$$\frac{C_p}{\theta^2} = f(K, \gamma) \tag{2.30}$$

We will find later that relations analogous to Eq. (2.30) abound in the theory of hypersonic flow over slender bodies.

2.4 Hypersonic Expansion-Wave Relations

Consider the centered Prandtl–Meyer expansion around a corner of deflection angle θ, as sketched in Fig. 2.4. An expansion fan consisting of an infinite number of Mach waves originates at the corner and spreads downstream. The Mach number and pressure upstream of the wave are M_1 and p_1, respectively; the corresponding quantities downstream of the wave are M_2 and p_2 respectively. From basic compressible flow (for example, see [4] and [5]), the relation between θ, M_1, and M_2 is given by

$$\theta = \nu(M_2) - \nu(M_1) \tag{2.31}$$

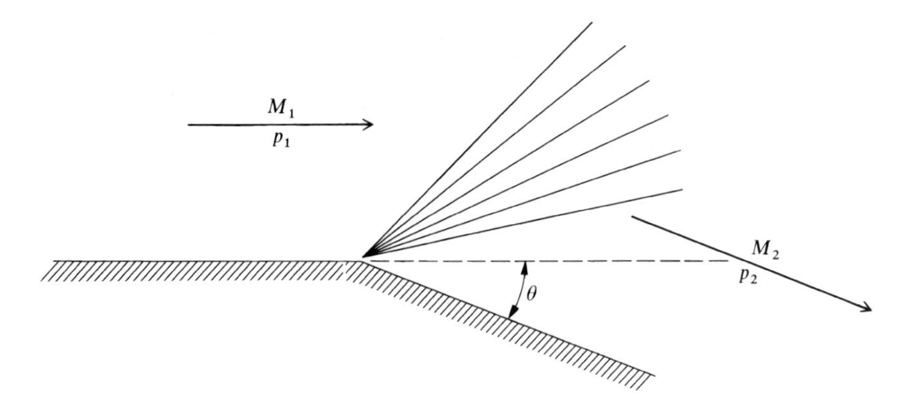

Fig. 2.4 Centered expansion wave.

where ν is the Prandtl–Meyer function

$$\nu(M) = \sqrt{\frac{\gamma+1}{\gamma-1}}\left[\tan^{-1}\sqrt{\frac{\gamma-1}{\gamma+1}(M^2-1)}\right] - \tan^{-1}\sqrt{M^2-1} \tag{2.32}$$

For large Mach numbers, $\sqrt{M^2-1} \approx M$. Hence, Eq. (2.32) can be written for hypersonic flow as

$$\nu(M) = \sqrt{\frac{\gamma+1}{\gamma-1}}\tan^{-1}\sqrt{\frac{\gamma-1}{\gamma+1}}M - \tan^{-1} M \tag{2.33}$$

Recalling the trigonometric identity

$$\tan^{-1} x = \frac{\pi}{2} - \tan^{-1}\left(\frac{1}{x}\right) \tag{2.34}$$

and the series expansion

$$\tan^{-1}\left(\frac{1}{x}\right) = \frac{1}{x} - \frac{1}{3x^3} + \frac{1}{5x^5} - \frac{1}{7x^7} + \cdots \tag{2.35}$$

we obtain by combining Eqs. (2.34) and (2.35),

$$\tan^{-1} x = \frac{\pi}{2} - \frac{1}{x} + \frac{1}{3x^3} - \frac{1}{5x^5} + \frac{1}{7x^7} + \cdots \tag{2.36}$$

Utilizing Eq. (2.36) to expand Eq. (2.33), we have

$$\nu(M) = \sqrt{\frac{\gamma+1}{\gamma-1}}\left(\frac{\pi}{2} - \sqrt{\frac{\gamma+1}{\gamma-1}}\frac{1}{M} + \cdots\right) - \left(\frac{\pi}{2} - \frac{1}{M} + \cdots\right) \tag{2.37}$$

At high Mach number, the higher-order terms associated with Eq. (2.37), that is, terms such as $1/3M^3$, $1/5M^5$, etc., can be ignored. For this case, Eq. (2.37) yields

$$\nu(M) = \sqrt{\frac{\gamma+1}{\gamma-1}}\frac{\pi}{2} - \left(\frac{\gamma+1}{\gamma-1}\right)\frac{1}{M} - \frac{\pi}{2} + \frac{1}{M} \tag{2.38}$$

Substituting Eq. (2.38) into (2.31), we obtain, for hypersonic Mach numbers,

$$\theta = \frac{1}{M_2} - \left(\frac{\gamma+1}{\gamma-1}\right)\frac{1}{M_2} - \frac{1}{M_1} + \left(\frac{\gamma+1}{\gamma-1}\right)\frac{1}{M_1}$$

or

$$\boxed{\theta = \frac{2}{\gamma-1} - \left(\frac{1}{M_1} - \frac{1}{M_2}\right)} \tag{2.39}$$

Equation (2.39) is the hypersonic relation for Prandtl–Meyer expansion waves; it is an approximate relation that becomes more accurate as M_1 and M_2 become larger. Recall that M increases through an expansion wave; hence, θ in Eq. (2.39) is a positive quantity. This is consistent with the sketch shown in Fig. 2.4, where the deflection angle θ is treated as a positive quantity.

The flow through an expansion wave is isentropic; hence the isentropic pressure relation holds as follows (again for example, see [4] and [5]):

$$\frac{p_2}{p_1} = \left[\frac{1 + (\gamma-1)/2\, M_1^2}{1 + (\gamma-1)/2\, M_2^2}\right]^{\gamma/(\gamma-1)} \tag{2.40}$$

For large Mach numbers, the hypersonic approximation for Eq. (2.40) becomes

$$\frac{p_2}{p_1} = \left(\frac{M_1}{M_2}\right)^{2\gamma/(\gamma-1)} \tag{2.41}$$

Rearranging Eq. (2.39), we obtain

$$\frac{M_1}{M_2} = 1 - \frac{\gamma-1}{2} M_1 \theta \tag{2.42}$$

Combining Eqs. (2.41) and (2.42), the pressure ratio across the expansion wave, at hypersonic speeds, becomes

$$\frac{p_2}{p_1} = \left(1 - \frac{\gamma-1}{2} M_1 \theta\right)^{2\gamma/(\gamma-1)} \tag{2.43}$$

Defining, as before, $M_1\theta$ as the hypersonic similarity parameter K, Eq. (2.43) can be written as

$$\frac{p_2}{p_1} = \left(1 - \frac{\gamma - 1}{2}K\right)^{2\gamma/(\gamma-1)} \tag{2.44}$$

Equation (2.44) is, for the expansion wave, the analog of Eq. (2.28) for the shock wave. In both cases, the pressure ratio p_2/p_1 is a function of K and γ. However, whereas Eq. (2.28) for the shock wave assumed both high Mach number and small angles, Eq. (2.44) for the expansion wave assumes only high Mach number; Eq. (2.44) is not restricted to small angles.

Finally, the pressure coefficient C_p is, from Eqs. (2.13) and (2.44),

$$C_p = \frac{2}{\gamma M_1^2}\left(\frac{p_2}{p_1} - 1\right) = \frac{2}{\gamma M_1^2}\left[\left(1 - \frac{\gamma - 1}{2}K\right)^{2\gamma/(\gamma-1)} - 1\right]$$

Multiplying and dividing the right-hand side by θ^2, we obtain

$$C_p = \frac{2\theta^2}{\gamma K^2}\left[\left(1 - \frac{\gamma - 1}{2}K\right)^{2\gamma/(\gamma-1)} - 1\right] \tag{2.45}$$

Equation (2.45) for the hypersonic expansion wave is analogous to Eq. (2.29) for the hypersonic shock wave. Indeed, analogous to Eq. (2.30), Eq. (2.45) gives the result, now becoming familiar, that

$$\frac{C_p}{\theta^2} = f(K, \gamma) \tag{2.46}$$

for the hypersonic expansion wave.

2.5 Summary and Comments

The conventional shock-wave and expansion-wave relations from basic compressible flow take on simplified but approximate forms at hypersonic Mach numbers. The more important of these forms are listed next.

2.5.1 Shock Waves

In the limit as $M_1 \to \infty$,

$$\frac{p_2}{p_1} = \frac{2\gamma}{\gamma + 1}M_1^2 \sin^2\beta \tag{2.2}$$

$$\frac{\rho_2}{\rho_1} = \frac{\gamma + 1}{\gamma - 1} \tag{2.4}$$

$$\frac{u_2}{V_1} = 1 - \frac{2\,\sin^2\beta}{\gamma + 1} \tag{2.7}$$

$$\frac{v_2}{V_1} = \frac{\sin 2\beta}{\gamma + 1} \tag{2.10}$$

$$C_p = \left(\frac{4}{\gamma + 1}\right)\sin^2\beta \tag{2.15}$$

In the combined limit, as $M_1 \to \infty$ and small angles,

$$\frac{\beta}{\theta} = \frac{\gamma + 1}{2} \tag{2.19}$$

Defining the hypersonic similarity parameter as $M_1\theta \equiv K$, we have, in the intermediate case of high but finite Mach number and small angles,

$$\frac{p_2}{p_1} = 1 + \frac{\gamma(\gamma + 1)}{4}K^2 + \gamma K^2\sqrt{\left(\frac{\gamma + 1}{4}\right)^2 + \frac{1}{K^2}} \tag{2.28}$$

$$C_p = 2\theta^2\left[\frac{\gamma + 1}{4} + \sqrt{\left(\frac{\gamma + 1}{4}\right)^2 + \frac{1}{K^2}}\right] \tag{2.29}$$

Note:

$$\frac{C_p}{\theta^2} = f(K, \gamma) \tag{2.30}$$

2.5.2 Expansion Waves

In the case of high but finite Mach numbers, we have

$$\theta = \frac{2}{\gamma - 1}\left(\frac{1}{M_1} - \frac{1}{M_2}\right) \tag{2.39}$$

where θ is the deflection angle and M_1 and M_2 are the Mach numbers upstream and downstream of the expansion wave. Also, for the same assumption

$$\frac{p_2}{p_1} = \left(1 - \frac{\gamma - 1}{2}K\right)^{2\gamma/(\gamma - 1)} \tag{2.44}$$

where $K = M_1\theta$

$$C_p = \frac{2\theta^2}{\gamma K^2}\left[\left(1 - \frac{\gamma - 1}{2}K\right)^{2\gamma/(\gamma - 1)} - 1\right] \tag{2.45}$$

Note:

$$\frac{C_p}{\theta^2} = f(K, \gamma) \tag{2.46}$$

Problem

2.1 Starting with the basic continuity, momentum, and energy equations for flow across an oblique shock wave (for example, see [4]), derive Eqs. (2.6) and (2.8). Note that u_2 and v_2 in these equations are the velocity components behind the shock parallel and perpendicular to the *upstream* velocity respectively—not parallel and perpendicular to the shock wave as is usually taken in most standard shock-wave derivations.

3
Local Surface Inclination Methods

Newton's ideas are as old as reason and as new as research.

J. C. Hunsaker, comments to the Royal Society,
Cambridge, England, at the occasion of
the Newton Tercentenary Celebration, July 1946

A striking difference between linear and nonlinear waves concerns the phenomenon of interaction: the principle of superposition holds for linear waves but not for nonlinear waves. As a consequence, for example, excess pressures of interfering sound waves are merely additive: in contrast to this fact, interaction and reflection of nonlinear waves may lead to enormous increases in pressure.

Richard Courant and K. O. Fredericks, 1948

Chapter Preview

The calculation of surface-pressure distributions over hypersonic bodies—that is the exclusive name of the game for this chapter. The resulting aerodynamic lift and wave drag are also treated here. This is our first opportunity in this book to calculate such pressure distribution and aerodynamic forces. And we do it in the quickest and easiest way possible by cutting directly to the body surface, calculating the pressure there, but nowhere else in the flowfield. All we need to know to calculate the pressure at a point on the surface of a body in a given hypersonic freestream is the local inclination angle the surface makes at that point with the freestream direction. Sounds very straightforward, does it not? But there is no free lunch here. We pay for this simplicity by sacrificing accuracy. How much? We will see. We will also see that the several methods discussed here are incredibly simple and straightforward.

These simple methods are grouped under the heading of "local surface inclination methods," for obvious reasons. In the early days of hypersonic aerodynamics—in the 1950s—these methods were about the only thing going. This is how aeronautical engineers in those days made estimates of surface pressures for the design of hypersonic vehicles. Today, in the world of modern computational fluid dynamics and very sophisticated calculations of complete flowfields around hypersonic bodies, the material in this chapter takes on the role of "back-of-the-envelope" calculations. This is the modern value of the methods discussed here—for you to carry them around in the pocket of your mind so that you can whip them out at any time to make a quick estimate of hypersonic pressure distributions, lift, and wave drag. These methods can be used as reality checks in the world of hypersonic vehicle design and performance.

This chapter is a great way to get off the ground in learning hypersonic aerodynamics. It is straightforward and to the point. It gives you some immediate tools to work with and allows you to make some fun calculations. So read on and have fun.

3.1 Introduction

Hypersonic flow is inherently *nonlinear*. This is intuitively obvious when we think of the important physical aspects of hypersonics discussed in Chapter 1—aspects such as high-temperature chemically reacting flows, viscous interaction, entropy layers, etc. It is hard to imagine that such complex phenomena could be described by simple linear relationships. Even without these considerations, the basic theory of inviscid compressible flow, when the Mach number becomes large, does not yield aerodynamic theories that are mathematically linear. This is in stark contrast to supersonic flow, which, for thin bodies at small angles of attack, can be described by a linear partial differential equation, leading to the familiar supersonic expression for pressure coefficient on a surface (or streamline) with local deflection angle θ:

$$C_p = \frac{2\theta}{\sqrt{M_\infty^2 - 1}} \tag{3.1}$$

In Eq. (3.1), M_∞ is the freestream Mach number. Equation (3.1) is a classic result from inviscid, linearized, two-dimensional, supersonic flow theory (for example, see [4] and [5]). It is simple and easy to apply. Unfortunately, it is not valid at hypersonic speeds, for reasons to be discussed in Chapter 4.

A virtue of Eq. (3.1) is that for a specified freestream Mach number it gives the pressure coefficient on the surface of a body strictly in terms of the *local deflection angle of the surface* θ. That is, within the framework of supersonic linearized theory, C_p at any point on a body does *not* depend on the details of the flowfield away from that point; thus, it does *not* require a detailed solution of the complete flowfield. In essence, Eq. (3.1) provides a *local surface inclination method* for the prediction of pressure distributions over two-dimensional supersonic bodies (restricted to thin bodies at small angles of attack). Such simplicity is always welcomed by practicing aerodynamicists who have to design flight vehicles. This leads to the question: although hypersonic aerodynamics is nonlinear, and hence Eq. (3.1) does not hold, are there other methods, albeit approximate, that allow the rapid estimate of pressure distributions over hypersonic bodies just in terms of the local surface inclination angle? In other words, is there a viable local surface inclination method for hypersonic applications? The answer is yes; indeed, there are several such methods that apply to hypersonic bodies. The purpose of this chapter is to present these methods.

Finally, examine the road map given in Fig. 1.24. Note that the material discussed in Chapter 2, as well as the present chapter, is itemized on the far left side of the road map. Keep in mind that we are still discussing inviscid hypersonic flow, where essentially we are examining the purely fluid-dynamic effect of large Mach numbers.

Also, a more detailed road map for the present chapter is given in Fig. 3.1. We begin with Newtonian flow, a classic fluid-dynamic theory postulated by Isaac Newton in 1687, which resulted in very poor accuracy for low-speed fluid-dynamic applications over the subsequent centuries. Only with the advent of modern hypersonic aerodynamic applications has Newtonian theory really come into its own. Newtonian theory provides the most straightforward and simplest prediction of surface pressure on a hypersonic body. That is why we start with it here. Moreover, we will explore several aspects of Newtonian flow as itemized in the subheadings in Fig. 3.1. Two other primary methods for the

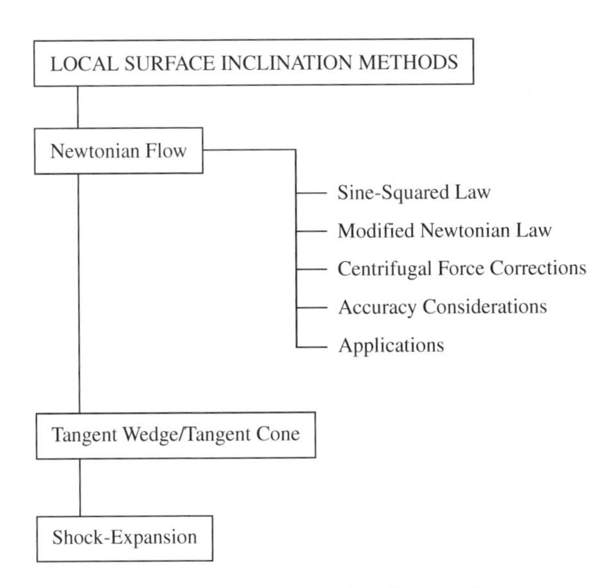

Fig. 3.1 Road map for Chapter 3.

direct calculation of surface pressure on a hypersonic body are also listed in Fig. 3.1, namely, the tangent-wedge/tangent-cone and the shock-expansion methods. These methods are slightly more elaborate than Newtonian, but provide inherently better accuracy. The road map for this chapter, Fig. 3.1, is relatively short and straightforward, but it is worthwhile to keep it in mind as you progress through the chapter.

3.2 Newtonian Flow

Three centuries ago, Isaac Newton established a fluid-dynamic theory that later was used to derive a "law" for the force on an inclined plane in a moving fluid. This law indicated that the force varies as the square of the sine of the deflection angle—the famous Newtonian sine-squared law. Experimental investigations carried out by d'Alembert more than a half-century later indicated that Newton's sine-squared law was not very accurate, and, indeed, the preponderance of fluid-dynamic experience up to the present day confirms this finding. The exception to this is the modern world of hypersonic aerodynamics. Ironically, Newtonian theory, developed 300 years ago for the application to low-speed fluid dynamics, has direct application to the prediction of pressure distributions on hypersonic bodies. What is the application and why? The answers are the subject of this section.

In propositions 34 and 35 of his *Principia*, first published in 1687, Newton modeled a fluid flow as a stream of particles in rectilinear motion, much like a stream of pellets from a shotgun blast, which, when striking a surface, would lose all of their momentum normal to the surface but would move tangentially to the surface without loss of tangential momentum. This picture is illustrated in Fig. 3.2, which shows a stream with velocity V_∞ impacting on a surface of area A inclined at the angle θ to the freestream. From this figure, we see that

$$\text{(Change in normal velocity)} = V_\infty \sin\theta$$

$$\{\text{Mass flux incident on a surface area } A\} = \rho_\infty V_\infty A \sin\theta$$

$$\begin{aligned}\{\text{Time rate of change of}\\ \text{momentum of this mass flux}\} &= (\rho_\infty V_\infty A \sin\theta)(V_\infty \sin\theta)\\ &= \rho_\infty V_\infty^2 A \sin^2\theta\end{aligned}$$

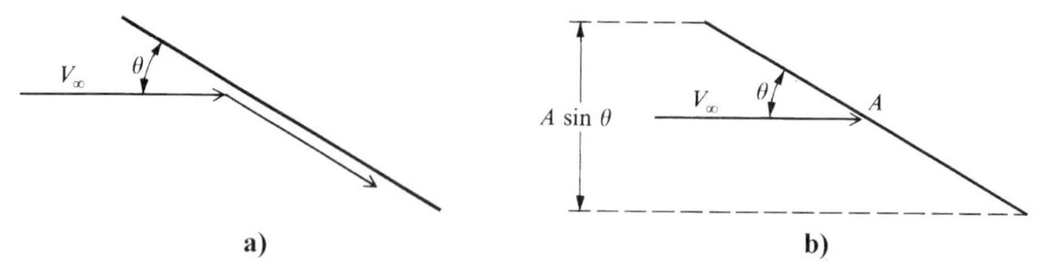

Fig. 3.2 Schematic for Newtonian impact theory.

From Newton's second law, the time rate of change of momentum is equal to the force F exerted on the surface

$$F = \rho_\infty V_\infty^2 A \sin^2 \theta$$

or

$$\frac{F}{A} = \rho_\infty V_\infty^2 \sin^2 \theta \tag{3.2}$$

The force F in Eq. (3.2) requires some interpretation. Newton assumed the stream of particles to be rectilinear, that is, he assumed that the individual particles do not interact with each other, and have no random motion. Because of this lack of random motion, F in Eq. (3.2) is a force associated only with the directed linear motion of the particles. On the other hand, modern science recognizes that the static pressure of a gas or liquid is a result of the purely *random* motion of the particles—motion not included in Newtonian theory. Hence, in Eq. (3.2), F/A, which has the dimensions of pressure, must be interpreted as the pressure *difference* above the freestream static pressure, namely,

$$\frac{F}{A} = p - p_\infty$$

where p is the surface pressure and p_∞ is the freestream static pressure. Hence, from Eq. (3.2)

$$p - p_\infty = \rho_\infty V_\infty^2 \sin^2 \theta$$

or

$$\frac{p - p_\infty}{\frac{1}{2}\rho_\infty V_\infty^2} = 2 \sin^2 \theta$$

or

$$\boxed{C_p = 2 \sin^2 \theta} \tag{3.3}$$

Equation (3.3) is the famous Newtonian sine-squared law for pressure coefficient.

What does the Newtonian pressure coefficient have to do with hypersonic flow? To answer this question, recall Fig. 1.13, which illustrated the shock wave and thin shock layer on a 15-deg wedge at Mach 36. An elaboration of this picture is given in Fig. 3.3, which shows the streamline pattern for the same Mach 36 flow over the same wedge. Here, upstream of the shock wave, we see straight, parallel streamlines in the horizontal freestream direction; downstream of the shock wave, the streamlines are also straight but parallel to the wedge surface inclined at a 15-deg angle. Now imagine that you examine

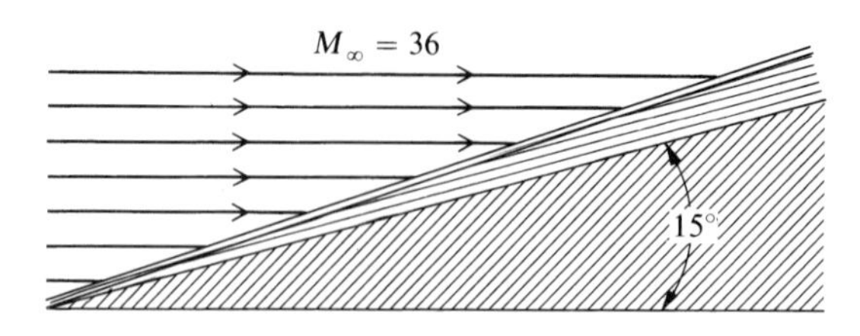

Fig. 3.3 Streamlines in the thin hypersonic shock layer.

Fig. 3.3 from a distance, say, from across the room. Because the shock wave lies so close to the surface at hypersonic speeds, Fig. 3.3 "looks" as if the incoming flow is directly impinging on the wedge surface and then is running parallel to the surface downstream—*precisely the picture Newton drew in 1687*. Therefore, the geometric picture of hypersonic flowfields has some characteristics that closely approximate Newtonian flow; Newton's model had to wait for more than two-and-a-half centuries before it came into own. By this reasoning, Eq. (3.3) should approximate the surface-pressure coefficient in hypersonic flow. Indeed, it has been used extensively for this purpose since the early 1950s.

In applying Eq. (3.3) to hypersonic bodies, θ is taken as the local deflection angle, that is, the angle between the tangent to the surface and the freestream. Clearly, Newtonian theory is a local surface inclination method, where C_p depends only on the local surface deflection angle; it does not depend on any aspect of the surrounding flowfield. To be specific, consider Fig. 3.4a, which shows an arbitrarily shaped two-dimensional body. Assume that we wish to estimate the pressure at point P on the body surface. Draw a line tangent to the body at point P; the angle between this line and the freestream is denoted by θ. Hence, from Newtonian theory the value of C_p at this point is given by $C_p = 2\sin^2\theta$. Now consider a three-dimensional body such as sketched in Fig. 3.4b. We

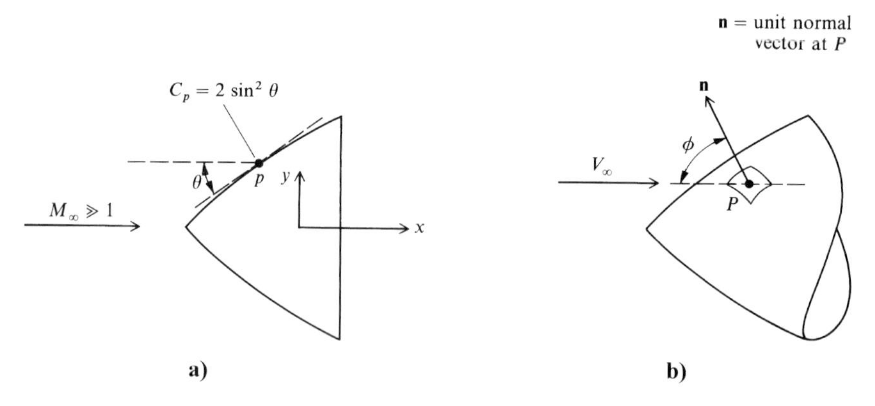

Fig. 3.4 Geometry for Newtonian applications in a) two-dimensional flow and b) three-dimensional flow.

wish to estimate the pressure at an arbitrary point P on this body. Draw a unit normal vector $\boldsymbol{n}$ to the surface at point P. Consider the freestream velocity as a vector $\boldsymbol{V}_\infty$. Then, by definition of the vector dot product, and using a trigonometric identity, we obtain

$$\boldsymbol{V}_\infty \cdot \boldsymbol{n} = |V_\infty| \cos \phi = |V_\infty| \sin\left(\frac{\pi}{2} - \phi\right) \tag{3.4}$$

where ϕ is the angle between $\boldsymbol{n}$ and $\boldsymbol{V}_\infty$. The vectors $\boldsymbol{n}$ and $\boldsymbol{V}_\infty$ define a plane, and in that plane the plane the angle $\theta = \pi/2 - \phi$ is the angle between a tangent to the surface and the freestream direction. Thus, from Eq. (3.4)

$$\boldsymbol{V}_\infty \cdot \boldsymbol{n} = |V_\infty| \sin \theta$$

or

$$\sin \theta = \frac{\boldsymbol{V}_\infty}{|V_\infty|} \cdot \boldsymbol{n} \tag{3.5}$$

The Newtonian pressure coefficient at point P on the three-dimensional body is then $C_p = 2 \sin^2 \theta$, where θ is given by Eq. (3.5).

In the Newtonian model of fluid flow, the particles in the freestream impact only on the frontal area of the body; they cannot curl around the body and impact on the backsurface. Hence, for that portion of a body that is in the "shadow" of the incident flow, such as the shaded region sketched in Fig. 3.5, no impact pressure is felt. Hence, over this shadow region it is consistent to assume that $p = p_\infty$, and therefore $C_p = 0$, as indicated in Fig. 3.5.

It is instructive to examine Newtonian theory applied to a flat plate, as sketched in Fig. 3.6. Here, a two-dimensional flat plat with chord length c is at

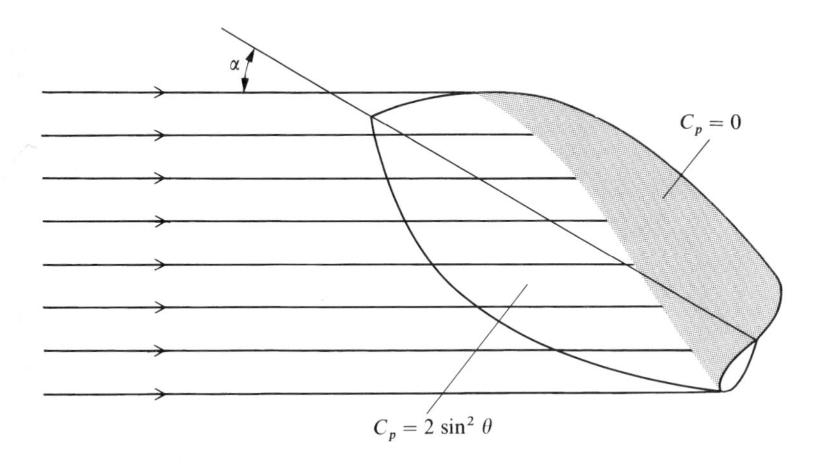

Fig. 3.5 Shadow region on the leeward side of a body, from Newtonian theory.

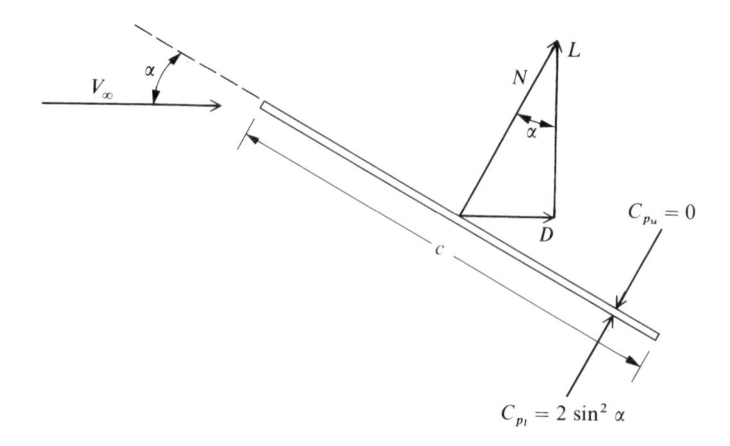

Fig. 3.6 Flat plate at angle of attack. Illustration of aerodynamic forces.

an angle of attack α to the freestream. Because we are not including friction and because surface pressure always acts normal to the surface, the resultant aerodynamic force is perpendicular to the plate, that is, in this case the normal force N is the resultant aerodynamic force. (For an infinitely thin flat plate, this is a general result that is not limited to Newtonian theory or even to hypersonic flow). In turn, N is resolved into lift and drag, denoted by L and D respectively, as shown in Fig. 3.6. According to Newtonian theory, the pressure coefficient on the lower surface is

$$C_{p_l} = 2\sin^2\alpha \tag{3.6}$$

and that on the upper surface, which is in the shadow region, is

$$C_{p_u} = 0 \tag{3.7}$$

Defining the normal-force coefficient as $c_n = N/q_\infty S$, where $S = (c)(1)$, we can readily calculate c_n by integrating the pressure coefficients over the lower and upper surfaces (for example, see the derivation given in [5]).

$$c_n = \frac{1}{c}\int_0^c (C_{p_l} - C_{p_u})\,\mathrm{d}x \tag{3.8}$$

where x is the distance along the chord from the leading edge. Substituting Eqs. (3.6) and (3.7) into (3.8), we obtain

$$c_n = \frac{1}{c}(2\sin^2\alpha)c$$

or

$$c_n = 2 \sin^2 \alpha \tag{3.9}$$

From the geometry of Fig. 3.6, we see that the lift and drag coefficients, defined as $c_l = L/q_\infty S$ and $c_d = D/q_\infty S$, respectively, where $S = (c)(1)$, are given by

$$c_l = c_n \cos \alpha \tag{3.10}$$

and

$$c_d = c_n \sin \alpha \tag{3.11}$$

Substituting Eq. (3.9) into Eqs. (3.10) and (3.11), we obtain

$$c_l = 2 \sin^2 \alpha \cos \alpha \tag{3.12}$$

and

$$c_d = 2 \sin^3 \alpha \tag{3.13}$$

Finally, from the geometry of Fig. 3.6, the lift-to-drag ratio is given by

$$\frac{L}{D} = \cot \alpha \tag{3.14}$$

[Note that Eq. (3.14) is a general result for inviscid flow over a flat plate. For such flows, the resultant aerodynamic force *is* the normal force N. From the geometry shown in Fig. 3.6, the resultant aerodynamic force makes the angle α with respect to lift, and clearly, from the right triangle between L, D, and N, we have $L/D = \cot \alpha$. Hence, Eq. (3.14) is not limited to just Newtonian theory.]

The results just obtained above for the application of Newtonian theory to an infinitely thin flat plate are plotted in Fig. 3.7. Here L/D, c_l, and c_d are plotted vs angle of attack α. From this figure, note the following aspects:

1) The value of L/D increases monotonically as α is decreased. Indeed, $L/D \to \infty$ as $\alpha \to 0$. However, this is misleading; when skin friction is added to this picture, D becomes finite at $\alpha = 0$, and then $L/D \to 0$ as $\alpha \to 0$.

2) The lift curve peaks at about $\alpha \approx 55$ deg. (To be exact it can be shown from Newtonian theory that maximum c_l occurs at $\alpha = 54.7$ deg; the proof of this is left as a homework problem.) It is interesting to note that $\alpha \approx 55$ deg for maximum lift is fairly realistic; the maximum lift coefficient for many practical hypersonic vehicles occurs at angles of attack in this neighborhood.

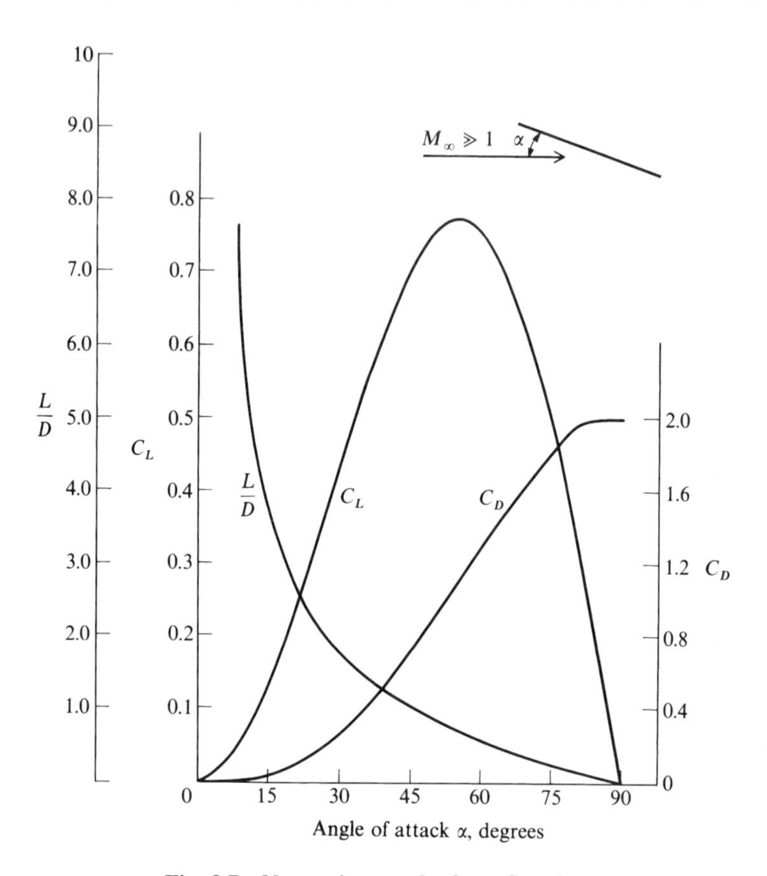

Fig. 3.7 Newtonian results for a flat plate.

3) Examine the lift curve at low angle of attack, say, in the range of α from 0 to 15 deg. Note that the variation of c_l with α is very *nonlinear*. This is in direct contrast to the familiar result for subsonic and supersonic flow, where for thin bodies at small α the lift curve is a linear function of α. (Recall, for example, that the theoretical lift slope from incompressible thin airfoil theory is 2π per radian). Hence, the nonlinear lift curve shown in Fig. 3.7 is a graphic demonstration of the nonlinear nature of hypersonic flow.

Consider two other basic aerodynamic bodies: the circular cylinder of infinite span and the sphere. Newtonian theory can be applied to estimate the hypersonic drag coefficients for these shapes; the results are as follows.

1) Circular cylinder of infinite span:

$$c_d = \frac{D}{q_\infty S}$$

$$S = 2R$$

where R = radius of cylinder,

$$c_d = \tfrac{4}{3} \qquad \text{(from Newtonian theory)}$$

2) Sphere:

$$C_D = \frac{D}{q_\infty S}$$

$$S = \pi R^2$$

where R = radius of sphere,

$$C_D = 1 \qquad \text{(from Newtonian theory)}$$

The derivations of these drag-coefficient values are left for homework problems.

The preceding results from Newtonian theory do not explicitly depend on Mach number. Of course, they implicitly assume that M_∞ is high enough for hypersonic flow to prevail; outside of that, the precise value of M_∞ does not enter the calculations. This is compatible with the *Mach-number independence principle*, to be discussed in Chapter 4. In short, this principle states that certain aerodynamic quantities become relatively independent of Mach number if M_∞ is made sufficiently large. Newtonian results are the epitome of this principle.

3.3 Modified Newtonian Law

Lester Lees [8] proposed a modification to Newtonian theory, writing Eq. (3.3) as

$$\boxed{C_p = C_{p_{\max}} \sin^2 \theta} \tag{3.15}$$

where $C_{p_{\max}}$ is the maximum value of the pressure coefficient, evaluated at a stagnation point behind a normal shock wave, that is,

$$C_{p_{\max}} = \frac{p_{O_2} - p_\infty}{\frac{1}{2}\rho_\infty V_\infty^2} \tag{3.16}$$

where p_{O_2} is the total pressure behind a normal shock wave at the freestream Mach number. From exact normal shock-wave theory, the Rayleigh pitot tube formula gives (see [5])

$$\frac{p_{O_2}}{p_\infty} = \left[\frac{(\gamma+1)^2 M_\infty^2}{4\gamma M_\infty^2 - 2(\gamma-1)}\right]^{\gamma/(\gamma-1)} \left[\frac{1-\gamma+2\gamma M_\infty^2}{\gamma+1}\right] \tag{3.17}$$

Noting that $\frac{1}{2}\rho_\infty V_\infty^2 = (\gamma/2)p_\infty M_\infty^2$, Eq. (3.16) becomes

$$C_{p_{\max}} = \frac{2}{\gamma M_\infty^2}\left[\frac{p_{O_2}}{p_\infty} - 1\right] \tag{3.18}$$

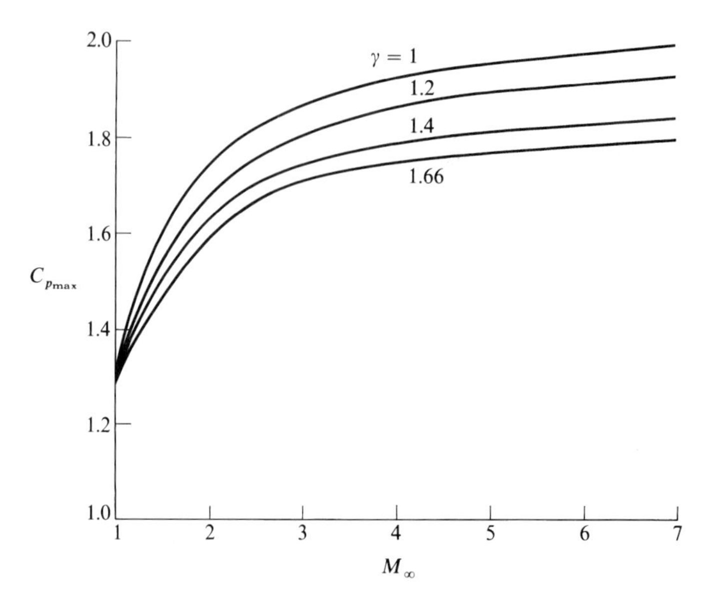

Fig. 3.8 Variation of stagnation-pressure coefficient with M_∞ and γ.

Combining Eqs. (3.17) and (3.18), we obtain

$$C_{p_{max}} = \frac{2}{\gamma M_\infty^2}\left\{\left[\frac{(\gamma+1)^2 M_\infty^2}{4\gamma M_\infty^2 - 2(\gamma-1)}\right]^{\gamma/(\gamma-1)}\left[\frac{1-\gamma+2\gamma M_\infty^2}{\gamma+1}\right]-1\right\} \tag{3.19}$$

This relation is plotted in Fig. 3.8. Note that, in the limit as $M \to \infty$, we have

$$\begin{aligned} C_{p_{max}} &\to \left[\frac{(\gamma+1)^2}{4\gamma}\right]^{\gamma(\gamma-1)}\left[\frac{4}{\gamma+1}\right] \\ &\to 1.839 \qquad \text{for} \quad \gamma = 1.4 \\ &\to 2.0 \qquad \text{for} \quad \gamma = 1 \end{aligned}$$

Equation (3.15), with $C_{p_{max}}$ given by Eq. (3.19), is called the *modified Newtonian law*. Note the following:

1) The modified Newtonian law is no longer Mach-number independent. The effect of a finite Mach number enters through Eq. (3.19).

2) As *both* $M_\infty \to \infty$ and $\gamma \to 1$, Eqs. (3.15) and (3.19) yield $C_p = 2\sin^2\theta$. That is, the straight Newtonian law is recovered in the limit as $M_\infty \to \infty$ and $\gamma \to 1$.

For the prediction of pressure distributions over blunt-nosed bodies, modified Newtonian, Eq. (3.15), is considerably more accurate than the straight Newtonian, Eq. (3.3). This is illustrated in Fig. 3.9, which shows the pressure distribution over a paraboloid at Mach 8. The solid line is an exact finite-difference

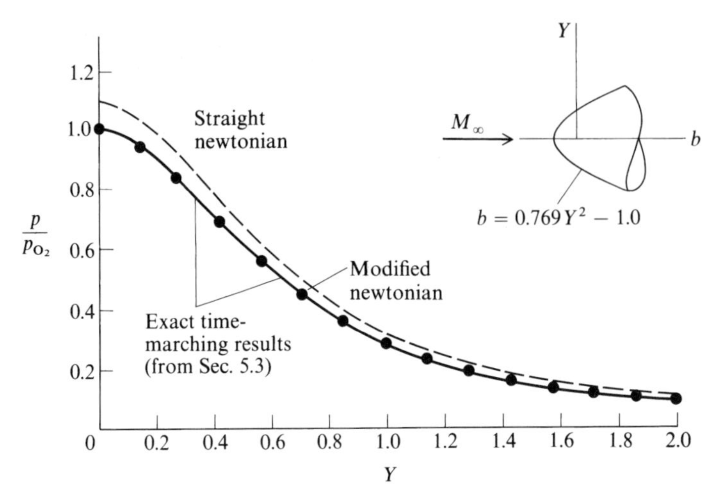

Fig. 3.9 Surface-pressure distribution over a paraboloid at $M_\infty = 8.0$; p_{O_2} is the total pressure behind a normal shock wave at $M_\infty = 8.0$.

solution of the blunt-body flowfield (to be discussed in Chapter 5); the solid symbols are the modified Newtonian results from Eqs. (3.15) and (3.19). Note the excellent agreement, particularly over the forward portion of the nose. The dashed line is the straight Newtonian result from Eq. (3.3); it lies 9% above the exact result. The inspiration for Lester Lee's modification to Newtonian theory appears obvious when examining Fig. 3.9. Clearly, from the proper physics of the flow, the pressure at the stagnation point on the body is equal to the stagnation pressure behind a normal shock wave, that is, the p_{O_2} given by Eq. (3.17); this yields the exact pressure coefficient at the stagnation point, given by Eq. (3.19). Therefore, it is rational to simply replace the coefficient 2 in Eq. (3.3) with the value $C_{p_{\max}}$, as shown in Eq. (3.15). This forces Newtonian theory to be exact at the stagnation point, and as can be seen in Fig. 3.9, the variation of C_p away from the stagnation point closely follows a sine-squared behavior.

3.4 Centrifugal Force Corrections to Newtonian Theory

In the derivation of the straight Newtonian law, Eq. (3.3), we considered flow over a flat surface, such as the model sketched in Fig. 3.1. However, we proceeded to apply Eq. (3.3) to curved surfaces, such as in Figs. 3.4, 3.5, and 3.9. Is this theoretically consistent? The answer is no; for flow over a curved surface, there is a centrifugal force acting on the fluid elements, which will affect the pressure on the surface. For an application of Newtonian theory to curved surfaces that is totally consistent with theoretical mechanics, we must modify the discussion in Sec. 3.2 to take into account the centrifugal force effects. This is the purpose of the present section. Moreover, the results of this section are needed to support the discussion in the next section on the real meaning of Newtonian theory.

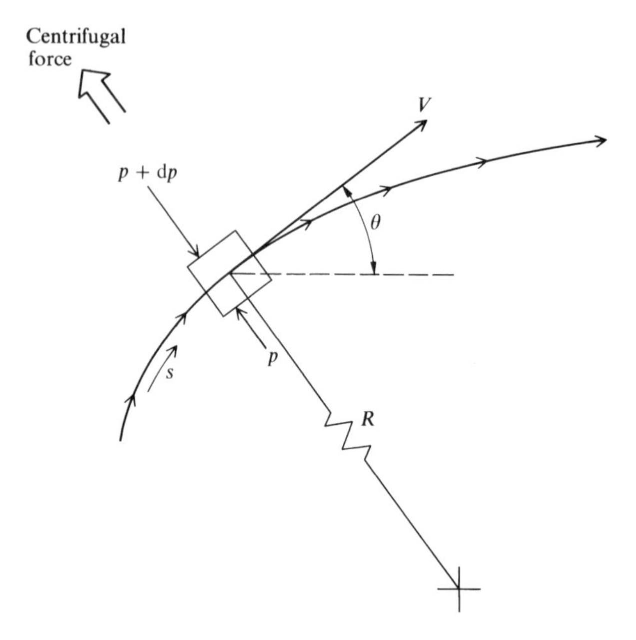

Fig. 3.10 Centrifugal force on a fluid element moving along a curved streamline.

To physically understand the nature of centrifugal force on a flowfield, consider a fluid element moving at velocity V along a curved streamline with radius of curvature R, as sketched in Fig. 3.10. The fluid element is experiencing a radial acceleration V^2/R with an attendant centrifugal force in the radial direction, as also shown in Fig. 3.10. To balance this centrifugal force and keep the fluid element moving along the streamline, the pressure $p + \mathrm{d}p$ on the top surface of the element must be larger than the pressure p on the bottom surface, that is, there must be a positive pressure gradient in the radial direction. One could then theorize that, in the flow over a convex surface, the pressure would decrease in a normal direction toward the surface. This is a general fluid dynamic trend, not just limited to Newtonian theory. However, it is especially true for the mechanics associated with the Newtonian model. For flow over a convex surface, we should expect the Newtonian pressure to be *decreased* as a result of the centrifugal effect. This is derived as follows.

Consider Fig. 3.11, which illustrates the Newtonian flow over a curved surface. Consistent with the Newtonian model, all particles that impact the surface subsequently move tangentially over the surface in an infinitely thin layer. For the time being, assume this layer to have small thickness Δn; later we will let $\Delta n \rightarrow 0$ consistent with the Newtonian approximation. Therefore, in Fig. 3.11 we are considering a thin layer of flow over the body, bounded by the dashed line and the body itself. (For clarity of presentation, the thickness of this layer is greatly magnified in Fig. 3.11.) Consider point i on the body surface. At point i we wish to calculate the pressure p_i. Through point i, consider

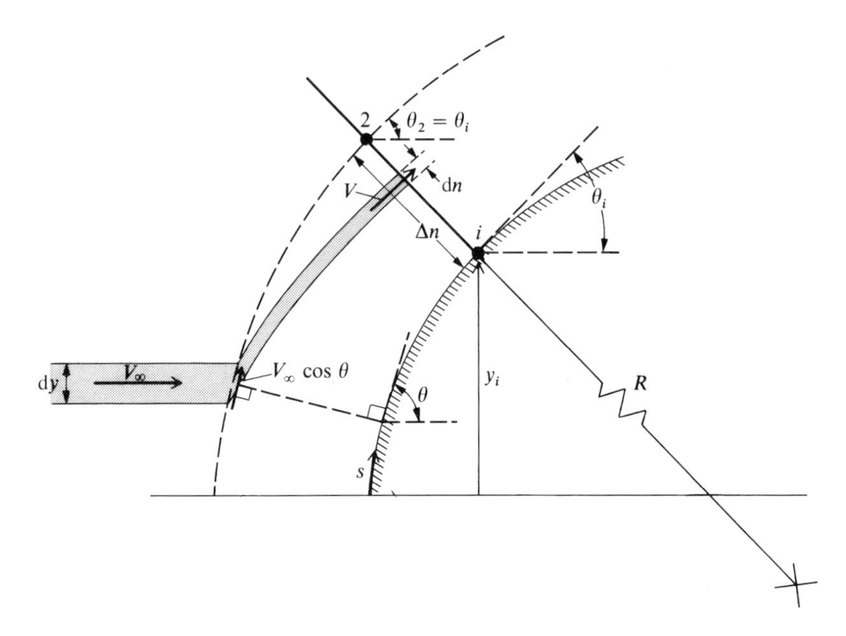

Fig. 3.11 Shock-layer model for centrifugal force corrections to Newtonian theory.

a streamline coordinate system, where s and n are coordinates locally tangential and perpendicular to the streamline. The radius of curvature of the streamline is R. The layer of flow over the body is so thin that we assume R is the same for all of the streamlines crossing the coordinate n drawn from point i over the distance Δn. As a result of this assumption, because the surface at point i is at the angle θ_i with respect to the freestream, then the angle at point 2 made by the outer edge of the layer (dashed line) with respect to the freestream is also θ_i. Now consider a streamtube within the layer, as shown by the shaded region in Fig. 3.11. In the freestream ahead of the layer, the height of this streamtube is dy, where y is the coordinate perpendicular to the freestream, and the velocity is V_∞. Immediately upon entering the layer, the flow direction is assumed to be θ, the local deflection angle of the body at that location, and the magnitude of the velocity is $V_\infty \cos\theta$—all consistent with the Newtonian model. Where the streamtube crosses the normal coordinate n drawn through point i, the thickness of the streamtube is dn, and the velocity is V. Concentrate on this part of the streamtube, that is, where it crosses n. At this location, Newton's second law written in streamline coordinates for the motion of a fluid element is, in the normal direction (for example, see [9]),

$$\frac{\partial p}{\partial n} = \frac{\rho V^2}{R} \tag{3.20}$$

Equation (3.20) states that the centrifugal force per unit volume of a fluid element $\rho V^2/R$ is exactly balanced by the normal pressure gradient $\partial p/\partial n$. Integrating

Eq. (3.20) across the layer from point i to point 2, we have

$$\int_{p_i}^{p_2} \mathrm{d}p = \int_0^{\Delta n} \frac{\rho V^2}{R} \mathrm{d}n \tag{3.21}$$

Assuming two-dimensional flow, the constant mass flow through the shaded streamtube dictates that

$$\rho_\infty V_\infty \,\mathrm{d}y = \rho V \,\mathrm{d}n \tag{3.22}$$

Substituting Eq. (3.22) into (3.21), we obtain

$$p_2 - p_i = \int_0^{y_i + \Delta n \cos\theta_i} \frac{\rho_\infty V_\infty}{R} V \,\mathrm{d}y \tag{3.23}$$

where the direction of integration now becomes the vertical coordinate y. Note that the vertical coordinates of points i and 2 are y_i and $y_i + \Delta n \cos\theta_i$, respectively. Recall that $\mathrm{d}y$ in Eqs. (3.22) and (3.23) is the incremental height of the streamtube measured in the freestream and that all of the mass flow through the section of the layer of thickness Δn above point i originates in the total vertical extent of the freestream from the bottom line up to point 2. Hence, in Eq. (3.23) the limits of integration are taken from $y = 0$ to $y = y_i + \Delta n \cos\theta_i$. Making the assumption of an infinitesimally thin layer, we let $\Delta n \to 0$ or, more correctly, $y_i \gg \Delta n \cos\theta_i$. In this limit, Eq. (3.23) becomes

$$p_2 - p_i = \int_0^{y_i} \frac{\rho_\infty V_\infty}{R} V \,\mathrm{d}y \tag{3.24}$$

We now make another assumption consistent with the Newtonian model. Because Newtonian theory assumes inelastic collisions of the particles with the surface wherein all of the normal momentum is lost but the tangential momentum is preserved, it is consistent to assume that the velocity of any given particle after collision is constant. Hence, in Fig. 3.11, we assume that the flow velocity along the shaded streamtube inside the layer is constant, that is, $V = V_\infty \cos\theta$ along the streamtube, including the section above point i. With this, and recalling that R is assumed constant for all streamlines crossing n above point i, Eq. (3.24) becomes

$$p_2 - p_i = \frac{\rho_\infty V_\infty^2}{R} \int_0^{y_i} \cos\theta \,\mathrm{d}y \tag{3.25}$$

Recall from the definition of radius of curvature that, at point i,

$$R = -\frac{1}{(\mathrm{d}\theta/\mathrm{d}s)_i} = -\frac{1}{(\mathrm{d}\theta/\mathrm{d}y)_i \sin\theta_i} \tag{3.26}$$

Combining Eqs. (3.25) and (3.26) and rearranging, we have

$$p_i = p_2 + \rho_\infty V_\infty^2 \left(\frac{d\theta}{dy}\right)_i \sin\theta_i \int_0^{y_i} \cos\theta \, dy \tag{3.27}$$

Subtracting p_∞ from both sides of Eq. (3.27) and dividing by q_∞, we obtain the pressure coefficient

$$C_{p_i} = C_{p_2} + 2\left(\frac{d\theta}{dy}\right)_i \sin\theta_i \int_0^{y_i} \cos\theta \, dy \tag{3.28}$$

Finally, at point 2 the flow is just entering the layer and is being deflected through the angle θ_i; there is no centrifugal effect at this point, and hence from Newtonian theory the pressure coefficient at point 2 must be interpreted as the straight Newtonian result given by Eq. (3.3), namely, $2\sin^2\theta_i$. With this, Eq. (3.28) is written as

$$\boxed{C_{p_i} = 2\sin^2\theta_i + 2\left(\frac{d\theta}{dy}\right)_i \sin\theta_i \int_0^{y_i} \cos\theta \, dy} \tag{3.29}$$

Equation (3.29) is the Newtonian pressure coefficient at point i on a curved *two-dimensional* surface taking into account the centrifugal force correction. The first term on the right-hand side is the straight Newtonian result; the second term is the theoretically consistent correction for centrifugal effects. An analogous equation for *axisymmetric bodies* is

$$\boxed{C_{p_i} = 2\sin^2\theta_i + 2\left(\frac{d\theta}{dy}\right)_i \frac{\sin\theta_i}{y_i} \int_0^{y_i} y\cos\theta \, dy} \tag{3.30}$$

Equation (3.30) can be written in terms of the local cross-sectional area $A = \pi y_2$.

$$C_{p_i} = 2\sin^2\theta_i + 2\left(\frac{d\theta}{dA}\right)_i \sin\theta_i \int_0^{A_i} \cos\theta \, dA \tag{3.31}$$

The derivations of Eqs. (3.30) and (3.31) are left as homework problems.

The results embodied in Eqs. (3.29–3.31) were first obtained by Adolf Busemann in 1933 [10], with analogous approaches given in [11] and [12]. For this reason, Newtonian theory as modified for centrifugal force effects is frequently called *Newtonian–Busemann theory*.

Note from Eqs. (3.29–3.31) that Newtonian theory with the centrifugal modification is not totally a local surface inclination result. The value of C_{pi} depends not only on the local inclination angle θ_i, but also on the shape of the body upstream of point i through the presence of the integral terms. In some sense, this is compatible with the true physical nature of steady supersonic and hypersonic flows where conditions at a given point are influenced by pressure waves from the upstream region but not from the downstream region. (Recall that information cannot propagate upstream in steady supersonic flow.) However, do not be misled; this aspect of Newtonian–Busemann theory has nothing to do with the true physical picture of the propagation of information via pressure waves—indeed, such propagation is not a part of the Newtonian model. Rather, the integral terms in Eqs. (3.29–3.31) are simply expressions associated with the *mass flow* through the layer immediately above point i in Fig. 3.11. This mass flow depends on the velocity profile along n, $V = V(n)$. In the Newtonian model shown in Fig. 3.11, recall that we assumed that the flow velocity is constant along a streamline inside the layer, and hence the value of V at a given n depends on the location (hence the local value of θ) where the streamline first enters the layer. This is how the dependence of C_{pi} on the shape of the body upstream of point i enters the formulation.

Equations (3.29) and (3.30) take on a particularly simple form for *slender bodies*, where θ is small. For small θ,

$$\sin\theta_i \to \theta_i$$

$$\int_0^{y_i} \cos\theta \, \mathrm{d}y \to y_i$$

Also, letting $\mathrm{d}s$ be an incremental length along the surface, $\mathrm{d}y = \sin\theta \, \mathrm{d}s$, and hence $\sin\theta_i(\mathrm{d}\theta/\mathrm{d}y)_i = (\mathrm{d}\theta/\mathrm{d}s)_i = \kappa_i$, where κ_i is the *curvature* of the surface at point i. Thus, Eqs. (3.29) and (3.30) become (dropping the subscript)

$$\boxed{C_p = 2(\theta^2 + \kappa y)\text{: for slender 2-D bodies}} \tag{3.32a}$$

$$\boxed{C_p = 2\theta^2 + \kappa y\text{: for slender bodies of revolution}} \tag{3.32b}$$

For flow over a blunt body, the centrifugal correction actually makes things worse. For example, Fig. 3.12 shows predictions for the pressure coefficient over a circular cylinder based on all three types of Newtonian-like flow: Newtonian, modified Newtonian, and Newtonian–Busemann. These results are compared with an exact numerical calculation carried out by Van Dyke for $M_\infty = \infty$ (see [14]). Note from Fig. 3.12 that Newtonian theory gives the correct qualitative variation, but is off by a constant percentage, and that modified Newtonian is quite accurate. However, the Newtonian–Busemann results are neither qualitatively nor quantitatively correct. A similar trend occurs for slender-body cases as shown in Fig. 3.13. Here, the pressure distribution over a 10% thick biconvex airfoil is predicted by both Newtonian and Newtonian–Busemann theories and compared with exact numerical results from the method of characteristics. For $\gamma = 1.4$, the Newtonian–Busemann is again worse than straight Newtonian. In Fig. 3.13, the method-of-characteristic results are obtained from [15], and the Newtonian results from [13] and [16].

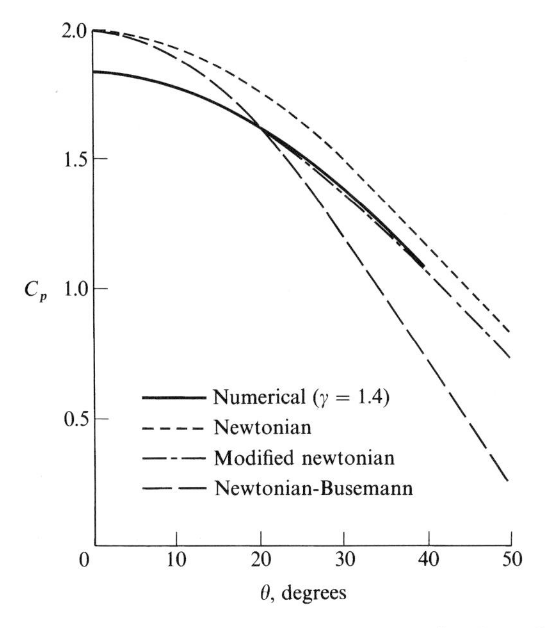

Fig. 3.12 Surface-pressure distributions for flow past a circular cylinder: $M_\infty = \infty$ and $\gamma = 1.4$ (from [13]).

In light of the results shown in Figs. 3.12 and 3.13, we conclude that the centrifugal force correction to Newtonian theory, although correct from the point or view of theoretical mechanics, is simply not valid for practical applications. For this reason, the centrifugal force corrections are rarely, if ever, seen in contemporary applications of Newtonian theory for hypersonic vehicle

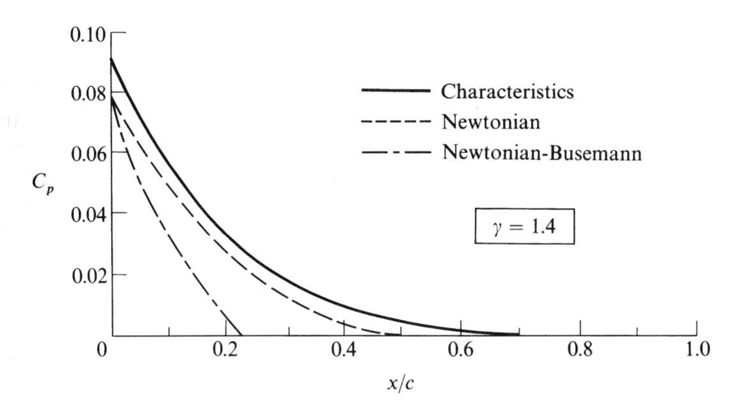

Fig. 3.13 Surface-pressure distribution over a 10% thick biconvex airfoil. Shape of the airfoil is shown in Fig. 3.14: $M_\infty = \infty$, and $\gamma = 1.4$ (from [15]).

design. Therefore, why have we spent an entire section of this book discussing such corrections? Is it only an academic exercise, at best? The answer is—not quite. This is the subject of the next section.

3.5 Newtonian Theory—What It Really Means

In Secs. 3.2–3.4, the theoretical basis of Newtonian theory was developed, including the centrifugal force effects. Given the Newtonian flow model, Eq. (3.3) for a flat surface and Eqs. (3.29) and (3.30) for curved surfaces are precise results, obtained by the rigorous application of theoretical mechanics to the postulated model. On the other hand, when we apply Newtonian theory to practical hypersonic flow problems in air we have seen in Secs. 3.3 and 3.4 that the best agreement with exact results is obtained *without* the centrifugal force corrections, which at first glance appears theoretically inconsistent. Indeed, straight Newtonian theory [Eq. (3.3), or Lee's modification given by Eq. (3.15)] frequently gives very acceptable results for pressure distributions over hypersonic bodies in air, whether or not these bodies have straight or curved surfaces. Therefore, is Newtonian theory just an approximation that fortuitously gives reasonable results for hypersonic flow? Is the frequently obtained good agreement between Newtonian and exact results just a fluke? The answer is *no*—Newtonian theory has true physical significance if, in addition to considering the limit of $M_\infty \to \infty$, we also consider the limit of $\gamma \to 1.0$. Let us examine this in more detail.

Temporarily discard any thoughts of Newtonian theory, and simply recall the exact oblique shock relation for C_p as given by Eq. (2.14), repeated next (with freestream conditions now denoted by a subscript ∞ rather than a subscript 1, as used in Chapter 2):

$$C_p = \frac{4}{\gamma + 1}\left(\sin^2 \beta - \frac{1}{M_\infty^2}\right) \tag{2.14}$$

Equation (2.15) gave the limiting value of C_p as $M_\infty \to \infty$ repeated here.

As $M_\infty \to \infty$:

$$C_p \to \frac{4}{\gamma + 1} \sin^2 \beta \tag{2.15}$$

Now take the additional limit of $\gamma \to 1.0$. From Eq. (2.15), in both limits as $M_\infty \to \infty$ and $\gamma \to 1.0$, we have

$$C_p \to 2 \sin^2 \beta \tag{3.33}$$

Equation (3.33) is a result from exact oblique shock theory; it has nothing to do with Newtonian theory (as yet). Keep in mind that β in Eq. (3.33) is the wave angle, not the deflection angle.

Let us go further. Consider the exact oblique shock relation for ρ_2/ρ_∞, given by Eq. (2.3) repeated here (again with subscript ∞ replacing the subscript 1):

$$\frac{\rho_2}{\rho_\infty} = \frac{(\gamma + 1)M_\infty^2 \sin^2 \beta}{(\gamma - 1)M_\infty^2 \sin^2 \beta + 2} \tag{2.3}$$

Equation (2.4) was obtained as the limit where $M_\infty \to \infty$, namely

As $M_\infty \to \infty$:

$$\frac{\rho_2}{\rho_\infty} \to \frac{\gamma + 1}{\gamma - 1} \tag{2.4}$$

In the additional limit as $\gamma \to 1$, we find that

As $\gamma \to 1$ and $M_\infty \to \infty$:

$$\boxed{\frac{\rho_2}{\rho_\infty} \to \infty} \tag{3.34}$$

That is, the density behind the shock is *infinitely large*. In turn, mass flow consideration then dictate that the *shock wave is coincident with the body surface*. This is further substantiated by Eq. (2.19), which is good for $M_\infty \to \infty$ and small deflection angles

$$\frac{\beta}{\theta} \to \frac{\gamma + 1}{2} \tag{2.19}$$

In the additional limit as $\gamma \to 1$, we have

As $\gamma \to 1$ and $M_\infty \to \infty$ and θ and β small:

$$\boxed{\beta = \theta}$$

That is, the shock wave lies on the body. In light of this result, Eq. (3.33) is written as

$$\boxed{C_p = 2 \sin^2 \theta} \tag{3.35}$$

Examine Eq. (3.35). It is a result from exact oblique shock theory, taken in the combined limit of $M_\infty \to \infty$ and $\gamma \to 1$. *However, it is also precisely the Newtonian result given by Eq. (3.3)*. Therefore, we make the following conclusion. The closer the actual hypersonic flow problem is to the limits $M_\infty \to \infty$ and $\gamma \to 1$, the closer it should be described physically by Newtonian

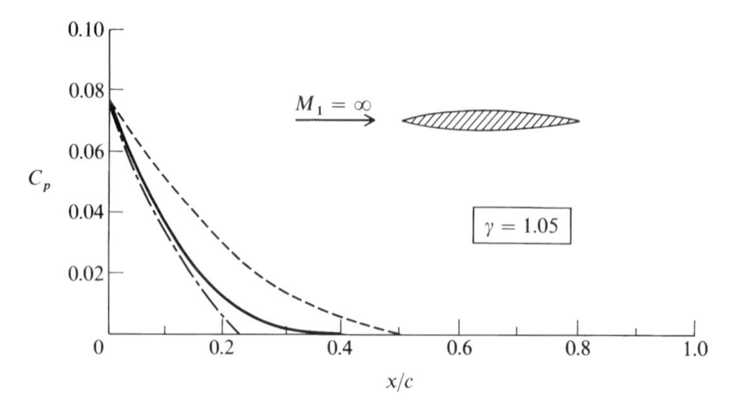

Fig. 3.14 Same as Fig. 3.13, except with $\gamma = 1.05$ (from [15]).

flow. Also in this combined limit, the centrifugal correction becomes physically appropriate, and the Newtonian–Busemann theory gives better results than straight Newtonian. For example, Fig. 3.14 illustrates the pressure coefficient over a 10% thick biconvex airfoil at $M_\infty = \infty$; this is the same type of comparison made in Fig. 3.13. However, Fig. 3.14 is for $\gamma = 1.05$, and clearly the Newtonian–Busemann theory gives much closer agreement with the exact method of characteristics than does the straight Newtonian. This is in direct contrast with the results for $\gamma = 1.4$, shown in Fig. 3.13. Therefore, we conclude that the application of Newtonian theory to hypersonic flow has some direct theoretical substance, becoming more accurate as $\gamma \rightarrow 1$. Furthermore, for hypersonic flows in air with $\gamma = 1.4$, we would not expect the full Newtonian theory (properly including centrifugal effects) to be accurate, and, as we have seen in Figs. 3.12 and 3.13, it is not. On the other hand, for air with $\gamma = 1.4$, agreement between exact results and the straight Newtonian theory (without centrifugal effects) does indeed appear to be rather fortuitous.

We might ask the rather academic question: if in the limit of $M_\infty \rightarrow \infty$ and $\gamma \rightarrow 1$, the shock-layer thickness goes to zero, then how can there be any centrifugal force felt over this zero thickness? The answer is, of course, that in the same limit the density becomes infinite, and although the shock layer approaches zero thickness, the infinite density felt over this zero thickness is an indeterminate form that yields a finite centrifugal force.

As a final note on our discussion of Newtonian theory, consider Fig. 3.15. Here, the pressure coefficients for a 15-deg half-angle wedge and a 15-deg half-angle cone are plotted vs freestream Mach number for $\gamma = 1.4$. The exact wedge results are obtained from oblique shock theory, and the exact cone results are obtained from the solution of the classical Taylor–Maccoll equation (for example, see [4]) as tabulated in [17] and [18]. Both sets of results are compared with Newtonian theory, $C_p = 2 \sin^2 \theta$, shown as the dashed line in Fig. 3.14. This comparison demonstrates two general aspects of Newtonian results:

1) The accuracy of Newtonian results results improves as M_∞ increases. This is to be expected from our preceding discussion. Note from Fig. 3.15 that below

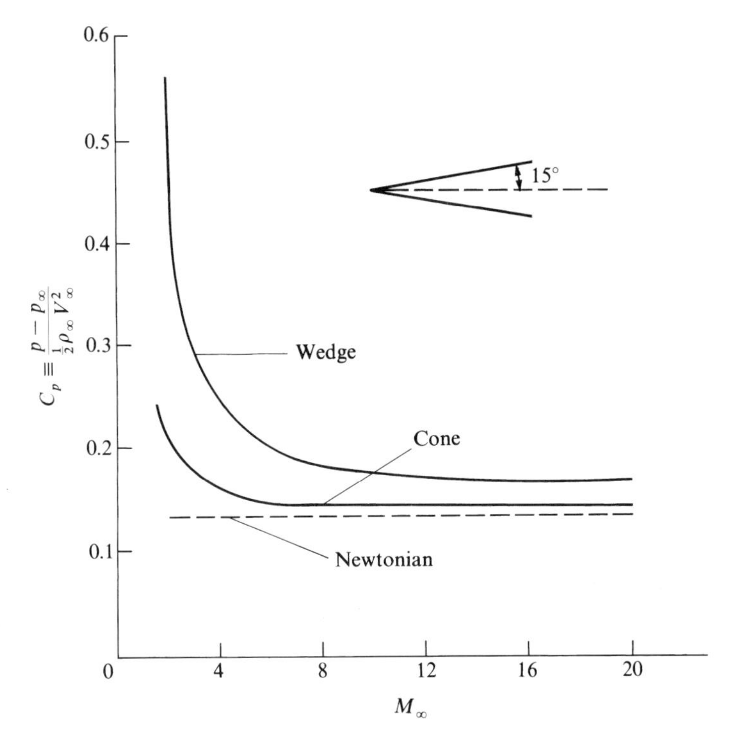

Fig. 3.15 Comparison between Newtonian and exact results for the pressure coefficient on a sharp wedge and a sharp cone.

$M_\infty = 5$ the Newtonian results are not even close, but the comparison becomes much closer as M_∞ increases above 5.

2) Newtonian theory is usually more accurate for three-dimensional bodies (e.g., the cone) than for two-dimensional bodies (e.g., the wedge). This is clearly evident in Fig. 3.15, where the Newtonian result is much closer to the cone results than to the wedge results.

These two trends are general conclusions that seem to apply to Newtonian results for hypersonic bodies in air. Furthermore, we are tempted to say that Newtonian results for blunt bodies should use the modified Newtonian formula [Eq. (3.15)], and that such results usually produce acceptable accuracy, as illustrated in Figs. 3.9 and 3.12. In contrast, we suggest that Newtonian results for slender bodies should use the straight Newtonian law [Eq. (3.3)], and we observe that its accuracy might not be totally acceptable in some cases. For example, for Fig. 3.15, at $M_\infty = 20$, the percentage error in using Newtonian results is 19 and 5% for the wedge and cone, respectively—not as accurate as might be required for some applications. If the modified Newtonian formula [Eq. (3.15)] had been used in Fig. 3.15, the errors would be even larger because $C_{p\max} < 2$. Therefore, we conclude that although Newtonian theory is very useful because of its simplicity, in some applications its accuracy leaves something to be desired.

As a parenthetical comment, Fig. 3.15 illustrates another trend that is characteristic of hypersonic flow. Note that, at low M_∞ the exact values of C_p for both the wedge and cone decrease rapidly with increasing Mach number. However, at higher values of M_∞ the pressure coefficient for each shape tends to seek a plateau, approaching a value that becomes rather independent of M_∞ at high Mach number. This is an example of the Mach-number independence principle, to be discussed in Chapter 4. There we will see that a number of properties in hypersonic flow, including C_p, lift coefficient, wave-drag coefficient, and moment coefficient become relatively independent of M_∞ at high Mach number.

3.5.1 *Worked Example: Comparison of Newtonian with Exact Theory*

The purpose of this worked example is to provide even better insight into the advantages and disadvantages of Newtonian theory, especially as applied to slender bodies at small angles of attack. Vehicles designed for efficient hypersonic flight for sustained periods of time in the atmosphere will most likely be slender shapes at small angles of attack. Is Newtonian theory a reasonable method for estimating the pressure distribution on such vehicles? Let us take a look by considering the most slender of vehicles, namely, a thin flat plate, at a moderate angle of attack.

Consider an infinitely thin flat plate at an angle of attack of 15 deg in a Mach 8 flow. Assume inviscid flow. Calculate the pressure coefficients on the top and bottom surface of the plate, the lift and drag coefficients, and the lift-to-drag ratio using a) exact shock-wave and expansion-wave theory and b) Newtonian theory. Compare the results.

Solution: exact theory. Consider the flat plate shown in Fig. 3.16. The flow over the top goes through an expansion wave, and the flow over the bottom goes through an oblique shock wave. First, consider the flow through the expansion wave. Using the terminology in Fig. 2.4, $M_1 = 8$, and from Eq. (2.32), the Prandtl–Meyer function is $\nu_1 = 95.62$ deg. From Eq. (2.31),

$$\nu_2 = \theta + \nu_1 = 15\,\text{deg} + 95.62\,\text{deg} = 110.62\,\text{deg}$$

With $\nu_2 = 110.62$ deg, Eq. (2.32) yields for M_2,

$$M_2 = 14.32$$

The flow through an expansion wave is isentropic; hence, the total pressure is constant through the wave. Let p_o denote the total pressure. Because the total pressure is constant,

$$p_{o_2} = p_{o_1}$$

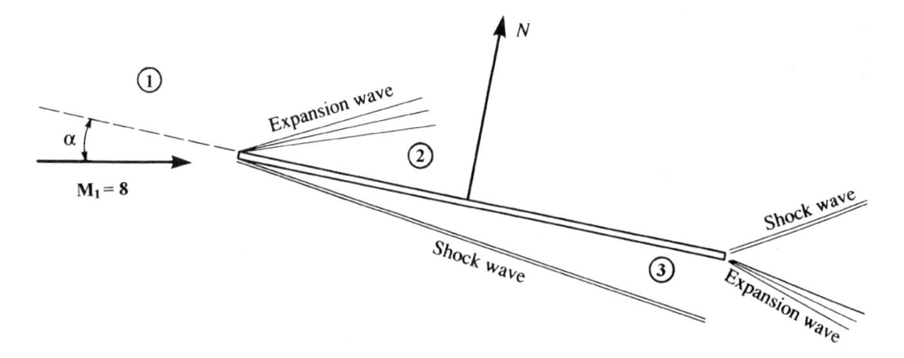

Fig. 3.16 Wave system on a flat plate in hypersonic flow.

and we can write the static-pressure ratio across the expansion wave as

$$\frac{p_2}{p_1} = \frac{p_{o_1}/p_1}{p_{o_2}/p_2}$$

For a calorically perfect gas, the ratio of total to static pressure at a point in the flow is a function of the local Mach number at that point (for example, see [4] and [5]), given by

$$\frac{p_o}{p} = \left[1 + \frac{\gamma - 1}{2} M^2\right]^{\frac{\gamma}{\gamma-1}}$$

Hence,

$$\frac{p_2}{p_1} = \frac{p_{o_1}/p_1}{p_{o_2}/p_2} = \left[\frac{1 + (\gamma - 1/2)M_1^2}{1 + (\gamma - 1/2)M_2^2}\right]^{\frac{\gamma}{\gamma-1}}$$

With $M_1 = 8$ and $M_2 = 14.32$, this gives

$$\frac{p_2}{p_1} = 0.0203$$

The pressure coefficient is given by Eq. (2.13)

$$C_{p_2} = \frac{2}{\gamma M_1^2}\left(\frac{p_2}{p_1} - 1\right)$$

Hence,

$$C_{p_2} = \frac{2}{(1.4)(8)^2}(0.0203 - 1) = \boxed{-0.0219}$$

To obtain the pressure coefficient on the bottom surface from oblique shock theory, Eq. (2.16) with $\theta = 15$ deg and $M_1 = 8$ yields a shock-wave angle of $\beta = 21$ deg. From Eq. (2.1) with $\beta = 21$ deg and $M_1 = 8$, and denoting the pressure behind the shock as p_3 consistent with Fig. 3.16, we have

$$\frac{p_3}{p_1} = 9.443$$

Hence, the pressure coefficient on the bottom surface is

$$C_{p_3} = \frac{2}{\gamma M_1^2}\left(\frac{p_3}{p_1} - 1\right)$$

$$C_{p_3} = \frac{2}{(1.4)(8)^2}(9.443 - 1) = \boxed{0.1885}$$

From Eq. (3.8), the normal-force coefficient is simply

$$c_n = C_{p_3} - C_{p_2} = 0.1885 - (-0.0219) = 0.2104$$

From Eq. (3.10)

$$c_\ell = c_n \cos\alpha = 0.2104 \cos 15\,\text{deg} = \boxed{0.2032}$$

and from Eq. (3.11),

$$c_d = c_n \sin\alpha = 0.2104 \cos 15\,\text{deg} = \boxed{0.0545}$$

Finally,

$$\frac{L}{D} = \frac{c_\ell}{c_d} = \frac{0.2032}{0.0545} = \boxed{3.73}$$

Note: For more details on how to make these exact shock-wave and expansion-wave calculations, for example, see [4] and [5]. As a case in point, in the preceding calculations, Eq. (2.16) was used to find the shock-wave angle, $\beta = 21$ deg. To do this arithmetically, Eq. (2.16) would have to be solved by trial and error. In reality, the value $\beta = 21$ deg was obtained from a graphical plot of Eq. (2.16) called the θ-β-M curves, as explained in [4] and [5].

Solution: Newtonian theory. From Eq. (3.7) for the upper surface of the plate,

$$\boxed{C_{p_2} = 0}$$

From Eq. (3.6) for the lower surface of the plate,

$$C_{p_3} = 2\sin^2\alpha = 2\sin^2 15\,\text{deg} = \boxed{0.134}$$

From Eq. (3.12)

$$c_\ell = 2\sin^2\alpha\cos\alpha = 2(\sin^2 15\,\text{deg})(\cos 15\,\text{deg}) = \boxed{0.1294}$$

and from Eq. (3.13),

$$c_d = 2\sin^3\alpha = 2\sin^3 15\,\text{deg} = \boxed{0.03468}$$

Finally,

$$\frac{L}{D} = \frac{c_\ell}{c_d} = \frac{0.1294}{0.03468} = \boxed{3.73}$$

3.5.2 Commentary

The preceding example provides some important comparisons between exact theory and Newtonian theory. First, compare the work effort to obtain the answers. Going through the exact calculations was much more work than using the simple Newtonian theory. Indeed, the Newtonian calculations were essentially "one-liners" to get the answers—nothing could be more simple. This comparison highlights the value of Newtonian theory—absolute simplicity. But this simplicity is obtained at a cost, namely, a loss of accuracy. For example, for the pressure coefficient on the bottom surface of the plate, we have the following.

Exact:

$$C_{p_3} = 0.1885$$

Newtonian:

$$C_{p_3} = 0.134$$

Newtonian theory *underpredicts* the pressure coefficient on the lower surface by 29%. For the top surface of the plate, we have the following.

Exact:

$$C_{p_2} = -0.0219$$

Newtonian:

$$C_{p_2} = 0$$

Newtonian theory *overpredicts* the pressure coefficient on the upper surface by a 100% of the exact value. For the lift coefficient, we have the following.

Exact:

$$c_\ell = 0.2032$$

Newtonian:

$$c_\ell = 0.1294$$

For the drag coefficient, we have the following.

Exact:

$$c_d = 0.0545$$

Newtonian:

$$c_d = 0.03468$$

The Newtonian results underpredict both c_ℓ and c_d by 36%. However, the values of L/D ratio compare exactly.

Exact:

$$\frac{L}{D} = 3.73$$

Newtonian:

$$\frac{L}{D} = 3.73$$

This is no surprise, for two reasons. First, the Newtonian values of c_ℓ and c_d are both underpredicted by the same percentage; hence, their ratio is not affected. Second, the value of L/D for inviscid flow over a flat plate, no matter what theory is used to obtain the pressures on the top and bottom surfaces, is simply a matter of *geometry* as discussed earlier in conjunction with Eq. (3.14). From this equation,

$$\frac{L}{D} = \cot\alpha = \cot 15\,\text{deg} = 3.73$$

A conclusion from the preceding example as well as from the comparison shown in Fig. 3.15 is that Newtonian theory when applied to hypersonic slender bodies at small to moderate angles of attack does not do a good job of estimating surface pressure or the resulting lift and wave drag. Newtonian theory is much more applicable to hypersonic blunt bodies; the Newtonian pressure distribution around the nose of a blunt body is reasonably accurate as illustrated in Fig. 3.9.

3.5.3 Lift-to-Drag Ratio for Hypersonic Bodies—Comment

The magnitude of the lift-to-drag ratio calculated in the preceding example is important to note—it is very low in comparison to those values for a flat plate in subsonic or even supersonic flow. Here is a harbinger of things to come. Hypersonic flight vehicles suffer from characteristically low values of L/D. This is not good news when designing hypersonic vehicles for sustained flight in the atmosphere. The lift-to-drag ratio of a body is a direct measure of its aerodynamic efficiency (for example, see [1]). For example, everything else being equal, the higher is the value of L/D, the larger is the range of the vehicle. The design of hypersonic vehicles with reasonable values of L/D is a challenge and a quest, as will be discussed later in this book.

3.6 Tangent-Wedge Tangent-Cone Methods

Referring again to the road map given in Fig. 1.24, we remind ourselves that we are discussing a class of hypersonic prediction methods based only on a knowledge of the local surface inclination relative to the freestream. The Newtonian theory discussed in Secs. 3.2–3.5 was one such example; the tangent-wedge/tangent-cone methods presented in this section are two others.

Let us consider first the tangent-wedge method, applicable to two-dimensional hypersonic shapes. Consider the two-dimensional body shown as the hatched area in Fig. 3.17. Assume that the nose of the body is pointed and that the

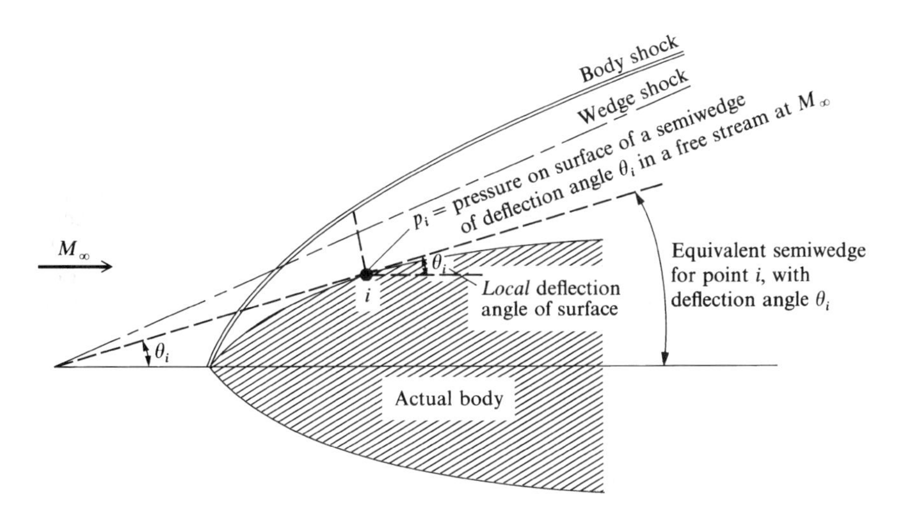

Fig. 3.17 Illustration of the tangent-wedge method.

local surface inclination angle θ at all points along the surface is less than the maximum deflection angle for the freestream Mach number. Consider point i on the surface of the body; we wish to calculate the pressure at point i. The local deflection angle at point i is θ_i. Imagine a line drawn tangent to the body at point i; this line makes an angle θ_i with respect to the freestream and can be imagined as the surface of an equivalent wedge with a half-angle of θ_i, as shown by the dashed line in Fig. 3.17. The tangent-wedge approximation assumes that the pressure at point i is the same as the surface pressure on the equivalent wedge at the freestream Mach number M_∞, that is, p_i is obtained directly from the exact oblique shock relations for a deflection angle of θ_i and a Mach number of M_∞.

The tangent-cone method for application to axisymmetric bodies is analogous to the tangent-wedge method and is illustrated in Fig. 3.18. Consider point i on the body: a line drawn tangent to this point makes the angle θ_i with respect to the freestream. Shown as the dashed line in Fig. 3.18, this tangent line can be imagined as the surface of an equivalent cone, with a semiangle of θ_i. The tangent-cone approximation assumes that the pressure at point i is the same as the surface pressure on the equivalent cone at a Mach number of M_∞, that is, p_i is obtained directly from the cone tables such as [17] and [18].

Both the tangent-wedge and tangent-cone methods are very straightforward. However, they are approximate methods, not based on any theoretical grounds. We cannot "derive" these methods from a model of the flow to which basic mechanical principles are applied, in contrast to the theoretical basis for Newtonian flow. Nevertheless, the tangent-wedge and tangent-cone methods frequently yield reasonable results at hypersonic speeds. Why? We can give an approximate, "hand-waving" explanation, as follows. First, consider a line drawn perpendicular to the body surface at point i, across the shock layer as sketched in Fig. 3.17. Note that the imaginary shock wave from the imaginary equivalent wedge crosses this line below the point where the actual shock wave from the body

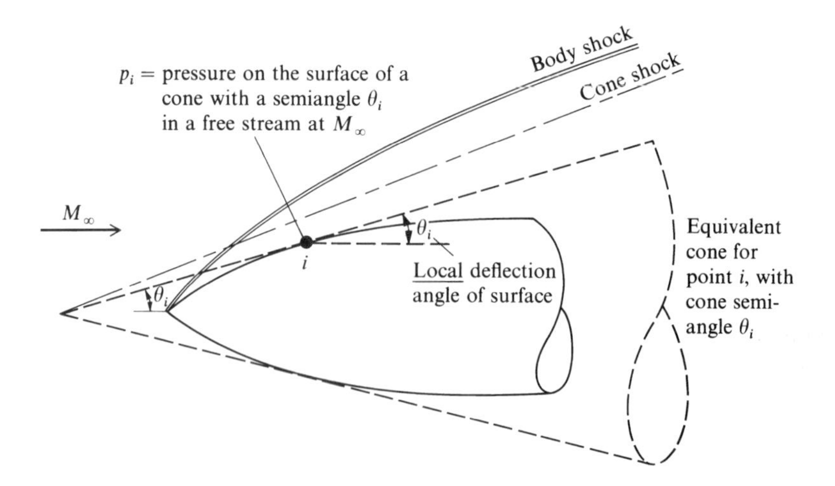

Fig. 3.18 Illustration of the tangent-cone method.

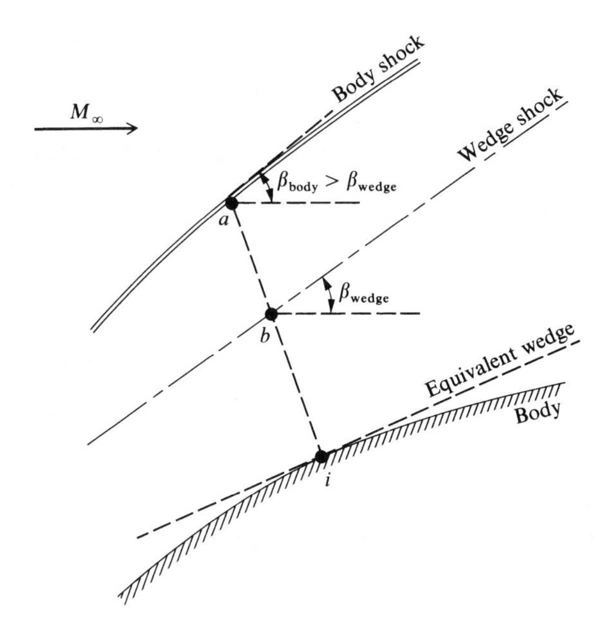

Fig. 3.19 Segment of a hypersonic shock layer, for use in partial justification of the tangent-wedge method.

crosses the line. The region around this line is isolated and magnified in Fig. 3.19. Now consider the following fact.

> *Fact*: In the hypersonic flow across an oblique shock wave on a slender body, the y component of the flow velocity v is changed much more strongly than the x component u.

This fact, which we will revisit several times in the following chapters, is proved by a combination of Eqs. (2.7), (2.10), and (2.19), which yields in the limit of $M_\infty \to \infty$ (referring to the shock geometry shown in Fig. 2.2)

$$\frac{\Delta u}{V_\infty} = \frac{V_\infty - u_2}{V_\infty} \to \frac{\gamma + 1}{2}\theta^2 \tag{3.36}$$

$$\frac{\Delta v}{V_\infty} = \frac{v_2}{V_\infty} \to \theta \tag{3.37}$$

In Eq. (3.36), Δu is the change in the x component of velocity across the oblique shock, and in Eq. (3.37) Δv is the change in the y component of velocity. Clearly, the change of the v velocity is considerably *smaller* (order of θ^2) than the change of the u velocity (order of θ). (Keep in mind that θ is a small angle in radians.) In turn, recalling Euler's equation $dp = -\rho V \mathrm{d}V$, this implies that the major pressure gradients are *normal to the flow*. Referring to Fig. 3.19, the principal change in pressure is therefore along the normal line *iab*; by comparison, changes in the flow direction are second order. Hence examining Fig. 3.19, the surface pressure on the body at point i is *dominated* by the *pressure behind the shock at point a*. Because of the centrifugal force effects, the pressure at point i, p_i, will be *less* than p_a. Now, in the tangent-wedge method $p_i = p_b$, where p_b is the pressure behind the imaginary wedge shock (at point b in Fig. 3.19). The pressure p_b is already less than p_a because the imaginary wedge shock angle at point b is less than the actual body-shock angle at point a ($\beta_{\text{body}} > \beta_{\text{wedge}}$). Thus we see that the wedge pressure p_b is a reasonable approximation for the surface pressure p_i because in the real flow picture the higher pressure p_a behind the body shock is mitigated by centrifugal effects as the pressure is impressed from the shock to the body at point i. The same reasoning holds for the tangent-cone method.

Results obtained with the tangent-cone method as applied to a pointed ogive are shown in Fig. 3.20, taken from [19]. Here, the surface-pressure distribution is plotted vs distance along the ogive. Four sets of results are presented, each for a different value of $K = M_\infty\,(d/l)$, where d/l is the slenderness ratio of the ogive. The solid line is an exact result obtained from the rotational method of characteristics, and the dashed line is the tangent-cone result. Very reasonable agreement is obtained, thus illustrating the usefulness of the tangent-cone method, albeit its rather tenuous foundations. The same type of agreement is typical of the tangent-wedge method. In Fig. 3.20, the parameter $K = M_\infty$ (d/l) is called the hypersonic similarity parameter. Its appearance in Fig. 3.20 is simply a precursor to our discussion of hypersonic similarity in Chapter 4.

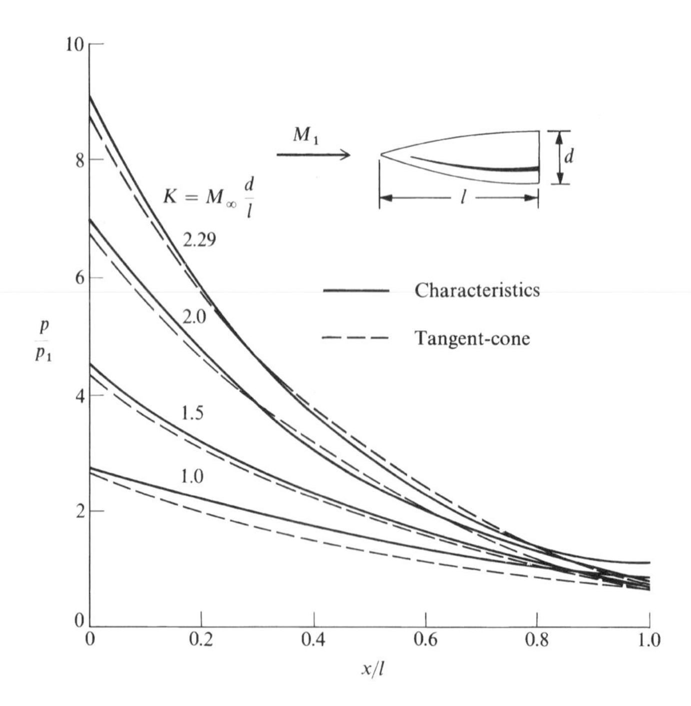

Fig. 3.20 Surface-pressure distributions for ogives of different slenderness ratio d/l (from [19]).

3.7 Shock-Expansion Method

Of the local surface inclination methods discussed so far, the Newtonian method can be applied to a body surface of any inclination angle, whereas the tangent-wedge/tangent-cone methods require a local surface angle less than the shock detachment angle for the given freestream Mach number. This is why Newtonian theory can be applied to blunt-nosed bodies, but the tangent-wedge/tangent-cone methods are limited to sharp-nosed bodies with attached shock waves. The method discussed in the present section—the shock-expansion method—is in the latter category. It assumes a sharp-nosed body with an attached shock wave. However, it has more theoretical justification than the tangent-wedge/tangent-cone methods, as described next.

Consider the hypersonic flow over a sharp-nosed two-dimensional body with an attached shock wave at the nose as sketched in Fig. 3.21. The deflection angle at the nose is θ_n. The essence of the shock-expansion theory is as follows:

1) Assume the nose is a wedge with semiangle θ_n. Calculate M_n and p_n behind the oblique shock at the nose by means of exact oblique-shock theory.

2) Assume a local Prandtl–Meyer expansion along the surface downstream of the nose. We wish to calculate the pressure at point i, p_i. To do this, we must first obtain the local Mach number at point i, M_i. This is obtained from the Prandtl–Meyer function, assuming an expansion through the deflection angle $\Delta\theta = \theta_n - \theta_i$.

$$\Delta\theta = \sqrt{\frac{\gamma+1}{\gamma-1}}\tan^{-1}\sqrt{\frac{\gamma-1}{\gamma+1}(M_n^2-1)} - \tan^{-1}\sqrt{\frac{\gamma-1}{\gamma+1}(M_i^2-1)} - \left[\tan^{-1}\sqrt{M_n^2-1} - \tan^{-1}\sqrt{M_l^2-1}\right] \tag{3.38}$$

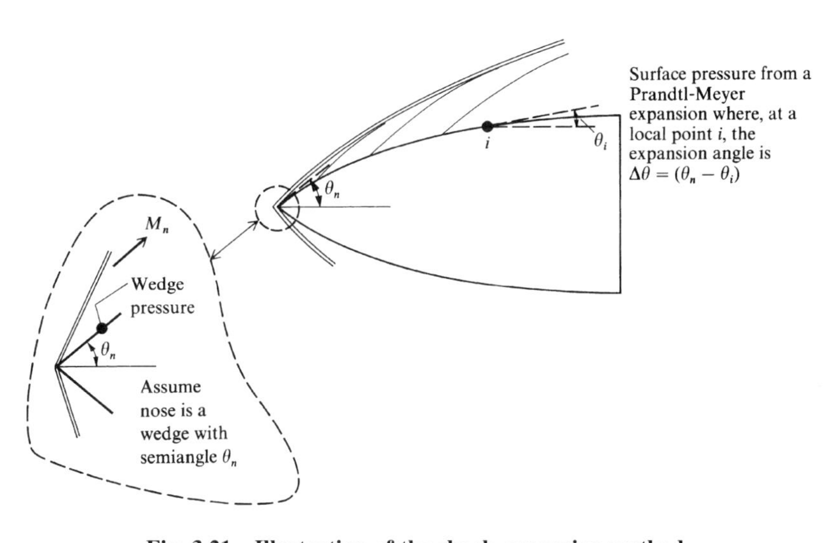

Fig. 3.21 Illustration of the shock-expansion method.

In Eq. (3.38), M_i is the only unknown; M_n is known from the preceding step 1, and $\Delta\theta = \theta_n - \theta_i$ is a known geometric quantity. Of course, for air with $\gamma = 1.4$ tables for the Prandtl–Meyer function abound (for example, see [4]), and in such a case the tables would be used to calculate M_i rather than attempting to solve Eq. (3.38) implicitly for M_i.

3) Calculate p_i from the isentropic flow relation

$$\frac{p_i}{p_n} = \left[\frac{1 + (\gamma - 1)/2M_n^2}{1 + (\gamma - 1)/2M_i^2}\right]^{\gamma/(\gamma-1)} \tag{3.39}$$

(Again, for air with $\gamma = 1.4$, the isentropic flow tables, such as found in [4], can be used to obtain p_i in a more convenient manner.)

Results from the shock-expansion method, obtained from [20] for flow over a 10%-thick biconvex airfoil are shown in Fig. 3.22, as compared with the exact method of characteristics. Excellent agreement is obtained. This is to be somewhat expected. After passing through the attached shock wave at the nose, the actual flow does indeed expand around the body, and this expansion process is approximated by the assumption of a local Prandtl–Meyer expansion. Why this is not a precisely exact calculation is discussed two paragraphs below.

The shock-expansion method can also be applied to bodies of revolution. The method is essentially the same as shown in Fig. 3.21, except now θ_n is assumed to be the semiangle of a cone, and M_n and p_n at the nose are obtained from the exact Taylor–Maccoll cone results. Then the Prandtl–Meyer expansion relations are applied locally downstream of the nose. This implies that the flow downstream of the nose is locally two dimensional, which assumes that the divergence of streamlines in planes tangential to the surface. For bodies of revolution at zero degree angle of attack, this condition is usually met. Results for the shock-expansion method applied to ogives at zero angle of attack are shown in

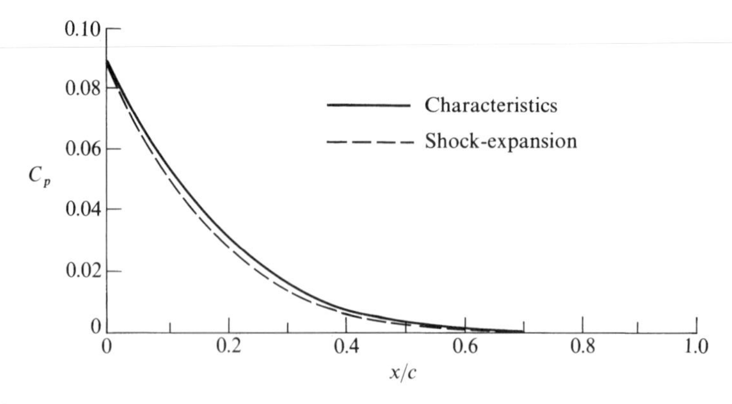

Fig. 3.22 Surface-pressure distribution over the same 10%-thick airfoil as shown in Fig. 3.14; comparison of the shock-expansion method with exact results from the method of characteristics: $M_\infty = \infty$ (from [20]).

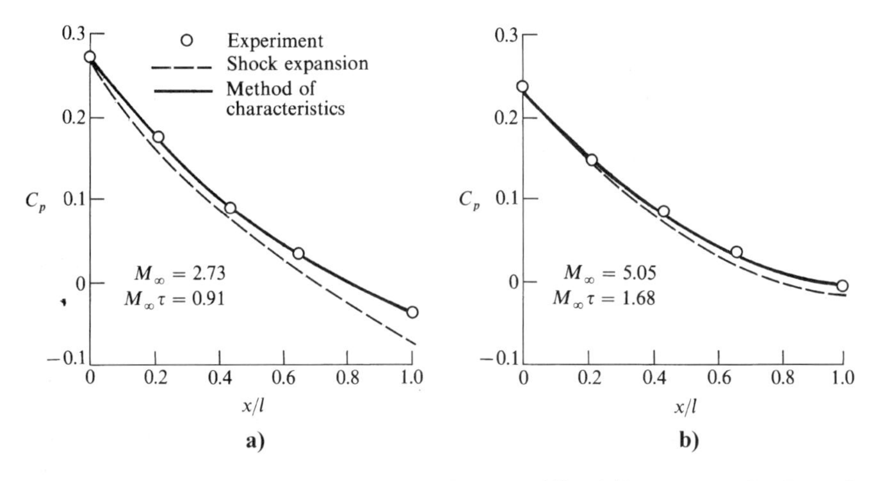

Fig. 3.23 Pressure distribution over an ogive with $d/l = 1/3$ at zero angle of attack with $\gamma = 1.4$ (from [21]): a) supersonic case and b) hypersonic case.

Fig. 3.23, obtained from [21]. The ogive has a slenderness ratio $d/l = 1/3$. In Fig. 3.23a, the results are for a supersonic Mach number $M_\infty = 2.73$, whereas in Fig. 3.23b, the results are for a slightly hypersonic case $M_\infty = 5.05$. The circles are experimental data, the solid line represents an exact result from the method of characteristics, and the dashed line is from the shock-expansion method. Note that, for the supersonic case the shock-expansion method yields poor agreement; however, for the hypersonic case the shock-expansion method is much closer to the hypersonic case, and the shock-expansion method is much closer to the exact result. There is a reason for this, as explained in the following.

Consider Fig. 3.24, which contains schematics of supersonic and hypersonic flows over a pointed body with an attached shock wave. Downstream of the

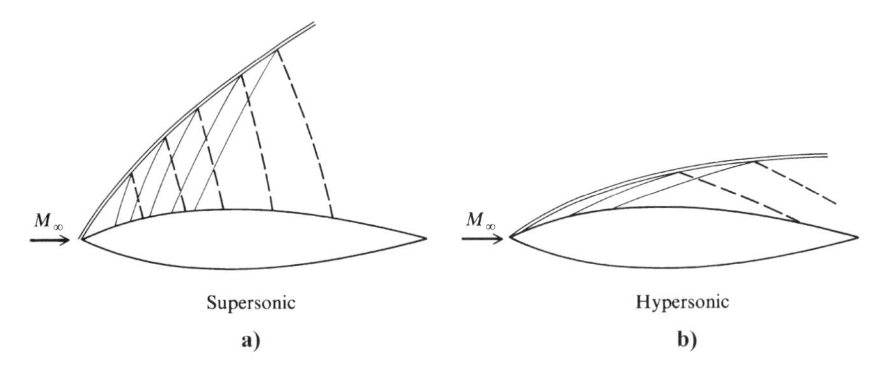

Fig. 3.24 Schematic of shock-wave and Mach-wave patterns: a) supersonic and b) hypersonic.

shock wave, expansion waves are generated at the surface of the body and propagate outward, eventually intersecting the bow shock wave. These expansion waves reflect from the shock wave; the reflected waves propagate back to the body surface, as shown by the dashed lines in Fig. 3.24. *Shock-expansion theory ignores the effect of these reflected waves on the body surface pressure.* Now consider just the supersonic case sketched in Fig. 3.24a. At supersonic Mach numbers, the shock angles and the incident and reflected wave angles are large. [The incident and reflected waves are essentially Mach waves with the Mach angle $\mu = \arcsin(1/M)$, where M is the local Mach number; at low Mach number, μ is large.] As a result, as seen in Fig. 3.24a, the reflected waves influence a considerable portion of the body surface, and this influence is not taken into account by the shock-expansion method. In contrast, for the hypersonic case shown in Fig. 3.24b, the shock and Mach angles are much smaller, and the reflected waves propagate much further downstream before they hit the body surface. As a result the reflected waves do not greatly influence the surface pressure, especially on the forward portion of the body. Therefore, the real hypersonic picture satisfies the assumption of shock-expansion theory more closely than the supersonic picture, and it is no surprise that shock-expansion theory yields better agreement at higher Mach numbers.

3.8 Summary and Comments

This chapter has dealt with hypersonic local surface inclination methods—such methods predict the local surface pressure as a function of the local surface inclination angle relative to the freestream direction θ. The methods discussed were 1) the straight Newtonian method, which yields

$$C_p = 2\sin^2\theta \tag{3.3}$$

2) the modified Newtonian method, which states

$$C_p = C_{p_{\max}} \sin^2\theta \tag{3.15}$$

3) the Newton–Busemann method, which takes into account the centrifugal force correction. For a two-dimensional body, this result is

$$C_{p_i} = 2\sin^2\theta_i + 2\left[\frac{\mathrm{d}\theta}{\mathrm{d}y}\right]_i \sin\theta_i \int_0^{y_i} \cos\theta \,\mathrm{d}y \tag{3.29}$$

4) the tangent-wedge method, where the pressure at point i on a two-dimensional body is assumed to be the same as a wedge with deflection angle θ_i; 5) the tangent-cone method, where the pressure at point i on an axisymmetric body is assumed to be the same as a cone with the semicone angle of θ_i; and 6) the shock-expansion method, where the pressure distribution downstream of the attached shock wave on a two- or three-dimensional body is assumed to be given by a local Prandtl–Meyer expansion. It is not possible to state with any certainty which of the preceding methods is the best for a given application. All of these

methods have their strengths and weaknesses, and some intuitive logic is required to choose one over the others for a given problem. For example, in the prediction of the pressure distribution over a hypersonic airplane any distinguishable portions of the fuselage might be treated with the tangent-cone method, whereas the wings might be better treated with the tangent-wedge method. Of course, for surfaces with large inclination angles (greater than the maximum deflection angle for an oblique shock wave at the given M_∞) the Newtonian method is appropriate. Within the confines of the Newtonian method itself, for blunt surfaces, where θ is very large, modified Newtonian is best, whereas straight Newtonian usually yields better results for slender bodies. In both cases, for $\gamma = 1.4$ the centrifugal force correction leads to poor results and should not be used. (Keep in mind that although the centrifugal force correction is theoretically consistent with mechanical principles, it is quantitatively correct only in the combined limit of $M_\infty \to \infty$ and $\gamma \to 1$.)

In regard to all of the local surface inclination methods discussed here, none of the preceding judgments on accuracy and applicability are totally definitive, and they all must be taken in the spirit of suggestions only. However, one definitive statement can be made about all of these methods, namely, that they are straightforward and easy to apply. For this reason, they are popular design tools for the investigation of large numbers of different hypersonic bodies. Indeed, *all* of the local surface inclination methods discussed in this chapter are embodied in an industry-standard computer program called the "Hypersonic Arbitrary Body Program" originally prepared by Gentry [22], and for this reason frequently referred to as the "Gentry program." This program has been in wide use throughout industry and government since the early 1970s. All of these methods discussed in this chapter are options within the Gentry program, which can be called at will for application to different portions of a hypersonic body. This program, and modified versions of it, is widely used in the preliminary design and analysis of hypersonic vehicles. It is mentioned here only to reinforce the engineering practicality of the methods discussed in this chapter.

Design Example 3.1

This is the first of a number of design examples in this book. The main thrust of this book is to present the *fundamental* aspects of hypersonic and high-temperature gas dynamics. We will from time to time, however, seize the moment to elaborate on the design applications of the fundamentals. This is such a moment.

The Hypersonic Arbitrary Body program (HABP), an elaborate computer program for predicting the surface-pressure, shear-stress, and heat-transfer distributions over hypersonic bodies of arbitrary shape, as well as their lift, drag, and moment coefficients, uses the local surface inclination methods discussed in this chapter. In fact, HABP offers a choice of 17 different pressure-prediction methods, but the primary ones of choice are Newtonian, modified Newtonian, tangent wedge, tangent cone, and shock expansion—all discussed in this chapter. The methods used for the prediction of shear stress and heat transfer are discussed in Chapter 6. The purpose of HABP is to provide an easy-to-use, fast, and reliable hypersonic aerodynamic prediction code for use in the preliminary design of hypersonic vehicles. Developed by the Douglas Aircraft Company in 1964, the code was greatly expanded in 1973, and then further updated in 1980. As mentioned earlier, this code is frequently referred to as the Gentry program after one of its originators [22]. The HABP is still in use today as a preliminary design tool.

An interesting and completely independent evaluation of HABP was made by Carren M. E. Fisher of British Aerospace P.L.C. In [225], Fisher presents calculations made with the Mark IV version of HABP for a variety of different vehicle configurations for which experimental data exist. In HABP, the vehicle surface is divided into a large number of flat surfaces (panels). For example, Fig. 3.25 shows the paneling used to represent the geometry of the space shuttle. Knowing the freestream Mach number and the angle of inclination of each panel relative to the freestream, a chosen local surface inclination method is used to obtain the pressure coefficient on each panel. Table 3.1, from [225], lists the available methods that can be chosen; there is a list for the windward side of the vehicle and a separate list for the leeward side.

Consider the tangent ogive-cylinder boattail shape given at the top of Fig. 3.26. Shown below the vehicle shape are results for the normal-force coefficient C_N, center-of-pressure location X_{cp}, moment coefficient about the nose C_m, and the axial-force coefficient C_A, respectively, as a function of angle of attack for a freestream Mach number of 4.63. The solid triangles are experimental data from Landrum ([226]). These data are compared with results from HABP using four different pressure-prediction methods labeled according to the numbering in Table 3.1. The curve labeled 1,1 ($K = 2.0$) pertains to item 1, modified Newtonian, for the windward side and item 1, Newtonian (i.e., $C_p = 0$), on the leeward side; K is the modified Newtonian correction factor, given by

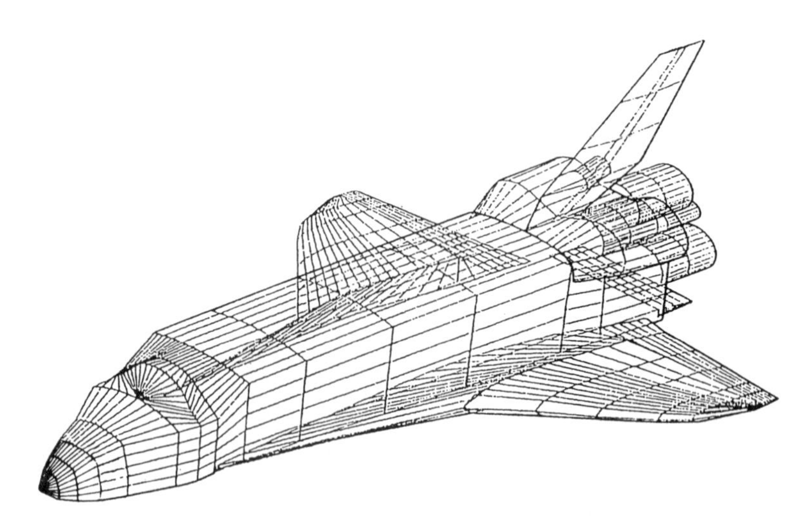

Fig. 3.25 Panel distribution over the space shuttle for an HAPB calculation (from Fisher [225]).

Table 3.1 List of options in HAPB[a]

Method no.	Impact method (applied to windward side of vehicle)	Shadow method (applied to leeward side of vehicle)
1	Modified Newtonian	Newtonian (i.e., $C_p = 0$)
2	Modified Newtonian and Prandtl–Meyer	Modified Newtonian and Prandtl–Meyer
3	Tangent-wedge (Using oblique shock)	Prandtl–Meyer expansion
4	Tangent-wedge empirical	Inclined-cone
5	Tanget-cone	Van Dyke unified
6	Inclined-cone	High Mach number base pressure ($C_p = -1/M^2$)
7	Van Dyke unified	Shock-expansion (using Strip theory)—Prandtl–Meyer expansion from freestream on first element of each stream-wise strip
8	Blunt-body skin-friction Shear-force	Input pressure coefficient
9	Shock-expansion (using Strip theory)	Free molecular flow
10	Free-molecular flow	Mirror Dahlem–Buck
11	Input pressure coefficient	ACM Empirical data
12	Hankey flat-surface Empirical	OSU Blunt body empirical
13	Delta-wing empirical	
14	Dahlem–Buck empirical	
15	Blast-wave pressure increments	
16	Modified tangent cone	
17	OSU Blunt body empirical	

[a]Impact methods 16 and 17 and shadow methods 10, 11, and 12 are recent updates to S/HABP.

$C_p = K \sin^2 \theta$. Here, $K = 2.0$; hence, $C_p = 2 \sin^2 \theta$, which is really the straight Newtonian results. The curve labeled 1,1 ($K = 1.81$) is the same set of methods except with a different value of K, where $C_p = 1.81 \sin^2 \theta$. The curve labeled 14, 1 uses the Dahlem–Buck empirical method (not described in this chapter) for the windward side and the Newtonian ($C_p = 0$) for the leeward side. And finally the fourth curve uses 6,4, inclined cone, for both the windward and leeward sides of the ogive portion of the vehicle, and 3,2 ($K = 1.81$), tangent wedge for the windward side and a combination of modified Newtonian with $K = 1.81$ and Prandtl–Meyer for the leeward side of the cylinder-boattail portion. Comparing the results shown in Fig. 3.26, this fourth curve gives

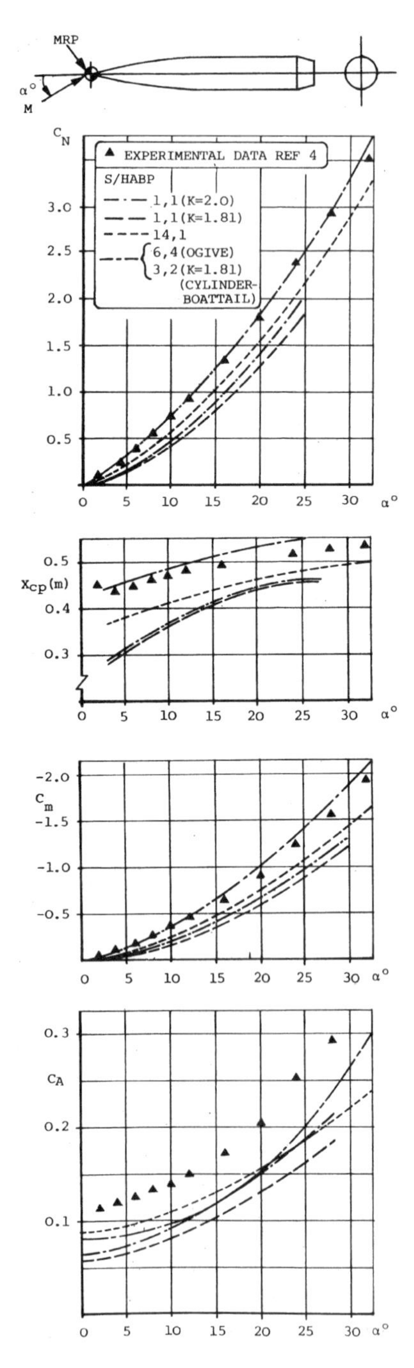

Fig. 3.26 Aerodynamic coefficients for a tangent ogive-cylinder boat-tail configuration, where $M_\infty = 4.63$ (from Fisher [225]).

the best agreement with experiment, illustrating the value of mixing and matching various options available in HABP for different parts of the vehicle. The normal-force coefficient C_N is accurately predicted, and the location of the center-of-pressure X_{cp} and the moment coefficient about the nose C_m are reasonably predicted. All of the pressure prediction methods underpredict the axial-force coefficient C_A because skin friction is not included in these particular results. Also, keep in mind that the results shown in Fig. 3.26 are for $M_\infty = 4.63$, a relatively low Mach number for applicability of the local surface inclination methods discussed in this chapter.

Return to Fig. 3.25, which shows the panel distribution for a space shuttle calculation. Predicted aerodynamic data for the space shuttle at Mach 13.5 obtained from HABP are compared with wind tunnel and flight data in Fig. 3.27. The predictions from HABP use only one combination, 1,1 ($K = 2.0$), that is, straight Newtonian. The wind-tunnel and flight data are from [227]. Figure 3.27 gives the variations of C_N, C_A, C_m, C_D, C_L, and L/D as functions of angle of attack. The wind-tunnel data span the complete angle-of-attack interval, whereas only one flight data point is shown on each graph because Mach 13.5 pertains to a specific point on the entry flight path, hence to only one specific space shuttle angle of attack. For the most part, the HABP calculations are markably close to the wind-tunnel data over the range of angle of attack. The flight data, given here from STS-1, for some of the coefficients deviate slightly from both the wind-tunnel data and the HABP calculations. But on the whole, we can see that HABP is a useful engineering prediction code for the space shuttle aerodynamics.

This statement is reinforced by the data shown in Fig. 3.28. Here the lift, drag, and moment coefficients, and the lift-to-drag ratio, for the space shuttle are given as a function of Mach number as the shuttle flies down its entry flight path. The flight data from STS-5 are given by the circles, and the preflight predictions from [227] are given by the triangles. Calculations using four different methods from HABP are also shown in Fig. 3.28. The discontinuities shown here are caused by shuttle maneuvers, which are not accounted for in the calculations. The two HABP methods labeled "with shielding" refer to the shielding option in HABP, which accounts for the reduction in pressure on an elemental panel that is shielded (hidden) from the freestream flow by another panel; for such shielded panels the pressure coefficient is set to zero. Clearly, from the results shown in Fig. 3.28 HABP gives a reasonable prediction of the space shuttle aerodynamic characteristics with the exception of the moment coefficient. The discrepancy in the moment coefficient calculations is caused by flowfield chemically reacting effects not included in HABP (and not reflected in the NASA preflight prediction data, which are largely based on cold-flow hypersonic wind-tunnel tests); at the end of Sec. 14.9, this matter is discussed at length, and the effect of chemically reacting flow on the moment coefficient is shown.

Very recently, Kinney [228] at the NASA Ames Research Center developed a new aerodynamic prediction code labeled CBAERO, the Configuration Based Aerodynamic software package, much in the same spirit as the earlier HABP. This new code makes pressure predictions using the modified Newtonian, tangent-wedge, and tangent-cone methods discussed in this chapter. In contrast to HABP, however, CBAERO uses panels on the body surface that are an unstructured mesh of triangles, such as shown in Fig. 3.29 for the surface of

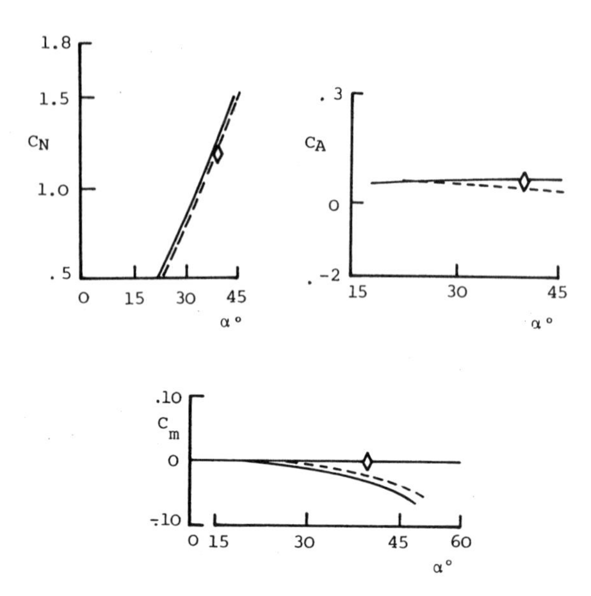

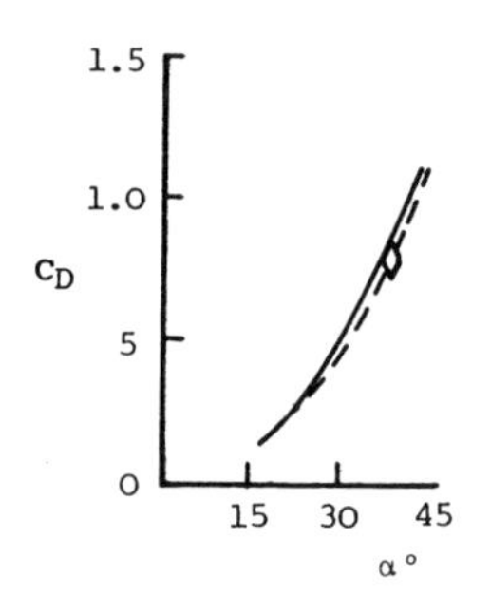

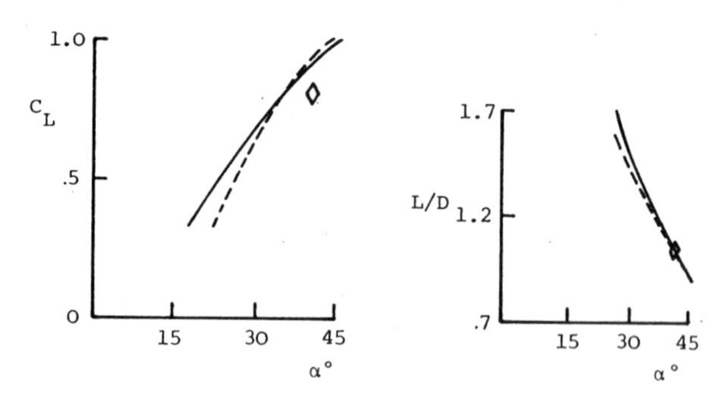

Fig. 3.27 Aerodynamic coefficient predictions using Newtonian theory, where $M_\infty = 13.5$ (from Fisher [225]).

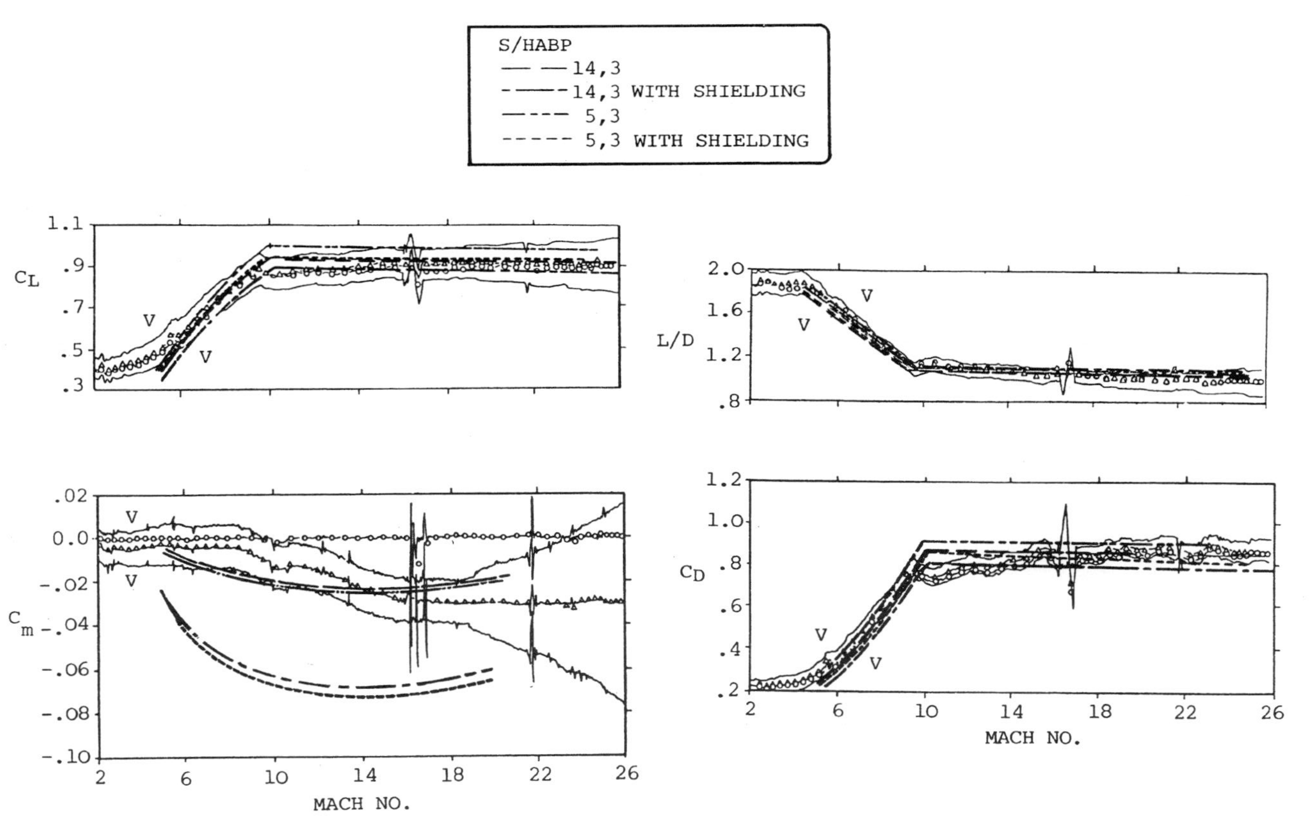

Fig. 3.28 STS-5 lift, drag, and moment variations during atmospheric entry. Predictions from HABP compared with flight data (from Fisher [225]).

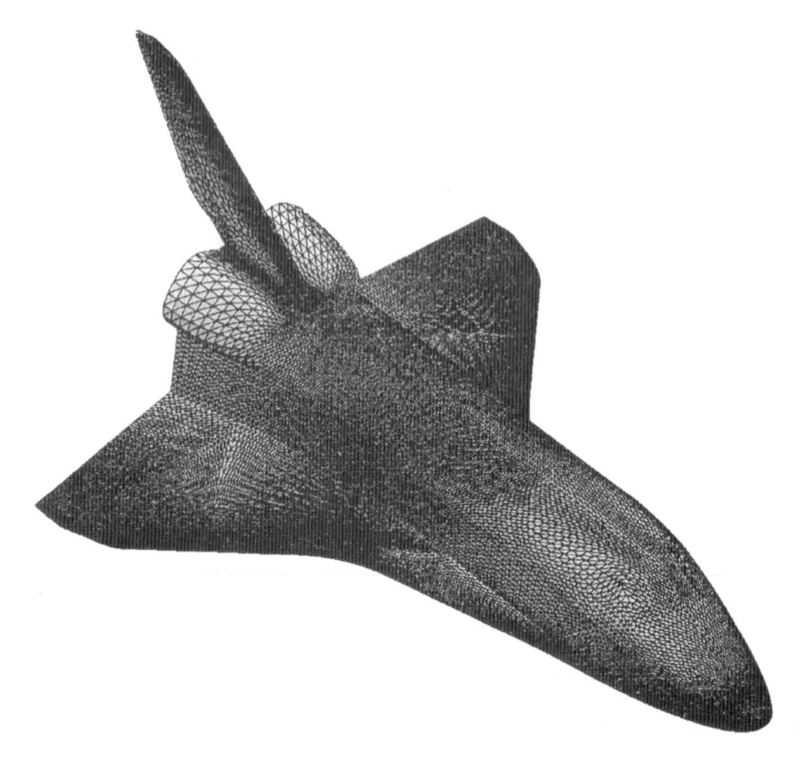

Fig. 3.29 Unstructured triangulated surface for the space shuttle (from Kinney and Garcia [229]).

the space shuttle. Compared to rectangular panels, this surface grid of triangles allows a more precise definition of complex vehicle geometries. Moreover, existing unstructured mesh-generation programs developed for use in computational fluid dynamics (CFD) can be used here for the body surface paneling. Indeed, this allows CBAERO results to be compared more directly with CFD results using the same mesh. In Fig. 3.29, the shuttle surface is covered by 103,104 triangles. Other modern features contained in CBAERO are the inclusion of high-temperature equilibrium chemically reacting flow thermodynamics (discussed in Chapter 14), the engineering prediction method of Tauber as well as the reference temperature method (discussed in Chapter 6) for convective aerodynamic heating, and an engineering method for stagnation radiative heat transfer (discussed in Chapter 18). This allows the use of CBAERO to predict the surface pressure, surface shear stress, convective heating, and radiative heating for vehicles operating in the severe aerothermal environments associated with flight Mach numbers as high as 36—that associated with Apollo–like atmospheric entries for a lunar mission.

Kinney and Garcia [229] computed surface pressures along the top and bottom centerlines of the space shuttle at Mach 24.87 and at angle of attack of 40.17 deg, as given in Fig. 3.30. Two sets of results are shown, one obtained from CBAERO and the other from a detailed CFD flowfield solution using a NASA Ames

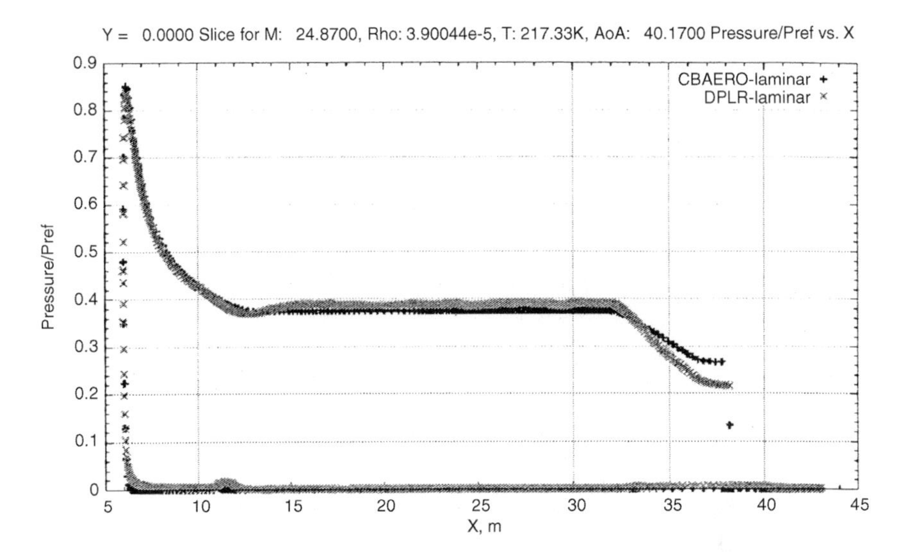

Fig. 3.30 Centerline pressure distribution for the space shuttle, where $M_\infty = 24.87$ and angle of attack 40 deg [229].

Research Center code labeled DPLR. The DPLR computational-fluid-dynamic results are the reference standard against which the CBAERO results are compared. The local surface inclination pressure prediction methods in CBAERO agree very well with the CFD results for both the windward surface (the upper curves) and the leeward surface (the lower curves).

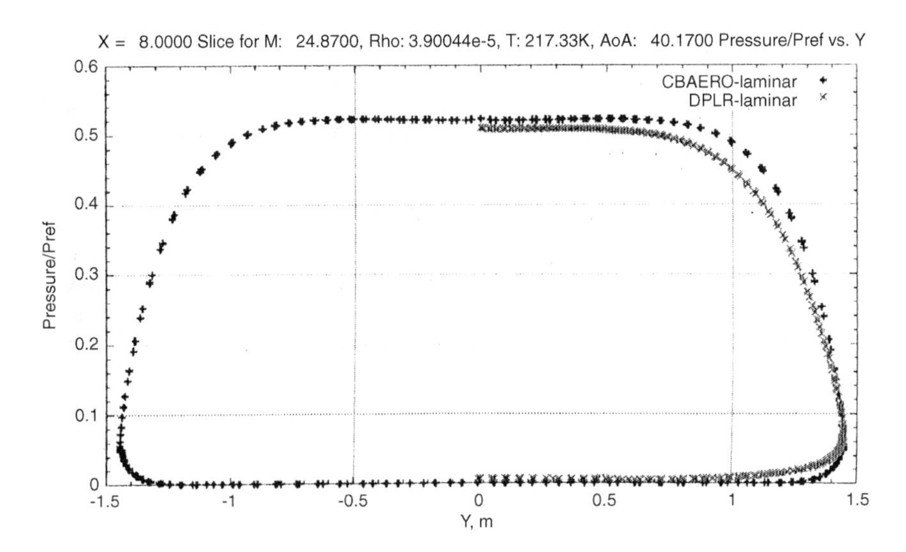

Fig. 3.31 Lateral pressure distribution for the space shuttle, where $M_\infty = 24.87$ and angle of attack 40 deg [229].

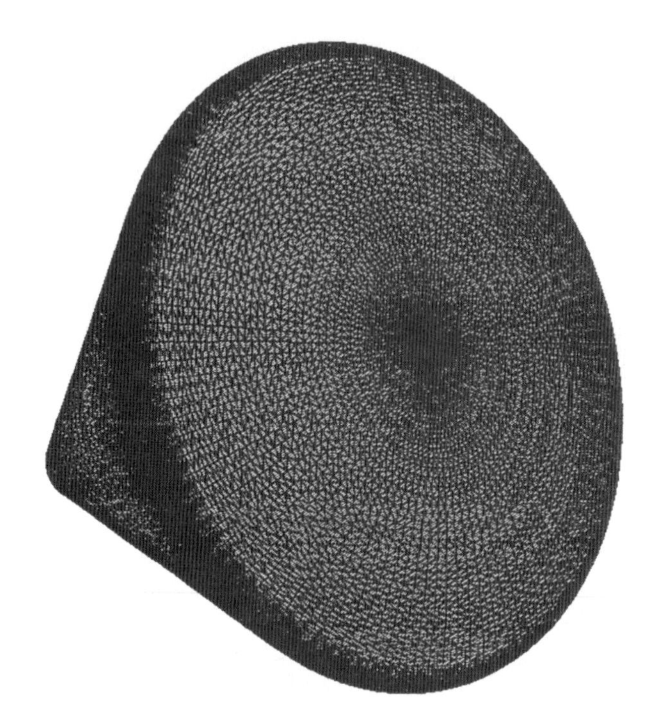

Fig. 3.32 Apollo command module with surface triangulation for CBAERO [229].

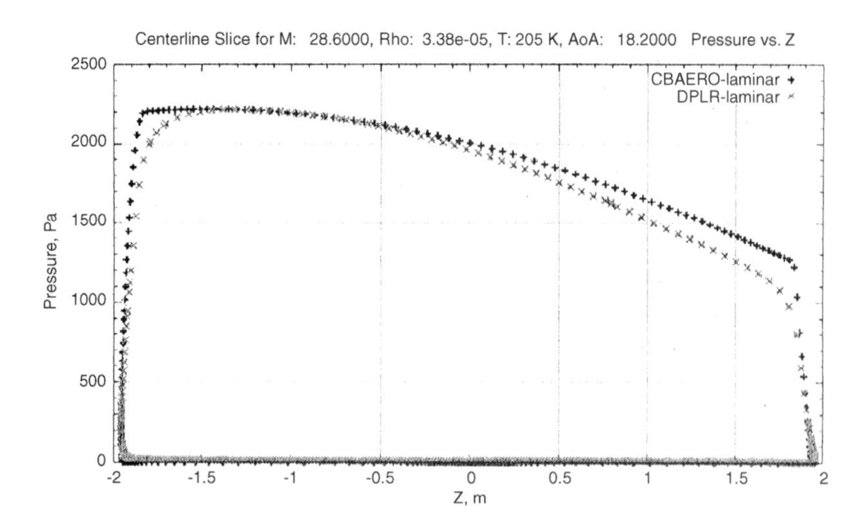

Fig. 3.33 Centerline pressure distribution for the Apollo command module, where M_∞ = 28.6 and angle of attack 18.2 deg [229].

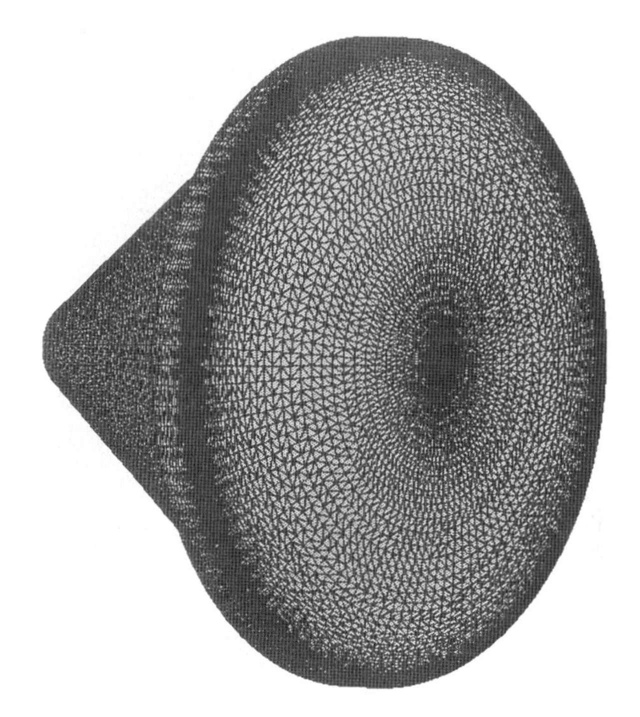

Fig. 3.34 Fire II test vehicle with surface triangulation for CBAERO [229].

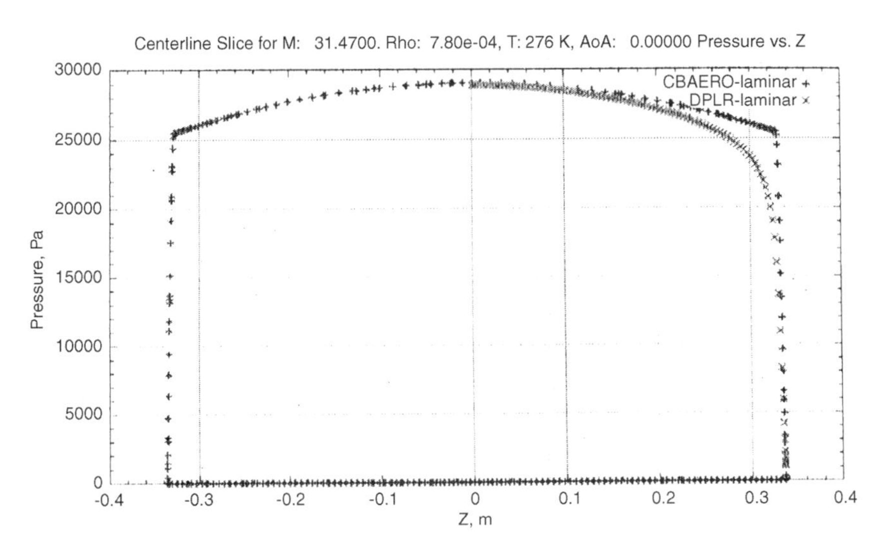

Fig. 3.35 Centerline pressure distribution for the Project Fire II test vehicle, where $M_\infty = 35.75$ and angle of attack 47 deg [229].

The lateral pressure distributions over the perimeter of a cross section of the space shuttle body are given in Fig. 3.31. The body cross section is located 8 m downstream of the nose. The pressure distribution is plotted vs the lateral coordinate y and forms a looped curve because at any given lateral location y there is a surface point on the windward side and another on the leeward side. The upper part of the loop (the higher pressures) corresponds to the windward side, and the lower part of the loop (the lower pressures) corresponds to the leeward side. Here we see that CBAERO slightly overpredicts the peak pressures on the windward side, especially along that side portion of the body with large lateral curvature, where three-dimensional flow effects are stronger.

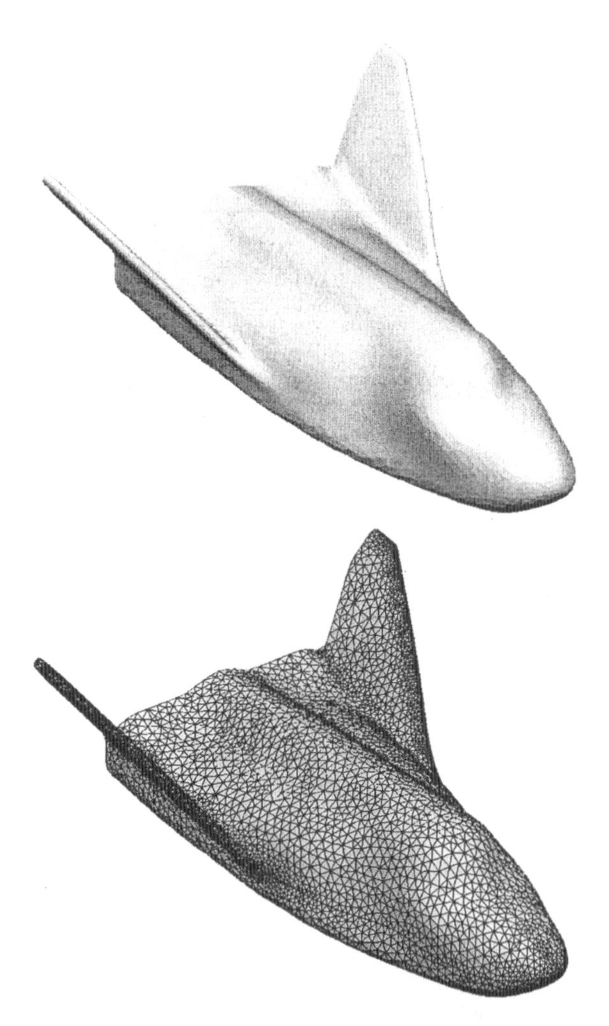

Fig. 3.36 HL-20 starting geometry and mesh for the optimization process using CBAERO (from Kinney [230]).

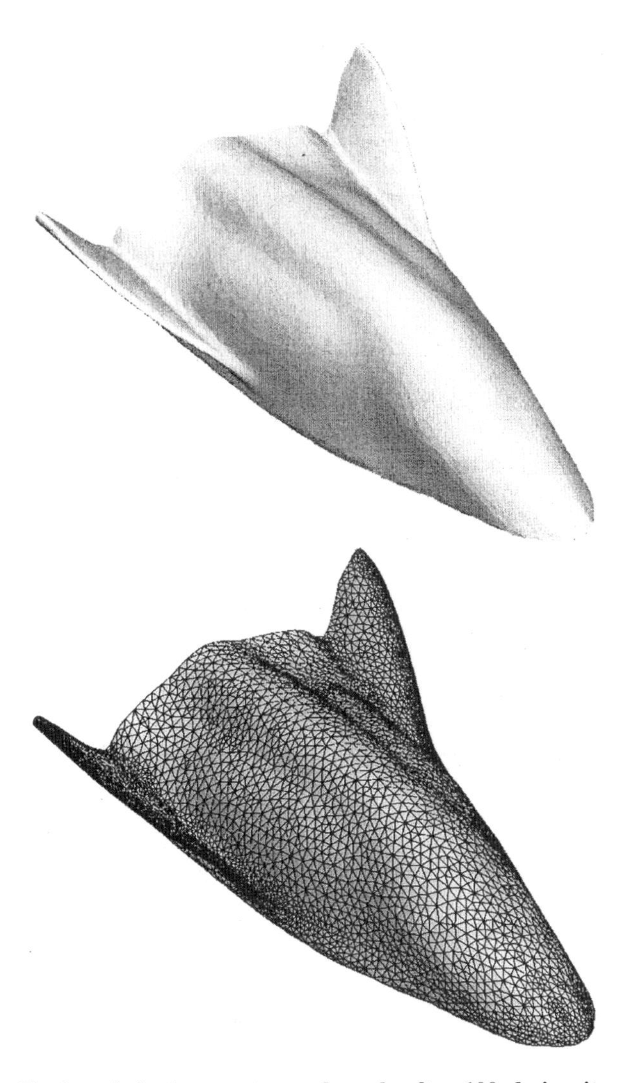

Fig. 3.37 Final optimized geometry and mesh after 100 design iterations using CBAERO [230].

Results obtained in a more severe aerothermal environment are given for the Apollo command module in Figs. 3.32 and 3.33. Figure 3.32 illustrates the triangular panel distribution over the surface of the module; here 29,000 triangles are used. The calculated centerline pressure distribution for the module at Mach 28.6, and an angle of attack of 18.2 deg is shown in Fig. 3.33, with results from CBAERO compared with the CFD results. In general, CBAERO does a reasonable job of predicting the Apollo Command Module pressure distribution.

Results obtained for an even more severe aerothermal environment are given for the Project Fire II test vehicle in Figs. 3.34 and 3.35. Figure 3.34 illustrates the triangular panel distribution over the vehicle surface; here 25,800 triangles are used. The calculated centerline pressure distribution for the test vehicle at Mach 35.75 and an angle of attack of 47 deg is shown in Fig. 3.35, with results from CBAERO compared with the CFD results. The CBAERO results compare very well with the detailed CFD calculations except in the shoulder region, where CBAERO slightly overpredicts the pressure.

We end this Design Example with an example of CBAERO used for the optimized design of hypersonic vehicles. This example is taken from the work of Kinney as described in [230]. Starting with the HL-20 crew transfer vehicle shown in Fig. 3.36 at Mach 20 and an angle of attack of 20 deg as a baseline, Kinney obtained the optimized shape shown in Fig. 3.37 at the same Mach number and angle of attack. An angle of attack 20 deg was chosen because the baseline HL-20 has a maximum value of L/D of 1.514 at this angle of attack. The objective of the optimization is to maximum L/D while holding the volume of the vehicle constant and constraining the pitching moment to zero. The optimized shape in Fig. 3.37 has $L/D = 2.78$, almost twice that of the baseline vehicle. Recall that in the optimization procedure the angle of attack is held constant at 20 deg. Interestingly enough, Kinney found that the resulting optimized shape in Fig. 3.37 actually has a *maximum* $L/D = 3.24$ and that it occurs at an angle of attack near 15 deg. Comparing the optimized shape in Fig. 3.37 with the baseline HL-20 in Fig. 3.36, Kinney observed that the optimization process drove the optimized geometry towards a wedge-like configuration on the forward portion of the body and that the windward side of the vehicle took on a waverider-like shape. (Waveriders are discussed in Chapters 5 and 6.) He also noted that the fins took on a smooth dihedral shape and were reduced in size.

In conclusion, at the time of writing the local surface inclination methods discussed in this chapter, although developed for hypersonic applications in the 1950s, are certainly alive and well today. Indeed, they are the basis of the modern Configuration Based Aerodynamics prediction code (CBAERO) highlighted here, as well as the well-established Hypersonic Arbitrary Body program (HABP) discussed earlier.

Problems

3.1 Consider the variation of lift with angle of attack for an infinitely thin flat plate. Using Newtonian theory, prove that maximum lift occurs at $\alpha = 54.7$ deg.

3.2 From Newtonian theory, prove that the drag coefficient for a circular cylinder of infinite span is $4/3$.

3.3 From Newtonian theory, prove that the drag coefficient for a sphere is 1.

3.4 In problems 3.1–3.3, are the results changed by using *modified* Newtonian theory? Explain.

3.5 Derive Eqs. (3.29) and (3.30) for the Newtonian pressure coefficient on an axisymmetric body including centrifugal effects.

3.6 The curves shown in Fig. 3.7 are changed when skin friction on the flat plate is included. In particular, the variation of L/D with α will peak at a low angle of attack and go to zero at $\alpha = 0$. (Why?) Let the drag coefficient caused by skin friction be assumed constant and denoted by C_{D_0}. Assuming a Newtonian pressure distribution, show that the maximum value of L/D is $0.667/C_{D_0}^{1/3}$ and occurs at an angle of attack (in radians) of $\alpha = C_{D_n}^{1/3}$. Furthermore, at $(L/D)_{\max}$ show that $C_{D_0} = \frac{1}{3} C_D$, where C_D is the total drag coefficient. [In other words, we can state that, at $(L/D)_{\max}$, wave drag is twice the friction drag.]

3.7 Using Newtonian theory, show that, at hypersonic speeds, stagnation pressure is about twice the dynamic pressure q_∞, where, by definition, $q_\infty = \frac{1}{2} \rho_\infty V_\infty^2$.

4
Hypersonic Inviscid Flowfields: Approximate Methods

No knowledge can be certain, if it is not based upon mathematics or upon some other knowledge which is itself based upon the mathematical sciences. Instrumental, or mechanical, science is the noblest and, above all others, the most useful.

Leonardo da Vinci (1425–1519)

Chapter Preview

This chapter is an intellectual excursion into the world of hypersonic aerodynamic *theory*. The material in this chapter is what people thought and what they did to calculate hypersonic flowfields before the advent and subsequent modern use of computational fluid dynamics. Today, a new practitioner learning hypersonic aerodynamics is frequently rushed into the use of massive computer programs, and the beauty and usefulness of the approximate flowfield analyses in this chapter are frequently overlooked. What a shame, and so this chapter presents some "golden nuggets" of hypersonic analyses that can really help the beginning student and the practitioner to better understand the nature of hypersonic flows. These golden nuggets are also very practical. They can help to reduce the amount of computational and experimental effort that might initially seem necessary for the solution of a given problem or the creation of a new design. For example, say that you have data, numerical or experimental, for the lift coefficient on a hypersonic body at Mach 10. Do you have to repeat the calculations or experiments to obtain the lift coefficient on the same, or a related body, at Mach 20? The principles of Mach-number independence and hypersonic similarity say NO! You might be wasting your time and money because by using these principles you can extract the lift coefficient at Mach 20 from the Mach 10 data. Pretty useful, say what? These principles are developed in the present chapter.

This chapter is full of useful hypersonic flow physics, around which the theory is wrapped. Did you know that if a slender hypersonic body, say in the shape of a ballpoint pen, went blasting past you at Mach 15 that the major change in the flow velocity would be in a direction perpendicular to the motion of the body and that the change in flow velocity parallel to the body would be practically nil? This physical variation is proven in the present chapter. Now you might say, "interesting, but so what?" Well, this interesting physical phenomenon leads directly to one of the most elegant and powerful theories in inviscid hypersonic flow, namely, the hypersonic equivalence principle and the resulting *blast wave theory* for calculating the surface-pressure distribution over a blunt-nosed slender body at hypersonic speeds. Blast wave theory leads to powerful, but simple, equations that allow fast, back-of-the envelope calculations. You need to know about all this, as well as much more in this chapter. So read on, allow yourself to enjoy the intellectual beauty of these approximate theoretical methods, and place the resulting practical equations on your analytical tool box for future use, both in your studies and in your practice.

4.1 Introduction

Examining the road map in Fig. 1.24, we note that our discussion of inviscid hypersonic aerodynamics started with the basic hypersonic shock and expansion relations (Chapter 2) and then carried on with local surface inclination methods for predicting pressure distributions on hypersonic bodies (Chapter 3). These discussions, which constitute the extreme left-hand branch in Fig. 1.24, have in common the need for only elementary mathematics; for the most part, the derivations and results involved only simple algebra. The reason for this is that straight oblique shock waves, expansion waves, and local surface inclination methods involve only localized phenomena—they do not require an integrated knowledge of whole regions of a flowfield. The material in Chapters 2 and 3 are about as far as we can proceed in this direction. For virtually all other considerations in hypersonic flow, we must examine the details of the complete flowfield. Therefore, we must now move to the second branch of our road map in Fig. 1.24, labeled "flowfield considerations." In so doing, our mathematical requirements increase because the details of any flowfield are governed by a system of conservation equations, which can be expressed in either integral or partial differential equation form. *Approximate* solutions of these equations for various hypersonic applications are the subject of the present chapter. *Exact* (numerical) solutions will be discussed in Chapter 5.

Another way to scope the material in this chapter is to establish the following philosophy. Up to as late as 1960, the history of the development of fluid mechanics had involved two dimensions: pure experiment and pure theory. With the advent of computational fluid dynamics after 1960, a new third dimension, namely, numerical computations, has been added, which complements the previous two. The science of fluid dynamics is now extended and applied by using all three dimensions in concert. The material in the present chapter is in the dimension of pure theory. The contributions of the other dimensions will be discussed in subsequent chapters. By its very nature, any hypersonic flowfield analysis before the advent of high-speed digital computers had to be in the dimension of pure theory. This was the only option for the analysis of hypersonic flows during the early development of the discipline. Many of these older analyses, all of which involved some approximations to allow the solution of the governing equations, are just as relevant to the modern hypersonics of today as they were in the 1950s. Moreover, they frequently have the advantage of illustrating more clearly than numerical solutions the effect of various parameters on the physical results. For these reasons, the present chapter is devoted to the discussion of approximate analyses of inviscid hypersonic flowfields. In so doing, we will begin to walk our way down the second branch of the road map in Fig. 1.24.

The local road map for this chapter is given in Fig. 4.1. As with any aerodynamic flowfield analysis, we start with the governing equations: the continuity, momentum, and energy equations. Using these equations, first we establish mathematically the Mach-number independence principle. Then we specialize these equations for the case of hypersonic flow over slender bodies at small angles of attack, obtaining the hypersonic small-disturbance equations, which are used in turn to develop a special aspect of flow similarity applicable to hypersonic slender bodies—the hypersonic similarity principle. One of the most important uses of the hypersonic small-disturbance equations is the demonstration of the hypersonic equivalence principle, which allows the application of blast wave theory to blunt-nosed slender hypersonic bodies. This application results in simple equations for the pressure distribution on the body surface downstream of the blunt nose, as well as the shape of the shock wave generated by the body. Finally, we apply the governing equations to the thin shock layer over a blunt-nosed hypersonic body, taking analytical advantage of the thinness of the shock layer and obtaining a simple method for calculating the pressure distribution over the nose and downstream surfaces of the body. We are reminded

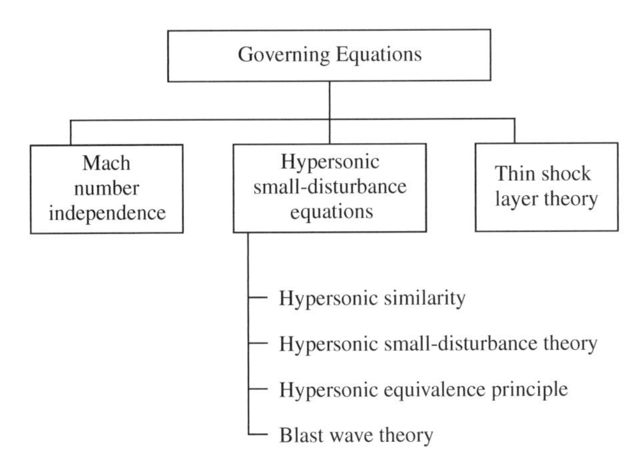

Fig. 4.1 Road map for Chapter 4.

again that the focus of the inviscid flow analyses in this chapter and in all others in Part 1 of this book is the calculation of pressure distributions on the surfaces of hypersonic bodies. Finally, we note that the road map in Fig. 4.1 has some mysterious-sounding destinations. By referring to this road map as you navigate through the chapter, hopefully all of these destinations will begin to make some sense to you, and when you reach the end of this chapter it will all hang together for you.

4.2 Governing Equations

Consider an inviscid, adiabatic (hence, isentropic) flowfield. The derivation of the governing conservation equations can be found in [4] and [5]; the results, written in Cartesian coordinates, are as follows.

Continuity:

$$\frac{\partial \rho}{\partial t} + \frac{\partial(\rho u)}{\partial x} + \frac{\partial(\rho v)}{\partial y} + \frac{\partial(\rho w)}{\partial z} = 0 \tag{4.1}$$

x momentum:

$$\rho\frac{\partial u}{\partial t} + \rho u\frac{\partial u}{\partial x} + \rho v\frac{\partial u}{\partial y} + \rho w\frac{\partial u}{\partial z} = -\frac{\partial p}{\partial x} \tag{4.2}$$

y momentum:

$$\rho\frac{\partial v}{\partial t} + \rho u\frac{\partial v}{\partial x} + \rho v\frac{\partial v}{\partial y} + \rho w\frac{\partial v}{\partial z} = -\frac{\partial p}{\partial y} \tag{4.3}$$

z momentum:

$$\rho\frac{\partial w}{\partial t} + \rho u\frac{\partial w}{\partial x} + \rho v\frac{\partial w}{\partial y} + \rho w\frac{\partial w}{\partial z} = -\frac{\partial p}{\partial z} \tag{4.4}$$

Energy:

$$\frac{\partial s}{\partial t} + u\frac{\partial s}{\partial x} + v\frac{\partial s}{\partial y} + w\frac{\partial s}{\partial z} = 0 \tag{4.5}$$

In the preceding, ρ is density; u, v, and w are the x, y, and z components of velocity, respectively; p is pressure; and s is entropy. Equations (4.1–4.5) are the well-known *Euler equations*, which govern inviscid flows. In reality, the preceding equations are a somewhat special form of the Euler equations, wherein body forces are neglected in Eqs. (4.2–4.4), and Eq. (4.5) is a specialized energy equation for an adiabatic, inviscid flow. In words, Eq. (4.1) is a statement that mass is conserved; Eqs. (4.2–4.4) are statements of Newton's second law, $F = ma$, in the x, y, and z directions, respectively, and Eq. (4.5) is a statement that the entropy is constant along a streamline for an inviscid, adiabatic flow. In some respects, Eq. (4.5) can be called the entropy equation, although it is fundamentally an energy equation. For an isentropic process in a calorically perfect gas (a perfect gas with constant specific heats), $p/\rho^\gamma =$ constant. Hence, if the entropy is constant along a streamline as stated by Eq. (4.5) then the quantity

p/ρ^γ is also constant along a streamline, and for a calorically perfect gas Eq. (4.5) can be replaced by

$$\boxed{\frac{\partial}{\partial t}\left(\frac{p}{\rho^\gamma}\right)+u\frac{\partial}{\partial x}\left(\frac{p}{\rho^\gamma}\right)+v\frac{\partial}{\partial y}\left(\frac{p}{\rho^\gamma}\right)+w\frac{\partial}{\partial z}\left(\frac{p}{\rho^\gamma}\right)=0} \tag{4.6}$$

The solution of the preceding equations for a given problem depends on the boundary and initial conditions for that problem. Discussions of the appropriate boundary and initial conditions will be made as appropriate in subsequent sections.

4.3 Mach-Number Independence

Return again to Fig. 3.15, where values of C_p for both a 15-deg half-angle wedge and cone are plotted vs Mach number. As noted at the end of Sec. 3.5, at low supersonic Mach numbers C_p decreases rapidly as M_∞ is increased. However, at hypersonic speeds the rate of decrease diminishes considerably, and C_p appears to reach a plateau as M_∞ becomes large, that is, C_p becomes relatively independent of M_∞ at high Mach numbers. This is the essence of the *Mach-number independence principle*; at high Mach numbers certain aerodynamic quantities such as pressure coefficient, lift, and wave-drag coefficients, and flowfield structure (such as shock wave shapes and Mach wave patterns) become essentially independent of Mach number. Indeed, straight Newtonian theory (discussed in Chapter 3) gives results that are *totally* independent of Mach number, as clearly demonstrated by Eq. (3.3). Modified Newtonian theory exhibits some Mach-number variation via $C_{p_{\max}}$ in Eq. (3.15); however, the variation of $C_{p_{\max}}$ with M_∞ in Fig. 3.8 exhibits a Mach-number independence at high M_∞. The hypersonic Mach-number independence principle is more than just an observed phenomena; it has a mathematical foundation, which is the subject of this section. We will examine the roots of this Mach-number independence more closely.

Let us nondimensionalize Eqs. (4.1–4.4) and (4.6) as follows. Define the nondimensional variables (the barred quantities) as

$$\bar{x}=\frac{x}{l} \quad \bar{y}=\frac{y}{l} \quad \bar{z}=\frac{z}{l}$$

$$\bar{u}=\frac{u}{V_\infty} \quad \bar{v}=\frac{v}{V_\infty} \quad \bar{w}=\frac{w}{V_\infty}$$

$$\bar{p}=\frac{p}{\rho_\infty V_\infty^2} \quad \bar{\rho}=\frac{\rho}{\rho_\infty}$$

where l denotes a characteristic length of the flow and ρ_∞ and V_∞ are the freestream density and velocity, respectively. Assuming steady flow ($\partial/\partial t=0$), we obtain from Eqs. (4.1–4.4) and (4.6)

$$\partial\frac{(\bar{\rho}\bar{u})}{\partial\bar{x}} + \partial\frac{(\bar{\rho}\bar{v})}{\partial\bar{y}} + \partial\frac{(\bar{\rho}\bar{w})}{\partial\bar{z}} = 0 \tag{4.7}$$

$$\bar{\rho}\bar{u}\frac{\partial\bar{u}}{\partial\bar{x}} + \bar{\rho}\bar{v}\frac{\partial\bar{u}}{\partial\bar{y}} + \bar{\rho}\bar{w}\frac{\partial\bar{u}}{\partial\bar{z}} = -\frac{\partial\bar{p}}{\partial\bar{x}} \tag{4.8}$$

$$\bar{\rho}\bar{u}\frac{\partial\bar{v}}{\partial\bar{x}} + \bar{\rho}\bar{v}\frac{\partial\bar{v}}{\partial\bar{y}} + \bar{\rho}\bar{w}\frac{\partial\bar{v}}{\partial\bar{z}} = -\frac{\partial\bar{p}}{\partial\bar{y}} \tag{4.9}$$

$$\bar{\rho}\bar{u}\frac{\partial\bar{w}}{\partial\bar{x}} + \bar{\rho}\bar{v}\frac{\partial\bar{w}}{\partial\bar{y}} + \bar{\rho}\bar{w}\frac{\partial\bar{w}}{\partial\bar{z}} = -\frac{\partial\bar{p}}{\partial\bar{z}} \tag{4.10}$$

$$\bar{u}\frac{\partial}{\partial\bar{x}}\left(\frac{\bar{p}}{\bar{\rho}^{\gamma}}\right) + \bar{v}\frac{\partial}{\partial\bar{y}}\left(\frac{\bar{p}}{\bar{\rho}^{\gamma}}\right) + \bar{w}\frac{\partial}{\partial\bar{z}}\left(\frac{\bar{p}}{\bar{\rho}^{\gamma}}\right) = 0 \tag{4.11}$$

Any particular solution of these equations is governed by the boundary conditions, which are discussed next.

The boundary condition for steady inviscid flow at a surface is simply the statement that the flow must be tangent to the surface. Let $\boldsymbol{n}$ be a unit normal vector at some point on the surface, and let $\boldsymbol{V}$ be the velocity vector at the same point. Then, for the flow to be tangent to the body

$$\boldsymbol{V}\cdot\boldsymbol{n} = 0 \tag{4.12}$$

[If there is any mass transfer through the surface, then $\boldsymbol{V}\cdot\boldsymbol{n} = v_T$, where v_T is the normal velocity of the fluid being transferred into or out of the surface. However, most inviscid flow problems do not involve mass transfer across the surface, and Eq. (4.12) is the pertinent boundary condition.] Let n_x, n_y, and n_z be the components of $\boldsymbol{n}$ in the x, y, and z directions, respectively. Then, Eq. (4.12) can be written as

$$un_x + vn_y + wn_z = 0 \tag{4.13}$$

Recalling the definition of direction cosines from analytic geometry, note in Eq. (4.13) that n_x, n_y, and n_z are also the direction cosines of $\boldsymbol{n}$ with respect to the x, y, and z axes, respectively. With this interpretation n_x, n_y, and n_z can be considered dimensionless quantities, and the nondimensional boundary condition at the surface is readily obtained from Eq. (4.13) as

$$\bar{u}n_x + \bar{v}n_y + \bar{w}n_z = 0 \tag{4.14}$$

Assume that we are considering the external flow over a hypersonic body, where the flowfield of interest is bounded on one side by the body surface and on the other side by the bow shock wave. Equation (4.14) gives the boundary condition on the body surface. The boundary conditions right behind the shock wave are given by the oblique shock properties expressed by Eqs. (2.1), (2.3), (2.6), and (2.8), repeated in the following for convenience (replacing the subscript 1 with subinfinity for freestream properties):

$$\frac{p_2}{p_\infty} = 1 + \frac{2\gamma}{\gamma+1}(M_\infty^2 \sin^2\beta - 1) \tag{2.1}$$

$$\frac{\rho_2}{\rho_\infty} = \frac{(\gamma+1)M_\infty^2 \sin^2\beta}{(\gamma-1)M_\infty^2 \sin^2\beta + 2} \tag{2.3}$$

$$\frac{u_2}{V_\infty} = 1 - \frac{2(M_\infty^2 \sin^2\beta - 1)}{(\gamma+1)M_\infty^2} \tag{2.6}$$

$$\frac{v_2}{V_\infty} = \frac{2(M_\infty^2 \sin\beta - 1)\cot\beta}{(\gamma+1)M_\infty^2} \tag{2.8}$$

In terms of the nondimensional variables, and noting that for a calorically perfect gas $p_2/p_\infty = \bar{p}_2(\rho_\infty V_\infty^2)/p_\infty = \bar{p}_2 V_\infty^2/RT_\infty = \bar{p}_2 \gamma V_\infty^2/a_\infty^2 = \bar{p}_2 \gamma M_\infty^2$, Eqs. (2.1), (2.3), (2.6), and (2.8) become

$$\bar{p}_2 = \frac{1}{\gamma M_\infty^2} + \frac{2}{\gamma+1}\left(\sin^2\beta - \frac{1}{M_\infty^2}\right) \tag{4.15}$$

$$\bar{\rho}_2 = \frac{(\gamma+1)M_\infty^2 \sin^2\beta}{(\gamma-1)M_\infty^2 \sin^2\beta + 2} \tag{4.16}$$

$$\bar{u}_2 = 1 - \frac{2(M_\infty^2 \sin^2\beta - 1)}{(\gamma+1)M_\infty^2} \tag{4.17}$$

$$\bar{v}_2 = \frac{2(M_\infty^2 \sin^2\beta - 1)\cot\beta}{(\gamma+1)M_\infty^2} \tag{4.18}$$

In the limit of high M_∞, as $M_\infty \to \infty$, Eqs. (4.15–4.18) go to [refer to Eqs. (2.2), (2.4), (2.7), and (2.10)]

$$\bar{p}_2 \to \frac{2\sin^2\beta}{\gamma+1} \tag{4.19}$$

$$\bar{\rho}_2 \to \frac{\gamma+1}{\gamma-1} \tag{4.20}$$

$$\bar{u}_2 \to 1 - \frac{2\sin^2\beta}{\gamma+1} \tag{4.21}$$

$$\bar{v}_2 \to \frac{\sin 2\beta}{\gamma+1} \tag{4.22}$$

Now consider a hypersonic flow over a given body. This flow is governed by Eqs. (4.7–4.11), with boundary conditions given by Eqs. (4.14–4.18).

Question: Where does M_∞ explicitly appear in these equations?

Answer: Only in the shock boundary conditions (4.15–4.18).

Now consider the hypersonic flow over a given body in the limit of large M_∞. The flow is again governed by Eqs. (4.7–4.11), but with boundary conditions given by Eqs. (4.14) and (4.19–4.22).

Question: Where does M_∞ explicitly appear in these equations?

Answer: No place!

Conclusion: At high M_∞, the solution is *independent* of Mach number.

Clearly, from this last consideration we can see that the Mach-number independence principle follows directly from the governing equations of motion with the appropriate boundary conditions written in the limit of high Mach number. Therefore, when the freestream Mach number is sufficiently high, the nondimensional dependent variables in Eqs. (4.7–4.11) become essentially independent of Mach number; this trend applies also to any quantities derived from these nondimensional variables. For example, C_p can be easily obtained as a function of $\bar{p}$ only; in turn, the lift and wave-drag coefficients for the body, C_L and C_{D_w}, respectively, can be expressed in terms of C_p integrated over the body surface (for example, see [5]). Therefore, C_p, C_L, and C_{D_w} also become independent of Mach number at high M_∞. This is demonstrated by the data shown in Fig. 4.2, obtained from [23–25], as gathered in [13]. In Fig. 4.2, the measured drag coefficients for spheres and for a large-angle cone-cylinder are plotted vs Mach number, cutting across the subsonic, supersonic, and hypersonic regimes. Note the large drag rise in the subsonic regime associated with the drag-divergence phenomena near Mach 1 and the decrease in C_D in the supersonic regime

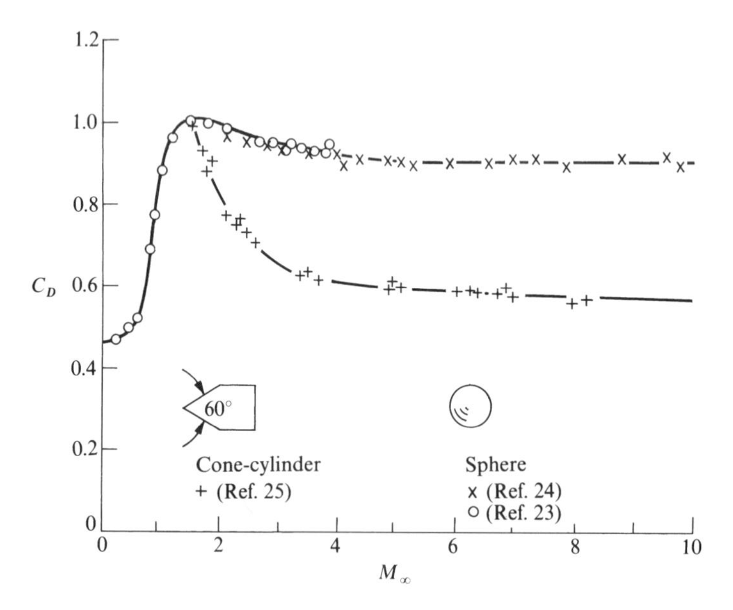

Fig. 4.2 Drag coefficient for a sphere and a cone-cylinder from ballistic range measurements; an illustration of Mach-number independence (from [13]).

beyond Mach 1. Both of these variations are expected and well understood for example, see [1] and [5]. For our purposes in the present section, note in particular the variation of C_D in the hypersonic regime; for both the sphere and cone-cylinder, C_D approaches a plateau and becomes relatively independent of Mach number as M_∞ becomes large. Note also that the sphere data appear to achieve Mach-number independence at lower Mach numbers than the cone-cylinder. This is to be expected, as follows. In Eqs. (4.15–4.18), the Mach number frequently appears in the combined form $M_\infty^2 \sin^2\beta$; for any given Mach number, this quantity is larger for blunt bodies (β large) than for slender bodies (β small). Hence blunt-body flows will tend to approach Mach-number independence at lower M_∞ than will slender bodies.

Finally, keep in mind from the preceding analysis that it is the *nondimensional* variables that become Mach-number independent. Some of the dimensional variables, such as p, are not Mach-number independent; indeed, $p \to \infty$ as $M_\infty \to \infty$.

4.4 Hypersonic Small-Disturbance Equations

The governing Euler equations discussed in Sec. 4.2 apply to the inviscid flow over a body of arbitrary shape—large or small, thick or thin, blunt or sharp. In applications involving low drag and/or high L/D hypersonic configurations, we are generally dealing with slender-body shapes; some examples are shown in Figs. 1.11 to 1.12. Therefore, a special, approximate form of the Euler equations, applicable to hypersonic slender bodies, is useful in studying the aerodynamic properties of such bodies. The purpose of this section is to obtain these equations, called the *hypersonic small-disturbance equations*.

We will follow an approach frequently employed in aerodynamic theory; instead of using the flow velocity itself as a dependent variable, we will deal with the *change* in velocity relative to the freestream, namely, the *perturbation* velocity. For example, consider the two-dimensional flow over the slender body shown in Fig. 4.3. At any given point in the flowfield, the vector velocity is $\boldsymbol{V}$. This is resolved into x and y components, u and v respectively. In turn, u and v can be expressed in terms of changes in velocity relative to the x and y components of the freestream velocity; these changes are denoted by u' and v', respectively, and are defined by

$$u = V_\infty + u'$$
$$v = v'$$

The preceding relations are written for the case where V_∞ is aligned with the x axis; hence, the y component of V_∞ is simply zero. The changes in velocity u' and v' are called perturbation velocities; in general, they do not have to be small.

In this section, we are considering the hypersonic flow over a slender body. In such a case, u' and v' are assumed to be small relative to V_∞, but not necessarily small relative to the freestream speed of sound. Hence, we will assume that we are dealing with small perturbations $u' \ll V_\infty$ and $v' \ll V_\infty$. To study the nature of these perturbations further, consider the velocity at a point on the surface of the body, as shown in Fig. 4.3. The body surface is given by

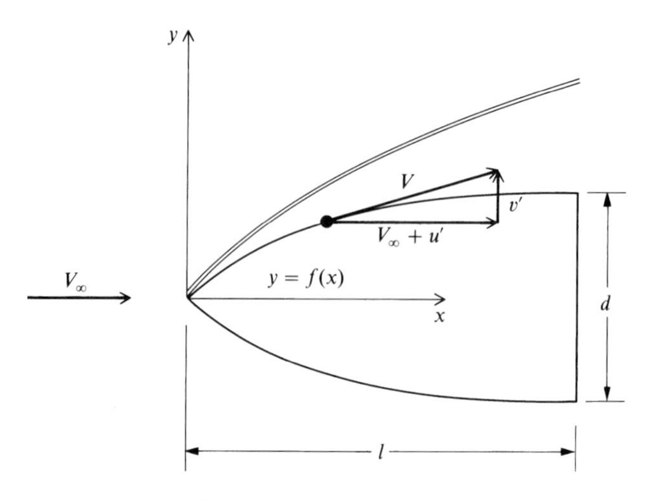

Fig. 4.3 Illustration of perturbation velocities.

$y = f(x)$; hence, the flow tangency condition dictates that

$$\frac{v'}{V_\infty + u'} = \frac{\mathrm{d}y}{\mathrm{d}x} \qquad \text{(on the body)} \tag{4.23}$$

However, examining Fig. 4.3 we see that

$$\frac{\mathrm{d}y}{\mathrm{d}x} = \mathcal{O}\left(\frac{d}{l}\right) \tag{4.24}$$

where the symbol $\mathcal{O}$ means "order of." Let us define

$$\frac{d}{l} = \tau = \text{slenderness ratio}$$

Then from Eqs. (4.23) and (4.24)

$$\frac{v'}{V_\infty + u'} = \frac{\mathrm{d}y}{\mathrm{d}x} = \mathcal{O}(\tau) \tag{4.25}$$

Because $u' \ll V_\infty$, then Eq. (4.25) is approximated by

$$\frac{v'}{V_\infty} = \mathcal{O}(\tau) \tag{4.26}$$

Let a_∞ = freestream speed of sound. From Eq. (4.26),

$$\frac{v'}{a_\infty} = \frac{V_\infty}{a_\infty}[\mathcal{O}(\tau)]$$

or

$$\frac{v'}{a_\infty} = \mathcal{O}(M_\infty \tau) \tag{4.27}$$

Clearly, from Eq. (4.27), the strength of the disturbance in the flow (relative to a_∞) is of the order of the parameter $M_\infty \tau$. This parameter will be identified as the hypersonic similarity parameter in the next section. However, for the time being, simply keep in mind the definition of the slenderness ratio $\tau = d/l$ and the fact that the product $M_\infty \tau$ is an indication of the strength of the disturbance created by the body in the flow, as expressed by v'/a_∞.

Let us now express the steady Euler equations in terms of the perturbation velocities u' and v', that is, in Eqs. (4.1–4.4) and (4.6), with zero time derivatives for steady flow, replace u with $V_\infty + u'$, v with v', and w with w', obtaining

$$\frac{\partial[\rho(V_\infty + u')]}{\partial x} + \frac{\partial(\rho v')}{\partial y} + \frac{\partial(\rho w')}{\partial z} = 0 \tag{4.28}$$

$$\rho(V_\infty + u')\frac{\partial(V_\infty + u')}{\partial x} + \rho v'\frac{\partial(V_\infty + u')}{\partial y} + \rho w'\frac{\partial(V_\infty + u')}{\partial z} = -\frac{\partial p}{\partial x} \tag{4.29}$$

$$\rho(V_\infty + u')\frac{\partial v'}{\partial x} + \rho v'\frac{\partial v'}{\partial y} + \rho w'\frac{\partial v'}{\partial z} = -\frac{\partial p}{\partial y} \tag{4.30}$$

$$\rho(V_\infty + u')\frac{\partial w'}{\partial x} + \rho v'\frac{\partial w'}{\partial y} + \rho w'\frac{\partial w'}{\partial z} = -\frac{\partial p}{\partial z} \tag{4.31}$$

$$(V_\infty + u')\frac{\partial}{\partial x}\left(\frac{p}{\rho^\gamma}\right) + v'\frac{\partial}{\partial y}\left(\frac{p}{\rho^\gamma}\right) + w'\frac{\partial}{\partial z}\left(\frac{p}{\rho^\gamma}\right) = 0 \tag{4.32}$$

Note in Eqs. (4.28–4.32) that only the velocities are expressed in terms of perturbations relative to the freestream values; the remaining flow quantities p and ρ are still carried as their whole values. [Sometimes, a perturbation analysis will also deal with changes in all the dependent variables relative to the freestream, i.e., a perturbation pressure p' and perturbation density ρ' would be defined as $p = p_\infty + p'$ and $\rho = \rho_\infty + \rho'$, respectively. This is not necessary in our present analysis; in Eqs. (4.28–4.32), p and ρ are the usual "whole" values of pressure and density.]

We wish to nondimensionalize Eqs. (4.28–4.32). Moreover, we wish to have nondimensional variables with an order of magnitude of unity, for reasons to be made clear later. To obtain a hint about reasonable nondimensionalizing quantities, consider the oblique shock relations in the limit as $M_\infty \to \infty$, obtained in Chapter 2. Also note that for a slender body at hypersonic speeds both the shock-wave angle β and the deflection angle θ are small; hence,

$$\sin\beta \approx \sin\theta \approx \theta \approx \frac{\mathrm{d}y}{\mathrm{d}x} \approx \tau$$

Thus, from Eq. (2.2), repeated here for convenience

$$\frac{p_2}{p_\infty} \to \frac{2\gamma}{\gamma+1} M_\infty^2 \sin^2\beta \tag{2.2}$$

we have the order-of-magnitude relationship

$$\frac{p_2}{p_\infty} \to \mathcal{O}[M_\infty^2 \tau^2] \tag{4.33}$$

This in turn implies that the pressure throughout the shock layer over the body will be on the order of $M_\infty^2 \tau^2 p_\infty$ and hence a reasonable definition for a nondimensional pressure, which would be on the order of magnitude of unity is $\bar{p} = p/\gamma\, M_\infty^2 \tau^2 p_\infty$. (The reason for the γ will become clear later.) In regard to density, consider Eq. (2.4), repeated here:

$$\frac{\rho_2}{\rho_\infty} \to \frac{\gamma+1}{\gamma-1} \tag{2.4}$$

For $\gamma = 1.4$, $\rho_2/\rho_\infty \to 6$, which for our purposes is on the order of magnitude near unity. Hence, a reasonable nondimensional density is simply $\bar{\rho} = \rho/\rho_\infty$. In regard to velocities, first consider Eq. (2.7), repeated here:

$$\frac{u_2}{V_\infty} \to 1 - \frac{2\sin^2\beta}{\gamma+1} \tag{2.7}$$

Define the change in the x component of velocity across the oblique shock as $\Delta u = V_\infty - u_2$. From Eq. (2.7), we have

$$\frac{\Delta u}{V_\infty} = \frac{V_\infty - u_2}{V_\infty} \to \frac{2\sin^2\beta}{\gamma+1} \to \mathcal{O}(\tau^2) \tag{4.34}$$

This implies that the nondimensional perturbation velocity $\bar{u}'$ (which is also a change in velocity in the x direction) should be defined as $\bar{u}' = u'/V_\infty \tau^2$ in order to be of an order of magnitude of unity. Finally, consider Eq. (2.10) repeated here:

$$\frac{v_2}{V_\infty} \to \frac{\sin 2\beta}{\gamma+1} \tag{2.10}$$

From Eq. (2.10), we have

$$\frac{\Delta v}{V_\infty} = \frac{v_2}{V_\infty} \to \frac{\sin 2\beta}{\gamma+1} \to \mathcal{O}(\tau) \tag{4.35}$$

This implies that the nondimensional perturbation velocity $\bar{v}'$ ought to be $\bar{v}' = v'/V_\infty\tau$, which is on the order of magnitude of one.

[We pause to observe an interesting physical fact evidenced by Eqs. (4.34) and (4.35). Because we are dealing with slender bodies, τ is a small number, much less than unity. Hence, by comparing Eqs. (4.34) and (4.35), we see that Δu, which varies as τ^2, is much smaller than Δv, which varies as τ. Therefore, we conclude in the case of hypersonic flow over a slender body that the change in v dominates the flow, that is, the changes in u and v are both small compared to V_∞, but that the change in v is large compared to the change in u. This fact was observed earlier, in Sec. 3.6, in conjunction with an argument that the major changes in properties in a hypersonic shock layer over slender bodies takes place across the flow rather than along the flow.]

Based on the preceding arguments, we define the following nondimensional quantities, all of which are on the order of magnitude of unity. Note that we add a third dimension in the z direction and that y and z in the thin shock layer are much smaller than x.

$$\bar{x} = \frac{x}{l} \quad \bar{y} = \frac{y}{l\tau} \quad \bar{z} = \frac{z}{l\tau}$$

$$\bar{u}' = \frac{u'}{V_\infty \tau^2} \quad \bar{v}' = \frac{v'}{V_\infty \tau} \quad \bar{w}' = \frac{w'}{V_\infty \tau}$$

$$\bar{p} = \frac{p}{\gamma M_\infty^2 \tau^2 p_\infty} \quad \bar{\rho} = \frac{\rho}{\rho_\infty}$$

(*Note:* The barred quantities here are *different* than the barred quantities used in Sec. 4.3, but because the present section is self-contained there should be no confusion.) In terms of the nondimensional quantities just defined, Eqs. (4.28–4.32) can be written as follows. From Eq. (4.28),

$$\frac{\partial}{\partial \bar{x}}\left[\bar{\rho}\left(\frac{1}{\tau^2} + \bar{u}'\right)\right][\rho_\infty V_\infty \tau^2] + \frac{\partial(\bar{\rho}\bar{v}')}{\partial \bar{y}}\left[\frac{\rho_\infty V_\infty \tau}{\tau}\right] + \frac{\partial(\bar{\rho}\bar{w}')}{\partial \bar{z}}\left[\frac{\rho_\infty V_\infty \tau}{\tau}\right] = 0 \tag{4.36}$$

From Eq. (4.29),

$$\bar{\rho}\left(\frac{1}{\tau^2} + \bar{u}'\right)\frac{\partial}{\partial \bar{x}}\left(\frac{1}{\tau^2} + \bar{u}'\right)[\rho_\infty V_\infty^2 \tau^4] + \bar{\rho}\,\bar{v}'\frac{\partial}{\partial \bar{y}}\left(\frac{1}{\tau^2} + \bar{u}'\right)[\rho_\infty V_\infty^2 \tau^3]$$
$$+\bar{\rho}\bar{w}'\frac{\partial}{\partial \bar{z}}\left(\frac{1}{\tau^2} + \bar{u}'\right)\left[\frac{\rho_\infty V_\infty^2 \tau^3}{\tau}\right] = -\frac{\partial \bar{p}}{\partial \bar{x}}[\gamma M_\infty^2 \tau^2 p_\infty]$$

or, noting that

$$\rho_\infty V_\infty^2 = \frac{\gamma p_\infty}{\gamma p_\infty} \rho_\infty V_\infty^2 = \gamma p_\infty \frac{V_\infty^2}{a_\infty^2} = \gamma p_\infty M_\infty^2$$

we have

$$\bar{\rho}(1 + \bar{u}'\tau^2)\frac{\partial \bar{u}'}{\partial \bar{x}} + \bar{\rho}\bar{v}'\frac{\partial \bar{u}'}{\partial \bar{y}} + \bar{\rho}\bar{w}'\frac{\partial \bar{u}'}{\partial \bar{z}} = -\frac{\partial \bar{p}}{\partial \bar{x}} \tag{4.37}$$

From Eq. (4.30)

$$\bar{\rho}\left(\frac{1}{\tau^2} + \bar{u}'\right)\frac{\partial \bar{v}'}{\partial \bar{x}}[\rho_\infty V_\infty^2 \tau^3] + \bar{\rho}\bar{v}'\frac{\partial \bar{v}'}{\partial \bar{y}}\left[\frac{\rho_\infty V_\infty^2 \tau^2}{\tau}\right]$$

$$+ \bar{\rho}\bar{w}'\frac{\partial \bar{v}'}{\partial \bar{z}}\left[\frac{\rho_\infty V_\infty^2 \tau^2}{\tau}\right] = -\frac{\partial \bar{p}}{\partial \bar{y}}\left[\frac{\gamma M_\infty^2 \tau^2 p_\infty}{\tau}\right]$$

or

$$\bar{\rho}(1 + \bar{u}'\tau^2)\frac{\partial \bar{v}'}{\partial \bar{x}} + \bar{\rho}\bar{v}'\frac{\partial \bar{v}'}{\partial \bar{y}} + \bar{\rho}\bar{w}'\frac{\partial \bar{v}'}{\partial \bar{z}} = -\frac{\partial \bar{p}}{\partial \bar{y}} \tag{4.38}$$

From Eq. (4.31), similarly we have

$$\bar{\rho}(1 + \bar{u}'\tau_2)\frac{\partial \bar{w}'}{\partial \bar{x}} + \bar{\rho}\bar{v}'\frac{\partial \bar{w}'}{\partial \bar{y}} + \bar{\rho}\bar{w}'\frac{\partial \bar{w}'}{\partial \bar{z}} = -\frac{\partial \bar{p}}{\partial \bar{z}} \tag{4.39}$$

From Eq. (4.32)

$$\left(\frac{1}{\tau^2} + \bar{u}'\right)\frac{\partial}{\partial \bar{x}}\frac{\bar{p}}{\bar{\rho}^\gamma}[V_\infty \tau^4 \gamma p_\infty M_\infty^2 \rho_\infty^\gamma] + \bar{v}'\frac{\partial}{\partial \bar{y}}\frac{\bar{p}}{\bar{\rho}^\gamma}\left[\frac{\gamma V_\infty \tau^3 p_\infty M_\infty^2 \rho_\infty^\gamma}{\tau}\right]$$

$$+ \bar{w}'\frac{\partial}{\partial \bar{z}}\frac{\bar{p}}{\bar{\rho}^\gamma}\left[\frac{V_\infty \tau^3 \gamma p_\infty M_\infty^2 \rho_\infty^\gamma}{\tau}\right] = 0$$

or

$$(1 + \tau^2 \bar{u}')\frac{\partial}{\partial \bar{x}}\left(\frac{\bar{p}}{\bar{\rho}^\gamma}\right) + \bar{v}'\frac{\partial}{\partial \bar{y}}\left(\frac{\bar{p}}{\bar{\rho}^\gamma}\right) + \bar{w}'\frac{\partial}{\partial \bar{z}}\left(\frac{\bar{p}}{\bar{\rho}^\gamma}\right) = 0 \tag{4.40}$$

Examine Eqs. (4.36–4.40) closely. Because of our choice of nondimensionalized variables, each term in these equations is of order of magnitude unity *except* for those mulitiplied by τ^2, which is very small. Therefore, the terms involving τ^2

can be ignored in comparison to the remaining terms, and Eqs. (4.36–4.40) can be written as

$$\frac{\partial \bar{\rho}}{\partial \bar{x}} + \frac{\partial(\bar{\rho}\bar{v}')}{\partial \bar{y}} + \frac{\partial(\bar{\rho}\bar{w}')}{\partial \bar{z}} = 0 \tag{4.41}$$

$$\bar{\rho}\frac{\partial \bar{u}'}{\partial \bar{x}} + \bar{\rho}\bar{v}'\frac{\partial \bar{u}'}{\partial \bar{y}} + \bar{\rho}\bar{w}'\frac{\partial \bar{u}'}{\partial \bar{z}} = -\frac{\partial \bar{p}}{\partial \bar{x}} \tag{4.42}$$

$$\bar{\rho}\frac{\partial \bar{v}'}{\partial \bar{x}} + \bar{\rho}\bar{v}'\frac{\partial \bar{v}'}{\partial \bar{y}} + \bar{\rho}\bar{w}'\frac{\partial \bar{v}'}{\partial \bar{z}} = -\frac{\partial \bar{p}}{\partial \bar{y}} \tag{4.43}$$

$$\bar{\rho}\frac{\partial \bar{w}'}{\partial \bar{x}} + \bar{\rho}\bar{v}'\frac{\partial \bar{w}'}{\partial \bar{y}} + \bar{\rho}\bar{w}'\frac{\partial \bar{w}'}{\partial \bar{z}} = -\frac{\partial \bar{p}}{\partial \bar{z}} \tag{4.44}$$

$$\frac{\partial}{\partial \bar{x}}\left(\frac{\bar{p}}{\bar{\rho}^{\gamma}}\right) + \bar{v}'\frac{\partial}{\partial \bar{y}}\left(\frac{\bar{p}}{\bar{\rho}^{\gamma}}\right) + \bar{w}'\frac{\partial}{\partial \bar{z}}\left(\frac{\bar{p}}{\bar{\rho}^{\gamma}}\right) = 0 \tag{4.45}$$

Equations (4.41–4.45) are the *hypersonic small-disturbance equations.* They closely approximate the hypersonic flow over slender bodies. They are limited to flow over slender bodies because we have neglected terms of order τ^2. They are also limited to hypersonic flow because some of the nondimensionalized terms are of order-of-magnitude unity only for high Mach numbers; we made certain of this in the argument that preceded the definition of the nondimensional quantities. Hence, the fact that each term in Eqs. (4.41–4.45) is of the order of magnitude unity [which is essential for dropping the τ^2 terms in Eqs. (4.36–4.40)] holds only for hypersonic flow.

Equations (4.41–4.45) exhibit an interesting property. Look for $\bar{u}'$ in these equations; you can find it only in Eq. (4.42). Therefore, in the hypersonic small-disturbance equations $\bar{u}'$ is *decoupled* from the system. In principle, Eqs. (4.41) and (4.43–4.45) constitute four equations for the four unknowns, $\bar{\rho}$, $\bar{p}$, $\bar{v}$, and $\bar{w}'$. After this system is solved, then $\bar{u}'$ follows directly from Eq. (4.42). This decoupling of $\bar{u}'$ from the rest of the system is another ramification of the fact already mentioned several times, namely, that the change in velocity in the flow direction over a hypersonic slender body is much smaller than the change in velocity perpendicular to the flow direction.

Equations (4.41–4.45) were obtained from the general nonlinear governing equations of motion (4.1–4.5). But in spite of containing the assumption of small perturbations, Eqs. (4.41–4.45) are still nonlinear. This is one of those distinctions that sets inviscid hypersonic flow apart from subsonic and supersonic flow. The small-disturbance equations for subsonic and supersonic flow are linear and lead to some straightforward solutions for subsonic and supersonic flows over slender bodies at small angles of attack (for example, see [4] and [5]). Not so for hypersonic flow. The hypersonic small-disturbance equations are a set of coupled, *nonlinear* partial differential equations for which no general analytical solution has yet been obtained. This circumstance is simply another ramification of the fact that there is nothing that is linear in hypersonic flow theory—hypersonic flow is inherently nonlinear.

The hypersonic small-disturbance equations, however, are used to obtain some practical information about hypersonic flows over slender bodies. The first such use will be made in the next section, dealing with hypersonic similarity.

(As a final, parenthetical comment, we now note the importance of obtaining the limiting hypersonic shock relations in Chapter 2. We have already used these relations several times for important developments. For example, they were used to help demonstrate Mach-number independence in Sec. 4.3, and they were instrumental in helping to define the proper nondimensional variables in the hypersonic small-disturbance equations obtained in this section. So the work done in Chapter 2 was more than just an academic exercise; the specialized forms of the oblique shock relations in the hypersonic limit are indeed quite useful.)

4.5 Hypersonic Similarity

The concept of flow similarity is well entrenched in fluid mechanics. In general, two or more different flows are defined to be dynamically similar when 1) the streamline shapes of the flows are geometrically similar and 2) the variation of the nondimensional flowfield properties is the same for the different flows when plotted in a nondimensional geometric space. Such dynamic similarity is ensured when 1) the body shapes are geometrically similar and 2) certain nondimensional parameters involving freestream properties and lengths, called *similarity parameters*, are the same between the different flows. See [5] for a more detailed discussion of flow similarity.

In the present section, we discuss a special aspect of flow similarity that applies to hypersonic flow over slender bodies. In the process, we will identify what is meant by hypersonic similarity and will define a useful quantity called the hypersonic similarity parameter.

Consider a slender body at hypersonic speeds. The governing equations are Eqs. (4.41–4.45). To these equations must be added the boundary conditions at the body surface and behind the shock wave. At the body surface, the flow tangency condition is given by Eq. (4.13), repeated here:

$$un_x + vn_y + wn_z = 0 \tag{4.13}$$

In terms of the perturbation velocities defined in Sec. 4.4, Eq. (4.13) becomes

$$(V_\infty + u')n_x + v'n_y + w'n_z = 0 \tag{4.46}$$

In terms of the nondimensional perturbation velocities defined in Sec. 4.4, Eq. (4.46) becomes

$$\left(\frac{1}{\tau^2} + \bar{u}'\right)(V_\infty \tau^2)n_x + \bar{v}'(V_\infty \tau)n_y + \bar{w}'(V_\infty \tau)n_z = 0$$

or

$$(1 + \tau^2\bar{u}')n_x + \bar{v}'\tau n_y + \bar{w}'\tau n_z = 0 \tag{4.47}$$

In Eq. (4.47), the direction cosines n_x, n_y, and n_z are in the (x, y, z) space; these values are somewhat changed in the transformed space $(\bar{x}, \bar{y}, \bar{z})$ defined in Sec. 4.4. Letting $\bar{n}_x$, $\bar{n}_y$, and $\bar{n}_z$ denote the direction cosines in the transformed space, we have (within the slender-body assumption)

$$n_x = \tau \bar{n}_x \quad n_y = \bar{n}_y \quad n_z = \bar{n}_z \tag{4.48}$$

The mathematical derivation of Eqs. (4.48) is left as a homework problem. However, the results are almost intuitively justified, as follows. For a slender body aligned along the x axis, the unit normal vector at the surface is almost perpendicular to the surface. This means that n_x is a small number, much less than unity, whereas n_y and n_z can be close to unity. In the transformed space, the slope of the body is increased by a factor $1/\tau$, and the unit normal vector in the transformed space is now more tilted with respect to the x axis by the factor $1/\tau$. Hence, the direction cosine with respect to the x axis is now $\bar{n}_x = n_x/\tau$. Moreover, in the transformed plane the unit normal vector is still close enough to being nearly perpendicular to the $\bar{x}$ axis to justify that $\bar{n}_y$ and $\bar{n}_z$ are still close to unity, just as in the case of n_y and n_z. Hence, we can say that $\bar{n}_y \approx n_y$ and $\bar{n}_z \approx n_z$. This is a justification for Eq. (4.48). With the relations given in Eq. (4.48), the boundary condition given by Eq. (4.47) becomes

$$(1 + \tau^2 \bar{u}')\tau \bar{n}_x + \bar{v}' \tau \bar{n}_y + \bar{w}' \tau \bar{n}_z = 0$$

or

$$(1 + \tau^2 \bar{u}')\bar{n}_x + \bar{v}' \bar{n}_y + \bar{w}' \bar{n}_z = 0 \tag{4.49}$$

Consistent with the derivation of the hypersonic small-disturbance equations in Sec. 4.4, we neglect the term of order τ^2 in Eq. (4.49), yielding the final result for the surface boundary condition:

$$\boxed{\bar{n}_x + \bar{v}' \bar{n}_y + \bar{w}' \bar{n}_z = 0} \tag{4.50}$$

The shock boundary conditions, consistent with the transformed coordinate system, can be obtained as follows. Consider Eq. (2.3) repeated here:

$$\frac{\rho_2}{\rho_\infty} = \bar{\rho}_2 = \frac{(\gamma + 1) M_\infty^2 \sin^2 \beta}{(\gamma - 1) M_\infty^2 \sin^2 \beta + 2} \tag{2.3}$$

or

$$\bar{\rho}_2 = \left(\frac{\gamma + 1}{\gamma - 1}\right) \left[\frac{M_\infty^2 \sin^2 \beta}{M_\infty^2 \sin^2 \beta + 2/(\gamma - 1)}\right] \tag{4.51}$$

For hypersonic flow over a slender body, β is small. Hence,

$$\sin \beta \approx \beta \approx \left(\frac{\mathrm{d}y}{\mathrm{d}x}\right)_s = \left(\frac{\mathrm{d}\bar{y}}{\mathrm{d}\bar{x}}\right)_s \tau$$

where $(d\bar{y}/d\bar{x})$ is the slope of the shock wave in the transformed space. Thus, Eq. (4.51) becomes

$$\bar{p}_2 = \left(\frac{\gamma+1}{\gamma-1}\right)\left\{\frac{(d\bar{y}/d\bar{x})_s^2}{(d\bar{y}/d\bar{x})_s^2 + 2/(\gamma-1)M_\infty^2\tau^2}\right\} \tag{4.52}$$

Repeating Eq. (2.2) here

$$\frac{p_2}{p_\infty} = 1 + \frac{2\gamma}{\gamma+1}(M_\infty^2 \sin^2\beta - 1) \tag{2.2}$$

and recalling that $\bar{p} = p/\gamma M_\infty^2 \tau^2 p_\infty$. Eq. (2.2) becomes

$$\frac{p_2}{\gamma M_\infty^2 \tau^2 p_\infty} = \frac{1}{\gamma M_\infty^2 \tau^2} + \frac{2\gamma}{\gamma+1}(M_\infty^2 \sin^2\beta - 1)\frac{1}{\gamma M_\infty^2 \tau^2}$$

$$\bar{p}_2 = \frac{1}{\gamma M_\infty^2 \tau^2} + \frac{2\gamma}{\gamma+1}\left[M_\infty^2\tau^2\left(\frac{d\bar{y}}{d\bar{x}}\right)_s^2 - 1\right]\frac{1}{\gamma M_\infty^2 \tau^2}$$

$$\bar{p}_2 = \frac{1}{\gamma M_\infty^2 \tau^2} + \frac{2(d\bar{y}/d\bar{x})_s^2}{\gamma+1} - \frac{2}{(\gamma+1)M_\infty^2\tau^2}$$

$$\bar{p}_2 = \frac{2(d\bar{y}/d\bar{x})_s^2}{\gamma+1} + \frac{(\gamma+1) - 2\gamma}{\gamma(\gamma+1)M_\infty^2\tau^2}$$

$$\bar{p}_2 = \frac{2}{\gamma+1}\left[\left(\frac{d\bar{y}}{d\bar{x}}\right)_s^2 + \frac{1-\gamma}{2\gamma M_\infty^2 \tau^2}\right] \tag{4.53}$$

Repeating Eq. (2.6)

$$\frac{u_2}{V_\infty} = 1 - \frac{2(M_\infty^2 \sin^2\beta - 1)}{(\gamma+1)M_\infty^2} \tag{2.6}$$

and recalling that $u_2 = V_\infty + u_2'$ and $\bar{u}_2' = u_2'/V_\infty\tau^2$, Eq. (2.6) becomes

$$1 + \frac{u_2'}{V_\infty} = 1 - \frac{2[M_\infty^2\tau^2(d\bar{y}/d\bar{x})_s^2 - 1]}{(\gamma+1)M_\infty^2}$$

$$\frac{u_2'}{V_\infty\tau^2} = \frac{-2[M_\infty^2\tau^2(d\bar{y}/d\bar{x})_s^2 - 1]}{(\gamma+1)M_\infty^2\tau^2}$$

$$\bar{u}_2' = -\frac{2}{\gamma+1}\left[\left(\frac{d\bar{y}}{d\bar{x}}\right)_s^2 - \frac{1}{M_\infty^2\tau^2}\right] \tag{4.54}$$

Repeating Eq. (2.8)

$$\frac{v_2}{V_\infty} = \frac{2(M_\infty^2 \sin^2\beta - 1)\cot\beta}{(\gamma+1)M_\infty^2} \tag{2.8}$$

and recalling that $v_2 = \bar{v}_2'$ and $\bar{v}_2 = \bar{v}_2'/V_\infty\tau$, Eq. (2.8) becomes

$$\frac{v_2'}{V_\infty\tau} = \frac{2}{\gamma+1}\left[\beta^2 - \frac{1}{M_\infty^2}\right]\frac{1}{\beta\tau}$$

$$\bar{v}_2' = \frac{2}{\gamma+1}\left[\left(\frac{\mathrm{d}\bar{y}}{\mathrm{d}\bar{x}}\right)_s^2 \tau^2 - \frac{1}{M_\infty^2}\right]\frac{1}{(\mathrm{d}\bar{y}/\mathrm{d}\bar{x})_s\tau^2}$$

$$\boxed{\bar{v}_2' = \frac{2}{\gamma+1}\left[\left(\frac{\mathrm{d}\bar{y}}{\mathrm{d}\bar{x}}\right)_s^2 - \frac{1}{M_\infty^2\tau^2}\right]\frac{1}{(\mathrm{d}\bar{y}/\mathrm{d}\bar{x})_s}} \tag{4.55}$$

Equations (4.52–4.55) represent boundary conditions immediately behind the shock wave in terms of the transformed variables. Note that these equations were obtained from the exact oblique shock relations, making only the one assumption of small wave angle; nothing was said about very high Mach numbers; hence, Eqs. (4.52–4.55) should apply to moderate as well as to large hypersonic Mach numbers.

Examine carefully the complete system of equations for hypersonic flow over a slender body—the governing flow equations (4.41–4.45), the surface boundary condition (4.50), and the shock boundary conditions (4.52–4.55). For this complete system, the freestream Mach number M_∞ and the body slenderness ratio τ appear only as the product $M_\infty\tau$, and this appears only in the shock boundary conditions. As first stated in Sec. 4.4, the product $M_\infty\tau$ is identified as the *hypersonic similarity parameter*, which we will denote by K.

Hypersonic similarity parameter:

$$K \equiv M_\infty\tau$$

The meaning of the hypersonic similarity parameter becomes clear from an examination of the complete system of equations. Because $M_\infty\tau$ and γ are the only parameters that appear in these nondimensional equations, then solutions for two different flows over two different but affinely related bodies (bodies that have essentially the same mathematical shape, but that differ by a scale factor on one direction, such as different values of thickness) will be the *same* (in terms of the nondimensional variables, $\bar{u}'$, $\bar{v}'$, etc.) *if* γ and $M_\infty\tau$ are the *same* between the two flows. This is the principle of *hypersonic similarity*.

For affinely related bodies at a small angle of attack α, the principle of hypersonic similarity holds as long as in addition to γ and $M_\infty\tau$, α/τ is also the same. For this case, the only modification to the preceding derivation occurs in the surface boundary condition, which is slightly changed; for small α, Eq. (4.50) is replaced by

$$\left(\bar{n}_x + \frac{\alpha}{\tau}\right) + \bar{v}'\bar{n}_y + \bar{w}'\bar{n}_z = 0 \tag{4.56}$$

The derivation of Eq. (4.56), as well as an analysis of the complete system of equations for the case of small α, is left to the reader as a homework problem. In summary, including the effect of angle of attack, the solution of the governing equations along with the boundary conditions takes the functional form

$$\bar{p} = \bar{p}\left(\bar{x}, \bar{y}, \bar{z}, \gamma, M_\infty\tau, \frac{\alpha}{\tau}\right)$$

$$\bar{\rho} = \bar{\rho}\left(\bar{x}, \bar{y}, \bar{z}, \gamma, M_\infty\tau, \frac{\alpha}{\tau}\right)$$

etc. Therefore, hypersonic similarity means that if γ, $M_\infty\tau$, and α/τ are the same for two or more different flows over affinely related bodies, then the variation of the nondimensional dependent variables over the nondimensional space $\bar{p} = \bar{p}(\bar{x}, \bar{y}, \bar{z})$, etc., is clearly the same between the different flows.

Consider the pressure coefficient, defined in Eq. (2.13), as

$$C_p = \frac{p - p_\infty}{\frac{1}{2}\rho_\infty V_\infty^2} = \frac{p - p_\infty}{(\gamma/2)p_\infty M_\infty^2}$$

This can be written in terms of $\bar{p}$ as

$$C_p = \frac{2(p - p_\infty)\tau^2}{\gamma p_\infty M_\infty^2 \tau^2} = 2\tau^2\left(\bar{p} - \frac{1}{\gamma M_\infty^2 \tau^2}\right) \tag{4.57}$$

Because $\bar{p} = \bar{p}(\bar{x}, \bar{y}, \bar{z}, \gamma, M_\infty\tau, \alpha/\tau)$, then Eq. (4.57) becomes the following functional relation:

$$\boxed{\frac{C_p}{\tau^2} = f_1\left(\bar{x}, \bar{y}, \bar{z}, \gamma, M_\infty\tau, \frac{\alpha}{\tau}\right)} \tag{4.58}$$

From Eq. (4.58), we see another aspect of hypersonic similarity, namely, that flows over affinely related bodies with the same values of γ, $M_\infty\tau$, and α/τ will have the same value of C_p/τ^2.

The viability of hypersonic similarity is reinforced by results that we have already obtained in Chapter 2. In Sec. 2.3, the hypersonic shock relations for large M_∞ and small deflection angles were obtained in terms $M_\infty\theta$, where θ is the flow deflection angle through the shock wave. There, we defined $M_\infty\theta \equiv K$ as the hypersonic similarity parameter; this is precisely the same as $M_\infty\tau$ because, for slender bodies, $\theta \approx \tan\theta \approx d/l = \tau$. Examine Eq. (2.29) and its functional form, namely, Eq. (2.30), repeated here:

$$\frac{C_p}{\tau^2} \approx \frac{C_p}{\theta^2} = f(K, \gamma) \tag{2.30}$$

This states that C_p/τ^2 for the flow behind an oblique shock (hence, over a wedge of slenderness ratio τ) is a function of γ and K only. Equations (2.45) and (2.46) obtained for the hypersonic expansion wave give analogous results. Hence, the results in Secs. 2.3 and 2.4 are precursors to the concept of hypersonic similarity discussed in the present section. It is recommended that, at this stage, you reread Secs. 2.3 and 2.4, keeping this point of view in mind.

Hypersonic similarity carries over to lift and wave-drag coefficients as well. Let us examine this in more detail. To begin with, assume a two-dimensional body of length l, hence a planform (or top-view) area per unit span of $(l)(1)$. The lift and wave-drag coefficients can be readily obtained by integrating the pressure coefficient over the surface of the body, resulting in (for example, see [5]).

$$c_l = \frac{1}{l}\int_0^l (C_{p_l} - C_{p_u})\, \mathrm{d}x \tag{4.59}$$

and

$$c_d = \frac{1}{l}\int_0^l (C_{p_l} + C_{p_u})\, \mathrm{d}y \tag{4.60}$$

In Eqs. (4.59) and (4.60), c_l and c_d are referenced to the *planform area*, and C_{p_l} and C_{p_u} are the pressure coefficients over the lower and upper surfaces, respectively. Equation (4.59), written in terms of $\bar{x}$, is

$$c_l = \int_0^1 (C_{p_l} - C_{p_u})\, \mathrm{d}\bar{x} \tag{4.61}$$

Dividing Eq. (4.61) by τ^2 and combining with Eq. (4.58), we obtain the following functional relation for c_l/τ^2:

$$\frac{c_l}{\tau^2} = \int_0^1 \left(\frac{C_{p_l}}{\tau^2} - \frac{C_{p_u}}{\tau^2}\right) \mathrm{d}\bar{x} = f_2\left(\gamma, M_\infty \tau, \frac{\alpha}{\tau}\right) \tag{4.62}$$

[Note that for a two-dimensional body $\bar{y} = \bar{y}(\bar{x})$, and there is no variation with $\bar{z}$; hence, the integral with respect to $\bar{x}$ in Eq. (4.62) takes care of the spatial variation of C_p with respect to $\bar{x}$, $\bar{y}$, $\bar{z}$ given in Eq. (4.58) resulting, after the integrations, in simply the functional variation shown by Eq. (4.62).] To obtain an analogous expression for the wave-drag coefficient, we write Eq. (4.60) in terms of $\bar{y}$ as follows:

$$c_d = \frac{1}{l}\int_0^1 (C_{p_l} + C_{p_u})\, \mathrm{d}\left(\frac{y}{l\tau}\right)(l\tau) = \tau \int_0^1 (C_{p_l} + C_{p_u})\, \mathrm{d}\bar{y} \tag{4.63}$$

Dividing Eq. (4.63) by τ^3 and combining with Eq. (4.58), we obtain the following functional relation for c_d/τ^3:

$$\frac{c_d}{\tau^3} = \int_0^1 \left(\frac{C_{p_l}}{\tau^2} + \frac{C_{p_u}}{\tau^2} \right) \mathrm{d}\bar{y} = f_3\left(\gamma,\, M_\infty \tau,\, \frac{\alpha}{\tau}\right) \tag{4.64}$$

Summarizing the preceding results, we have

$$\boxed{\begin{aligned} \frac{c_l}{\tau^2} &= f_2\left(\gamma,\, M_\infty \tau,\, \frac{\alpha}{\tau}\right) \\ \frac{c_d}{\tau^3} &= f_3\left(\gamma,\, M_\infty \tau,\, \frac{\alpha}{\tau}\right) \end{aligned}} \qquad \text{referenced to planform area}$$

Let us repeat the preceding arguments, except now for a three-dimensional body. The considerations are only slightly more involved, as follows. Consider Fig. 4.4, which shows an arbitrary body in an x-y-z coordinate system. In an inviscid flow, the net aerodynamic force is caused by the integration of the surface pressure distribution over the body. Consider an elemental force $p\,\mathrm{d}S$ caused by the pressure acting on the element of surface area $\mathrm{d}S$, as shown in Fig. 4.4. The component of this force in the z direction is $p\,\mathrm{d}x\,\mathrm{d}y$, where $(\mathrm{d}x\,\mathrm{d}y)$ is the projection of $\mathrm{d}S$ into the x-y plane. Hence the lift L is

$$L = \iint_S p(x, y, z)\,\mathrm{d}x\,\mathrm{d}y \tag{4.65}$$

In terms of the transformed variables, Eq. (4.65) becomes

$$L = \left[\iint_S \bar{p}(\bar{x}, \bar{y}, \bar{z})\,\mathrm{d}\bar{x}\,\mathrm{d}\bar{y} \right] (\gamma p_\infty M_\infty^2 \tau^2)(\tau) \tag{4.66}$$

We define the lift coefficient for the three-dimensional body as $C_L = L/q_\infty S$, where $q_\infty = (\gamma/2) p_\infty M_\infty^2$ and the area S is taken to be the *base area* (in contrast

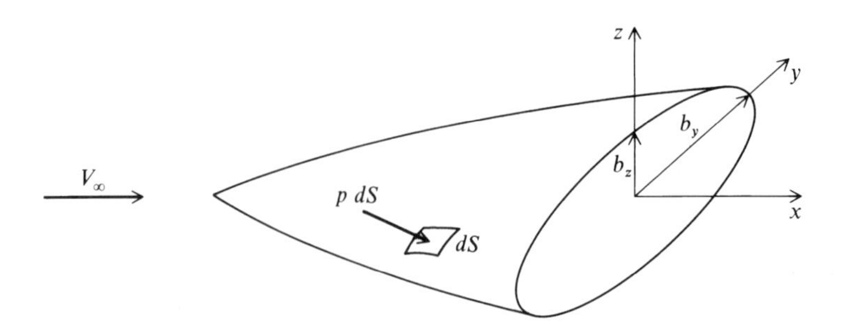

Fig. 4.4 Arbitrary body.

to the planform area, used for our preceding two-dimensional case). Letting b_y and b_z be the half-width and half-height of the base respectively (as shown in Fig. 4.4), then $S \propto b_y b_z = \bar{b}_y \bar{b}_z(\tau^2)$. Note that S is proportional to τ^2. Hence, from Eq. (4.66),

$$C_L \propto \frac{2}{\gamma p_\infty M_\infty^2 \tau^2}\left[\int\int \bar{p}(\bar{x}, \bar{y}, \bar{z})\, \mathrm{d}\bar{x}\, \mathrm{d}\bar{y}\right](\gamma p_\infty M_\infty^2 \tau^2)(\tau) \tag{4.67}$$

Recall that $\bar{p}(\bar{x}, \bar{y}, \bar{z})$ is obtained from the solution of the hypersonic small-disturbance equations for a given γ, $M_\infty \tau$, and α/τ. Therefore, the surface integral given in Eq. (4.67) depends only on γ, $M_\infty \tau$, and α/τ. With this in mind, Eq. (4.67) leads to the functional relation

$$\frac{C_L}{\tau} = F_1\left(\gamma, M_\infty \tau, \frac{\alpha}{\tau}\right) \tag{4.68}$$

Returning to Fig. 4.4, the component, of $p\,\mathrm{d}S$ in the x direction is $p\,\mathrm{d}y\,\mathrm{d}z$. Hence, the drag D is

$$D = \int\int_S p(x, y, z)\, \mathrm{d}y\, \mathrm{d}z$$

$$D = \left[\int\int_S \bar{p}(\bar{x}, \bar{y}, \bar{z})\, \mathrm{d}\bar{y}\, \mathrm{d}\bar{z}\right](\gamma p_\infty M_\infty^2 \tau^2)(\tau^2)$$

$$C_D = \frac{D}{q_\infty S} \propto \frac{2}{\gamma p_\infty M_\infty^2 \tau^2}\left[\int\int \bar{p}(\bar{x}, \bar{y}, \bar{z})\, \mathrm{d}\bar{y}\, \mathrm{d}\bar{z}\right](\gamma p_\infty M_\infty^2 \tau^2)(\tau^2)$$

or

$$\frac{C_D}{\tau^2} = F_2(\gamma, M_\infty \tau, \alpha/\tau) \tag{4.69}$$

Summarizing the preceding results, we have

$$\boxed{\begin{aligned} \frac{C_L}{\tau} &= F_1(\gamma, M_\infty \tau, \alpha/\tau) \\ \frac{C_D}{\tau^2} &= F_2(\gamma, M_\infty \tau, \alpha/\tau) \end{aligned}} \qquad \text{referenced to base area}$$

Examine the results summarized in the preceding two boxes, namely, the results for c_l and c_d for a two-dimensional flow, and C_L and C_D for a three-dimensional flow. From these results, the principle of hypersonic similarity states that affinely related bodies with the same values of γ, $M_\infty \tau$, and α/τ will have 1) the same values of c_l/τ^2 and c_d/τ^3 for two-dimensional flows, when referenced to planform area and 2) the same values of C_L/τ and C_D/τ^2 for three-dimensional flows when referenced to base area.

The validity of the hypersonic similarity principle is verified by the results shown in Figs. 4.5 and 4.6, obtained from the work of Neice and Ehret [26]. Consider first Fig. 4.5a, which shows the variation C_p/τ^2 as a function of distance downstream of the nose of a slender ogive-cylinder (as a function of $x = x/l$, expressed in percent of nose length). Two sets of data are presented, each for a different M_∞ and τ, but such that the product $K \equiv M_\infty \tau$ is the same value, namely, 0.5. The data are exact calculations made by the method of characteristics. Hypersonic similarity states that the two sets of data should be identical, which is clearly the case shown in Fig. 4.5a.

A similar comparison is made in Fig. 4.5b, except for a higher value of the hypersonic similarity parameter, namely, $K = 2.0$. The conclusion is the same; the data for two different values of M_∞ and τ, but with the same K, are identical. An interesting sideline is also shown in Fig. 4.5b. Two different methods of characteristics calculations are made—one assuming irrotational flow (the solid line) and the other treating rotational flow (the dashed line). There are substantial differences in implementing the method of characteristics for these two cases (for example, see [4] for more details). In reality, the flow over the ogive-cylinder is rotational because of the slightly curved shock wave over the nose. The effect of rotationality is to increase the value of C_p, as shown in Fig. 4.5b. This effect is noticeable for the high value of $K = 2$ in Fig. 4.5b. However, Neice and Ehret state that no significant differences between the rotational and irrotational calculations resulted for the low value of $K = 0.5$ in Fig. 4.5a, which is why only one curve is shown. One can conclude from this comparison the almost intuitive fact that the effects of rotationality become more important as M_∞, τ, or both are progressively increased. However, the main reason for bringing up the matter of rotationality is to ask the question: would we expect hypersonic similarity to hold for rotational flows? The question is rhetorical, because the answer is obvious. Examining the governing flow equations upon which hypersonic similarity is based, namely, Eqs. (4.41–4.45), we note that they contain no assumption of irrotational flow—they apply to both cases. Hence, the principle of hypersonic similarity holds for *both* irrotational and rotational flows. This is clearly demonstrated in Fig. 4.5b, where the data calculated for irrotational flow for two different values of M_∞ and τ (but the same K) fall on the same curve, and the data calculated for rotational flow for the two different values of M_∞ and τ (but the same K) also fall on the same curve (but a different curve than the irrotational results).

Figures 4.5a and 4.5b contain results at zero angle of attack. For the case of bodies at angle of attack, our similarity analysis has indicated that α/τ is an additional similarity parameter. This, as well as the general principle of hypersonic similarity, is *experimentally* verified by the wind-tunnel data shown in Fig. 4.6. Neice and Ehret [26] reported some experimental pressure distributions over two sharp, right-circular cones at various angles of attack obtained in the NACA Ames 10- by 14-in. supersonic wind tunnel. The freestream Mach numbers were 4.46 and 2.75, and the cones had different slenderness ratios such that $K = 0.91$ for both cases. Because the flow was conical, the values of C_p on the surface were constant along a given ray from the nose, but because

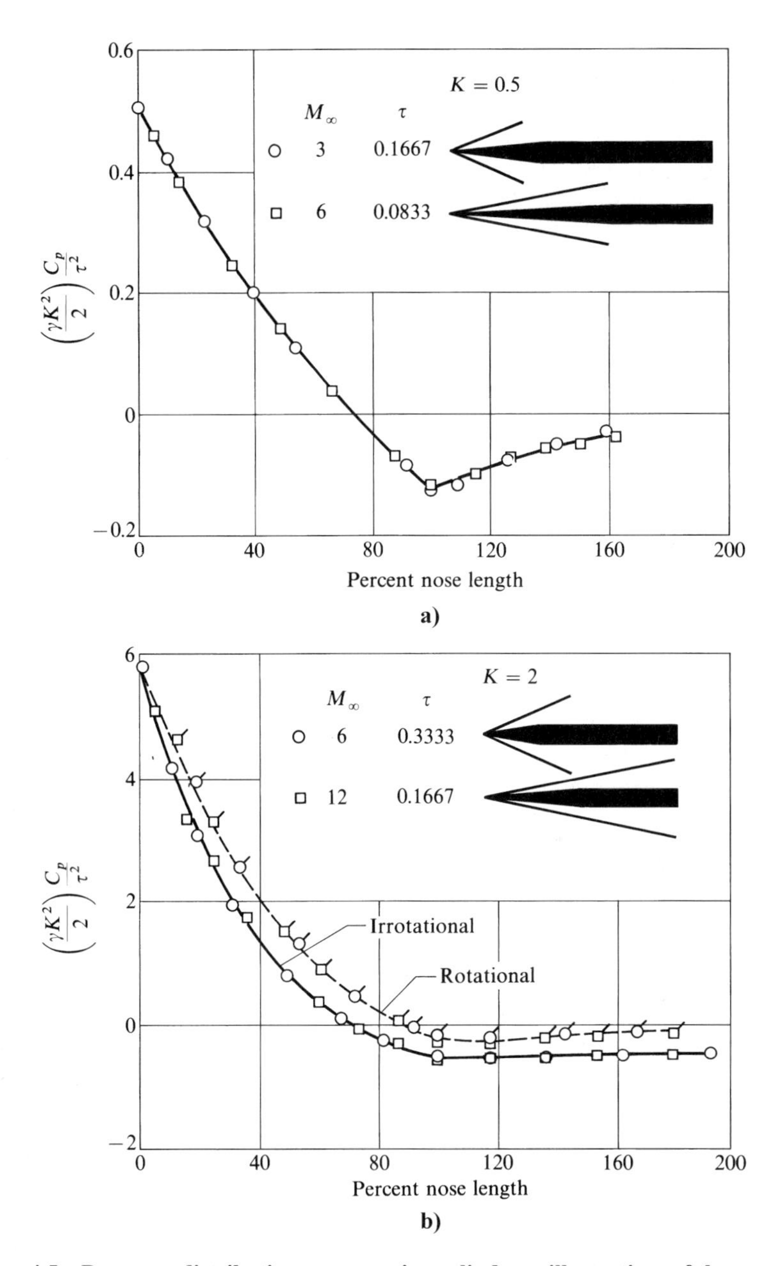

Fig. 4.5 Pressure distributions over ogive-cylinders, illustration of hypersonic similarity: a) $K = 0.5$ and b) $K = 2.0$ (from [26]).

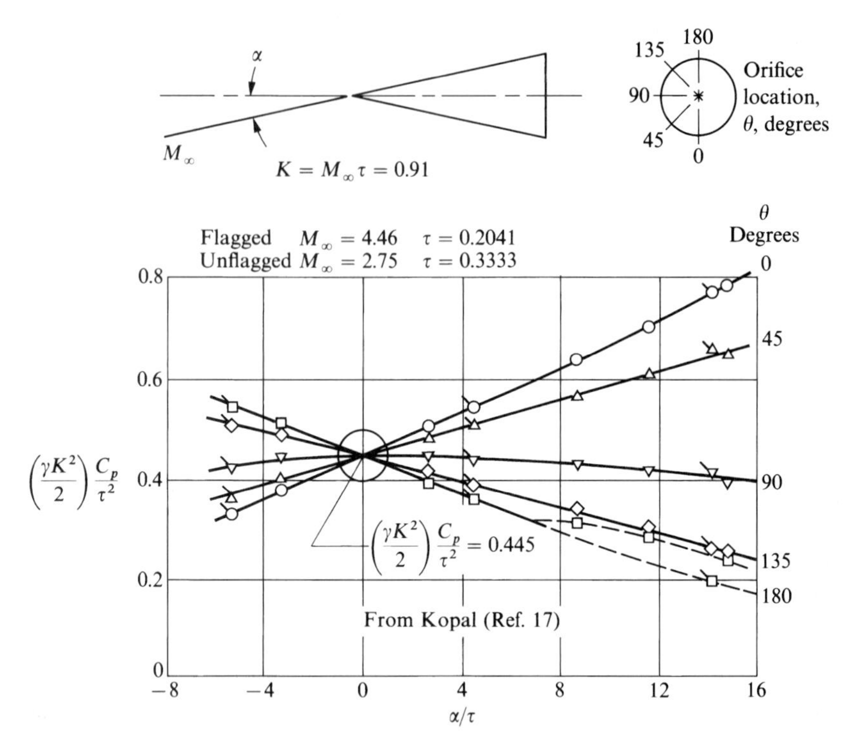

Fig. 4.6 Cone pressure at angle of attack, correlated by hypersonic similarity (from [26]).

of the angle of attack C_p varied from one ray to another around the cone as a function of angular location. Note in Fig. 4.6 that the data along any given ray for the two different values of M_∞ and τ (but both such that $K = 0.91$) fall on the same curve when plotted vs α/τ. Hence, the data in Fig. 4.6 are a direct experimental verification of hypersonic similarity for bodies at angle of attack. (Note that at $\alpha = 0$ all of the curves pass through the value of C_p predicted from exact cone theory, as tabulated by Kopal in [17].)

Hypersonic similarity appears to hold even at very moderate hypersonic Mach numbers. (The data in Fig. 4.6 even show some correlation at supersonic Mach numbers.) Indeed, Van Dyke [27] has pointed out a combined supersonic-hypersonic similarity rule that replaces $M_\infty\tau$ with $\tau\sqrt{M_\infty^2 - 1}$, which closely approximates $M_\infty\tau$ at high values of M_∞. By replacing $M_\infty\tau$ with $\tau\sqrt{M_\infty^2 - 1}$, a single similarity rule holds for the entire Mach-number regime starting just above the transonic range and going to an infinite Mach number. See [27] for more details.

Question: Over what range of values of $K \equiv M_\infty\tau$ does hypersonic similarity hold?

The answer cannot be made precisely. However, many results show that for very slender bodies (such as a 5-deg half-angle cone), hypersonic similarity holds for values of K ranging from less than 0.5 to infinitely large. On the other hand, for less slender bodies (say, a 20-deg half-angle cone), the data do not correlate well until $K > 1.5$. Homework problems 4.4 and 4.5 are very instructive in this regard. However, always keep in mind that hypersonic similarity is based on the hypersonic small-disturbance equations, and we would expect the results to become more tenuous as the thickness of the body is increased.

An important historical note is in order here. The concept of hypersonic similarity was first developed by H. S. Tsien in 1946 and published in [28]. In this paper, Tsien treated a two-dimensional potential (hence irrotational) flow. This work was further extended by Hayes [29], who showed that Tsien's results applied to rotational flows as well. (As noted earlier, the development of hypersonic similarity in the present chapter started right from the beginning with the governing equations for rotational flow. There is no need to limit ourselves to the special case treated by Tsien.) However, of equal (or more) historical significance, Tsien's 1946 paper seems to be the source that coined the word *hypersonic*. After an extensive search of the literature, the present author could find no reference to the word hypersonic before 1946. Then, in his 1946 paper—indeed, in the *title* of the paper—Tsien makes liberal use of the word hypersonic, without specifically stating that he is coining a new word. In this sense, the word hypersonic seems to have entered our vocabulary with little or no fanfare.

4.6 Hypersonic Small-Disturbance Theory: Some Results

Return to our road map in Fig. 1.24. We are presently working under the general heading of flowfield considerations, and we have, so far, treated both the concepts of Mach-number independence and hypersonic similarity under this heading. Also, return to our chapter road map in Fig. 4.1. Recall that we have discussed the general partial differential equations for an inviscid flow (Sec. 4.2), situated at the top center of Fig. 4.1. From these governing equations, we have proven mathematically the existence of Mach-number independence (left box in Fig. 4.1). We have also obtained the hypersonic small-disturbance equations (Sec. 4.4) represented by the center box in Fig. 4.1. It is important to note that, in our discussions of both Mach-number independence and hypersonic similarity, we have only *examined* the appropriate equations—*we have not solved them*. Specifically, our examination of a nondimensional form of the Euler equations and the boundary conditions in Sec. 4.2 clearly demonstrated the mathematical justification for Mach-number independence. Similarly, our examination of the hypersonic small-disturbance equations and the boundary conditions in Sec. 4.4 led to the important conclusions dealing with hypersonic similarity. But in both cases, we did not actually solve the governing equations. This is as far as we can proceed in such a fashion: for the remainder of the items listed under flowfield considerations in Fig. 1.24, we will deal with actual *solutions* of the governing equations for specific cases. This will constitute the remainder of the present chapter (on approximate methods) as well as all of Chapter 5 (on exact methods).

Consider again the hypersonic small-disturbance equations given by Eqs. (4.41–4.45). The purpose of the present section is to discuss how these equations can be solved for the hypersonic flow over slender bodies. The material in this section is a representative sample of a bulk of solutions generated over the past 35 years, all originating with Eqs. (4.41–4.45). Such solutions come under the general description of hypersonic small-disturbance theory. This theory was first developed in some detail by Milton Van Dyke [30], and we will partly follow his approach in this section.

To begin with, consider the hypersonic small-disturbance equations written for two-dimensional flow, and recall that the x-momentum equation is decoupled from the remaining equations in the system. For this case, from Eqs. (4.41), (4.43), and (4.45), we have

$$\frac{\partial \bar{\rho}}{\partial \bar{x}} + \frac{\partial(\bar{\rho}\bar{v}')}{\partial \bar{y}} = 0 \tag{4.70}$$

$$\bar{\rho}\frac{\partial \bar{v}'}{\partial \bar{x}} + \bar{\rho}\bar{v}'\frac{\partial \bar{v}'}{\partial \bar{y}} = -\frac{\partial \bar{p}}{\partial \bar{y}} \tag{4.71}$$

$$\frac{\partial}{\partial \bar{x}}\left(\frac{\bar{p}}{\bar{\rho}^{\gamma}}\right) + \bar{v}'\frac{\partial}{\partial \bar{y}}\left(\frac{\bar{p}}{\bar{\rho}^{\gamma}}\right) = 0 \tag{4.72}$$

which are three equations to be solved for the three unknowns $\bar{v}$, $\bar{p}$, and $\bar{\rho}$. However, this system can be reduced to just one equation in terms of one unknown by introducing a stream function ψ, defined as

$$\frac{\partial \psi}{\partial \bar{y}} = \bar{\rho} \tag{4.73}$$

and

$$\frac{\partial \psi}{\partial \bar{x}} = -\bar{\rho}\bar{v}' \tag{4.74}$$

To be a valid stream function, ψ must satisfy the continuity equation. Substitution of Eqs. (4.73) and (4.74) into (4.70) yields

$$\frac{\partial}{\partial \bar{x}}\left(\frac{\partial \psi}{\partial \bar{y}}\right) + \frac{\partial}{\partial \bar{y}}\left(-\frac{\partial \psi}{\partial \bar{x}}\right) = 0$$

or

$$\frac{\partial^2 \psi}{\partial \bar{x}\,\partial \bar{y}} - \frac{\partial^2 \psi}{\partial \bar{x}\,\partial \bar{y}} \equiv 0$$

that is, ψ as defined in Eqs. (4.73) and (4.74) does indeed satisfy the continuity equation. Using the subscript notation for partial derivatives, Eqs. (4.73) and (4.74) become

$$\bar{\rho} = \psi_{\bar{y}} \tag{4.75}$$

and

$$\bar{v}' = -\frac{\psi_{\bar{x}}}{\bar{\rho}} = -\frac{\psi_{\bar{x}}}{\psi_{\bar{y}}} \tag{4.76}$$

Also, denote $\bar{p}/\bar{\rho}^\gamma$ by ω, where ω is a function of ψ only. This is true because, for an isentropic flow, $\bar{p}/\bar{\rho}^\gamma$ is constant along a streamline, and by definition of a stream function (for example, see [5]) ψ is also constant along a streamline. Hence,

$$\frac{\bar{p}}{\bar{\rho}^\gamma} = \omega(\psi) \tag{4.77}$$

or

$$\bar{p} = \omega\bar{\rho}^\gamma = \omega(\psi_{\bar{y}})^\gamma \tag{4.78}$$

From Eq. (4.76),

$$\frac{\partial \bar{v}'}{\partial \bar{x}} = \frac{-\psi_{\bar{y}}\psi_{\bar{x}\bar{x}} + \psi_x\psi_{\bar{x}\bar{y}}}{(\psi_{\bar{y}})^2} \tag{4.79}$$

and

$$\frac{\partial \bar{v}'}{\partial \bar{y}} = \frac{-\psi_{\bar{y}}\psi_{\bar{x}\bar{y}} + \psi_{\bar{x}}\psi_{\bar{y}\bar{y}}}{(\psi_{\bar{y}})^2} \tag{4.80}$$

From Eq. (4.78),

$$\frac{\partial \bar{p}}{\partial \bar{y}} = \omega\gamma(\psi_{\bar{y}})^{\gamma-1}\psi_{\bar{y}\bar{y}} + (\psi_{\bar{y}})^\gamma\frac{\partial \omega}{\partial \bar{y}} \tag{4.81}$$

Because

$$\frac{\partial \omega}{\partial \bar{y}} = \left(\frac{\partial \omega}{\partial \psi}\right)\frac{\partial \psi}{\partial \bar{y}} = \omega'\psi_{\bar{y}}$$

then Eq. (4.81) becomes

$$\frac{\partial \bar{p}}{\partial \bar{y}} = \gamma\omega(\psi_{\bar{y}})^{\gamma-1}\psi_{\bar{y}\bar{y}} + \omega'(\psi_{\bar{y}})^{\gamma+1} \tag{4.82}$$

Substitute Eqs. (4.75), (4.76), (4.79), (4.80), and (4.82) into the y-momentum equation (4.71).

$$\psi_{\bar{y}}\left[\frac{-\psi_{\bar{y}}\psi_{\bar{x}\bar{x}} + \psi_{\bar{x}}\psi_{\bar{x}\bar{y}}}{(\psi_y)^2}\right] + (-\psi_{\bar{x}})\left[\frac{-\psi_{\bar{y}}\psi_{\bar{x}\bar{y}} + \psi_x\psi_{\bar{y}\bar{y}}}{(\psi_{\bar{y}})^2}\right]$$
$$= -\gamma\omega(\psi_{\bar{y}})^{\gamma-1}\psi_{\bar{y}\bar{y}} - \omega'(\psi_{\bar{y}})^{\gamma+1}$$

or

$$\boxed{(\psi_{\bar{y}})^2\psi_{\bar{x}\bar{x}} - 2\psi_x\psi_{\bar{y}}\psi_{\bar{x}\bar{y}} + (\psi_{\bar{x}})^2\psi_{\bar{y}\bar{y}} = (\psi_{\bar{y}})^{\gamma+1}[\gamma\omega\psi_{\bar{y}\bar{y}} + \omega'(\psi_{\bar{y}})^2]} \tag{4.83}$$

Equation (4.83) is a single equation for a single unknown, namely, ψ, based on the hypersonic small-disturbance assumptions. Note that in the development of this equation, no additional assumptions were made (other than that of two-dimensional flow); hence, Eq. (4.83) is of the same order of accuracy as the original hypersonic small-disturbance equations.

Equation (4.83) holds for two-dimensional planer flow; hence, it can be applied to two-dimensional shapes such as airfoils. On the other hand, for axisymmetric bodies a cylindrical coordinate system (x, r, ϕ) is more convenient, where x and r are the coordinates parallel and perpendicular respectively to the body centerline, and ϕ is the familiar azimuthal angle. For an axisymmetric body at zero angle of attack, the flowfield is independent of ϕ and depends on x and r only. For this case, the governing hypersonic small-perturbation equations become

$$\frac{\partial \bar{\rho}}{\partial \bar{x}} + \frac{\partial(\bar{\rho}\bar{v}')}{\partial \bar{r}} + \frac{\bar{\rho}\bar{v}'}{\bar{r}} = 0 \tag{4.84}$$

$$\bar{\rho}\frac{\partial \bar{v}'}{\partial \bar{x}} + \bar{\rho}\bar{v}'\frac{\partial \bar{v}'}{\partial \bar{r}} = -\frac{\partial \bar{p}}{\partial \bar{r}} \tag{4.85}$$

$$\frac{\partial}{\partial \bar{x}}\left(\frac{\bar{p}}{\bar{\rho}^{\gamma}}\right) + \bar{v}'\frac{\partial}{\partial \bar{r}}\left(\frac{\bar{p}}{\bar{\rho}^{\gamma}}\right) = 0 \tag{4.86}$$

These are the same as Eqs. (4.70–4.72), except for the additional term in Eq. (4.84). In the preceding, $\bar{x} = x/l$, $\bar{r} = r/\tau l$, $\bar{v}'$ is the nondimensional perturbation velocity in the $\bar{r}$ direction, and all of the other quantities are the same as before. For the axisymmetric flow described by Eqs. (4.84) and (4.85), a stream function ψ can be defined as

$$\frac{\partial \psi}{\partial \bar{r}} = \bar{r}\bar{\rho} \tag{4.87}$$

$$\frac{\partial \psi}{\partial \bar{x}} = -\bar{r}\bar{\rho}\bar{v}' \tag{4.88}$$

A derivation similar to that for Eq. (4.83) leads to the following equation for axisymmetric flow (the derivation is left to the reader as homework problem 4.6):

$$\begin{aligned}(\psi_{\bar{r}})^2\psi_{\bar{x}\bar{x}} - 2\psi_{\bar{x}}\psi_{\bar{r}}\psi_{\bar{x}\bar{r}} + (\psi_{\bar{x}})^2\psi_{\bar{r}\bar{r}} \\ = \frac{(\psi_{\bar{r}})^{\gamma+1}}{\bar{r}^{\gamma-1}}\left[\gamma\omega\left(\psi_{\bar{r}\bar{r}} - \frac{\psi_{\bar{r}}}{\bar{r}}\right) + \omega'(\psi_{\bar{r}})^2\right]\end{aligned} \tag{4.89}$$

Equation (4.89) is the axisymmetric analog to Eq. (4.83). As before, it is a single equation in terms of one unknown, namely, ψ. In principle, Eq. (4.89) is easier to solve than the original coupled system of three equations, namely, Eqs. (4.84–4.86).

We will illustrate a solution of Eq. (4.89) for the case of flow over a slender right-circular cone at zero angle of attack. For this case, we take advantage of

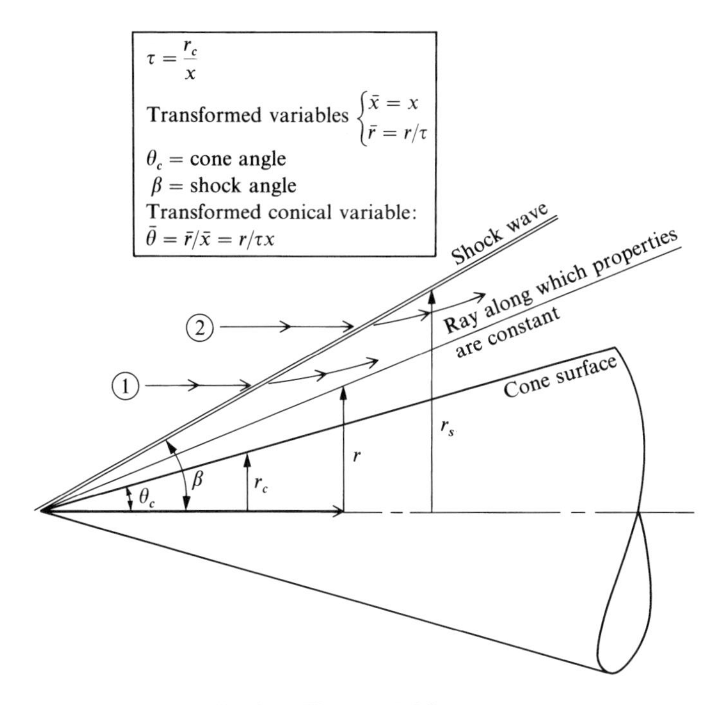

Fig. 4.7 Flow model for a cone.

the nature of conical flow, namely, that flow properties are constant along any ray emanating from the cone vertex. Consider the sketch shown in Fig. 4.7. Along any ray making a slope r/x with respect to the centerline, the flow properties are constant. For this ray, we define a conical variable $\bar{\theta}$ such that

$$\bar{\theta} \equiv \frac{\bar{r}}{\bar{x}} = \frac{r}{\tau x} \tag{4.90}$$

In addition, for conical flow the stream function $\psi(\bar{x}, \bar{r})$ can be expressed as a function of $\bar{x}$ and $\bar{\theta}$ through Eq. (4.90), where $\bar{r} = \bar{x}\bar{\theta}$. A proper form for $\psi = \psi(\bar{x}, \bar{\theta})$ applicable to conical flow is

$$\psi = \bar{x}^2 f(\bar{\theta}) \tag{4.91}$$

An intuitive justification for Eq. (4.91) can be obtained from Fig. 4.7. Recall that for two-dimensional flow the difference in ψ between two streamlines is equal to the mass flow between these streamlines; for an axisymmetric flow the difference in ψ between two stream surfaces (designated 1 and 2 in Fig. 4.7) is equal to the mass flow between these surfaces. This mass flow is proportional to the circular ring of area between stream surfaces 1 and 2, which in turn is proportional to r^2 and thus to $\bar{r}^2$. Hence, it makes sense to define the stream function as $\psi = \bar{r}^2 g(\bar{\theta})$, where $\bar{r}^2$ is proportional to the area, and $g(\bar{\theta})$ yields the flow properties necessary to complete the mass flow expression. However, because $\bar{r} = \bar{x}\,\bar{\theta}$

from Eq. (4.90), then $\psi = \bar{r}^2 g(\bar{\theta}) = \bar{x}^2\bar{\theta}^2 g(\bar{\theta}) = \bar{x}^2 f(\bar{\theta})$, which is Eq. (4.91). We wish to substitute this expression for ψ into Eq. (4.89). To do so, we need expressions for the derivatives of ψ, constructed in the following. In the process, keep in mind that $\psi = \bar{x}^2 f(\bar{\theta})$ and that we are essentially transforming from one set of independent variables, $\bar{x}$ and $\bar{r}$, into another set, $\bar{x}$ and $\bar{\theta}$, where $\bar{x} = \bar{x}$ and $\bar{\theta} = \bar{r}/\bar{x}$. For example, from the chain rule

$$\left(\frac{\partial\psi}{\partial\bar{r}}\right)_{\bar{x}} = \left(\frac{\partial\psi}{\partial\bar{\theta}}\right)_{\bar{x}}\left(\frac{\partial\bar{\theta}}{\partial\bar{r}}\right)_{\bar{x}} + \left(\frac{\partial\psi}{\partial\bar{x}}\right)_{\bar{\theta}}\left(\frac{\partial\bar{x}}{\partial\bar{r}}\right)_{\bar{x}} \tag{4.92}$$

where the subscripts are added to remind the reader what independent variable is being held constant for each of the partial differentiations. From Eq. (4.91),

$$\left(\frac{\partial\psi}{\partial\bar{\theta}}\right)_{\bar{x}} = \bar{x}^2 f'(\bar{\theta}) \tag{4.93}$$

where $f'(\bar{\theta}) \equiv \mathrm{d}f/\mathrm{d}\bar{\theta}$. Also, from Eq. (4.90),

$$\left(\frac{\partial\bar{\theta}}{\partial\bar{r}}\right)_{\bar{x}} = \frac{1}{\bar{x}} \tag{4.94}$$

Because $\bar{x}$ is being held constant in $(\partial\bar{x}/\partial\bar{r})_{\bar{x}}$, then

$$\left(\frac{\partial\bar{x}}{\partial\bar{r}}\right)_{\bar{x}} \equiv 0 \tag{4.95}$$

Substituting Eqs. (4.93–4.95) into Eq. (4.92), and using the subscript notation for partial derivatives, we have

$$\psi_{\bar{r}} = \bar{x} f'(\bar{\theta}) \tag{4.96}$$

Similarly, from the chain rule applied to $\psi_{\bar{r}}$,

$$\psi_{\bar{r}\bar{r}} \equiv \left(\frac{\partial\psi_{\bar{r}}}{\partial\bar{r}}\right)_{\bar{x}} = \left(\frac{\partial\psi_{\bar{r}}}{\partial\bar{\theta}}\right)_{\bar{x}}\left(\frac{\partial\bar{\theta}}{\partial\bar{r}}\right)_{\bar{x}} + \left(\frac{\partial\psi_{\bar{r}}}{\partial\bar{x}}\right)_{\bar{\theta}}\left(\frac{\partial\bar{x}}{\partial\bar{r}}\right)_{\bar{x}}$$

or

$$\psi_{\bar{r}\bar{r}} = f''(\bar{\theta}) \tag{4.97}$$

Also, from the chain rule

$$\psi_{\bar{r}\bar{x}} \equiv \left(\frac{\partial\psi_{\bar{r}}}{\partial\bar{x}}\right)_{\bar{r}} = \left(\frac{\partial\psi_{\bar{r}}}{\partial\bar{\theta}}\right)_{\bar{x}}\left(\frac{\partial\bar{\theta}}{\partial\bar{x}}\right)_{\bar{r}} + \left(\frac{\partial\psi_{\bar{r}}}{\partial\bar{x}}\right)_{\bar{\theta}}\left(\frac{\partial\bar{x}}{\partial\bar{x}}\right)_{\bar{r}} \tag{4.98}$$

From Eq. (4.90),

$$\left(\frac{\partial\bar{\theta}}{\partial\bar{x}}\right)_{\bar{r}} = -\frac{\bar{r}}{\bar{x}^2} \tag{4.99}$$

Noting that $(\partial\bar{x}/\partial\bar{r})_r = 1$, and utilizing Eq. (4.99), Eq. (4.98) becomes

$$\psi_{\bar{r}\bar{x}} = [\bar{x}f''(\bar{\theta})]\left(-\frac{\bar{r}}{\bar{x}^2}\right) + f''(\bar{\theta})$$

or

$$\psi_{\bar{r}\bar{x}} = -\frac{\bar{r}}{\bar{x}}f''(\bar{\theta}) + f'(\bar{\theta}) = -\bar{\theta}f''(\bar{\theta}) + f'(\bar{\theta}) \tag{4.100}$$

Similarly, from the chain rule,

$$\psi_{\bar{x}} \equiv \left(\frac{\partial\psi}{\partial\bar{x}}\right)_{\bar{r}} = \left(\frac{\partial\psi}{\partial\bar{\theta}}\right)_{x}\left(\frac{\partial\bar{\theta}}{\partial\bar{x}}\right)_{\bar{r}} + \left(\frac{\partial\psi}{\partial\bar{x}}\right)_{\bar{\theta}}\left(\frac{\partial\bar{x}}{\partial\bar{x}}\right)_{\bar{r}}$$

or

$$\psi_{\bar{x}} = [\bar{x}^2 f'(\bar{\theta})]\left(-\frac{\bar{r}}{\bar{x}^2}\right) + 2\bar{x}f(\bar{\theta}) = -\bar{r}f'(\bar{\theta}) + 2\bar{x}f(\bar{\theta})$$

or

$$\psi_{\bar{x}} = -\bar{x}\bar{\theta}f'(\bar{\theta}) + 2\bar{x}f(\bar{\theta}) \tag{4.101}$$

Similarly, from the chain rule,

$$\psi_{\bar{x}\bar{x}} \equiv \left(\frac{\partial\psi_{\bar{x}}}{\partial\bar{x}}\right)_{r} = \left(\frac{\partial\psi_{\bar{x}}}{\partial\bar{\theta}}\right)_{\bar{x}}\left(\frac{\partial\bar{\theta}}{\partial\bar{x}}\right)_{\bar{r}} + \left(\frac{\partial\psi_{\bar{x}}}{\partial\bar{x}}\right)_{\bar{\theta}}\left(\frac{\partial\bar{x}}{\partial\bar{x}}\right)_{\bar{r}} \tag{4.102}$$

From Eq. (4.101),

$$\begin{aligned}\left(\frac{\partial\psi_{\bar{x}}}{\partial\bar{\theta}}\right)_{\bar{x}} &= -\bar{x}\bar{\theta}f''(\bar{\theta}) - \bar{x}f'(\bar{\theta}) + 2\bar{x}f'(\bar{\theta})\\ &= -\bar{x}\bar{\theta}f''(\theta) + \bar{x}f'(\bar{\theta})\end{aligned} \tag{4.103}$$

and

$$\left(\frac{\partial\psi_{\bar{x}}}{\partial\bar{x}}\right)_{\bar{\theta}} = -\bar{\theta}f'(\bar{\theta}) + 2f(\bar{\theta}) \tag{4.104}$$

Substituting Eqs. (4.103) and (4.104) into (4.102), we have

$$\psi_{\bar{x}\bar{x}} = [-\bar{x}\bar{\theta}f''(\bar{\theta}) + \bar{x}f'(\bar{\theta})]\left(-\frac{\bar{r}}{\bar{x}^2}\right) + 2f(\bar{\theta}) - \bar{\theta}f'(\bar{\theta})$$

$$= \frac{\bar{r}}{\bar{x}}\bar{\theta}f''(\bar{\theta}) - \frac{\bar{r}}{\bar{x}}f'(\bar{\theta}) + 2f(\bar{\theta}) - \bar{\theta}f'(\bar{\theta})$$

$$= \bar{\theta}^2 f''(\bar{\theta}) - \bar{\theta}f'(\bar{\theta}) + 2f(\bar{\theta}) - \bar{\theta}f'(\bar{\theta})$$

or

$$\psi_{\bar{x}\bar{x}} = \bar{\theta}^2 f''(\bar{\theta}) - 2\bar{\theta}f'(\bar{\theta}) + 2f(\bar{\theta}) \tag{4.105}$$

We have now completed all of our derivative transformations. Substituting Eqs. (4.96), (4.97), (4.100), (4.101), and (4.105) into Eq. (4.89) and noting that $\omega' \equiv d\omega/d\bar{\theta} = 0$ because the entropy, hence ω, is constant between the shock wave and the body, we have

$$\bar{x}^2(f')^2(\bar{\theta}^2 f'' - 2\bar{\theta}f' + 2f) - 2(-\bar{x}\bar{\theta}f' + 2\bar{x}f)(\bar{x}f')(-\bar{\theta}f'' + f')$$
$$+ (-\bar{x}\bar{\theta}f' + 2\bar{x}f)^2 f'' = \frac{(\bar{x})^{\gamma+1}(f')^{\gamma+1}}{(\bar{r})^{\gamma-1}}\left[\gamma\omega\left(f'' - \frac{\bar{x}f'}{\bar{r}}\right)\right]$$

Dividing by $\bar{x}^2$ and grouping coefficients of like powers of $\bar{\theta}$, this becomes

$$\bar{\theta}^2[(f')^2 f'' - 2(f')^2 f'' + (f')^2 f''] + \bar{\theta}[-2(f')^3 + 2(f')^3 + 4ff'f'' - 4ff'f'']$$
$$+ 2(f')^2 f - 4f(f')^2 + 4f^2 f'' = \frac{(\bar{x})^{\gamma-1}(f')^{\gamma+1}}{(\bar{r})^{\gamma-1}}\left[\gamma\omega\left(f'' - \frac{\bar{x}}{\bar{r}}f'\right)\right]$$

Recalling that $\bar{r}/\bar{x} = \bar{\theta}$ and noting that the terms within the square brackets cancel each other, the preceding equation becomes

$$4(f)^2 f'' - 2f(f')^2 = \gamma\omega\frac{(f')^{\gamma+1}}{(\bar{\theta})^{\gamma-1}}\left(f'' - \frac{f'}{\bar{\theta}}\right)$$

or, rearranging

$$\boxed{f'' - \frac{f'}{\bar{\theta}} = \frac{2}{\gamma\omega}\frac{(\bar{\theta})^{\gamma-1}f}{(f')^{\gamma+1}}[2ff'' - (f')^2]} \tag{4.106}$$

Equation (4.106) is the governing equation for hypersonic flow over a slender cone. It was obtained from the system of hypersonic small-disturbance equations; indeed, it replaces that system with a single ordinary differential equation in terms of one unknown, namely, $f(\bar{\theta})$. When a system of partial differential equations is replaced by one or more ordinary differential equations in terms of one independent variable (in this case $\bar{\theta}$), then the solution is said to be *self-similar*. Such is the case here. However, this should be no surprise; the Taylor–Maccoll equation for the *exact* solution of conical flows (for example, see [4]) is also an ordinary differential equation. Hence, Eq. (4.106) can be viewed as the approximate counterpart of the Taylor–Maccoll equation, applicable to hypersonic flow over slender cones.

Question: Why have we gone to such length to obtain Eq. (4.106), when we could more easily use the exact Taylor–Maccoll results to obtain hypersonic (as well as supersonic) flow over cones, as tabulated, for example, in [17] and [18]?

The answer lies in the fact that, in the present section, we are demonstrating an actual solution of the hypersonic small-disturbance equations, and we have chosen to treat the case of a cone specifically because an exact solution exists. In this fashion, by comparing the results we can obtain some feeling for just how accurate this small-disturbance theory is. Moreover, we will also demonstrate how the hypersonic small-disturbance theory leads to a closed-form analytical solution for flows over cones—an advantage not to be enjoyed by the exact numerical Taylor–Maccoll results. Therefore let us proceed to solve Eq. (4.106).

Our next step in treating Eq. (4.106) is to recall that $\omega \equiv \bar{p}/\bar{\rho}^{\gamma}$ and to recognize that ω is a constant for the isentropic conical flow, equal to its value behind the oblique shock wave. An expression for ω can be obtained directly from the hypersonic shock-wave relations derived in Sec. 4.5. Examine Eqs. (4.52) and (4.53) for $\bar{p}$ and $\bar{\rho}$, respectively. For flow over a cone, the shock wave is a straight oblique surface, with a constant transformed slope, that is, in Eqs. (4.52) and (4.53), $(\mathrm{d}\bar{y}/\mathrm{d}\bar{x})_s$ is constant. Moreover,

$$\left(\frac{\mathrm{d}\bar{y}}{\mathrm{d}\bar{x}}\right)_s = \frac{1}{\tau}\frac{\mathrm{d}y}{\mathrm{d}x} \tag{4.107}$$

But for hypersonic flow over a slender body

$$\frac{\mathrm{d}y}{\mathrm{d}x} = \tan\beta \approx \tan\theta_c \approx \tau \tag{4.108}$$

where θ_c is the cone angle. Combining Eqs. (4.107) and (4.108), we see that

$$\left(\frac{\mathrm{d}\bar{y}}{\mathrm{d}\bar{x}}\right)_s \approx 1 \tag{4.109}$$

Inserting the results of Eq. (4.109) into Eqs. (4.52) and (4.53) and returning to the definition of ω, we obtain

$$\omega = \frac{\bar{p}}{\bar{\rho}^{\gamma}} = \frac{2}{\gamma+1}\left(1+\frac{1-\gamma}{2\gamma M_\infty^2\tau^2}\right)\left(\frac{\gamma-1}{\gamma+1}\right)^{\gamma}\left[1+\frac{2}{(\gamma-1)M_\infty^2\tau^2}\right]^{\gamma}$$

Noting that $M_\infty\tau = K$, the hypersonic similarity parameter, the preceding equation can be written as

$$\omega = \frac{2}{\gamma+1}\left(\frac{\gamma-1}{\gamma+1}\right)^{\gamma}\left(1+\frac{1-\gamma}{2\gamma K^2}\right)\left[1+\frac{2}{(\gamma-1)K^2}\right]^{\gamma} \tag{4.110}$$

Clearly, $\omega = \omega(K)$; it is a constant for a given flow with a given value of K. Reflecting again on Eq. (4.106), the solution to the flowfield in the form of $f(\bar{\theta})$ will depend on K as a parameter because ω in Eq. (4.106) is a function of K. This is yet another example of the close relationship between solutions of the hypersonic small-disturbance equations, hypersonic similarity, and the hypersonic similarity parameter K.

The solution of Eq. (4.106) must satisfy boundary conditions at the body and behind the shock wave. Let us address these boundary conditions by first noting the values of $\bar{\theta}$ on the body and at the shock wave. At the body (the surface of the cone with semiangle θ_c)

$$\bar{\theta} = \bar{\theta}_c = \frac{\bar{r}_c}{\bar{x}} = \frac{r_c}{\tau x}$$

However, from Fig. 4.7, r_c/x is precisely τ. Thus, from the preceding equation,

$$\text{At the body} \qquad \bar{\theta} = \bar{\theta}_c = 1$$

At the shock wave, with wave angle β,

$$\bar{\theta} = \frac{\bar{r}_s}{\bar{x}} = \frac{1}{\tau}\frac{r_s}{x} \tag{4.111}$$

From Fig. 4.7, and noting that, for hypersonic flow over a slender body, β is small,

$$\frac{r_s}{x} = \tan\beta \approx \beta \tag{4.112}$$

Thus, combining Eqs. (4.111) and (4.112),

$$\bar{\theta} = \frac{1}{\tau}\tan\beta = \frac{\beta}{\tau} \tag{4.113}$$

From Fig. 4.7, for a slender cone, we note that

$$\tau = \frac{r_c}{x} = \tan\theta_c \approx \theta_c$$

Hence, Eq. (4.113) can be written as

$$\text{At the shock} \qquad \bar{\theta} = \bar{\theta}_s = \frac{\beta}{\tau} = \frac{\beta}{\theta_c} = \frac{\text{shock angle}}{\text{cone angle}} \tag{4.113a}$$

The boundary conditions for Eq. (4.106) are the known values of $f(1)$ at the surface and $f(\beta/\tau)$ and $f'(\beta/\tau)$ at the shock wave. These values are known as follows. First, at the surface we know (by definition of the stream function) that $\psi = 0$. From Eq. (4.91) applied at the surface, we have $\psi = \bar{x}^2 f(\bar{\theta}) = \bar{x}^2 f(1) = 0$. Thus, the body boundary condition is

$$\boxed{\text{At the body} \qquad f(1) = 0} \tag{4.114}$$

At the shock wave, we can obtain values of both f and f' by using Eqs. (4.96) and (4.101) as follows. From Eq. (4.96),

$$f' = \frac{\psi_{\bar{r}}}{\bar{x}} \tag{4.115}$$

Substituting Eq. (4.115) into Eq. (4.101), we have

$$\psi_{\bar{x}} = -\bar{r}f' + 2\bar{x}f = -\frac{\bar{r}}{\bar{x}}\psi_{\bar{r}} + 2\bar{x}f \tag{4.116}$$

Substituting Eqs. (4.87) and (4.88) into Eq. (4.116), we obtain

$$-\bar{r}\bar{\rho}\bar{v}' = -\frac{\bar{r}}{\bar{x}}(\bar{r}\bar{\rho}) + 2\bar{x}f$$

Solving for f,

$$f = \frac{1}{2}\left[\left(\frac{\bar{r}}{\bar{x}}\right)^2 \bar{\rho} - \left(\frac{\bar{r}}{\bar{x}}\right)\bar{\rho}\bar{v}'\right] \tag{4.117}$$

At the shock wave, Eq. (4.117) becomes

$$f\left(\frac{\beta}{\tau}\right) = \frac{\bar{\rho}}{2}\left[\left(\frac{\bar{r}_s}{\bar{x}}\right)^2 - \left(\frac{\bar{r}_s}{\bar{x}}\right)\bar{v}'\right] \tag{4.118}$$

However, as noted in Eq. (4.113a),

$$\frac{\bar{r}_s}{\bar{x}} = \bar{\theta}_s = \frac{\beta}{\tau}$$

Hence, Eq. (4.118) becomes

$$\boxed{\text{At the shock} \qquad f\left(\frac{\beta}{\tau}\right) = \frac{\bar{\rho}}{2}\left[\left(\frac{\beta}{\tau}\right)^2 - \left(\frac{\beta}{\tau}\right)\bar{v}'\right]} \tag{4.119}$$

To obtain f' at the shock, substitute Eq. (4.87) into Eq. (4.96):

$$f' = \frac{\psi_r}{\bar{x}} = \frac{\bar{r}\bar{\rho}}{\bar{x}}$$

At the shock, this becomes

$$f'\left(\frac{\beta}{\tau}\right) = \frac{\bar{r}_s}{\bar{x}}\bar{\rho} = \left(\frac{\beta}{\tau}\right)\bar{\rho}$$

Thus,

$$\boxed{\text{At the shock} \qquad f'\left(\frac{\beta}{\tau}\right) = \left(\frac{\beta}{\tau}\right)\bar{\rho}} \tag{4.120}$$

In both Eqs. (4.119) and (4.120), the values of $\bar{\rho}$ and $\bar{v}'$ are those values immediately behind the shock wave, given by Eqs. (4.52) and (4.55), respectively. Recalling from Eq. (4.109) that $(\mathrm{d}\bar{y}/\mathrm{d}\bar{x})_s = 1$ and noting that $M_\infty\tau = K$, then Eq. (4.52) becomes

$$\bar{\rho} = \frac{\gamma+1}{\gamma-1}\left\{\frac{1}{1+2/[(\gamma-1)K^2]}\right\} \tag{4.121}$$

and Eq. (4.55) becomes

$$\bar{v}' = \frac{2}{\gamma+1}\left(1-\frac{1}{K^2}\right) \tag{4.122}$$

In summary, the boundary conditions for Eq. (4.106) are given by Eq. (4.114) at the body and Eqs. (4.119) and (4.120) at the shock wave, wherein the values of $\bar{\rho}$ and $\bar{v}'$ in Eqs. (4.119) and (4.120) are given by Eqs. (4.121) and (4.122).

We are now in a position to set up a straightforward numerical solution to Eq. (4.106) for the hypersonic flow over a slender cone. In most practical cases, we are interested in the flow over a cone of specified angle θ_c (or equivalently specified slenderness ratio τ) with a specified M_∞. However, keep in mind that, within the framework of hypersonic small-disturbance theory, M_∞ or τ individually are not germane; the solutions depend only on the *product*, $M_\infty\tau = K$. Scan over the equations we are dealing with, namely, Eqs. (4.106), (4.110), (4.114), and (4.19–4.122); note that K is the parameter that appears, not M_∞ or τ by themselves. (τ also appears in the ratio β/τ, which is one of the unknowns of the problem—to be obtained as part of the solution.) Therefore, let us *specify the value of* K, and set up a numerical solution for this value of K as follows:

1) *Assume* a value of β/τ (a suggested value might be 1.1 for $\gamma = 1.4$). Note that this establishes an *assumed* value for the shock-wave angle β.

2) Starting at $\bar{\theta} = \beta/\tau$, that is, starting at the shock wave, with boundary values of $f(\beta/\tau)$ and $f'(\beta/\tau)$ given by Eqs. (4.119) and (4.120), respectively, numerically integrate Eq. (4.106) in steps of $(-\Delta\bar{\theta})$, that is, in the direction of decreasing $\bar{\theta}$, that is, starting at the shock wave, integrate Eq. (4.106) in the direction toward the body. This integration can be carried out by any standard

numerical technique for a nonlinear ordinary differential equation, such as the Runge–Kutta method.

3) Continue this integration until $\bar{\theta}$ reaches the value $\bar{\theta} = 1$. Then check to see if the body boundary condition (4.114) is satisfied; that is, is the relation $f(1) = 0$ satisfied by the numerical integration? If not, assume a new value of β/τ, and repeat steps 2 and 3. Repeat this process until the proper value of β/τ is found such that $f(1) = 0$.

4) We have now arrived at the final result. For the specified value of K, we have found the ratio of wave angle to cone angle β/τ, and we have obtained numerical values of f and f' between the shock (where $\bar{\theta} = \beta/\tau$) and the body (where $\bar{\theta} = 1$).

After the preceding numerical procedure is completed, the conventional flowfield variables can be obtained from f and f'. For example, from Eqs. (4.77), (4.78), and (4.96), we can obtain the pressure as

$$\bar{p} = \omega\bar{\rho}^{\gamma} = \omega\left(\frac{\psi_{\bar{r}}}{\bar{r}}\right)^{\gamma} = \omega\left(\frac{\bar{x}f'}{\bar{r}}\right)^{\gamma} = \omega\left(\frac{f'}{\bar{\theta}}\right)^{\gamma} \tag{4.123}$$

In turn, the pressure coefficient can be obtained from

$$C_p = \frac{2}{\gamma M_\infty^2}\left(\frac{p}{p_\infty} - 1\right) = \frac{2}{\gamma M_\infty^2}\left[\frac{p}{p_\infty \gamma M_\infty^2 \tau^2}(\gamma M_\infty^2 \tau^2) - 1\right]$$

$$= \frac{2}{\gamma M_\infty^2}[\bar{p}(\gamma M_\infty^2 \tau^2) - 1]$$

Dividing by τ^2, and noting that $K = M_\infty \tau$, the preceding equation becomes

$$\frac{C_p}{\tau^2} = \frac{2}{\gamma K^2}(\gamma K^2 \bar{p} - 1)$$

Substituting Eq. (4.123) into the preceding, we obtain

$$\frac{C_p}{\tau^2} = \frac{2}{\gamma K^2}\left[\gamma K^2 \omega\left(\frac{f'}{\bar{\theta}}\right)^{\gamma} - 1\right] \tag{4.124}$$

The pressure coefficient on the cone surface can be obtained by inserting $\bar{\theta} = 1$ and the numerically obtained value of $f'(1)$ into Eq. (4.124).

A numerical solution to the preceding problem was first obtained by Van Dyke [30]. Van Dyke's formulation differs from our preceding derivation in that he defines K as $M_\infty\beta$ and utilizes β instead of τ in the nondimensional variables. This has an advantage in the numerical solution of Eq. (4.106) because his conical coordinate is defined as $r/\beta x$ [in contrast to our $r/\tau x$, from Eq. (4.90)]. In turn, $r/\beta x$ at the shock wave is unity, and hence $f(1)$ and $f'(1)$ denote values at the shock wave, in contrast to our formulation where $f(\beta/\tau)$ and $f'(\beta/\tau)$ denote values at the shock wave. Because β/τ is an unknown, we were led to an *iterative* numerical solution, assuming values of β/τ until we converged on the proper body boundary condition. In Van Dyke's approach, no iteration is necessary; starting with $f(1)$ and $f'(1)$ at the shock, he simply integrates until $f = 0$ (the body boundary condition). The value of his conical coordinate at $f = 0$ yields the ratio of wave angle to cone angle. For pedagogical reasons, we have deliberately chosen not to follow Van Dyke in this regard; instead, we maintained a consistent usage of the *body* slenderness ratio τ (instead of β), and $K = M_\infty \tau$, throughout our development, because such usage was introduced right from the beginning of this chapter having to do with hypersonic similarity, where K was initially defined as $M_\infty \tau$ (not $M_\infty \beta$). Moreover, in practical applications, involving a given body, we know τ, while β is usually an unknown; hence, the practical hypersonic similarity parameter is $M_\infty \tau$, not $M_\infty \beta$. Of course, in the final solution the flowfield results are the same, no matter which approach is taken.

Figure 4.8 shows the final results for C_p/τ^2 on the surface of the cone, as reported in [30]. Note from our numerical solution that a specific value of C_p/τ^2 on the cone corresponds to the specified value of K. When another value of K is chosen, another value of C_p/τ^2 is obtained from the solution, that is, C_p/τ^2 is a function of K, as known from our previous work. This function is given by the numerical results shown in Fig. 4.8, where C_p/τ^2 is plotted vs $K = M_\infty \tau$. The upper line is the present numerical solution; the two lower lines are exact conical flow results from Kopal [17] for cones of 10- and 15-half-angles. (The solid circles in Fig. 4.8 correspond to a closed-form analytical expression, to be discussed subsequently.) The value of Fig. 4.8 is that it illustrates the degree of accuracy of the hypersonic small-disturbance theory when compared with exact results; reasonable accuracy is indeed obtained over a wide range of values of K. The agreement is better for the more slender cone, as expected. Recall our earlier statement that the application of hypersonic small-disturbance

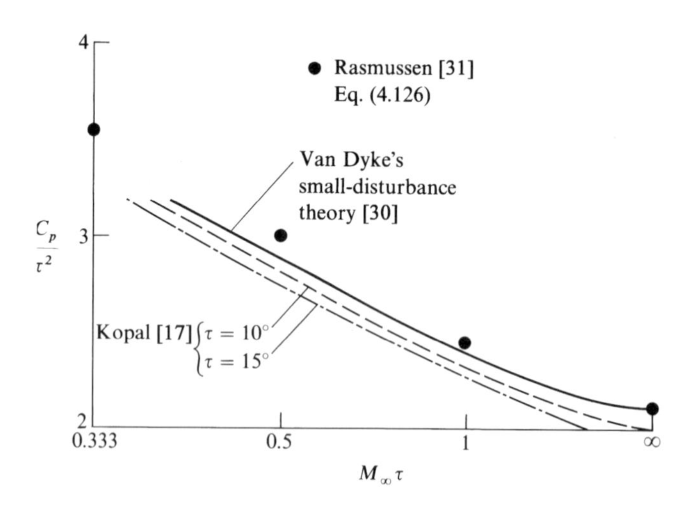

Fig. 4.8 Cone surface pressure: comparison between exact theory [17], hypersonic small-disturbance theory [30], and analytical formula [31].

theory to the flow over a cone is given here partly as an academic exercise—an exercise to demonstrate for a relatively simple flow what the hypersonic small-disturbance theory is all about.

There is another reason for treating the case of a cone. For this case, the hypersonic small-disturbance theory leads to a *closed-form analytical result* for C_p and β for a given τ and M_∞. This now becomes much more than just an academic exercise because a closed-form analytic result for C_p for hypersonic flow over cones allows some very practical engineering calculations. For example, the tangent-cone method discussed in Sec. 3.6 becomes even simpler and more useful if we have a *formula* for C_p on a cone, rather than constantly having to look up values in the Kopal tables [17]. Moreover, for certain optimization studies of hypersonic vehicles using the calculus of variations, a closed-form expression for C_p is absolutely necessary. Therefore, we will end this section by discussing such closed-form results, thus illustrating one of the most useful advantages of hypersonic small-disturbance theory.

Starting with Eq. (4.106), Rasmussen [31] integrated twice from the shock wave, obtaining an integral equation for $f(\theta)$. By successive approximation, this led to closed-form analytical expressions for both $f(\bar{\theta})$ and $f'(\theta)$ as functions of $\bar{\theta}$. The details are described in [31], which the reader is encouraged to examine; hence, no further elaboration will be given here. Utilizing the fact that $f(\bar{\theta}) = 0$ at the body surface, Rasmussen obtained the following closed-form expression for the shock-wave angle:

$$\boxed{K_\beta = K\sqrt{\frac{\gamma+1}{2} + \frac{1}{K^2}}} \tag{4.125}$$

where $K_\beta = M_\infty \beta$. Furthermore, by substituting his closed-form result for $f'(\bar{\theta})$ into Eq. (4.124) Rasmussen obtained the following expression for the pressure coefficient on a cone:

$$\boxed{\frac{C_p}{\theta_c^2} = 1 + \frac{(\gamma+1)K^2+2}{(\gamma-1)K^2+2}\,\ell n\left(\frac{\gamma+1}{2} + \frac{1}{K^2}\right)} \tag{4.126}$$

In his analysis, Rasmussen approximated $\tau = \tan\theta_c$ by θ_c itself; hence, in Eq. (4.126), $K = M_\infty \theta_c$. Results from Eq. (4.126) are plotted as the solid circles in Fig. 4.8. Note that Eq. (4.126) agrees well with the numerical results of Van Dyke when $K > 1$. Rasmussen observed that Eq. (4.126) agrees well with the exact cone results (say, from Kopal [17]) when τ is small and M_∞ is large; however, better agreement for larger values of τ is obtained when $\tau \approx \theta_c$ is replaced by $\sin\theta_c$. These results are shown in Fig. 4.9, where $C_p/\sin^2\theta_c$ is plotted vs $M_\infty \sin\theta_c$. The open symbols are exact results from Kopal, and the solid line is from Eq. (4.126), with θ_c replaced by $\sin\theta_c$. Excellent agreement is obtained, even for a reasonably large cone semi-angle of 30 deg.

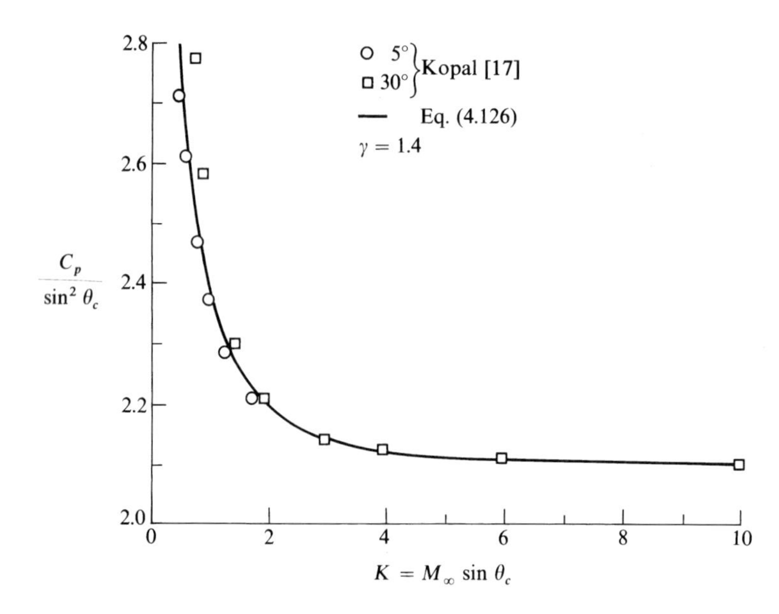

Fig. 4.9 Cone surface pressure: comparison of Rasmussen's formula [31] with exact results [17].

Doty and Rasmussen [32] extended this work to include angle-of-attack effects. Defining the normal-force coefficient as

$$C_N = \frac{N}{\frac{1}{2}\rho_\infty V_\infty^2 A}$$

where A = base area, a closed-form expression for the slope of the moment coefficient curve $dC_N/d\alpha$ was obtained in the following form:

$$\frac{1}{\cos^2\theta}\left(\frac{dC_N}{d\alpha}\right) = \frac{1}{\varepsilon_0(1-\varepsilon_0)}\frac{8}{\gamma+1} + (1+g)\left[\varepsilon_0 - \frac{\gamma+5}{\gamma+1}\right] - \left(\frac{1-g}{2}\right)(1-\varepsilon_0)^2\left[1 + \frac{1+\varepsilon}{\varepsilon_0^{1/2}}\ell n\left(\frac{1+\varepsilon_0^{1/2}}{\sqrt{1-\varepsilon_0}}\right)\right] \tag{4.127}$$

where

$$\varepsilon_0 = \frac{2+(\gamma-1)M_\infty^2\sin^2\theta}{2+(\gamma+1)M_\infty^2\sin^2\theta}$$

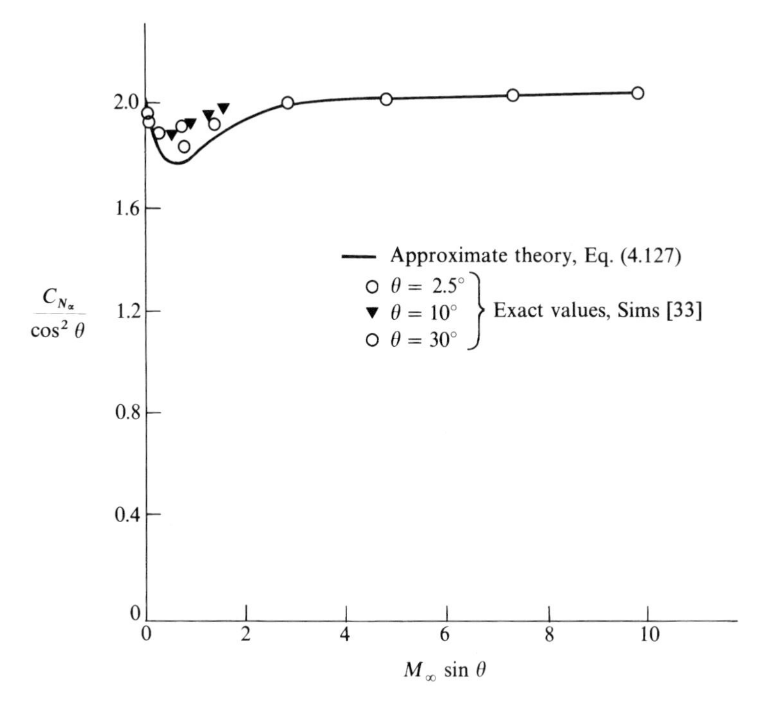

Fig. 4.10 Slope of the normal-force coefficient for slender cones. Comparison between Rasmussen's formula and exact results.

and

$$g = \frac{g_N}{g_D}$$

$$g_N = (5 - \varepsilon_0)(1 - \varepsilon_0) - 2(2 - \varepsilon_0)\frac{\varepsilon_0 - (\gamma - 3)}{(\gamma + 1)} - \ell n \frac{1 + \varepsilon_0^{1/2}}{(1 - \varepsilon_0)^{1/2}} \frac{(1 - \varepsilon_0)^3}{\varepsilon_0^{1/2}}$$

$$g_D = 5(1 - \varepsilon_0)^2 + 2(2 - \varepsilon_0)\frac{\varepsilon_0 - (\gamma + 5)}{(\gamma + 1)} - \ell n \frac{1 + \varepsilon_0^{1/2}}{(1 - \varepsilon_0)^{1/2}} \frac{(1 - \varepsilon_0)^3}{\varepsilon_0^{1/2}}$$

The results for $dC_N/d\alpha$ are shown in Fig. 4.10, where Eq. (4.127) is compared with the results of Sims [33].

A further extension to elliptic cones at angle of attack is made in [34], which should be consulted for details.

4.7 Comment on Hypersonic Small-Disturbance Theory

Small-disturbance (small-perturbation) theories abound in aerodynamics. In the areas of subsonic and supersonic aerodynamics, the small-perturbation approach leads to linear theories, with correspondingly simple results (for example, see [4] and [5]). In contrast, hypersonic flow is inherently nonlinear—even the small-perturbation theory for hypersonic flow is nonlinear. As a consequence, the hypersonic small-disturbance theory is more elaborate and leads to more complex results. For proof, just compare the lengthy discussions we have presented in this chapter with the analogous simple discussions for subsonic and supersonic flow that you can find in standard textbooks, such as [4] and [5]. However, in spite of its nonlinearity, hypersonic small-disturbance theory does provide useful results for the analysis of hypersonic flow over slender bodies—witness the principle of hypersonic similarity (Sec. 4.5), the self-similar solutions obtained in Sec. 4.6, as well as the closed-form analytical expressions presented at the end of Sec. 4.6. For these reasons, hypersonic small-disturbance theory occupies a relatively high status within the general class of approximate flowfield solutions for hypersonic flow.

Referring to our road map in Fig. 1.24, we now leave this subject, and for the remainder of this chapter we move on to two other approximate hypersonic flowfield methods, namely, blast-wave theory and thin shock-layer theory. Also, in regard to our chapter road map in Fig. 4.1, we continue to travel down the center section.

4.8 Hypersonic Equivalence Principle and Blast-Wave Theory

Return to the Euler equations given in Sec. 4.2, namely, Eqs. (4.1–4.5) and (4.6). Let us write these equations for an *unsteady, two-dimensional flow* in the y-z plane. (Note that in our previous work, the x axis is in the freestream direction; hence, the y-z plane is *perpendicular* to the freestream direction.) Because we are dealing with flow in the y-z plane only, $u = 0$ in Eqs. (4.1–4.4) and (4.6), yielding

$$\frac{\partial \rho}{\partial t} + \frac{\partial(\rho v)}{\partial y} + \frac{\partial(\rho w)}{\partial z} = 0 \tag{4.128}$$

$$\rho\frac{\partial v}{\partial t} + \rho v\frac{\partial v}{\partial y} + \rho w\frac{\partial v}{\partial z} = -\frac{\partial p}{\partial y} \tag{4.129}$$

$$\rho\frac{\partial w}{\partial t} + \rho v\frac{\partial w}{\partial y} + \rho w\frac{\partial w}{\partial z} = -\frac{\partial p}{\partial z} \tag{4.130}$$

$$\frac{\partial}{\partial t}\left(\frac{p}{\rho^\gamma}\right) + v\frac{\partial}{\partial y}\left(\frac{p}{\rho^\gamma}\right) + w\frac{\partial}{\partial z}\left(\frac{p}{\rho^\gamma}\right) = 0 \tag{4.131}$$

Let us nondimensionalize these equations as follows. Let

$$\tilde{\rho} = \frac{\rho}{\rho_\infty} \quad \tilde{v} = \frac{v}{V_\infty} \quad \tilde{w} = \frac{w}{V_\infty} \quad \tilde{p} = \frac{p}{\rho_\infty V_\infty^2}$$

$$\tilde{t} = \frac{t}{(l/V_\infty)} \quad \tilde{y} = \frac{y}{l} \quad \tilde{z} = \frac{z}{l}$$

In the preceding, ρ_∞ and V_∞ can be treated as reference quantities. You might ask what physical meaning they have in terms of an unsteady two-dimensional flow in the y-z plane. Indeed, there is no physical meaning necessary here; both ρ_∞ and V_∞ are just reference quantities. However, we will use ρ_∞ and V_∞ as we have before, namely, ρ_∞ is some freestream density and V_∞ is some freestream velocity in the x direction, and their physical connection with the unsteady two-dimensional flow in the y-z plane will be made later. Then Eqs. (4.128–4.131) become

$$\frac{\partial\tilde{\rho}}{\partial\tilde{t}}+\frac{\partial(\tilde{\rho}\tilde{v})}{\partial\tilde{y}}+\frac{\partial(\tilde{\rho}\tilde{w})}{\partial\tilde{z}}=0 \tag{4.132}$$

$$\tilde{\rho}\frac{\partial\tilde{v}}{\partial\tilde{t}}+\tilde{\rho}\tilde{v}\frac{\partial\tilde{v}}{\partial\tilde{y}}+\tilde{\rho}\tilde{w}\frac{\partial\tilde{v}}{\partial\tilde{z}}=-\frac{\partial\tilde{p}}{\partial\tilde{y}} \tag{4.133}$$

$$\tilde{\rho}\frac{\partial\tilde{w}}{\partial\tilde{t}}+\tilde{\rho}\tilde{v}\frac{\partial\tilde{w}}{\partial\tilde{y}}+\tilde{\rho}\tilde{w}\frac{\partial\tilde{w}}{\partial\tilde{z}}=-\frac{\partial\tilde{p}}{\partial\tilde{z}} \tag{4.134}$$

$$\frac{\partial}{\partial\tilde{t}}\left(\frac{\tilde{p}}{\tilde{\rho}^\gamma}\right)+\tilde{v}\frac{\partial}{\partial\tilde{y}}\left(\frac{\tilde{p}}{\tilde{\rho}^\gamma}\right)+\tilde{w}\frac{\partial}{\partial\tilde{z}}\left(\frac{\tilde{p}}{\tilde{\rho}^\gamma}\right)=0 \tag{4.135}$$

Now, with Eqs. (4.132–4.135) in sight, turn back to Eqs. (4.41–4.45), and compare these two sets of equations; note that, other than slightly different symbols, they are *identical sets of partial differential equations*. On one hand, Eqs. (4.41–4.45) are the hypersonic small-disturbance equations, which govern the *steady, three-dimensional flow* over a hypersonic slender body. On the other hand, Eqs. (4.132–4.135) govern an *unsteady, two-dimensional flow*. However, because the sets of equations for these two cases are identical, there obviously is an equivalence between these two types of flow. This is the mathematical justification for the *hypersonic equivalence principle*, which can be traced back to Hayes in [29]. Simply stated, we have: *the hypersonic equivalence principle: the steady hypersonic flow over a slender body is equivalent to an unsteady flow in one less space dimension.* Furthermore, examining these two sets of equations further, note that the symbols $\bar{x}$ in Eqs. (4.41–4.45) and $\tilde{t}$ in Eqs. (4.132–4.135) are equivalent, that is,

$$\bar{x}=\frac{x}{l}=\tilde{t}=\frac{tV_\infty}{l} \tag{4.136}$$

Thus, from Eq. (4.136), we have

$$\boxed{x=V_\infty t} \tag{4.137}$$

Equation (4.137) is useful in the physical interpretation of the hypersonic equivalence principle, to be discussed next.

The preceding equivalence was established mathematically. It can be established on a *physical basis*, as well. To see this, consider the sketch shown in Fig. 4.11. Visualize a fixed (y-z) plane perpendicular to the page, as illustrated by the vertical lines at the left. A hypersonic body moving at velocity V_∞ penetrates this plane. (In Fig. 4.11, the body is shown as a body of revolution, but, in general, the body can have an arbitrary cross section.) The trace of the body and its shock wave on the y-z plane at three separate times is shown at the right of

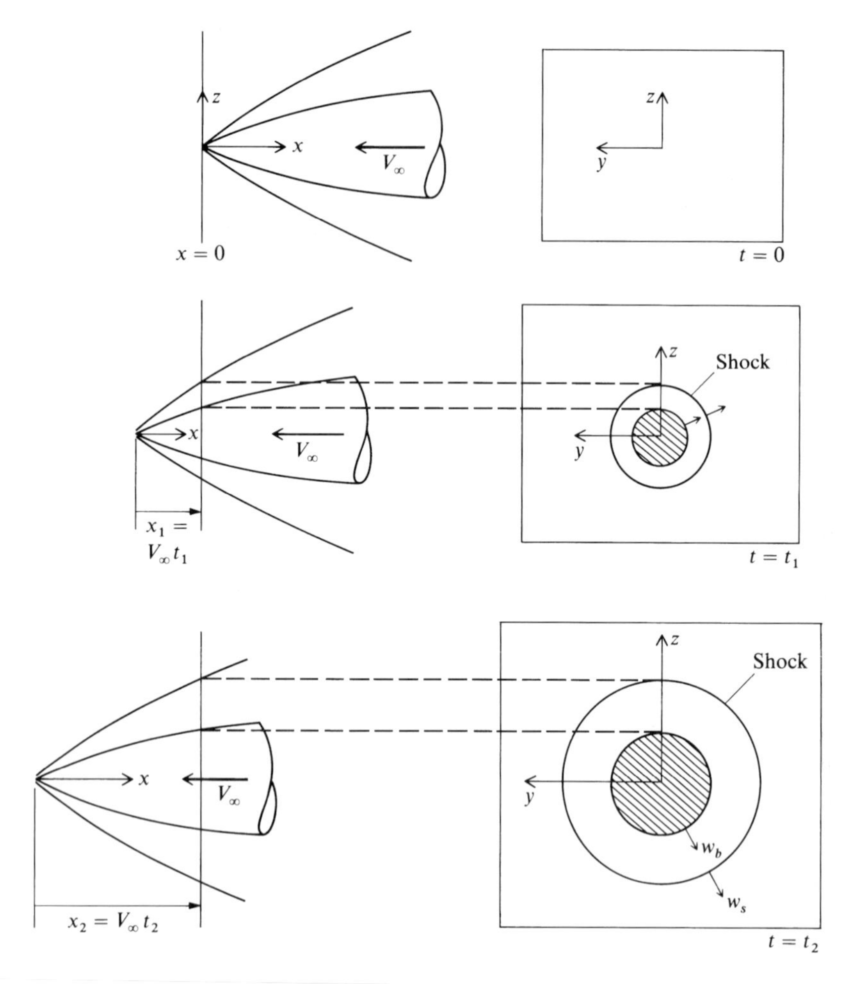

Fig. 4.11 Illustration of the hypersonic equivalence principle; three-dimensional steady flow and an equivalent two-dimensional unsteady flow.

Fig. 4.11. In the y-z plane, the changing body shape looks like an expanding cylindrical piston moving at velocity w_b, driving a cylindrical shock outward at velocity w_s. Because of the hypersonic equivalence principle, the unsteady flow in the y-z plane at the right of Fig. 4.11 shown at various times $t = t_1$, t_2, etc., gives the corresponding steady flow results in the y-z planes located at various corresponding values of $x = x_1$, x_2, etc., shown at the left, where $x = V_\infty t$. Therefore, we see how the steady hypersonic flow over a body (the left-hand side of Fig. 4.11) can be constructed from an unsteady flow in one less space dimension (the right-hand side of Fig. 4.11).

Question: What is the practical advantage of this equivalence?

The answer lies in the fact that solutions of the unsteady one-dimensional flow driven by a moving flat-faced piston and the unsteady two-dimensional flow driven by a radially expanding circular piston (the case shown at the right in Fig. 4.11) exist in the classical literature. An excellent source for these classical solutions is the book by Sedov [35]. These solutions are carried out by self-similar methods, wherein Eqs. (4.132–4.135) are reduced to a simpler set of ordinary differential equations. We will not go into the lengthy details here; see [35] for a discussion of these self-similar solutions. The important point here is that a solution to the unsteady flow shown at the right of Fig. 4.11 does indeed exist in the literature, and because of the hypersonic equivalence principle this solution can be carried over directly to the hypersonic steady flow shown at the left of Fig. 4.11. Moreover, solutions to the classical unsteady flow problem existed *before* the advent of major interest in hypersonic aerodynamics in the 1950s and therefore were waiting there, in the literature, to be of help to hypersonic aerodynamicists when the time came.

To further illustrate the hypersonic equivalence principle, consider a simpler case, for example, the flow over a two-dimensional airfoil with chord length c, as shown at the left of Fig. 4.12 (obtained from [8]). As the airfoil penetrates the fixed vertical plane (fixed vertical slit), the body motion acts like a one-dimensional piston moving in the z direction. This piston motion is shown in the z–t wave diagram at the right of Fig. 4.12. Note that, as the airfoil passes through the vertical plane, the equivalent piston motion is first toward increasing z, reaching a maximum z (corresponding to the maximum airfoil thickness), and then retreating toward decreasing z. The resulting unsteady shock and Mach waves are shown in the z–t wave diagram on the right of Fig. 4.12. These waves are directly equivalent to the steady shock and Mach waves over the airfoil, on the left side of Fig. 4.12, where again $x = V_\infty t$. As before, the known, classical solution of the unsteady one-dimensional flow shown on the

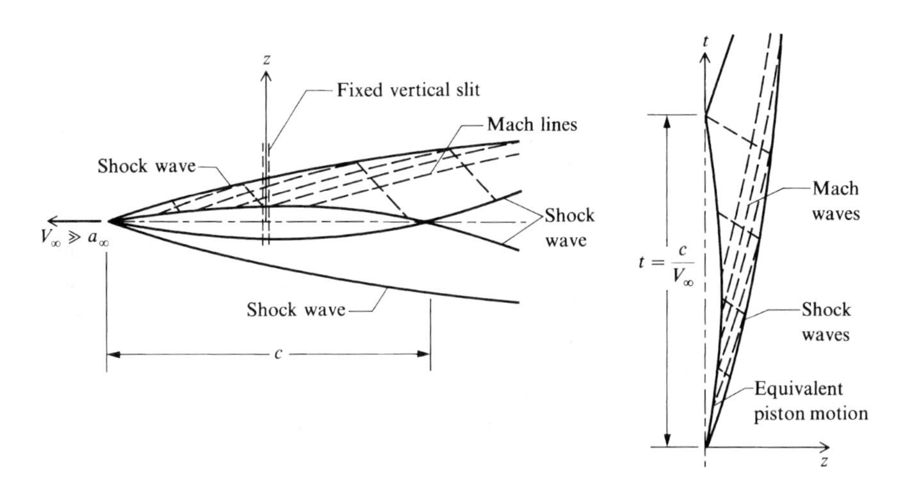

Fig. 4.12 Illustration of the hypersonic equivalence principle; two-dimensional steady flow and an equivalent one-dimensional unsteady flow [8].

right can be carried over directly to construct the steady two-dimensional flow on the left. Note that if the airfoil shape is given by

$$z = z_{\max} f\left(\frac{x}{c}\right)$$

then the equivalent piston motion is

$$z = z_{\max} f\left(\frac{t}{t_0}\right)$$

where t is measured from the instant the leading edge of the airfoil contacts the vertical plane and t_0 is the duration of piston motion, $t_0 = c/V_\infty$. Also, in the steady flow picture shown at the left, let w_b be the value of the vertical component of the flow velocity on the body surface where the slope of the body is $\mathrm{d}z/\mathrm{d}x$:

$$w_b = V_\infty \frac{\mathrm{d}z}{\mathrm{d}x} \tag{4.138}$$

Through the equivalence principle, this is exactly the same as the flow velocity in the z direction adjacent to the face of the piston in the unsteady flow picture w_p, where

$$w_p = \frac{\mathrm{d}z}{\mathrm{d}t} \tag{4.139}$$

Because $x = V_\infty t$, we have from Eqs. (4.138) and (4.139),

$$w_b = V_\infty \frac{\mathrm{d}z}{\mathrm{d}x} = \frac{V_\infty}{V_\infty} \frac{\mathrm{d}z}{\mathrm{d}t} = w_p$$

which is consistent with the equivalence principle, namely, the piston velocity is the same as the vertical velocity of the body surface as seen from the fixed vertical plane penetrated by the body. If we divide Eq. (4.138) by the freestream speed of sound a_∞, we obtain

$$\frac{w_b}{a_\infty} = \frac{V_\infty}{a_\infty}\left(\frac{\mathrm{d}z}{\mathrm{d}x}\right) = M_\infty \tan\theta \tag{4.140}$$

where θ is the local inclination angle of the surface. For small θ, $\tan\theta = \theta$. Moreover, the order of $\theta_{\max}$ is on the order of $z_{\max}/c$. From Eq. (4.140), we obtain

$$\left(\frac{w_b}{a_\infty}\right)_{\max} = \left(\frac{w_p}{a_\infty}\right)_{\max} = M_\infty \theta_{\max} = \mathcal{O}\left(M_\infty \frac{z_{\max}}{c}\right) \tag{4.141}$$

If we consider $z_{\max}/c$ as a measure of the slenderness ratio of the airfoil, that is, $z_{\max}/c = \tau$, then, from Eq. (4.141),

$$\left(\frac{w_b}{a_\infty}\right)_{\max} = \left(\frac{w_p}{a_\infty}\right)_{\max} = \mathcal{O}[K] \tag{4.142}$$

where $K = M_\infty \tau$ is the familiar hypersonic similarity parameter. Equation (4.142) indicates two points:

1) The conditions for the hypersonic equivalence principle are the same as those for hypersonic similarity, which makes absolute sense considering that the hypersonic small-disturbance equations are the basis for both lines of thought.

2) The hypersonic similarity parameter K can be given some physical significance on its own, namely, that it is on the same order as the maximum-disturbance Mach number in the shock layer. (This has already been demonstrated for all practical purposes in our previous discussions involving K; the present development is simply a reinforcement.)

An important variation on the hypersonic equivalence principle is the application of *blast-wave theory*. Returning to the right side of Fig. 4.11, note that the unsteady shock-wave motion and ensuing flowfield are driven mechanically by an expanding piston. A similar unsteady flow can also be driven by the instantaneous release of energy at the origin, as sketched on the right of Fig. 4.13. Here, at time $t = 0$, a large amount of energy is released at a point in the y-z plane. A strong cylindrical shock wave propagates from the point of energy release. It can be argued that the unsteady two-dimensional flow shown at the right of Fig. 4.13 is equivalent to the steady three-dimensional flow over a blunt-nosed slender

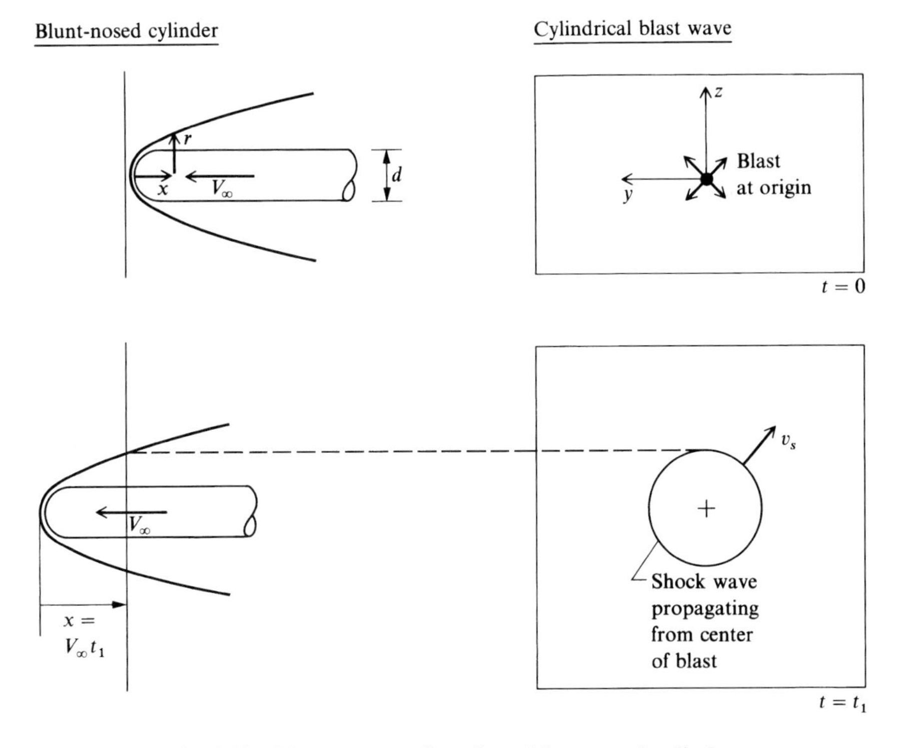

Fig. 4.13 Blast-wave analogy for a blunt-nosed cylinder.

body, where the blunt nose, in "blasting through" the fixed vertical plane, provides the equivalent instantaneous energy release shown on the right. Results obtained from this equivalency are called *blast-wave results*. Such results have been used to estimate the pressure distribution on axisymmetric blunt-nosed cylinders at hypersonic speeds, with the cylindrical axis aligned in the direction of the flow as sketched on the left in Fig. 4.13. Blast-wave results have also been used to estimate the pressure distribution on two-dimensional slabs with blunt leading edges in hypersonic flow; such a body is sketched on the left of Fig. 4.14. Here, the blunt nose, in blasting through the vertical y-z plane, represents a concentrated line of energy release, which drives planar shock waves in both the upward and downward directions, as sketched on the right of Fig. 4.14. The shock waves shown on the right of both Figs. 4.13 and 4.14 are called blast waves because they are created in both cases by the instantaneous release of large amounts of energy, as would be the case of a concentrated explosion, or blast. In these applications, the blast-wave results provide pressure distributions on the flat surface *downstream* of the blunt nose as well as shock-wave shapes in the same region; the pressure distribution and flowfield in the nose region itself is quite another problem and is *not* provided by blast-wave theory. (The detailed blunt-body flowfield is discussed in Chapter 5.)

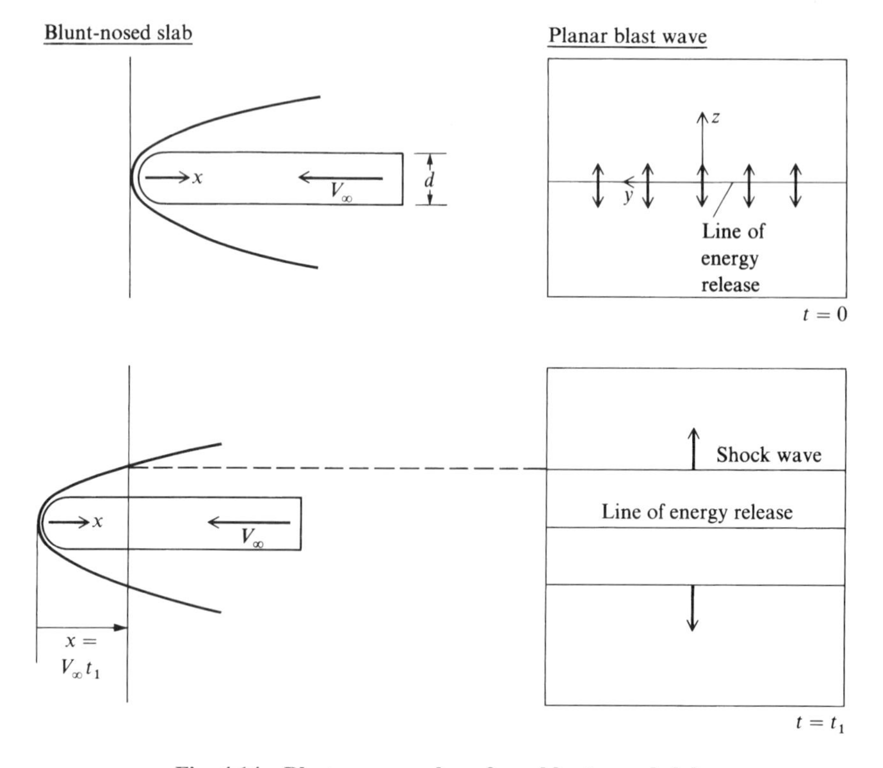

Fig. 4.14 Blast-wave analogy for a blunt-nosed slab.

In the blast-wave analogy, the energy that is released (the right side of Figs. 4.13 and 4.14) is related to the wave drag of the nose, as follows. Consider the blunt-nosed flat plate shown in Fig. 4.15. Let D be the wave drag of the nose per unit span. The plate moves through a slab of air, which has thickness $\mathrm{d}x$ in the direction of flight and has unit length in the spanwise direction. Drag is the force exerted on the body by the air; in turn, because of Newton's third law, the body exerts a force on the air in the equal and opposite direction, namely, D. Hence, the body does *work* on the air equal to $D\,\mathrm{d}x$. Because work is energy, then the amount of energy per unit span deposited in the air is $\mathrm{d}E$, where

$$\mathrm{d}E = D\,\mathrm{d}x \tag{4.143}$$

If we let the body move a unit distance in the x direction, then from Eq. (4.143) the energy released to the air is

$$E = D(1) = D \tag{4.143a}$$

From Eq. (4.143), we see that the nose drag is equal to E. In turn, from Fig. 4.15 considering a unit span and a unit length in the x direction, we see that E is the energy released over a horizontal plane of unit area; that is, E is the energy release per unit area. In the one-dimensional unsteady blast-wave problem sketched on the right of Fig. 4.14, we visualize that the line of energy release shown is in reality an infinite *sheet* of energy release, where the sheet is

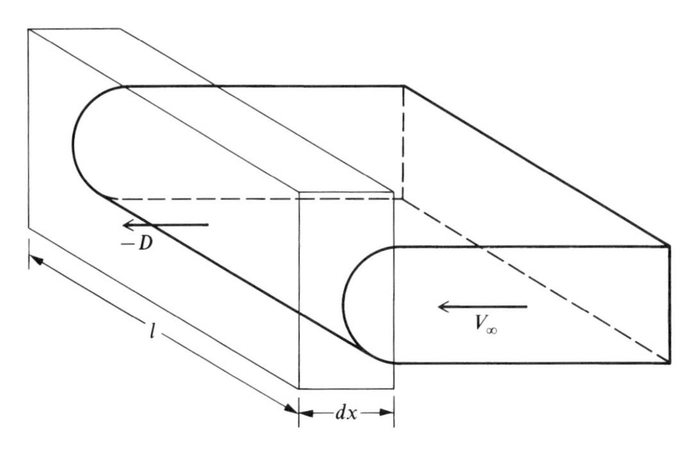

$D\,dx = dE$ where D = nose drag per unit span of body

Energy per unit area:

$$\frac{dE}{dx} = D$$

Fig. 4.15 Equivalence between nose drag and blast-wave energy; blunt-nosed slab.

perpendicular to the page. In turn, the blast waves in this picture are planar waves perpendicular to the page, of infinite extent, and propagating both upward and downward. Hence, in this picture, from Eq. (4.143), E represents the energy released per unit area of this sheet, and in turn $E = D$, where D is the nose drag of the body per unit span. For the case of the blunt-nosed cylinder shown in Fig. 4.13, the nose drag and the energy release are also related, as follows. Consider the axisymmetric cylinder moving in the x direction, as shown in Fig. 4.16. The body moves through a cylindrical slab of air of thickness $\mathrm{d}x$. From the same argument as just shown, the nose drag D of the body adds energy to this cylindrical slab, equal to

$$\mathrm{d}E = D\,\mathrm{d}x$$

Thus, when the body moves a unit length in the x direction, the energy released to the air is

$$E = D(1) = D \tag{4.143b}$$

So once again we see that the energy release is equal to the nose drag; however, here E represents the energy release per unit length along the x axis (in contrast to the energy release per unit area in the case of the two-dimensional slab). Returning to the blunt-nosed cylinder shown in Fig. 4.13, we visualize that the blast at the origin shown at the right is in reality a blast concentrated along an *infinite line* perpendicular to the page and that E is the energy release per unit length along this line. The shock wave generated by this energy release is a cylindrical blast wave, propagating outward in the radial direction, and extending to an infinite extent perpendicular to the page. From the preceding arguments, we have shown that the nose drag D is equal to the energy release per unit length E.

Return to Figs. 4.13 and 4.14. The advantage of the blast-wave analogy is that solutions to the unsteady blast-wave problem (the right sides of Figs. 4.13 and 4.14) can be found in the classical literature and hence can be immediately

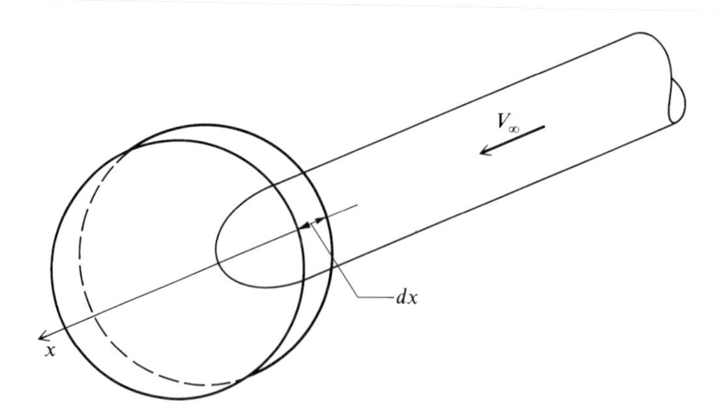

Fig. 4.16 Sketch for the blunt-nosed cylinder.

transferred to the steady hypersonic flows shown at the left of Figs. 4.13 and 4.14. As in our earlier discussion involving the unsteady piston problem, such blast-wave solutions can be obtained from self-similar solutions involving ordinary differential equations. A detailed presentation of these solutions is given in Chapter 4 of Sedov [35], which the reader is encouraged to examine. In [35] treatments are given for spherical, cylindrical, and planer blast waves; only the latter two are germane to Figs. 4.13 and 4.14, respectively. It is beyond the scope of the present book to go into the lengthy details of these unsteady blast-wave solutions. However, in the case of very intense explosions, where the pressure ahead of the blast wave can be neglected in comparison with the pressure behind the wave, analytic, asymptotic formulas for velocity, density, and pressure near the center of the explosion can be obtained. Of most interest to us is the pressure, given in [35] for the *cylindrical blast wave* (Fig. 4.13) as

$$p = k_1 \rho_\infty \left(\frac{E}{\rho_\infty}\right)^{1/2} t^{-1} \tag{4.144a}$$

where

$$k_1 = \frac{\gamma^{[2(\gamma-1)/(2-\gamma)]}}{2^{[(4-\gamma)/(2-\gamma)]}} \tag{4.144b}$$

and for the *planar blast wave* (Fig. 4.14) as

$$p = k_2 \rho_\infty \left(\frac{E}{\rho_\infty}\right)^{2/3} t^{-2/3} \tag{4.145a}$$

where

$$k_2 = \frac{2^{7/3}(2\gamma-1)^{[(5\gamma-4)/3(2-\gamma)]}}{9(\gamma+1)^{[2(\gamma+1)/3(2-\gamma)]}} \tag{4.145b}$$

Equations (4.144a) and (4.145a) give the pressure near the center of the blast as a function of time t, with the energy release E as a parameter. In addition, let the coordinate of the shock wave be denoted by r; for the cylindrical blast wave r is the radial coordinate of the wave, whereas for the planer blast wave r is the vertical coordinate of the wave. From [35] we find that, for the cylindrical blast wave,

$$r = \left(\frac{E}{\rho_\infty}\right)^{1/4} t^{1/2} \tag{4.146}$$

and for the planar blast wave,

$$r = \left(\frac{E}{\rho_\infty}\right)^{1/3} t^{2/3} \tag{4.147}$$

Therefore, we can look upon Eqs. (4.144–4.147) as solutions to the flows shown at the right of Figs. 4.13 and 4.14. Let us now obtain the equivalent results for the steady hypersonic flows shown on the left of Figs. 4.13 and 4.14.

First, consider the cylindrical case shown in Fig. 4.13. In Eq. (4.144a), E is the energy release per unit length along the axis of the cylindrical shock wave; as shown by Eq. (4.143b), $E = D$. Because D is the nose drag, let us define a nose-drag coefficient C_D as $C_D = D/q_\infty S$, where $q_\infty = \frac{1}{2}\rho_\infty V_\infty^2$ and $S = \pi d^2/4$. Here, ρ_∞ and V_∞ are the freestream density and velocity, respectively, for the body shown at the left of Fig. 4.12, and d is the base diameter of the body. Thus, from Eq. (4.143b) we have

$$E = D = \frac{1}{2}\rho_\infty V_\infty^2 C_D \frac{\pi d^2}{4} \tag{4.148}$$

Also, from the equivalence principle as embodied in Eq. (4.137), we have

$$t = \frac{x}{V_\infty} \tag{4.149}$$

Substituting Eqs. (4.148) and (4.149) into (4.144a), we obtain

$$p = k\rho_\infty \sqrt{\frac{\pi}{8}} V_\infty d \sqrt{C_D} \frac{V_\infty}{x}$$

Recalling the perfect-gas equation of state, namely, $p_\infty = \rho_\infty R T_\infty$, the preceding equation can be written as

$$p = k \frac{\gamma p_\infty}{\gamma R T_\infty} \sqrt{\frac{\pi}{8}} V_\infty^2 \sqrt{C_D} \left(\frac{x}{d}\right)^{-1}$$

Recognizing $\gamma R T_\infty = a_\infty^2$, where a_∞ is the freestream speed of sound, and noting that $V_\infty / a_\infty = M_\infty$, the preceding equation becomes, for $\gamma = 1.4$,

$$\frac{p}{p_\infty} = 0.8773 k M_\infty^2 \sqrt{C_D} \left(\frac{x}{d}\right)^{-1} \tag{4.150}$$

From Eq. (4.144b), for $\gamma = 1.4$, $k = 0.07768$. Thus, Eq. (4.150) becomes the blunt cylinder:

$$\boxed{\frac{p}{p_\infty} = 0.0681 M_\infty^2 \frac{\sqrt{C_D}}{(x/d)}} \tag{4.151}$$

Inserting Eqs. (4.148) and (4.149) into Eq. (4.146), we have

$$r = \left[\frac{1}{2} V_\infty^2 C_D \frac{\pi d^2}{4}\right]^{1/4} \left(\frac{x}{V_\infty}\right)^{1/2}$$

or

$$\frac{r}{d} = \left(\frac{\pi}{8}\right)^{1/4} C_D^{1/4} \sqrt{\frac{x}{d}}$$

or blunt cylinder:

$$\boxed{\frac{r}{d} = 0.792 C_D^{1/4} \sqrt{\frac{x}{d}}} \tag{4.152}$$

Again examining the left side of Fig. 4.13, we note that the pressure distribution downstream of the nose of the blunt-nosed cylinder is given by Eq. (4.151) as a function of x. Moreover, the shape of the shock wave is given by Eq. (4.152) as a function of x. Equations (4.151) and (4.152) are *blast-wave results*, applied to the steady flow over the blunt-nosed cylinder via the hypersonic equivalence principle.

Next, consider the planar case shown in Fig. 4.14. In Eq. (4.145a), E is the energy release per unit area of the plane perpendicular to the page; as shown by Eq. (4.143a), $E = D$. For the blunt-nosed slab shown on the left of Fig. 4.14, let us define a nose-drag coefficient as $C_D = D/q_\infty S$, where, as before $q_\infty = \frac{1}{2}\rho_\infty V_\infty^2$, but now $S = d(1) = d$, namely, the base area per unit span. Thus, from Eq. (4.143a),

$$E = D = \tfrac{1}{2}\rho_\infty V_\infty^2 d C_D \tag{4.153}$$

Using Eqs. (4.153) and (4.149) along with Eqs. (4.145a), (4.145b), and (4.147), we obtain, for $\gamma = 1.4$ (the details are left as a homework problem), the following.

Blunt slab:

$$\boxed{\frac{p}{p_\infty} = 0.127 M_\infty^2 C_D^{2/3} \left(\frac{x}{d}\right)^{-2/3}} \tag{4.154}$$

and blunt slab:

$$\boxed{\frac{r}{d} = 0.794 C_D^{1/3} \left(\frac{x}{d}\right)^{2/3}} \tag{4.155}$$

Examining the preceding results further, we note the following, for the blunt-nosed cylinder:

1) From Eq. (4.151), the pressure distribution varies inversely with x.

2) From Eq. (4.152), the shock-wave shape varies as $x^{1/2}$, that is, it is a *parabolic shape*.

Also, for the blunt-nosed slab, we note the following:

1) From Eq. (4.154), the pressure distribution varies inversely as $x^{2/3}$.

2) From Eq. (4.155), the shock-wave shape varies as $x^{2/3}$.

Also observe that p/p_∞ for both cases varies with the square of the freestream Mach number and that all of the results depend on C_D to some power. Recall again that C_D is the nose-drag coefficient. We can estimate the values of C_D from Newtonian theory as given in Sec. 3.2. Specifically, as noted in Sec. 3.2 for a hemicylindrical nose (the blunt slab case) $C_D = \frac{4}{3}$, and for a hemispherical nose (the blunt-nosed cylinder case) $C_D = 1$. Also, recall again that the preceding blast-wave results are to be applied downstream of the nose of the body, although x in the preceding equations is measured from the tip of the body. Blast-wave theory cannot be applied to obtain detailed results on the nose itself; this region must be analyzed by detailed numerical solutions, such as to be discussed in Chapter 5.

Question: How accurate is blast-wave theory as applied to hypersonic bodies?

One of the most definitive answers to this is given by Lukasiewicz [36]. Utilizing the blast-wave results of Sakuri [37] and [38], Lukasiewicz compared theory with wind-tunnel data for blunt-nosed flat plates and cylinders. Sakurai [37] and [38] obtained two sets of blast-wave results, identified as first and second approximations. The first approximation, which ignores the freestream pressure ahead of the blast wave, gives the results described by Eqs. (4.151), (4.152), (4.154), and (4.155). The second approximation takes into account a finite pressure ahead of the blast wave. These results, as applied by Lukasiewicz [36], are listed next, note that the first approximation results are the same as obtained earlier, with only negligible differences in the leading coefficients. From [36], we have the following.

Blunt-nosed flat plate (first approximation):

$$\frac{p}{p_\infty} = 0.121 M_\infty^2 \left(\frac{C_D}{x/d}\right)^{2/3} \tag{4.156}$$

$$\frac{r}{d} = 0.774 C_D^{1/3} \left(\frac{x}{d}\right)^{2/3} \tag{4.157}$$

Blunt-nosed flat plate (second approximation):

$$\frac{p}{p_\infty} = 0.121 M_\infty^2 \left(\frac{C_D}{x/d}\right)^{2/3} + 0.56 \tag{4.158}$$

$$\left(\frac{r}{d}\right) \Big/ (M_\infty^2 C_D) = \frac{0.774}{M_\infty^2 [C_D/(x/d)]^{2/3} - 1.09} \tag{4.159}$$

Blunt-nosed cylinder (first approximation):

$$\frac{p}{p_\infty} = 0.067 M_\infty^2 \frac{\sqrt{C_D}}{(x/d)} \tag{4.160}$$

$$\frac{r}{d} = 0.795 C_D^{1/4} \left(\frac{x}{d}\right)^{1/2} \tag{4.161}$$

Blunt-nosed cylinder (second approximation):

$$\frac{p}{p_\infty} = 0.067 M_\infty^2 \frac{\sqrt{C_D}}{(x/d)} + 0.44 \tag{4.162}$$

$$\frac{r/d}{M_\infty C_D^{1/2}} = 0.795 \sqrt{\frac{(x/d)}{M_\infty^2 C_D^{1/2}} \left[1 + 3.15 \frac{(x/d)}{(M_\infty^2 C_D^{1/2})}\right]} \tag{4.163}$$

where

x = distance measured from the nose, in the flow direction
C_D = wave drag coefficient of the nose
d = plate thickness or cylinder diameter
r = value of z at the shock wave

Lukasiewicz [36] compared the preceding equations (4.156–4.163) with experimental data obtained at the Arnold Engineering Development Center (AEDC) and with more exact theoretical results based on the method of characteristics. Some of his comparisons are shown in Figs. 4.17–4.24. In Fig. 4.17, results are given for the pressure distribution over a fiat plate with a cylindrical leading edge. Note that the first approximation, Eq. (4.156), compares more favorably with the wind-tunnel data than the second approximation, Eq. (4.158). However, as shown by the solid curve in Fig. 4.17, the second approximation can be brought into close agreement with the data if the origin of the x axis, namely, the point at which $x = 0$, is taken not at the nose of the body, but rather at a location $\frac{2}{3}d$ *upstream* of the nose. In Fig. 4.18, the data in Fig. 4.17 are plotted vs the blast analogy parameter $(x/d)^{2/3}/M_\infty^2 C_D^{2/3}$, along with additional results obtained from the method of characteristics as described in [39] and [40]. Also shown as the dashed line is a simple correlation of the

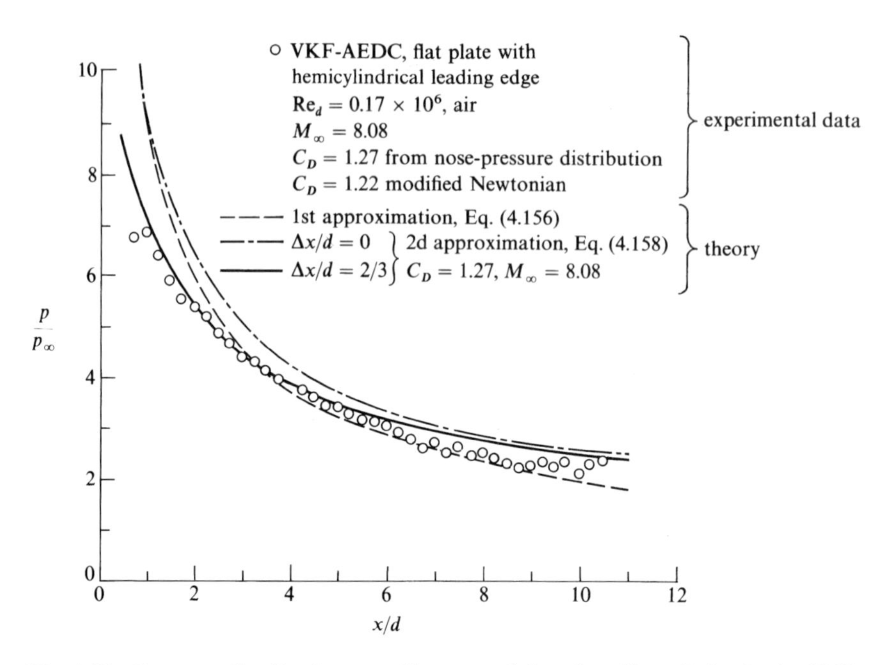

Fig. 4.17 Pressure distribution on a blunt-nosed flat plate (from Lukasiewicz [36]).

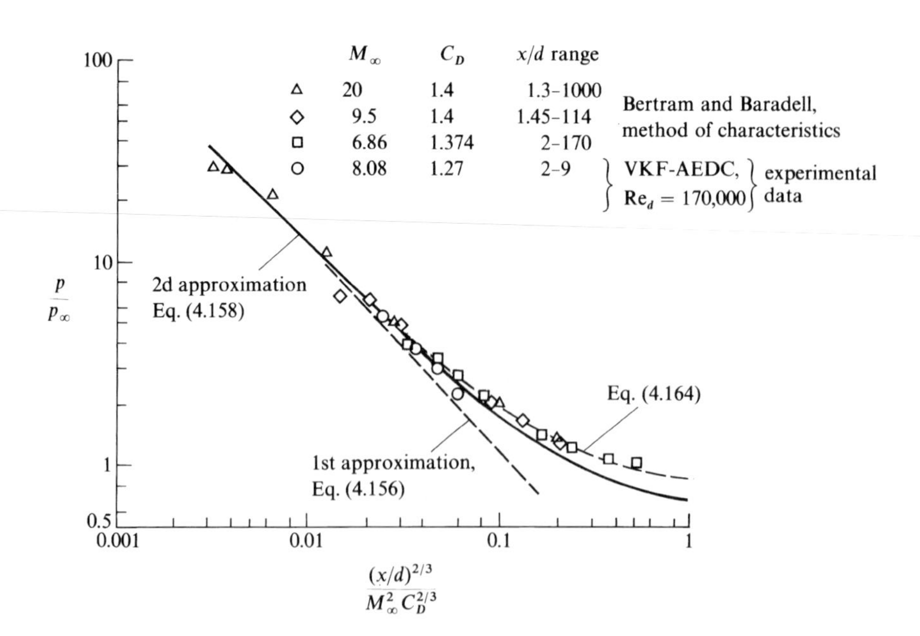

Fig. 4.18 Correlation of pressure distribution for a blunt-nosed flat plate (from [36]).

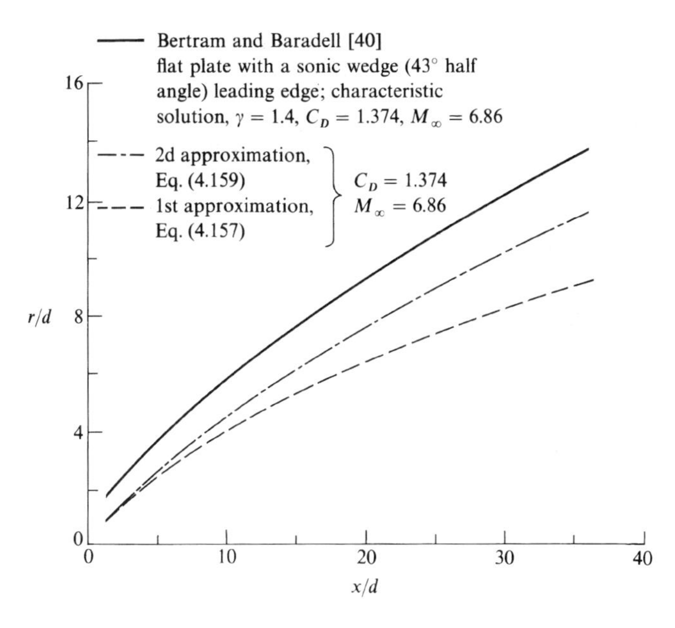

Fig. 4.19 Shock-wave shape calculated by the method of characteristics and by blast-wave theory; blunt-nosed flat plate (from [36]).

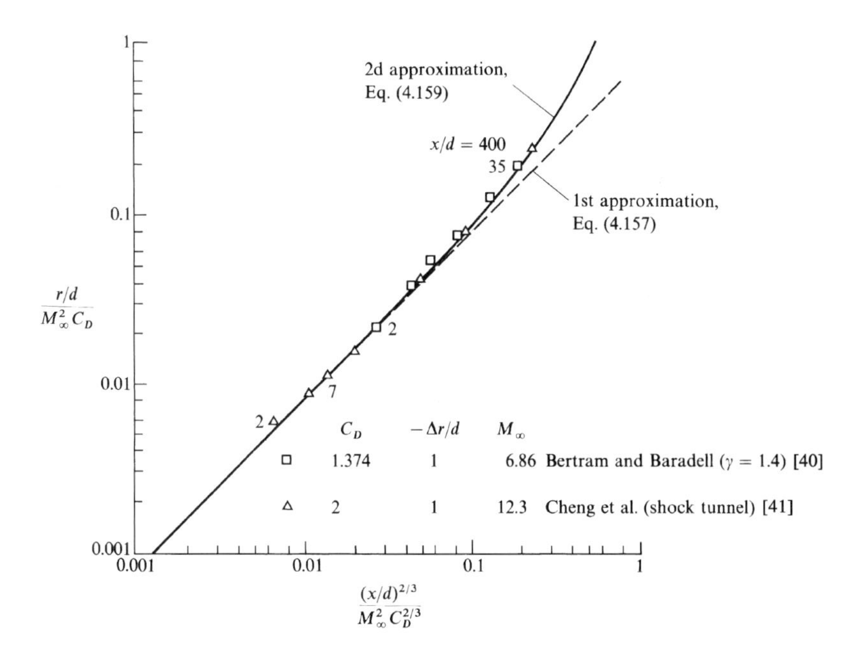

Fig. 4.20 Correlation of shock-wave shapes; blunt-nosed flat plate (from [36]).

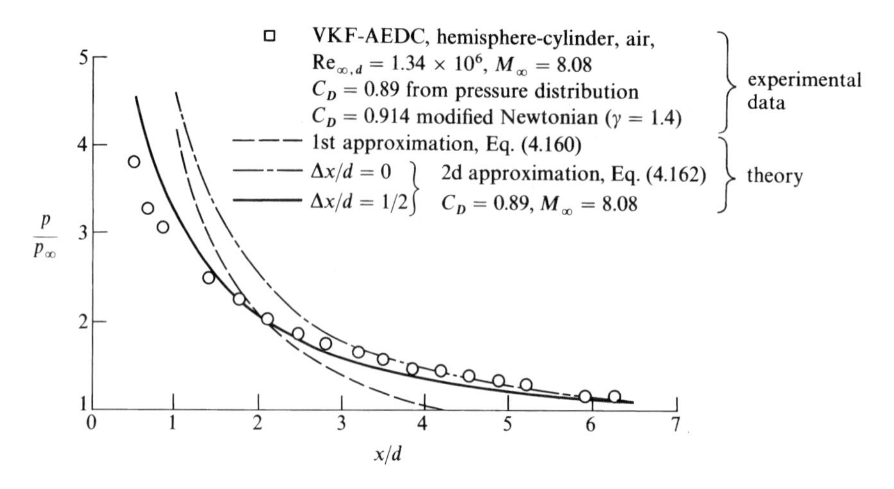

Fig. 4.21 Pressure distributions on a hemisphere cylinder (from [36]).

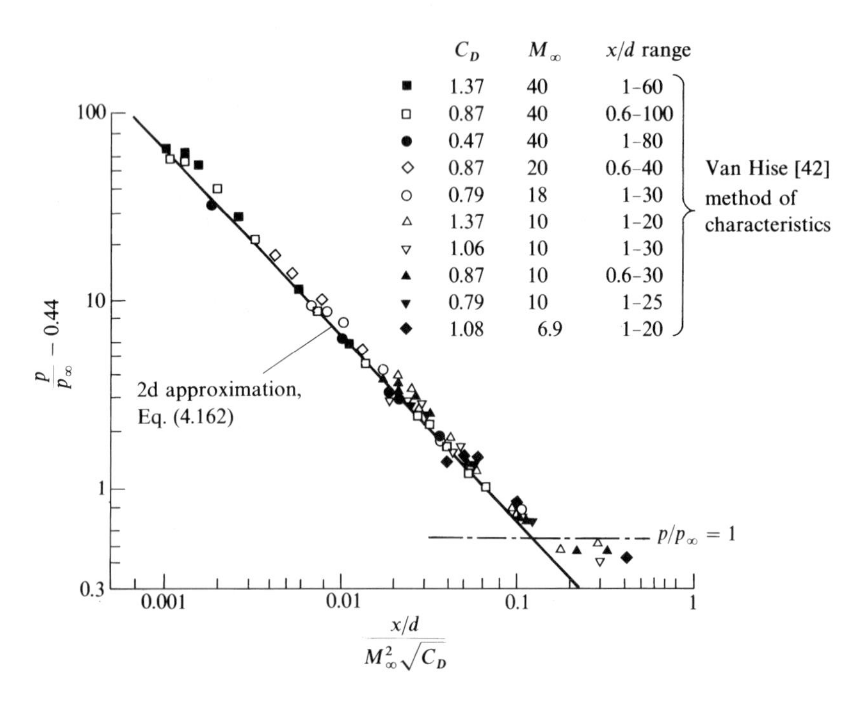

Fig. 4.22 Correlation of pressure distributions for a blunt-nosed cylinder (from [36]).

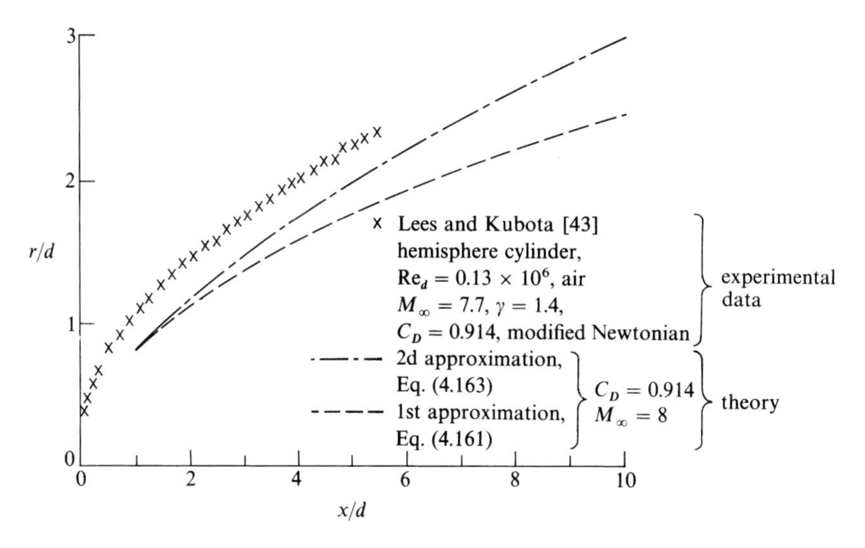

Fig. 4.23 Shock-wave shape around a blunt-nosed cylinder (from [36]).

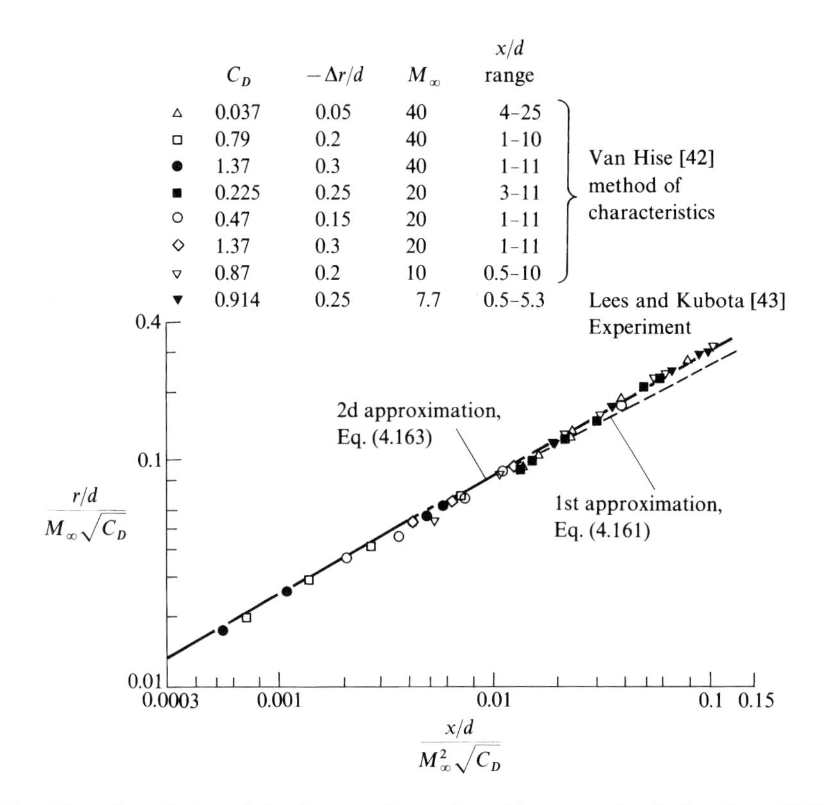

Fig. 4.24 Correlation of shock-wave shapes for a blunt-nosed cylinder (from [36]).

method of characteristic results from [39] and [40], given as

$$\frac{p}{p_\infty} = 0.117 M_\infty^2 C_D^{2/3} \left(\frac{x}{d}\right)^{-2/3} + 0.732 \tag{4.164}$$

For the theoretical results shown in Fig. 4.18, the origin is again shifted by the amount $\Delta x = \frac{2}{3}d$; with this shift, the second approximation [Eq. (4.158)] is seen to give good agreement with both wind-tunnel data and the method of characteristics. Moreover, the results in Fig. 4.18 clearly show that the trends which are predicted by blast-wave theory are confirmed by experiment, namely, that, for a *blunt-nosed flat plate*, the pressure distribution p/p_∞ varies

1) directly as M_∞^2,
2) directly as $C_D^{2/3}$, and
3) inversely as $(x/d)^{2/3}$.

Results for shock-wave shapes are shown in Fig. 4.19, where blast-wave theory is compared with the exact method of characteristics, from [40]. Clearly, the first approximation, Eq. (4.147), gives poorer agreement than the second approximation, Eq. (4.159), and neither of the blast-wave results is particularly good. However, Eq. (4.159) appears to predict the proper shape, but it is simply shifted from the exact results. If the results for both the first and second approximations are shifted upward by the amount $\Delta r = d$, much better agreement is obtained, as shown in Fig. 4.20. Here, comparison is also made with the shock-tunnel data of Cheng et al. [41]. The results of Fig. 4.20 tend to confirm that, for a *blunt-nosed flat plate*, the shock-wave shape varies

1) directly as $C_D^{1/3}$, and
2) directly as $(x/d)^{2/3}$.

Results for the pressure distribution over a hemisphere cylinder are shown in Fig. 4.21. The first and second approximation blast-wave results are obtained from Eqs. (4.160) and (4.162), respectively. Note that the best agreement with wind-tunnel data at Mach 8 is obtained with the second approximation, with the origin shifted upstream of the nose by $\Delta x = \frac{1}{2}d$. Other data for higher M_∞ from [42] are plotted in Fig. 4.22 vs the blast analogy parameter $(x/d)/(M_\infty^2 C_D^{1/2})$. Here, the origin is shifted by $\Delta x = d$. Once again we see the general confirmation of the trends established in blast-wave theory, namely that, for a *blunt-nosed cylinder*, the pressure distribution p/p_∞ varies

1) directly as M_∞^2,
2) directly as $C_D^{1/2}$, and
3) inversely as (x/d).

The shock-wave shape for a hemisphere cylinder is given in Fig. 4.23; the blast-wave results are compared with experimental data from Lees and Kubota [43]. Once again we see that the shock-wave location predicted by blast-wave theory is not accurate. However, if the predicted shock wave is shifted by an amount Δr, as indicated in Fig. 4.24, then good agreement is obtained. Moreover, Fig. 4.24 demonstrates that blast-wave theory predicts properly the following (noting that the slope of the curve on the logarithmic plot is $\frac{1}{2}$): for the *blunt-nosed cylinder*, the shock-wave shape varies

1) directly as $C_D^{1/4}$ and
2) directly as $(x/d)^{1/2}$.

A comment is in order here. The analogy between unsteady blast-wave flowfields and the steady hypersonic flowfields over slender, blunt-nosed bodies is somewhat tenuous on physical grounds, mainly because of the assumption in the classical blast-wave solutions of *instantaneous* energy release at a *point* or *line* in space. A hypersonic body does not add energy to the flow instantly, nor is this energy addition precisely at a point or along a line. This is why better agreement is obtained in some of the previous plots by shifting the virtual origin of x. However, there is good physical reasoning behind the analogy between the steady flow over a hypersonic slender body and the unsteady flow in one less space dimension (the hypersonic equivalence principle) because in the steady flowfield the disturbance velocities v' and w' perpendicular to the body axis are truly much larger than the disturbance velocity in the flow direction u'. [Review, for example, Eqs. (4.34) and (4.35).] In the final analysis, for whatever reason, blast-wave theory does provide some relatively accurate predictions of the pressure distributions and shock-wave shapes on blunt-nosed slender bodies where the Reynolds number is high enough such that viscous interaction effects are negligible. *Moreover, these blast-wave results are in the form of analytic formulas, which are extremely handy for quick, approximate estimates*—an advantage not to be ignored.

As a final note, the preceding results from blast-wave theory can readily be expressed in terms of the pressure coefficient C_p.

$$C_p \equiv \frac{p - p_\infty}{\frac{1}{2}\rho_\infty V_\infty^2} = \frac{2}{\gamma M_\infty^2}\left(\frac{p}{p_\infty} - 1\right) \tag{4.165}$$

At very high values of M_∞, $p/p_\infty \gg 1$; hence, Eq. (4.165) can be approximated by

$$C_p = \frac{2}{\gamma M_\infty^2}\left(\frac{p}{p_\infty}\right) \tag{4.166}$$

Combining Eq. (4.166) with Eqs. (4.156) and (4.160), we have, for $\gamma = 1.4$, the following.

Blunt-nosed plate:

$$C_p = \frac{0.173 C_D^{2/3}}{(x/d)^{2/3}} \tag{4.167}$$

Blunt-nosed cylinder:

$$C_p = \frac{0.096 C_D^{1/2}}{(x/d)} \tag{4.168}$$

Note that C_p is independent of M_∞; blast-wave theory is another example of Mach-number independence at hypersonic speeds.

In [3], a combination of blast-wave theory and straight Newtonian was used to predict the pressure distribution along the windward centerline of the space shuttle. Let l denote the length of the shuttle and d the thickness of the fuselage near the canopy. For the shuttle, the fineness ratio is $l/d = 7$. Moreover, the drag

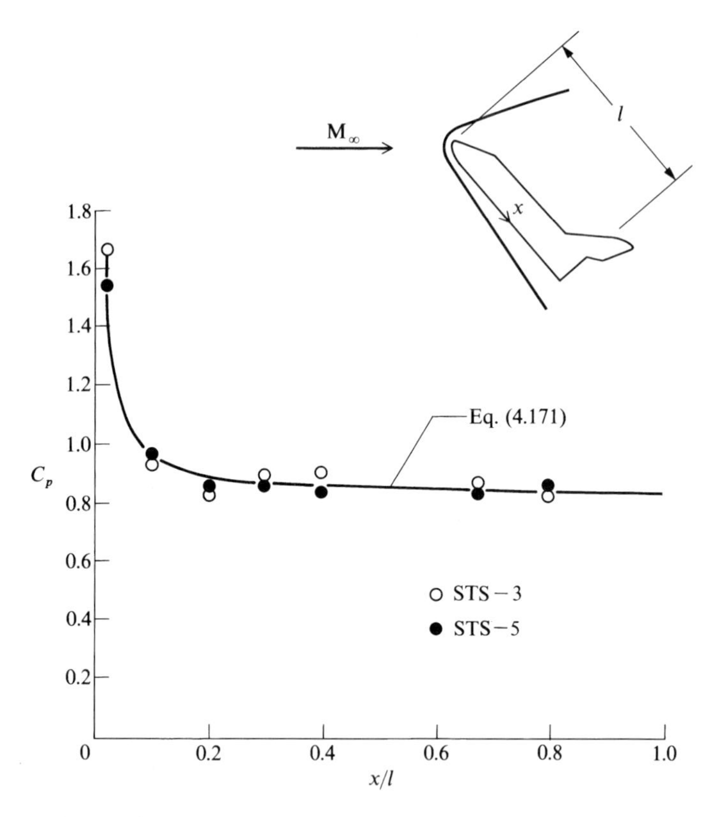

Fig. 4.25 Comparison of pressure coefficients obtained with combined blast-wave/ Newtonian theory [Eq. (4.171)] with flight data for the space shuttle: windward centerline, $M_\infty = 21.6$, and $\alpha = 40$ deg.

coefficient of a hemisphere from straight Newtonian theory is $C_D = 1$. Substituting these values into Eq. (4.168), written as

$$C_p = 0.096 C_D^{1/2} \left(\frac{x}{l}\right)^{-1} \left(\frac{l}{d}\right)^{-1}$$

we obtain

$$C_p = \frac{0.0137}{x/l} \tag{4.169}$$

Equation (4.169) holds for zero-degree angle of attack. To take angle of attack, α into effect, let us simply add the Newtonian contribution $2\sin^2\alpha$ to Eq. (4.169), obtaining

$$C_p = \frac{0.0137}{x/l} + 2\sin^2\alpha \tag{4.170}$$

Let us choose a point, on the shuttle trajectory corresponding to $\alpha = 40$ deg and $M_\infty = 21.6$. For $\alpha = 40$ deg, Eq. (4.170) becomes

$$C_p = \frac{0.0137}{x/l} + 0.826 \tag{4.171}$$

Results from Eq. (4.171) are plotted as the solid curve in Fig. 4.25, obtained from [3]. These results are compared with actual flight data from the STS-3 (open circles) and STS-5 (solid circles) shuttle missions; these flight data are obtained from [44]. The agreement between theory and flight data in Fig. 4.25 is quite remarkable, especially when considering that the theoretical curve can be calculated in a few minutes by hand. This clearly demonstrates the value of both blast-wave theory and Newtonian results.

This ends our discussion of blast-wave theory and the general idea of the hypersonic equivalence principle. Our purpose has been to describe these ideas, to make them plausible on a physical and mathematical basis, and to show some practical results. Keep in mind that all of these results are limited to slender bodies at hypersonic speeds. In the next section, we will discuss a class of approximate inviscid flow theory that can be applied to blunt as well as to slender bodies.

4.9 Thin Shock-Layer Theory

We have already discussed that shock layers over hypersonic bodies are *thin* (refer again to Fig. 1.13 and the related discussion in Chapter 1). In the limit as $M_\infty \to \infty$ and $\gamma \to 1$, we have shown that $\beta \to \theta$, and the shock layer becomes infinitely thin and infinitely dense. In such a limit, we can consider the shock shape, the body shape, and the streamline shapes in between to be all of the same. Such approximations, or variations of them, are the basis of *thin shock-layer theory.* An interesting discussion of thin shock-layer theory can be found in [45]; additional discussion is given in [46].

In this section, the analysis developed by Maslen [47] will be outlined as an example of a theory based on the assumption of a thin shock layer. Maslen's method is chosen here because of its simplicity and because of its frequent application—even today—for the approximate analysis of hypersonic inviscid shock layers. Moreover, Maslen's method gives results for the flowfield over blunt as well as slender bodies, and therefore it makes a nice intellectual as well as chronological bridge between the recently discussed classical material in this chapter and the more modern, computationally based blunt-body solutions to be discussed in Chapter 5.

Consider the curvilinear coordinate system shown in Fig. 4.26, where x and y, respectively, are parallel and perpendicular to the shock, and u and v are the corresponding components of velocity. For simplicity we will assume a two-dimensional flow; however, Maslen's method also applies to axisymmetric flow (see [47] for details). Now assume that the shock layer is thin, and hence the streamlines are essentially parallel to the shock wave. In a streamline-based

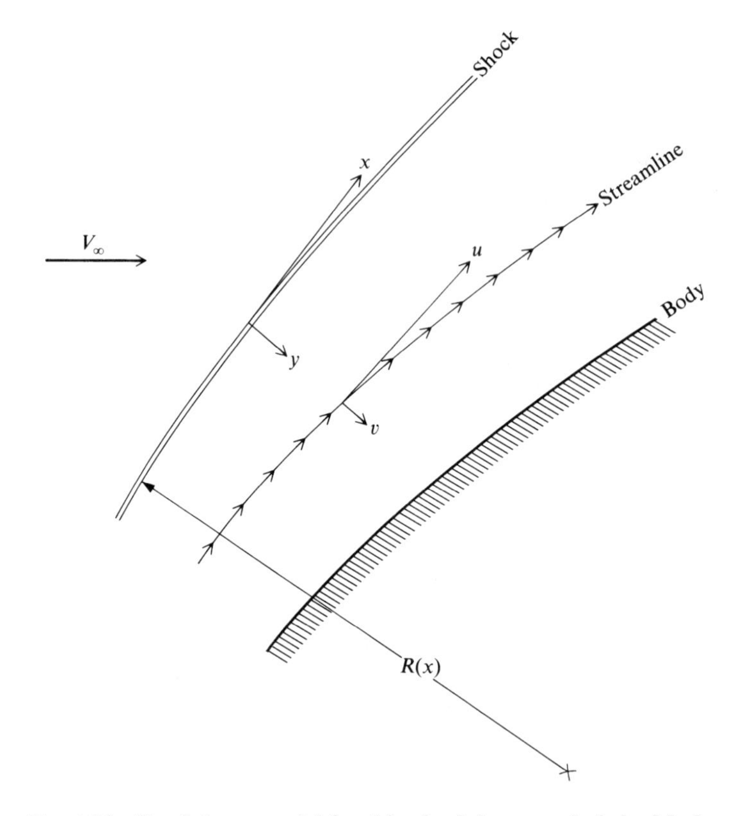

Fig. 4.26 Shock-layer model for thin shock-layer analysis by Maslen.

coordinate system, the momentum equation normal to a streamline is

$$\rho \frac{u^2}{R} = \frac{\partial p}{\partial n} \tag{4.172}$$

where R is the local streamline radius of curvature. For the preceding assumptions, Eq. (4.172) becomes

$$\frac{\rho u^2}{R_s} = \frac{\partial p}{\partial y} \tag{4.173}$$

where R_s is the shock radius of curvature. Define a stream function ψ such that

$$\rho u = \frac{\mathrm{d}\psi}{\mathrm{d}y} \tag{4.174}$$

and replace y in Eq. (4.173) with ψ [i.e., introduce a von Mises transformation such that the independent variables are (x, ψ) rather than (x, y)]:

$$\rho \frac{u^2}{R_s} = \frac{\partial p}{\partial \psi}(\rho u)$$

or

$$\frac{\partial p}{\partial \psi} = \frac{u}{R_s} \tag{4.175}$$

Again, consistent with a thin shock layer where all of the streamlines are essentially parallel to the shock, $u \approx u_s$, the velocity just behind the shock. Thus, Eq. (4.175) becomes

$$\frac{\partial p}{\partial \psi} = \frac{u_s}{R_s} \tag{4.176}$$

Integrating Eq. (4.176) between a point in the shock layer where the value of the stream function is ψ and just behind the shock wave where $\psi = \psi_s$, we have

$$\boxed{p(x, \psi) = p_s(x) + \frac{u_s(x)}{R_s(x)}[\psi - \psi_s(x)]} \tag{4.177}$$

Equation (4.177) is the crux of Maslen's method. The flowfield solution progresses as follows:

1) *Assume* a shock-wave shape, as shown in Fig. 4.27. In this sense, Maslen's method is an *inverse method*, where a shock wave is assumed and the body that supports this shock is calculated.

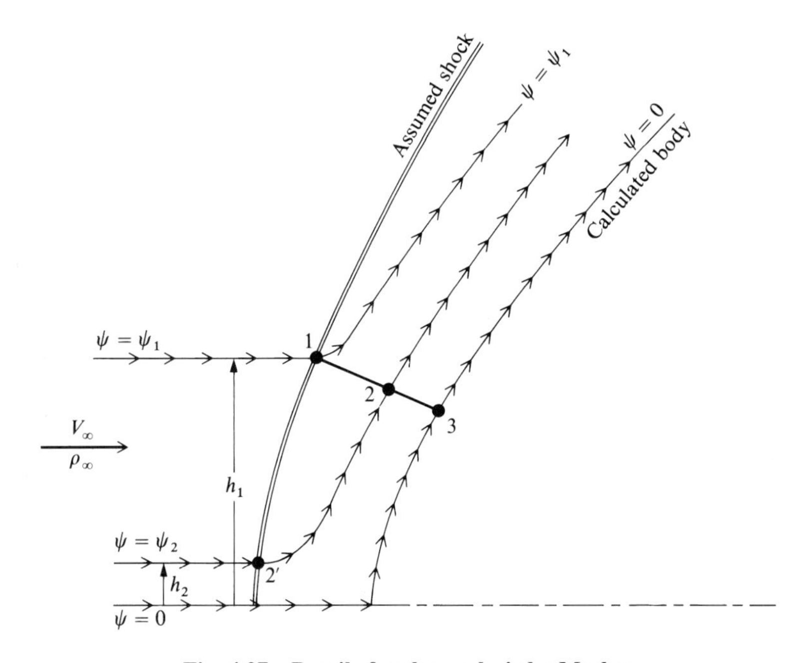

Fig. 4.27 Details for the analysis by Maslen.

2) Hence, all flow quantities are known at point 1 (Fig. 4.27) just behind the shock, from the oblique shock relations. The value of $\psi = \psi_1$ at point 1 is known from

$$\psi_1 = \rho_\infty V_\infty h_1$$

3) Choose a value of ψ_2, where $0 < \psi_2 < \psi_1$. This identifies a point 2 inside the flowfield along the y axis, as shown in Fig. 4.27, where $\psi = \psi_2$. (The precise value of the physical coordinate y_2 will be found in a subsequent step.)

4) Calculate the pressure at point 2 from Eq. (4.177)

$$p_2 = p_1 + \frac{u_1}{(R_s)}(\psi_2 - \psi_1)$$

5) The entropy at point 2, s_2 is *known* because the streamline at point 2, corresponding to $\psi = \psi_2$, has come through that point on the shock wave, point 2′, where $\psi_{2'} = \psi_2$, and where

$$\psi_{2'} = \psi_2 = \rho_\infty V_\infty h_2$$

or

$$h_2 = \frac{\psi_2}{\rho_\infty V_\infty} \tag{4.178}$$

Therefore, h_2 is obtained from Eq. (4.178), which locates point 2′ on the shock. In turn, $s_{2'}$ is known from the oblique shock relations, and because the flow is isentropic along any given streamline $s_2 = s_{2'}$.

6) Calculate the enthalpy h_2 and density ρ_2 from the thermodynamic equations of state

$$h_2 = h(s_2, p_2)$$
$$\rho_2 = \rho(s_2, p_2)$$

7) Calculate the velocity at point 2 from the adiabatic energy equation (total enthalpy is constant). That is,

$$h_0 = h_\infty + \frac{V_\infty^2}{2}$$

where h_0 is the total enthalpy, which is constant throughout the adiabatic flowfield. In turn,

$$h_0 = h_2 + \frac{u_2^2}{2} \qquad \text{(ignoring } v_2\text{)}$$

or

$$u_2 = \sqrt{2(h_0 - h_2)}$$

8) All of the flow quantities are now known at point 2. Referring to Figs. 4.26 and 4.27 repeat the preceding steps for all points along the y axis between the shock (point 1) and the body (point 3). The body surface is defined by $\psi = 0$.

9) The physical coordinate y, which corresponds to a particular value of ψ, can now be found by integrating the definition of the stream function (which is really the continuity equation). Because

$$\frac{\mathrm{d}\psi}{\mathrm{d}y} = \rho u$$

then

$$y = \int_{\psi}^{\psi_s} \frac{\mathrm{d}\psi}{\rho u} \tag{4.179}$$

where ρ and u are known as a function of ψ from the preceding steps. This also locates the body coordinate, where

$$y_b = \int_{0}^{\psi_s} \frac{\mathrm{d}\psi}{\rho u}$$

10) This procedure is repeated for any desired number of points along the specified shock wave, hence generating the flowfield and body shape which supports that shock.

Again, remember that the preceding assumed a two-dimensional flow. Extension to an axisymmetric body is straightforward (see [47]).

Some results calculated by Maslen, taken from [47], are shown in Figs. 4.28–4.30. In Fig. 4.28, a paraboloidal shock is assumed, and the calculated body shape is shown. (Note that Maslen's method is an inverse method, where the shock-wave shape is assumed, and the body shape that supports the assumed shock, as well as the flowfield between the shock and body, are calculated.) Maslen's calculations of the body shape are shown as the triangles and are compared with more exact calculations. For example, the left side of Fig. 4.28 shows results in the nose region, which are compared with the exact numerical calculations by Van Dyke [14]. Excellent agreement is achieved. On the right side of Fig. 4.28, results extending far downstream of the nose are shown. Here, Maslen's results are compared with the theories of Yakura [48] and Sychev [49], based on hypersonic small-disturbance theory and blast-wave analysis. The corresponding surface-pressure distributions (in terms of p/p_0, where p_0 is

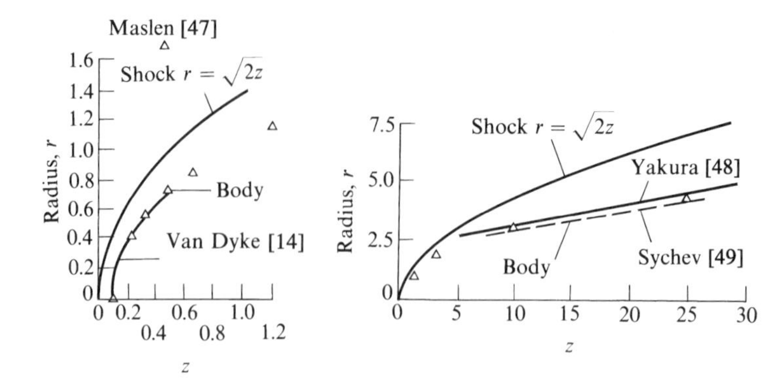

Fig. 4.28 Body associated with paraboloidal shock wave (from Maslen [47]): $M_\infty = \infty$, and $\gamma = 1.4$.

the stagnation point pressure), are shown in Fig. 4.29. Note that only a few points representing Maslen's results are shown. There is good reason for this; all of the results given by Maslen in [47] were calculated by him over the space of a few days using a *hand calculator*, which naturally limited the number of calculated points. (However, this illustrates the practicality of Maslen's method, namely, that it is indeed simple enough to be carried out using a hand calculator.)

The inverse method can be used, in an iterative sense, to calculate the flow over a given body, that is, the shock is assumed, and the supporting body shape is calculated. This body shape is compared with the given body, and a new shock is assumed that will produce results closer to the given body. This iteration is repeated until the calculated body matches the given body closely enough. Using this approach, Maslen calculated the pressure distribution and shock shape for a hemisphere cylinder, shown in Fig. 4.30. Good agreement is obtained with the experimental data of Kubota [50]. Also shown are the numerical results of Inouye and Lomax [51], based on an inverse blunt-body solution (iterated) and the method of characteristics. Again, Maslen's method gives

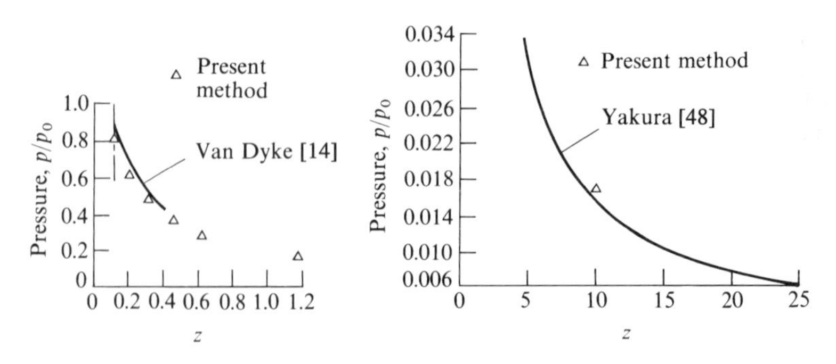

Fig. 4.29 Surface pressure on body supporting a paraboloidal shock (from Maslen [47]): $M_\infty = \infty$, and $\gamma = 1.4$.

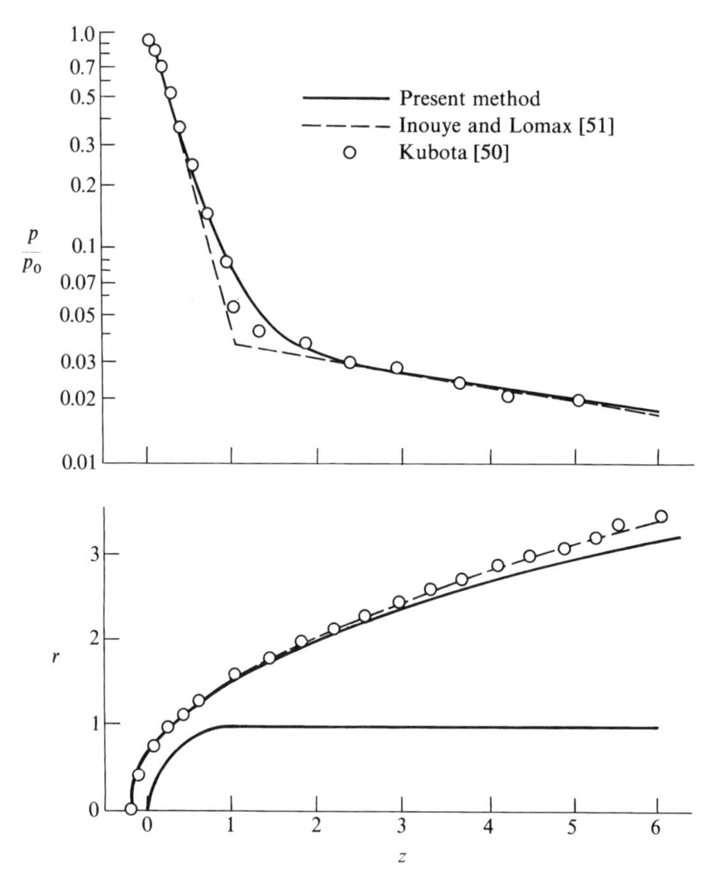

Fig. 4.30 Shock-wave shape and surface pressure for a hemisphere cylinder (from Maslen [47]): $M_\infty = \infty$, and $\gamma = 1.4$.

reasonable results. Therefore, in light of Figs. 4.28–4.30. Maslen's method, which is relatively straightforward to apply, can be considered an excellent example of the general class of thin shock-layer methods. Also, note that Maslen's method applies to blunt as well as to slender bodies. Because of its accuracy and simplicity, Maslen's method has found frequent applications in inviscid hypersonic flow analysis, including up to the present day.

4.10 Summary and Comments

Hypersonic aerodynamics is highly nonlinear; even the assumption of small perturbations, which in subsonic and supersonic flows leads to simple linear theories, does not yield a system of linear equations for hypersonic flow. In spite of this, various approximate methods have been successfully developed for the analysis of inviscid hypersonic flows. We have discussed several of these methods in the present chapter, some of them predicated upon the hypersonic small-disturbance equations, given by Eqs. (4.41–4.45). In particular, the following is noted:

1) The mathematical basis of *hypersonic similarity* rests upon these equations; from them, we have shown that flows over affinely related bodies with the same values of γ, $M_\infty\tau$, and α/τ will have the same values of C_p/τ^2, c_l/τ^2, c_d/τ^3 (where c_l and c_d are referenced to planform area). Here,

$$M_\infty\tau = \text{hypersonic similarity parameter}$$

For a three-dimensional body where base area is used as the reference for C_L and C_D, hypersonic similarity states that affinely related bodies with the same values of γ, $M_\infty\tau$, and α/τ will have the same values of C_L/τ and C_D/τ^2. Because hypersonic similarity stems from the hypersonic small-disturbance equations, this concept applies only to slender bodies at small angles of attack.

2) The hypersonic small-disturbance equations themselves can be directly solved for the flowfield between the shock wave and body; a particular application to cones was used here to illustrate such a solution based on hypersonic small-disturbance theory. Although such solutions usually require a numerical treatment at some stage, the results can sometimes lead to closed-form analytical formulas, such as are repeated next for the pressure coefficient on a slender cone at hypersonic speeds,

$$\frac{C_p}{\theta_c^2} = 1 + \frac{(\gamma+1)K^2+2}{(\gamma-1)K^2+2}\ln\left(\frac{\gamma+1}{2}+\frac{1}{K^2}\right) \tag{4.126}$$

as well as for the cone shock-wave angle β,

$$K_\beta = K\sqrt{\frac{\gamma+1}{2}+\frac{1}{K^2}} \tag{4.125}$$

where, in the preceding, $K = M_\infty\theta_c$ and $K_\beta = M_\infty\beta$. The hypersonic small-disturbance equations also lead to the *hypersonic equivalence principle*, which states that the steady hypersonic flow over a slender body is equivalent to an unsteady flow in one less space dimension. A corollary to this principle is *blast-wave theory*, which allows the self-similar solutions to the unsteady flowfield generated by an instantaneous release of energy along a line or plane to be carried over to the steady flow downstream of the nose of hypersonic blunt-nosed slender bodies. The results, as documented in [36], are as follows.

Blunt-nosed flat plate (first approximation).

$$\frac{p}{p_\infty} = 0.121M_\infty^2\left(\frac{C_D}{x/d}\right)^{2/3} \tag{4.156}$$

$$\frac{r}{d} = 0.774C_D^{1/3}\left(\frac{x}{d}\right)^{2/3} \tag{4.157}$$

Blunt-nosed flat plate (second approximation):

$$\frac{p}{p_\infty} = 0.121 M_\infty^2 \left(\frac{C_D}{x/d}\right)^{2/3} + 0.56 \tag{4.158}$$

$$\left(\frac{r}{d}\right)(M_\infty^2 C_D) = \frac{0.774}{M_\infty^2 [C_D/(x/d)]^{2/3} - 1.09} \tag{4.159}$$

Blunt-nosed cylinder (first approximation):

$$\frac{p}{p_\infty} = 0.067 M_\infty^2 \frac{\sqrt{C_D}}{x/d} \tag{4.160}$$

$$\frac{r}{d} = 0.795 C_D^{1/4} \left(\frac{x}{d}\right)^{1/2} \tag{4.161}$$

Blunt-nosed cylinder (second approximation):

$$\frac{p}{p_\infty} = 0.067 M_\infty^2 \frac{\sqrt{C_D}}{x/d} + 0.44 \tag{4.162}$$

$$\frac{r/d}{M_\infty C_D^{1/2}} = 0.795 \sqrt{\frac{(x/d)}{M_\infty^2 C_D^{1/2}} \left[1 + 3.15 \frac{(x/d)}{(M_\infty^2 C_D^{1/2})}\right]} \tag{4.163}$$

Note that, for a blunt-nosed flat plate, the pressure distribution p/p_∞ downstream of the nose varies directly as M_∞^2 and $C_D^{2/3}$, and inversely as $(x/d)^{2/3}$; the shock-wave shape varies directly as $C_D^{1/3}$ and $(x/d)^{2/3}$. For a blunt-nosed cylinder, the pressure distribution p/p_∞ downstream of the nose varies directly as M_∞^2 and $C_D^{1/2}$ and inversely as x/d; the shock-wave shape varies directly as $C_D^{1/4}$ and $(x/d)^{1/2}$, that is, it is parabolic.

Methods discussed in this chapter that are not predicated on the hypersonic small-disturbance equations and that therefore are not restricted to slender bodies at small angle of attack are as follows.

1) *Concept of Mach-number independence*: Here, we observe both experimentally and from the governing Euler equations (4.1–4.5) with the appropriate boundary conditions that certain nondimensional aerodynamic quantities such as C_L, C_D, and C_p become relatively independent of Mach number above a sufficiently higher value of M_∞, which for blunt bodies can be as low as $M_\infty = 5$.

2) *Thin shock-layer methods*: These methods make use of approximations based on the thinness of hypersonic shock layers. As an example, Maslen's method is a straightforward application of the thin-shock-layer assumption, leading to a closed-form equation for the variation of pressure across the shock

layer as

$$p(x, \psi) = p_s(x) + \frac{u_s(x)}{R_s(x)}[\psi - \psi_s(x)] \tag{4.177}$$

which in turn allows the solution for all other flow variables within the shock layer.

Problems

4.1 Referring to Sec. 4.3, *prove* that the Mach-number independence principle applies to the pressure coefficient C_p, the lift coefficient C_L, and the wave-drag coefficient C_{D_w}.

4.2 Derive Eqs. (4.48) for the transformed direction cosines.

4.3 The condition that two or more different flows over different affinely related bodies satisfy hypersonic similarity is that γ, $M_\infty \tau$, and α/τ be the same between these flows. Show how the derivation of the principle of hypersonic similarity, carried out in Sec. 4.5 for zero angle of attack, is modified to include small angle of attack.

4.4 The purpose of this problem is to demonstrate the degree of validity of hypersonic similarity by plotting data for wedges. Proceed as follows:
(a) From exact oblique shock theory, tabulate C_p vs M_∞ for wedges of $\theta = 5$-, 10-, 15-, 20-, and 30-deg half-angles.
(b) Plot these data for all five wedges on the same piece of graph paper in the form of $C_p/\tan^2\theta$ vs $M_\infty \tan\theta$. (Note that, within the framework of hypersonic small-disturbance theory, $C_p/\tan^2\theta = C_p/\tau^2 \approx C_p/\theta^2$ and $M_\infty \tan\theta = M_\infty \tau = K \approx M_\infty \theta$).
(c) On the same graph, plot Eq. (2.29). Finally, after observing the results shown on the graph, make some statements about: the accuracy and range of validity of hypersonic similarity.

4.5 The purpose of this problem is to demonstrate the degree of validity of hypersonic similarity by plotting data for cones. Proceed as follows:
(a) From exact cone results (such as from [17] or [18]), tabulate C_p vs M_∞ for cones of $\theta_c = 5$-, 10-, 15-, 20-, and 30-deg half-angle.
(b) Plot these data for all five cones on the same piece of graph paper in the form of $C_p/\tan^2\theta_c = C_p/\tau^2$ and $M_\infty \tan\theta_c = M_\infty \tau$.
(c) On the same graph, plot Eq. (4.126). After observing the results, make some statements about the accuracy and range of validity of hypersonic similarity. Are these conclusions any different than those made in problem 4.4 for wedges?

4.6 Derive Eq. (4.89).

4.7 For the solution of hypersonic flow over a slender cone, this problem demonstrates how the flowfield variables can be obtained after a solution of Eq. (4.106) is carried out. For example, Eq. (4.123) gives an equation from which $\bar{p}$ can be obtained in terms of f'. Derive the analogous equations for $\bar{v}'$ and $\bar{\rho}$ as function of f and f'. Finally, show how $\bar{u}'$ can be obtained.

4.8 Derive Eqs. (4.154) and (4.155).

5
Hypersonic Inviscid Flowfields: Exact Methods

Regarding computing as a straightforward routine, some theoreticians still tend to underestimate its intellectual value and challenge, while practitioners often ignore its accuracy and overrate its validity.

C. K. Chu, Columbia University, 1978

Chapter Preview

Aeronautical history was made on 27 March 2004 when the X-43 Hyper-X test vehicle, shown in Fig. 5.1, achieved sustained flight for 11 s at Mach 6.9 powered by a supersonic combustion ramjet engine (scramjet). This was the first time that an airbreathing scramjet engine had successfully powered a vehicle for sustained light in the atmosphere. On 16 November, another scramjet-powered X-43 achieved sustained flight at nearly Mach 10, making it the fastest airplane in history to date. A three view of this unmanned vehicle is shown in Fig. 5.2; the vehicle is about 12 ft in length. A photograph of the X-43 resting on supports in the laboratory is given in Fig. 5.3.

Imagine that you are given the job of calculating the flow field and aerodynamic characteristics of the X-43 in hypersonic flight. The approximate methods discussed in the preceding chapters would be good for back-of-the-envelope calculations and for gaining insight helpful during the preliminary conceptual stages of design. But the X-43 is designed right on the margin of its required performance, and absolutely the best accuracy is required when calculating its detailed aerodynamics. The approximate techniques discussed earlier are simply not good enough for this purpose. Part of obtaining the best accuracy is to use the full continuity, momentum, and energy equations for fluid flow, with no geometric simplifications. If you are satisfied at first with the assumption of an inviscid nonchemically reacting flow, then Eqs. (4.1–4.5) are the ones you want to use. But the only way to solve these "exact" equations, especially for the geometry of the X-43, is *numerically.*

That is what this chapter is all about—exact methods for calculating hypersonic inviscid flowfields. Exact means that the full conservation equations (4.1–4.5) are being used with no geometric simplification. In turn, all of the exact methods are numerical methods—they have to be. So this chapter deals with a sequence of numerical methods for calculating inviscid hypersonic flow. In many respects for the calculation of hypersonic flows, this is where the rubber hits the road. When you finish this chapter, you will be much closer to the calculation of the hypersonic flow over a vehicle such as the X-43 in Figs. 5.1–5.3. That is what is exciting about this chapter. So read on—with excitement.

Fig. 5.1 X-43 side view (NASA).

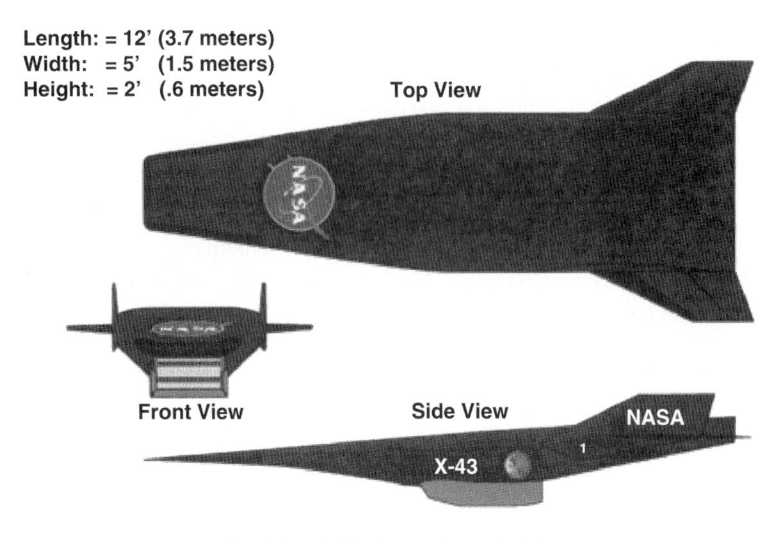

Fig. 5.2 X-43 three view (NASA).

Fig. 5.3 X-43 rear view (NASA).

5.1 General Thoughts

Chapter 4 covered material that is in the realm of "classical" hypersonic aerodynamics, that is, theoretical analyses developed in the 1950s and 1960s before the widespread use of high-speed digital computers. Indeed, before the age of the computer there was no other recourse; for the analysis of inviscid hypersonic flows, the exact nonlinear Euler equations (4.1–4.5) do not yield closed-form theoretical results of a general nature; hence, various physical approximations had to be applied to these exact equations in order to obtain a system of approximate equations more tractable to theoretical analyses. For hypersonic flows, even these approximate governing equations are still nonlinear, but, as we have seen in Chapter 4, various approximate methods have been successfully developed to obtain useful solutions. The point here is that such solutions are indeed *approximate*, either because the governing equations themselves are reduced to simpler form as a result of approximations about the physics of the flow (e.g., small perturbations), or during the course of solution of the exact equations various limiting cases are taken (for example, $M_\infty \to \infty$). This is why Chapter 4 was subtitled "Approximate Methods." However, even though such methods are classical, keep in mind that the results are frequently very practical and useful, and indeed many of these classical methods are used extensively today in the engineering analysis of hypersonic flows.

In contrast, the present chapter is subtitled "Exact Methods." This requires some definition. Here, we will deal with the exact, governing Euler equations for inviscid flow without any subsequent reduction of these equations based on physical approximations. Because we are using the full equations (4.1–4.5) without any approximations, and because these are the exact equations for inviscid flow (over all parts of the flight spectrum, from subsonic to hypersonic), we will label the subsequent solutions of these equations as "exact." This is a slight misnomer, however, because all of the exact solutions are numerical, and any numerical solution is subject to numerical error. For example, we will see that the exact, governing partial differential equations can be replaced by finite difference equations; these difference equations are numerically and theoretically different than the original partial differential equations because of the *truncation error* that is always present in the finite difference formulation. Moreover, during the course of the numerical solution of these difference equations computer *round-off errors* are incurred. Finally, there is sometimes a lack of preciseness brought about by the numerical treatment of boundary conditions. So, strictly speaking, even numerical solutions are not truly exact solutions of the governing equations. However, with this slight proviso in mind, we will proceed to label such numerical solutions as exact solutions because they begin with the exact governing equations. Stated slightly differently, once we choose to work with the full Euler equations, then the only type of solution with any generality must be *numerical*; hence, the terms exact solutions and numerical solutions are used here synonymously.

In more modern terms, this chapter deals with the application of computational fluid dynamics (CFD) to inviscid hypersonic flows. This chapter could not have been written 40 years ago; indeed, CFD is a newly emerging dimension in fluid dynamics, which now complements the more classical dimensions of pure experiment and pure theory. Applications of CFD are impacting aerodynamic research and development across the entire flight spectrum, from subsonic to hypersonic speeds. The impact of CFD is particularly strong on hypersonic aerodynamics because the availability of hypersonic wind tunnels and other hypersonic ground-test facilities is severely limited, both in regard to number of facilities as well as the practical flight range of Mach number, Reynolds number, and temperature levels attainable in such facilities. Thus, in the modern world of hypersonics CFD serves as a powerful tool for research, development, and design.

In this chapter, there is no intent to give a detailed presentation of the fundamentals of CFD; such an endeavor justifies a book on that subject alone. There are many such books. The interested reader is encouraged to study [52], which is an introduction to CFD at the advanced senior/first-year graduate level. On the other hand, to understand and appreciate some of the inviscid-flow calculations discussed here some of the general ideas and methodology of CFD must be understood. The assumption is made here that the reader has not had formal education or experience in CFD, and therefore, as the case demands, we will present various details of the computational techniques in a fashion that will be reasonably self-contained.

As a final introductory note, advances in CFD now make possible the inviscid, three-dimensional, unsteady flowfield solution over complete flight-vehicle configurations. Moreover, there are a variety of different computational techniques, ranging from the method of characteristics (which is in reality an older classical technique now made very practical by high-speed digital computers), to finite difference, finite volume, and finite element methods. In the present chapter, only a representative selection of solutions and solution procedures will be given. The choice is based on the author's experience and bias—10 different authors would most likely make 10 different choices, all justified in their own way. The result, however, would be the same, that is, the absolute appreciation that the Euler equations are now made solvable by a seemingly endless variety of numerical techniques. The purpose of this chapter is to give the reader the flavor of such solutions as applied to hypersonic inviscid flows.

The road map for this chapter is given in Fig. 5.4. Two different numerical methods for the solution of the exact Euler equations for inviscid hypersonic

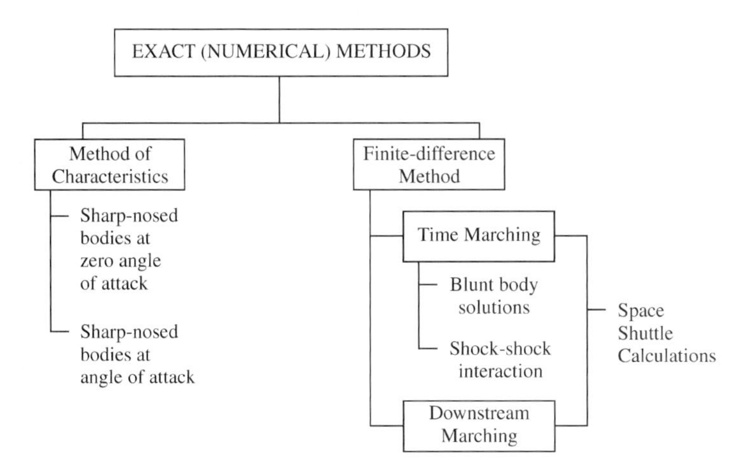

Fig. 5.4 Road map for Chapter 5.

flows are discussed: the method of characteristics and the finite difference method. In reality there is no one "method of characteristics" and no one "finite difference method"; rather, the two primary boxes at the top of the road map represent two separate and different generic methods for each of which there are numerous different detailed numerical approaches. See [4] for more details on the method of characteristics and [52] for an extensive presentation of various finite difference methods. As stated earlier, our purpose here is just to give you the flavor of these methods as applied to hypersonic inviscid flow. Along the way, following down the left column in Fig. 5.4, we will look at some method-of-characteristics solutions for the flow over sharp-nosed two-dimensional and axisymmetric bodies at zero angle of attack and the flow around an axisymmetric body at angle of attack. The latter application involves the three-dimensional method of characteristics—a rigorous challenge. Then we move to the right-hand column in Fig. 5.4 and consider two different finite difference approaches: time marching and downstream marching. The time-marching approach will be applied to the flow over a blunt body and to the interaction of an external shock wave impinging on the curved blow shock ahead of a blunt body. The downstream-marching approach will be applied to sharp-nosed bodies and to the locally supersonic and hypersonic flow regions over the space shuttle. I suggest that you make frequent references to this road map as you progress through this chapter.

5.2 Method of Characteristics

In 1929 in Germany, Ludwig Prandtl and Adolf Busemann were the first to apply the method of characteristics to a problem in supersonic flow; they utilized a graphical approach to construct the contour of a supersonic nozzle. Since then, the method of characteristics has become a classical technique for the solution of inviscid supersonic and hypersonic flows, both internal and external. Because it is classical and widely known, and because it is usually part of a basic course in compressible flow, the assumption is made that the reader has some familiarity with the method of characteristics; hence, the method will not be developed in detail here. Rather, some of the general considerations will be reviewed, and some applications to hypersonic flows will be shown. For the interested reader, an excellent and detailed presentation of the method of characteristics applied to various aerodynamic problems is given in [53]. See also [4] for an introductory presentation.

The method of characteristics is useful when the system of governing partial differential equations is hyperbolic. For this case, the problem is mathematically well posed by starting from an initial data surface and calculating the flow along the characteristic directions. Steady, inviscid, supersonic (and hypersonic) flow is one such case; the governing Euler equations are hyperbolic, and hence, starting with an initial data surface situated downstream of the limiting characteristics, the supersonic flow can be calculated by marching downstream along the characteristic lines (if the flow is two dimensional) or characteristic surface (if the flow is three dimensional). The governing Euler equations for two- or three-dimensional steady flow [Eqs. (4.1–4.5) with $\partial/\partial t = 0$] exactly reduce to simpler differential equations in one less space dimension, called the compatibility equations, along the characteristics. For two-dimensional flow, these compatibility equations become ordinary differential equations, which are more readily solved than the original partial differential equations. The solution of the compatibility equations and the evolution of the characteristic directions are in general computed simultaneously as the solution procedure marches downstream from the initial data surface. In this manner, the entire supersonic and hypersonic inviscid flowfield can be calculated in an exact fashion.

There is a hierarchy of solutions involving the method of characteristics. The simplest application of the method involves a steady, two-dimensional, irrotational flow. This is frequently the first application encountered by the student when first studying the method. For this application, there are two characteristics at each point in the flow—the left- and right-running Mach waves. Moreover, the compatibility equations that hold along these characteristics are algebraic relations; this leads to a particularly straightforward solution of the flow. For a steady, axisymmetric, irrotational flow, the characteristics at each point are still the left- and right-running Mach waves, but the compatibility equations are now ordinary differential equations, which are readily solved numerically along the characteristics. On the other hand, for a steady two-dimensional or axisymmetric *rotational* flow, there are three characteristics lines through each point—the left- and right-running Mach waves and the streamline. The compatibility equations are appropriate ordinary differential equations that hold along these characteristics. Finally, the most complex application of the method of characteristics is to three-dimensional flows, where the characteristics are Mach surfaces and stream surfaces, and the compatibility equations are, in general, partial differential equations in two space dimensions.

For the application of the method of characteristics to external hypersonic flow, the aspect of *rotationality* is of primary concern. To understand this better, recall that, for a given shaped body, as the supersonic or hypersonic freestream Mach number increases the strength of the shock wave increases. If these strong shock waves have curvature, as sketched in Fig. 5.5 for both slender and blunt bodies, then the entropy increase across the shock wave will be different from one streamline to the next. For example, in both Figs. 5.5a and 5.5b, streamline 1 has a higher entropy than streamline 2 because it has come through a stronger portion of the shock wave. This is particularly acute for the blunt body in Fig. 5.5b, where the bow shock wave is highly curved, and an intense region of large entropy gradients is produced in the nose region. This is the source of the entropy layer discussed in Sec. 1.3.2, and sketched in Fig. 1.14, which should be reviewed at this stage. Such entropy gradients occur behind curved shock waves at any supersonic Mach number, but because the shocks are stronger and usually more highly curved at hypersonic Mach numbers then the entropy gradients become more severe. In turn, this introduces a large amount of rotationality into inviscid flows over hypersonic bodies, as can be quantitatively obtained from Crocco's theorem (see [4]), written as follows:

$$T\nabla s = \nabla h_0 - \mathbf{V} \times (\nabla \times \mathbf{V}) \tag{5.1}$$

Here, ∇h_0 is the gradient in total enthalpy at a point in the flow; for the steady, adiabatic flows considered here, h_0 is constant, and hence $\nabla h_0 = 0$. Also

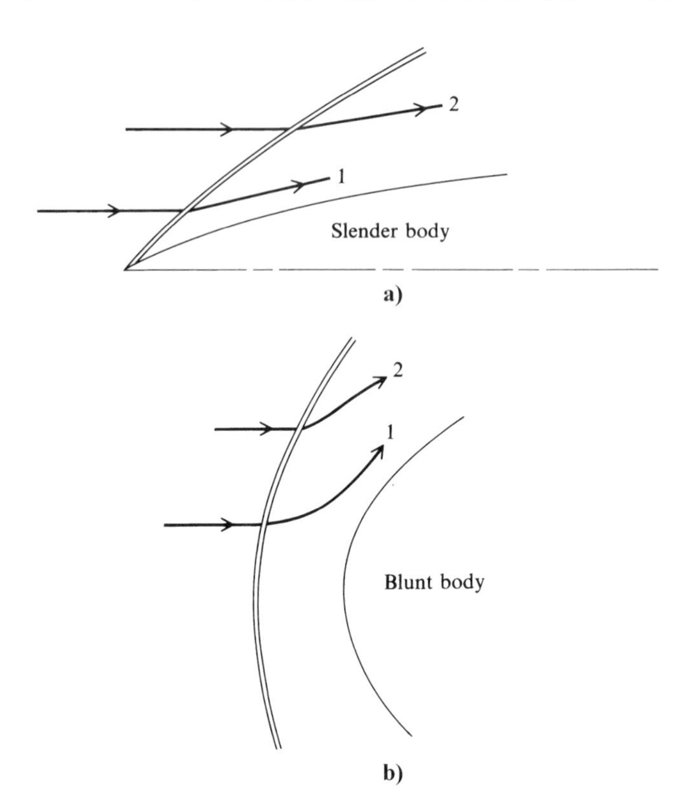

Fig. 5.5 Bodies with curved shock waves.

in Eq. (5.1), $\nabla \times \boldsymbol{V}$ is the vorticity; if the vorticity is finite, then by definition the flow is rotational. From Eq. (5.1), with $\nabla h_0 = 0$ we see that an entropy gradient, ∇s directly produces vorticity in the flow, hence making such flows rotational. For hypersonic applications, these flows can be *highly* rotational. This is illustrated in Fig. 5.6 (obtained from [13]), which gives the variation of vorticity behind a parabolic bow shock wave as a function of the local wave angle β. The different curves correspond to different values of M_∞. Note that 1) the vorticity peaks at a large value in the vicinity of the sonic point on the shock and 2) the magnitude of the vorticity increases with Mach number. The point here is that any application of the method of characteristics to calculate the inviscid flow over a *hypersonic* body should definitely utilize the *rotational* method of characteristics.

Historically, the first major application of the rotational method of characteristics was made by Antonio Ferri in 1946, as described in [54]. (Ferri was an ebullient Italian aerodynamicist who developed a pioneering supersonic aerodynamic laboratory in Guidonia near Rome during the 1930s and who was brought to the United States toward the end of World War II to help foster research in supersonic flows at the NACA Langley Memorial Laboratory in Virginia.) For example, Ferri's work was the basis of the characteristic calculations shown in Figs. 3.13,

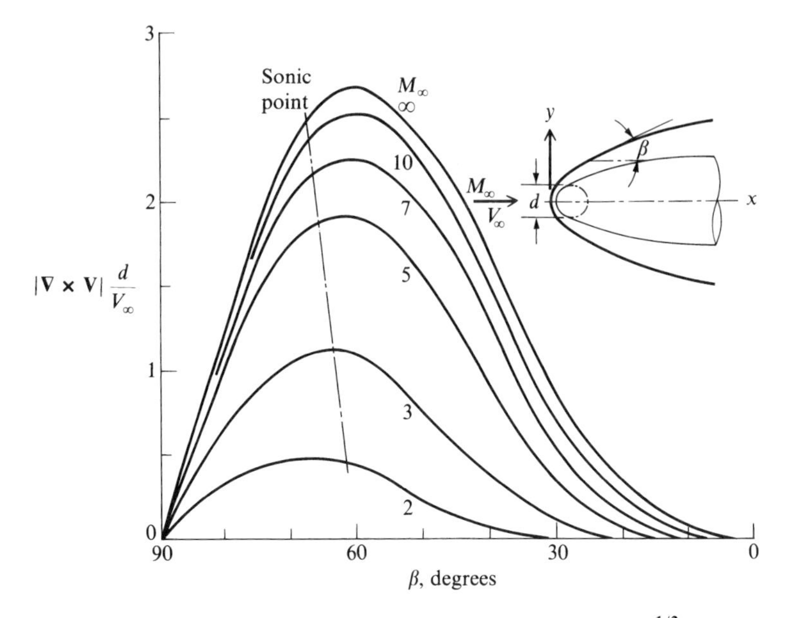

Fig. 5.6 Vorticity behind a parabolic shock wave: $y/d = (x'/d)^{1/2}$, and $\gamma = 1.4$ (from [13]).

3.14, 3.20, 3.22, 3.23, and 4.5. For technical details in a more modern setting, make certain to read the sections on the rotational method of characteristics in Zucrow and Hoffman [53].

To this point in the present section, we have been reviewing some general considerations about the method of characteristics, relying somewhat upon prior familiarity with the method on the part of the reader. In the interest of being slightly more precise, the following is a brief outline of the application of the rotational method of characteristics to a two-dimensional or axisymmetric external flow:

1) The method must be started from an initial data line that lies totally within the supersonic portion of the flow and in particular should be downstream of the limiting characteristics (for example, see [4]). The flow properties on this initial data line must be obtained from another, independent calculation. The usual methods of obtaining such initial data are as follows:

a) If the body has a pointed nose with an attached shock wave, such as shown in Fig. 5.5a, then the flow in the immediate vicinity of the nose is totally supersonic, and it can be closely approximated by wedge flow (in the two-dimensional case) or conical flow (in the axisymmetric case). Then, as indicated in Fig. 5.7a, all flow properties are known along the initial data line from the exact oblique shock solution, or from the Taylor–Maccoll conical solution.

b) If the body has a blunt nose with a detached bow shock wave, an appropriate blunt-body solution must be carried out (blunt-body solutions are discussed in Sec. 5.3). The initial data line must be taken along or downstream of the limiting characteristic, as shown in Fig. 5.7b.

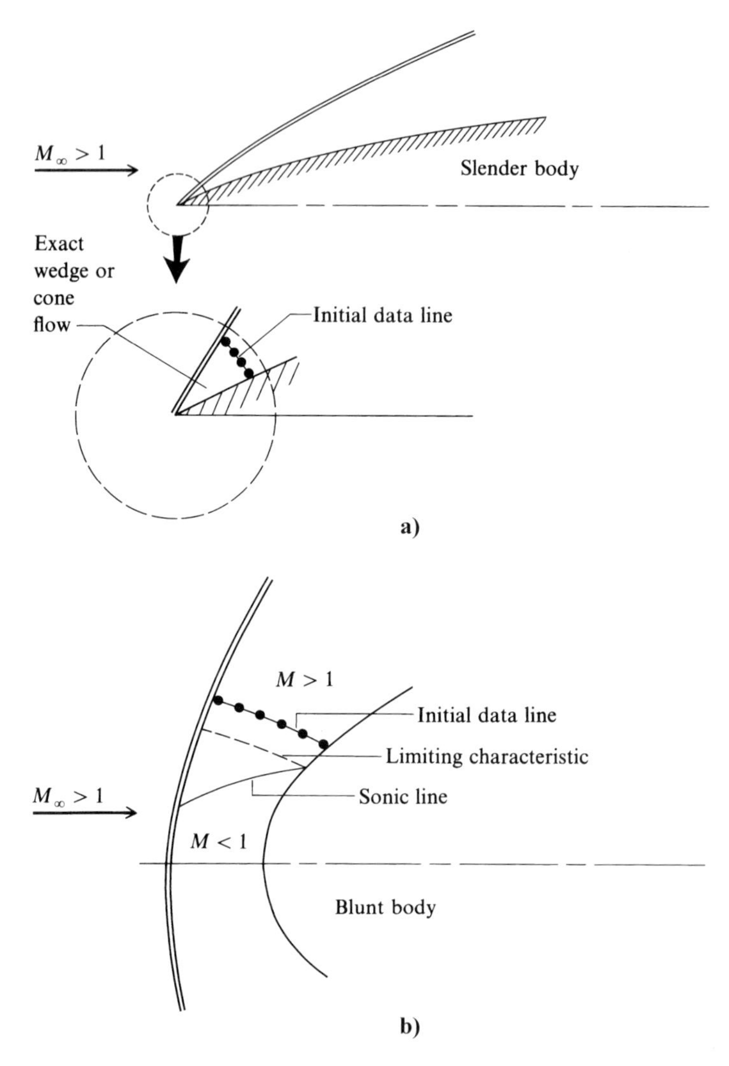

Fig. 5.7 Initial data lines for slender and blunt-body flows.

In both of the preceding cases, the flow must be totally supersonic along the initial data line; moreover, care should be taken not to use a characteristic line as the initial data line.

2) Starting from the initial data line, the solution progresses by marching downstream along the characteristic lines. A single element of this process, called a unit process, is illustrated in Fig. 5.8. Here, points 1 and 2 are two points on the initial data line; all flow properties, including the streamline angles θ_1 and θ_2, are known at these points. From the known Mach numbers, hence the known Mach angles μ_1 and μ_2 at these points, construct the left- and right-running Mach waves (designated by C_+ and C_-, respectively) at

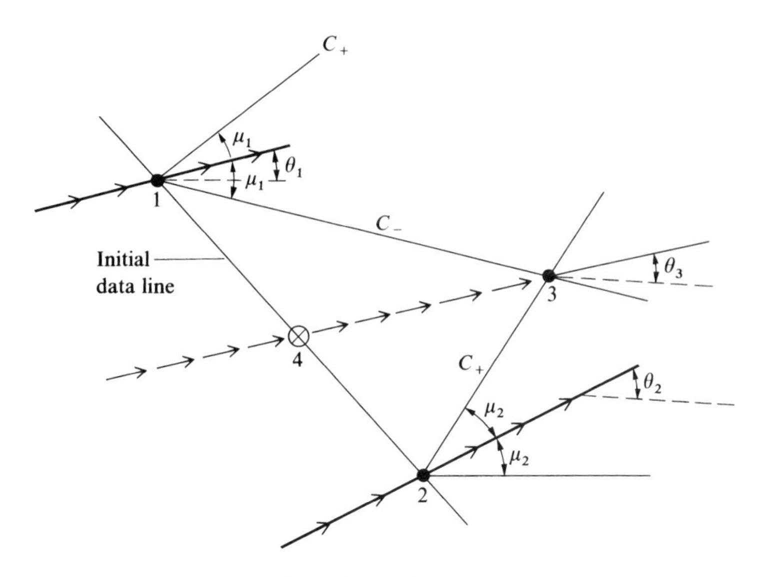

Fig. 5.8 Unit process.

both points, as shown in Fig. 5.8. (Because the unit process involves only small distances in the flow, all constructed Mach lines and streamlines are drawn as straight lines between adjacent points.) The Mach waves are characteristic lines, hence C_+ and C_- are appropriate designations for these lines. Note that the C_- characteristic from point 1 and the C_+ characteristic from point 2 intersect at point 3, thus locating point 3 in space. We wish to calculate the flow properties at point 3.

3) Assume that the streamline through point 3 is at an angle θ_3, where θ_3 is an average between the known θ_1 and θ_2. (This assumption is made more accurate by iterating the unit process, as will be noted later.) Trace the streamline at point 3 backward until it intersects the initial data line at point 4, as shown in Fig. 5.8. Recall that, for rotational flow, this streamline is also a characteristic line.

4) Calculate the flow properties at point 3 by solving the compatibility equations along the characteristic lines. These equations are obtained from the governing Euler equations (4.1–4.5) after considerable manipulation (see [4] and [53]) and are as follows.

Along Mach lines:

$$\frac{\mathrm{d}p}{\rho V^2 \tan\mu} \pm \mathrm{d}\theta + \frac{j \sin\theta \sin\mu}{\sin(\theta \pm \mu)} \frac{\mathrm{d}y}{y} = 0 \tag{5.2}$$

Along streamlines:

$$\mathrm{d}s = 0 \tag{5.3}$$

In Eq. (5.2), the plus and minus signs correspond to the C_+ and C_- characteristics, respectively, and $j = 0$ or 1 for two-dimensional and axisymmetric flow, respectively. Equation (5.3) is simply a statement that the entropy is constant along a streamline in an inviscid, adiabatic flow—a statement already contained in Eq. (4.5). Equation (5.3) can be replaced by two relations that depend on constant entropy along a streamline, namely,

Along streamlines:

$$\mathrm{d}p = -\rho V \,\mathrm{d}V \tag{5.4}$$

Along streamlines:

$$\mathrm{d}p = a^2 \,\mathrm{d}\rho \tag{5.5}$$

Equation (5.4) is the familiar Euler equation that holds along a streamline in a rotational flow; it can be readily obtained by a suitable manipulation of Eqs. (4.2–4.4) along with the definition of a streamline, as shown in [5]. Equation (5.5) is simply based on the definition of the speed of sound, $a^2 = (\partial p/\partial \rho)_s$; because s is constant along the streamline, then any change in pressure along the streamline $\mathrm{d}p$ is related to a corresponding change in density along the streamline $\mathrm{d}\rho$, through the relation $a^2 = (\mathrm{d}p/\mathrm{d}\rho)$. Please note that whereas $\mathrm{d}p$ in Eqs. (5.4) and (5.5) denotes a change in pressure along a streamline, $\mathrm{d}p$ in Eq. (5.2) denotes a change in pressure along the Mach lines. Returning to Fig. 5.8, we now have a system of compatibility equations, namely, Eqs. (5.2), (5.4), and (5.5), which hold along the characteristic lines. The integration of these equations along the characteristics can be carried out in a variety of ways. For simplicity, let us replace the differentials in Eqs. (5.2), (5.4), and (5.5) with forward differences. For example, Eq. (5.2) written along the C_- characteristic through point 1 in Fig. 5.8 is

$$\frac{p_3 - p_1}{\rho_1 V_1^2 \tan \mu_1} - (\theta_3 - \theta_1) + \frac{j \sin \theta_1 \sin \mu_1}{\sin(\theta_1 - \mu_1)} \frac{(y_3 - y_1)}{y_1} = 0 \tag{5.6}$$

Equation (5.2) written along the C_+ characteristic through point 2 is

$$\frac{p_3 - p_2}{\rho_2 V_2^2 \tan \mu_2} + (\theta_3 - \theta_2) + \frac{j \sin \theta_2 \sin \mu_2}{\sin(\theta_2 + \mu_2)} \frac{(y_3 - y_2)}{y_2} = 0 \tag{5.7}$$

Equation (5.4) written along the streamline is

$$p_3 - p_4 = -\rho_4 V_4 (V_3 - V_4) \tag{5.8}$$

Equation (5.5) written along the streamline is

$$p_3 - p_4 = a_4^2 (\rho_3 - \rho_4) \tag{5.9}$$

In Eqs. (5.6–5.9), all conditions at points 1, 2, and 4 are known. (Conditions at point 4 are interpolated between points 1 and 2.) The locations of points 3 and 4 are known, from steps 2 and 3; hence, y_3 is known. Thus, Eqs. (5.6–5.9) are four algebraic equations that can be solved for the four unknowns p_3, θ_3, V_3, and ρ_3.

5) Repeat steps 2–4, where now the slopes of the C_- and C_+ characteristics are based on an *average* of θ_1, μ_1, θ_3, and μ_3, and the streamline at point 3 is traced back using the value of θ_3 obtained in step 4. Iterate until convergence is obtained. At the completion of this iteration, point 3 is now accurately located in space, and the flow properties at point 3 are accurately obtained.

The preceding unit process is carried out from point to point in a sequential fashion, marching downstream from the initial data line. Slight modifications to the unit process are made to satisfy the boundary conditions at the shock wave and the body: see [53] for details. In this fashion, the complete inviscid flowfield between the body and the shock wave can be numerically constructed. Again, emphasis is made that this is an exact solution; the compatibility equations (5.2) and (5.3) are obtained directly from the exact Euler equations without any mathematical approximations or further physical simplifications, and any errors introduced in the solution are numerical errors involved with the finite difference representation of the compatibility equations. In this vein, the method of characteristics becomes truly exact in the limit of an infinitely fine characteristics mesh.

An example of a characteristics mesh for the calculation of the rotational flow over a two-dimensional body is shown in Fig. 5.9, obtained from [53]. For clarity, only the Mach lines are shown here, although keep in mind that the streamlines are also characteristics. This calculation is made for a supersonic freestream with

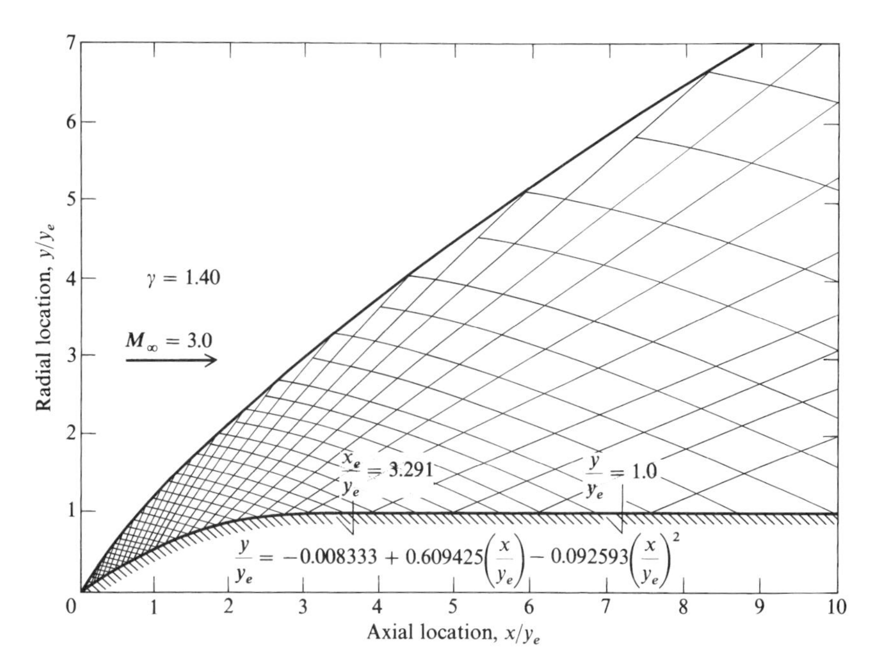

Fig. 5.9 Typical characteristics mesh (from Zucrow and Hoffman [53]).

$M_\infty = 3$; at hypersonic Mach numbers, the shock wave would be much closer to the surface, and because the Mach angles would be smaller the characteristics mesh would be more highly skewed. The pressure distributions behind the shock wave and along the body corresponding to the case in Fig. 5.9 are shown in Fig. 5.10. (Note that the pressure behind the shock is higher than that along the body. This is another example of the normal pressure gradient necessary to balance the centrifugal force along curved streamlines; for the convex streamlines over the body in Figs. 5.9 and 5.10, the pressure quite naturally is going to decrease from the shock to the body, as originally discussed in conjunction with Fig. 3.9.) The results shown in Figs. 5.9 and 5.10 are included here only to illustrate the use of the rotational method of characteristics for a two-dimensional body and to emphasize again that the method is a reasonable approach for the calculation of two-dimensional and axisymmetric inviscid hypersonic flows.

The vast majority of practical aerodynamic problems involve three-dimensional flows. The method of characteristics can also be applied to such flows (for example, see [55–59]); however, the three-dimensional method of characteristics requires considerably more effort than its two-dimensional counterpart. In steady, three-dimensional rotational flow, the characteristics are surfaces, namely, Mach cones and stream surfaces. In general, the compatibility equations that hold along these characteristic surfaces are partial differential

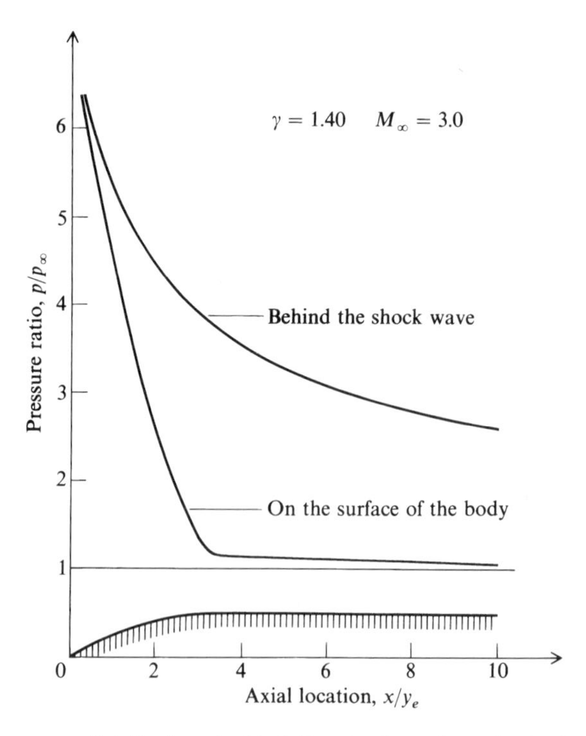

Fig. 5.10 Pressure distributions behind the shock and on the body for the case shown in Fig. 5.9 (from [53]).

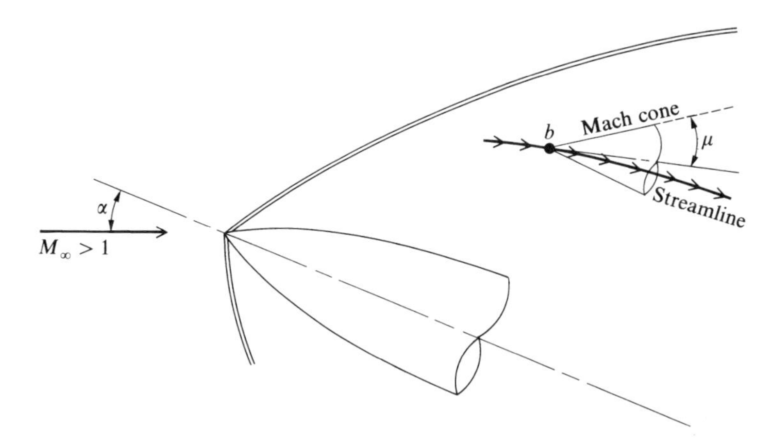

Fig. 5.11 Three-dimensional characteristic surfaces.

equations in two space dimensions. To examine this in more detail, consider Fig. 5.11, which illustrates a general supersonic or hypersonic three-dimensional flowfield. Point b is an arbitrary point in the flow. Through this point, the characteristic directions generate two sets of three-dimensional surfaces—a Mach cone with its vertex at point b and with a half-angle equal to the local Mach angle μ and a stream surface through point b. The intersections of these surfaces establish a complex three-dimensional network of grid points. Moreover, as if this were not complicated enough, the compatibility equations along arbitrary rays of the Mach cone (called bicharacteristics) are partial differential equations that contain cross derivatives which have to be evaluated in directions not along the characteristics. Nevertheless, such solutions can be obtained; see [53] for a detailed discussion of these matters.

Although a detailed presentation of the three-dimensional method of characteristics is beyond the scope of this book, it is important to note some results obtained with this method. In particular, we will examine the calculations of Rakich [58] and [59]; in this work, Rakich utilized a modification to the general philosophy of the three-dimensional method of characteristics, which somewhat simplifies the calculations. In this approach, which is sometimes labeled "semicharacteristics," or the "reference plane method," the three-dimensional flowfield is divided into an arbitrary number of reference planes containing the centerline of the body. This is sketched in Fig. 5.12 for the case of an axisymmetric body at angle of attack; Fig. 5.12 shows a front view of the body and shock wave. Each reference plane is identified by its angular location $\mathbf{\Phi}_1$, $\mathbf{\Phi}_2$, etc. One of these planes, say, $\mathbf{\Phi} = \mathbf{\Phi}_2$, is projected on Fig. 5.13. In this particular reference plane, a series of grid points are established along arbitrarily spaced straight lines locally perpendicular to the body surface. These grid points are not the intersections in a systematic characteristic mesh (such as shown in Fig. 5.9 for the two-dimensional case); rather, they are placed along arbitrary straight lines in much the same way that a finite difference grid is established (to be discussed in subsequent sections). This represents a

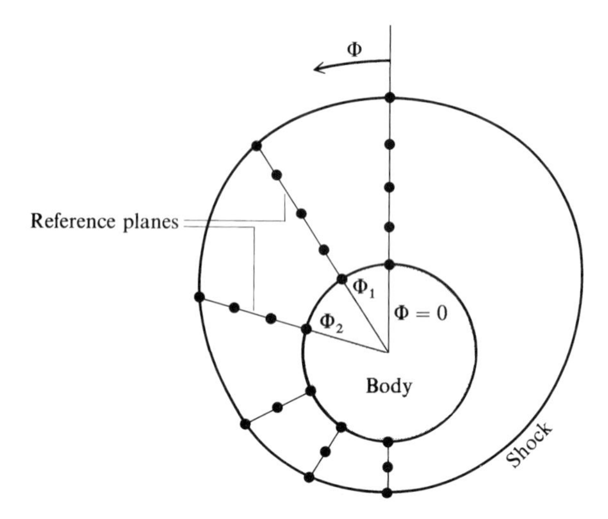

Fig. 5.12 Grid network in a cross-sectional plane for an axisymmetric body at angle of attack: three-dimensional method of characteristics.

major simplification over the general three-dimensional method of characteristics. In Fig. 5.13, assume that the flowfield properties are known at the grid points denoted by solid circles along the straight line *ab*. Furthermore, arbitrarily choose point 1 on the next downstream line *cd*. Let C_+, C_-, and S denote the projection in the reference plane of the Mach cone and streamline through point 1. Extend these characteristics upstream until they intersect the data line *ab* at the cross marks. Data at these cross marks are obtained by interpolating

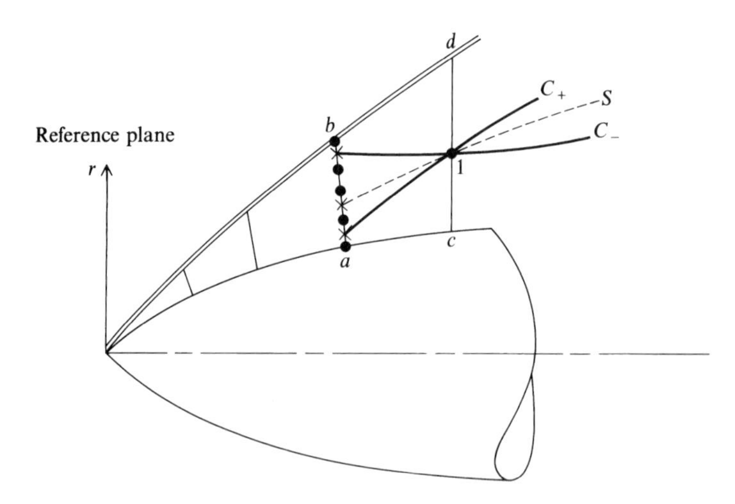

Fig. 5.13 Gird network in the meridional plane for an axisymmetric body at angle of attack; three-dimensional method of characteristics.

between the known data at the solid circles. Then, the flowfield properties at point 1 are obtained by solving the following compatibility equations along C_+, C_-, and S (see [58] for a derivation of these equations):

$$\frac{\beta}{\rho V^2}\frac{\partial p}{\partial C_+} + \cos\phi \frac{\partial \theta}{\partial C_+} = (f_1 + \beta f_2)\sin\mu^* \tag{5.10}$$

$$\frac{\beta}{\rho V^2}\frac{\partial p}{\partial C_-} - \cos\phi \frac{\partial \theta}{\partial C_-} = (f_1 - \beta f_2)\sin\mu^* \tag{5.11}$$

$$\frac{\partial \phi}{\partial S} = f_3 \tag{5.12}$$

where $\beta = \sqrt{M^2 - 1}$, μ^* is the angle between S and C_+ or C_- (i.e., the projection of the Mach angle μ onto the reference plane), θ is the flow angle from the x axis in the meridional plane (that is, $\theta = \tan^{-1} v/u$, where u and v are the velocity components in the x and r directions, respectively), ϕ is the crossflow angle (that is, $\phi = \sin^{-1} w/v$, where w is the velocity component in the $\mathbf{\Phi}$ direction), and f_1 and f_2 are expressions containing the cross derivatives (see [58]).

The preceding limited description is intended only to give the flavor of Rakich's method and to set the stage for the presentation of some results illustrating the use of the three-dimensional method of characteristics. Such results are given in Figs. 5.14–5.18, obtained from [59]. In these figures, results are given for the hypersonic flow over a blunt-nosed, 15-deg cone at angle of attack; theoretical results for the inviscid flow obtained by Rakich using the three-dimensional method of characteristics are directly compared with experimental wind-tunnel results. For example, Fig. 5.14 shows the calculated and measured

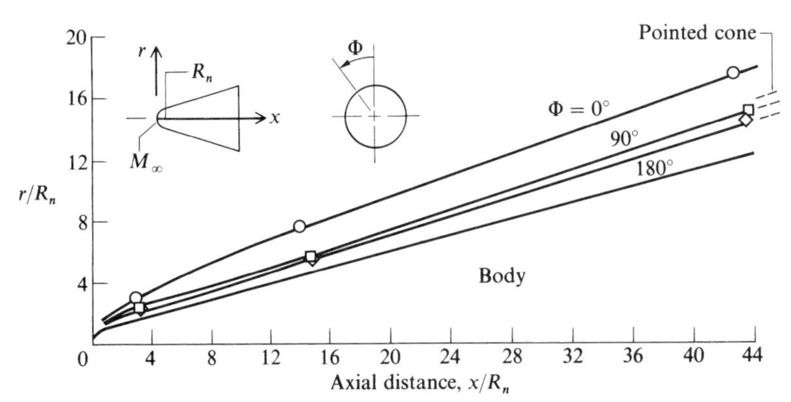

Fig. 5.14 Variation of shock-wave shape: comparison between theory and experiment, where θ_c = 15 deg, α = 10 deg, M_∞ = 10, and γ = 1.4 (from Rakich and Cleary [59]).

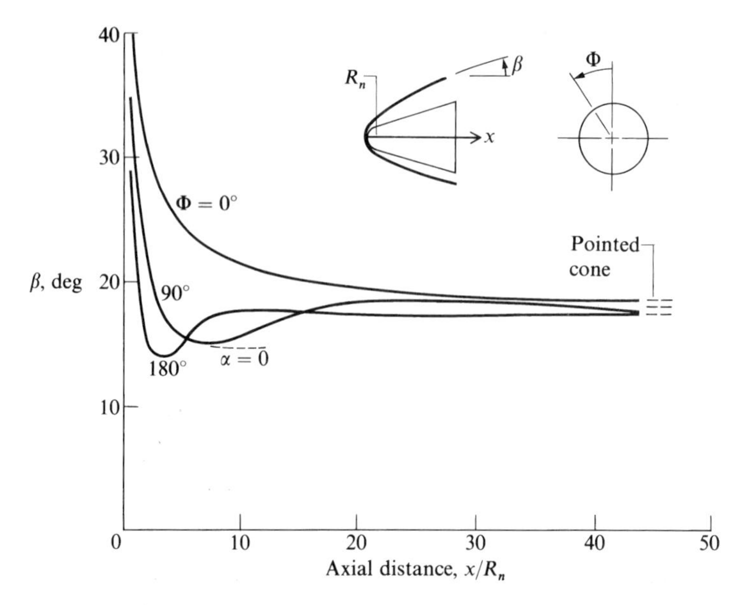

Fig. 5.15 Variation of shock-wave angle; calculations from the three-dimensional method of characteristics: $\theta_c = 15$ deg, $\alpha = 10$ deg, $M_\infty = 10$, and $\gamma = 1.4$ (from [59]).

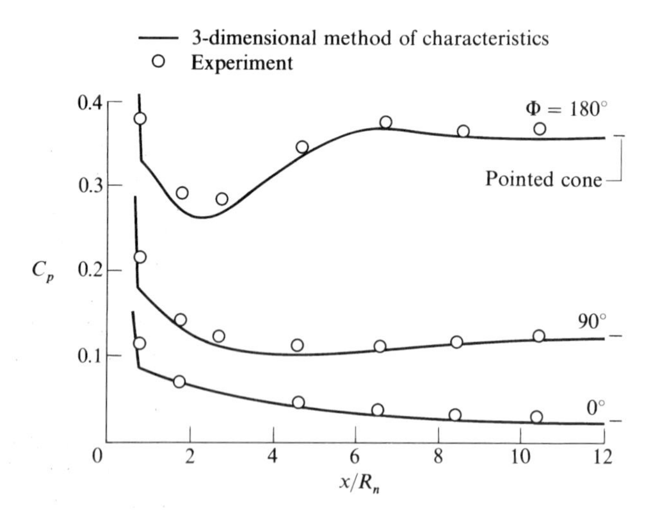

Fig. 5.16 Pressure distribution over a blunt-nosed cone; comparison between theory and experiment: $\theta_c = 15$ deg, $\alpha = 10$ deg, $Re = 0.6 \times 10^6$, $M_\infty = 10$, and $\gamma = 1.4$ (from [59]).

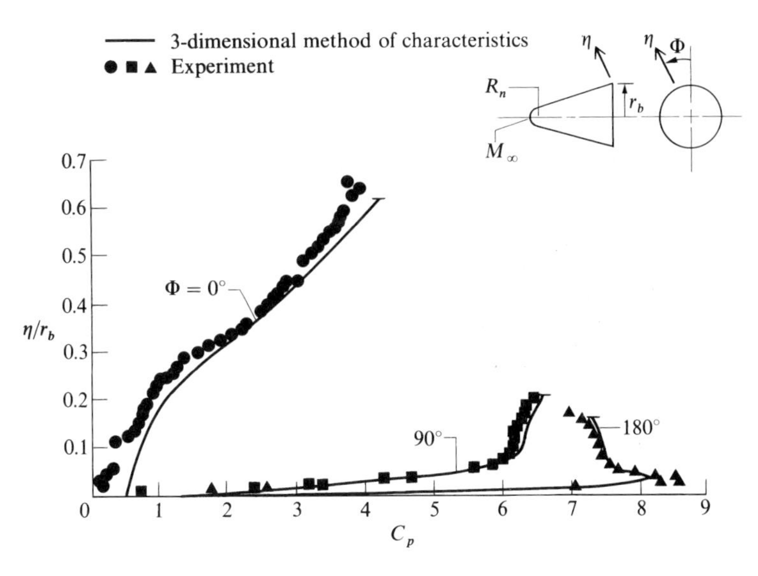

Fig. 5.17 Pitot-pressure variations from the body to the shock: $\theta_c = 15$ deg, $\alpha = 10$ deg, $M_\infty = 10$, $\gamma = 1.4$, and $Re = 0.6 \times 10^6$ deg (from [59]).

shock-wave shape in air ($\gamma = 1.4$) at three meridional planes, $\Phi = 0$, 90, and 180 deg, for an angle of attack of 10 deg at Mach 10. For the wind-tunnel data, the Reynolds number referenced to the cone base radius was 0.6×10^6. Agreement between the three-dimensional method of characteristics and experiment is excellent for this case. This figure also illustrates the effect of nose bluntness on the shock shape. The shock location for a sharp-nosed cone at 15-deg angle of attack is shown as the dashed lines at the right of Fig. 5.14; note that nose bluntness displaces the leeward portion of the shock ($\Phi = 0$ deg) outward, whereas the windward portion of the shock ($\Phi = 180$ deg) is not noticeably displaced. The calculated variation of the shock-wave angle β with axial distance x is shown in Fig. 5.15. It is well known that shock waves around blunted cones at zero angle of attack exhibit a local minimum in the wave angle, that is, as the strong bow shock wave progresses from the normal shock at the nose to the weaker shock downstream, β first decreases, reaches a local minimum, and then increases, finally approaching the sharp cone result far downstream. For the conditions shown in Fig. 5.15, this zero-angle-of-attack case results in the local minimum β occurring at about $x/R_n = 10$, as shown in Fig. 5.15 by the dashed line labeled $\alpha = 0$. For the angle-of-attack case, the method-of-characteristics results indicate a similar trend, except with the minimum β occurring at different axial locations, as shown by the solid lines in Fig. 5.15. Note that the wave angles for the blunted cone eventually approach the sharp-cone results at large distances downstream of the nose, as seen at the right of Fig. 5.15. Surface-pressure distributions are shown in Fig. 5.16. Again, excellent agreement is obtained between the three-dimensional method of characteristics and experiment. In analogy with the shock angle, note that the pressure goes

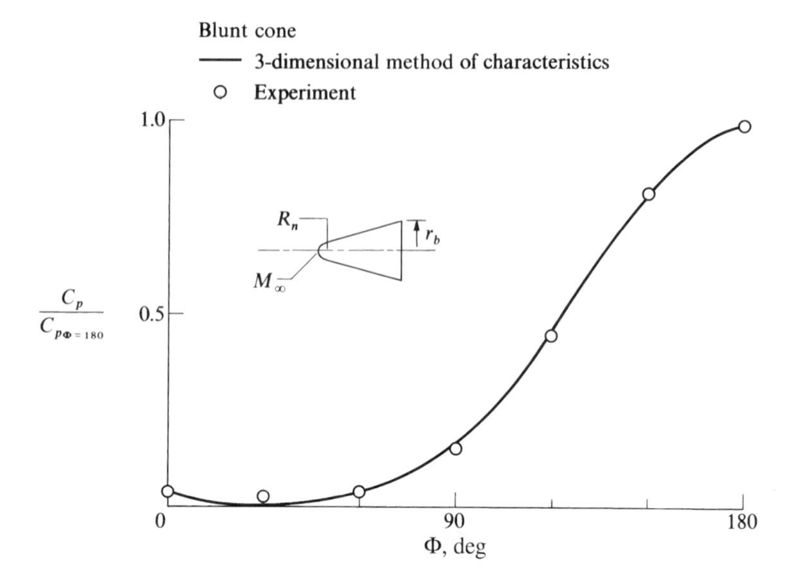

Fig. 5.18 Circumferential surface-pressure distribution at $x/R_n = 8$; comparison between theory and experiment in helium: $\theta_c = 15$ deg, $\alpha = 20$ deg, $M_\infty = 14.9$, $\gamma = 1.667$, and $Re = 0.86 \times 10^6$ (from [59]).

through a local minimum as illustrated in Fig. 5.16. In expanding over the blunt nose, the pressure overexpands downstream of the shoulder, falling below the sharp cone result, and then recompresses to the sharp cone result far downstream. (We note here a weakness of the blast-wave theory discussed in Sec. 4.8; blast-wave theory is incapable of predicting the type of overexpansion and recompression shown in Fig. 5.16.) Figure 5.17 shows the variation of pitot pressure in the flowfield, namely, along lines locally perpendicular to the body, and extending from the body to the shock wave. A comparison with experiment of detailed flowfield information throughout the shock layer (as opposed to just along the body surface) is always a good test of any flowfield theory; in Fig. 5.17, the comparison between the wind-tunnel data and the method-of-characteristics calculation is quite good. For the leeward section of the flowfield ($\Phi = 0$ deg), some lack of agreement between theory and experiment occurs near the body surface; this is because of the thick viscous boundary layer on the leeward side, which is not taken into account by the inviscid theory. On the windward section ($\Phi = 180$ deg), the strong variation of pitot pressure within the entropy layer (see Sec. 1.3) is very apparent. Pitot pressure reaches a peak just outside the entropy layer and then decreases towards the shock wave. This is because, outside the entropy layer and boundary layer on the surface, the local supersonic Mach number increases toward the shock wave, hence resulting in a progressively lower pitot pressure. Finally, to emphasize the three-dimensional nature of this flowfield, Fig. 5.18 illustrates the circumferential surface-pressure distribution around the blunt cone at angle of attack. Unlike the previous data, Fig. 5.18 pertains to hypersonic flow of helium ($\gamma = 1.67$) at $M_\infty = 14.9$. Once again, excellent agreement between experiment and the three-dimensional method of characteristics is obtained.

On this note, we end this discussion on the method of characteristics and its application to hypersonic inviscid flows. We have seen that the method of characteristics is a viable approach toward exact solutions of such flow and indeed has been used extensively for such cases, especially in the time period before 1970. However, the method of characteristics is sometimes tedious to set up and program (in the days before high-speed digital computers, the method of characteristics was carried out by hand calculations—the ultimate in tediousness), with particular complexity in the three-dimensional case. For this reason, in more recent times simpler finite difference solutions have supplanted the method of characteristics in many applications. Modern finite difference methods are treated in the remainder of the chapter.

5.3 Time-Marching Finite Difference Method: Application to the Hypersonic Blunt-Body Problem

Let us return to the road map in Fig. 1.24 and scan over the items discussed so far in this book. Starting with the basic hypersonic shock relations in Chapter 2, we have covered all of the left branch of Fig. 1.24 and most of the second branch, down to and including the method of characteristics. These sections of the road map, and hence all of the preceding discussion in this book, pertain to the state of the art in hypersonic aerodynamics prior to 1966. In fact, if this book were being written in 1966, our discussion of inviscid hypersonic flow would be essentially finished at this point, except for some mention of the calculation of the flow over a blunt hypersonic body. However, this discussion would have been inhibited by the then-existing severe difficulties in obtaining blunt-body solutions. This is emphasized in a statement made in 1966 by Hayes and Probstein [60], to the effect that "in spite of the amount of effort that has gone into this problem in recent years, at present no single method has been agreed on as being the best for calculating the hypersonic flow past general blunt shapes." This situation changed rapidly in 1966 when the first practical blunt-body solution was published by Moretti and Abbett [61]. This solution was obtained by means of a time-marching finite difference technique that greatly simplified the calculation of flows over blunt hypersonic bodies. Indeed, the general idea of using time-marching methods to calculate steady flowfields for a whole host of different problems is now a major endeavor in computational fluid dynamics. The situation has changed so rapidly that the hypersonic blunt-body problem, which in the 1950s and early 1960s was one of the major research problems of the day—with millions of dollars and the efforts of scores of researchers spent on its solution, is today an extended homework problem in several university courses in computational fluid dynamics. Because of the importance of the blunt body in hypersonic aerodynamic applications and because of the efficiency and power of the time-marching technique used to solve such blunt-body flows, both will be discussed at length in this section. Also with this section we move to the right-hand column of our chapter road map in Fig. 5.4.

On a practical basis, the blunt body is a particularly important shape in hypersonic aerodynamics because all hypersonic vehicles have blunt noses to reduce aerodynamic heating. Such heating is a driving design factor for most types of hypersonic vehicles (as we will soon see in Part 2 of this book). Indeed, in Chapter 6 we will demonstrate that stagnation-point aerodynamic heating varies inversely as the square root of nose radius; hence, the larger the nose radius, the lower the aerodynamic heating. This fact was not always recognized. In the 1940s and early 1950s, hypersonic aerodynamic practice was viewed as a high-speed extension of supersonic aerodynamic practice, where slender bodies with sharp leading edges were employed to produce the weakest possible shock waves with an attendent low wave drag. However, as M_∞ increases, aerodynamic heating becomes a major factor, and the heat transfer to a sharp-nosed vehicle becomes severe. (If a hypersonic vehicle in flight does employ a sharp leading edge, nature will soon blunt it by melting away the surface via intense aerodynamic heating.) The desirability of a blunt nose to reduce aerodynamic heating was first advanced by H. Julian Allen in the mid-1950s. Some simple reasons for this, and some of the historical background, are given in Chapter 1 of [5] and Chapter 8 of [1], which should be consulted for more details. However, on a heuristic basis, we can demonstrate the viability of a blunt body in reducing aerodynamic heating as follows. Consider a hypersonic vehicle at high altitude and high velocity, hence with large values of potential and kinetic energy. Imagine the vehicle returns to the ground at zero velocity; hence, the potential and kinetic energies are now both zero. Where has all of the energy gone? The answer is into the air and into the body. The mechanism for heating the air is in part the temperature increase across the shock wave. On one hand, if the body were slender with a sharp nose the shock wave would be weak; hence, less energy would go into heating the air and more into heating the body. On the other hand, if the body had a blunt nose, then the bow shock wave would be strong; hence, more energy would go into heating the air and less would be available to heat the body. On this physical argument alone, we can see why a blunt nose reduces the aerodynamic heating to a body. The point here is that blunt-body flowfields are an important part of the study of hypersonic aerodynamics. Clearly, a detailed knowledge of the flow in the blunt-nose region is essential to the accurate prediction of the heat-transfer distribution around the nose, as well as to the detailed structure of the entropy layer created in the nose region. In turn, the properties of the blunt-body shock layer, as well as the shape of the shock wave in the nose region, can have a strong impact on the body surface conditions far downstream of the nose; recall, for example, the blunt-nosed cone results discussed in Sec. 5.2. Furthermore, recall that the method-of-characteristics solutions over the blunt cone as seen in Sec. 5.2 must be started from an initial data line obtained from a blunt-body solution; thus, the accuracy of such blunt-body solutions is critical to the accuracy of the method-of-characteristics solutions downstream. For all these reasons, and more, the blunt-body problem discussed in the present section is an essential aspect of hypersonic flow. Here, we will treat the *inviscid* blunt-body flow, which is particularly important for the prediction of surface-pressure distribution, shock-wave shape, entropy-layer structure, and for the calculation of properties at the edge of the boundary layer. Finally *viscous* blunt-body flows will be treated in Part 2 of this book.

What made the hypersonic blunt-body problem originally so hard to solve, and why is it an almost routine calculation today? To answer this question, let us examine some physical aspects of the blunt-body shock layer. Consider the steady flow over a blunt body moving at supersonic or hypersonic speeds, as shown in Fig. 5.19. The shock wave in front of this body is detached and curved, ranging from a normal shock wave right at the nose, and becoming a weak Mach wave at large distances from the body. Hence, this single shock wave represents all possible oblique shock solutions for the given upstream Mach number M_∞ with the wave angle ranging from $\beta = \pi/2$ to $\beta = \mu$, where μ is the Mach angle. Behind the normal, and nearly normal, portions of the shock wave, the flow is subsonic, whereas behind the more oblique portion of the shock wave the flow is supersonic. (See [4] for a general description of shock-wave phenomena.) Hence, the blunt-body shock layer is a mixed subsonic–supersonic flow, where the subsonic and supersonic regions are divided by sonic lines, shown as the dashed lines in Fig. 5.19. In the steady, subsonic regions the governing Euler equations (4.1–4.5) with $\partial/\partial t = 0$ are

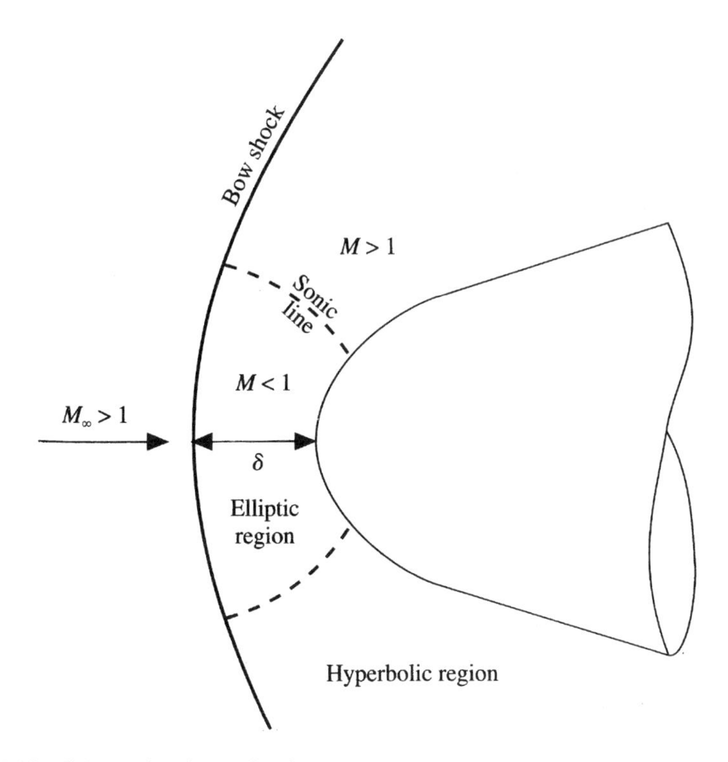

Fig. 5.19 Schematic of the flowfield over a blunt body moving at supersonic or hypersonic speeds.

mathematically *elliptic*, whereas in the supersonic regions these same equations are mathematically *hyperbolic*. (For a description of these mathematical classifications, and their impact on the fluid-dynamic equations, see [4].) The same Euler equations obviously apply in all regions of the flowfield. However, their elliptic nature in the subsonic region means that the flow at any given point depends simultaneously on the properties at all other points in the subsonic region and in particular on the conditions along the total boundary of the subsonic region. In contrast, their hyperbolic nature in the supersonic region means that the flow at any given point depends only on the properties at other points that are contained within the domain of dependence, bounded by Mach lines reaching up-stream from the given point. This situation is a partial answer to the question posed at the beginning of this paragraph. Any theoretical or numerical technique suitable for the exact solution of the subsonic region is improperly posed and hence falls apart in the supersonic region, and vice versa. As just described, in the early days of hypersonics, this mixed nature of the blunt-body flowfield made a consistent exact analysis, valid for both the subsonic and supersonic regions exceptionally difficult to obtain. This state of affairs was nicely reviewed by Van Dyke in 1958 (see [62]). Indeed, it can be said that until 1966 no practical blunt-body solution existed for routine operation, which carried the flow far enough downstream of the sonic line (at least downstream of the limiting characteristics) to provide valid initial conditions for a method of characteristics solution in the supersonic region.

This situation changed dramatically in 1966 when Moretti and Abbett [61] published the first truly practical supersonic blunt-body solution. This approach utilizes a *time-marching* (sometimes called *time-dependent*) *finite difference* solution of the governing *unsteady* Euler equations, starting from arbitrarily assumed initial conditions, and calculating the steady flowfield as an asymptotic limit at large times. The unsteady Euler equations (4.1–4.5) are *hyperbolic* with respect to time, no matter whether the flow is locally subsonic or supersonic. Hence, a time-marching approach starting from assumed initial conditions is a properly posed mathematical problem in all regions of the flow and allows the solution of both the subsonic and supersonic regions simultaneously with the same numerical technique. Today, the time-marching approach is always used for the exact solution of blunt-body flowfields; the calculations are considered "routine," and every major aeronautical company and laboratory has one or more versions of their "standard" blunt-body computer program for this purpose. Because of the importance of these time-marching solutions to *modern* hypersonic aerodynamics, the general procedure is outlined next.

Here, we will follow the philosophy as originally set forth by Moretti and Abbett [61]. However, in [61] the Lax–Wendroff finite difference technique was employed, which later was superseded by a simpler version developed by MacCormack [63]. MacCormack's explicit, predictor-corrector finite difference method was widely used throughout the 1970s and 1980s. Today it remains the most "graduate-student friendly time-marching method available because of its simplicity," and hence it will be utilized here for our solution to the blunt-body problem.

For simplicity, assume a two-dimensional flow. The governing unsteady Euler equations are, from Eqs. (4.1–4.3) and (4.6),

$$\frac{\partial \rho}{\partial t} = -\left[\frac{\partial(\rho u)}{\partial x} + \frac{\partial(\rho v)}{\partial y}\right] \tag{5.13}$$

$$\frac{\partial u}{\partial t} = -\left[u\frac{\partial u}{\partial x} + v\frac{\partial u}{\partial y} + \frac{1}{\rho}\frac{\partial p}{\partial x}\right] \tag{5.14}$$

$$\frac{\partial v}{\partial t} = -\left[u\frac{\partial v}{\partial x} + v\frac{\partial v}{\partial y} + \frac{1}{\rho}\frac{\partial p}{\partial y}\right] \tag{5.15}$$

$$\frac{\partial}{\partial t}\left(\frac{p}{\rho^\gamma}\right) = -\left[u\frac{\partial}{\partial x}\left(\frac{p}{\rho^\gamma}\right) + v\frac{\partial}{\partial y}\left(\frac{p}{\rho^\gamma}\right)\right] \tag{5.16}$$

To solve these equations for the blunt-body flowfield, the following steps can be followed:

1) We are considering a *given* body shape. Hence, this is a *direct* solution, that is, we are calculating the flowfield and shock-wave shape for a given body.

2) Assume the shock-wave shape and shock-detachment distance. Cover the flowfield between the shock and body with a series of discrete grid points, as shown in Fig. 5.20a. In this figure, the body shape is specified as $b = b(y)$, independent of time. The shock-wave shape, which is initially assumed at time $t = 0$, will change with time, and is given by $s = s(y, t)$. Here, b and s are the x coordinates of the body and shock, respectively.

3) Assume the flowfield variables ρ, u, v, p at each of the grid points shown in Fig. 5.20a. This assumed flowfield will be considered as initial conditions at time $t = 0$.

4) Calculate the flowfield at the next step in time by means of an appropriate finite difference solution of Eqs. (5.13–5.16). As mentioned earlier, during the 1970s and 1980s, the most popular finite difference technique for this purpose had been the explicit predictor-corrector approach of MacCormack, first described in [63], and discussed in an introductory sense in [4] and [5]. This technique will be followed here. (You should be aware, however, that today a number of more sophisticed CFD algorithms are available for the time-marching solution of blunt-body flow fields, one of which is discussed in Sec. 5.5.) Because the finite difference quotients should be formed in a rectangular grid, the curvilinear physical space shown in Fig. 5.20a can be transformed into a rectangular space shown in Fig. 5.20b via

$$\zeta = \frac{x - b}{\delta} \tag{5.17}$$

where δ is the local shock-detachment distance, $\delta = s - b$. In this transformed space, the body ($x = b$) is obtained from Eq. (5.17) as $\zeta = 0$. The shock ($x = s$) is also obtained from Eq. (5.17) as $\zeta = 1$. Hence in Fig. 5.20b, the left side of the rectangular space, $\zeta = 0$, represents all of the grid points along the

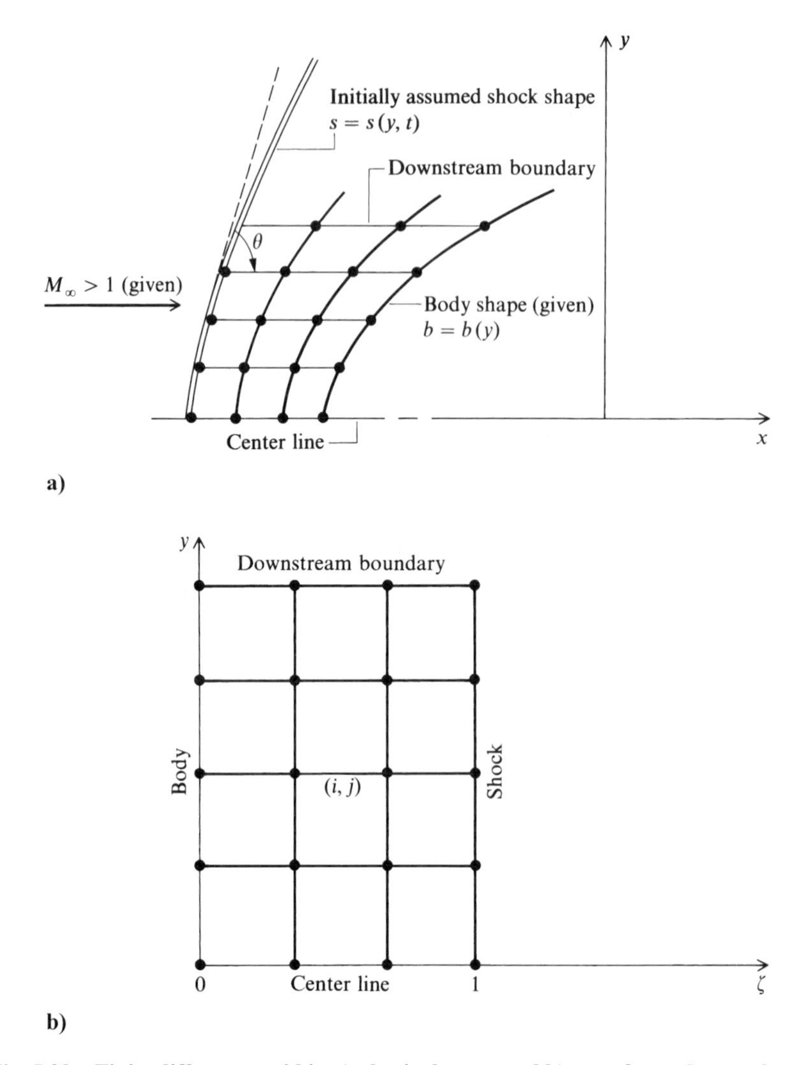

Fig. 5.20 Finite difference grid in a) physical space and b) transformed space for the blunt-body problem.

body, and the right side, $\zeta = 1$, represents all of the grid points along the shock. Because the y coordinate remains the same in both the physical and the transformed space, then the top and bottom of the rectangular space represent the downstream boundary and centerline, respectively. In this fashion, the curvilinear grid in the physical space (Fig. 5.20a) is transformed to the rectangular grid in the transformed space (Fig. 5.20b). Because the finite difference calculations are performed on this rectangular grid, Fig. 5.20b is also called the computational space.

5) For convenience, Moretti and Abbett also transformed the dependent variables as

$$P = \ell n\, p$$
$$R = \ell n\, \rho$$
$$\psi = \ell n\, p - \gamma \ell n\, \rho = P - \gamma R$$

Also define

$$C \equiv (\zeta - 1)\frac{db}{dy} - \zeta \cot \theta$$

$$W = \frac{ds}{dt} = x \text{ component of the shock-wave velocity}$$

$$B = \frac{u - W\zeta + vC}{\delta}$$

In terms of the just transformed dependent and independent variables, Eqs. (5.13–5.16) become the following.

Continuity:

$$\frac{\partial R}{\partial t} = -\left[B\frac{\partial R}{\partial \zeta} + \frac{1}{\delta}\frac{\partial u}{\partial \zeta} + \frac{C}{\delta}\frac{\partial v}{\partial \zeta} + \frac{\partial v}{\partial y} + v\frac{\partial R}{\partial y}\right] \tag{5.18}$$

x Momentum:

$$\frac{\partial u}{\partial t} = -\left[B\frac{\partial u}{\partial \zeta} + v\frac{\partial u}{\partial y} + \frac{p}{\rho\delta}\frac{\partial p}{\partial \zeta}\right] \tag{5.19}$$

y Momentum:

$$\frac{\partial v}{\partial t} = -\left[B\frac{\partial v}{\partial \zeta} + v\frac{\partial v}{\partial y} + \frac{pC}{\rho\delta}\frac{\partial P}{\partial \zeta} + \frac{p}{\rho}\frac{\partial P}{\partial y}\right] \tag{5.20}$$

Energy:

$$\frac{\partial \psi}{\partial t} = -\left[B\frac{\partial \psi}{\partial \zeta} + v\frac{\partial \psi}{\partial y}\right] \tag{5.21}$$

(The derivation of these transformed equations is left as a homework problem.) Note that these equations have been written with the time derivatives on the left side and all of the spatial derivatives on the right side, for reasons that will be clear shortly. Also note that the transformed equations (5.18–5.21) are to be evaluated in the computational space, Fig. 5.20b. Once the flowfield variables are obtained at the grid points in this computational space, then the results can be directly carried to the corresponding grid points in the physical space, Fig. 5.20a. Note that, whereas the computational space is fixed, independent of time [by virtue of the transformation, Eq. (5.17)], the shock-layer thickness in the physical space is varying with time because the shock wave is moving, constantly changing the local shock-detachment distance δ. This means that the grid points in the physical space are moving. Only in the steady state, obtained at large times, do the shock wave and grid network in the physical space become stationary. The movement of the shock wave, that is, the varying shock-layer thickness with time, is accounted for in Eqs. (5.18–5.21) via the term B, which contains the local shock-wave velocity W. In the steady state, W becomes zero.

6) For illustration of the calculation of the flowfield, let us pick the x component of velocity u. All of the other flow variables are calculated in an analogous fashion. Consider a given grid point in the computational space, denoted by (i, j), where i is the point index in the ζ direction and j is the point index in the y direction; $i = 1, 2, \ldots, N$, and $j = 1, 2, \ldots, M$, where N and M are the number of grid points along a given ζ and y coordinate line, respectively. In Fig. 5.20b, $N = 4$, and $M = 5$, for purposes of illustration. At this grid point, $u(t)$ is the known velocity from the earlier time step; we wish to calculate $u(t + \Delta t)$ at the next step in time, where Δt is the time interval between steps. Calculate $u(t + \Delta t)$ at grid point (i, j) denoted by $u_{i,j}^{t+\Delta t}$, from $u(t)$, denoted by $u_{i,j}^{t}$, using

$$u_{i,j}^{t+\Delta t} = u_{i,j}^{t} + \left(\frac{\partial u}{\partial t}\right)_{\text{ave}} \Delta t \tag{5.22}$$

where $(\partial u/\partial t)_{\text{ave}}$ is an average time derivative of u between t and $t + \Delta t$. This average time derivative is evaluated by means of a predictor-corrector philosophy as follows.

7) Calculate a value of $(\partial u/\partial t)_{i,j}^{t}$ from Eq. (5.19), using *forward differences* for the spatial derivatives. These spatial derivatives are *known* at times t. (Remember that we are trying to calculate the value of u at time $t + \Delta t$ from a known flow at time t.) So, from Eq. (5.19),

$$\begin{aligned}\left(\frac{\partial u}{\partial t}\right)_{i,j}^{t} = -\Bigg[&B_{i,j}^{t}\left(\frac{u_{i+1,j}^{t} - u_{i,j}^{t}}{\Delta \zeta}\right) + v_{i,j}^{t}\left(\frac{u_{i,j+1}^{t} - u_{i,j}^{t}}{\Delta y}\right) \\ &+ \left(\frac{p}{\rho\delta}\right)_{i,j}^{t}\left(\frac{p_{i+1,j}^{t} - p_{i,j}^{t}}{\Delta \zeta}\right)\Bigg]\end{aligned} \tag{5.23}$$

8) Calculate a *predicted* value of velocity from the first two terms in a Taylor's series

$$\bar{u}_{i,j}^{t+\Delta t} = u_{i,j}^{t} + \left(\frac{\partial u}{\partial t}\right)_{i,j}^{t} \Delta t \tag{5.24}$$

where the bar denotes *predicted* values. Carry out the same process itemized in steps 6–8 to obtain predicted values of the other dependent-flow variables, namely, $\overline{v}_{i,j}^{t+\Delta t}$, $\overline{R}_{i,j}^{t+\Delta t}$, and $\overline{\psi}_{i,j}^{t+\Delta t}$, but now using Eqs. (5.20), (5.18), and (5.21), respectively.

9) As a corrector step, calculate a value of the time derivative by inserting the predicted quantities obtained in step 8 into Eqs. (5.18–5.21), but using *rearward* spatial derivatives. For Example, from Eq. (5.19),

$$\left(\frac{\overline{\partial u}}{\partial t}\right)_{i,j}^{t+\Delta t} = -\left[\overline{B}_{i,j}^{t+\Delta t}\left(\frac{\overline{u}_{i,j}^{t+\Delta t} - \overline{u}_{i-1,j}^{t+\Delta t}}{\Delta\zeta}\right) + \overline{v}_{i,j}^{t+\Delta t}\left(\frac{\overline{u}_{i,j}^{t+\Delta t} - \overline{u}_{i,j-1}^{t+\Delta t}}{\Delta y}\right)\right.$$
$$\left. + \left(\frac{\bar{p}}{\bar{\rho}\delta}\right)_{i,j}^{t+\Delta t}\left(\frac{\bar{P}_{i,j}^{t+\Delta t} - \bar{P}_{i-1,j}^{t+\Delta t}}{\Delta\zeta}\right)\right] \tag{5.25}$$

10) Calculate the *average* time derivative that appears in Eq. (5.22) by

$$\left(\frac{\partial u}{\partial t}\right)_{\text{ave}} = \frac{1}{2}\left[\underbrace{\left(\frac{\partial u}{\partial t}\right)_{i,j}^{t}}_{\substack{\text{from}\\ \text{Eq. (5.23)}}} + \underbrace{\left(\frac{\overline{\partial u}}{\partial t}\right)_{i,j}^{t+\Delta t}}_{\substack{\text{from}\\ \text{Eq. (5.25)}}}\right] \tag{5.26}$$

11) Calculate the final corrected value of $u_{i,j}^{t+t\Delta t}$ from Eq. (5.22) repeated here:

$$u_{i,j}^{t+\Delta t} = u_{i,j}^{t} + \left(\frac{\partial u}{\partial t}\right)_{\text{ave}} \Delta t \tag{5.22}$$

12) Repeat steps 7–11 for a large number of time steps. The variation of u (and the other flow variables) from one time step to the next will initially be large. However, after a sufficient number of steps are taken, $u^{t+\Delta t} \approx u^{t}$, that is, *a steady state will be approached in the limit of large times.* This steady state is the desired result; the time-dependent approach is simply a means to that end.

Before proceeding further, examine steps 6–12 once again; these steps are the essence of MacCormack's predictor-corrector method. In this manner you will begin to appreciate how straightforward and strikingly simple the method is. Furthermore, we will have use for this method in subsequent applications in this book, so make certain that you feel comfortable with the approach. In regard to the numerical accuracy of this method (something that workers in computational fluid dynamics are always sensitive to—for example, see, [52]), although first-order forward and rearward differences are used on the predictor and corrector steps respectively, the combination of the two steps via Eq. (5.25) results in a second-order-accurate technique. Second-order accuracy is usually sufficient for most applications in computational fluid dynamics and is certainly sufficient for the inviscid blunt-body problem being discussed here.

In terms of the blunt-body problems steps 1–12 outline a solution procedure for the *interior* points in the flow, that is, for the points in Fig. 5.20 that are not on any of the four boundaries. The calculation of the flowfield variables at the boundary points is especially important and requires some special attention. Indeed, in the general theoretical context of the solution of the Euler equations the only way that the governing equations, can recognize one type of application from another is through the different boundary conditions imposed by each application. Hence, the boundary conditions are a powerful influence in determining the solution for a given problem, and any numerical solution must have an appropriate method for properly treating these boundary conditions. Thus, in the following paragraphs we will sequentially examine the shock, body, and downstream and centerline boundary conditions.

In the present discussion, we are treating the shock wave as a discontinuity, across which the usual shock-wave relations (sometimes called the Rankine–Hugoniot relations) hold. Because the shock wave is moving, the flow velocities in front of and behind the shock that appear in the shock relations must be interpreted as velocities *relative* to the shock wave itself. (For a discussion of the governing relations for a moving shock wave, see Chapter 7 of [4].) For example, in the basic normal shock case Eq. (2.1) holds for a moving shock wave as long as M_1 is interpreted as the Mach number of the flow ahead of the wave relative to the wave. Also, in the present hypersonic blunt-body solution the *exact* oblique shock relations are used, such as Eqs. (2.1), (2.3), and (2.16); because we are working with an exact solution of the blunt-body problem, it is neither necessary nor appropriate to utilize the limiting, approximate, hypersonic shock expressions developed in Chapter 2. In [61], the flow properties at each of the grid points along the shock (along $\zeta = 1$ in Fig. 5.20b) are obtained as follows. Consider a given grid point on the shock wave. The flow properties and wave velocity $W(t)$ at this point are known at time t from the previous time step. To obtain the flow properties and wave velocity at this grid point at time $t + \Delta t$, first *assume* a value for $W(t + \Delta t)$. Also, set up a localized, one-dimensional, unsteady method of characteristics calculation written in a direction locally perpendicular to the shock wave at the given grid point, reaching back into the internal part of the shock layer (see [61] for details). For the assumed $W(t + \Delta r)$, the Rankine–Hugoniot shock relations predict the flow properties immediately behind the shock at the given grid point. Alternatively, the localized one-dimensional method of characteristics method, via the solution of the appropriate compatibility equation along the normal direction, propagates information from the neighboring internal flow at time t to the shock grid point at time $t + \Delta t$. Do these two sets of flowfield results at the given shock grid point agree? If not, assume another value of $W(t + \Delta t)$, and try again. In this manner, an iterative process results, which, after a number of iterations, will finally match the Rankine–Hugoniot shock properties with the properties predicted from the unsteady one-dimensional method of characteristics from the internal flowfield. When the iteration is complete, then $W(t + \Delta t)$ is known, as well as the flow properties at the given shock grid point at time $t + \Delta t$. To better understand this approach, see [61] for an extended discussion of the idea, a well as for a presentation of the appropriate compatibility equation. We will not elaborate any further here because there is a simpler method of handling the shock points that, in the author's experience, works just as well as the preceding approach. This simpler method is as follows. Return to Fig. 5.20b and again consider a given point on the shock boundary. Calculate the flow properties at this grid point at time $t + \Delta t$ by employing the *internal* flow algorithm outlined earlier in steps 6–11, with one modification. We cannot employ a forward difference as called for in the predictor step (step 7) because there are no points to the right of the shock in Fig. 5.20b. Hence, at the shock grid points a rearward difference must be used on both the predictor and corrector steps, that is, the forward differences in equations such as Eq. (5.23) must be replaced with rearward differences. This is called a "one-sided" difference approach. When step 11 is finished, the flow properties at the shock grid point are now obtained at time $t + \Delta t$. In particular, the pressure at time $t + \Delta t$, $p(t + \Delta t)$, is now obtained. In turn, from this pressure (the pressure immediately behind the shock), and the known freestream conditions, the value of $W(t + \Delta t)$ is immediately fixed by the exact oblique shock relations (the Rankine–Hugoniot relations), as long as, for this part of the calculation only, we assume the wave angle at the grid point at time $t + \Delta t$ to be the same as the known value at time t. To understand this more clearly, recall that, from exact oblique shock theory, only two quantities are needed to fix the strength of a shock wave. Here, we are using the calculated static-pressure ratio p_2/p_1 [where $p_2 = p(t + \Delta t)$ and $p_1 = p_\infty$], and the wave angle β to define the specific shock wave: from this, the Mach number of the flow upstream of the shock relative to the shock M_1 is directly obtained from the shock relations. Because the shock wave is moving, M_1 is not the same as the freestream Mach number M_∞ in Fig. 5.20a. However, knowing M_1 and M_∞, as well as the speed of sound in the freestream, W is immediately obtained as $W = a_\infty(M_\infty - M_1)$. [Here, keep in mind that W is the shock velocity *relative to the laboratory*, treated positive when the shock is moving to the right in Fig. 5.20a, and hence the velocity of the flow ahead of the wave *relative to wave* is $V_\infty - W$, as sketched in Fig. 5.21. Thus, the Mach number of the flow ahead of the wave relative to the wave is $M_1 = (V_\infty - W)/a_\infty$, which in turn yields the wave velocity $W = a_\infty(M_\infty - M_1)$.] From the value of $W(t + \Delta t)$ just obtained, the shock

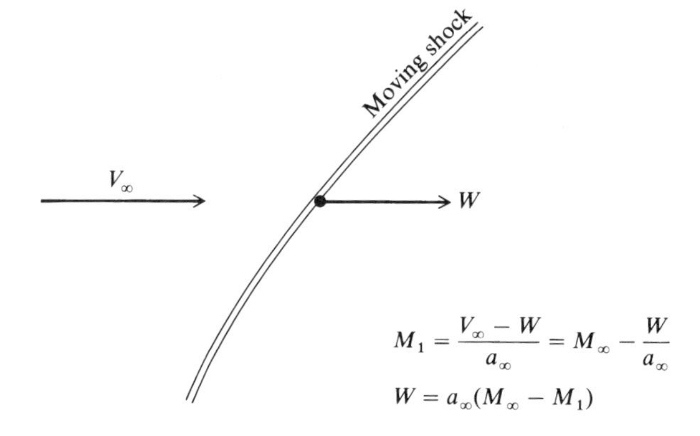

Fig. 5.21 Schematic of a moving shock wave.

wave at the given grid point is moved a distance Δs in the x direction, where Δs is based on an average velocity between times t and $t + \Delta t$, that is, $\Delta s = \frac{1}{2} [W(t + \Delta t) + W(t)] \Delta t$. This can be taken as the new location in physical space (Fig. 5.20a) of the shock wave at time $t + \Delta t$ at the given grid point, and the value just calculated for $W(t + \Delta t)$ is the appropriate shock velocity at time $t + \Delta t$. Finally, given this value of $W(t + \Delta t)$, the other flow properties at the shock grid point, such as $\rho(t + \Delta t)$, $T(t + \Delta t)$, etc., are obtained from the Rankine–Hugoniot shock relations. In short, what we have done here is to use the internal flow algorithm (the MacCormack predictor-corrector method) with one-sided differences to obtain the pressure behind the shock and then using this calculated pressure in conjunction with the freestream properties to uniquely define $W(t + \Delta t)$ from the shock relations. Once $W(t + \Delta t)$ is known, the other flow variables at the shock grid point are obtained from the exact oblique shock relations. Because, in applying these shock relations, we assumed that the wave angle β was the value at time t, the accuracy of this approach can be improved slightly by repeating the shock calculation, now using an improved β based on the predicted new location of the shock at time $t + \Delta t$. This now concludes our discussion of the numerical treatment of the shock boundary condition.

The boundary condition along the body ($\zeta = 0$ in Fig. 5.20b) is the usual inviscid flow condition that the velocity must be tangent to the surface, that is, $\boldsymbol{V} \cdot \boldsymbol{n} = 0$, where $\boldsymbol{n}$ is a unit vector normal to the surface. To implement this boundary condition within the context of the blunt-body problem, Moretti and Abbett [61] used a local, unsteady, one-dimensional method-of-characteristics approach written in the local normal direction at the body much along the lines of their treatment of the shock boundary condition as described earlier (except now the boundary is stationary—the body is fixed). See [61] for more details. Here, we will describe an alternate and simpler treatment at the body surface, which, in the author's experience, works just as well. Consider a given grid point on the body. Calculate the velocity at this point using the internal flow algorithm, that is, using MacCormack's technique as outlined in steps 6–11. Once again, we will have to use one-sided differences, in this case forward differences on both the predictor and corrector steps. For example, in Eq. (5.25) the rearward differences have to be replaced with forward differences. At the end of step 11, both the x and y components of velocity u and v will be obtained at time $t + \Delta t$ at the given grid point on the body (labeled as point 1 in Fig. 5.22). These components add vectorally to yield the vector velocity $\boldsymbol{V}$ at point 1 on the surface, as sketched in Fig. 5.22. In general, $\boldsymbol{V}$ will not be tangent to the surface, that is, the boundary condition will be violated, and we have to modify the boundary calculation to force $\boldsymbol{V}$ to be tangent to the surface. Another way to state this is to consider the component of $\boldsymbol{V}$ normal to the surface, namely, V_n in Fig. 5.22; in general, V_n will be some finite value obtained by the process in steps 6–11, and we need to make $V_n = 0$ in order to satisfy the body boundary conditions. To accomplish this, let us send a local, finite, one-dimensional, isentropic expansion or compression wave away from the surface at point 1 of sufficient strength to cancel V_n. (See Chapter 7 of [4] for a discussion of general, unsteady, finite wave motion.) Note that at the end of step 11, in addition to the velocity, values of pressure, density, etc. at point 1 will also be obtained at time $t + \Delta t$. For example, let us designate

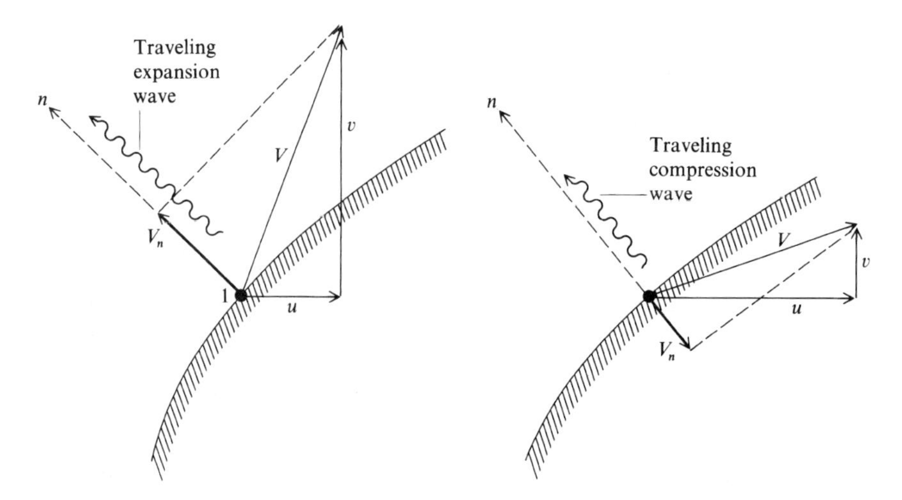

Fig. 5.22 Illustration of boundary condition at the wall.

the resulting pressure at grid point 1 at time $t + \Delta t$ by p_{old}, as obtained from steps 6–11. However, by sending a traveling, finite wave along $\boldsymbol{n}$ to cancel V_n, all of the other flow properties at point 1 will be slightly changed by the traveling wave, that is, the pressure at point 1 after the wave interaction will be denoted by p_{new}. Similar definitions hold for T_{old}, T_{new}, ρ_{old}, ρ_{new}, etc. at point 1. Return now to our imaginary finite wave traveling away from the surface in Fig. 5.22. In the case shown in Fig. 5.22a where V_n is directed away from the surface, the finite wave should be an expansion wave because the mass motion induced by an unsteady expansion wave is in the opposite direction to the propagation of the wave, hence canceling V_n. After the expansion wave does its job, the new pressure at point 1, denoted by p_{new}, is less than p_{old} because the pressure decreases through an expansion wave. Examining Fig. 5.22b, if $\boldsymbol{V}$ were directed into the surface as shown, and hence V_n were into the surface, the finite wave should be a compression wave because the mass motion induced by an unsteady compression wave is in the same direction as the propagation of the wave, thus canceling V_n. After the compression wave does its job, the new pressure at point 1 p_{new} is greater than p_{old} because the pressure increases through a compression wave. To quantify these arguments, recall the relations for pressure ratio and temperature ratio through an unsteady, isentropic, one-dimensional, finite wave for example, see Chapter 7 of [4]. Written in terms of the standard nomenclature for unsteady waves, we have

$$\frac{p}{p_4} = \left[1 \pm \frac{\gamma - 1}{2}\left(\frac{u'}{a_4}\right)\right]^{2\gamma/(\gamma-1)} \tag{5.27}$$

and

$$\frac{T}{T_4} = \left[1 \pm \frac{\gamma - 1}{2}\left(\frac{u'}{a_4}\right)\right]^{2} \tag{5.28}$$

where p_4, a_4, and T_4 are the pressure, speed of sound, and temperature in front of the propagating wave; u' is the induced mass motion at an arbitrary point inside the wave; and p and T are the corresponding pressure and temperature at that point. The plus and minus signs correspond to a compression wave and an expansion wave, respectively; with the plus and minus nomenclature, the velocity u' is taken as positive in both cases. Applied to our discussion here, Eqs. (5.27) and (5.28) are written as

$$\frac{p_{\text{new}}}{p_{\text{old}}} = \left[1 \pm \frac{\gamma - 1}{2}\left(\frac{V_n}{a_{\text{old}}}\right)\right]^{2\gamma/(\gamma - t)} \tag{5.29}$$

$$\frac{T_{\text{new}}}{T_{\text{old}}} = \left[1 \pm \frac{\gamma - 1}{2}\left(\frac{V_n}{a_{\text{old}}}\right)\right]^{2} \tag{5.30}$$

where V_n is taken as a positive number in both the cases shown in Fig. 5.22, the plus sign corresponds to Fig. 5.22b, and the minus sign corresponds to Fig. 5.22a. In summary, the flow properties at the body can be calculated from the internal flow algorithm using one-sided differences, giving p_{old}, T_{old}, etc.; then the precise flow-tangency condition at the body is enforced by expanding or compressing the flow through a finite unsteady wave of strength just sufficient to cancel any finite component of velocity perpendicular to the wall. This yields slightly modified flow values at the wall, namely, p_{new}, T_{new}, etc. In turn, these are the final values of the flowfield variables at the wall at time $t + \Delta t$, that is, $p(t + \Delta t) = p_{\text{new}}$, $T(t + \Delta t) = T_{\text{new}}$, etc. This approach to the wall boundary condition is an unsteady analog to the familiar "Abbett's" boundary treatment (see [64]) used for steady flows, to be discussed in Sec. 5.5. This completes our discussion of the numerical treatment of the wall boundary condition.

Returning to Fig. 5.20b, the downstream and centerline grid points (the top and bottom of the rectangle in Fig. 5.20b) are easily treated, as follows. At the downstream boundary, the flow properties at the boundary grid points are simply obtained from linear extrapolation from the values at the adjacent internal grid points. This is sufficient as long as the downstream boundary is taken far enough downstream to be supersonic all along the boundary; this is an important consideration because extrapolation (of any order) is a properly posed supersonic boundary condition but an improperly posed subsonic boundary condition. Hence, if any of the grid points along the downstream boundary are subsonic, and extrapolation is used to obtain the flow properties at these points, numerical instabilities are usually encountered. In regard to the centerline boundary condition, for a two-dimensional or axisymmetric flow at zero angle of attack, the centerline is a line of symmetry. In such a case, the usual symmetry conditions are employed, namely, $\partial p/\partial y = \partial T/\partial y = \partial u/\partial y = 0$. In terms of our numerical calculations, these conditions are writtens as (referring to the nomenclature in Fig. 5.23)

$$p_{j+1} = p_{j-1}; \quad T_{j+1} = T_{j-1}; \quad u_{j+1} = u_{j-1}$$

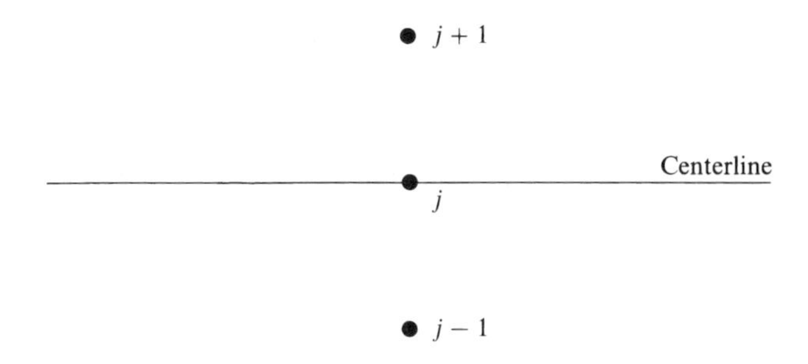

Fig. 5.23 Grid points above and below a centerline.

Because the y component of velocity v changes sign as the centerline is crossed, the symmetry boundary condition on v is

$$v_{j+1} = -v_{j-1}$$

These symmetry conditions are sufficient to form the forward and rearward differences at grid points along the centerline, thus allowing the use of the usual internal flow algorithm to calculate properties along the centerline, that is, to allow the calculation of p_j, T_j, u_j, etc.

A final aspect of the time-marching approach is the value of Δt, which appears in Eqs. (5.22) and (5.24). The finite difference technique discussed in this section is an explicit method, and therefore Δt is subject to a stability criterion. (See [52] for an in-depth discussion of both explicit and implicit finite difference methods and the governing stability considerations; an introductory discussion of such matters is given in Chapters 11 and 12 of [4].) In the present method, Δt cannot exceed a certain value in order to maintain a numerically stable solution. The stability criterion on Δt is

$$\Delta t \leq \min(\Delta t_x, \Delta t_y) \tag{5.31}$$

where

$$\Delta t_x = \frac{\Delta x}{u + a} \tag{5.32}$$

$$\Delta t_y = \frac{\Delta y}{v + a} \tag{5.33}$$

Equations (5.31–5.33) constitute a version of the Courant–Friedrichs–Lewy (or CFL) criterion, which governs the stability of explicit methods dealing with hyperbolic equations [65]. On a physical basis, Δt_x is the time it takes a sound wave to travel between two adjacent grid points in the x direction, and Δt_y is the similar time in the y direction. Equation (5.31) states that the allowable time step in the explicit method is less than, or at best equal to, the minimum of these two times. The CFL criterion was first derived on the basis of linear partial differential equations; therefore, for the nonlinear Euler equations Eqs. (5.31–5.33) are to be interpreted as a guideline only and not as precise condition. Hence, in practice, Δt is chosen such that

$$\Delta t = K[\min(\Delta t_x, \Delta t_y)] \tag{5.34}$$

where K is less than unity, typically on the order of 0.5 to 0.8. A particular value of K suited to a particular application is usually determined by trial and error.

Let us examine some typical results for hypersonic blunt-body flows obtained by means of the time-marching procedure. Such results are given in Figs. 5.24–5.28 for the flow over a parabolic cylinder at zero angle of attack from [4] (with the exception of Fig. 5.27, which is for an axisymmetric paraboloid). In particular Figs. 5.24 and 5.25 illustrate the time-marching mechanism. In Fig. 5.24, the unsteady bow shock-wave motion is shown for the case where $M_\infty = 4.0$; the fixed, parabolic cylinder is shown at the right, and four different shock shapes and locations are shown, corresponding with four different times during the calculation. The shock labeled $0\Delta t$ is the assumed shock-wave shape and location at time $t = 0$ (part of the assumed initial conditions). The shock labeled $100\Delta t$ is the shock shape and location after executing the preceding time-marching technique for 100 time steps. The shock waves for 200, 300, and $500\Delta t$ are also shown. Note that, at early times, the shock wave moves rapidly, but after 300 time steps the wave motion has decreased considerably, and the shock is essentially steady; the shock waves for 300, 400, and 500 time steps are virtually the same, as shown in Fig. 5.24. The result shown at $500\Delta t$ is essentially the final, steady-state shock-wave shape and location—that is, the desired result. The time-marching behavior is further illustrated in Fig. 5.25, which gives the time variation of the pressure at the stagnation point. Note that the pressure changes very rapidly at early times during the time-marching procedure, but at large times it asymptotically approaches the steady-state value. Again, emphasis is made that

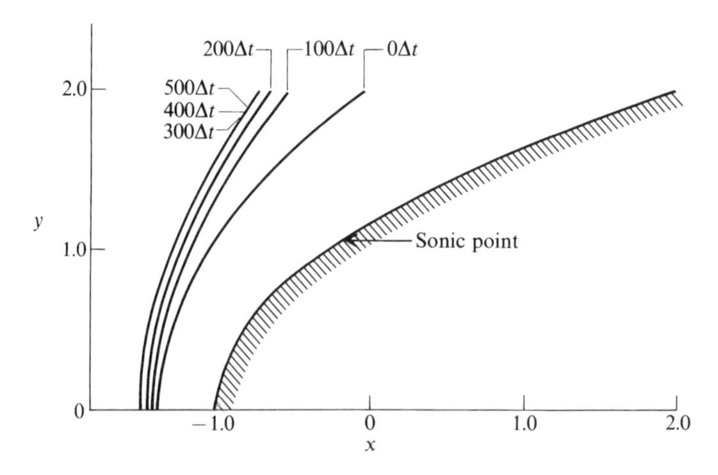

Fig. 5.24 Time-marching shock-wave motion, parabolic cylinder, where $M_\infty = 4$.

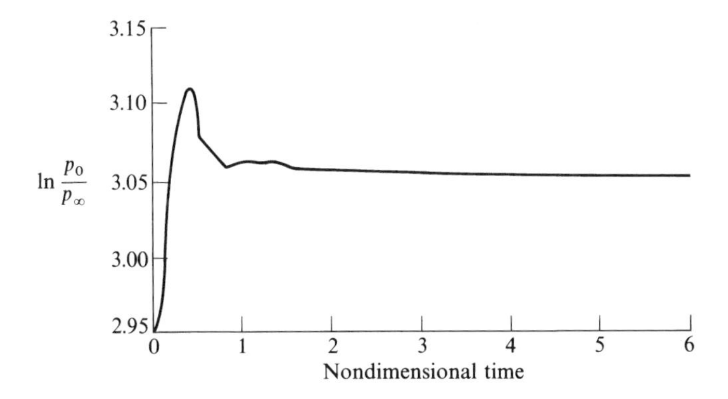

Fig. 5.25 Time variation of stagnation-point pressure, parabolic cylinder, where $M_\infty = 4$.

we desire a solution to the steady-state flowfield, and the time-marching procedure is simply a means to that end. (In carrying out such time-marching solutions, my students used to generate large amounts of computer printout for a given case; I sometimes jokingly told them to tear off the last sheet, keep it, and throw out the rest because the last sheet contains the solution to the problem. Today the output appears graphically on a computer screen.) Some steady-state results are shown in Figs. 5.26–5.28. In Fig. 5.26, the steady-state surface-pressure distributions are shown for $M_\infty = 4$ and 8. The exact time-marching finite difference results are shown as the solid curves; also, for the sake of comparison, the symbols give the modified Newtonian prediction [from Eq. (3.15)]. Note that, as already discussed in Chapter 3, the Newtonian results are not very accurate for a blunt, two-dimensional body; we see in Fig. 5.26 that Newtonian results underpredict the exact numerical results downstream of the immediate nose region. This is not the case for an axisymmetric body, as shown in Fig. 5.27. Here, the surface-pressure distribution is given for

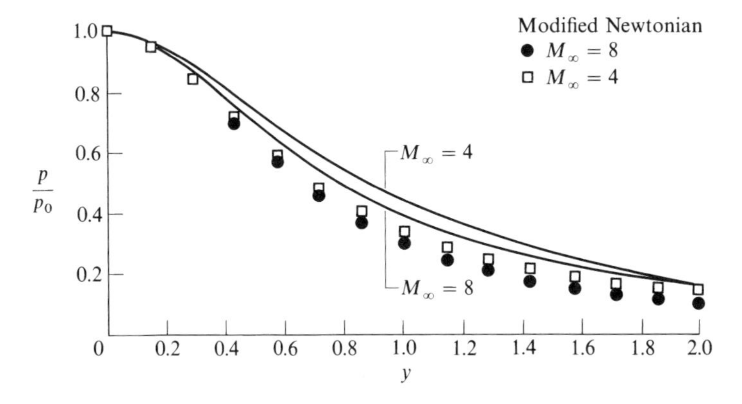

Fig. 5.26 Surface-pressure distribution, parabolic cylinder.

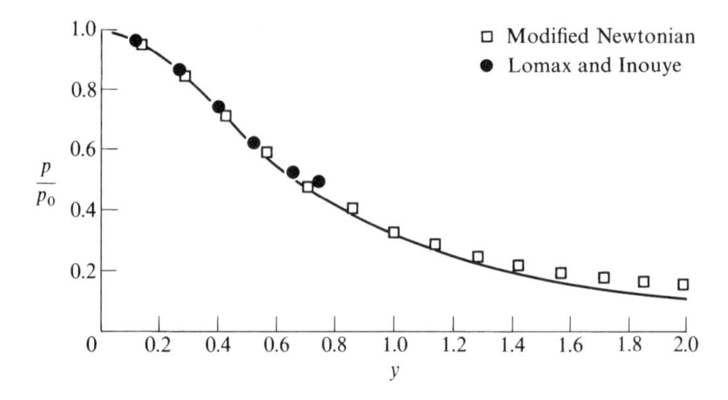

Fig. 5.27 Surface-pressure distribution, paraboloid, where $M_\infty = 4$.

an axisymmetric paraboloid (with the same meridional shape as the parabolic cylinder shown in Fig. 5.24). The solid curve gives the exact numerical results, and the open squares are from modified Newtonian. Here, agreement between the exact results and Newtonian is quite good, again emphasizing that Newtonian theory appears to be more applicable to three-dimensional rather than two-dimensional bodies. Figure 5.27 is similar to Fig. 3.9, used in Chapter 3 to demonstrate the viability of Newtonian theory. However, in Fig. 5.27, some additional data are shown, namely, the results of Lomax and Inouye [66], which were obtained from a numerical, steady-flow inverse blunt-body solution. These data are shown here to emphasize a particular advantage of the time-marching method. To see this, recall that for $\gamma = 1.4$ sonic flow on the surface occurs when $p/p_0 = 0.528$; examining Fig. 5.27, we note that the inverse blunt-body solution is discontinued in the vicinity of the sonic point—a problem encountered by all steady-flow blunt-body techniques prior to 1966. In contrast, the time-marching procedure gives results far downstream of the body sonic

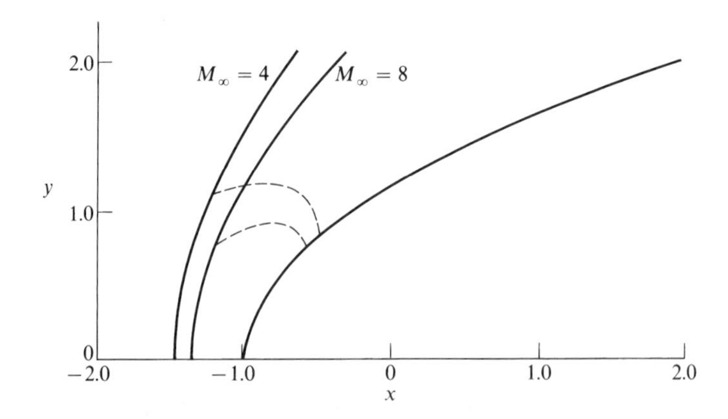

Fig. 5.28 Shock shapes and sonic lines, parabolic cylinder.

point—indeed, as far as one wants to go downstream. As a final example of the present technique, Fig. 5.28 shows the steady-state shock shapes and sonic lines for a parabolic cylinder at Mach 4 and 8, obtained by means of the time-marching procedure. Note that, as M_∞ increases, the shock wave moves closer to the body, and the sonic points on both the shock and the body move closer to the center-line—all standard physical behavior for blunt-body flows. Furthermore, observe that, as M_∞ increases, the sonic point on the shock moves down faster than the sonic point on the body, and thus the sonic line actually rotates in a counterclockwise fashion as the Mach number increases.

Some interesting details on the physical aspects of the sonic line behavior are given by Hayes and Probstein [46] and [60] and are summarized in Fig. 5.29, taken from those references. In Fig. 5.29, qualitative results are sketched for two cases, namely, the flow over a two-dimensional circular cylinder and the flow over an axisymmetirc sphere; although the shapes are the same, the behavior of the sonic lines is not. For example, in Fig. 5.29a, the sonic line is shown for both the cylinder and the sphere at low supersonic Mach number. The sonic point on the shock is much higher than on the body, and the angle made by the sonic line at the body (ω_b in Fig. 5.29) is acute. For the cylinder, as the Mach number increases, the sonic points on both the shock and the body move closer to the centerline, and the sonic line becomes more curved, as shown in Fig. 5.29b. The sketch shown in Fig. 5.29b pertains to a Mach number of approximately 2 and greater. For the cylinder, the angle ω_b always remains acute, no matter how high the Mach number. (Note that the sonic lines at the body in Fig. 5.28 for a two-dimensional parabolic cylinder are consistent with this fact.) Figure 5.29b also pertains to the case of a sphere, but only for the limited Mach-number range approximately between 2 and 3. At higher Mach numbers, as shown in Fig. 5.29c for the sphere, ω_b becomes obtuse. Note that Fig. 5.29 also illustrates the limiting characteristics and how they change with Mach number. By definition, the limiting characteristic is the locus of points, each of which has only one point of the sonic line in its zone of action. For

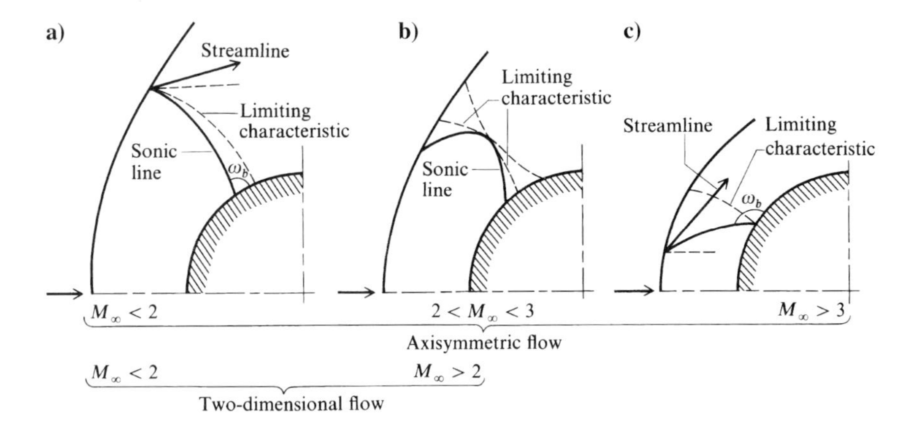

Fig. 5.29 General sonic line and limiting characteristic behavior as Mach number increases (from [46] and [61]).

example, in Fig. 5.29, the flow is locally supersonic at each point downstream of the sonic line. However, in Fig. 5.29a, imagine a left-running characteristic line (Mach wave) initiated at some point on the body that lies between the sonic line and the limiting characteristic. This left-running characteristic will propagate upward and to the left and will intersect that sonic line somewhere between the body and the shock. When we move downstream to the limiting characteristic itself, the left-running characteristic will only intersect the sonic line at the shock point; only when we move downstream of the limiting characteristic will the left-running characteristics no longer intersect the sonic line. The physical implication of this is that, although the flow region between the sonic line and the limiting characteristic is totally supersonic, disturbances produced in this region will propagate to the sonic line and can affect the entire subsonic portion of the flow. Similar arguments hold for the cases shown in Figs. 5.29b and 5.29c. This is why, in Sec. 5.2, repeated warnings were given that the initial data line for a method of characteristics solution over a blunt-nosed body must be taken downstream of the limiting characteristic, *not* just downstream of the sonic line. An extended, but introductory discussion of limiting characteristics can be found in Chapter 12 of [4].

Another interesting physical aspect of hypersonic blunt-body flows is the location of the stagnation point and the point of maximum entropy. For a symmetric body at zero angle of attack, the stagnation streamline and the stagnation point are along the centerline, as sketched in Fig. 5.30a. This streamline crosses the bow shock at precisely the point where $\beta = \pi/2$, that is, it crosses a normal shock, and hence the entropy of the stagnation streamline in the shock layer is the maximum value. In contrast, consider the *asymmetric* cases shown in Figs. 5.30b and 5.30c, an asymmetric flow can be produced by a nonsymmetric body, an angle of attack, or both. In these cases, the shape and location of the stagnation streamline, and hence of the stagnation point, are not known in advance; they must be obtained as part of the solution. Moreover, the stagnation streamline does not pass through the normal portion of the shock wave, and hence it is not the maximum entropy streamline. The relative locations of the stagnation streamline and the maximum entropy streamline for two nose shapes are shown in Figs. 5.30b and 5.30c. Note that the stagnation streamline is always attracted to that portion of the body with maximum curvature, whereas the maximum entropy streamline will turn in the direction of decreasing body curvature. More details on this matter can be found in [60].

A further interesting point concerning entropy, and one with particular consequence to the time-marching procedure, is as follows. Consider the entropy equation (4.5) repeated here:

$$\frac{\partial s}{\partial t} + u\frac{\partial s}{\partial x} + v\frac{\partial s}{\partial y} + w\frac{\partial s}{\partial z} = 0 \tag{4.5}$$

When applied at a stagnation point, where $u = v = w = 0$, Eq. (4.5) yields

$$\frac{\partial s}{\partial t} = 0 \tag{5.35}$$

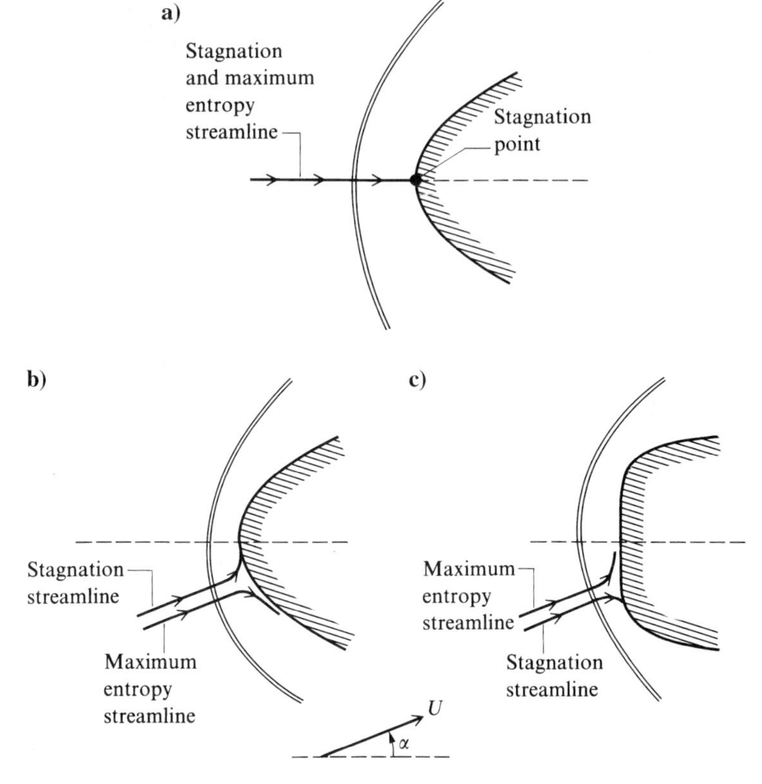

Fig. 5.30 Stagnation and maximum entropy streamlines.

that is, at a stagnation point in an unsteady, inviscid flow, the entropy remains constant, independent of time. Return to Fig. 5.20a, which shows the physical plane for a symmetric blunt body at zero angle of attack. Consider the stagnation point, which occurs on the centerline. Equation (5.35) dictates that, at the stagnation point, the initial conditions at time $t = 0$ for a time-marching solution cannot be chosen arbitrarily. Indeed, the proper steady-state value of entropy must be used because it will remain constant at the stagnation point throughout the time-marching procedure. However, this is no problem for the symmetric case; we know in advance that the steady-state conditions at the stagnation point are identical to the stagnation conditions behind a normal shock wave, which are easily calculated from the normal shock relations. Therefore, the proper initial conditions at time $t = 0$ at the stagnation point on the blunt body in Fig. 5.20a are simply the stagnation conditions behind a normal shock wave. This is demonstrated in Fig. 5.25, where the initial value of p_0 at time $t = 0$ was indeed chosen as the proper steady-state value. After going through the massive variations shown in Fig. 5.25, p_0 finally approaches, in the limit of large times, the value it started with at $t = 0$. For the asymmetric case, where the location of the stagnation point is not known in advance, chances are that none of the chosen grid points will correspond to the stagnation point, and thus the problem is not encountered.

The example chosen in this section to describe and illustrate the time-marching solution of hypersonic blunt-body flow was a two-dimensional body at zero angle of attack. This was done for simplicity, as well as to underscore the basic ideas and philosophy of the method without cluttering our discussion with tedious details. The extension to three-dimensional flows is straightforward, although the amount of detail and tedious computation increases by almost an order of magnitude. Among the first extension of the time-marching idea to blunt bodies at angle of attack was the work of Moretti and Bleich [67]. An example of a three-dimensional, inviscid, blunt-body calculation is the work of Weilmuenster [68], who solved the flowfield over a space-shuttle-like vehicle at large angle of attack. Weilmuenster utilized the explicit MacCormack predictor-corrector scheme, just as we have described here, except extended to three-dimensional flow. The governing three-dimensional Euler equations (4.1–4.5) were solved in a time-marching fashion, just as outlined earlier in this section. The three-dimensional shock wave was treated as a discontinuity and moved in space during the time-marching procedure. In the physical space, a spherical coordinate system was used in the blunt-nose region of the body, matched to a cylindrical coordinate system for the remainder of the flowfield. The physical grid is presented in Fig. 5.31, which shows both the symmetry plane and the crossflow plane. This physical plane was transformed to a three-dimensional rectangular box, analogous to the transformation shown in Fig. 5.20,

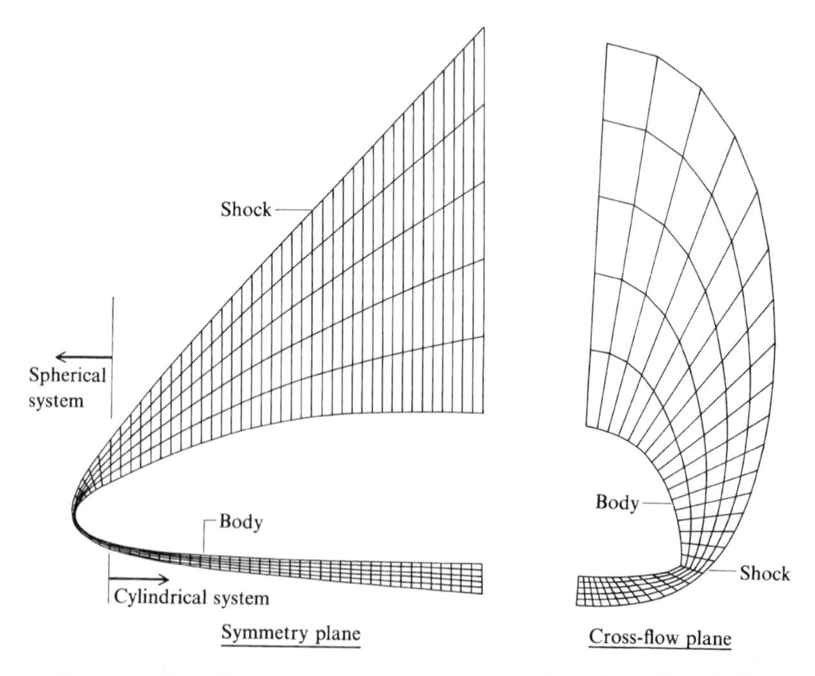

Fig. 5.31 Coordinate system for space shuttle calculations (from [68]).

Fig. 5.32 Calculated three-dimensional shock-wave shape on a shuttle-like configuration (from Weilmuenster [68]).

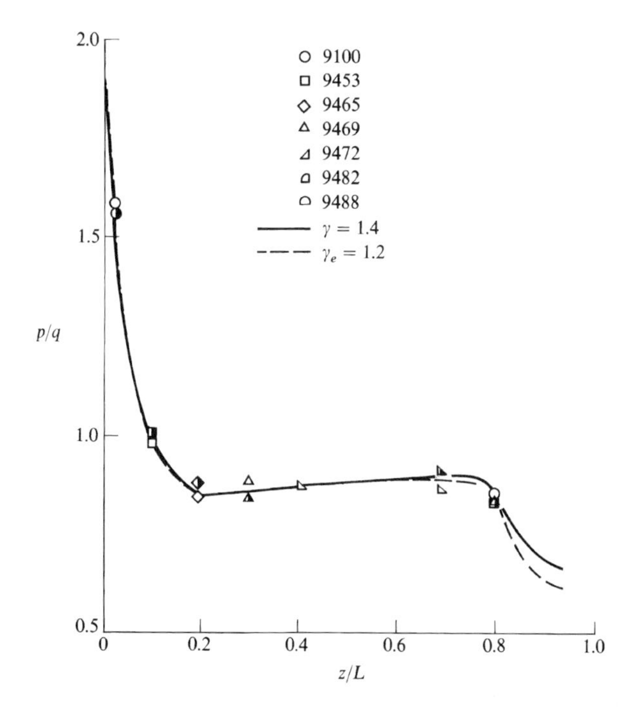

Fig. 5.33 Calculated pressure distribution on the space shuttle windward centerline; $M_\infty = 16.25$, and $\alpha = 39.8$ deg; comparison with flight data (from [68]).

for the finite difference calculations, along with the appropriate transformed equations. A total of 84,825 grid points were used in the calculation, which was carried out on a CDC Cyber 203 supercomputer. As a sample of the results, Fig. 5.32 illustrates the final, steady-state shock wave at $M_\infty = 16.25$ at an angle of attack of 39.8 deg. Figure 5.33 gives the centerline pressure distribution over the bottom surface for the same flight conditions. In Fig. 5.33, the solid and dashed lines are calculations for $\gamma = 1.4$ and 1.2, respectively; the symbols are flight-test data from the space shuttle itself. Excellent agreement is obtained. (Note that the pressure distribution is relatively insensitive to changes in γ.) These results are presented here as the epitome of time-marching solutions to inviscid, hypersonic blunt-body flows.

On this note, we conclude our discussion of exact solutions to hypersonic blunt-body flows. The time-marching solutions discussed here represent a substantial milestone in the progress of aerodynamic theory, not only for hypersonics, but for the whole spectrum of aerodynamics.

5.4 Correlations for Hypersonic Shock-Wave Shapes

As a corollary to our discussion on exact solutions of the hypersonic blunt-body problem in Sec. 5.3, in the present section we provide some simple engineering correlations of blunt-body shock-wave shapes. Such correlations are quite useful for rapid engineering analysis of blunt-body aerodynamic properties. Here, we present the results of Billig [69], which are based on experimental data. The correlations hold for sphere-cone and circular cylinder-wedge bodies and assume a hyperbolic shock shape give by the equation:

$$x = R + \delta - R_c \cot^2 \beta \left[\left(1 + \frac{y^2 \tan^2 \beta}{R_c^2} \right)^{1/2} - 1 \right] \tag{5.36}$$

The nomenclature in Eq. (5.36) is illustrated in Fig. 5.34; R is the radius of the nose, R_c is the radius of curvature of the shock wave at the vertex of the hyperbola, δ is the shock detachment distance, x and y are Cartesian coordinates, and β is the angle of the shock wave in the limit of an infinite distance away from the nose. If the body downstream of the blunt nose is a cone of angle θ_c, then β is the wave angle for an attached shock wave on a sharp cone of angle θ_c. Similarly, if the body downstream of the nose is a wedge of angle θ, then β is the wave angle for an attached shock wave on a sharp wedge of angle θ. If, in the axisymmetric case, the downstream body is a cylinder (aligned with the flow) or if, in the two-dimensional case, the downstream body is a flat slab (where in both cases the downstream body surface is parallel to the freestream), then β is a Mach wave. In Eq. (5.36), the values of δ and R_c are correlated from experimental data as

$$\frac{\delta}{R} = \begin{cases} 0.143 \ \exp[3.24/M_\infty^2] & \text{sphere-cone} \\ 0.386 \ \exp[4.67/M_\infty^2] & \text{cylinder-wedge} \end{cases} \tag{5.37}$$

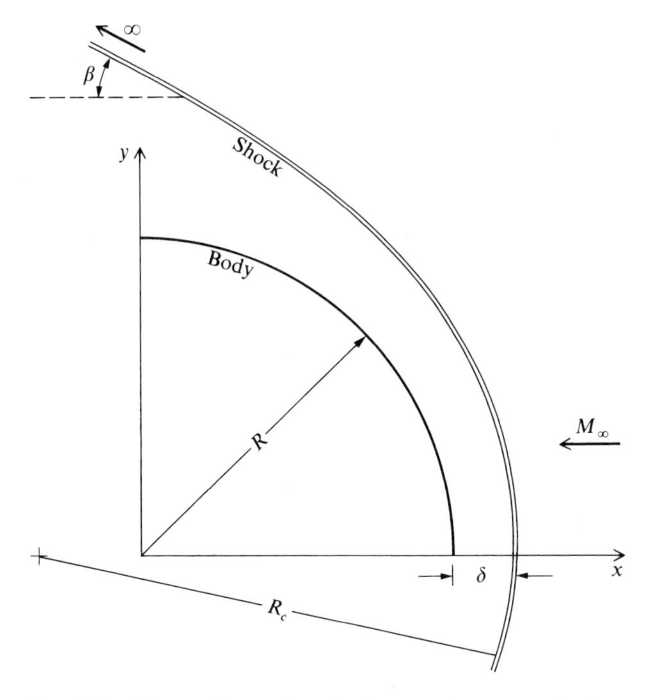

Fig. 5.34 Nomenclature for shock-wave shape correlations.

and

$$\frac{R_c}{R} = \begin{cases} 1.143 \ \exp[0.54/(M_\infty - 1)^{1.2}] & \text{sphere-cone} \\ 1.386 \ \exp[1.8/(M_\infty - 1)^{0.75}] & \text{cylinder-wedge} \end{cases} \tag{5.38}$$

In Eqs. (5.37) and (5.38), M_∞ is the freestream Mach number.

In [70], Billig's correlations are compared with numerical results obtained by means of the exact, time-marching method described in Sec. 5.3. The comparison is shown in Figs. 5.35 and 5.36, obtained from [70]. (The details of the numerical calculations are given in [71].) Figure 5.35 gives steady-state shock-wave shapes at Mach 4 and 8 for a sphere cone. The solid lines are the exact time-marching results, and the open symbols are from Billig's correlation; excellent agreement is obtained. Figure 5.36 gives the shock-wave shape for a cylinder-wedge at Mach 8; the solid curves are shock shapes obtained at various time steps by means of the time-marching method, with the steady-state shock wave identified by 300–500Δt. Billig's correlation is given by the open circles; again, excellent agreement is obtained for the steady-state shock shape. From the comparisons shown in Figs. 5.35 and 5.36, we conclude that the shock correlations given by Eqs. (5.36–5.38) are quite accurate.

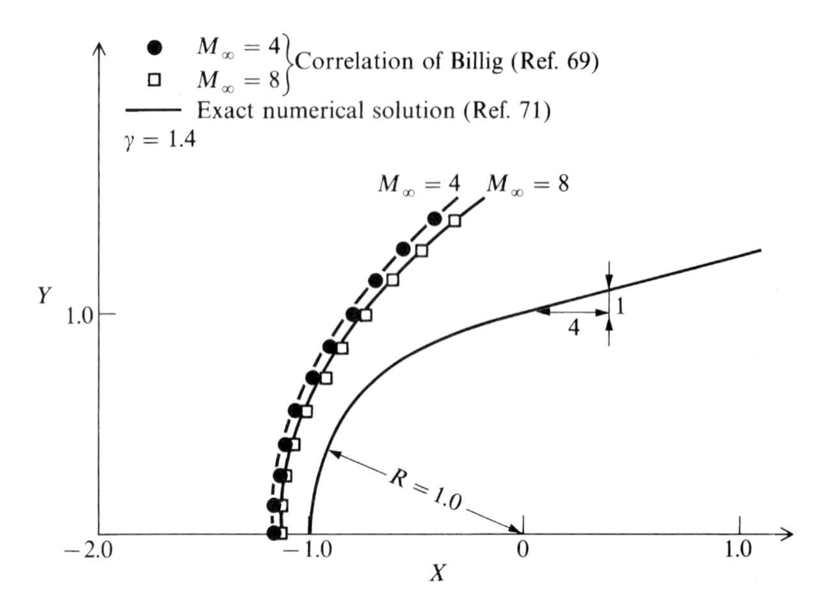

Fig. 5.35 Steady-state shock-wave shapes for a sphere-cone.

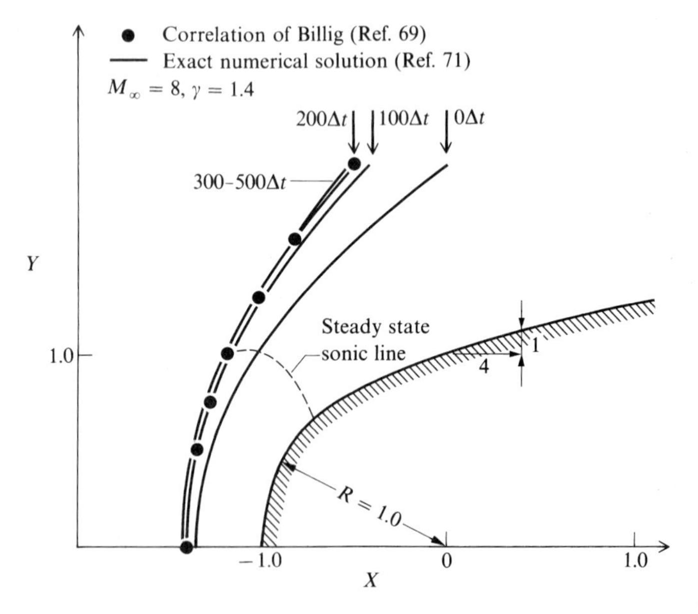

Fig. 5.36 Transient and steady-state shock-wave shapes for a cylinder-wedge (from [70]).

As a parenthetical comment, Eqs. (5.36–5.38) are very useful for constructing *initial conditions* for a time-marching numerical blunt-body solution. Suggestions for constructing the initial conditions are as follows:

1) Assume a shock-wave shape and location as given by Eqs. (5.36–5.38), and assume that the wave velocity W at all grid points is initially zero.

2) The initial flow conditions at the shock grid points (see Fig. 5.16) are then obtained from the exact oblique shock equations.

3) Assume a Newtonian pressure distribution along the body.

4) Interpolate between the body and the shock wave to obtain pressures at the internal grid points.

5) Assume a linear velocity variation along the body surface, starting with zero at the stagnation point and assigning a sharp cone value, wedge value, or freestream value (whichever makes the most sense for the given body) at the last downstream body point.

6) Interpolate between the body and the shock wave to obtain velocities at the internal grid points.

7) Obtain the temperature at each point from the adiabatic relation

$$C_p T + \frac{V^2}{2} = C_p T_\infty + \frac{V_\infty^2}{2} \tag{5.39}$$

where T_∞ and V_∞ are known freestream values. [Note that Eq. (5.39), which states that the total enthalpy is constant throughout the flowfield, is only valid for a *steady flow*. It cannot be used as part of the unsteady, time-marching procedure. However, here we are discussing the construction of initial conditions, which are somewhat arbitrary in the first place.]

8) Obtain the density at each grid point from the equation of state, $p = \rho RT$.

Although in theory the initial conditions can be purely arbitrary, in practice it is helpful that they be somewhat near the proper steady-state solution because in such a case 1) the number of time steps required to obtain the steady state is less, hence reducing the required computer time; and 2) the stability behavior of the numerical solution will be enhanced.

5.5 Shock–Shock Interactions

In some supersonic and hypersonic flowfields, shock waves impinge on other shock waves—*shock–shock interactions*. Some of these interactions are straightforward, such as those involving straight oblique shock waves that can be calculated by algebraic methods from classical compressible flow as described in [4]. Others are more complex, involving a mixture of straight and curved shock waves producing mixed subsonic-supersonic flows that must be calculated using time-marching numerical methods. In all cases, the interaction pattern is driven by inviscid flow phenomena. Such shock–shock interactions, therefore, are appropriate for discussion in this chapter on numerical solutions of hypersonic inviscid flows.

There is a very practical reason for this discussion. Some hypersonic flight vehicles are plagued with shock–shock interactions that, if not properly taken into account, can destroy parts of the vehicle. Consider for example the hypersonic flow over the X-43 shown in three view in Fig. 5.2. The oblique shock wave created at the nose propagates downstream underneath the vehicle and can impinge on the leading edge of the cowl of the SCRAMjet engine. A generic illustration of this shock pattern is given in Fig. 5.37; here the shock wave from the nose is labeled the inlet bow shock because it is designed to impinge on the inlet cowl of the engine for efficient containment of the flow that enters the engine. The inlet cowl has a blunt leading edge to reduce aerodynamic heating, and therefore a detached curved shock wave exists just upstream of the cowl leading edge similar to the blunt-body flows discussed in Secs. 5.3 and 5.4. The shock from the nose of the vehicle does not impinge directly on the surface of the cowl, but rather on the cowl shock wave, setting up a rather complex shock-shock interaction that can change the nature of both shocks and the surrounding flow field. One possibility is sketched in the inset in Fig. 5.37, which illustrates a type-IV shock-shock interaction. (The different types of shock–shock interactions are discussed in the following paragraphs.) The type-IV interaction is particularly interesting and important because it results in a supersonic jet that impinges on the engine cowl as shown in Fig. 5.37, creating large peaks in pressure and heat transfer on the cowl that can do damage. Designers of hypersonic airbreathing vehicles, therefore, are particularly concerned about the type IV interaction. But there are other types of shock–shock interactions that can occur. Indeed, Edney [219] has identified six types of shock–shock interactions; a summary of his work can be found in [220]. Edney's classification for the six types has become standard in the literature.

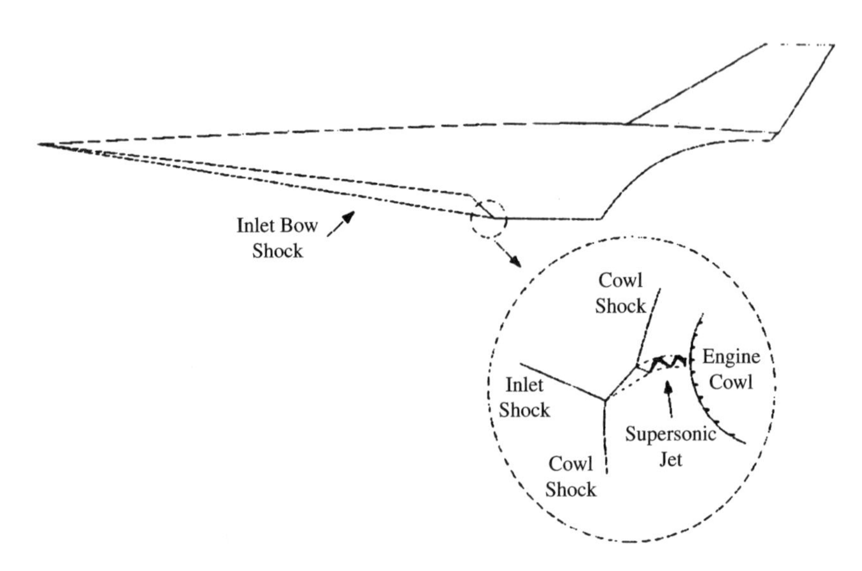

Fig. 5.37 Generic hypersonic vehicle showing shock interaction in the vicinity of an engine cowling (Lind and Lewis [222]).

The six types of shock–shock interactions are summarized in Fig. 5.38, taken from Lind [221]. At the middle right of Fig. 5.38, there is a picture of a blunt body with a curved bow shock labeled BS. The sonic points on the body are identified by SP. No shock–shock interaction is illustrated in this blunt-body picture; the flow is that calculated in Sec. 5.3. However, if an impinging shock IS, also shown at the middle right of Fig. 5.38, intersects the bow shock wave, the flowfield will change. The resulting shock–shock interaction pattern depends on where the impinging shock strikes the curved bow shock wave. The bow shock is divided

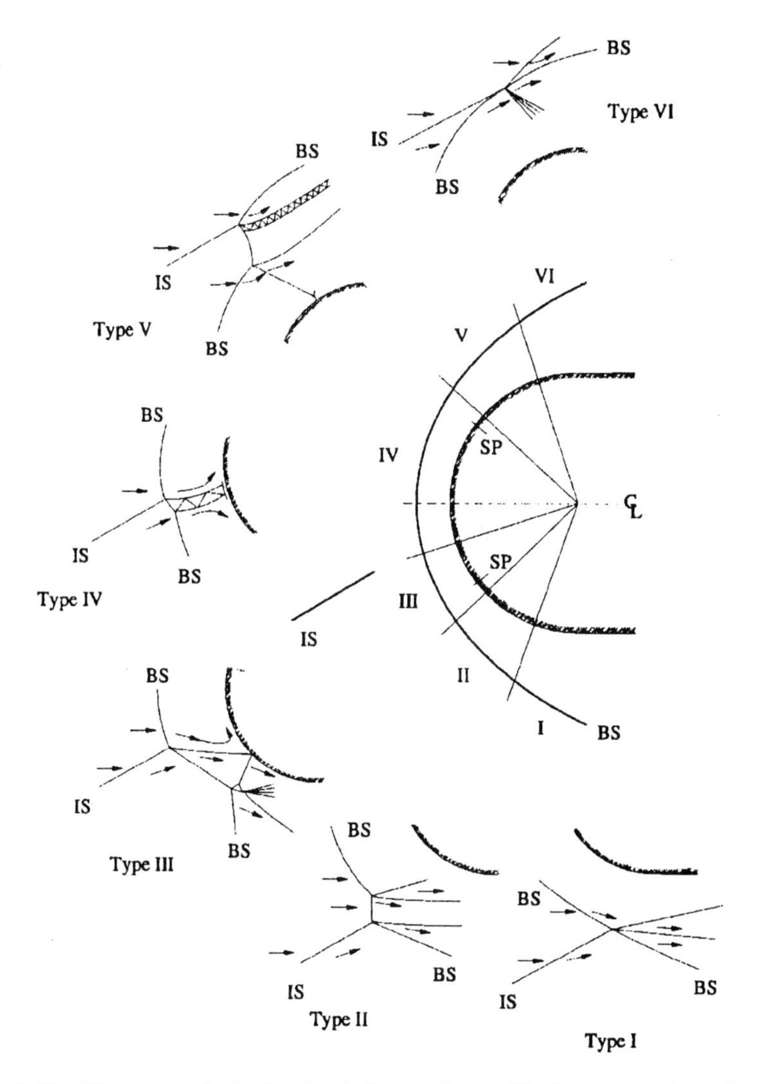

Fig. 5.38 Six types of shock–shock interactions: SP denotes sonic point; IS, impinging shock; and BS, bow shock (Lind [221]).

into six sectors labeled I-VI. If the impinging shock strikes the bow shock in sector I, the resulting shock-shock interaction is called a type-I interaction. If the impinging shock strikes the bow shock in sector IV, the resulting interaction is called a type-IV interaction and so forth. Qualitative sketches of the six different interaction patterns are arrayed around the periphery of Fig. 5.38 starting with the type-I interaction at the bottom and progressing clockwise to the type-VI interaction shown at the top.

Consider the sketch for the type-I interaction shown at the bottom of Fig. 5.38. This is the straightforward intersection of two straight oblique shocks of opposite families discussed in most compressible flow texts (for example, see [4]). The left-running shock IS intersects the right-running shock BS, resulting in two transmitted shocks and a slip line trailing downstream from the intersection point. The strength and angles of the transmitted shocks and the direction of the slip line can be calculated by standard algebraic methods. The flow is locally supersonic everywhere in this pattern.

Consider the sketch for the type-II interaction shown in Fig. 5.38. Here the shock IS impinges on the bow shock BS at a location just downstream of the sonic point behind shock BS. The curved bow shock is stronger at this location, the local flow Mach number is lower, and the simple type-I pattern cannot occur because the flow deflection angles exceed the maximum allowable by straight oblique shock theory. Instead, the two intersecting shock waves are connected by a nearly normal shock, essentially a Mach reflection, and an imbedded core of subsonic flow trails downstream from the intersection region. As a result, the type-II shock–shock interaction pattern is a mixed subsonic supersonic flowfield that must be calculated by a time-marching numerical technique.

Continuing clockwise around the pheriphy of Fig. 5.38, the type-III interaction occurs when the shock IS impinges on the bow shock BS at a location just upstream of the sonic point behind shock BS, that is, where the flow behind shock BS is locally subsonic, albeit at a fairly high subsonic Mach number. A slip line trails downstream from the intersection point and impinges on the body; this slip line separates the locally subsonic flow above it from the locally supersonic flow behind it. The locally supersonic flow below the slip line is turned by the body surface through a rather complex wave pattern, the details of which are not completely shown in Fig. 5.38 (see Fig. 12 of [220] for the details).

The type-IV interaction occurs when the shock IS impinges on the stronger part of the bow shock BS, at a location usually far upstream of the sonic point behind the shock BS, and where the turning angle of the flow by the body is too large for the supersonic flow to be deflected downstream through an attached shock (as is allowed in the type-III interaction). Hence, the flow pattern of the type-IV interaction is completely different than that of the type-III interaction. A sketch of the type-IV interaction flowfield is shown in Fig. 5.39. Behind the impingement point of the impinging shock IS and the bow shock BS, a supersonic jet SJ is formed, bounded on both sides by slip lines SL that separate the supersonic and subsonic regions. The supersonic jet penetrates the flowfield towards the body and terminates in a normal shock NS close to the body surface. There are huge spikes in the pressure distribution and the local aerodynamic heating on the body surface in this region. Indeed, pressure and heating rates up to 30 times the noninteracting case have been measured and calculated [219–222].

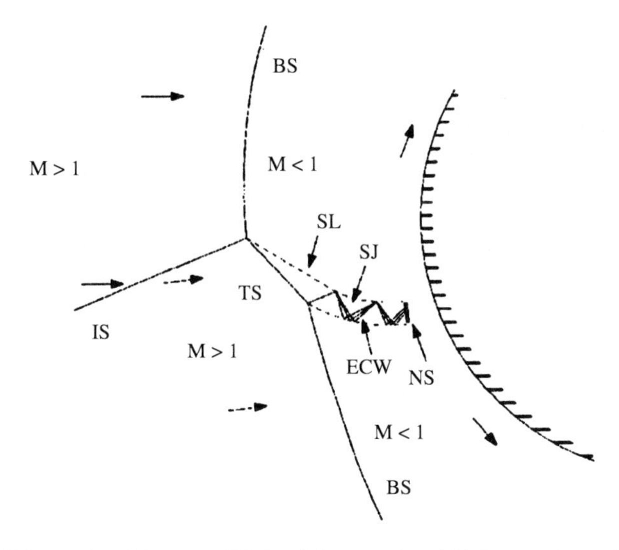

Fig. 5.39 Schematic of the type-IV shock interaction: BS denotes bow shock; ECW, expansion/compression waves; IS, impinging shock; NS, normal shock; SL, shear layer; SJ, supersonic jet; and TS, transmitted shock (Lind [221]).

Computational-fluid-dynamic results for the surface-pressure distribution obtained from a time-marching solution by Lind [221] are given in Fig. 5.40. Here, the surface pressure nondimensionalized by the total pressure behind a normal shock in the freestream is plotted vs angular location along the body, with zero deg being the stagnation-point location in the noninteracting blunt-body flow. The freestream Mach number is 8.144, and the wave angle of the impinging shock (relative to the freestream) is 19 deg. Results are shown for three slightly different impinging shock locations θ_s, where θ_s is a polar angle measured counterclockwise from the horizontal centerline (CL in Fig. 5.38) stretching to the right, for example, the symmetry point of the noninteraction bow shock would be $\theta_s = 180$ deg. For the case $\theta_s = 176.5$ deg, the shock IS impinges the bow shock BS slightly above the noninteracting symmetry point, and for the cases $\theta_s = 181.3$ and 183.5 deg the shock IS impinges the bow shock BS slightly below the noninteracting symmetry point. Note the large peaks in surface pressure induced by the type-IV supersonic jet impinging on the body surface. Because local heat-transfer rates are nearly proportional to the pressure (see Chapters 6 and 7), similar peaks in aerodynamic heating also occur. These pressure and heat-transfer peaks make the type-IV shock interaction a serious consideration for hypersonic vehicle design. Unfortunately, for the type of airbreathing hypersonic vehicle shown in Figs. 5.1–5.3 and sketched in Fig. 5.37, where the on-design condition calls for the inlet bow shock emanating from the vehicle nose to impinge on the engine cowl, the type-IV interaction can be a problem. Moreover, the type-IV interaction flowfield can be unsteady, with interaction frequencies on the order of 1–32 kHz (see [221] and [222]). This unsteadiness by itself requires the calculation of the flow to be time-marching

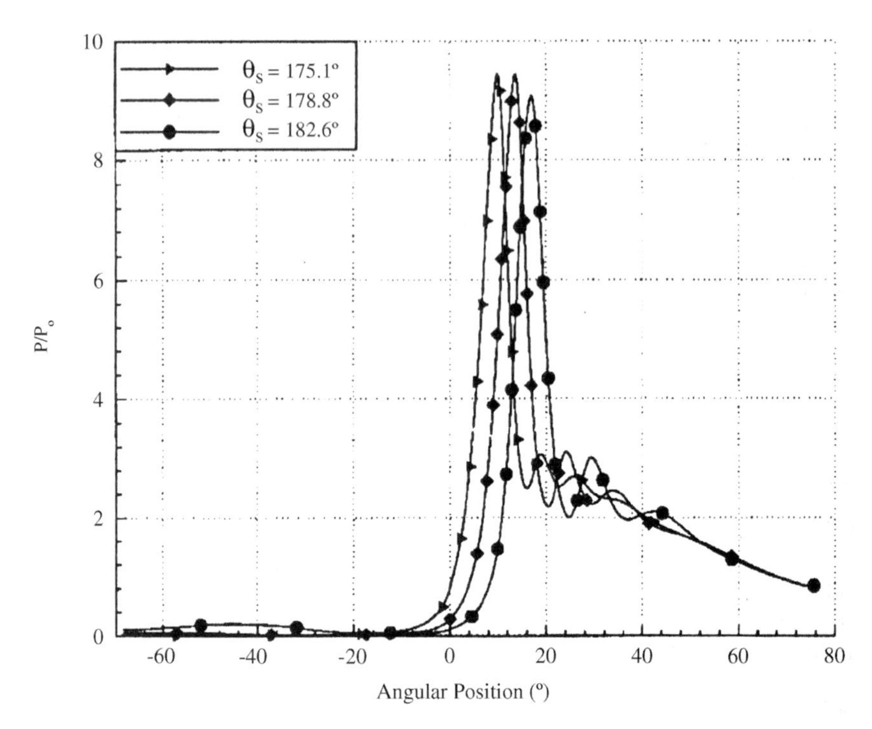

Fig. 5.40 Calculated surface-pressure distributions over a blunt nose for three different locations of the impinging shock for a type-IV interaction (Lind [221]).

solution. Not shown in Fig. 5.39 are the transient vortices that are formed in the unsteady flowfield (see [222] for more details).

Returning to Fig. 5.38, and continuing clockwise around the periphery, the type-V shock–shock interaction involves the impinging IS and the bow shock BS to be of the same family; in Fig. 5.38 they are both left-running shock waves. The impingement point is just downstream of the sonic point behind the bow shock BS, and therefore the type-V interaction is somewhat like the type-II interaction. In the type-V interaction, the two intersecting shock waves are connected by a nearly normal shock, essentially a Mach reflection as with the type-II interaction. An imbedded core of subsonic flow trails downstream from the intersection region, bounded on one side by a slip line and on the other side by a very thin supersonic jet, so thin that the jet basically serves as another slip line.

The type-VI shock-shock interaction as shown in Fig. 5.38 involves two intersection left-running shock waves, where the impingement point is far downstream of the sonic point behind the bow shock. Analogous to the type-I interaction, the type-VI shock–shock interacting is the classic picture of two shock waves of the same family intersecting at a point, coalescing into a single transmitted shock and a slip line trailing downstream from the intersection point. This interaction can be calculated algebraically using classic methods from compressible flow, such as described in [4].

The beginning of Sec. 7.5 dealing with shock-wave/boundary-layer interaction describes a serious incident on one of the X-15 hypersonic test airplane flights wherein shock waves from a dummy ramjet nacelle mounted on a pylon underneath the airplane impinged on the pylon, burning a hole in the pylon surface. Moreover, the bow shock wave from the pylon itself burned a potentially fatal hole through the bottom surface of the airplane. Although some of this damage was caused by a shock-wave/boundary-layer interaction as described in Sec. 7.5, the damage to the pylon itself was most likely caused by a type-IV shock–shock interaction. This incident reinforces the importance of the shock–shock interactions discussed in the present section. Moreover, such interactions can be particularly important in hypersonic flowfields because shock waves can become particularly strong at high Mach numbers, as discussed in Chaper 2.

5.6 Space-Marching Finite Difference Method: Additional Solutions of the Euler Equations

In the present chapter, we are dealing with exact solutions of hypersonic inviscid flows. Although not intentional, the presentation in this chapter has been chronological, starting with the classical method of characteristics (dating from 1928 in terms of its application to supersonic flow), and then discussing the time-marching technique, applied with much success to the hypersonic blunt-body problem in 1966. In the present section, we continue this chronological development by presenting a space-marching finite difference procedure for the solution of steady hypersonic flows—a procedure that has been widely applied since the early 1970s. This space-marching finite difference method applies only to flowfields that are totally supersonic or hypersonic (for example, it cannot be used for the mixed subsonic-supersonic flow in the blunt-nose region); in this fashion, it is analogous to the method of characteristics. But the analogy ends there, because the finite difference method is usually easier to set up and apply than the characteristics method (this is especially true for three-dimensional flows) and is just as accurate. For this reason, downstream-marching finite difference solutions today have all but supplanted the method of characteristics for solutions of purely supersonic and hypersonic inviscid flowfields. However, please keep in mind that *all* of the approaches discussed in this chapter are used today, to some degree or more, for the solution of hypersonic inviscid flows, and therefore represent the modern world of hypersonics.

To introduce the general idea of the downstream-marching procedure, consider the two-dimensional or axisymmetric steady flow over a sharp-nosed body, as sketched in Fig. 5.41a. The general governing Euler equations are given by Eqs. (4.1–4.5). Writing these equations in a form suitable for two-dimensional or axisymmetric steady flow, we have the following.

Continuity:

$$\frac{\partial(\rho u)}{\partial x} + \frac{\partial(\rho v)}{\partial y} + \frac{j\rho v}{y} = 0 \tag{5.40}$$

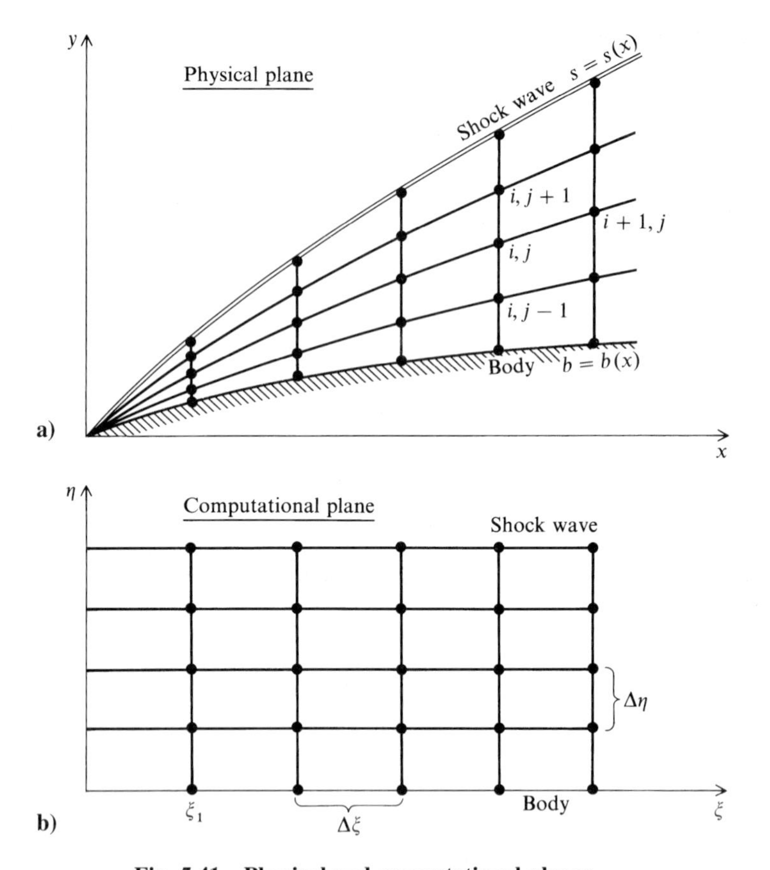

Fig. 5.41 Physical and computational planes.

x Momentum:

$$\rho u \frac{\partial u}{\partial x} + \rho v \frac{\partial u}{\partial y} = -\frac{\partial p}{\partial x} \tag{5.41}$$

y Momentum:

$$\rho u \frac{\partial v}{\partial x} + \rho v \frac{\partial v}{\partial y} = -\frac{\partial p}{\partial y} \tag{5.42}$$

where $j = 0$ or 1 for two-dimensional or axisymmetric flow, respectively. Because the flowfield is steady and adiabatic, the total enthalpy is constant; therefore, the partial differential energy equation [in the form of Eq. (4.5) or (4.6)] can be replaced by the algebraic relation

$$h + \frac{V^2}{2} = h_\infty + \frac{V_\infty^2}{2} = h_0 \tag{5.43}$$

where h_0 is the known total enthalpy. For a calorically perfect gas,

$$h = C_p T = \frac{\gamma RT}{\gamma - 1} = \frac{\gamma}{\gamma - 1}\left(\frac{p}{\rho}\right)$$

Hence, Eq. (5.43) can be written as

$$\frac{\gamma}{\gamma - 1}\left(\frac{p}{\rho}\right) + \frac{u^2 + v^2}{2} = h_0 \tag{5.44}$$

Equations (5.40–5.42) and (5.44) constitute four equations with four unknowns, namely, p, ρ, u, and v. Let us write these equations in a slightly different form as follows. Multiplying Eq. (5.40) by u, and adding the result to Eq. (5.41), we have

$$u\frac{\partial(\rho u)}{\partial x} + \rho u\frac{\partial u}{\partial x} + u\frac{\partial(\rho v)}{\partial y} + \rho v\frac{\partial u}{\partial y} + \frac{j\rho uv}{y} = -\frac{\partial p}{\partial x}$$

or

$$\frac{\partial(\rho u^2)}{\partial x} + \frac{\partial(\rho uv)}{\partial y} + \frac{j\rho uv}{y} = -\frac{\partial p}{\partial x}$$

or

$$\frac{\partial}{\partial x}(p + \rho u^2) + \frac{\partial(\rho uv)}{\partial y} + \frac{j\rho uv}{y} = 0 \tag{5.45}$$

Similarly, multiplying Eq. (5.40) by v, and adding the result to Eq. (5.42), we obtain

$$\frac{\partial(\rho uv)}{\partial x} + \frac{\partial}{\partial y}(p + \rho v^2) + \frac{j\rho v^2}{y} = 0 \tag{5.46}$$

Examine Eqs. (5.40), (5.45), and (5.46) closely; they can be written in the general form

$$\frac{\partial E}{\partial x} + \frac{\partial F}{\partial y} + H = 0 \tag{5.47}$$

where E, F, and H are column vectors

$$E = \begin{Bmatrix} \rho u \\ p + \rho u^2 \\ \rho u v \end{Bmatrix} \quad F = \begin{Bmatrix} \rho v \\ \rho u v \\ p + \rho v^2 \end{Bmatrix} \quad H = \frac{j}{y} \begin{Bmatrix} \rho v \\ \rho u v \\ \rho v^2 \end{Bmatrix}$$

Equation (5.47), with the quantities for E, F, and H as just given, is a form of the Euler equation called the "strong-conservation form." Various classifications of the governing equations have grown out of the computational-fluid-dynamics literature in recent years. Depending on the manner in which the equations are written, they can be classified as nonconservation form, weak-conservation form, or strong-conservation form. The distinction between these forms is described in [52] and is discussed in detail in [72]. Because the emphasis in the present chapter is hypersonic aerodynamics and not the details of computational fluid dynamics, no further elaboration will be given here. Let us simply state that for the application discussed here, involving the hypersonic flow over a body with a distinct shock wave treated as a discontinuity, the particular form of the Euler equations used is not important. We have just chosen to express the governing equations in strong conservation form [Eq. (5.47)] to illustrate that such a form is used in some analyses. For the purposes of this section, we could just as well use the form of the equations expressed by Eqs. (5.40–5.42), where Eq. (5.40) is in conservation form, but Eqs. (5.41) and (5.42) are in nonconservation form. There are instances, however, where the form of the equations used for a particular computation is important; this will be discussed at the end of the present section.

Continuing with the Euler equations in the form of Eq. (5.47), we wish to calculate the hypersonic flow between the body and the shock wave, as sketched in Fig. 5.41a, where the shape and location of the shock wave are also obtained as part of the solution. Because the grid in Fig. 5.41 is curvilinear, a transformation to a rectangular grid in the computational plane is necessary. This can be accomplished by the following transformation:

$$\xi = x \tag{5.48a}$$

$$\eta = \frac{y - b}{\delta} \tag{5.48b}$$

where δ is the local shock-layer thickness $\delta = s - b$, s is the local ordinate of the shock $s = s(x)$, and b is the local ordinate of the body $b = b(x)$. Equations (5.48a) and (5.48b) transform the curvilinear grid in the physical plane (Fig. 5.41a) to the rectangular grid in the computational plane (Fig. 5.41b). Here, $\eta = 0$ is the body, and, $\eta = 1$ is the shock wave. The derivative transformation can be obtained from the chain rule of differentiation as follows:

$$\frac{\partial}{\partial x} = \left(\frac{\partial}{\partial \xi}\right)\left(\frac{\partial \xi}{\partial x}\right) + \left(\frac{\partial}{\partial \eta}\right)\left(\frac{\partial \eta}{\partial x}\right) \tag{5.49a}$$

$$\frac{\partial}{\partial y} = \left(\frac{\partial}{\partial \xi}\right)\left(\frac{\partial \xi}{\partial y}\right) + \left(\frac{\partial}{\partial \eta}\right)\left(\frac{\partial \eta}{\partial y}\right) \tag{5.49b}$$

where, from Eqs. (5.48a) and (5.48b)

$$\frac{\partial \xi}{\partial x} = 1 \quad \frac{\partial \xi}{\partial y} = 0$$

$$\frac{\partial \eta}{\partial x} = \frac{\delta(-\mathrm{d}b/\mathrm{d}x) - (y-b)\,\mathrm{d}\delta/\mathrm{d}x}{\delta^2} = \frac{1}{\delta}\left(-\eta\frac{\mathrm{d}\delta}{\mathrm{d}x} - \frac{\mathrm{d}b}{\mathrm{d}x}\right)$$

$$\frac{\partial \eta}{\partial y} = \frac{1}{\delta}$$

Substituting the preceding results into Eqs. (5.49a) and (5.49b), we have the following derivative transformation:

$$\frac{\partial}{\partial x} = \frac{\partial}{\partial \xi} + \frac{1}{\delta}\left(-\eta\frac{\mathrm{d}\delta}{\mathrm{d}x} - \frac{\mathrm{d}b}{\mathrm{d}x}\right)\left(\frac{\partial}{\partial \eta}\right) \tag{5.50a}$$

$$\frac{\partial}{\partial y} = \frac{1}{\delta}\left(\frac{\partial}{\partial \eta}\right) \tag{5.50b}$$

Using Eqs. (5.50a) and (5.50b), the transformed version of Eq. (5.47) is

$$\frac{\partial E}{\partial \xi} + \frac{1}{\delta}\left(\eta\frac{\mathrm{d}\delta}{\mathrm{d}x} - \frac{\mathrm{d}b}{\mathrm{d}x}\right)\frac{\partial E}{\partial \eta} + \frac{1}{\delta}\frac{\partial F}{\partial \eta} + H = 0$$

Writing the preceding equation with the ξ derivative on the left and the η derivatives on the right, we have

$$\boxed{\frac{\partial E}{\partial \xi} = -H - \frac{1}{\delta}\left(-\eta\frac{\mathrm{d}\delta}{\mathrm{d}x} - \frac{\mathrm{d}b}{\mathrm{d}x}\right)\frac{\partial E}{\partial \eta} - \frac{1}{\delta}\frac{\partial F}{\partial \eta}} \tag{5.51}$$

Equation (5.51) is reminiscent of Eqs. (5.18–5.21) used for the time-marching solution of the blunt-body problem; in Eqs. (5.18–5.21) the time derivatives are on the left sides of the equations, and all of the spatial derivatives are on the right sides. However, in the case of Eq. (5.51) the ξ derivative is on the left, and the η derivatives are on the right. This suggests a marching procedure in steps of ξ, that is, a *spatial-marching procedure* in the downstream direction. Indeed, MacCormack's predictor-corrector method, used for the time-marching solutions in Sec. 5.3, can also be used here for the spatial marching. Such a downstream-marching approach is mathematically valid, because, for a supersonic or hypersonic inviscid flow, Eq. (5.47) and, hence, the transformed version Eq. (5.51), is a hyperbolic partial differential equation. Hence, starting with an initial data line at some ξ station, the downstream-marching procedure is mathematically well posed.

In light of the preceding, the following is an outline of the application of MacCormack's method to the solution of the flowfield at the internal grid points as shown in Fig. 5.41:

1) Begin with an initial data line at some value of ξ, say ξ_1. For a pointed body, the properties along this initial data line can be obtained from exact wedge flow (for a two-dimensional body) or from exact cone flow (for an axisymmetric body). For a blunt-nosed body, the initial data line is obtained from a blunt-body solution, such as described in Sec. 5.3. The preceding comments about the generation of data for an initial data line are exactly the same as made in conjunction with the method of characteristics, which also required an initial data line (recall Sec. 5.2). In short, referring to Fig. 5.41b, all properties are considered known along the initial data line, $\xi = \xi_1$.

2) Knowing properties along $\xi = \xi_1$ (or any other line of constant ξ), the flow properties at the next downstream location $\xi + \Delta\xi$ can be found from

$$E_{i+1,j} = E_{i,j} + \left(\frac{\partial E}{\partial \xi}\right)_{\text{ave}} \Delta\xi \tag{5.52}$$

where $E_{i+1,j}$ is the column vector of properties, ρu, $p + \rho u^2$, and ρuv at grid point $(i+1, j)$, and the value of $(\partial E/\partial \xi)_{\text{ave}}$ is obtained from MacCormack's predictor-corrector method, as described next. In other words, the notation in Eq. (5.52) represents three individual equations, one each for the flow quantities, ρu, $p + \rho u^2$, and ρuv. Note here that the unknowns are not directly p, ρ, u, and v (called the "primitive variables"), but rather the "flux" quantities ρu, $p + \rho u^2$, and ρuv. The process described here will produce numerical values for ρu, $p + \rho u^2$, and ρuv at the given grid point; in turn, the primitive variables (p, ρ, u, and v) at the grid point can be extracted from these numbers and from Eq. (5.44) by simultaneous solution of the algebraic equations

$$\begin{aligned} \rho u &= c_1 \\ p + \rho u^2 &= c_2 \\ \rho uv &= c_3 \\ \frac{\gamma}{\gamma - 1}\left(\frac{p}{\rho}\right) + \frac{u^2 + v^2}{2} &= h_0 \end{aligned}$$

where c_1, c_2, and c_3 are the known values from the computation at the grid point and h_0 is the known total enthalpy.

3) The first step in obtaining $(\partial E/\partial \xi)_{\text{ave}}$, which appears in Eq. (5.52), is the predictor step of MacCormack. Therefore, calculate a predicated value of E at

grid point $(i+1, j)$ denoted by $\bar{E}_{i+1,j}$, from

$$\bar{E}_{i+1,j} = E_{i,j} + \left(\frac{\partial E}{\partial \xi}\right)_{i,j} \Delta\xi \tag{5.53}$$

where $E_{i,j}$ is the known value at the given ξ and $\bar{E}_{i+1,j}$ is the predicted value at $\xi + \Delta\xi$. In Eq. (5.53), $(\partial E/\partial \xi)_{i,j}$ comes from Eq. (5.51), where the right-hand side contains only known values at ξ and where the derivatives are obtained from forward differences, that is,

$$\left(\frac{\partial E}{\partial \xi}\right)_{i,j} = -H_{i,j} - \frac{1}{\delta_{ij}}\left(-\eta\frac{\mathrm{d}\delta}{\mathrm{d}x} - \frac{\mathrm{d}b}{\mathrm{d}x}\right)_{i,j}\left(\frac{E_{i,j+1} - E_{i,j}}{\Delta\eta}\right) - \frac{1}{\delta_{ij}}\left(\frac{F_{i,j+1} - F_{i,j}}{\Delta\eta}\right) \tag{5.54}$$

Knowing $\overline{E}_{i+1,j}$ from Eq. (5.53), predicted values of the primitive flow variables $\bar{p}$, $\bar{\rho}$, $\bar{u}$, and $\bar{v}$ can be obtained (as described in step 2), which in turn yields predicted values for F and H, namely, $\overline{F}_{i+1,j}$ and $\overline{H}_{i+1,j}$.

4) On the corrector step, insert the predicted quantities into Eq. (5.51), using rearward differences

$$\left(\frac{\overline{\partial E}}{\partial \xi}\right)_{i+1,j} = -\overline{H}_{i+1,j} - \frac{1}{\delta_{i+1,j}}\left(-\eta\frac{\mathrm{d}\delta}{\mathrm{d}x} - \frac{\mathrm{d}h}{\mathrm{d}x}\right)_{i+1,j}\left(\frac{\overline{E}_{i+1,j} - \overline{E}_{i+1,j-1}}{\Delta\eta}\right) - \frac{1}{\delta_{i+1,j}}\left(\frac{\overline{F}_{i+1,j} - \overline{F}_{i+1,j-1}}{\Delta\eta}\right) \tag{5.55}$$

5) Obtain the average derivative that appears in Eq. (5.52) by

$$\left(\frac{\partial E}{\partial \xi}\right)_{\text{ave}} = \frac{1}{2}\left[\underbrace{\left(\frac{\partial E}{\partial \xi}\right)_{i,j}}_{\substack{\text{obtained from}\\ \text{Eq. (5.54)}}} + \underbrace{\left(\frac{\overline{\partial E}}{\partial \xi}\right)_{i+1,j}}_{\substack{\text{obtained from}\\ \text{Eq. (5.55)}}}\right] \tag{5.56}$$

6) Calculate the final, corrected value of $E_{i+1,j}$ from Eq. (5.52), repeated here:

$$E_{i+1,j} = E_{i,j} + \left(\frac{\partial E}{\partial \xi}\right)_{\text{ave}} \Delta\xi \tag{5.52}$$

Evaluation of Eq. (5.52) via steps 3–6 at each of the $i+1$ grid points for the jth column results in the complete determination of the internal part of the flowfield at $\xi + \Delta\xi$. The entire procedure (steps 2–6) is then repeated in order to progressively march downstream from the initial data line.

The boundary condition at the shock wave is handled in an analogous fashion as described in Sec. 5.3, except now the flow is steady, that is, there is no moving shock wave. In this respect, the application of the shock boundary condition is simpler. For the present downstream-marching procedure, the flow properties at the shock grid points (the upper boundary in Fig. 5.41b) as well as the shock-wave angle can be obtained as follows:

1) Consider the shock grid points labeled 1 and 2 in Fig. 5.42. We wish to calculate the flow properties and wave angle β at point 2. The flow conditions and wave angle at point 1 have already been obtained from the preceding downstream-marching step. Initially calculate the flow properties at point 2 using the internal flow algorithm as outlined in the preceding steps 2–6, except using one-sided differences, that is, use rearward differences in both Eqs. (5.54) and (5.55).

2) Among the flow properties obtained in the preceding step is the pressure at point 2, p_2. This pressure along with the freestream pressure and Mach number provide two known quantities about the shock at point 2, namely, p_2/p_∞ and M_∞. Recall that the strength of an oblique shock wave (for a calorically perfect gas) is uniquely defined by two quantities, such as the preceding two. Hence, the oblique shock wave, including the wave angle β_2, is now determined at point 2.

3) Although all of the flow properties at point 2 were originally calculated from the internal flow algorithm as stated in step 1, our main interest was in the pressure in order to establish the strength of the shock wave, as described in step 2. Now reset the values of ρ_2, T_2, u_2, and V_2 at point 2 to be equal to the proper values behind the calculated oblique shock wave, as determined by the exact oblique shock relations. This now finalizes the flowfield properties at the shock grid point.

4) Construct the shock-wave shape and location at point 2 by drawing a straight line from point 1 with the angle $\frac{1}{2}(\beta_1 + \beta_2)$.

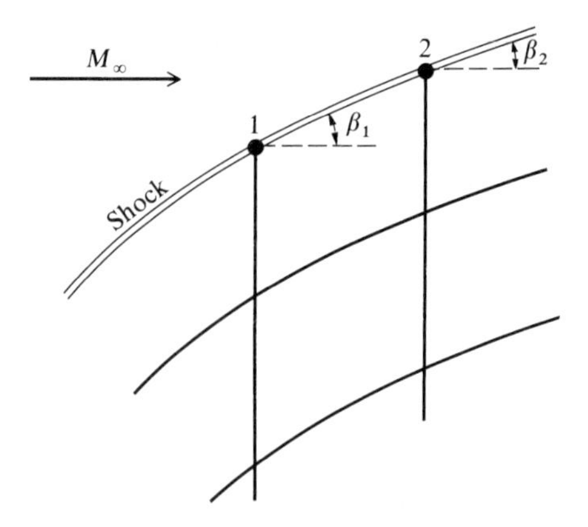

Fig. 5.42 Shock boundary for the downstream-marching procedure.

The boundary condition on the body is also handled in an analogous fashion as treated in Sec. 5.3, except now, because the flow is steady, an ordinary steady Prandtl–Meyer expansion or compression is used at the surface. For the present downstream-marching procedure, the flow properties at the body grid points (the lower boundary in Fig. 5.41b) can be obtained as follows:

1) Consider the body grid points labeled 1 and 2 in Fig. 5.43. All properties at point 1 are known from the previous calculation, and in the downstream-marching sequence we wish to calculate the properties at point 2. Initially calculate these properties using the internal flow algorithm at point 2, except using one-sided differences, that is, use forward differences in both Eqs. (5.54) and (5.55).

2) The values of u and v at point 2 obtained from the preceding step will, in general, result in a velocity that is not tangent to the surface. This velocity is denoted by $\boldsymbol{V}_{\text{old}}$, shown in Fig. 5.43 making an angle θ with the tangent to the surface at point 2. To satisfy the flow tangency condition, this velocity vector must be "rotated" through the angle θ, such that the resulting velocity, denoted by V_{new} in Fig. 5.43, is tangent to the surface. This rotation is accomplished by a Prandtl–Meyer expansion through the angle θ. The flowfield values at point 2 obtained from preceding step 1 are denoted as "old" values, p_{old}, ρ_{old}, u_{old}, v_{old}, etc. These are assumed to represent the flowfield upstream of the local Prandtl–Meyer expansion. After expansion through the angle θ, the flowfield calculated downstream of the Prandtl–Meyer expansion (using the Prandtl–Meyer function and the isentropic flow relations—for example, see [4] and [5]) is denoted as p_{new}, ρ_{new}, u_{new}, v_{new}, etc. These "new" values are now assigned as the final flowfield values at point 2, satisfying the flow tangency condition. The treatment of the wall boundary condition described here was first suggested by Abbett [64] and therefore is frequently called Abbett's method (see also Chapter 11 of [4]).

Because the downstream-marching technique described here is an explicit, finite difference method, it must satisfy the Courant–Friedrichs–Lewy stability criterion applied to steady flow. This criterion is applied in the physical plane shown in Fig. 5.41a. In essence, it states the following. Consider grid point (i, j), $(i, j+1)$, and $(i, j-1)$, located at a given x station. The next neighboring

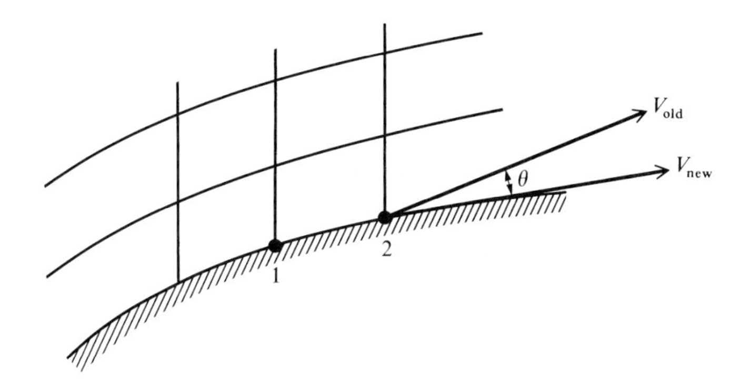

Fig. 5.43 Body boundary for the downstream-marching procedure.

point downstream is point $(i+1, j)$ as shown in Fig. 5.41a. The spacing between points (i, j) and $(i+1, j)$ is denoted by Δx. The CFL criterion states that Δx must be small enough such that point $(i+1, j)$ falls *upstream* of the left-running characteristic (left-running Mach line) through point $(i, j-1)$ and upstream of the right-running characteristic through point $(i, j+1)$. On a quantitative basis, this criterion is given by

$$\Delta x \leq \frac{\Delta y}{|\tan(\theta \pm \mu)|_{\max}}$$

where θ and μ are the streamline direction and Mach angle respectively at either point $(i, j-1)$ or $(i, j+1)$. See [4] for more details.

This completes our description of the downstream-marching finite difference method. Some results of this method, applied to an axisymmetric three-quarter power-law body, are shown in Figs. 5.44 and 5.45. In Fig. 5.44, the given body shape, the calculated shock-wave shape, and grid in the physical plane are shown for a case at Mach 5. Pressure coefficient distributions as a function of the downstream distance x are shown in Fig. 5.45 for $M_\infty = 5$, 10, and 15. Note that little difference exists between these results—another demonstration of the Mach-number independence principle.

The preceding description and results are for a two-dimensional or axisymmetric body. For such applications, the method of characteristics (Sec. 5.2) and the downstream-marching finite difference method (decribed in the present section) are competing techniques. The choice is up to the user as to which technique is employed. However, the choice most often made today is the finite difference approach, caused primarily by its relative simplicity. This is particularly true in the case of three-dimensional flow, where the method of characteristics becomes very tedious and where the finite difference method is still, relatively speaking, straightforward.

One of the first three-dimensional, downstream-marching, inviscid hypersonic flow calculations was carried out by Kutler et al. [73]. Here, the flow over a

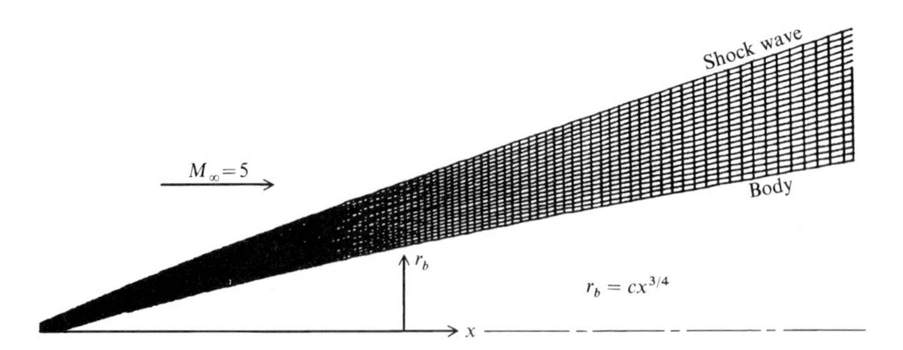

Fig. 5.44 Shock-wave and finite difference grid for a downstream-marching solution (courtesy of Stephen Corda, University of Maryland).

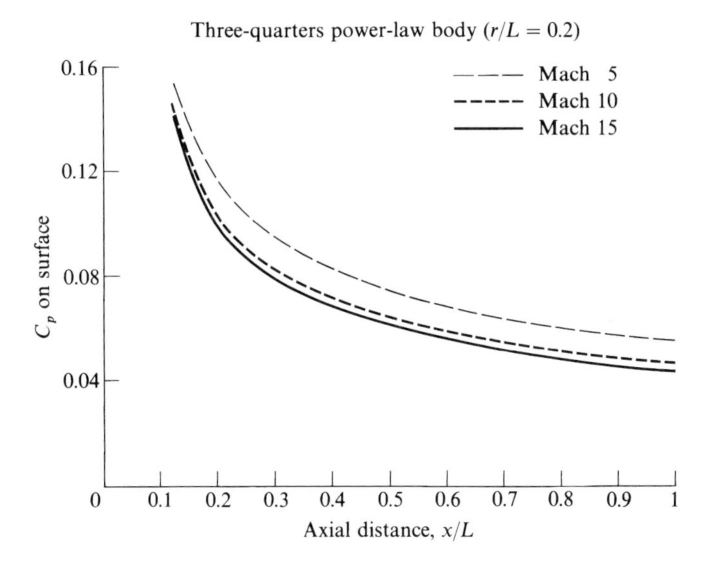

Fig. 5.45 Pressure distributions obtained for the body shown in Fig. 5.44 (calculations made by Stephen Corda, University of Maryland).

space-shuttle-like vehicle is calculated at Mach 7.4. This work used the cylindrical coordinate system illustrated in Fig. 5.46, where r, ϕ, and z are the usual cylindrical coordinates. The axis of the body is taken along the z axis, which is at an angle of attack α to the freestream. The flowfield in the initial data plane is obtained from an independent blunt-body calculation, which today is almost always a time-marching calculation such as described in Sec. 5.3. Starting

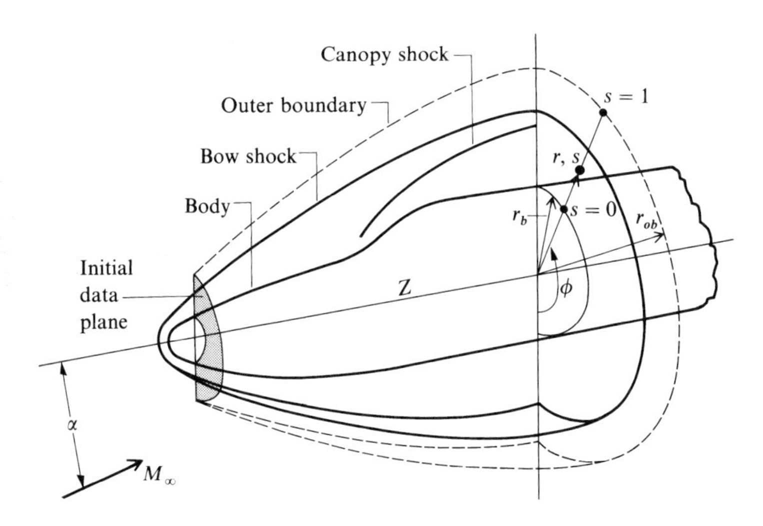

Fig. 5.46 Coordinate systems for a three-dimensional body (from Kutler et al. [73]).

from the initial data plane, the finite difference calculations are marched downstream in the z direction, using the same type of philosophy described earlier in the present section. Consult [73] for details.

In Fig. 5.46, a transformed coordinate s is also displayed, which is defined in such a manner that $s = 0$ is the body surface and $s = 1$ is the outer-flow boundary of the computation. Note that the outer-flow boundary is taken *outside* the shock wave (the outer flow boundary is in the freestream), and hence the shock wave itself is handled differently than described earlier. (Elaboration on this will be made in the next paragraph.) For the finite difference calculations, the physical space shown in Fig. 5.46 is transformed to a rectangular box, much in the same spirit as described earlier for the two-dimensional and axisymmetric cases.

In [73], the shock wave is calculated differently than described in Sec. 5.3, or to this point in the present section. In these sections, the shock was treated as a *discontinuity*, and only the flowfield between the shock and body was calculated, using the oblique shock equations to determine properties behind the shock. *Such a philosophy is called shock fitting.*

In contrast, in Fig. 5.46, the outer boundary of the coordinate system is *outside* the bow shock wave. Here, the shock comes naturally out of the finite difference calculations, showing up as a rapid change of flow properties across several grid points. It is *not* treated explicitly as a discontinuity, and the oblique shock relations are *not* used. *Such a philosophy is called shock capturing.*

The relative merits of using a shock-fitting or a shock-capturing approach are a matter of continued discussion within the computational-fluid-dynamics community and are beyond the scope of the present book. For further information on these matters, see [4] and [72].

Results from the calculations of Kutler et al. are shown in Figs. 5.47–5.51. In Fig. 5.47, the shock locations are shown in both the planform and side views. The solid lines are experimental results obtained from [74]. The squares and circles pertain to the downstream-marching calculation; the squares are a second-order-accurate calculation using the MacCormack technique described earlier, and the circles are a related finite difference formulation but of third-order accuracy. (Again, see [73] for details.) Note the excellent accuracy between calculation and experiment shown in Fig. 5.47. Also, on a physical basis, note that a shock wave is generated at the nose of the vehicle and that this bow shock interacts with a second shock wave generated by the canopy, as seen in the side view. A slip surface is generated by the interaction of the bow and canopy shocks and flows downstream. The computed and experimentally measured slip surfaces agree very well. Also, note from the plan view that another shock wave is generated by the wing leading edge and interacts with the bow shock wave. Observe that the calculations are not carried further downstream of the interaction of the bow and wing shock. This is because a pocket of locally subsonic flow was encountered in the interaction region. In a steady flow, such a subsonic region is mathematically elliptic, and hence the downstream-marching solution (which applies to hyperbolic and parabolic regions only) becomes invalid (it will usually "blow up" in the subsonic region). The only way to overcome this problem is to calculate such subsonic regions by a time-marching procedure and resume the downstream-marching technique in the region where the flow

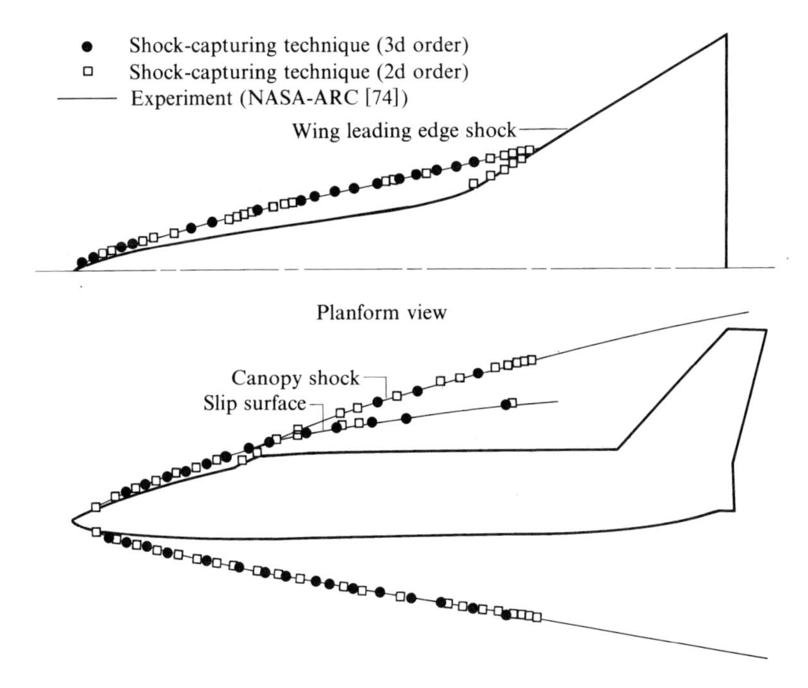

Fig. 5.47 Shock locations for shuttle-like configuration obtained from second- and third-order downstream-marching finite difference techniques: $M_\infty = 7.4$, and $\alpha = 0$ deg (from [73]).

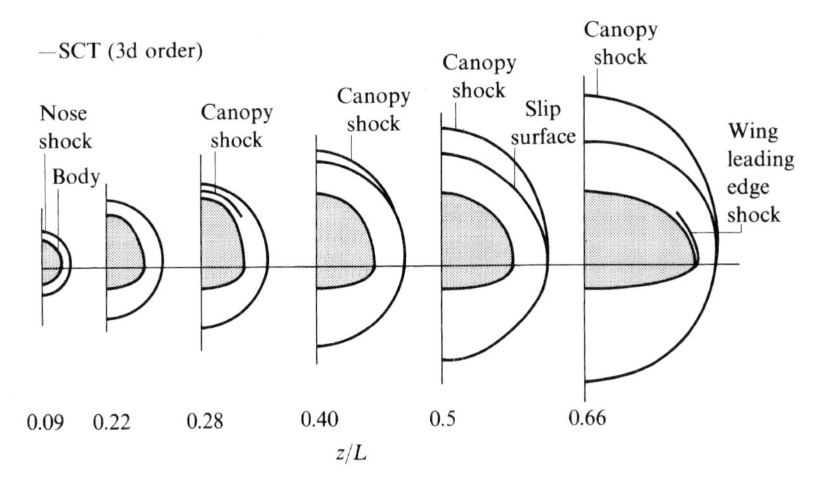

Fig. 5.48 Cross-sectional shock-wave shapes at various streamwise stations of the shuttle-like configuration shown in Fig. 5.47, $M_\infty = 7.4$, and $\alpha = 0$ deg (from [73]).

becomes supersonic again. In most modern downstream-marching computer solutions, a provision is made to switch to a time-marching solution for those local pockets of subsonic flow. (Such a provision was not available for the calculations of [73].) To further illustrate the three-dimensional nature of these calculations. Fig. 5.48 shows the calculated development of the shock shapes and slip surface in the crossflow plane at various axial locations along the body. In Fig. 5.49, pressure coefficient distributions are shown as a function axial distance z for various azimuthal angles around the body, starting with the bottom of the vehicle ($\phi = 0$) and concluding with the top of the vehicle ($\phi = 180$ deg). Note that, for this case, the pressures are higher on the top than on the bottom of the vehicle; this is because the angle of attack is zero, and, noting the shape of the vehicle as shown in the side view in Fig. 5.47, the top surface at $\alpha = 0$ is more of a compression surface than the bottom of the vehicle. Also note the sharp spike in pressure for $\phi = 180$ deg; this is because of the canopy shock wave on the top surface. Calculated streamline shapes on the bottom surface are shown in Fig. 5.50; these are given here just to emphasize the many different types of data that can be obtained in such flowfield calculations. Finally, the calculated and measured shock-wave shapes for an angle-of-attack case ($\alpha = 15.3$ deg) are given in Fig. 5.51. Again, excellent agreement is obtained. Also, at this stage the reader is cautioned that downstream-marching calculations must be limited to low enough angle-of-attack applications so as not to have large regions of subsonic flow over the bottom surface. For cases at high angle of attack with large regions of subsonic flow, a time-marching three-dimensional solution must be employed, such as described in [68] and illustrated in Figs. 5.32 and 5.33.

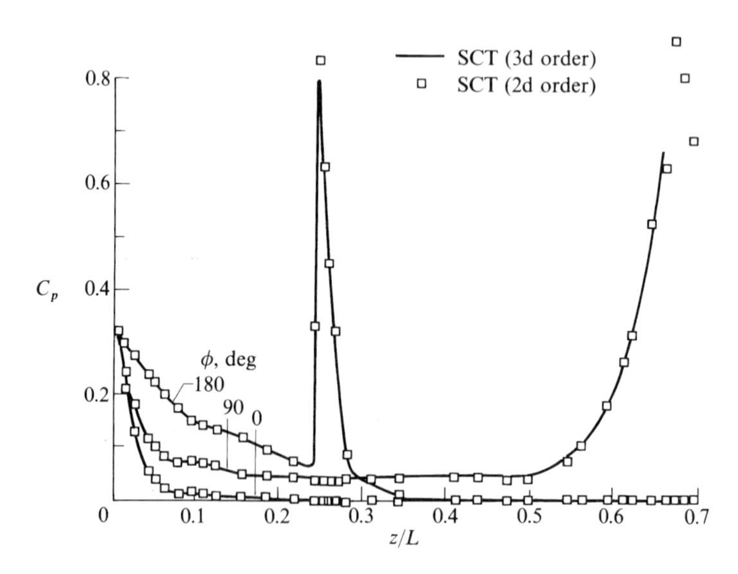

Fig. 5.49 Longitudinal surface-pressure distributions for the 0, 90, and 180 meridians of the shuttle-like configuration: $M_\infty = 7.4$, and $\alpha = 0$ deg (from [73]).

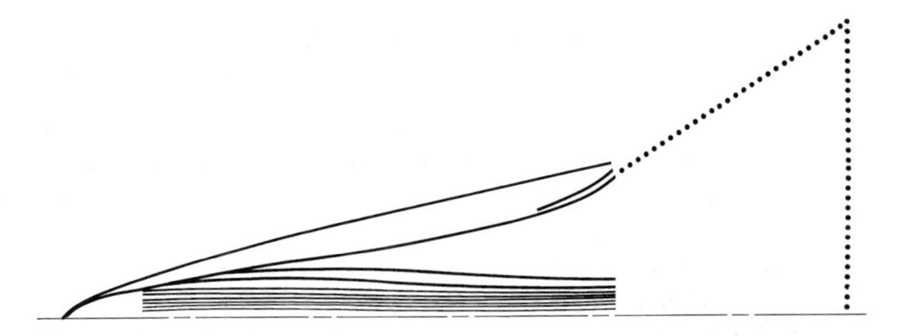

Fig. 5.50 Surface streamline distribution on bottom of shuttle-like configuration: $M_\infty = 7.4$, and $\alpha = 0$ deg (from [73]).

A more recent example of downstream-marching, three-dimensional, hypersonic flow solutions is the work of Maus et al. [75] wherein inviscid flowfields over the space shuttle are calculated for both a calorically perfect gas and an equilibrium chemically reacting gas. (Chemically reacting flows are the subject of Part III of this book.) Results from [75] are given in Fig. 5.52, which shows calculated pressure distributions on the windward centerline of the space shuttle for angles of attack of 20 and 30 deg. The calculations are made at two Mach numbers, $M_\infty = 8$ and 23. Note that, at a given angle of attack, the C_p results for both Mach 8 and 23 are almost identical—yet another demonstration of the Mach-number independence principle.

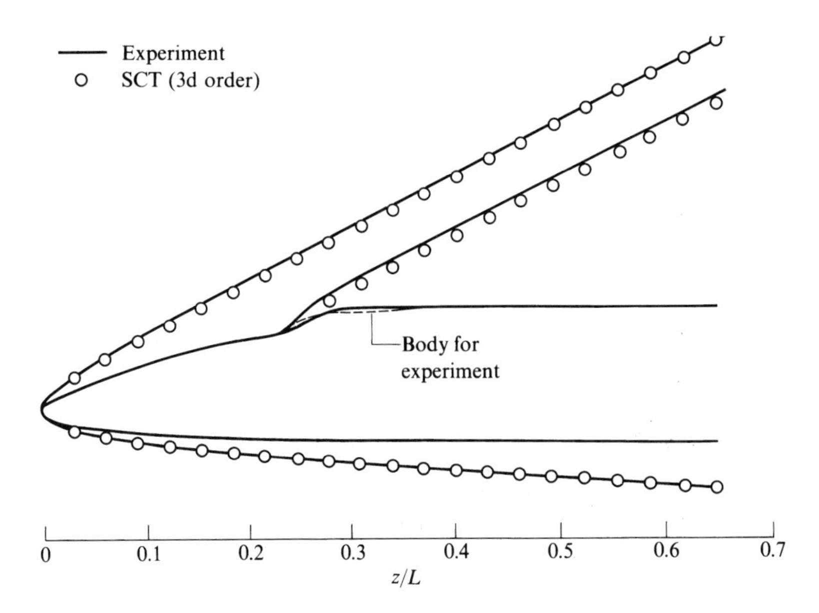

Fig. 5.51 Shock location for a shuttle-like configuration: $M_\infty = 7.4$, and $\alpha = 15.30$ deg (from [73]).

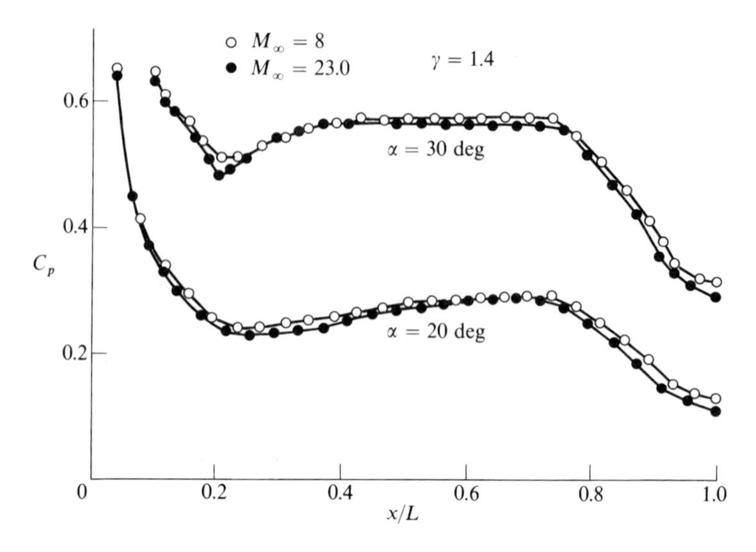

Fig. 5.52 Pressure coefficient distribution along the windward centerline on the bottom of the space shuttle; illustration of Mach-number independence. Downstream-marching finite difference calculations by Maus et al. [75].

5.7 Comments on the State of the Art

There are no general, closed-form, analytical solutions to hypersonic inviscid flows. There are, however, approximate theoretical solutions based on simplified forms of the exact governing equations, as discussed in Chapter 4. On the other hand, the exact governing equations (the Euler equations) can be solved numerically, as demonstrated in the present chapter. Indeed, the power of modern computational fluid dynamics gives us the ability to obtain exact solutions of hypersonic inviscid flows for virtually any arbitrary geometry, including complex, three-dimensional configurations.

However, do not be lulled into a false sense of security by these statements. Even though computational fluid dynamics gives us the ability to make such exact calculations, the actual carrying out of such calculations is frequently tedious, sometimes difficult, and laced with details that have to be handled properly in order to obtain accurate and stable solutions. It is not within the scope of the present book to elaborate on computational fluid dynamics. Indeed, the purpose of the present book is to provide an educational experience for the reader in the areas of hypersonic and high-temperature gas dynamics, and only enough computational fluid dynamics is discussed to give the reader a flavor of its application to these areas. Before embarking on serious work on multidimensional hypersonic flow calculations, the reader is encouraged to study the introductory discussions on computational fluid dynamics in [4] and [72] and in particular the thorough treatment in [52].

The reader is also encouraged to examine, and keep current with, the contemporary literature in computational fluid dynamics, and its applications to hypersonic flows. The CFD state of the art is dynamically changing, particularly at the present time of writing. One example is the current work on upwind differencing.

In the present chapter we have utilized MacCormack's finite difference method, which is basically a central difference method. In the presence of strong shock waves, this method can produce spatial oscillations both upstream and downstream of the shock. Because hypersonic shock waves are usually strong, these oscillations can become a very undesirable aspect of some hypersonic flow calculations. Therefore, much current work is being devoted to the development of "upwind" schemes, that is, numerical schemes that pay attention to the domain of dependence of a given grid point in supersonic and hypersonic flow, and which utilize data only from the upstream locations within the domain of dependence. Such upwind schemes have captured shock waves that are crisply defined over only one (or at most two) grid point, and with little or no oscillations. For example, see [76–78] for more details. And to become even more general, we have to mention that finite difference schemes do not have a monopoly on hypersonic flowfield calculations; finite volume and finite element techniques have applications in hypersonics as well. It is not feasible for us to elaborate on such matters here. See [223] and [224] for thorough discussions on modern computational fluid dynamics.

In final perspective, the present chapter makes one important statement: exact solutions of the governing equations of hypersonic inviscid flow for general problems can be obtained if one is willing to accept the methods of computational fluid dynamics as supplying such solutions. This is an aspect of the modern hypersonics; indeed, the bulk of this chapter could not have been written before 1966. We have given examples of exact solutions for hypersonic inviscid flows from 1) the method of characteristics (a "classical" method), 2) a time-marching finite difference method, and 3) a space-marching finite difference method. The methods and results presented here are intended to provide only the flavor of such work.

5.8 Summary and Comments

This brings to an end our discussion of inviscid hypersonic flows, wherein the purely fluid mechanical effect of high Mach number was illustrated. Part 1 of this book has concentrated on such flows, both from classical and modern points of view. In the modern hypersonic aerodynamics of today, it is still useful to be aware of the classical theory and engineering approaches described in the earlier sections of Part 1. Also, we must recognize that computational fluid dynamics dominates the analysis of modern hypersonic problems. Before proceding to Part 2 return again to the road map in Fig. 1.24, and scan over the items listed under the general heading of inviscid flows, namely, the two left-hand branches. Make certain that you feel comfortable with the material contained within each of the items and that you appreciate how each item is related to the general scheme of hypersonic inviscid flows.

Design Example 5.1

The method of characteristics (Sec. 5.2) is alive and well in modern hypersonic vehicle design. Based on the method of characteristics, Fred Billig [232] developed a method of tracing streamlines in a flowfield and along with Ajay

Kothari applied it to the design of hypersonic vehicles. Based on this technique, Kothari et al. [231] discuss a novel inlet design for airbreathing scramjet-powered hypersonic vehicles that uses an inward turning geometry. An inward-turning inlet is compared to the more common two-dimensional inlet in Fig. 5.53, from [233]. Although still in the research stage at the time of writing, the inward-turning inlet offers some performance gains that are bringing it some attention within the hypersonic design community. As noted by Dissel et al. in [233], the inward-turning inlet offers the following advantages over a two-dimensional inlet:

1) It has less wetted area in the high heating regions at the end of the inlet, through the combustor, and the entrance to the nozzle. This can result in about a 35% reduction in the amount of active cooling required and an overall 50% reduction in heat transfer.

2) It has a single combustor flowpath, which reduces the complexity and the amount of actuators and seals compared to the six to eight combustor flowpaths of the two-dimensional engine.

3) The reduced cooling loads and combustor provisions result in lighter engine and thermal protection weights.

4) The reduced viscous losses and smaller cooling requirements result in a higher engine specific impulse, enabling the inward-turning vehicle to accelerate to a higher Mach number before scramjet turnoff.

The root source of all of these advantages is a modern application of the method of characteristics to advanced hypersonic vehicle design.

This chapter deals in part with time-marching and downstream-marching numerical solutions of hypersonic inviscid flows using computational-fluid-dynamic methods. The practical output of these solutions for hypersonic vehicle design is the surface-pressure distributions and the resulting lift and wave drag. Also, such inviscid solutions are used to set the flow conditions at the outer edge of the boundary layer for boundary-layer calculations of surface shear stress and aerodynamic heating distributions (discussed in Chapter 6).

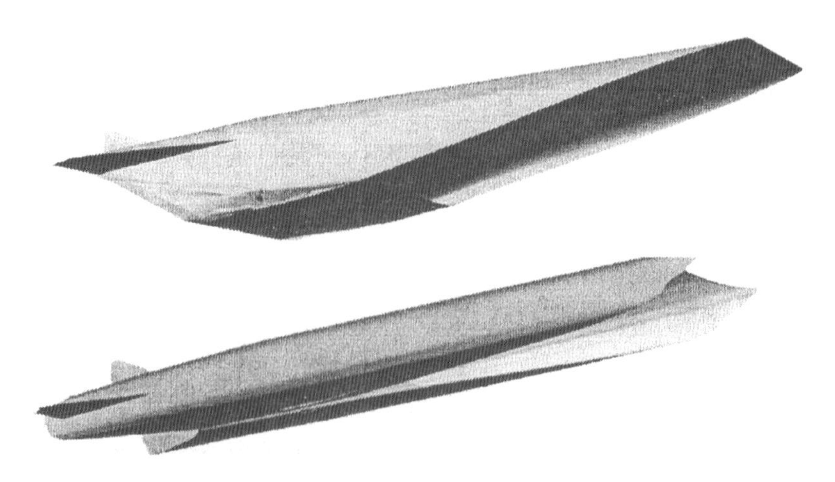

Fig. 5.53 Two conceptual single-stage-to-orbit vehicles, one with a two-dimensional inlet (upper) and the other with an inward-turning inlet (lower) (Dissel et al. [233]).

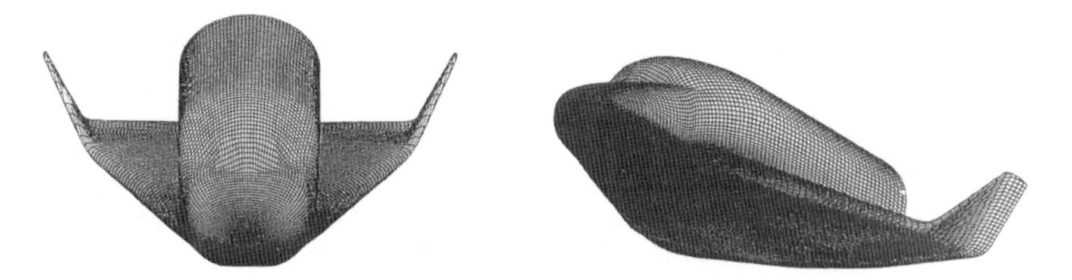

Fig. 5.54 Geometry of the HERMES space shuttle (Rieger [234]).

This combination of CFD inviscid flow calculations combined with a boundary-layer analysis constitutes a relatively fast engineering package for vehicle design in comparison with the more detailed and complex CFD solutions of fully viscous flowfield as discussed in Chapter 8. Examples of some of the first inviscid CFD solutions over hypersonic vehicles have already been given in Secs. 5.3–5.5. An example of more recent inviscid flow calculations in support of the European HERMES space shuttle design is given in Figs. 5.54–5.56, from the work of H. Rieger of Dornier in Germany [234]. The HERMES configuration is shown in Fig. 5.54, with its surface covered by part of the mesh for a finite volume calculation. The total mesh used about 300,000 points. Inviscid flow results for the Mach-number distribution throughout the flowfield over the forward portion of

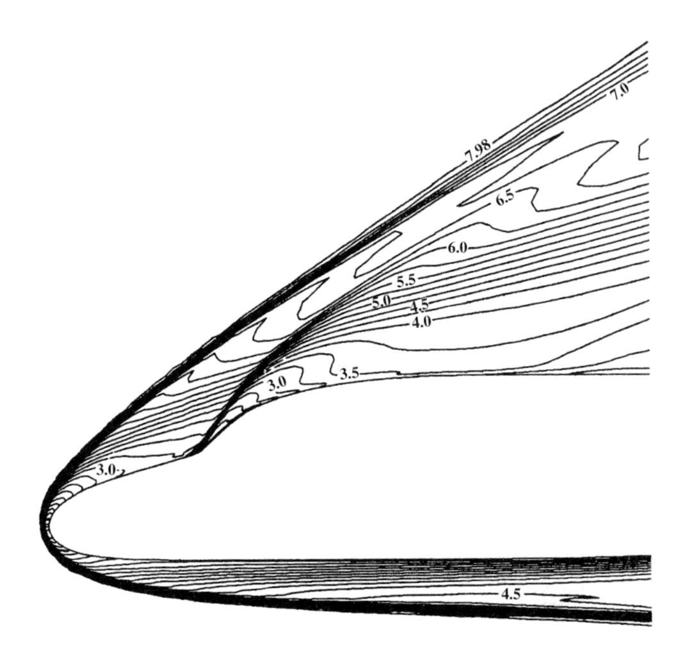

Fig. 5.55 Computed inviscid flow Mach-number distribution in the symmetry plane around the HERMES space shuttle, where $M_\infty = 8$ and angle of attack = 20 deg [234].

Fig. 5.56 Computed inviscid flow Mach-number distributions in the cross section for $x = 11.2$ m for the HERMES space shuttle, where $M_\infty = 8$ and angle of attack = 20 deg [234].

the HERMES at Mach 8 and a 20-deg angle of attack are given in Figs. 5.55 and 5.56. The Mach-number distribution in the symmetric plane is seen in Fig. 5.55, and that in the cross-section plane located at 11.2 m downstream of the nose is given in Fig. 5.56. The shock-wave pattern is nicely captured in these results, including the crossflow shocks above the wing and the upper fuselage as seen in Fig. 5.56.

Today, CFD solutions of the completely viscous flowfield over complex configurations are available (see Chapter 8). Such Navier–Stokes solutions, however, are computer intensive and are used for design purposes only on selective and specialized instances. We will examine some of these instances in Chapter 8. In contrast, CFD solutions of an inviscid flowfield are simpler and quicker to run on the computer and therefore are more suited for design applications. We have seen several instances in the present chapter.

Design Example 5.2: Hypersonic Waveriders—Part 1

The maximum lift-to-drag ratio $(L/D)_{max}$ for a flight vehicle is a measure of its aerodynamic efficiency. Unfortunately, for supersonic and hypersonic flight vehicles, as the freestream Mach number increases, $(L/D)_{max}$ decreases rather dramatically. This is just a fact of nature, brought about by the rapidly increasing shock-wave strength as Mach number increases (see Chapter 2), with consequent

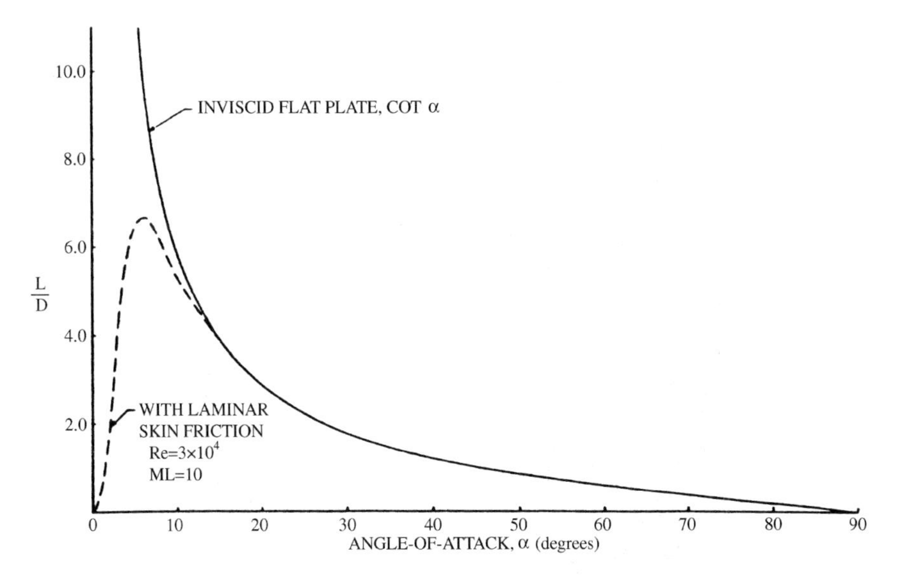

Fig. 5.57 Newtonian results for a flat plate.

large increases in wave drag. Return to Fig. 3.7, and note the variation of L/D with angle of attack for a hypersonic flat plate as calculated from Newtonian theory. The results in Fig. 3.7 assume an inviscid flow, and as a consequence L/D theoretically approaches infinity as the angle of attack α approaches zero. The Newtonian variation of L/D for a flat plate is repeated in Fig. 5.57 and is labeled as the inviscid flat-plate result given by cot α. In reality, the viscous shear stress acting on the plate surface causes L/D to peak at a low value of α and to go to zero as $\alpha \to 0$. This is illustrated by the dashed curve in Fig. 5.57, which shows the variation of L/D modified by skin friction as predicted by a reference temperature method (to be discussed in Sec. 6.9). The skin-friction calculation is for laminar flow at Mach 10 and a Reynolds number of 3×10^6. Note that $(L/D)_{\max}$ for the flat plate is about 6.5. By comparison, $(L/D)_{\max}$ for a Boeing 707 at normal cruising conditions near Mach 1 is about 20. So the $(L/D)_{\max}$ for a hypersonic flat plate, as shown in Fig. 5.57, is a low value, reflecting the characteristically low lift-to-drag ratios generated by hypersonic vehicles. And the infinitely thin flat plate is the most efficient lifting surface aerodynamically compared to other hypersonic shapes with finite thickness. *Conclusion*: The L/D value of vehicles at hypersonic Mach numbers are low.

There is a class of hypersonic vehicle shapes, however, that generates higher value of L/D than other shapes—*waveriders*. A waverider is a supersonic or hypersonic vehicle that has an *attached* shock wave all along its leading edge, as sketched in Fig. 5.58a. Because of this, the vehicle appears to be riding on top of its shock wave, hence the term "waverider." This is in contrast to a more conventional hypersonic vehicle, where the shock wave is usually detached from the leading edge, as sketched in Fig. 5.58b. The aerodynamic advantage of the waverider in Fig. 5.58a is that the high pressure behind the shock wave under

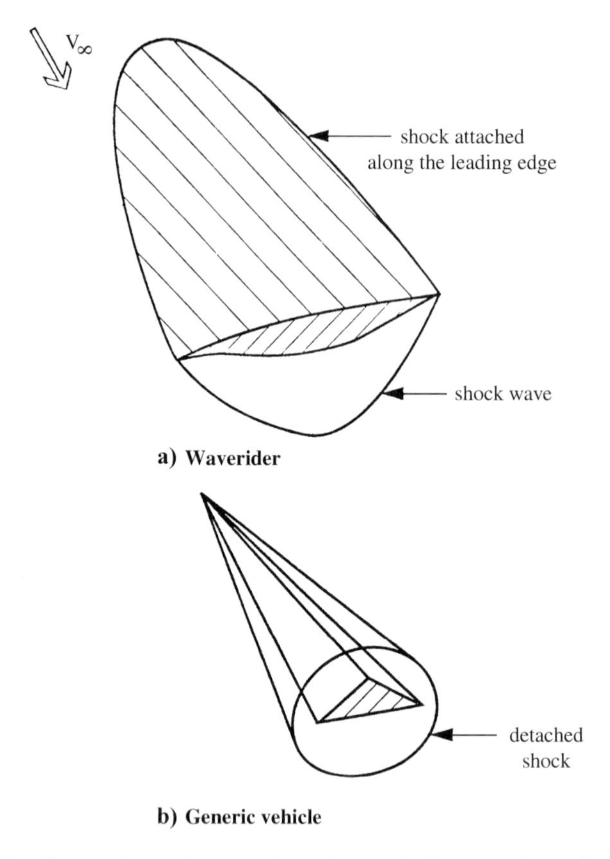

Fig. 5.58 Comparison of waverider and generic hypersonic configurations.

the vehicles does not "leak" around the leading edge to the top surface; the flowfield over the bottom surface is contained, and the high pressure is preserved, therefore generating more lift on the vehicle. In contrast, for the vehicle shown in Fig. 5.58b, there is communication between the flows over the bottom and top surfaces; the pressure tends to leak around the leading edge, and the general integrated pressure level on the bottom surface is reduced, resulting in less lift. Because of this, the generic vehicle in Fig. 5.58b must fly at a larger angle of attack α to produce the same lift as the waverider in Fig. 5.58a. This is illustrated in Fig. 5.59, where the lift curves (L vs α) are sketched for the two vehicles in Fig. 5.58. Note that the lift curve for the waverider is considerably higher because of the pressure containment as compared to that for the generic vehicle. At the same lift, points 1a and 1b in Fig. 5.59 represent the waverider and generic vehicles, respectively. Also shown in Fig. 5.59 are typical variations of L/D vs α, which for slender hypersonic vehicles are not too different for the shapes in Figs. 5.58a and 5.58b. (Although the lift of the waverider at a given angle of attack is increased by the pressure containment on the bottom surface,

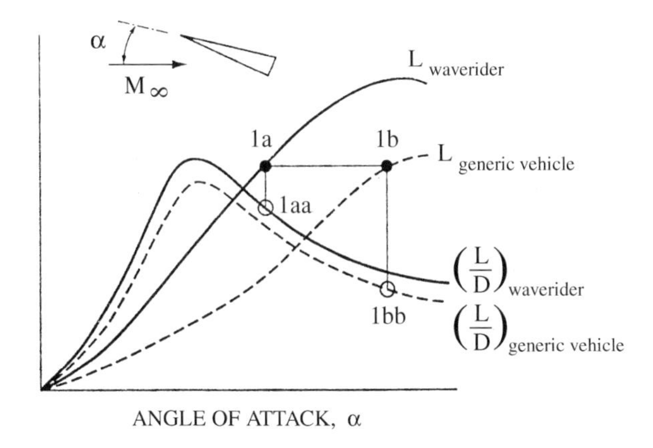

Fig. 5.59 Comparison of lift and L/D curves between a waverider and a generic vehicle.

so is the wave drag; hence, the L/D ratio *at a given angle of attack* for the waverider is better, but not greatly so, than that for the generic vehicle.) However, note that because the waverider generates the same lift *at a smaller* α (point 1a in Fig. 5.59) than does the generic vehicle, which must fly at a large α (point 1b in Fig. 5.59), the L/D for the waverider is considerably higher (point 1aa) than that for the generic shape (point 1bb). Therefore, for sustained hypersonic cruising flight in the atmosphere the waverider configuration has a definite advantage.

> *Question*: How do you design a vehicle shape such that the shock wave is attached all along its leading edge, that is, how do you design a waverider?

One answer is as follows. Consider the simple flowfield generated by a wedge in a supersonic or hypersonic freestream. Imagine that the top surface of the wedge is parallel to the freestream, and hence the only wave in the flow is the planar shock wave propagating below the wedge, as sketched at the top of Fig. 5.60. Now imagine two straight lines arbitrarily traced on the surface of the shock wave, coming to a point at the front of the shock. Consider all of the streamlines of the flow behind the shock that emanate from these arbitrarily traced lines. Taken together, these streamlines form a streamsurface that can be considered the surface of a vehicle with its leading edges defined by the two arbitrarily traced lines on the shock wave. Because the flowfield behind a planar shock wave is uniform with parallel streamlines, these streamsurfaces are flat surfaces that trace out a vehicle shape with a caret cross section as shown in Fig. 5.60, named after the "caret" dictionary symbol of the same name. If you now mentally strip away the imaginary generating flowfield shown at the top of Fig. 5.60, you have left the caret-shaped vehicle shown at the bottom of Fig. 5.60. Concentrating on the vehicle shape at the bottom of Fig. 5.60, the planar surfaces on the bottom of the vehicle are streamsurfaces that exist behind a planar oblique shock wave—streamsurfaces that are generated by streamlines that begin on the shock surface itself. Hence, the shock wave is, by

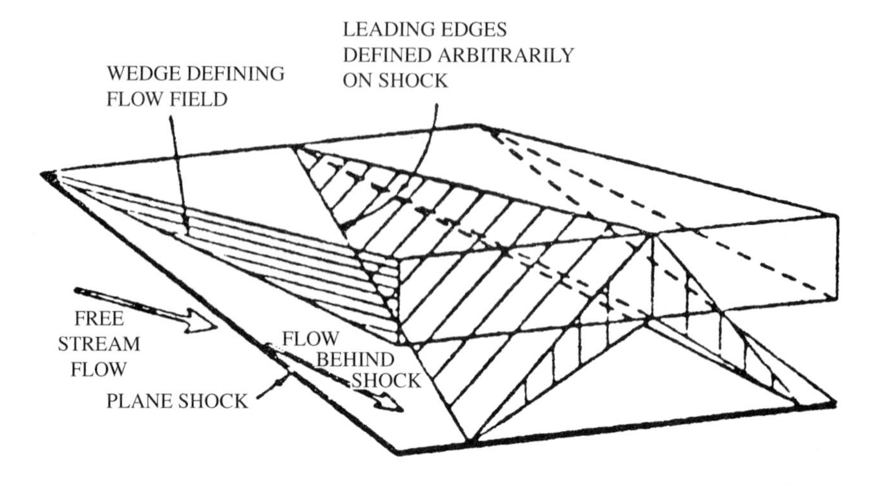

CONSTRUCTION FROM KNOWN FLOW FIELD

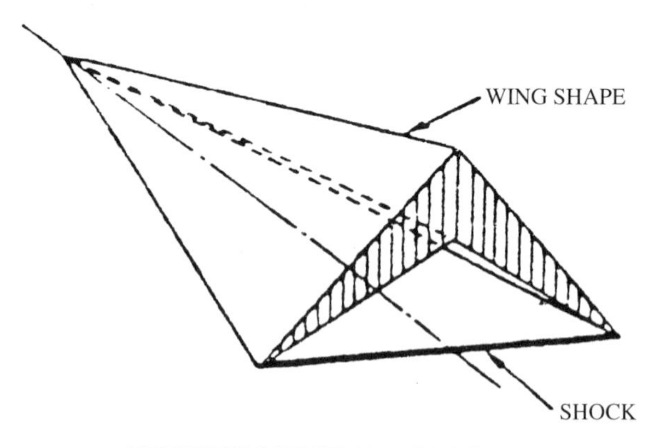

RESULTING WING AND SHOCK

Fig. 5.60 Nonweiler or "caret" wing.

definition, attached to the leading edge of the vehicle; this planar attached shock is shown stretching between the two straight leading edges of the vehicle sketched at the bottom of Fig. 5.60. By definition, therefore, this vehicle is a waverider. *Caution*: The waverider is in principle a point-designed vehicle. The generating oblique shock sketched at the top of Fig. 5.60 pertains to a given freestream Mach number M_∞ and a given flow deflection angle of the imaginary wedge that generates the oblique shock. Nevertheless, if you construct the vehicle shape shown at the bottom of Fig. 5.60 and put it in a freestream at the given M_∞ and at an angle of attach such that the flow deflection angle of the vehicle bottom surface is the same as that of the imaginary generating wedge, then nature will make certain that the shock wave is attached all along the vehicle's leading edge, that is, the vehicle will be a waverider. Note that in Fig. 5.60 we have oriented the imaginary generating wedge such that its top surface is parallel to the freestream; hence, there is no wave over the top surface of the wedge. Consequently, the top surfaces of the resulting caret waverider shown at the bottom of Fig. 5.60 are aligned with the freestream, and there is no wave above the waverider.

In principle any shape can be used for the imaginary body producing the flowfield from which a waverider shape is carved. The simplest case is to use a wedge for the imaginary body as just described. This has the advantage that a wedge produces a simple *known* flowfield that is easily calculated, as treated in Chapter 2. You do not need a CFD solution for this flow. The flow over a cone at zero angle of attack in a supersonic or hypersonic flow is similarly a known flowfield that can be used to generate waverider shapes. Because this conical flowfield is quasi-three-dimensional, it provides more flexibility in the generation of waverider shapes. The idea is the same. Consider the supersonic or hypersonic conical flowfield over a right-circular cone at zero angle of attack as sketched at the top of Fig. 5.61. The solution of this flowfield using the hypersonic small-disturbance equations was discussed in Sec. 4.6. The exact numerical solution of this flow was first obtained by Taylor and Maccoll [235]; it can be found

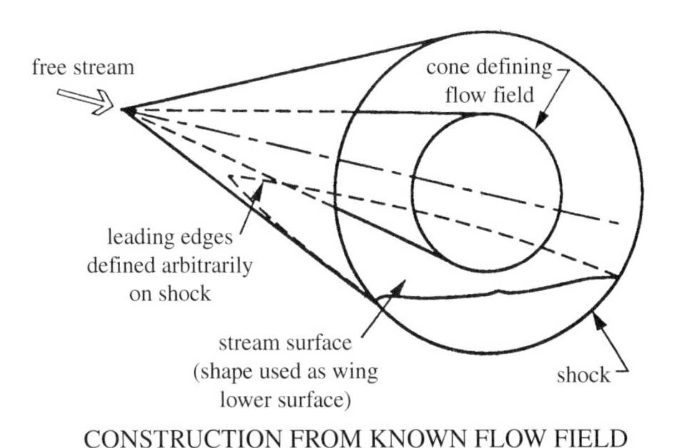

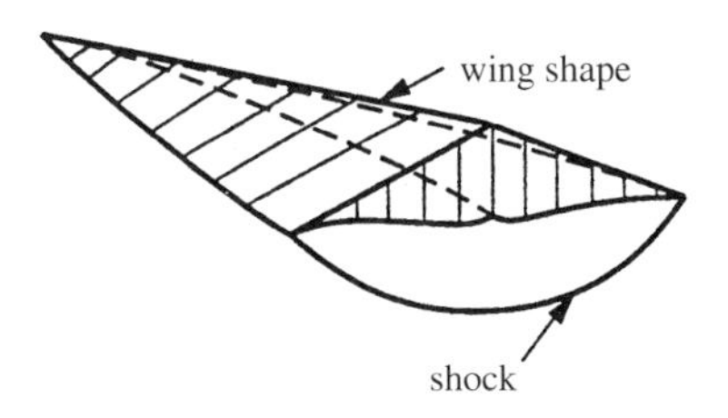

Fig. 5.61 Cone flow wing.

discussed at length in [4]. Tabulated results are given in [17] and [18]. In short, this is a known flowfield. At the top of Fig. 5.61, we see a conical shock wave attached at the vertex of the right-circular cone. This cone is simply the imaginary body generating the flowfield.

Consider the dashed curve drawn on the bottom surface of the conical shock wave as sketched at the top of Fig. 5.61. All of the streamlines flowing through this dashed curve constitute a streamsurface. In turn, this streamsurface defines the bottom surface of a waverider with a leading edge traced out by the dashed curve, as sketched at the bottom of Fig. 5.61. Any curve can be traced on the conical shock; hence, any streamsurface of this conical flowfield downstream of the shock can be used as the surface of a waverider. In so doing, the shock wave will be attached all along the leading edge of the waverider, as shown in Fig. 5.61. Moreover, the attached shock wave on this resulting waverider will, of course, be a segment of the conical shock wave shown at the top of Fig. 5.61.

The waverider concept was first introduced by Nonweiler [236] in 1959, who generated caret-shaped waveriders from the two-dimensional flowfield behind a planar oblique shock wave generated by a wedge, as described earlier. Nonweiler was interested in such waveriders as lifting atmospheric entry bodies. The first extension of Nonweiler's concept to the use of a conical flow as a generating flowfield was by Jones [237] in 1963, and further extensions to other axisymmetic generating flows are discussed by Jones et al. in [238]. An excellent and authoritative survey of waverider research up to 1979 is given by Townend in [239]. In the early 1980s Rasmussen and his colleagues at the University of Oklahoma (for example, see [240–242]) utilized hypersonic small-disturbance theory to design waveriders from flowfields over right-circular cones as well as elliptic cones. (Some of Rasmussen's small-disturbance analyses of conical flows are discussed in Sec. 4.6.) Consistent with his use of analytical solutions of the waverider flows, Rasmussen was also able to use the classic calculus of variations to optimize the waverider shapes utilizing the inviscid properties of the flow.

In the work just described, the waverider configurations were designed (and sometimes optimized) on the basis of inviscid flowfields, not including the effect of skin-friction drag. In turn, the drag predicted by such inviscid analyses was simply wave drag, and the resulting values of the inviscid L/D looked promising. However, waveriders tend to have large wetted surface areas, and the skin-friction drag, always added to the waverider aerodynamics after the fact, tended to greatly decrease the predicted inviscid lift-to-drag ratio. This made the waverider a less interesting prospect and led to a periodic lack of interest, indeed outright skepticism by researchers and vehicle designers in the waverider as a viable hypersonic configuration. Beginning in 1987, the author and his students at the University of Maryland took a different tact. New families of waveriders were generated wherein the skin-friction drag was *included* within an optimization routine to calculate waveriders with maximum L/D. In this fashion, the tradeoffs between wave drag and friction drag were accounted for during the optimization process, and the resulting family of waveriders had a shape and wetted surface area so as to optimize L/D. This family of waveriders is called *viscous-optimized hypersonic waveriders*, and subsequent CFD calculations and wind-tunnel tests have proven their viability, thus greatly enhancing modern interest in the waverider concept. But this is a story to be continued in Chapter 6 in Design Example 6.2: Hypersonic Waveriders—Part 2.

Problems

5.1 Starting with Eqs. (5.13–5.16) and using the transformation of both the independent and dependent variables as given in Sec. 5.3, derive Eqs. (5.18–5.21).

5.2 (a) Consider the bow shock wave over a cylinder-wedge in air, where the wedge half-angle is 20 deg. Draw this body on a piece of graph paper. Using the shock-wave shape correlations given in Sec. 5.4, plot on the *same graph* the shock shapes on the cylinder-wedge for $M_\infty = 2, 4, 6, 10, 15, 20$, and 25. Comment on these results as an illustration of the Mach-number independence principle.

(b) On another piece of graph paper, repeat part a; except for a 20-deg sphere-cone.

(c) Comment on the Mach-number range at which Mach-number independence for the shock-wave shape is reasonably obtained for the two dimensional shape in part a as compared to the axisymmetric shape in part b.

Part 2
Viscous Hypersonic Flow

In Part 2, we emphasize the effects of viscosity and thermal conduction in combination with high Mach numbers, and we will label such flows as hypersonic viscous flow. The effects of high temperature and diffusion will be covered in Part 3. In dealing with inviscid hypersonic flow in Part 1, we examined the question: What happens to the fluid dynamics of an inviscid flow when the Mach number is made very large? In Part 2 we take the next logical step and address the question: What happens in a high-Mach-number flow when the transport phenomena of viscosity and thermal conduction are included? The answer to this question leads to many practical results regarding the prediction of skin friction and aerodynamic heating in hypersonic flow.

6
Viscous Flow: Basic Aspects, Boundary Layer Results, and Aerodynamic Heating

Two major problems encountered today in aeronautics are the determination of skin friction and skin temperatures of high-speed aircraft.

E. R. Van Driest, 1950

Chapter Preview

With this chapter we leave the ideal world of inviscid flow and enter the real world of viscous flow. What physical phenomena determine the skin-friction drag and aerodynamic heating of hypersonic vehicles? We will find out here. How can we calculate the skin-friction drag and aerodynamic heating of hypersonic vehicles? We will find out here. If you are a practical person looking for practical methods to deal with skin-friction drag and aerodynamic heating, then you will eventually feel at home in this chapter. To get to that point, however, we have to wade through some rather interesting theoretical considerations. We have to start with the governing continuity, momentum, and energy equations for a viscous flow—the Navier–Stokes equations—and then specialize these equations for the case of flow in a boundary layer—the boundary-layer equations. Both of these systems of equations are coupled nonlinear partial differential equations, the difference being that the Navier–Stokes equations have no known analytical solutions, whereas the boundary-layer equations, being simpler, lend themselves to a partially analytical approach that provides practical results for skin-friction drag and heat transfer. If you are a person who enjoys analysis and working with equations, you will immediately feel at home with the first part of this chapter.

In short, this chapter is where the rubber hits the road, where some of the most important aspects of hypersonic aerodynamics come into play. In this chapter you are entering a different world, a world where friction and thermal conduction reign. This is not an easy world, but if you treat it with seriousness and respect it will reward you with brand new vistas in hypersonic aerodynamics.

6.1 Introduction

As noted in the preceding quotation by the well-known American aerodynamicist, E. R. Van Driest, the practical impact of viscous flow on hypersonic vehicles was recognized as early as 1950. The matter of aerodynamic heating (hence skin temperature) and shear stress (hence skin-friction drag) is an extremely important aspect of hypersonic vehicle design. This has never been more true than in the modern hypersonic applications of today. For example, consider the concept of an aerospace plane, designed to take off horizontally from an existing runway, and then literally blast its way into orbit mainly on the strength of airbreathing propulsion. A sketch of such a concept is shown in Fig. 1.8. It will be necessary for such a vehicle to acquire enough kinetic energy *within the sensible atmosphere* to "coast" into low Earth orbit. At such speeds (approximately Mach 25) within the atmosphere, aerodynamic heating will be extremely severe. For example, Tauber and Menees [79] have made engineering estimates of the aerodynamic heating to an aerospace plane for both ascent and reentry and compared these results with the space shuttle reentry. These results are summarized in the bar chart shown in Fig. 6.1, which gives both the maximum heat-transfer rate (in W/cm^2) and the total heat transfer (in kJ/cm^2) at the stagnation point. Here we see the striking result that the aerospace plane reentry stagnation-point heating is three times larger than the reentry heating of the space shuttle, and even more striking the *ascent* heating of the aerospace plane is an *order of magnitude* larger than reentry heating of the space shuttle. Hence, because of the requirement of the aerospace plane to achieve essentially orbital velocity within the atmosphere, the aerodynamic heating during ascent dominates its design. Another example, this time emphasizing the role of skin-friction drag, is given in Fig. 6.2. Here, a hypersonic waverider designed to optimize the lift/drag ratio is shown, as obtained from [80]. Such waveriders are promising hypersonic cruise vehicle configurations, wherein a high value of lift/drag is necessary for efficient, long-range cruising conditions. (Hypersonic waveriders are discussed in Design Examples at the end of Chapters 5, 6, 7, and 15.) The hypersonic transport shown in Fig. 1.11 is another example of a hypersonic vehicle designed for relatively high lift/drag. For these types of vehicles, skin-friction drag at hypersonic speeds is a dominant concern because, unlike a blunt body (where the drag is mostly wave drag due to the high pressures behind the strong bow shock wave), the slender configurations shown in Figs. 1.11 and 6.2 experience considerable skin-friction drag. In [80], it was observed that the magnitudes of wave drag and skin-friction drag for the optimized hypersonic waverider were approximately the same, never differing by more than a factor of two. The important

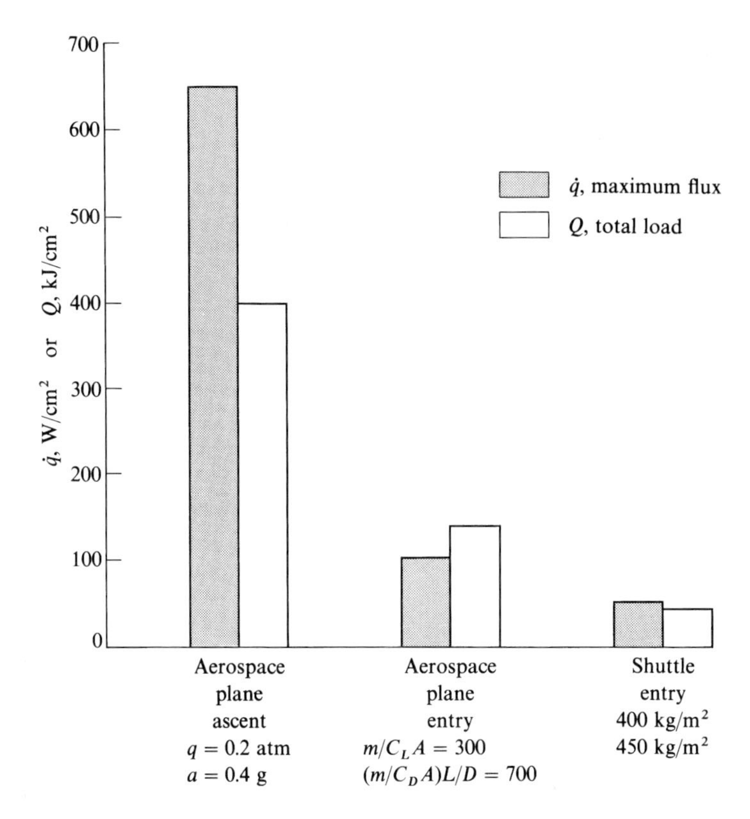

Fig. 6.1 Comparison between ascent and reentry stagnation-point aerodynamic heating for an aerospace plane and the reentry stagnation-point heating of the space shuttle; calculations by Tauber and Menees [79].

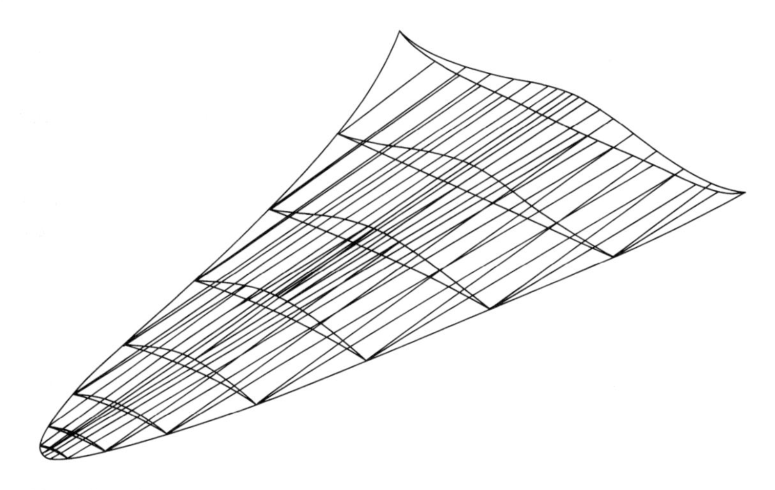

Fig. 6.2 Viscous-optimized hypersonic wave rider, by Bowcutt et al. [80].

point here is that skin-friction drag has a major impact on the design of slender hypersonic vehicles.

In light of the preceding, we repeat that aerodynamic heating and skin friction are very important aspects of practical hypersonic aerodynamics. In turn, the understanding and accurate prediction of these aspects is a vital part of the study of hypersonic viscous flows. In Part 2, and especially in the present chapter, we will emphasize these aspects. The introductory discussion in the preceding paragraph is given simply to motivate our subsequent discussions. As we progress in our study of hypersonic viscous flow, always keep in mind the preceding practical reasons for our interest.

Let us continue to examine the importance of hypersonic viscous flow, but from a slightly different point of view emphasizing a more purely fluid-dynamic aspect. Consider Fig. 6.3, which is a velocity-altitude map showing several lifting reentry trajectories from orbit, each with different values of the lift parameter m/C_LS (see Sec. 1.5). The shaded portion corresponds to the reentry of the space shuttle. Superimposed on this velocity-altitude map are lines of constant Reynolds number per meter, obtained from [81]. Note that the higher-altitude portions of the flight trajectories experience combined conditions of high Mach number and low Reynolds number—conditions that accentuate the effects of hypersonic viscous flows. Indeed, for most of the reentry trajectory a hypersonic vehicle is going to experience important Reynolds-number effects. Also note that a purely arbitrary transition Reynolds number of 10^6 is assumed, so that regions of purely laminar flow and of turbulent flow for a 10-m-long vehicle are identified on the right of Fig. 6.3. The main thrust of Fig. 6.3 is to indicate that viscous effects are important in hypersonic flight; such viscous effects are the subject of Part 2. Again, emphasis is made that only the purely viscous effects of viscosity and thermal conduction are highlighted in Part 2; the effects of high

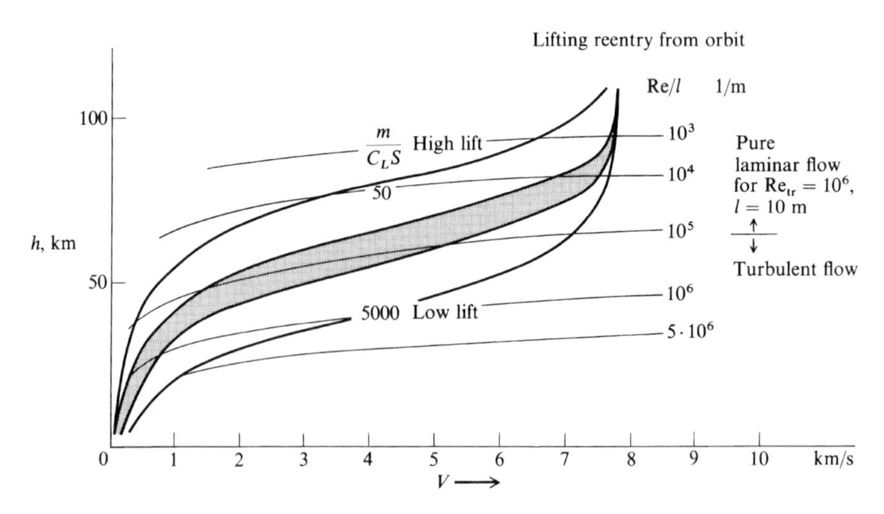

Fig. 6.3 Velocity-altitude map, with superimposed lines of constant unit Reynolds number (from Koppenwallner [81]).

temperatures and diffusion, which so frequently accompany hypersonic viscous flow, are treated in Part 3.

In the present chapter, some basic aspects of viscous flows will be discussed, including the full, governing equations (the Navier–Stokes equations), the boundary-layer equations and how they are affected by hypersonic conditions, and important results from the boundary-layer equations. Throughout Part 2 of this book, the assumption is made that the reader has been previously introduced to some elementary concepts of viscous flow, at least to the extent covered in chapter 15 of [5]. It is strongly recommended that the reader review this preliminary material before progressing further.

Return for a moment to the general road map in Fig. 1.24. We have completed our discussions on inviscid flows listed under the left-hand column in Fig. 1.24. We now move to the center column labeled viscous flows. This column represents the material in Part 2 of this book. Indeed, in the present chapter we will work our way through the first five items in the center column. A more detailed road map for this chapter is given in Fig. 6.4. The Navier–Stokes equations are the general governing equations for viscous flow, and we start with them. From these equations we extract the similarity parameters for compressible viscous flow—one of the most fundamental of considerations. We also examine the boundary conditions that necessarily go along with the

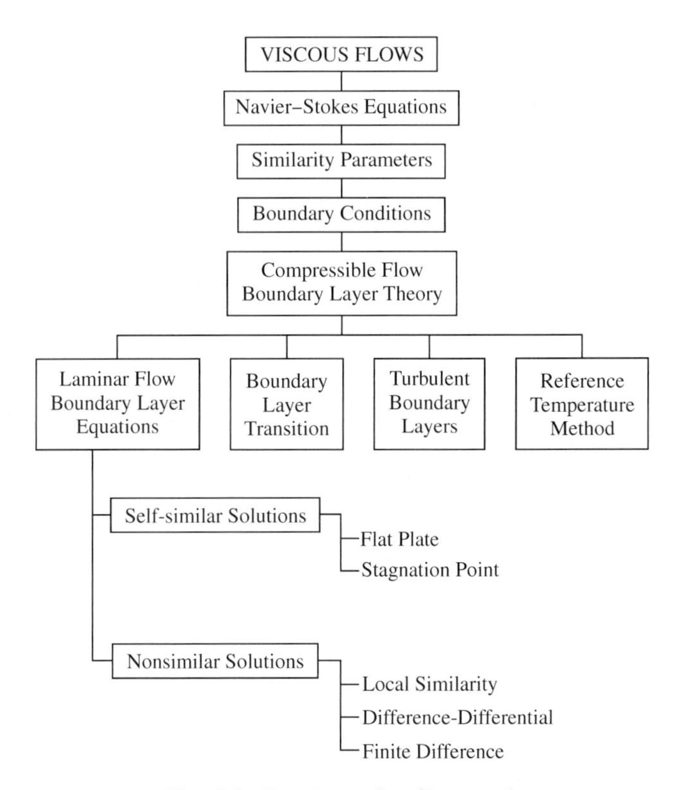

Fig. 6.4 Road map for Chapter 6.

Navier–Stokes equations. Then we introduce the concept of a boundary layer—one of the most important intellectual concepts in fluid dynamics. Conceived by Ludwig Prandtl in 1904, the boundary layer is the region adjacent to a surface where the effects of friction and thermal conduction are dominant. The flow external to the boundary layer, however, is not greatly affected by such effects and can be treated as inviscid. This division of a fluid flow into two distinct regions, the viscous boundary layer adjacent to the surface and the inviscid flow external to the boundary layer, is arguably the most important intellectual advancement in fluid dynamics since the beginning of the 20th century. (See the recent article on Prandtl and the boundary-layer concept in [243].) As shown in Fig. 6.4, our discussion on boundary layers then divides into four headings. The first is the derivation of the boundary-layer equations for a laminar compressible flow, followed by discussions of two classes of solutions of these equations, self-similar solutions (giving results for flow over a flat plate and around a stagnation point) and the more general nonsimilar solutions for general configurations. Then we move to the second heading and discuss the phenomena of transition from a laminar to a turbulent boundary layer, followed by the third heading, namely, turbulent boundary layers. Finally, the fourth heading, the reference temperature method, provides a simple engineering approach to calculating the surface skin-friction and aerodynamic heating for compressible boundary layers using classic results from incompressible flow. For all subject items in the road map in Fig. 6.4, the special aspects of hypersonic flow will be highlighted and discussed.

6.2 Governing Equations for Viscous Flow: Navier–Stokes Equations

In Sec. 4.2, we presented the governing equations for an inviscid flow, namely, the Euler equations (4.1–4.5). These equations are, in reality, a special form of the general governing equations of fluid dynamics wherein the viscous terms have been deleted. Another way of stating this is that the Euler equations are the limiting form of the general viscous flow equations in the limit of infinite Reynolds number. Indeed, it is frequently convenient to think of inviscid flow as a flow that results from the Reynolds number approaching infinity.

In the general equations of motion for a fluid flow, viscous effects do not influence the basic principle of mass conservation; hence, the continuity equation is the same as we presented in Sec. 4.2 [namely, Eq. (4.1)]. However, visualizing a moving fluid element, the shear and normal viscous stresses on the surface of the element result in stress terms that appear in both the momentum and energy equations. Moreover, thermal conduction across the surface of the element provides an additional mode of energy transfer, which appears in the energy equation. The resulting governing equations, called the *Navier–Stokes equations*, are derived (for example) in chapter 15 of [5]. Therefore, no details will be given here. These equations are given here.

Continuity equation:

$$\frac{\partial \rho}{\partial t} + \nabla \cdot (\rho \mathbf{V}) = 0 \tag{6.1}$$

x Momentum:

$$\rho \frac{\mathrm{D}u}{\mathrm{D}t} = -\frac{\partial p}{\partial x} + \frac{\partial \tau_{xx}}{\partial x} + \frac{\partial \tau_{yx}}{\partial y} + \frac{\partial \tau_{zx}}{\partial z} \tag{6.2}$$

y Momentum:

$$\rho \frac{\mathrm{D}v}{\mathrm{D}t} = -\frac{\partial p}{\partial y} + \frac{\partial \tau_{xy}}{\partial x} + \frac{\partial \tau_{yy}}{\partial y} + \frac{\partial \tau_{zy}}{\partial z} \tag{6.3}$$

z Momentum:

$$\rho \frac{\mathrm{D}w}{\mathrm{D}t} = -\frac{\partial p}{\partial z} + \frac{\partial \tau_{xz}}{\partial x} + \frac{\partial \tau_{yz}}{\partial y} + \frac{\partial \tau_{zz}}{\partial z} \tag{6.4}$$

Energy:

$$\begin{aligned}\rho \frac{\mathrm{D}(e + V^2/2)}{\mathrm{D}t} &= \rho\dot{q} + \frac{\partial}{\partial x}\left(k\frac{\partial T}{\partial x}\right) + \frac{\partial}{\partial y}\left(k\frac{\partial T}{\partial y}\right) + \frac{\partial}{\partial z}\left(k\frac{\partial T}{\partial z}\right) - \nabla \cdot p\mathbf{V} \\ &+ \frac{\partial(u\tau_{xx})}{\partial x} + \frac{\partial(u\tau_{yx})}{\partial y} + \frac{\partial(u\tau_{zx})}{\partial z} + \frac{\partial(v\tau_{xy})}{\partial x} + \frac{\partial(v\tau_{yy})}{\partial y} \\ &+ \frac{\partial(v\tau_{zy})}{\partial z} + \frac{\partial(w\tau_{xz})}{\partial x} + \frac{\partial(w\tau_{yz})}{\partial y} + \frac{\partial(w\tau_{zz})}{\partial z}\end{aligned} \tag{6.5}$$

where

$$\tau_{xy} = \tau_{yx} = \mu\left(\frac{\partial v}{\partial x} + \frac{\partial u}{\partial y}\right) \tag{6.6a}$$

$$\tau_{yz} = \tau_{zy} = \mu\left(\frac{\partial w}{\partial y} + \frac{\partial v}{\partial z}\right) \tag{6.6b}$$

$$\tau_{zx} = \tau_{xz} = \mu\left(\frac{\partial u}{\partial z} + \frac{\partial w}{\partial x}\right) \tag{6.6c}$$

$$\tau_{xx} = \lambda(\nabla \cdot \mathbf{V}) + 2\mu\frac{\partial u}{\partial x} \tag{6.6d}$$

$$\tau_{yy} = \lambda(\nabla \cdot \mathbf{V}) + 2\mu\frac{\partial v}{\partial y} \tag{6.6e}$$

$$\tau_{zz} = \lambda(\nabla \cdot \mathbf{V}) + 2\mu\frac{\partial w}{\partial z} \tag{6.6f}$$

The preceding equations are written for an unsteady, compressible, viscous, three-dimensional flow in Cartesian coordinates. In addition to the familiar symbols from Chapter 4, we now have the shear stresses τ_{xy}, τ_{yz}, etc., and the normal viscous stresses τ_{xx}, τ_{yy}, and τ_{zz}, which are related to velocity gradients in the flow via Eqs. (6.6a–6.6f). Also, μ is the viscosity coefficient, k is the thermal conductivity, and λ is the bulk viscosity coefficient (where the usual Stokes hypothesis is $\lambda = -\frac{2}{3}\mu$). In the energy equation (6.5) e is the internal energy per unit mass, $\dot{q}$ represents the volumetric heating that might occur, say, by the absorption or emission of radiation by the gas, and the temperature gradient terms $(\partial/\partial x)[k(\partial T/\partial x)]$, etc. represent energy transfer across a surface caused by the thermal conduction. More details concerning the physical significance of all of these terms can be found in [5].

A comment on nomenclature is made here. Historically, the term "Navier–Stokes equations" identified only the momentum equations (6.2–6.4) because these were the very equations derived by the Frenchman Claude Louis M. H. Navier in 1827 and independently by the Englishman George Stokes in 1845. However, in recent times, particularly with the advent of computational fluid dynamics, most citations in the literature referring to "solutions of the Navier–Stokes equations" denote solutions of the *complete system of equations*, namely, Eqs. (6.1–6.5). We will follow this modern trend here and will label the complete system of equations for viscous flow, Eqs. (6.1–6.5), as the Navier–Stokes equations.

Just as in the case of the Euler equations (4.1–4.5), there is no general analytic solution to the complete Navier–Stokes equations. However, in analogy with the approximate solutions of the Euler equations given in Chapter 4, we can simplify the Navier–Stokes equations via an appropriate set of assumptions and obtain approximate viscous flow results. Such a simplification involves the boundary-layer equations, to be discussed in Sec. 6.4. Also, in analogy with the "exact" solutions of the Euler equations given in Chapter 5, there are numerical solutions of the exact Navier–Stokes equations, to be discussed in Chapter 8.

6.3 Similarity Parameters and Boundary Conditions

As a precursor to the boundary-layer equations, to be discussed in the next section, and as a means to highlight the important similarity parameters for a viscous flow, it is useful to have a nondimensional form of the Navier–Stokes equations. To reduce the number of operations and terms, without loss of instructional value, we will consider a two-dimensional steady flow, and we will ignore the normal stresses τ_{xx} and τ_{yy}. Let us introduce the following dimensionless variables:

$$\bar{\rho} = \frac{\rho}{\rho_\infty} \qquad \bar{u} = \frac{u}{V_\infty} \qquad \bar{v} = \frac{v}{V_\infty} \qquad \bar{p} = \frac{p}{p_\infty} \qquad \bar{e} = \frac{e}{c_v T_\infty}$$

$$\bar{\mu} = \frac{\mu}{\mu_\infty} \qquad \bar{x} = \frac{x}{c} \qquad \bar{y} = \frac{y}{c} \qquad \bar{k} = \frac{k}{k_\infty} \qquad \frac{T}{T_\infty} = \bar{T}$$

where ρ_∞, V_∞, p_∞, μ_∞, k_∞, and T_∞ are reference values (for example, say free-stream values) and c is a reference length (say, the chord of an airfoil). In terms of these dimensionless variables, Eqs. (6.1–6.5) become (for two-dimensional, steady flow, and neglecting volumetric heating)

$$\frac{\partial(\bar{\rho}\bar{u})}{\partial\bar{x}} + \frac{\partial(\bar{\rho}\bar{v})}{\partial\bar{y}} = 0 \tag{6.7}$$

$$\bar{\rho}\bar{u}\frac{\partial\bar{u}}{\partial\bar{x}} + \bar{\rho}\bar{v}\frac{\partial\bar{u}}{\partial\bar{y}} = -\frac{1}{\gamma M_\infty^2}\frac{\partial\bar{p}}{\partial\bar{x}} + \frac{1}{Re_\infty}\frac{\partial}{\partial\bar{y}}\left[\bar{\mu}\left(\frac{\partial\bar{v}}{\partial\bar{x}} + \frac{\partial\bar{u}}{\partial\bar{y}}\right)\right] \tag{6.8}$$

$$\bar{\rho}\bar{u}\frac{\partial\bar{v}}{\partial\bar{x}} + \bar{\rho}\bar{v}\frac{\partial\bar{v}}{\partial\bar{y}} = -\frac{1}{\gamma M_\infty^2}\frac{\partial\bar{p}}{\partial\bar{y}} + \frac{1}{Re_\infty}\frac{\partial}{\partial\bar{x}}\left[\bar{\mu}\left(\frac{\partial\bar{v}}{\partial\bar{x}} + \frac{\partial\bar{u}}{\partial\bar{y}}\right)\right] \tag{6.9}$$

$$\begin{aligned}\bar{\rho}\bar{u}\frac{\partial\bar{e}}{\partial\bar{x}} + \bar{\rho}\bar{v}\frac{\partial\bar{e}}{\partial\bar{y}} = {} & -\frac{\gamma(\gamma-1)}{2}M_\infty^2\left[\bar{\rho}\bar{u}\frac{\partial}{\partial\bar{x}}(\bar{u}^2+\bar{v}^2) + \bar{\rho}\bar{v}\frac{\partial}{\partial\bar{y}}(\bar{u}^2+\bar{v}^2)\right] \\ & + \frac{\gamma}{\mathrm{Pr}_\infty Re_\infty}\left[\frac{\partial}{\partial\bar{x}}\left(\bar{k}\frac{\partial\bar{T}}{\partial\bar{x}}\right) + \frac{\partial}{\partial\bar{y}}\left(\bar{k}\frac{\partial\bar{T}}{\partial\bar{y}}\right)\right] - (\gamma-1)\left[\frac{\partial(\bar{u}\bar{p})}{\partial\bar{x}} + \frac{\partial(\bar{v}\bar{p})}{\partial\bar{y}}\right] \\ & + \gamma(\gamma-1)\frac{M_\infty^2}{Re_\infty}\left\{\frac{\partial}{\partial\bar{x}}\left[\bar{\mu}\bar{v}\left(\frac{\partial\bar{v}}{\partial\bar{x}} + \frac{\partial\bar{u}}{\partial\bar{y}}\right)\right] + \frac{\partial}{\partial\bar{y}}\left[\bar{\mu}\bar{u}\left(\frac{\partial\bar{v}}{\partial\bar{x}} + \frac{\partial\bar{u}}{\partial\bar{y}}\right)\right]\right\}\end{aligned} \tag{6.10}$$

The derivation of Eqs. (6.7–6.10) is left as homework Problem 6.1. Note that several parameters have emerged in Eqs. (6.8–6.10).

Ratio of specific heats:

$$\gamma = \frac{c_p}{c_v}$$

Mach number:

$$M_\infty = \frac{V_\infty}{a_\infty}$$

Reynolds number:

$$Re = \frac{\rho_\infty V_\infty c}{\mu_\infty}$$

Prandtl number:

$$Pr = \frac{\mu c_p}{k}$$

These four dimensionless parameters are called *similarity parameters* and are very important in determining the nature of a given viscous-flow problem. Indeed, a formal method for identifying the similarity parameters in any mechanical system is to nondimensionalize the governing equations; the dimensionless constants that appear in front of the derivative terms are the governing similarity parameters. The significance of flow similarity, and the meaning of the similarity parameters, is discussed in detail in chapter 15 of [5]. We will make only a few brief comments here, in the way of a remainder. First of all, from our experience with inviscid flows in Part 1 it is no surprise that γ and M_∞ carry over as similarity parameters for viscous flows. Thermodynamic properties, as reflected through γ, are important for any high-speed flow problem. A combination of thermodynamics and flow kinetic energy can be found in M_∞; indeed, it can readily be shown (for example, see [4]) that

$$M_\infty^2 \propto \frac{\text{flow kinetic energy}}{\text{flow internal energy}}$$

For the Reynolds number, we have (for example, see [82])

$$Re \propto \frac{\text{inertia force}}{\text{viscous force}}$$

The Prandtl number, introduced via the energy equation, is an index that is proportional to the ratio of energy dissipated by friction to the energy transported by thermal conduction, that is,

$$Pr \propto \frac{\text{frictional dissipation}}{\text{thermal conduction}}$$

In the study of compressible, viscous flow, Pr is just as important as γ, Re, or M. For air at standard conditions, $Pr = 0.71$. Note that Pr is a property of the gas; for different gases Pr is different. Also, like μ, k, and c_p, Pr for a nonreacting gas is a function of temperature only. (For a chemically reacting gas, Pr is also dependent on the local chemical composition, which in turn depends on the local temperature and pressure for an equilibrium flow and on the history of the upstream conditions for a nonequilibrium flow; these ideas will be introduced in Part 3.)

An important difference between inviscid and viscous flows not seen explicitly in the Navier–Stokes equations is the wall boundary conditions. In Part 1, we utilized the flow tangency condition at the wall—the usual boundary condition for an inviscid flow (with no mass transfer through the wall). This boundary condition changes drastically for a viscous flow. Because of the existence of friction, the flow can no longer "slip along the wall" at a finite value. Rather, for a continuum viscous flow we have the *no-slip* boundary condition at the wall, namely, the velocity is *zero* at the wall.

Wall boundary condition:

$$u = v = 0 \tag{6.11}$$

If there is mass transfer at the wall (as a result of ablation or transpiration cooling, for example), then Eq. (6.11) is modified as follows.

Wall boundary condition with mass transfer:

$$u = 0$$

$$v = v_w$$

where v_w is the specified velocity normal to the surface. In addition, because of energy transport by thermal conduction, we require an additional boundary condition at the wall involving internal energy (or more usually temperature). If the wall is at constant temperature, then the boundary condition is simple.

Constant wall-temperature boundary condition:

$$T = T_w \tag{6.12}$$

where T_w denotes the *specified* wall temperature. As is more usually the case, the wall will not be at constant temperature. If we know a priori the distribution of temperature along the surface, then Eq. (6.12) is slightly modified as follows.

Variable wall-temperature boundary condition:

$$T = T_w(s) \tag{6.13}$$

where $T_w(s)$ is the specified wall-temperature variation as a function of distance along the surface s. Unfortunately, in a high-speed flow problem, the wall temperature is usually one of the unknowns, and we cannot utilize either Eq. (6.12) or (6.13). Instead, the more general condition on temperature at the wall is given by Fourier's law of heat conduction.

Heat-transfer wall boundary condition:

$$q_w = -k\left(\frac{\partial T}{\partial n}\right)_w \tag{6.14}$$

where q_w is the heat transfer (energy per second per unit area) into or out of the wall, n is the coordinate normal to the wall, and $(\partial T/\partial n)_w$ is the normal temperature gradient existing in the gas immediately at the wall. In general, the wall heat transfer (and hence the wall-temperature gradient) are unknowns of the problem, and, therefore, in the most general case the wall boundary condition [Eq. (6.14)] must be matched to a separate heat-conduction analysis describing the heat distribution within the surface material itself, and both the flow problem and the surface material problem must be solved in a coupled fashion. A special case of Eq. (6.14) is the *adiabatic wall condition*, wherein by definition the heat transfer to the wall is *zero*. From Eq. (6.14), we have in this case

Adiabatic wall condition:

$$\left(\frac{\partial T}{\partial n}\right)_w = 0 \tag{6.15}$$

Note that here the boundary condition is not on the wall temperature itself, but rather on the *temperature gradient*, namely, a specified *zero* gradient at the wall. The resulting wall temperature (which comes out as part of the solution) is defined as the *adiabatic wall temperature* T_{aw}, sometimes called the "equilibrium" temperature.

Although the choice of an appropriate boundary condition for temperature (or temperature gradient) at the wall appears somewhat open ended from the preceding discussion, the majority of high-speed viscous-flow calculations assume one of the two extremes, that is, they either treat a uniform, constant-temperature wall [Eq. (6.12)] or an adiabatic wall [Eq. (6.15)]. However, for a detailed and accurate solution of many practical problems Eq. (6.14) must be employed along with a coupled solution of the heat-conduction problem in the surface material itself. Such detailed, coupled solutions are beyond the scope of this book.

The temperature boundary condition adds another similarity parameter to our viscous-flow analysis. In Eq. (6.10), the dimensionless internal energy is defined as $\bar{e} = e/c_v T_\infty$. The value of $\bar{e}$ at the wall is $\bar{e}_w$, which for a constant-temperature wall is a specified constant value. Moreover, for a calorically perfect gas $e = c_v T$. Hence, at the wall

$$\bar{e}_w = \frac{e_w}{c_v T_\infty} = \frac{c_v T_w}{c_v T_\infty} = \frac{T_w}{T_\infty}$$

Therefore, to achieve flow similarity in the solution of Eqs. (6.7–6.10), not only are γ, M_∞, Re_∞, and Pr_∞ similarity parameters, but the wall-to-freestream temperature ratio T_w/T_∞ is also a similarity parameter.

As a final note in this section, we observe that the nondimensional Navier–Stokes equations (6.7–6.10), in the limit of $Re \to \infty$, reduce to the nondimensional Euler equations, thus supporting our earlier statement that an inviscid flow can be thought of as a limiting case of a viscous flow when the Reynolds number becomes infinite.

6.4 Boundary-Layer Equations for Hypersonic Flow

Until the advent of computational fluid dynamics, exact solutions of the complete Navier–Stokes equations for practical problems were virtually nonexistent. Even today, numerical solutions of these equations (to be discussed later) are not easy and generally require a lot of computer power, as well as human resources to generate the computer solutions. Therefore, reasons exist for simpler viscous-flow solutions. By a suitable order-of-magnitude reduction of the Navier–Stokes equations, a simpler set of equations—*the boundary-layer equations*—can be obtained. The compressible boundary-layer equations are the same, whether the flow is subsonic, supersonic, or hypersonic (neglecting high-temperature effects). However, there is one aspect of the standard boundary-layer equations that becomes rather tenuous at hypersonic speeds and that is not always recognized. For this reason, the following excerpts from [5] on the derivation of the boundary-layer equations are given next. Please keep in mind that it is not the purpose of this book to "rehash" basic fluid mechanics, which is assumed to be part of the reader's background. However, in the present case regarding the derivation of the boundary-layer equations, the following review material is important to our subsequent comments on hypersonic boundary layers.

Considering two-dimensional, steady flow, the nondimensionalized form of the x-momentum equation (one of the Navier–Stokes equations) was given by Eq. (6.8):

$$\bar{\rho}\bar{u}\frac{\partial \bar{u}}{\partial \bar{x}} + \bar{\rho}\bar{v}\frac{\partial \bar{u}}{\partial \bar{y}} = -\frac{1}{\gamma M_\infty^2}\frac{\partial \bar{p}}{\partial \bar{x}} + \frac{1}{Re_\infty}\frac{\partial}{\partial \bar{y}}\left[\bar{\mu}\left(\frac{\partial \bar{v}}{\partial \bar{x}} + \frac{\partial \bar{u}}{\partial \bar{y}}\right)\right] \tag{6.8}$$

Let us now reduce Eq. (6.8) to an approximate form that holds reasonably well within a boundary layer.

Consider the boundary layer along a flat plate of length c. The basic assumption of boundary-layer theory is that a boundary layer is very thin in comparison with the scale of the body, that is,

$$\boxed{\delta \ll c} \tag{6.16}$$

where δ is the boundary-layer thickness. Consider the continuity equation for a steady, two-dimensional flow, which in terms of the nondimensional variables is given by Eq. (6.7):

$$\frac{\partial(\bar{\rho}\bar{u})}{\partial \bar{x}} + \frac{\partial(\bar{\rho}\bar{v})}{\partial \bar{y}} = 0 \tag{6.7}$$

Because $\bar{u}$ varies from 0 at the wall to 1 at the edge of the boundary layer, let us say that $\bar{u}$ is of the order of magnitude equal to 1, symbolized by 0(1). Similarly, $\bar{\rho} = 0(1)$. Also, because x varies from 0 to c, $\bar{x} = 0(1)$. However, because y varies from 0 to δ, where $\delta < c$, then $\bar{y}$ is of the smaller order of magnitude, denoted by $\bar{y} = 0(\delta/c)$. Without loss of generality, we can assume that c is a unit length. Therefore $\bar{y} = 0(\delta)$. Putting these orders of magnitude in Eq. (6.7), we have

$$\frac{[0(1)][0(1)]}{0(1)} + \frac{[0(1)][\bar{v}]}{0(\delta)} = 0 \tag{6.17}$$

Hence, from Eq. (6.17) clearly $\bar{v}$ must be of an order of magnitude equal to δ, that is, $v = 0(\delta)$. Now examine the order of magnitude of the terms in Eq. (6.8). We have

$$\bar{\rho}\bar{u}\frac{\partial \bar{u}}{\partial \bar{x}} = 0(1) \qquad \bar{\rho}\bar{v}\frac{\partial \bar{u}}{\partial \bar{y}} = 0(1) \qquad \frac{\partial \bar{p}}{\partial \bar{x}} = 0(1)$$

$$\frac{\partial}{\partial \bar{y}}\left(\bar{\mu}\frac{\partial \bar{v}}{\partial \bar{x}}\right) = 0(1) \qquad \frac{\partial}{\partial \bar{y}}\left(\bar{\mu}\frac{\partial \bar{u}}{\partial \bar{y}}\right) = 0\left(\frac{1}{\delta^2}\right)$$

Hence, the order-of-magnitude equation for Eq. (6.8) can be written as

$$0(1) + 0(1) = -\frac{1}{\gamma M_\infty^2}0(1) + \frac{1}{Re_\infty}\left[0(1) + 0\left(\frac{1}{\delta^2}\right)\right] \tag{6.18}$$

Let us now introduce another assumption of boundary-layer theory, namely, that the Reynolds number is large, indeed large enough such that

$$\boxed{\frac{1}{Re_\infty} = 0(\delta^2)} \tag{6.19}$$

Then, Eq. (6.18) becomes

$$0(1) + 0(1) = -\frac{1}{\gamma M_\infty^2} 0(1) + 0(\delta^2)\left[0(1) + 0\left(\frac{1}{\delta^2}\right)\right] \tag{6.20}$$

In Eq. (6.20), there is one term with an order of magnitude that is much smaller than the rest, namely, the product $0(\delta^2)[0(1)] = 0(\delta^2)$. This term corresponds to $(1/Re_\infty)\partial/\partial\bar{y}(\bar{\mu}\ \partial\bar{v}/\partial\bar{x})$ in Eq. (6.8). Hence, neglect this term in comparison to the remaining terms in Eq. (6.8). We obtain

$$\bar{\rho}\bar{u}\frac{\partial \bar{u}}{\partial \bar{x}} + \bar{\rho}\bar{v}\frac{\partial \bar{u}}{\partial \bar{y}} = -\frac{1}{\gamma M_\infty^2}\frac{\partial \bar{p}}{\partial \bar{x}} + \frac{1}{Re_\infty}\frac{\partial}{\partial \bar{y}}\left(\bar{\mu}\frac{\partial \bar{u}}{\partial \bar{y}}\right) \tag{6.21}$$

In terms of dimensional variables, Eq. (6.21) is

$$\rho u\frac{\partial u}{\partial x} + \rho v\frac{\partial u}{\partial y} = -\frac{\partial p}{\partial x} + \frac{\partial}{\partial y}\left(\mu\frac{\partial u}{\partial y}\right) \tag{6.22}$$

Equation (6.22) is the approximate x-momentum equation, which holds for flow in a thin boundary layer at high Reynolds number.

Consider the y-momentum equation for two-dimensional, steady flow, obtained in terms of the nondimensional variables as Eq. (6.9)

$$\bar{\rho}\bar{u}\frac{\partial \bar{v}}{\partial \bar{x}} + \bar{\rho}\bar{v}\frac{\partial \bar{v}}{\partial \bar{y}} = -\frac{1}{\gamma M_\infty^2}\frac{\partial \bar{p}}{\partial \bar{y}} + \frac{1}{Re_\infty}\frac{\partial}{\partial \bar{x}}\left[\bar{\mu}\left(\frac{\partial \bar{v}}{\partial \bar{x}} + \frac{\partial \bar{u}}{\partial \bar{y}}\right)\right] \tag{6.23}$$

The order-of-magnitude equation for Eq. (6.23) is

$$0(\delta) + 0(\delta) = -\frac{1}{\gamma M_\infty^2}\frac{\partial \bar{p}}{\partial \bar{y}} + 0(\delta^2)\left[0(\delta) + 0\left(\frac{1}{\delta}\right)\right] \tag{6.24}$$

From Eq. (6.24), we see that $\partial\bar{p}/\partial\bar{y} = 0(\delta)$ or smaller, assuming that $\gamma M_\infty^2 = 0(1)$. Because δ is very small, this implies that $\partial\bar{p}/\partial\bar{y}$ is very small. Therefore, from the y-momentum equation specialized to a boundary layer, we have

$$\frac{\partial p}{\partial y} = 0 \tag{6.25}$$

Equation (6.25) is important; it states that at a given x station the pressure is constant through the boundary layer in a direction normal to the surface. This implies that the pressure distribution at the outer edge of the boundary layer is impressed directly to the surface without change. Hence, throughout the boundary layer $p = p(x) = p_e(x)$, where $p_e(x)$ is the pressure distribution at the outer edge of the boundary layer (determined from inviscid-flow calculations).

This leads to a major point concerning hypersonic boundary layers and the reason for reviewing the preceding order-of-magnitude analysis of the boundary-layer equations. Consider again Eq. (6.24), but now in the case of *large hypersonic Mach numbers*. In Eq. (6.24), if M_∞^2 is very large, then $\partial\bar{p}/\partial\bar{y}$ does not have to be small. For example, if M_∞ were large enough such that $1/\gamma M_\infty^2 = 0(\delta)$, then $\partial\bar{p}/\partial\bar{y}$ could be as large as $0(1)$, and Eq. (6.24) would still be satisfied. Thus, *for very large hypersonic Mach numbers, the assumption that p is constant in the normal direction through a boundary layer is not always valid*. This aspect of hypersonic boundary layers is not frequently discussed or widely appreciated, and hence some emphasis is being made here. Also, it is important to properly interpret the preceding statement; it is a fluid-dynamic result which states that the normal pressure gradient through a hypersonic boundary layer need not be zero. However, this does not preclude the pressure gradient from being zero or nearly zero; it is simply saying that, within the conventional boundary-layer assumptions resulting in Eq. (6.24), $\partial\bar{p}/\partial\bar{y}$ *does not have to be zero*. Since the 1950s, a large number of hypersonic boundary-layer calculations have been made with the conventional boundary-layer assumption that $\partial\bar{p}/\partial\bar{y} = 0$, and in many applications this is justified. However, we should not expect this to always be the case. Indeed, the question concerning the possible existence of a finite normal pressure gradient adds more support to carrying out hypersonic viscous-flow calculations by going *beyond* the usual boundary-layer calculations and instead dealing with the entire shock layer as fully viscous from the body to the shock wave. In such viscous shock-layer analyses, the normal pressure gradient is calculated as part of the solution to the problem, thus circumventing the uncertainty as to whether $\partial\bar{p}/\partial\bar{n}$ is zero or finite. Such viscous shock-layer calculations will be discussed in Chapter 8.

As a corollary to the preceding discussion, return to the x-momentum equation (6.21) and its order-of-magnitude comparison given by Eq. (6.20). If M_∞ is large, then the pressure-gradient term $(1/\gamma M_\infty^2)(\partial\bar{p}/\partial\bar{x})$ can be small; in such a case, the hypersonic boundary layer will not be greatly influenced by the axial pressure gradient $\partial\bar{p}/\partial\bar{x}$. This is vaguely analogous to the inviscid-flow result discussed in Part 1, namely, that for hypersonic flow over slender bodies most of the flowfield changes take place in the y direction, and only small changes take place in the x direction.

Keeping the preceding considerations in mind, we will proceed in the present chapter with a discussion of the conventional boundary-layer equations and results based upon these equations. To round out our presentation of the boundary-layer equations, we must consider the general energy equation given in nondimensional form for two-dimensional, steady flow by Eq. (6.10). Inserting $e = h - p/\rho$ into this equation, subtracting the momentum equation multiplied by velocity, and performing an order-of-magnitude analysis similar to those in the preceding, we can obtain the boundary-layer equation as

$$\rho u \frac{\partial h}{\partial x} + \rho v \frac{\partial h}{\partial y} = \frac{\partial}{\partial y}\left(k \frac{\partial T}{\partial y}\right) + u \frac{\partial p}{\partial x} + \mu \left(\frac{\partial u}{\partial y}\right)^2 \tag{6.26}$$

The details are left to you.

In summary, by making the combined assumptions of $\delta < c$ and $Re \geq 1/\delta^2$, the complete Navier–Stokes equations given in Sec. 6.2 can be reduced to simpler forms that apply to a boundary layer. These boundary-layer equations are as follows.

$$\boxed{\text{Continuity:} \quad \frac{\partial(\rho u)}{\partial x} + \frac{\partial(\rho v)}{\partial y} = 0} \tag{6.27}$$

$$\boxed{x\ \text{Momentum:} \quad \rho u \frac{\partial u}{\partial x} + \rho v \frac{\partial u}{\partial y} = -\frac{\mathrm{d}p_e}{\mathrm{d}x} + \frac{\partial}{\partial y}\left(\mu \frac{\partial u}{\partial y}\right)} \tag{6.28}$$

$$\boxed{y\ \text{Momentum:} \quad \frac{\partial p}{\partial y} = 0} \tag{6.29}$$

$$\boxed{\text{Energy:} \quad \rho u \frac{\partial h}{\partial x} + \rho v \frac{\partial h}{\partial y} = \frac{\partial}{\partial y}\left(k \frac{\partial T}{\partial y}\right) + u \frac{\mathrm{d}p_e}{\mathrm{d}x} + \mu \left(\frac{\partial u}{\partial y}\right)^2} \tag{6.30}$$

Note that, as in the case of the Navier–Stokes equations, the boundary-layer equations are nonlinear. However, the boundary-layer equations are simpler and therefore are more readily solved. Also, because $p = p_e(x)$ the pressure gradient expressed as $\partial p/\partial x$ in Eq. (6.22) is reexpressed as $\mathrm{d}p_e/\mathrm{d}x$ in Eqs. (6.28) and (6.30). In the preceding equations, the unknowns are u, v, ρ, and h; p is known from $p = p_e(x)$, and μ and k are properties of the fluid that vary with temperature. To complete the system, we have for a calorically perfect gas

$$p = \rho R T \tag{6.31}$$

and

$$h = c_p T \tag{6.32}$$

Hence, Eqs. (6.27), (6.28), and (6.30–6.32) are five equations for the five unknowns u, v, ρ, T, and h.

The boundary conditions for the preceding equations are as follows.
At the wall:

$$y = 0, \quad u = 0, \quad v = 0, \quad T = T_w$$

or adiabatic wall:

$$\left(\frac{\partial T}{\partial n}\right)_w = 0$$

At the boundary-layer edge:

$$y \to \infty, \quad u \to u_e, \quad T \to T_e$$

Note that because the boundary-layer thickness is not known a priori, the boundary condition at the edge of the boundary layer is given at large y, essentially y approaching infinity.

The boundary-layer equations just given apply to compressible flow; they are equally applicable to subsonic and supersonic flows with no distinction made for such cases. They can be (and have been) applied to hypersonic flows. However, when using Eqs. (6.27–6.32) for hypersonic flows, keep in mind our earlier discussion concerning $\partial\bar{p}/\partial\bar{y}$. Also, if M_∞ is high enough, viscous dissipation within the boundary layer creates high temperatures, which in turn causes chemical reactions within the boundary layer. In such a case, the system of equations given by Eqs. (6.27–6.32) is not totally applicable; diffusion of chemical species and energy changes caused by chemical reactions must be included. The subject of hypersonic chemically reacting boundary layers will be treated in Part 3. Nevertheless, the application of Eqs. (6.27–6.32) to relatively moderate hypersonic conditions yields useful results. Also, many hypersonic wind-tunnel tests are conducted in "cold flows," flows where the total enthalpy is low enough to ignore high-temperature effects. Hence, there are many hypersonic applications where the governing boundary-layer equations for a calorically perfect gas in the form of Eqs. (6.27–6.32) are appropriate. Thus, we will pursue various aspects of these equations throughout the remainder of this chapter.

This ends our introductory discussion of the basic aspects of hypersonic viscous flow. It is instructive at this stage to return to the road map given in Fig. 1.24. We are now located on the second major branch of hypersonic flows, namely, hypersonic viscous flows. We have just finished the item labeled "basic aspects" and are now ready to move on to discussions of hypersonic boundary-layer theory. Under this category, our discussions will first cover some aspects of self-similar boundary layers and then examine some approaches for nonsimilar boundary layers. (What is meant by similar and nonsimilar boundary layers will be explained in the next section.) Also, return to the chapter road map in Fig. 6.4. We have progressed about halfway down the map and have just finished the left-hand box labeled "laminar flow boundary-layer equations." We are ready to move to the first subitem under this box, namely, self-similar solutions. The material we will discuss is somewhat classical in nature. It will be, for all practical purposes, a discussion of compressible boundary-layer theory not limited to just hypersonic flow. However, our interest here is in the application of the results of this classical theory to high-Mach-number problems.

6.5 Hypersonic Boundary-Layer Theory: Self-Similar Solutions

Although the title of this section involves the word "hypersonic," in reality we will be dealing with compressible boundary-layer theory, and the results will apply to both subsonic and supersonic, as well as hypersonic conditions. As a reminder, a major aspect that distinguishes hypersonic boundary-layer theory from the subsonic and supersonic cases is the intense viscous dissipation, resulting in high-temperature, chemically reacting flow. This aspect will be considered in Part 3. In contrast, in the present section as well as throughout Part 2, we are assuming a calorically perfect gas, that is, we are highlighting only the fluid-dynamic effect of viscosity and thermal conductivity in combination with high Mach numbers. In this regard, the present section is classical in its scope; the material discussed here has evolved since the 1940s, when interest in compressible boundary layers began to emerge under the impetus of high-speed, subsonic, and supersonic flight. Furthermore, we will assume some slight familiarity on the part of the reader, at least to the extent of the material covered in chapters 15 and 16 in [5]. For an excellent discussion of classical compressible boundary-layer theory, see the book by White [83].

The concept of self-similar boundary layers is illustrated in Fig. 6.5. In general, the variation of flow properties throughout a two-dimensional boundary layer is a function of both x and y. This is sketched in the physical plane shown at the left of Fig. 6.5, where two velocity profiles are shown at different x locations, x_1 and x_2, along the surface. In general, the profiles are different, that is, $u(x, y) \neq u(x_2, y)$. However, for certain cases under the appropriate independent variable

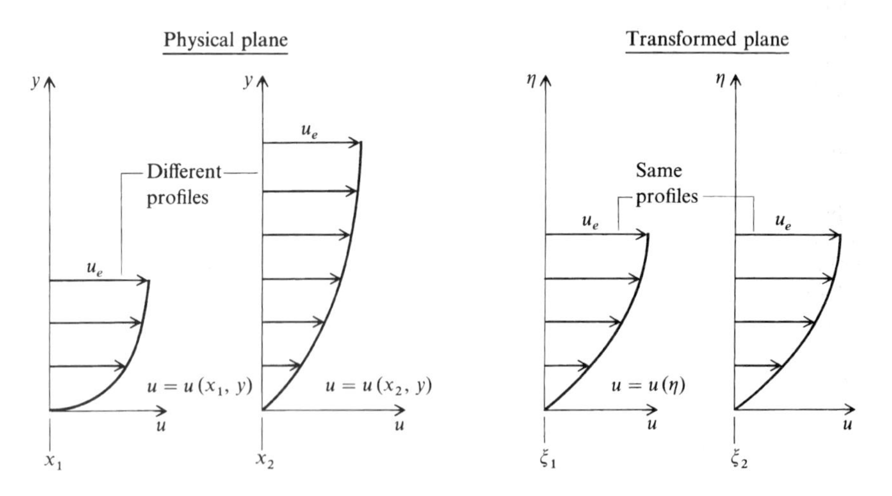

Fig. 6.5 Illustration of the concept of self-similarity.

transformation from (x, y) to (ξ, η), the flowfield profiles become independent of location along the surface. This is sketched at the right of Fig. 6.5, which shows a transformed plane wherein the velocity profile is independent of the transformed surface distance ξ; that is, the same velocity profiles exist at different values of ξ, say, ξ_1 and ξ_2. Thus, in the transformed plane the velocity profile is given by $u = u(\eta)$, independent of ξ. Boundary layers that exhibit this property are called *self-similar boundary layers*, and solutions for these boundary layers are called *self-similar solutions*—the subject of this section. Self-similar solutions to boundary layers have been investigated since the original incompressible flat-plate solution obtained by Blasius in 1908; the fact that the flow might be hypersonic does not preclude the occurrence of self-similar solutions, as we will see.

Let us now transform the boundary-layer equations (6.27–6.30) from physical (x, y) space to a transformed (ξ, η) space and examine the possibility of self-similar solutions. The appropriate transformation is based on work initiated in the 1940s by Illingworth [84], Stewartson [85], Howarth [86], and Dorodnitsyn [87], and put in a more useful form by Levy [88] and Lees [89]. The transformation is

$$\xi = \int_0^x \rho_e u_e \mu_e \, \mathrm{d}x \tag{6.33}$$

$$\eta = \frac{u_e}{\sqrt{2\xi}} \int_0^y \rho \, \mathrm{d}y \tag{6.34}$$

where ρ_e, u_e, and μ_e are the density, velocity, and viscosity coefficients, respectively, at the edge of the boundary layer. Because ρ_e, u_e, and μ_e are functions of x only, then $\xi = \xi(x)$. The transformation given by Eqs. (6.33) and (6.34) has been identified by various names in the literature, with some inconsistency caused by the number of researchers contributing to its development (for example, see the names associated with [84]–[89]). If for no other reason, it seems appropriate to recognize the chronological first [87] and last [89] of the references given, and hence the transformation given by Eqs. (6.33) and (6.34) will be called the *Lees–Dorodnitsyn* transformation in this book.

Let us now apply the preceding transformation to the boundary-layer equations (6.27–6.30). In the process, we will also transform the dependent variables as well, resulting in a system of partial differential equations describing the boundary-layer flow that looks completely different than Eqs. (6.27–6.30), but which are easier to analyze and solve. Although the following transformations might at first look involved, they are in reality quite straightforward. There are four basic steps to the transformation process, leading to the final transformed equations; these four steps will be clearly identified next, for the convenience of the reader.

Step 1—Transformation of the independent variables: The independent variable transformation given by Eqs. (6.33) and (6.34) must be couched in terms of derivatives because terms involving x and y in the original boundary-layer equations (6.27–6.30) are derivative terms. From the chain rule of differential calculus, we have

$$\frac{\partial}{\partial x} = \left(\frac{\partial}{\partial \xi}\right)\left(\frac{\partial \xi}{\partial x}\right) + \left(\frac{\partial}{\partial \eta}\right)\left(\frac{\partial \eta}{\partial x}\right) \tag{6.35}$$

$$\frac{\partial}{\partial y} = \left(\frac{\partial}{\partial \xi}\right)\left(\frac{\partial \xi}{\partial y}\right) + \left(\frac{\partial}{\partial \eta}\right)\left(\frac{\partial \eta}{\partial y}\right) \tag{6.36}$$

From Eqs. (6.33) and (6.34), keeping in mind that $\xi = \xi(x)$ only, we have

$$\frac{\partial \xi}{\partial x} = \rho_e u_e \mu_e \tag{6.37a}$$

$$\frac{\partial \xi}{\partial y} = 0 \tag{6.37b}$$

$$\frac{\partial \eta}{\partial y} = \frac{u_e \rho}{\sqrt{2\xi}} \tag{6.37c}$$

(As we will soon see, we do not need an explicit expression for $\partial\eta/\partial x$). Substituting Eqs. (6.37a–6.37c) into Eqs. (6.35) and (6.36), we obtain the following *derivative transformations*:

$$\frac{\partial}{\partial x} = \rho_e u_e \mu_e \frac{\partial}{\partial \xi} + \left(\frac{\partial \eta}{\partial x}\right)\frac{\partial}{\partial \eta} \tag{6.38}$$

$$\frac{\partial}{\partial y} = \frac{u_e \rho}{\sqrt{2\xi}} \frac{\partial}{\partial \eta} \tag{6.39}$$

At this stage, it is convenient (but not necessary) to introduce the stream function ψ defined, as usual, by

$$\frac{\partial \psi}{\partial y} = \rho u \tag{6.40a}$$

$$\frac{\partial \psi}{\partial x} = -\rho v \tag{6.40b}$$

In terms of ψ, the x-momentum boundary-layer equation (6.28) becomes

$$\frac{\partial \psi}{\partial y}\frac{\partial u}{\partial x} - \frac{\partial \psi}{\partial x}\frac{\partial u}{\partial y} = -\frac{\mathrm{d}p_e}{\mathrm{d}x} + \frac{\partial}{\partial y}\left(\mu \frac{\partial u}{\partial y}\right) \tag{6.41}$$

Introducing the derivative transformations given by Eqs. (6.38) and (6.39) into Eq. (6.41), we have

$$\left(\frac{u_e\rho}{\sqrt{2\xi}}\frac{\partial\psi}{\partial\eta}\right)\left[\rho_e u_e \mu_e \frac{\partial u}{\partial\xi} + \left(\frac{\partial\eta}{\partial x}\right)\frac{\partial u}{\partial\eta}\right] - \left[\rho_e u_e \mu_e \frac{\partial\psi}{\partial\xi} + \left(\frac{\partial\eta}{\partial x}\right)\frac{\partial\psi}{\partial\eta}\right]\frac{u_e\rho}{\sqrt{2\xi}}\frac{\partial u}{\partial\eta}$$
$$= -\rho_e u_e \mu_e \frac{\mathrm{d}p_e}{\mathrm{d}\xi} + \frac{u_e\rho}{\sqrt{2\xi}}\frac{\partial}{\partial\eta}\left(\frac{u_e\rho\mu}{\sqrt{2\xi}}\frac{\partial u}{\partial\eta}\right) \tag{6.42}$$

Multiplying Eq. (6.42) by $\sqrt{2\xi}/u_e\rho$, we obtain

$$\frac{\partial\psi}{\partial\eta}\left[\rho_e u_e \mu_e \frac{\partial u}{\partial\xi} + \left(\frac{\partial\eta}{\partial x}\right)\frac{\partial u}{\partial\eta}\right] - \left[\rho_e u_e \mu_e \frac{\partial\psi}{\partial\xi} + \left(\frac{\partial\eta}{\partial x}\right)\frac{\partial\psi}{\partial\eta}\right]\frac{\partial u}{\partial\eta}$$
$$= -\sqrt{2\xi}\frac{\rho_e}{\rho}\mu_e \frac{\mathrm{d}p_e}{\mathrm{d}\xi} + \frac{\partial}{\partial\eta}\left(\frac{u_e\rho\mu}{\sqrt{2\xi}}\frac{\partial u}{\partial\eta}\right) \tag{6.43}$$

This is the end of step 1; Eq. (6.43) represents the boundary-layer x-momentum equation in terms of the transformed independent variables.

Step 2—Transformation of the dependent variables: Let us define a function of ξ and η, $f(\xi,\eta)$ such that

$$\frac{u}{u_e} = \frac{\partial f}{\partial\eta} \equiv f' \tag{6.44}$$

where the prime denotes (for the time being) the *partial* derivative with respect to η. Recalling that the velocity at the edge of the boundary layer is a function of x (hence ξ) only, that is, $u_e = u_e(\xi)$, the derivatives of u follow from Eq. (6.44) as

$$\frac{\partial u}{\partial\xi} = f'\frac{\mathrm{d}u_e}{\mathrm{d}\xi} + u_e\frac{\partial f'}{\partial\xi} \tag{6.45}$$

$$\frac{\partial u}{\partial\eta} = u_e f'' \tag{6.46}$$

where f'' denotes $(\partial^2 f/\partial\eta^2)$.

Step 3—Identification of f with ψ: The new dependent variable $f(\xi,\eta)$, defined by Eq. (6.44), is essentially a stream function in its own right and is indeed related to ψ as follows. From Eq. (6.40a), written in terms of the transformation given in Eqs. (6.39) and (6.44), we have

$$\frac{u_e\rho}{\sqrt{2\xi}}\frac{\partial\psi}{\partial\eta} = \rho u = \rho f' u_e$$

or

$$\frac{\partial \psi}{\partial \eta} = \sqrt{2\xi} f' \tag{6.47}$$

Integrating Eq. (6.47) with respect to η, we have

$$\psi = \sqrt{2\xi} f + F(\xi) \tag{6.48}$$

where $F(\xi)$ is an arbitrary function of ξ. However, recall from the general properties of the stream function that, with no mass injection at the wall, the value of ψ at the wall is zero, that is, $\psi(\xi, 0) = 0$. In Eq. (6.48) applied at the wall, the only way to ensure that $\psi = 0$ at each point along the wall is for each term of Eq. (6.48) to be zero, that is, both $f = 0$ and $F(\xi) = 0$. Hence, the arbitrary function $F(\xi)$ in Eq. (6.48) must be zero, and we have

$$\psi = \sqrt{2\xi} f \tag{6.49}$$

Clearly, from Eq. (6.49), f is a stream function related to ψ. Finally, from Eq. (6.49) we have

$$\frac{\partial \psi}{\partial \xi} = \sqrt{2\xi} \frac{\partial f}{\partial \xi} + \frac{1}{\sqrt{2\xi}} f \tag{6.50}$$

Step 4—Obtaining the final transformed equation: Substituting Eqs. (6.45–6.47) and (6.50) into Eq. (6.43), we obtain

$$\begin{aligned} &\sqrt{2\xi} f' \left[\rho_e u_e \mu_e \left(f' \frac{\mathrm{d}u_e}{\mathrm{d}\xi} + u_e \frac{\partial f'}{\partial \xi} \right) + \left(\frac{\partial \eta}{\partial x} \right) u_e f'' \right] \\ &\quad - \left[\rho_e u_e \mu_e \left(\sqrt{2\xi} \frac{\partial f}{\partial \xi} + \frac{1}{\sqrt{2\xi}} f \right) + \left(\frac{\partial \eta}{\partial x} \right) \sqrt{2\xi} f' \right] u_e f'' \\ &= -\sqrt{2\xi} \frac{\rho_e}{\rho} \mu_e \frac{\mathrm{d}p_e}{\mathrm{d}\xi} + \frac{\partial}{\partial \eta} \left(\frac{u_e^2 \rho \mu}{\sqrt{2\xi}} f'' \right) \end{aligned} \tag{6.51}$$

From Euler's equation, which governs the inviscid flow at the boundary-layer edge,

$$\mathrm{d}p_e = -\rho_e u_e \, \mathrm{d}u_e \tag{6.52}$$

Inserting Eq. (6.52) into Eq. (6.51) and multiplying terms, we obtain

$$\begin{aligned}&\sqrt{2\xi}\rho_e u_e \mu_e (f')^2 \frac{\mathrm{d}u_e}{\mathrm{d}\xi} + \sqrt{2\xi} f' \rho_e u_e^2 \mu_e \frac{\partial f'}{\partial \xi} + \sqrt{2\xi}\, u_e f' f'' \left(\frac{\partial \eta}{\partial x}\right)\\ &- \rho_e u_e^2 \mu_e \sqrt{2\xi} f'' \frac{\partial f}{\partial \xi} - \frac{\rho_e u_e^2 \mu_e}{\sqrt{2\xi}} f f'' - \sqrt{2\xi} u_e f' f'' \left(\frac{\partial \eta}{\partial x}\right)\\ &= \sqrt{2\xi} \frac{(\rho_e)^2}{\rho} u_e \mu_e \frac{\mathrm{d}u_e}{\mathrm{d}\xi} + \frac{\partial}{\partial \eta}\left(\frac{u_e^2 \rho \mu}{\sqrt{2\xi}} f''\right)\end{aligned} \tag{6.53}$$

Note that the third and sixth terms (involving $\partial\eta/\partial x$) in Eq. (6.53) cancel; this is why we never bothered to find an explicit expression for $\partial\eta/\partial x$. Dividing Eq. (6.53) by $\sqrt{2\xi}\rho_e u_e^2 \mu_e$, we have

$$\frac{1}{u_e}(f')^2 \frac{\mathrm{d}u_e}{\mathrm{d}\xi} + f' \frac{\partial f'}{\partial \xi} - \frac{\partial f}{\partial \xi} f'' - \frac{1}{2\xi} f f'' = \frac{\rho_e}{\rho}\frac{1}{u_e}\frac{\mathrm{d}u_e}{\mathrm{d}\xi} + \frac{\partial}{\partial \eta}\left(\frac{1}{2\xi}\frac{\rho\mu}{\rho_e \mu_e} f''\right) \tag{6.54}$$

Denote the "rho-mu" ratio in Eq. (6.54) by $C = \rho\mu/\rho_e\mu_e$. Grouping terms in Eq. (6.54), we finally obtain

$$\boxed{(Cf'')' + ff'' = \frac{2\xi}{u_e}\left[(f')^2 - \frac{\rho_e}{\rho}\right]\frac{\mathrm{d}u_e}{\mathrm{d}\xi} + 2\xi\left(f' \frac{\partial f'}{\partial \xi} - \frac{\partial f}{\partial \xi} f''\right)} \tag{6.55}$$

Equation (6.55) is the transformed boundary-layer x-momentum equation for a two-dimensional, compressible flow.

The boundary-layer y-momentum equation, namely, Eq. (6.29), stating that $\partial p/\partial y = 0$ becomes in the transformed space

$$\boxed{\frac{\partial p}{\partial \eta} = 0} \tag{6.56}$$

The boundary-layer energy equation given by Eq. (6.30) can also be transformed. Defining a nondimensional static enthalpy as

$$g = g(\xi, \eta) = \frac{h}{h_e} \tag{6.57}$$

where h_e is the static enthalpy at the boundary-layer edge, and utilizing the same transformation as before, Eq. (6.30) becomes

$$\boxed{\left(\frac{C}{\mathrm{Pr}} g'\right)' + fg' = 2\xi\left[f' \frac{\partial g}{\partial \xi} + \frac{f' g}{h_e}\frac{\partial h_e}{\partial \xi} - g' \frac{\partial f}{\partial \xi} + \frac{\rho_e u_e}{\rho h_e} f' \frac{\mathrm{d}u_e}{\mathrm{d}\xi}\right] - C\frac{u_e^2}{h_e}(f'')^2} \tag{6.58}$$

where $Pr = \mu c_p/k$ and, as before, $C = \rho\mu/\rho_e\mu_e$. (In some of the literature, C is called the Chapman–Rubesin factor.) The derivation of Eq. (6.58) is left as homework problem 6.2 for the reader.

Examine Eqs. (6.55), (6.56), and (6.58); they are transformed compressible boundary-layer equations. They are still *partial* differential equations, where both f and g are functions of ξ and η. They contain no further approximations or assumptions beyond those associated with the original boundary-layer equations, namely, Eqs. (6.27–6.30). However, they are certainly in a less recognizable, somewhat more complicated-looking form than the original equations. But do not be disturbed by this; in reality, Eqs. (6.55), (6.56), and (6.58) are in a form that will prove to be practical and useful in the following discussion. Indeed, transformed equations like Eqs. (6.55), (6.56), and (6.58) will occur frequently in our presentation of hypersonic viscous flow, not only in Part 2, but also in our discussion of high-temperature chemically reacting flows in Part 3. Thus, it is important to understand and feel comfortable with these equations.

The preceding transformed boundary-layer equations must be solved subject to the following boundary conditions. The physical boundary conditions were given immediately following Eqs. (6.26–6.32); the corresponding transformed boundary conditions are as follows.

At the wall for fixed wall temperature:

$$\eta = 0, \quad f = f' = 0, \quad g = g_w$$

or adiabatic wall:

$$g' = 0$$

At the boundary-layer edge:

$$\eta \to \infty, \quad f' = 1, \quad g = 1$$

In general, solutions of Eqs. (6.55), (6.56), and (6.58) along with the appropriate boundary conditions yield variations of velocity and enthalpy throughout the boundary layer, via $u = u_e f''(\xi, \eta)$ and $h = h_e g(\xi, \eta)$. The pressure throughout the boundary layer is known because the known pressure distribution (or equivalently the known velocity distribution) at the edge of the boundary is given by $p_e = p_e(\xi)$, and this pressure is impressed without change through the boundary layer in the locally normal direction via Eq. (6.56), which says that p = constant in the normal direction at any ξ location. (This is the usual boundary-layer result. Keep in mind that here we are ignoring the possibility, discussed earlier, that a finite normal pressure gradient can occur in a hypersonic boundary layer.) Finally, knowing h and p throughout the boundary layer, equilibrium thermodynamics provides the remaining variables through the appropriate equations of state, for example, $T = T(h, p)$ $\rho = \rho(h, p)$, etc. For convenience, it is useful to visualize solutions of Eqs. (6.55) and (6.58) displayed as profiles through the boundary layer at various ξ locations, as qualitatively sketched in Fig. 6.6. At the top of Fig. 6.6, velocity profiles (η as the ordinate and u as the abscissa) are shown at three different stations along the surface, denoted by ξ_1, ξ_2, and ξ_3. At the bottom of Fig. 6.6, static enthalpy profiles (η as the ordinate and h as the abscissa) are sketched at three different stations. In general, even though these profiles are calculated in the transformed $\xi - \eta$ space, they will be *different profiles* at each different value of ξ. Boundary layers that exhibit this behavior, which is the case in general, are called *nonsimilar* boundary layers. This is in contrast to the concept of a self-similar boundary layer illustrated earlier in Fig. 6.5, and which is a special case that will be discussed in subsequent paragraphs.

Included in the general boundary-layer solution such as sketched in Fig. 6.6 are the velocity and enthalpy gradients *at the wall*, given by $f''(\xi, 0)$ and $g'(\xi, 0)$. From the point of view of applied problems, this is the real payoff from a boundary-layer solution because the local surface skin-friction coefficient c_f is related to $f''(\xi, 0)$ and the local heat-transfer rate at the surface is related to $g'(\xi, 0)$. These relations are obtained as follows. The local skin-friction coefficient c_f is defined as

$$c_f = \frac{\tau_w}{\frac{1}{2}\rho_e u_e^2} \tag{6.59}$$

where τ_w is the local shear stress at the wall, given by

$$\tau_w = \left[\mu\left(\frac{\partial u}{\partial y}\right)\right]_w \tag{6.60}$$

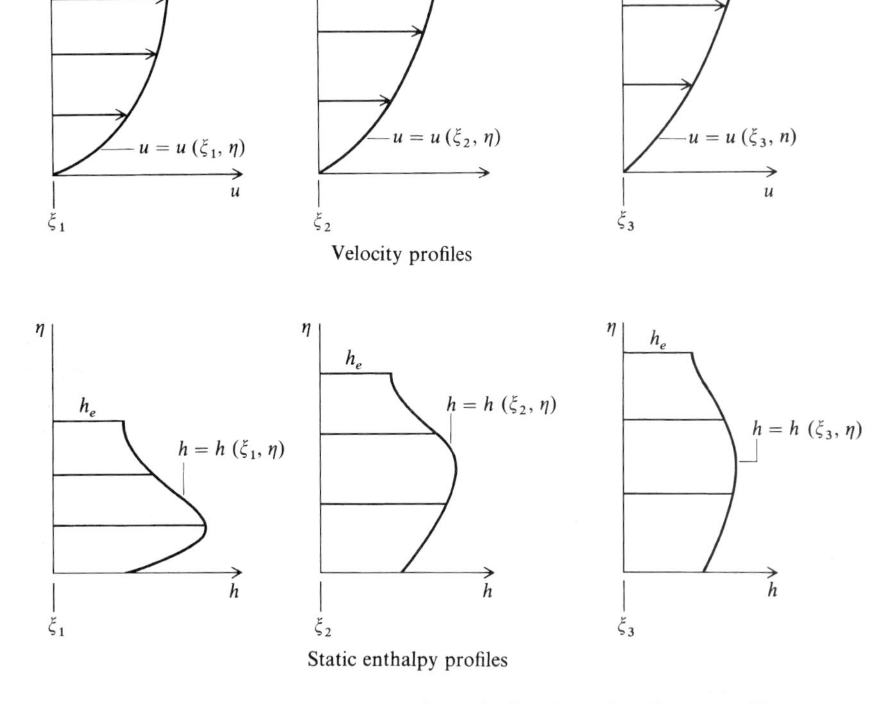

Fig. 6.6 Qualitative sketches of nonsimilar boundary-layer profiles.

Combining Eqs. (6.59) and (6.60) and utilizing the transformations given by Eqs. (6.39) and (6.44), we obtain

$$c_f = \left(\frac{2}{\rho_e u_e^2}\right)\mu_w\left(\frac{\partial u}{\partial y}\right)_w = \left(\frac{2}{\rho_e u_e^2}\right)\mu_w\frac{u_e\rho_w}{\sqrt{2\xi}}\left(\frac{\partial u}{\partial \eta}\right)_w$$

$$= \left(\frac{2}{\rho_e u_e^2}\right)\mu_w\frac{u_e^2\rho_w}{\sqrt{2\xi}}f''(\xi, 0)$$

or

$$\boxed{c_f = \frac{2\mu_w\rho_w}{\rho_e\sqrt{2\xi}}f''(\xi, 0)} \tag{6.61}$$

The local heat-transfer coefficient can be expressed by any one of several defined parameters, such as the Nusselt number Nu, or the Stanton number C_H, defined as follows:

$$Nu = \frac{q_w x}{k_e(T_{\text{aw}} - T_w)} \tag{6.62}$$

$$C_H = \frac{q_w}{\rho_e u_e(h_{\text{aw}} - h_w)} \tag{6.63}$$

where q_w is the local heat-transfer rate (energy per second per unit area) at the wall, x is the distance along the wall measured from the leading edge, k_e is the thermal conductivity at the edge of the boundary layer, and T_{aw} and h_{aw} are the adiabatic wall temperature and adiabatic wall enthalpy, respectively. (By definition, T_{aw} and h_{aw} are the temperature and enthalpy respectively at the wall when the heat transfer to the wall q_w is zero. Sometimes T_{aw} is referred to as the "equilibrium" temperature or "equilibrium" enthalpy, respectively.) From the definitions given by Eqs. (6.62) and (6.63) and noting that $h = c_p T$ for a calorically perfect gas, we write

$$Nu \equiv \frac{q_w x}{k_e(T_{\text{aw}} - T_w)} = \left[\frac{q_w}{\rho_e u_e c_p(T_{\text{aw}} - T_w)}\right]\left[\frac{\rho_e u_e x}{\mu_e}\right]\left[\frac{\mu_e c_p}{k_e}\right]$$

or

$$\boxed{Nu = C_H Re\, Pr} \tag{6.64}$$

Let us concentrate on the Stanton number because Nu is related to C_H via Eq. (6.64). From Eq. (6.63) and the Fourier equation for heat conduction, namely,

$$q_w = \left[k\frac{\partial T}{\partial y}\right]_w$$

we have

$$C_H = \frac{q_w}{\rho_e u_e (h_{aw} - h_w)} = \frac{1}{\rho_e u_e (h_{aw} - h_w)} \left[k \frac{\partial T}{\partial y} \right]_w = \frac{1}{\rho_e u_e (h_{aw} - h_w)} \left[\frac{k}{c_p} \frac{\partial h}{\partial y} \right]_w$$

Using the transformations given by Eqs. (6.39) and (6.57), this becomes

$$C_H = \frac{1}{\rho_e u_e (h_{aw} - h_w)} \left[\frac{k}{c_p} \frac{u_e \rho h_e}{\sqrt{2\xi}} \right]_w g'(\xi, 0)$$

or

$$C_H = \frac{1}{\sqrt{2\xi}} \frac{k_w}{c_{p_w}} \frac{\rho_w}{\rho_e} \frac{h_e}{(h_{aw} - h_w)} g'(\xi, 0) \tag{6.65}$$

In summary, Eqs. (6.61) and (6.65) express the skin-friction coefficient and the heat-transfer coefficient at the wall in terms of $f''(\xi, 0)$ and $g'(\xi, 0)$. In turn, $f''(\xi, 0)$ and $g'(\xi, 0)$ are obtained from the solution of Eqs. (6.55) and (6.58) for the *complete* flowfield within the boundary layer, taking into account the proper boundary conditions. There is no way of obtaining f'' and g' at the wall directly; only a complete solution of the boundary layer will provide the results at the wall—the main practical results of any boundary-layer analysis.

The actual solution of Eqs. (6.55) and (6.58) for a general, nonsimilar boundary layer requires the solution of coupled, nonlinear partial differential equations as a two-point boundary-value problem, the two boundaries being $\eta = 0$ and $\eta \to \infty$ (or η at least large enough to be outside the boundary layer). These matters will be discussed in Sec. 6.6. In the remainder of the present section, a simpler approach will be considered. We will examine cases where the boundary layer is self-similar, that is, where the picture shown in Fig. 6.5 holds. We will consider two classic aerodynamic problems—flow over a flat plate and flow around a stagnation point.

6.5.1 *Flat-Plate Case*

The inviscid flow over a flat plate at zero angle of attack is characterized by constant properties, that is,

$$u_e = \text{const} \qquad T_e = \text{const} \qquad \rho_e = \text{const}$$

Furthermore, let us assume either a constant-temperature wall

$$T_w = \text{const}$$

or an adiabatic wall

$$\left(\frac{\partial T}{\partial y}\right)_w = 0$$

Examine Eqs. (6.55) and (6.58) for this case. Here, $du_e/d\xi = 0$, and u_e, ρ_e, and h_e are constant values, independent of ξ. Under these conditions, Eqs. (6.55) and (6.58) become

$$(Cf'')' + ff'' = 2\xi\left(f'\frac{\partial f'}{\partial \xi} - \frac{\partial f}{\partial \xi}f''\right) \tag{6.66}$$

and

$$\left(\frac{C}{\Pr}g'\right)' + fg' = 2\xi\left(f'\frac{\partial g}{\partial \xi} - g'\frac{\partial f}{\partial \xi}\right) - C\frac{u_e^2}{h_e}(f'')^2 \tag{6.67}$$

Equations (6.66) and (6.67) are still *partial* differential equations. Let us now go through a thought experiment. Let us *assume* that f and g are functions of η only, that is, assume that f and g are independent of ξ. Insert this assumption into Eqs. (6.66) and (6.67). If in the resulting equations *all* dependency upon ξ drops out, then the equations become ordinary differential equations, and we have a verification that the assumption is correct. For the flat-plate case, this is indeed true because when the assumptions of $f = f(\eta)$ and $g = g(\eta)$ are inserted into Eqs. (6.66) and (6.67), they become

$$(Cf'')' + ff'' = 0 \tag{6.68}$$

and

$$\left(\frac{C}{Pr}g'\right)' + fg' + C\frac{u_e^2}{h_e}(f'')^2 = 0 \tag{6.69}$$

Examining Eqs. (6.68) and (6.69), we see no ξ dependency; indeed, these equations are now *ordinary differential equations* in terms of the single independent variable η. Equations (6.68) and (6.69) are the governing equations for a compressible boundary layer over a flat plate with constant wall conditions, and they demonstrate that such a boundary layer is self-similar. Along with the boundary conditions, these equations represent a two-point boundary-value problem for coupled, ordinary differential equations. Also note in these equations both $C = \rho\mu/\rho_e\mu_e$ and $Pr = \mu c_p/k$ are the *local values* at each point within the boundary layer, and in general are variables, that is, $C = C(\eta)$ and $Pr = Pr(\eta)$. Indeed, the variation of C across the boundary layer can be quite large for hypersonic boundary layers, ranging over an order of magnitude or more. On the other hand, the variation of Pr across the boundary layer is usually no more than 20 or 30%.

The actual numerical solution of Eqs. (6.68) and (6.69) frequently takes the form of a "shooting" technique, as described next. Equation (6.68) is a third-order equation, and Eq. (6.69) is second order. Therefore, in numerically integrating the equations using a standard technique such as the Runge–Kutta method starting at the wall and marching across the boundary layer to the outer edge, five boundary conditions at $\eta = 0$ must be specified. In terms of the problem itself, we have specified only three conditions at the wall, namely,

$$f(0) = 0 \quad f'(0) = 0 \quad g(0) = g_w$$

Thus, to integrate the equations, we must assume two additional conditions at the wall, that is, we must assume values for $f''(0)$ and $g'(0)$. With this in mind, the straightforward shooting technique is carried out as follows:

1) Assume values for $f''(0)$ and $g'(0)$. Numbers on the order of 0.5 to 1.0 are usually good assumptions.

2) Numerically integrate Eqs. (6.68) and (6.69) across the boundary layer, going to large enough values of η such that $f'(\eta)$ and $g(\eta)$ become relatively constant with η. This would correspond to conditions outside the boundary layer.

3) Do the resulting values of $f'(\eta)$ and $g(\eta)$ at large η approach $f'(\eta) = 1$ and $g(\eta) = 1$, which are the appropriate boundary conditions at the edge of the boundary layer? If not, return to step 1, and assume new values for $f''(0)$ and $g'(0)$.

4) Repeat steps 1–3 until the proper values for $f''(0)$ and $g'(0)$ are assumed at the wall such that the integration of Eqs. (6.68) and (6.69) produces the proper results at large η, namely, $f'(\eta) = 1$ and $g(\eta) = 1$.

At the end of step 4, all aspects of the compressible, laminar boundary layer on a flat plate are known, including the skin friction and heat transfer determined from the final, converged values of $f''(0)$ and $g'(0)$ via simplified versions of Eqs. (6.61) and (6.65). These simplified forms are obtained as follows. For a flat plate, where ρ_e, u_e, and μ_e are constant, Eq. (6.33) gives

$$\xi = \rho_e u_e \mu_e x \tag{6.70}$$

Inserting Eq. (6.70) into (6.61) and replacing $f''(\xi, 0)$ with simply $f''(0)$, we have

$$c_f = \frac{2\mu_w \rho_w f''(0)}{\rho_e \sqrt{2\rho_e u_e \mu_e x}} = \sqrt{2}\frac{\mu_w \rho_w}{\mu_e \rho_e}\frac{f''(0)}{\sqrt{\rho_e u_e x/\mu_e}}$$

or

$$\boxed{c_f = \sqrt{2}\frac{\rho_w \mu_w}{\rho_e \mu_e}\frac{f''(0)}{\sqrt{Re_x}}} \tag{6.71}$$

where Eq. (6.71) is applicable to the flat-plate case only. Note from Eq. (6.71) that $c_f \propto 1/\sqrt{Re_x}$. It is interesting to compare this result with the familiar result for incompressible flow over a flat plate (for example, see [5]), given by

$$c_f(\text{incompressible}) = \frac{0.664}{\sqrt{Re_x}} \tag{6.72}$$

Working further with Eq. (6.71), we note from the equation of state $p = \rho RT$ that

$$\frac{\rho_w}{\rho_e} = \frac{T_e}{T_w} \tag{6.73}$$

Also, recall that μ for a gas is a function of temperature only; if we assume an exponential variation $\mu \propto T^n$, then

$$\frac{\mu_w}{\mu_e} = \left(\frac{T_w}{T_e}\right)^n \tag{6.74}$$

Combining Eqs. (6.71), (6.73), and (6.74), we have

$$c_f = \sqrt{2}\left(\frac{T_w}{T_e}\right)^{n-1} \frac{f''(0)}{\sqrt{Re_x}} \tag{6.75}$$

The value of $f''(0)$ itself is a function of M_e, Pr, and γ through the solution of Eqs. (6.68) and (6.69). This is because Pr appears explicitly in Eq. (6.69), and the term ${u_e}^2/h_e$ is proportional to M_e^2 and $(\gamma - 1)$, that is [noting that for a calorically perfect gas, $c_p = \gamma R/(\gamma - 1)$ and the freestream speed of sound is $a_e = \sqrt{\gamma RT_e}$],

$$\frac{u_e^2}{h_e} = \frac{u_e^2}{c_p T_e} = (\gamma - 1)\frac{u_e^2}{\gamma RT_e} = (\gamma - 1)\frac{u_e^2}{a_e^2} = (\gamma - 1)\,M_e^2$$

Thus, $f''(0)$ is a function of M_e, Pr, and γ through Eqs. (6.68) and (6.69); it also depends on the wall-to-freestream-temperature (or enthalpy) ratio $T_w/T_e = h_w/h_e$ through the boundary condition. The net result is that we can express the coefficient of $1/\sqrt{Re_x}$ in Eq. (6.75) with the functional expression $F(M_e, Pr, \gamma, T_w/T_e)$, writing Eq. (6.75) as

$$c_f(\text{compressible}) = \frac{F(M_e, Pr, \gamma, T_w/T_e)}{\sqrt{Re_x}} \tag{6.76}$$

Thus, comparing Eqs. (6.72) and (6.76), our compressible boundary-layer theory demonstrates that the familiar coefficient 0.664 in the incompressible result is replaced by another number, which is a function of M_e, Pr, γ, and T_w/T_e. The form of Eq. (6.76) is certainly to be expected because we identified M_e, Pr, γ, T_w/T_e, and Re in Sec. 6.3 as the governing similarity parameters for a compressible viscous flow. The point here is that compressible boundary-layer theory, just as in the familiar incompressible case, demonstrates that for laminar flow over flat plate c_f is inversely proportional to $\sqrt{Re_x}$; however, the constant of proportionality, which is 0.664 for the familiar incompressible case, becomes for the compressible case a number that is a function of the compressible flow similarity parameters M_e, Pr, γ, and T_w/T_e. This number is obtained from the boundary layer solution just discussed. In regard to heat transfer to a flat plate, Eq. (6.65) is combined with Eq. (6.70), resulting in

$$C_H = \frac{1}{\sqrt{2}}\frac{1}{\sqrt{\rho_e u_e x/\mu_e}}\frac{1}{\mu_e}\frac{k_w}{c_{pw}}\frac{\rho_w}{\rho_e}\frac{h_e}{(h_{\text{aw}} - h_w)}g'(0)$$

or

$$\boxed{C_H = \frac{1}{\sqrt{2}}\frac{1}{\mu_e}\frac{k_w}{c_{pw}}\frac{\rho_w}{\rho_e}\frac{h_e}{(h_{\text{aw}} - h_w)}\frac{g'(0)}{\sqrt{Re_x}}} \tag{6.77}$$

where Eq. (6.77) is applicable to the flat-plate case only. Note from Eq. (6.77) that $C_H \propto 1/\sqrt{Re_x}$. It is interesting to compare this result with the familiar result for incompressible flow over a flat plate given by

$$C_H(\text{incompressible}) = \frac{0.332}{\sqrt{Re_x}}Pr^{-2/3} \tag{6.78}$$

Working further with Eq. (6.77), and using the calorically perfect-gas relation $h = c_p T$, we obtain

$$C_H = \frac{1}{\sqrt{2}}\frac{\rho_w \mu_w}{\rho_e \mu_e}\frac{1}{Pr_w}\frac{1}{(T_{\text{aw}}/T_e - T_w/T_e)}\frac{g'(0)}{\sqrt{Re_x}} \tag{6.79}$$

Recalling Eqs. (6.73) and (6.74), we obtain from Eq. (6.79)

$$C_H = \frac{1}{\sqrt{2}}\left(\frac{T_w}{T_e}\right)^{n-1}\frac{1}{Pr_w}\frac{1}{(T_{\text{aw}}/T_e - T_w/T_e)}\frac{g'(0)}{\sqrt{Re_x}} \tag{6.80}$$

In Eq. (6.80), T_{aw}/T_e can be found from a solution of Eqs. (6.68) and (6.69), using the adiabatic wall boundary condition that $(\partial T/\partial \eta) = 0$. In turn, the solution of Eqs. (6.68) and (6.69) depends on M_e, γ, and Pr. Thus, in Eq. (6.80), T_{aw}/T_e is a function of M_∞, γ, and Pr, and therefore the entire factor multiplying $1/\sqrt{Re_x}$ in Eq. (6.80) is simply a function of the similarity parameters

$$C_H(\text{compressible}) = \frac{G(M_e, Pr, \gamma, T_w/T_e)}{\sqrt{Re_x}} \tag{6.81}$$

Thus, comparing Eqs. (6.78) and (6.81), our compressible boundary-layer theory demonstrates that the familiar coefficient 0.332 $Pr^{-2/3}$ in the incompressible result is replaced by another number, which is a function of M_e, Pr, γ, and T_w/T_e. Finally, we note that Reynolds analogy linking c_f and C_H, which for the incompressible case is [from Eqs. (6.72) and (6.78)]

$$\frac{C_H}{C_f}\ (\text{incompressible}) = \tfrac{1}{2}Pr^{-2/3} \tag{6.82}$$

now for the compressible case becomes [from Eqs. (6.76) and (6.81)]

$$\frac{C_H}{C_f}\ (\text{compressible}) = \frac{G}{F} = f\left(M_e,\ Pr,\ \gamma,\ \frac{T_w}{T_e}\right) \tag{6.83}$$

Emphasis is again made that, in Eqs. (6.76), (6.81), and (6.83), the values of F and G are obtained by solving the boundary-layer equations (6.68) and (6.69) with the appropriate boundary conditions. There is no exact method that can give an a priori answer for F and G.

A word about the viscosity coefficient and thermal conductivity is in order here. For a pure, nonreacting gas, the viscosity coefficient is dependent only on temperature. An engineering approximation is to assume an exponential temperature variation, such as already given in Eq. (6.74), where in the literature the exponent n seems to vary from 0.5 to 1.0, depending on the nature of the particular gas. However, perhaps the most commonly used expression for μ is Sutherland's law

$$\frac{\mu}{\mu_{\text{ref}}} = \left(\frac{T}{T_{\text{ref}}}\right)^{3/2} \frac{T_{\text{ref}} + S}{T + S} \tag{6.84a}$$

where for air $\mu_{\text{ref}} = 1.789 \times 10^{-5}$ kg/m s, $T_{\text{ref}} = 288$ K, $S = 110$ K, and μ and T are in units of kg/m s and K, respectively. Sutherland's law is accurate for air over a range of several thousand degrees and is certainly appropriate for hypersonic viscous-flow calculations under the assumptions considered in Part 2 of this book. Moreover, under these same assumptions the thermal conductivity k can be obtained from μ and the Prandtl number as

$$Pr = \frac{\mu c_p}{k}$$

hence

$$k = \frac{\mu c_p}{Pr} \tag{6.84b}$$

For air at standard conditions, $Pr = 0.71$.

The preceding discussion has presented the *theory* of laminar, compressible boundary layers over a flat plate; it was given here to provide the reader with the flavor of such boundary-layer solutions. Let us now consider some representative results, particularly at high Mach numbers. Various studies have addressed

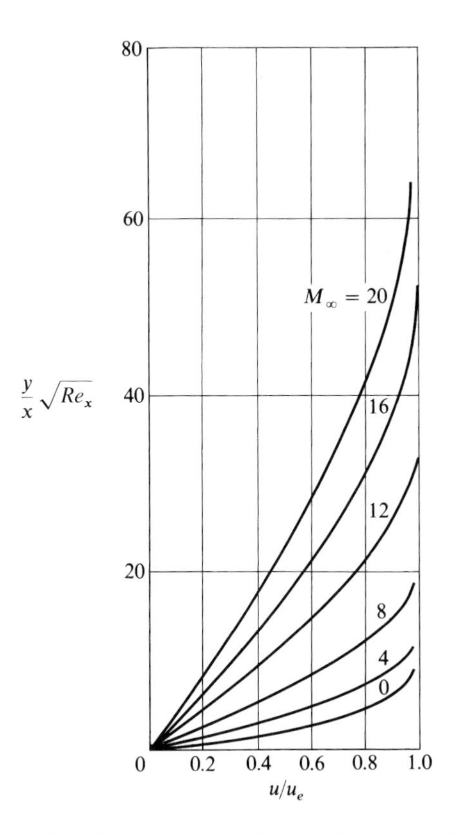

Fig. 6.7 Velocity profiles in a compressible laminar boundary layer over an insulated flat plate (from Van Driest [90]).

the laminar, compressible boundary layer. Most notably, Van Driest [90] calculated flows over a flat plate, and Cohen and Reshotko [91] addressed the entire spectrum of possible self-similar solutions. Figures 6.7 and 6.8 contain results for an insulated flat plate (zero heat transfer) obtained by Van Driest [90] using Sutherland's law for μ and assuming a constant $Pr = 0.75$. The velocity profiles are shown in Fig. 6.7 for different Mach numbers ranging from 0 (incompressible flow) to the large hypersonic value of 20. Note that at a given x station at a given Re_x the boundary-layer thickness increases markedly as M_e is increased to hypersonic values. *This clearly demonstrates one of the most important aspects of hypersonic boundary layers, namely, that the boundary-layer thickness becomes large at large Mach numbers.* Indeed, in Chapter 7 we will easily demonstrate that the laminar boundary-layer thickness varies approximately as M_e^2. Figure 6.8 illustrates the temperature profiles for the same case as Fig. 6.7. Note the obvious physical trend that, *as M_e increases to large hypersonic values, the temperatures increase markedly.* Also note in Fig. 6.8 that at the wall ($y = 0$) $(\partial T/\partial y)_w = 0$, as it should be for an insulated

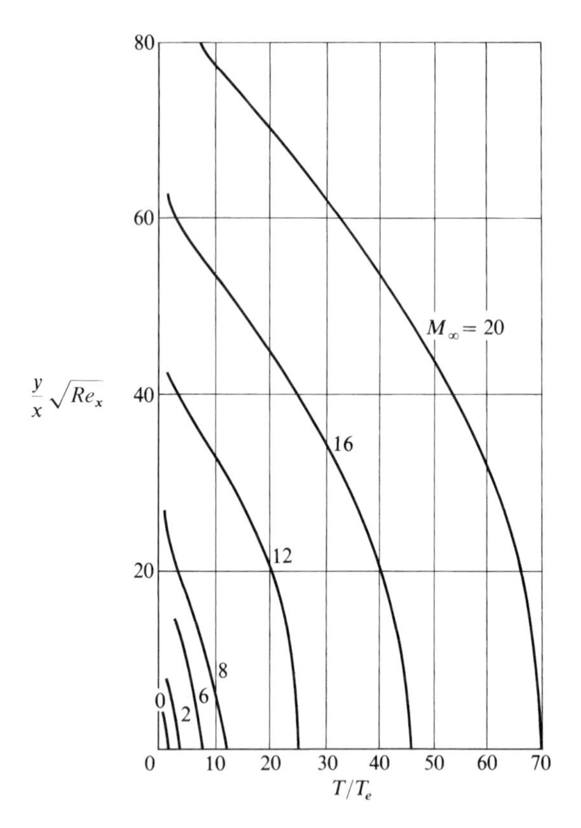

Fig. 6.8 Temperature profiles in a compressible laminar boundary layer over an insulated flat plate (from [90]).

surface ($q_w = 0$). Figures 6.9 and 6.10 also contain results by Van Driest [90], but now for the case of heat transfer to the wall. Such a case is called a "cold-wall" case because $T_w < T_{aw}$. (The opposite case would be a "hot wall," where heat is transferred from the wall into the flow; in this case, $T_w > T_{aw}$.) For the results shown in Figs. 6.9 and 6.10, $T_w/T_e = 0.25$ and $Pr = 0.75 =$ constant. Figure 6.9 shows velocity profiles for various different values of M_e, again demonstrating the rapid growth in boundary-layer thickness with increasing M_e. In addition, the effect of a cold wall on the boundary-layer thickness can be seen by comparing Figs. 6.7 and 6.9. For example, consider the case of $M_e = 20$ in both figures. For the insulated wall at Mach 20 (Fig. 6.7), the boundary-layer thickness reaches out beyond a value of $(y/x)\sqrt{Re_x} = 60$, whereas for the cold wall at Mach 20 (Fig. 6.9) the boundary-layer thickness is slightly above $(y/x)\sqrt{Re_x} = 30$. *This illustrates the general fact that the effect of a cold wall is to reduce the boundary-layer thickness.* This trend is easily explainable on a physical basis when we examine Fig. 6.10, which illustrates the temperature profiles through the boundary layer for the cold-wall case.

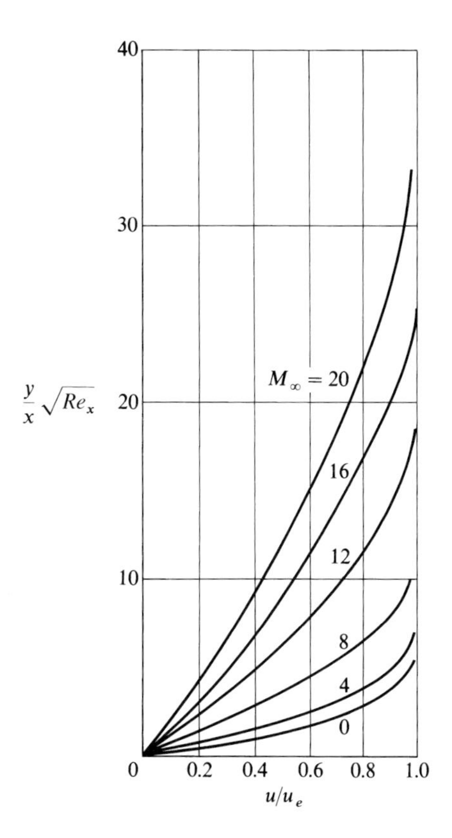

Fig. 6.9 Velocity profiles in a laminar, compressible boundary layer over a cold plate [90].

Comparing Figs. 6.8 and 6.10, we note that, as expected, the temperature levels in the cold-wall case are considerably lower than in the insulated case. In turn, because the pressure is the same in both cases, we have from the equation of state $p = \rho RT$ that the *density in the cold-wall case is much higher*. If the density is higher, the mass flow within the boundary layer can be accommodated within a smaller boundary-layer thickness; hence, the effect of a cold wall is to *thin* the boundary layer. Also note in Fig. 6.10 that, starting at the outer edge of the boundary layer and going toward the wall, the temperature first increases, reaches a peak somewhere within the boundary layer, and then decreases to its prescribed cold-wall value T_w. The peak temperature inside the boundary layer is an indication of the amount of viscous dissipation occurring within the boundary layer. Figure 6.10 clearly demonstrates the rapidly growing effect of this viscous dissipation as M_e increases—yet another basic aspect of hypersonic boundary layers.

Carefully study the boundary-layer profiles shown in Figs. 6.7–6.9. They are an example of the detailed results that emerge from a solution of Eqs. (6.68)

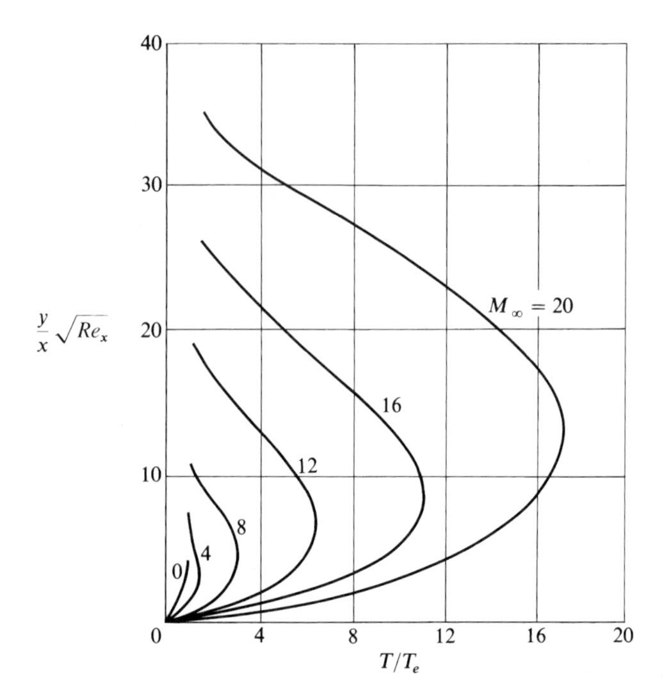

Fig. 6.10 Temperature profiles in a laminar, compressible boundary layer over a cold flat plate [90].

and (6.69); indeed, these figures are graphical representations of Eqs. (6.68) and (6.69), with the results cast in the physical (x, y) space (rather than in terms of the transformed variable η). In turn, the surface values of c_f and C_H can be obtained from these solutions, as given by Eqs. (6.71) and (6.77), respectively. These results are given in Figs. 6.11 and 6.12. In particular, c_f is shown in Fig. 6.11 as a function of M_e. Note from this figure the following important trends:

1) The effect of *increasing* M_e is to *decrease* c_f. For an insulated flat plate, c_f is reduced by approximately a factor of two in going from $M_e = 0$ to 20. Do not be misled by this, however. We see that c_f decreases as M_e increases, but this does *not* mean that the actual shear stress at the wall, τ_w decreases. Keep in mind that $\tau_w = \frac{1}{2}\rho_e u_e^2 c_f = \frac{1}{2}\gamma p_e M_e^2 c_f$. Hence, although c_f decreases gradually as M_e increases, τ_w increases considerably as M_e increases because of the M_e^2 variation just shown.

2) The effect of *cooling* the wall is to *increase* c_f. This makes good physical sense in light of our preceding discussion on the effects of cooling on the boundary-layer thickness δ. A cold wall decreases δ, as we have already seen. In turn, the velocity gradient at the wall is increased when δ decreases, that is, $(du/dy)_w = 0(u_e/\delta)$. Because $\tau_w = \mu(du/dy)_w$, then τ_w will increase. Because $c_f = \tau_w/\frac{1}{2}\rho_e u_e^2$, then c_f will also increase, which confirms the trend shown in Fig. 6.11.

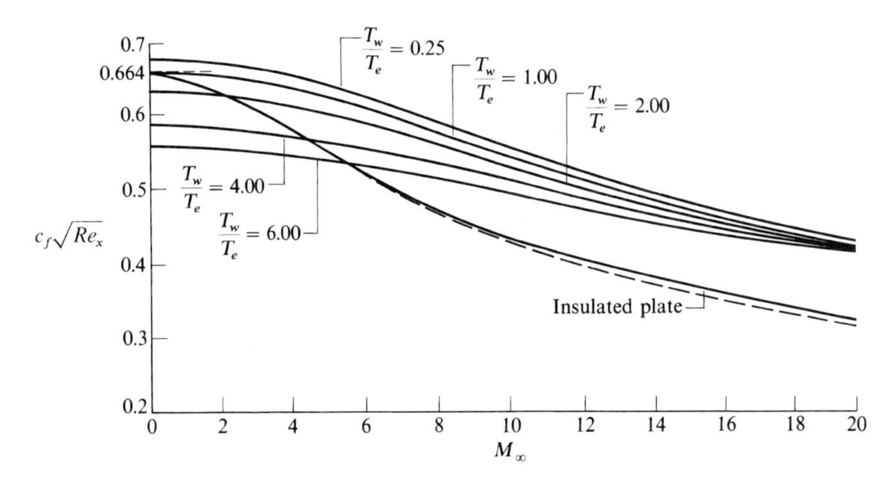

Fig. 6.11 Flat-plate skin-friction coefficients [90].

3) For the insulated wall, as $M_e \rightarrow 0$, $c_f\sqrt{Re_x} \rightarrow 0.664$. This is the familiar result for incompressible flow, as noted in Eq. (6.72).

Heat-transfer results are given in Fig. 6.12. Here, C_H [as calculated from Eq. (6.77)] is plotted vs M_e. The trends shown here are identical to the trends shown for c_f in Fig. 6.11. This is simply a demonstration of Reynolds analogy [Eq. (6.83)], which states the direct relation between c_f and C_H. Also note that Fig. 6.12 gives finite values of C_H for the insulated-wall case. Recall the

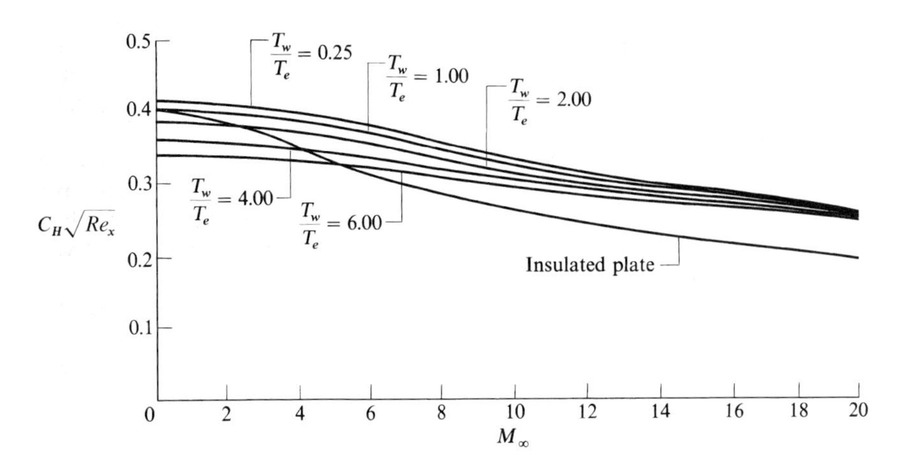

Fig. 6.12 Flat-plate Stanton numbers [90].

definition of C_H from Eq. (6.63), namely,

$$C_H = \frac{q_w}{\rho_e u_e (h_{\text{aw}} - h_w)}$$

For an insulated wall, by definition $q_w = 0$ and $h_w = h_{aw}$. Thus, for the insulated-wall case C_H becomes an indefinite form expressed as

$$C_H = \frac{0}{0}$$

which clearly has a finite value, as shown in Fig. 6.12.

The physical results shown in Figs. 6.7–6.12 are so important that we summarize them here.

1) Boundary thickness δ increases rapidly with M_e.

2) Temperature inside the boundary layer increases rapidly with M_e as a result of viscous dissipation.

3) Cooling the wall reduces δ.

4) Both c_f and C_H decrease as M_e increases.

5) Both c_f and C_H increase as the wall is cooled.

Let us consider some further aspects of aerodynamic heating at hypersonic speeds. Return to the definition of C_H given by Eq. (6.63). Note from this definition that aerodynamic heating to the surface is given by

$$\boxed{q_w = \rho_e u_e C_H (h_{\text{aw}} - h_w)} \tag{6.85}$$

This equation is important because it emphasizes that the "driving potential" for aerodynamic heating to the surface is the enthalpy difference $(h_{\text{aw}} - h_w)$. We will find this to be the case for virtually all cases in aerodynamic heating to high-speed vehicles, even in the chemically reacting cases discussed in Part 3. In turn, the calculation of the adiabatic-wall enthalpy h_{aw} is an important consideration before Eq. (6.85) can be used to obtain q_w. An exact solution for h_{aw} for the flat plate can be obtained by solving Eqs. (6.68) and (6.69) along with the insulated-wall boundary condition $(\partial T/\partial y)_w = 0$. However in most engineering-related calculations, the value of h_{aw} (and of $T_{\text{aw}} = h_{\text{aw}}/c_p$) is expressed in terms of the *recovery factor r*, defined as

$$h_{\text{aw}} = h_e + r\frac{u_e^2}{2} \tag{6.86}$$

At the outer edge of the boundary layer, we have

$$h_0 = h_e + \frac{u_e^2}{2} \tag{6.87}$$

where h_0 is the total enthalpy in the inviscid flow outside the boundary layer. Substituting Eq. (6.87) into (6.86), we have

$$h_{\text{aw}} = h_e + r(h_0 - h_e)$$

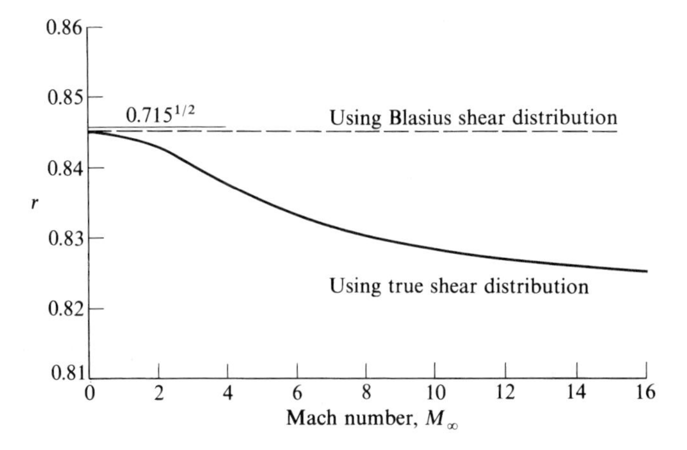

Fig. 6.13 Comparison of exact and approximate recovery factor for laminar flow over a flat plate [83].

$$\boxed{r = \frac{h_{\text{aw}} - h_e}{h_0 - h_e}} \tag{6.88}$$

For a calorically perfect gas, where $h = c_p T$, Eq. (6.88) can be written as

$$r = \frac{T_{\text{aw}} - T_e}{T_0 - T_e} \tag{6.89}$$

For incompressible flow, the value of r is related to the Prandtl number as $r = \sqrt{Pr}$. Exact results for r for compressible flow are shown in Fig. 6.13, obtained from [83], and are compared with the $\sqrt{Pr} = \sqrt{0.715} = 0.845$. Note that r decreases as M_e increases through the hypersonic regime. However, also note that the ordinate is an expanded scale, showing that r decreases by only 2.4% from $M_e = 0$ to 16. Hence, for all practical purposes we can assume for laminar hypersonic flow over a flat plate that

$$\boxed{r = \sqrt{Pr}} \tag{6.90}$$

With Eqs. (6.90) and (6.88), we can readily estimate h_{aw} for use in Eq. (6.85). To complete an engineering analysis of q_w using Eq. (6.85), we must obtain an estimate of C_H. Again, in an exact solution C_H would be obtained by solving Eqs. (6.68) and (6.69) for the specified wall temperature T_w. However, we can estimate C_H using Reynolds analogy. The general, exact value for Reynolds analogy is expressed by Eq. (6.83); numerical solutions are given in Fig. 6.14, obtained from [90]. Note that the ratio c_f/C_H decreases as M_e increases across the hypersonic regime. However, again note that the ordinate is an expanded

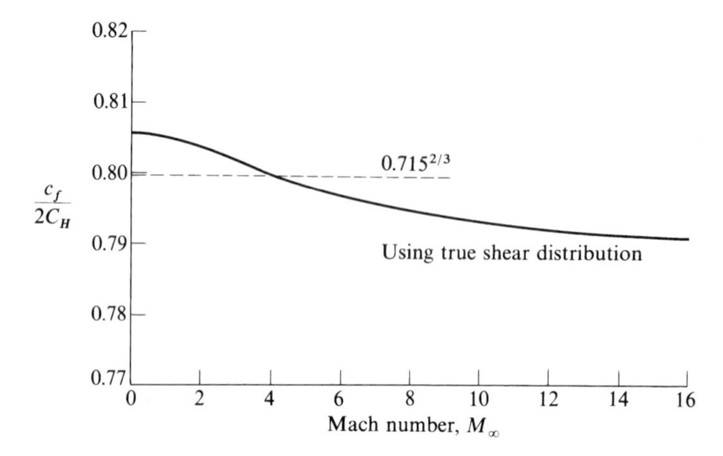

Fig. 6.14 Comparison of exact and approximate Reynolds analogy factor for laminar flow over a flat plate [90].

scale, and c_f/C_H decreases by only 2% from $M_e = 0$ to 16. Thus, the incompressible result given by Eq. (6.82) is a reasonable approximation at hypersonic speeds, namely,

$$\frac{C_H}{c_f} = \frac{1}{2} Pr^{-2/3} \tag{6.91}$$

This brings to an end our discussion of hypersonic flat-plate laminar boundary layers. Although the results have been obtained for the special case of a flat plate, their value goes far beyond that special case. For example, the physical trends just listed hold for hypersonic boundary layers over general aerodynamic shapes. Moreover, the actual flat-plate results are frequently applied to slender three-dimensional shapes in a "localized" sense, following a given streamline over a thin, three-dimensional body. Therefore, before progressing further, review all of the material in the present section until you feel comfortable with the ideas and results.

Example 6.1

Consider a flat plate at zero angle of attack in an airflow at standard sea-level conditions ($p_\infty = 1.01 \times 10^5$ N/m^2 and $T_\infty = 288$ K). The chord length of the plate (distance from the leading edge to the trailing edge) is 2 m. The planform area of the plate is 40 m^2. At standard sea-level conditions, $\mu_\infty = 1.7894 \times 10^{-5}$ kg/(m)(s). Assume the wall temperature is the adiabatic-wall temperature T_{aw}. Assuming laminar flow, calculate the local shear stress on the plate at the location 0.5 m downstream from the leading edge when the freestream velocity is 3402 m/s.

Solution: The freestream density is

$$\rho_\infty = \frac{p_\infty}{RT_\infty} = \frac{1.01 \times 10^5}{(287)(288)} = 1.22 \text{ kg/m}^3$$

The speed of sound is

$$a_\infty = \sqrt{\gamma R T_\infty} = \sqrt{(1.4)(287)(288)} = 340.2 \text{ m/s}$$

Hence, for $V_\infty = 3402$ m/s,

$$M_\infty = \frac{V_\infty}{a_\infty} = \frac{3402}{340.2} = 10$$

From Fig. 6.11, for the insulated-plate results at Mach 10,

$$c_f \sqrt{Re_x} = 0.43$$

From Eq. (6.59),

$$\tau_w = \frac{1}{2} \rho_e u_e^2 c_f$$

For the flat plate, the properties at the edge of the boundary layer are the free-stream properties; hence,

$$\frac{1}{2} \rho_e u_e^2 = \frac{1}{2} \rho_\infty V_\infty^2 = q_\infty = \frac{1}{2} (1.22)(3402)^2 = 7.06 \times 10^6 \text{ N/m}^2$$

Thus,

$$\tau_w = q_\infty c_f = \frac{0.43\, q_\infty}{\sqrt{Re_x}}$$

At $x = 0.5$ m,

$$Re_x = \frac{\rho_\infty V_\infty x}{\mu_\infty} = \frac{(1.22)(3402)(0.5)}{1.7894 \times 10^{-5}} = 1.16 \times 10^8$$

Thus,

$$\tau_w = \frac{0.43 q_\infty}{\sqrt{Re_x}} = \frac{0.43(7.06 \times 10^6)}{\sqrt{1.16 \times 10^8}} = \boxed{281.9 \text{ N/m}^2}$$

Note: Because this problem deals with flight at Mach 10 at sea level, the Reynolds number is high, and the assumption of laminar flow in reality might be suspect. Matters of hypersonic boundary-layer transition and turbulent flow are discussed in Secs. 6.7 and 6.8.

Example 6.2

For the conditions in Example 6.1, calculate the skin-friction drag for the whole plate. Assume the plate is elevated in the air so that air flows over both the top and bottom surfaces, that is, the total friction drag is caused by shear stress acting on both the top and bottom surfaces.

Solution: The overall skin-friction drag coefficient is conventionally defined as

$$C_f \equiv \frac{D_f}{q_\infty S}$$

where D_f is the skin-friction drag on *one surface* of the plate and S is the reference area taken as the surface area of one side of the plate. We first obtain an expression for C_f. To do this, note that the drag per unit span of the plate is simply the integral of the local shear stress along the plate from the leading edge, where $x = 0$, to the trailing edge, where $x = c$.

$$D = \int_0^c \tau_w \, \mathrm{d}x \qquad \text{(per unit span)}$$

From Example 6.1,

$$\tau_w = \frac{0.43\, q_\infty}{\sqrt{\rho_\infty V_\infty x/\mu_\infty}}$$

Hence,

$$D = \frac{0.43\ q_\infty}{\sqrt{\rho_\infty V_\infty/\mu_\infty}} \int_0^c \frac{\mathrm{d}x}{x^{1/2}} = \frac{0.43\ q_\infty}{\sqrt{\rho_\infty V_\infty/\mu_\infty}} \left[2x^{\frac{1}{2}}\right]_0^c$$

$$= \frac{0.86\ q_\infty c^{\frac{1}{2}}}{\sqrt{\rho_\infty V_\infty/\mu_\infty}} = \frac{0.86\, q_\infty c}{\sqrt{\rho_\infty V_\infty c/\mu_\infty}} = \frac{0.86\ q_\infty c}{\sqrt{Re_c}}$$

Recall that D is the drag per unit span. Thus, using D in the definition of C_f, the reference area must be per unit span, that is, $S = c(1)$. Thus,

$$C_f = \frac{D}{q_\infty S} = \frac{D}{q_\infty c(1)} = \frac{0.86}{q_\infty c(1)} \frac{q_\infty c}{\sqrt{Re_c}}$$

or

$$C_f = \frac{0.86}{\sqrt{Re_c}}$$

(*Note*: The classical incompressible flow result is

$$C_f = \frac{1.328}{\sqrt{Re_c}}$$

The result for Mach 10 has a smaller number in the numerator.)

The friction drag D_f on one side of the plate is given by

$$D_f = q_\infty \, S \, C_f$$

where S is now the total surface area of one side of the plate and $S = 40\ \text{m}^2$ as given in Example 6.1. Because $c = 2$ m (given in Example 6.1),

$$Re_c = \frac{\rho_\infty V_\infty c}{\mu_\infty} = \frac{(1.22)(3402)(2)}{1.7894 \times 10^{-5}} = 4.639 \times 10^8$$

Thus,

$$C_f = \frac{0.86}{\sqrt{Re_c}} = \frac{0.86}{\sqrt{4.639 \times 10^8}} = 3.99 \times 10^{-5}$$

and

$$D_f = q_\infty S C_f = 7.06 \times 10^6 (40)(3.99 \times 10^{-5})$$
$$D_f = 11{,}276\ \text{N}$$

The total drag, accounting for both the top and bottom surfaces, is

$$D = 2(11276) = \boxed{22552\ \text{N}}$$

We note an interesting fact. From Example 6.1, the local shear stress at $x = 0.5$ m, that is, the quarter-chord location, is $\tau_w = 281.9\ \text{N/m}^2$. If the shear stress were assumed constant over the surface of the plate, which of course it is *not*, then the total skin-friction drag on one surface would be τ_w $S = (281.9)\ (40) = 11{,}276$ N. This is the precise drag value already obtained for one surface of the plate. This is not a fluke, but rather a geometric feature of the inverse square-root variation of τ_w for a laminar boundary layer. If we let $y = x^{-1/2}$, then the area under the curve for $x = 0$ to c is simply given by

$$A = \int_0^c y\ \mathrm{d}x = \int_0^c x^{-1/2}\ \mathrm{d}x = \left[2x^{1/2}\right]_0^c = 2\sqrt{c}$$

Now consider a rectangle of length c and height equal to y evaluated at $x = c/4$. At $x = c/4$,

$$y = \frac{1}{\sqrt{x}} = \frac{1}{\sqrt{c/4}} = \frac{2}{\sqrt{c}}$$

The area of the rectangle $c(2/\sqrt{c}) = 2\sqrt{c}$, which is precisely the area under the curve of $y = x^{-1/2}$. This result tells us that one way to calculate the total laminar skin-friction drag on one surface of a flat plate is to evaluate the local shear stress τ_w at the quarter-chord location and multiply that shear stress by the surface area of the plate.

Example 6.3

For the same flow conditions treated in Examples 6.1 and 6.2, consider a flat plate at zero angle of attack with a constant wall temperature $T_w = 288$ K. Calculate the local heat-transfer rate at the quarter-chord location.

Solution: From Example 6.1, at the quarter-chord location,

$$Re_x = 1.16 \times 10^8$$

From Fig. 6.12, at Mach 10, and a wall-to-freestream ratio of $T_w/T_e = 288/288 = 1$, we have

$$C_H \sqrt{Re_x} = 0.33$$

or

$$C_H = \frac{0.33}{\sqrt{Re_x}} = \frac{0.33}{\sqrt{1.16 \times 10^8}} = \frac{0.33}{1.077 \times 10^4} = 3.06 \times 10^{-5}$$

From Eq. (6.63)

$$C_H = \frac{q_w}{\rho_e u_e (h_{\text{aw}} - h_w)}$$

To evaluate the local heating rate q_w from Eq. (6.63), we need the adiabatic-wall enthalpy h_w or equivalently the adiabatic-wall temperature T_w. Consistent with our presentations in Parts 1 and 2 of this book, we assume a calorically perfect gas, where $h = c_p\, T$, and $\gamma = c_p/c_v =$ constant. For a calorically perfect gas at a freestream Mach number M_∞, the ratio of total to static temperature in the freestream is given by (for example, see [4] and [5]),

$$\frac{T_0}{T_\infty} = 1 + \frac{\gamma - 1}{2} M_\infty^2 = 1 + \frac{1.4 - 1}{2} (10)^2 = 1 + 0.2(100) = 21$$

$$T_0 = 21\ T_\infty = 21(288) = 6048\ \text{K}$$

To find the adiabatic-wall temperature, we use the recovery factor expressed by Eq. (6.88), which for a calorically perfect gas becomes

$$r = \frac{h_{\text{aw}} - h_e}{h_0 - h_e} = \frac{T_{\text{aw}} - T_e}{T_0 - T_e}$$

where the recovery factor is given by Eq. (6.90)

$$r = \sqrt{Pr} = \sqrt{0.715} = 0.845$$

Thus,

$$T_{\text{aw}} = r(T_0 + T_e) + T_e = 0.845(6048 - 288) + 288 = 5155\ \text{K}$$

Returning to Eq. (6.63) written in the form of Eq. (6.85),

$$q_w = \rho_e u_e C_H (h_{aw} - h_w)$$

or

$$q_w = \rho_e u_e C_H c_p (T_{aw} - T_w)$$

The relation between c_p and the specific gas constant R is (for example, see [4] and [5])

$$c_p = \frac{\gamma R}{\gamma - 1}$$

The specific gas constant for standard air is 287 J/kg K. Hence,

$$c_p = \frac{(1.4)(287)}{0.4} = 1004.5 \, \frac{\text{J}}{\text{kg K}}$$

The local aerodynamic heating rate is therefore

$$q_w = \rho_e u_e C_H c_p (T_{aw} - T_w) = (1.22)(3402)(3.06 \times 10^{-5})(1004.5)(5155 - 288)$$

$$q_w = 6.21 \times 10^5 \, \frac{J}{\text{sec m}^2} = 6.21 \times 10^5 \, \frac{\text{W}}{\text{m}^2} = \boxed{62.1 \; W/\text{cm}^2}$$

For flat-plate heat-transfer rates, this is a very large value. It compares to the approximately 50 W/cm^2 stagnation-point heating rate of the space shuttle shown in Fig. 6.1. The preceding calculation was made for Mach 10 flight *at sea level*, and it illustrates why we do not generally fly hypersonic vehicles at sea level—the aerothermal environment is too severe.

Example 6.4

In Design Example 5.2: Hypersonic Waveriders—Part 1 at the end of Chapter 5, the lift-to-drag ratio of a hypersonic flat plate was shown in Fig. 5.57 assuming a Newtonian pressure distribution. In particular, the dashed curve in Fig. 5.57 includes the effect of skin friction and shows L/D increasing as the angle of attack α decreases, then reaching a maximum value at some low angle of attack, and then going to zero as $\alpha \to 0$. The dashed curve is plotted for $M_\infty = 10$ and $Re = 3 \times 10^6$. In homework problem 3.6, you are asked to prove that for the flat plate with a Newtonian pressure distribution $(L/D)_{\max} = 0.667/C_{D_o}^{1/3}$ and that it occurs at an angle of attack (in radians) of $\alpha = C_{D_o}^{1/3}$, where C_{D_o} is the total skin-friction drag coefficient for the flat plate. Using these expressions from homework problem 3.6, and assuming an adiabatic wall, calculate the value of $(L/D)_{\max}$ and the angle of attack at which it occurs. Compare with the results shown in Fig. 5.57.

Solution: From Fig. 6.11 for $M_\infty = 10$ and an adiabatic wall, we have

$$c_f\sqrt{Re_x} = 0.43 \qquad \text{(same as in Example 6.1)}$$

In Example 6.2, using the preceding result for c_f, we proved that the overall skin-friction coefficient for the drag on one side of the plate C_f is

$$C_f = \frac{0.86}{\sqrt{Re_c}}$$

The Reynolds number based on the chord of the plate is given as $Re_c = 3 \times 10^6$. Hence,

$$C_f = \frac{0.86}{\sqrt{3 \times 10^6}} = 4.965 \times 10^{-4}$$

This is the skin-friction coefficient for one side of the plate. The total skin-friction drag coefficient for the plate including the shear stress acting on both the top and bottom surfaces is

$$C_{D_0} = 2C_f = 9.93 \times 10^{-4}$$

From the expressions given in homework problem 3.6, we have

$$(L/D)_{\max} = \frac{0.667}{C_{D_0}^{1/3}} = \frac{0.667}{(9.93 \times 10^{-4})^{1/3}} = \boxed{6.74}$$

and the angle of attack at which it occurs is

$$\alpha = C_{D_0}^{1/3} = 0.0998 \text{ rad} = \boxed{5.7 \text{ deg}}$$

Examining Fig. 5.57, we see that these values agree with the results shown by the dashed curve.

6.5.2 Stagnation-Point Case

We now discuss the second of the two classical problems considered in this section, namely, the laminar boundary layer at a stagnation point. Consider the stagnation region on a blunt body, as sketched in Fig. 6.15. The boundary-layer thickness is finite at the stagnation point. As before, x is the distance measured along the surface, and u_e is the velocity in the x direction at the outer edge of the boundary layer. For the time being, we will consider two-dimensional flow; hence, Fig. 6.1.5 represents a blunt, two-dimensional cylindrical body with infinite span perpendicular to the page. The local surface radius of curvature at the stagnation point is R.

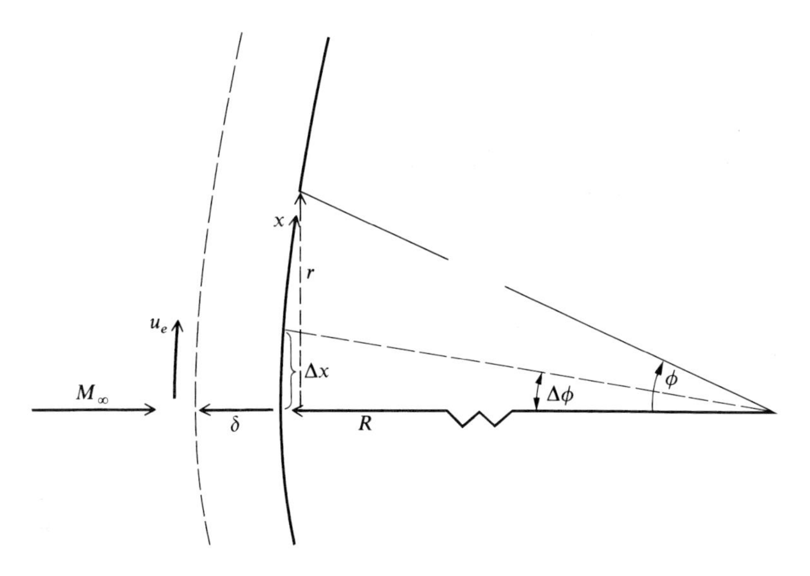

Fig. 6.15 Stagnation region geometry.

Let us consider the possibility of a self-similar solution to the governing boundary-layer equations (6.55) and (6.58) for the stagnation-point case. As before, we make the assumption that f and g are functions of η only; hence, $\partial f'/\partial\xi = \partial f/\partial\xi = \partial g/\partial\xi = 0$ in Eqs. (6.55) and (6.58). With these assumptions, Eqs. (6.55) and (6.58) become, respectively,

$$(Cf'')' + ff'' = \frac{2\xi}{u_e}\left[(f')^2 - \frac{\rho_e}{\rho}\right]\frac{\mathrm{d}u_e}{\mathrm{d}\xi} \tag{6.92}$$

and

$$\left(\frac{C}{Pr}g'\right)' + fg' = 2\xi\left[\frac{f'g}{h_e}\frac{\partial h_e}{\partial\xi} + \frac{\rho_e u_e}{\rho h_e}f'\frac{\mathrm{d}u_e}{\mathrm{d}\xi}\right] - C\frac{u_e^2}{h_e}(f'')^2 \tag{6.93}$$

Equations (6.92) and (6.93) still exhibit ξ dependency. However, consider the following aspects associated with the stagnation point. First, in the stagnation region u_e is very small, and $h_e = h_0$ (stagnation enthalpy) is very large. Hence

$$\frac{u_e^2}{h_e} \approx 0 \tag{6.94}$$

Next, we observe that the flow velocity is so low in the stagnation region that we can assume almost incompressible flow conditions to exist in the inviscid flow outside the boundary layer. Thus, we use a result from incompressible, inviscid flow at a stagnation point, namely,

$$u_e = \left(\frac{\mathrm{d}u_e}{\mathrm{d}x}\right)_s x \tag{6.95}$$

where $(\mathrm{d}u_e/\mathrm{d}x)_s$ is the velocity gradient at the stagnation point external to the boundary layer. Substitute Eq. (6.95) into (6.33),

$$\xi = \int_0^x \rho_e u_e \mu_e \,\mathrm{d}x = \int_0^x \rho_e \mu_e \left(\frac{\mathrm{d}u_e}{\mathrm{d}x}\right)_s x \,\mathrm{d}x$$

or

$$\xi = \rho_e \mu_e \left(\frac{\mathrm{d}u_e}{\mathrm{d}x}\right)_s \frac{x^2}{2} \tag{6.96}$$

Also, consider the term $\mathrm{d}u_e/\mathrm{d}\xi$. This can be expressed as

$$\frac{\mathrm{d}u_e}{\mathrm{d}\xi} = \left(\frac{\mathrm{d}u_e}{\mathrm{d}x}\right)\left(\frac{\mathrm{d}x}{\mathrm{d}\xi}\right) = \frac{(\mathrm{d}u_e/\mathrm{d}x)}{(\mathrm{d}\xi/\mathrm{d}x)} \tag{6.97}$$

From Eq. (6.33)

$$\frac{\mathrm{d}\xi}{\mathrm{d}x} = \rho_e u_e \mu_e \tag{6.98}$$

Substituting Eq. (6.98) into (6.97), we have

$$\frac{\mathrm{d}u_e}{\mathrm{d}\xi} = \frac{1}{\rho_e u_e \mu_e} \frac{\mathrm{d}u_e}{\mathrm{d}x} \tag{6.99}$$

Substituting Eq. (6.95) into (6.99), we have at the stagnation point

$$\left(\frac{\mathrm{d}u_e}{\mathrm{d}\xi}\right)_s = \frac{1}{\rho_e \mu_e (\mathrm{d}u_e/\mathrm{d}x)_s x}\left(\frac{\mathrm{d}u_e}{\mathrm{d}x}\right)_s$$

or

$$\left(\frac{\mathrm{d}u_e}{\mathrm{d}\xi}\right)_s = \frac{1}{\rho_e \mu_e x} \tag{6.100}$$

Consider the term $(2\xi/u_e)\,\mathrm{d}u_e/\mathrm{d}\xi$, which appears in Eq. (6.92). Using Eqs. (6.95), (6.96), and (6.100), we obtain

$$\frac{2\xi}{u_e}\frac{\mathrm{d}u_e}{\mathrm{d}\xi} = \frac{2[\rho_e \mu_e (\mathrm{d}u_e/\mathrm{d}x)_s (x^2/2)]}{(\mathrm{d}u_e/\mathrm{d}x)_s x}\left(\frac{1}{\rho_e \mu_e x}\right) = 1 \tag{6.101}$$

Consider the term $2\xi(\rho_e u_e/\rho h_e)(\mathrm{d}u_e/\mathrm{d}\xi)$, which appears in Eq. (6.93). Using Eqs. (6.95), (6.96), and (6.100), this becomes

$$2\xi\frac{\rho_e u_e}{\rho h_e}\frac{\mathrm{d}u_e}{\mathrm{d}\xi} = 2\frac{\rho_e}{\rho h_e}\left[\rho_e\mu_e\left(\frac{\mathrm{d}u_e}{\mathrm{d}x}\right)_s\frac{x^2}{2}\right]\left[\left(\frac{\mathrm{d}u_e}{\mathrm{d}x}\right)_s x\right]\left[\frac{1}{\rho_e\mu_e x}\right]$$

$$= \frac{\rho_e}{\rho h_e}\left(\frac{\mathrm{d}u_e}{\mathrm{d}x}\right)_s^2 x^2$$

Because, at the stagnation point, $x = 0$, then the preceding becomes

$$2\xi\frac{\rho_e u_e}{\rho h_e}\frac{\mathrm{d}u_e}{\mathrm{d}\xi} = 0 \tag{6.102}$$

Also, note in Eq. (6.92) that the term ρ_e/ρ can be expressed, for a calorically perfect gas, as

$$\frac{\rho_e}{\rho} = \frac{p_e}{p}\frac{T}{T_e} = \frac{p_e}{p}\frac{h}{h_e} = \frac{h}{h_e} \equiv g \tag{6.103}$$

where we have recognized for a boundary layer that $p_e = p$. Substituting Eqs. (6.94) and (6.101)–(6.103) into Eqs. (6.92) and (6.93), we have

$$\boxed{(Cf'')' + ff'' = (f')^2 - g} \tag{6.104}$$

and

$$\boxed{\left(\frac{C}{Pr}g'\right)' + fg' = 0} \tag{6.105}$$

Equations (6.104) and (6.105) are the governing equations for a compressible, stagnation-point boundary layer. Examining these equations, we see no ξ dependency. Hence, the stagnation-point boundary layer is a self-similar case.

Numerical solutions to Eqs. (6.104) and (6.105) can be obtained by the "shooting technique" as described earlier in the flat-plate case. There is nothing to be gained in going through the details at this stage of our discussion; rather, such details will be deferred until Part 3, where we will discuss at length a solution of the stagnation-point problem in a dissociating and ionizing gas. Instead, we simply state the result of solving Eqs. (6.104) and (6.105), correlated in the following expression obtained from [92].

Cylinder:

$$q_w = 0.57\,Pr^{-0.6}(\rho_e\mu_e)^{1/2}\sqrt{\frac{\mathrm{d}u_e}{\mathrm{d}x}}(h_{aw} - h_w) \tag{6.106}$$

If we had considered an *axisymmetric* body, the original transformation given by Eqs. (6.33) and (6.34) would have been slightly modified as follows:

$$\xi = \int_0^x \rho_e u_e \mu_e r^2 \, dx \tag{6.107}$$

and

$$\eta = \frac{u_e r}{\sqrt{2\xi}} \int_0^y \rho \, dy \tag{6.108}$$

where r is the vertical coordinate measured from the centerline, as shown in Fig. 6.15. Equations (6.107) and (6.108) lead to equations for the axisymmetric stagnation point almost identical to Eq. (6.104) and (6.105), namely,

$$(Cf'')' + ff'' = \tfrac{1}{2}[(f')^2 - g] \tag{6.109}$$

and

$$\left(\frac{C}{Pr} g'\right)' + fg' = 0 \tag{6.110}$$

The derivation of Eqs. (6.109) and (6.110) is left as a homework problem. In turn, the resulting heat-transfer expression is [92] as follows.

Sphere:

$$q_w = 0.763\, Pr^{-0.6} (\rho_e \mu_e)^{1/2} \sqrt{\frac{du_e}{dx}} (h_{aw} - h_w) \tag{6.111}$$

Compare Eq. (6.106) for the two-dimensional cylinder with Eq. (6.111) for the axisymmetric sphere. The equations are the same except for the leading coefficient, which is higher for the sphere. Everything else being the same, this demonstrates that stagnation-point heating to a sphere is larger than to a two-dimensional cylinder. Why? The answer lies in a basic difference between two- and three-dimensional flows. In a two-dimensional flow, the gas has only two directions to move when it encounters a body—up or down. In contrast, in an axisymmetric flow, the gas has three directions to move—up, down, and side-ways, and hence the flow is somewhat "relieved," that is, in comparing two- and three-dimensional flows over bodies with the same longitudinal section (such as a cylinder and a sphere), there is a well-known three-dimensional relieving effect for the three-dimensional flow. As a consequence of this relieving effect, the boundary-layer thickness δ at the stagnation point is smaller for the sphere than for the cylinder. In turn, the temperature gradient at the wall $(\partial T/\partial y)_w$, which is $\mathcal{O}(T_e/\delta)$, is larger for the sphere. Because $q_w = k(\partial T/\partial y)_w$, then q_w is larger for the sphere. This confirms the comparison between Eqs. (6.106) and (6.111).

The preceding results for aerodynamic heating to a stagnation point have a stunning impact on hypersonic vehicle design, namely, they impose the requirement for the vehicle to have a blunt, rather than a sharp, nose. To see this, consider the velocity gradient $\mathrm{d}u_e/\mathrm{d}x$, which appears in Eqs. (6.106) and (6.111). From Euler's equation applied at the edge of the boundary layer,

$$\mathrm{d}p_e = -\rho_e u_e \,\mathrm{d}u_e \tag{6.112}$$

we have

$$\frac{\mathrm{d}u_e}{\mathrm{d}x} = -\frac{1}{\rho_e u_e}\frac{\mathrm{d}p_e}{\mathrm{d}x} \tag{6.113}$$

Assuming a Newtonian pressure distribution over the surface, we have from Eq. (3.2)

$$C_p = 2\sin^2\theta$$

where θ is defined as the angle between a tangent to the surface and the freestream direction. If we define ϕ as the angle between the *normal* to the surface and the freestream, then Eq. (3.2) can be written as

$$C_p = 2\cos^2\phi \tag{6.114}$$

From the definition of C_p, Eq. (6.114) becomes

$$\frac{p_e - p_\infty}{q_\infty} = 2\cos^2\phi$$

or

$$p_e = 2q_\infty \cos^2\phi + p_\infty \tag{6.115}$$

Differentiating Eq. (6.115), we obtain

$$\frac{\mathrm{d}p_e}{\mathrm{d}x} = -4q_\infty \cos\phi\sin\phi\frac{\mathrm{d}\phi}{\mathrm{d}x} \tag{6.116}$$

Combining Eqs. (6.113) and (6.116), we have

$$\frac{\mathrm{d}u_e}{\mathrm{d}x} = \frac{4q_\infty}{\rho_e u_e}\cos\phi\sin\phi\frac{\mathrm{d}\phi}{\mathrm{d}x} \tag{6.117}$$

Equation (6.117) is a general result that applies at all points along the body. Now consider the stagnation-point region, as sketched in Fig. 6.15. In this region, let Δx be a small increment of surface distance above the stagnation point, corresponding to the small change in ϕ, $\Delta\phi$. From Eq. (6.95)

$$u_e = \left(\frac{\mathrm{d}u_e}{\mathrm{d}x}\right)_s \Delta x \tag{6.118}$$

Also, in the stagnation region ϕ is small; hence, from Fig. 6.15,

$$\cos\phi \approx 1 \tag{6.119a}$$

$$\sin\phi \approx \phi \approx \Delta\phi \approx \frac{\Delta x}{R} \tag{6.119b}$$

$$\frac{\mathrm{d}\phi}{\mathrm{d}x} = \frac{1}{R} \tag{6.119c}$$

where R is the local radius of curvature of the body at the stagnation point. Finally, at the stagnation point, Eq. (6.114) becomes

$$C_p = 2 = \frac{p_e - p_\infty}{q_\infty}$$

or

$$q_\infty = \tfrac{1}{2}(p_e - p_\infty) \tag{6.120}$$

Substituting Eqs. (6.118–6.120) into Eq. (6.117), we have

$$\left(\frac{\mathrm{d}u_e}{\mathrm{d}x}\right)^2 = \frac{2(p_e - p_\infty)}{\rho_e \Delta x}\left(\frac{\Delta x}{R}\right)\left(\frac{1}{R}\right)$$

or

$$\frac{\mathrm{d}u_e}{\mathrm{d}x} = \frac{1}{R}\sqrt{\frac{2(p_e - p_\infty)}{\rho_e}} \tag{6.121}$$

Examine Eqs. (6.106) and (6.111) in light of Eq. (6.121). We see that

$$\boxed{q_w \propto \frac{1}{\sqrt{R}}} \tag{6.122}$$

This states that *stagnation-point heating varies inversely with the square root of the nose radius; hence, to reduce the heating, increase the nose radius.* This is the reason why the nose and leading-edge regions of hypersonic vehicles are blunt; otherwise, the severe aerothermal conditions in the stagnation region would quickly melt a sharp leading edge. Indeed, for Earth entry bodies, such as the Mercury and Apollo space vehicles (see Fig. 1.7), the only viable design to overcome aerodynamic heating is a very blunt body. The derivation leading to Eq. (6.122) is quantitative proof of the need for blunt bodies in hypersonic applications. There is also an important qualitative rationale for hypersonic blunt bodies, which as presented in [1] and [5] and hence will not be repeated here. The reader is encouraged to examine these qualitative discussions in [1] and [5], in order to acquire a more in-depth understanding of the hypersonic aerodynamic heating differences between slender and blunt bodies. The fact that q_w is inversely proportional to $\sqrt{R}$ is experimentally verified in Fig. 6.16, obtained from [81]. Here, various sets of experimental data for C_H at the stagnation point are plotted vs Reynolds number based on nose diameter; the abscissa is essentially proportional R. This is a log-log plot, and the data exhibit a slope of -0.5, hence verifying that $q_w \propto 1/\sqrt{R}$.

As a corollary to the preceding discussion on stagnation-point heating, we note that for a laminar flow *around* a cylindrical or spherical nose, q_w drops considerably with distance from the stagnation point. This is graphically demonstrated in Fig. 6.17, taken from [81]. Here, the heat-transfer distribution around a circular cylinder is given in terms of $q_w(\phi)/q_w(0)$, where $q_w(0)$ is the stagnation-point heat transfer and ϕ is the angle shown in Fig. 6.15. Figure 6.17 displays experimental data recently obtained by Koppenwallner at Germany's DFVLR and reported in [81]. The solid curve in Fig. 6.17 is simply a fairing of the data. Note the rapid drop in q_w as ϕ increases. The local values of q_w vary approximately as $\cos^{3/2}\phi$. Indeed, Beckwith and Gallagher [93] have given the following curve fit for heat-transfer data around an unswept circular cylinder:

$$Nu = Nu_s\,(0.7\cos^{3/2}\phi + 0.3)$$

where Nu_s is the Nusselt number at the stagnation point. [Recall from Eq. (6.64) that $Nu = C_H\,Re\,Pr$.]

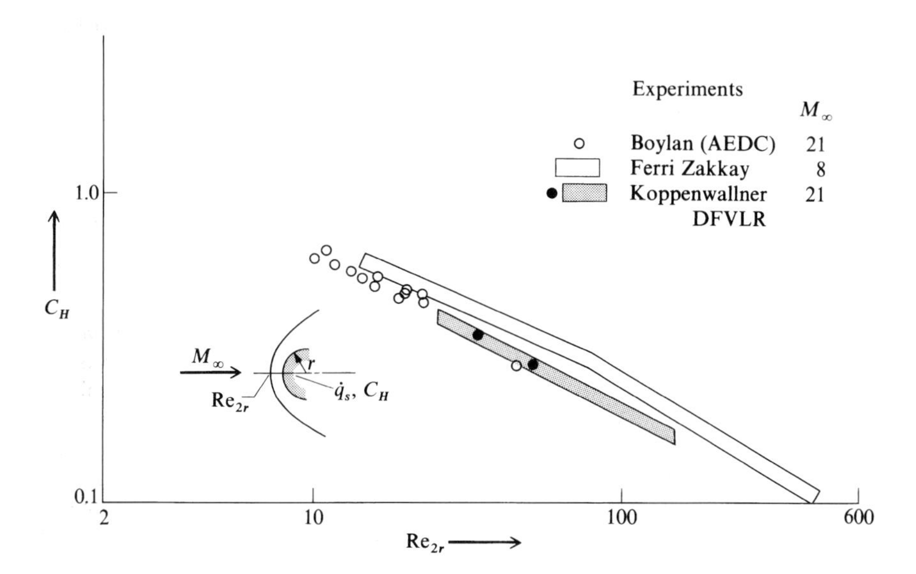

Fig. 6.16 Stagnation-point Stanton number vs *Re* based on nose radius (from Koppenwallner [81]).

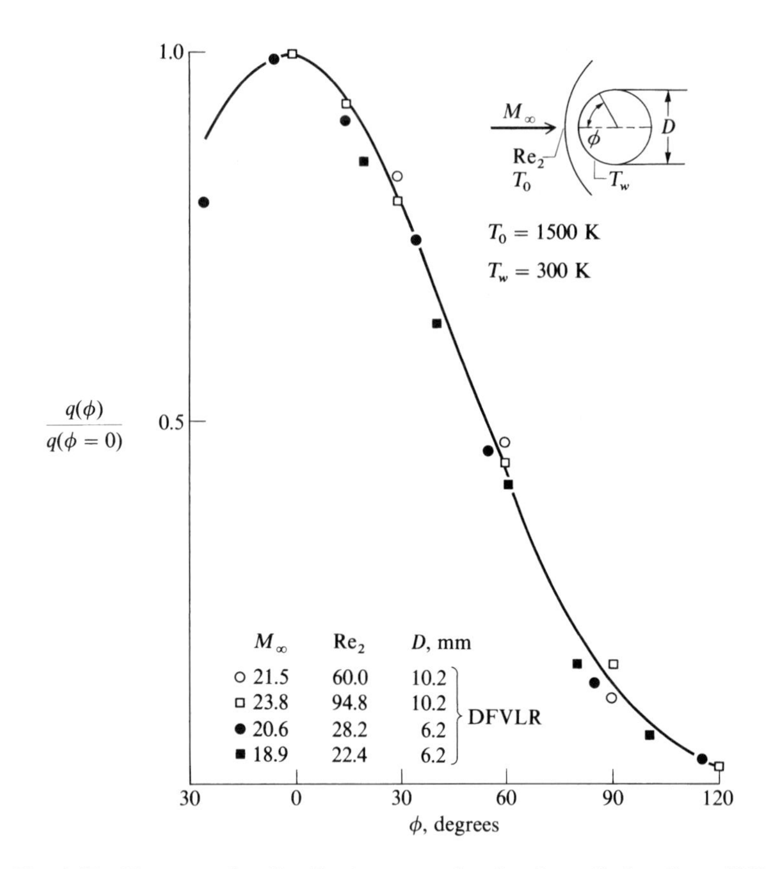

Fig. 6.17 Heat-transfer distribution around a circular cylinder (from [81]).

6.5.3 Summary

This concludes the present section on similar solutions of hypersonic laminar boundary layers. We have seen that the governing boundary-layer equations, which are partial differential equations (6.27–6.30), reduce to a system of ordinary differential equations for the special cases of the flat plate and the stagnation point. Hence these special cases are examples of self-similar solutions. There are other cases where self-similar solutions apply, for example, supersonic and hypersonic flow over a right-circular cone, where the inviscid flow at the edge of the boundary layer is constant, independent of distance from the nose tip. In addition, Cohen and Reshotko [91] give self-similar results for a whole spectrum of external flows generated by

$$u_e = cx^m$$

where c is a constant. Defining a pressure gradient parameter as

$$\beta = \frac{2m}{m+1}$$

Reference [91] tabulates results for $0.326 \leq \beta \leq 2.0$, where $\beta = 0$ and 1 represent the special cases of the flat plate and stagnation point, respectively. For cases other than the preceding, self-similar solutions are not possible. Indeed, the vast majority of hypersonic boundary-layer applications involve general situations where the boundary layers are *nonsimilar*. Such nonsimilar boundary layers are discussed in the next section. However, the time we spent on similar solutions in the present section was in no way wasted. Quite the contrary, the flat-plate and stagnation-point cases represent fundamental applications in hypersonic aerodynamics. They are useful in two regards: 1) as the source of engineering formulas for predicting aerodynamic heating and 2) as a clear demonstration of the basic behavior of hypersonic boundary layers—behavior that is indicative of all hypersonic boundary layers, similar or nonsimilar.

6.6 Nonsimilar Hypersonic Boundary Layers

Prior to 1960, the everyday world of boundary-layer applications emphasized *approximate* solutions of the boundary-layer equations; the only *exact* solutions that were available were the self-similar solutions discussed in Sec. 6.5. Many of the approximate solutions involved the assumption of polynomial profiles for u and h across the boundary layer and the application of the integral forms of the governing equations (in contrast to the partial differential equation form presented in Sec. 6.3). Such integral solutions of the boundary layer are well known and extensively presented elsewhere (for example, see [83]); hence, they will not be considered here.

In the modern world of hypersonic aerodynamics today, exact solutions of the boundary-layer equations (6.27–6.30) can be obtained numerically for arbitrary pressure (hence, velocity) gradients external to the boundary layer. Here, the word "exact" is being used in the same computational-fluid-dynamic sense as in Chapter 5, that is, the exact boundary-layer equations are used, but errors are introduced via the numerical solution in the form of truncation and round-off errors. Hence, we can readily state that the numerical solution of arbitrary nonsimilar boundary layers is a fairly common practice in hypersonic aerodynamics today. The purpose of the present section is to provide the flavor of such nonsimilar solutions; the existing literature in this field is so expansive that we can only highlight some of the more important developments. A thorough review of numerical boundary-layer solutions is given by Blottner in [94], which should be consulted for more details.

In this section, we will introduce three separate methods for solving general, nonsimilar boundary layers: 1) local similarity, 2) difference-differential approach, and 3) finite difference solutions. The first two are somewhat historical in the sense that they are no longer in widespread use today, but they constitute interesting ideas with which any student of hypersonic viscous flows should be acquainted. The last item, finite difference solutions, represents today's state of the art. Finally, in regard to the chapter road map in Fig. 6.4, we are now temporarily at the bottom of the map, that is, the last of the items shown under the box at the left side of the map, labeled "laminar flow boundary-layer equations."

6.6.1 Local Similarity Method

The method of local similarity is not a precisely exact solution for general nonsimilar boundary layers, but it is an important bridge between the exact self-similar technique discussed in Sec. 6.4 and the exact nonsimilar solutions in the present section. The concept of local similarity is outlined next.

1) Consider a boundary layer with properties at the outer edge and at the wall that have an arbitrary variation with x, as sketched in Fig. 6.18.

2) Apply the general transformed boundary-layer equations (6.55) and (6.58) to a *small slice* of the boundary layer located at some local value of x, say, $x = x_1$; this small slice is shown as the shaded region located at x_1 in Fig. 6.18. Take the thickness of this slice Δx to be small enough such that the variations of T_w, u_e, h_e, p_e, etc., over Δx are small. Indeed, assume T_w, u_e, h_e, etc. to be equal to their *local* values at x_1. This includes the gradient $\mathrm{d}u_e/\mathrm{d}\xi$, which is taken to be a numerical value in Eqs. (6.55) and (6.58) equal to its local value at x_1.

3) In Eqs. (6.55) and (6.58), assume that all of the partial derivatives with respect to ξ, namely, $\partial f'/\partial\xi$, $\partial f/\partial\xi$, and $\partial g/\partial\xi$ are small and can be neglected. This is why the local similarity method is an approximate method. In a truly self-similar solution, these derivatives are precisely zero. Here, they are finite, but we assume them to be small enough so that they can be neglected in Eqs. (6.55) and (6.58).

4) Under these assumptions, Eq. (6.55) becomes [recalling Eq. (6.103)]

$$(Cf'')' + ff'' = \frac{2\xi}{u_e}[(f')^2 - g]\frac{\mathrm{d}u_e}{\mathrm{d}\xi} \tag{6.123}$$

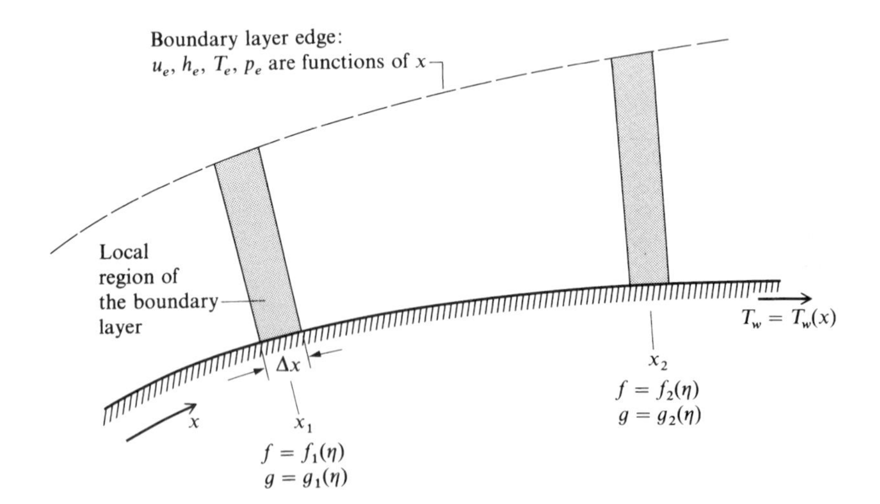

Fig. 6.18 Schematic for the concept of local similarity.

and Eq. (6.58) becomes

$$\left(\frac{C}{Pr}g'\right)' + fg' = 2\xi\frac{\rho_e u_e}{\rho h_e}f'\frac{du_e}{d\xi} - C\frac{u_e^2}{h_e}(f'')^2 + 2\xi\frac{f'g}{h_e}\frac{\partial h_e}{\partial \xi} \tag{6.124}$$

Equations (6.55) and (6.58) are the exact governing, transformed, boundary-layer equations, whereas Eqs. (6.123) and (6.124) are *approximate* forms. At any given x (or ξ) station, $du_e/d\xi$, ρ_e, u_e, and h_e are the *local* values and hence enter Eqs. (6.123) and (6.124) as specific numerical values. Equations (6.55) and (6.58) are partial differential equations; in contrast, Eqs. (6.123) and (6.124) are ordinary differential equations, in the same spirit as in Sec. 6.5. Hence, these equations can be solved (say, by the shooting technique described earlier) at the local value of x. The solution gives $f = f(\eta)$ and $g = g(\eta)$ for the slice of the boundary layer located at x_1 (the first shaded region of Fig. 6.18).

5) Pick another slice of the boundary layer at another value of x, say, $x = x_2$, and repeat the preceding process. This is shown schematically in Fig. 6.18, where two locations are denoted by x_1 and x_2. The preceding locally similar solution is carried out at each value of x, resulting in $f_1(\eta)$ and $g_1(\eta)$ at x_1 and $f_2(\eta)$ and $g_2(\eta)$ at x_2. In general,

$$f_1(\eta) \neq f_2(\eta)$$
$$g_1(\eta) \neq g_2(\eta)$$

Thus, the "locally similar" solution is a solution of the nonsimilar boundary layer, albeit in an approximate sense.

6) After application to many values of x, the preceding procedure yields the skin friction [via $f''(0)$] and heat transfer [via $g'(0)$] as functions of x (numerically).

One of the best examples of the application of local similarity is the work by Kemp et al. [95], which treated the boundary layer over a hemisphere-cylinder as sketched at the top right of Fig. 6.19. This work treated chemically reacting, dissociating air, which is the purview of Part 3. However, these results are presented here to demonstrate the viability of the local similarity method. Figure 6.19 gives the variation of q_w as a function of angular distance ϕ away from the stagnation point, as measured in a shock tube. The freestream conditions were such as to simulate the pressure and enthalpy levels associated with free flight in the atmosphere with a velocity of 18,000 ft/s at an altitude of 70,000 ft. [The actual shock-tube freestream conditions were different than the preceding velocity and altitude; recall only low supersonic Mach numbers can be produced in a shock tube (for example, see [4]), but that the actual stagnation enthalpy and pressure can be directly simulated.] The symbols in Fig. 6.19 denote shock-tube data, and the curve represents the local similarity calculation. Considering the scatter of the shock-tube data, the local similarity method compared fairly well with experiment. The numerical results in [95]

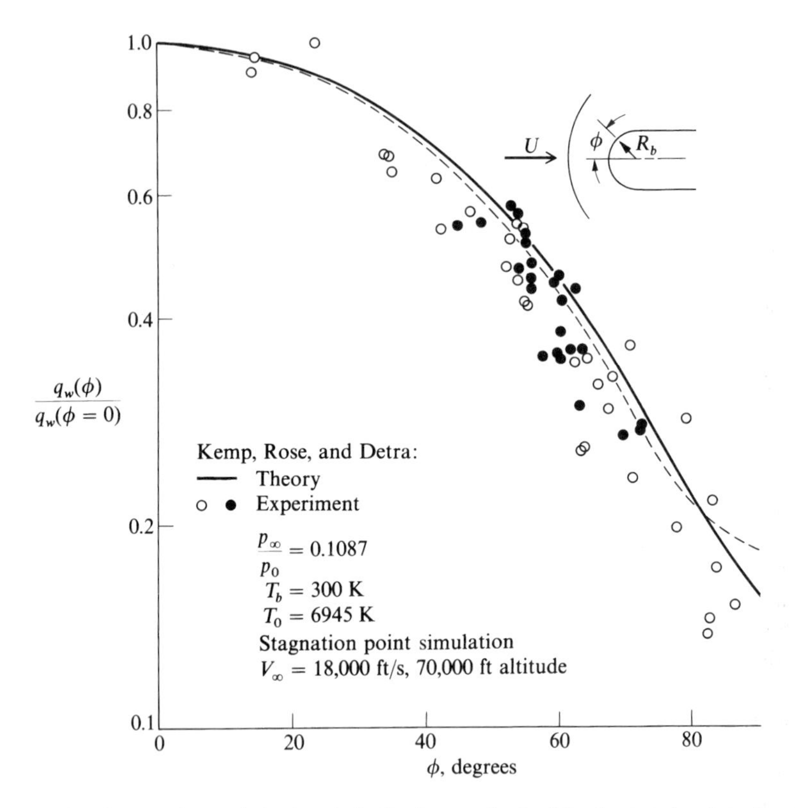

Fig. 6.19 Comparison of the local similarity method with shock-tube data for the heat-transfer distribution over a hemisphere-cylinder (from [95]).

indicate that, for ordinary blunt bodies, the calculated variation of $g'(0)$ as a function of x around the body is very weak. To be precise, for the conditions treated in [95],

$$0.96 \leq \frac{g'(0) \text{ as a function of } x}{g'(0) \text{ at the stagnation point}} \leq 1.03$$

that is, the *transformed* enthalpy gradient at the wall is essentially the same at all points around the body. Because q_w is obtained from Eqs. (6.63) and (6.65), then the strong variation of q_w with ϕ shown in Fig. 6.19 is caused by the variation of h_e, ρ_e, and u_e in those equations.

Note that the local similarity method blocks out any effect of the upstream properties within the boundary layer. The calculation at any given x does not utilize values of the boundary-layer profiles upstream of x. Therefore, the "history" of the development of the boundary layer from the leading edge to the local station x is not properly accounted for in the local similarity method.

However, the upstream history of the *inviscid* flow is transmitted to the local similarity solutions insofar as the local values of u_e, h_e, etc., are influenced by this history. This physical defect does not appear to be serious, probably for the following reason. The boundary-layer equations are parabolic partial differential equations; for such equations, information is readily transmitted across the flow, normal to the surface. At the same time, information is carried downstream; however, the history of this information "damps out" quickly with distance downstream. That is to say, the real, physical boundary layer at any given location x is mainly dominated by the local wall and edge conditions at x, and these conditions are the driving parameters in the local similarity method.

6.6.2 Difference-Differential Method

Unlike the approximate local similarity method just discussed, the difference-differential method is inherently an exact solution of the general boundary-layer equations. The general idea was originated in 1937 by Hartree and Womersley [96], but was not applied in a practical sense until the work of A. M. O. Smith in the 1960s. Smith utilized the difference-differential method extensively and with success; a typical example of his work is represented by [97]. The idea is as follows. Consider the general, transformed boundary-layer equations (6.55) and (6.58). These equations are to be solved for arbitrary flow conditions at the boundary-layer edge and arbitrary wall conditions. This arbitrary boundary layer is sketched in Fig. 6.20. Also shown is a grid network at four different ξ (or x) stations, namely, $(i-2)$, $(i-1)$, i and $(i+1)$. Assume that we wish to calculate the boundary-layer profiles at the station denoted by i. In the difference-differential method, the ξ derivatives in Eqs. (6.55) and (6.58) are replaced by finite difference quotients. For second-order accuracy, Smith used the following three-point one-sided difference:

$$\left(\frac{\partial f}{\partial \xi}\right)_i = \frac{3f_i - 4f_{(i-1)} + f_{(i-2)}}{2\Delta\xi} \qquad \text{at a given } j \tag{6.125}$$

Identical expressions are used for $\partial f'/\partial\xi$ and $\partial g/\partial\xi$. We assume that the boundary-layer profiles have already been solved at locations $(i-1)$ and $(i-2)$; hence, $f_{(i-1)}$ and $f_{(i-2)}$ in Eq. (6.125) are *known numbers.* The only unknown in Eq. (6.125) is f_i. When the difference expressions such as Eq. (6.125) are substituted into Eqs. (6.55) and (6.58), the only derivatives that appear are η derivatives. This can easily be seen by displaying, for example. Eq. (6.55) as follows:

$$(Cf'')' + ff'' = \frac{2\xi}{u_e}\left[(f')^2 - \frac{\rho_e}{\rho}\right]\frac{\mathrm{d}u_e}{\mathrm{d}\xi} + 2\xi\left\{f'\left[\frac{3f' - 4f'_{(i-1)} + f'_{(i-2)}}{2\Delta\xi}\right]\right.$$
$$\left. - f''\left[\frac{3f - 4f_{(i-1)} + f_{(i-2)}}{2\Delta\xi}\right]\right\} \tag{6.126}$$

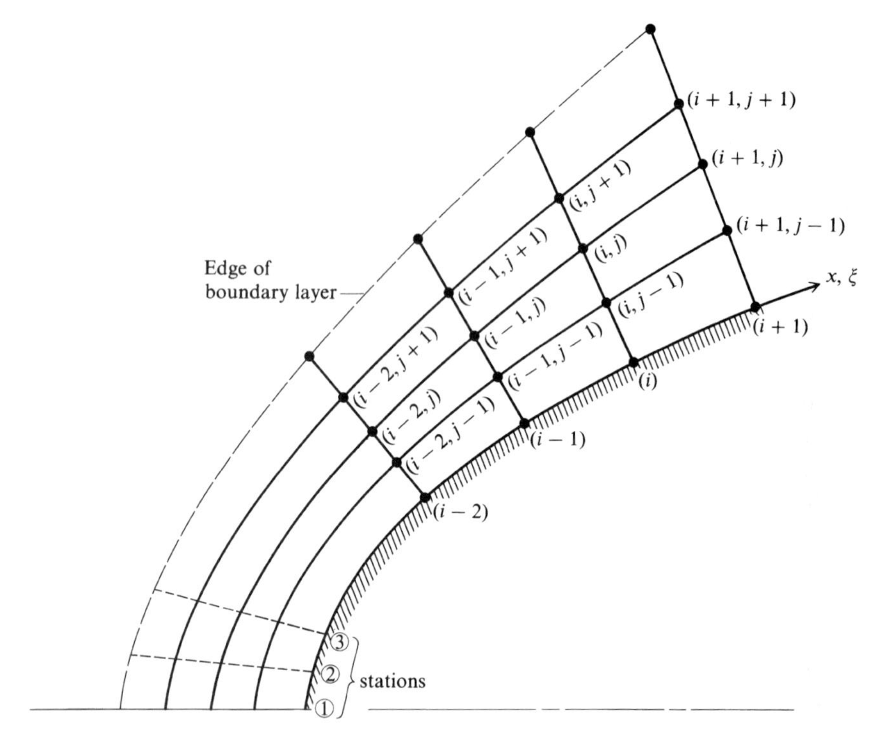

Fig. 6.20 Schematic for finite difference solution of the boundary layer.

where the subscript i has been dropped to emphasize that f_i is simply the unknown f at the given station denoted by i. Equation (6.126) can be rewritten as

$$(Cf'')' + ff'' = \frac{2\xi}{u_e}\left[(f')^2 - \frac{\rho_e}{\rho}\right]\frac{du_e}{d\xi} + \frac{\xi}{\Delta\xi}[3(f')^2 + Af' - 3f''f + Bf''] \quad (6.127)$$

where A and B are simply known numbers, obtained from the previous boundary-layer solutions at stations $(i-1)$ and $(i-2)$. [*Note*: These numbers will change with j as the integration is carried out across the boundary layer at a given station i.] Carrying out the same substitution in Eqs. (6.58), we obtain for the energy equation

$$\left(\frac{C}{Pr}g'\right)' + fg = \frac{\xi}{\Delta\xi}(3f'g + Ef' - 3fg' + Fg') + 2\xi\frac{\rho_e u_e}{\rho h_e}f'\frac{du_e}{d\xi} - C\frac{u_e^2}{h_e}(f'')^2 + \frac{f'g}{h_e}\frac{dh_e}{d\xi} \quad (6.128)$$

where E and F are known numbers. Examine Eqs. (6.127) and (6.128) closely; the only derivatives that appear are η derivatives, denoted by the prime. (Recall that

$du_e/d\xi$ is a known number, obtained from the known velocity variation at the outer edge of the boundary layer.) Hence, Eqs. (6.127) and (6.128) are *ordinary* differential equations that can be integrated across the boundary layer at station i in the same manner as described earlier, that is, using an iterative shooting technique to match the boundary conditions at the wall and the outer edge. Note that Eqs. (6.127) and (6.128) are still the exact boundary-layer equations; no simplifying physical assumptions have been made in going from Eqs. (6.55) and (6.58) to Eqs. (6.127) and (6.128). Therefore, the difference-differential method is an exact method for the solution of general nonsimilar boundary layers.

Returning to Fig. 6.20, recall that the solution at location i is dependent on previous solutions upstream of i, namely, at $(i-1)$ and $(i-2)$. The entire boundary-layer solution must be *started* at some location, say, a leading edge or at a stagnation point. In Fig. 6.20, station 1 denotes the stagnation point. The boundary-layer solution at this location can be obtained from the self-similar solution discussed in Sec. 6.4. Moving to the next downstream location, station 2 in Fig. 6.20, the difference-differential method can be implemented, but with two-point one-sided differences for the ξ derivatives, that is,

$$\frac{\partial f}{\partial \xi} = \frac{f_2 - f_1}{\Delta \xi} \qquad \text{etc.}$$

Moving to the next downstream location, station 3, the full method as just described can now be implemented. In this fashion, the complete nonsimilar boundary layer over the whole body can be calculated.

A word of caution is noted here. In comparison to the usual self-similar equations [such as Eqs. (6.68) and (6.69) for the flat plate], Eqs. (6.127) and (6.128) contain a number of extra terms. These terms act as "forcing functions" and cause the numerical solution of these equations to be much "stiffer" than in the flat-plate or stagnation-point cases, that is, the solution tends to become numerically unstable unless fairly accurate guesses for $f''(0)$ and $g'(0)$ are made at the wall to start the iterative shooting technique. As Smith and Clutter state in [97] "meeting boundary conditions efficiently has been the most difficult part of the entire problem."

Nevertheless the difference-differential method is an exact solution to general, nonsimilar boundary layers, and it has been applied with success to hypersonic problems. An example is given by Fig. 6.21, which shows the heat-transfer distribution over a flat-faced cylinder at Mach numbers between 7.4 and 9.6 obtained from [97]. The open symbols represent experimental data, the dashed line is from the local similarity method as calculated in [95], and the solid curve is from the difference-differential method as calculated in [97]. Note that, as is to be expected, the difference-differential method gives better agreement with experiment than with local similarity. The flow over a flat-faced cylinder is a good test case for any theory because the boundary layer is highly nonsimilar, especially in the rapid expansion region at the corner. Note also the physical trends shown in Fig. 6.21. Here is a case where the stagnation point is *not* the location of maximum heating; rather, the peak heat transfer occurs in the corner region. The physical explanation for this is as follows. The rapid

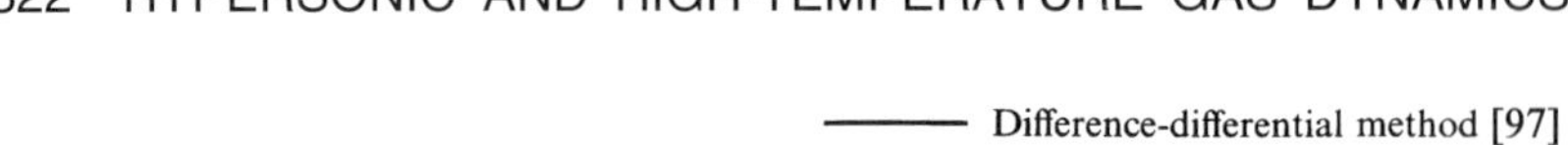

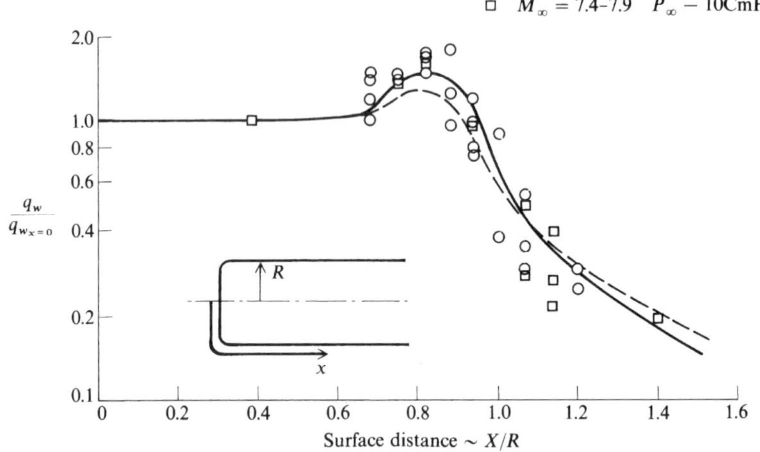

Fig. 6.21 Heat-transfer distribution over a flat-faced cylinder (from [97]).

expansion of the inviscid flow around the corner imposes an extremely large favorable pressure gradient on the boundary layer, which results in an actual reduction of the boundary-layer thickness. In turn, the temperature gradients within the boundary layer, including at the wall, are increased because they are inversely proportional to the boundary-layer thickness, $(\partial T/\partial y)_y = \mathcal{O}(T_e/\delta)$. Because $q_w = k(\partial T/\partial y)_w$, we therefore expect the local heat transfer to increase. This trend is clearly demonstrated in Fig. 6.21.

6.6.3 Finite Difference Method

In the difference-differential method just discussed, the ξ derivatives are replaced by finite differences. The next logical step is to replace *both* the ξ and η derivatives by finite differences. Such finite difference solutions are discussed here; they represent the current state of the art in hypersonic boundary-layer solutions.

Consider again the general, transformed boundary-layer equations given by Eqs. (6.55) and (6.58). Assume that we wish to calculate the boundary layer at station $(i + 1)$ in Fig. 6.20. As discussed in Chapter 5, the general philosophy of finite difference approaches is to evaluate the governing partial differential equations at a given grid point by replacing the derivatives by finite difference quotients at that point. Consider, for example, the grid point (i, j) in Fig. 6.20. At this point, replace the derivatives in Eqs. (6.55) and (6.58) by finite difference expressions of the form:

$$\frac{\partial f}{\partial \xi} = \frac{f_{i+1,j} - f_{i,j}}{\Delta \xi} \tag{6.129}$$

$$\frac{\partial f}{\partial \eta} = \frac{\theta(f_{i+1,j+1} - f_{i+1,j-1})}{2\Delta \eta} + \frac{(1-\theta)(f_{i,j+1} - f_{i,j-1})}{2\Delta \eta} \tag{6.130}$$

$$\frac{\partial^2 f}{\partial \eta^2} = \frac{\theta(f_{i+1,j+1} - 2f_{i+1,j} + f_{i+1,j-1})}{(\Delta \eta)^2} + \frac{(1-\theta)(f_{i,j+1} - 2f_{i,j} + f_{i,j-1})}{(\Delta \eta)^2} \tag{6.131}$$

$$f = \theta f_{i+1,j} + (1-\theta) f_{i,j} \tag{6.132}$$

where θ is a parameter that adjusts Eqs. (6.129–6.132) to various finite difference approaches (to be discussed next). Similar relations for the derivatives of g are employed. When Eqs. (6.129–6.132) are inserted into Eqs. (6.55) and (6.58), along with the analogous expressions for g, two algebraic equations are obtained. If $\theta = 0$, the only unknowns that appear are $f_{i+1,j}$ and $g_{i+1,j}$, which can be obtained directly from the two algebraic equations. This is an explicit approach. Using this approach, the boundary-layer properties at grid point $(i+1, j)$ are solved explicitly in terms of the known properties at points $(i, j+1)$, (i, j), and $(i, j-1)$. [Recall that the boundary-layer solution is a downstream-marching procedure; we are calculating the boundary-layer profiles at station $(i+1)$ only after the flow at the previous station (i) has been obtained.]

When $0 < \theta \leq 1$, then $f_{i+1,\ j+1}$, $f_{i+1,\ j}$, $f_{i+1,\ j-1}$, $g_{i+1,\ j+1}$, $g_{i+1,\ j}$, and $g_{i+1,j-1}$ appear as unknowns in Eqs. (6.55) and (6.58). We have six unknowns and only two equations. Therefore, the finite difference forms of Eqs. (6.55) and (6.58) must be evaluated at *all* of the grid points through the boundary layer at station $(i+1)$ *simultaneously*, leading to an implicit formulation for the unknowns. In particular, if $\theta = 1/2$, the scheme becomes the well-known Crank-Nicolson implicit procedure, and if $\theta = 1$ the scheme is called fully implicit. These implicit schemes result in large systems of simultaneous algebraic equations, the coefficients of which constitute block tridiagonal matrices.

Already the reader can sense that implicit solutions are more elaborate than explicit solutions. Indeed, we remind ourselves that the subject of this book is hypersonics, and it is beyond our scope to go into great computational-fluid-dynamic detail. Therefore, we will not elaborate any further. Our purpose here is only to give the flavor of the finite difference approach to boundary-layer solutions. Chapter 7 of [224] contains a detailed discussion of such matters, and [94] is a thorough survey of the subject. The reader is strongly encouraged to consult these references. We emphasize that modern hypersonic boundary-layer solutions (of an exact nature) are predominately finite difference solutions. They are inherently faster and more accurate solutions than any of the methods discussed before. We will revisit such finite difference solutions in Part 3, when we discuss the analysis of chemically reacting boundary layers.

At this stage, return to the original boundary-layer equations in physical coordinates, Eqs. (6.27–6.30). The finite difference schemes already mentioned can be applied *directly* to these equations; there is no compelling need to deal with the transformed equations. In this case, the derivatives in Eqs. (6.27–6.30) are replaced by difference quotients such as

$$\frac{\partial u}{\partial x} = \frac{u_{i+1,j} - u_{i,j}}{\Delta x}$$

$$\frac{\partial u}{\partial y} = \frac{\theta(u_{i+1,j+1} - u_{i+1,j-1})}{2\Delta y} + \frac{(1-\theta)(u_{i,j+1} - u_{i,j-1})}{2\Delta y}$$

etc.

In this case, the real physical variables are the unknowns, such as $u_{i+1,\,j+1}$, $u_{i+1,\,j}$, $u_{i+1,\,j-1}$, etc. However, when the computations are carried out in physical (x, y) space, the grid spacing in the y direction must be very small; this is because the boundary-layer properties change rapidly near the wall, and the grid must be fine enough to accurately define these changes. Therefore, the transformation to ξ-η space given by Eqs. (6.33) and (6.34) is still useful here because the Lees–Dorodnitsyn transformation stretches the grid in the normal direction, especially near the wall, that is, a uniformly spaced grid in terms of η is equivalent in physical space to fine spacing near the wall, and coarse spacing near the boundary-layer edge, a desirable arrangement for efficient boundary-layer calculations. Therefore, it is frequently recommended to carry out finite-difference calculations using the transformed ξ-η space.

In summary, a finite difference solution of a general, nonsimilar boundary layer proceeds as follows:

1) The solution must be started from a given solution at the leading edge, or at a stagnation point (say, station 1 in Fig. 6.20). As stated earlier, this can be obtained from appropriate self-similar solutions.

2) At station 2, the next downstream station, the finite difference procedure reflected by Eqs. (6.129–6.132) yields a solution of the flowfield variables across the boundary layer.

3) Once the boundary-layer profiles of u and T are obtained, the skin friction and heat transfer at the wall are determined from

$$\tau = \left[\mu\left(\frac{\partial u}{\partial y}\right)\right]_w$$

and

$$q = \left(k\frac{\partial T}{\partial y}\right)_w$$

Here, the velocity gradients can be obtained from the known profiles of u and T by using one-sided differences, such as

$$\left(\frac{\partial u}{\partial y}\right)_w = \frac{-3u_1 + 4u_2 - u_3}{2\Delta y} \tag{6.133}$$

$$\left(\frac{\partial T}{\partial y}\right)_w = \frac{-3T_1 + 4T_2 - T_3}{2\Delta y} \tag{6.134}$$

In Eqs. (6.133) and (6.134), the subscripts 1, 2, and 3 denote the wall point and the next two adjacent grid points above the wall. Of course, because of the specified boundary conditions of no velocity slip and a fixed wall temperature, $u_1 = 0$ and $T_1 = T_w$ in Eqs. (6.133) and (6.134).

4) The preceding steps are repeated for the next downstream location, say, station 3 in Fig. 6.20. In this fashion, by repeating applications of these steps, the complete boundary layer is computed, marching downstream from a given initial solution.

An example of results obtained from such finite difference boundary-layer solutions is given in Figs. 6.22 and 6.23, obtained by Blottner [94]. These are

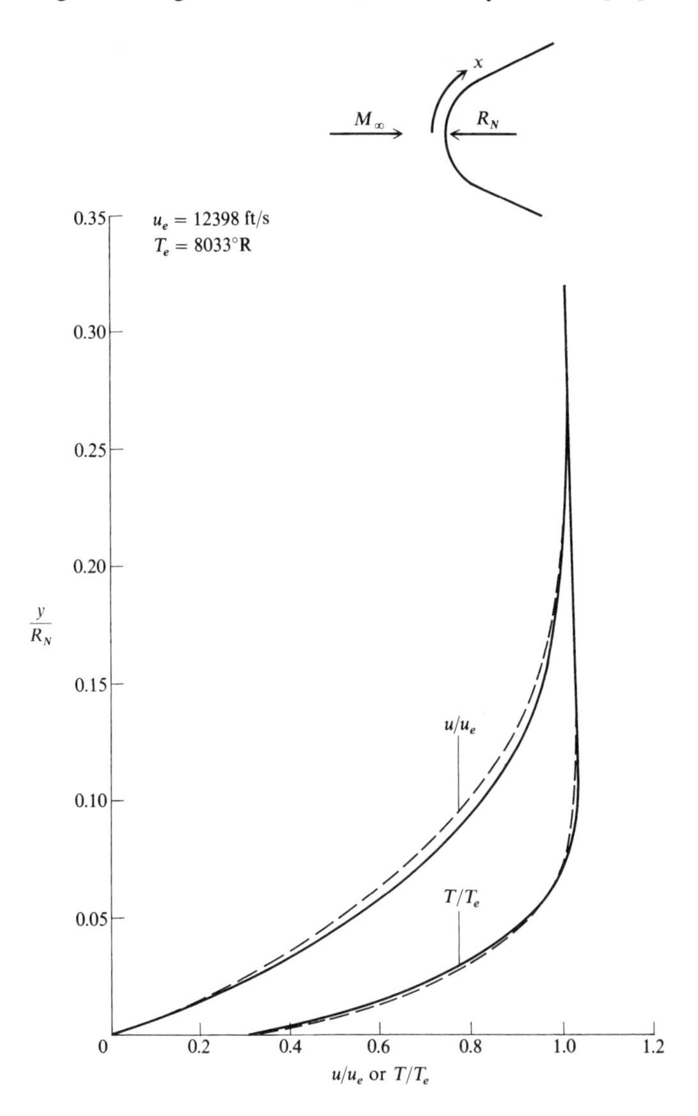

Fig. 6.22 Velocity and temperature profiles across the boundary layer at $x/R_N = 50$ on an axisymmetric hyperboloid (from Blottner [94]).

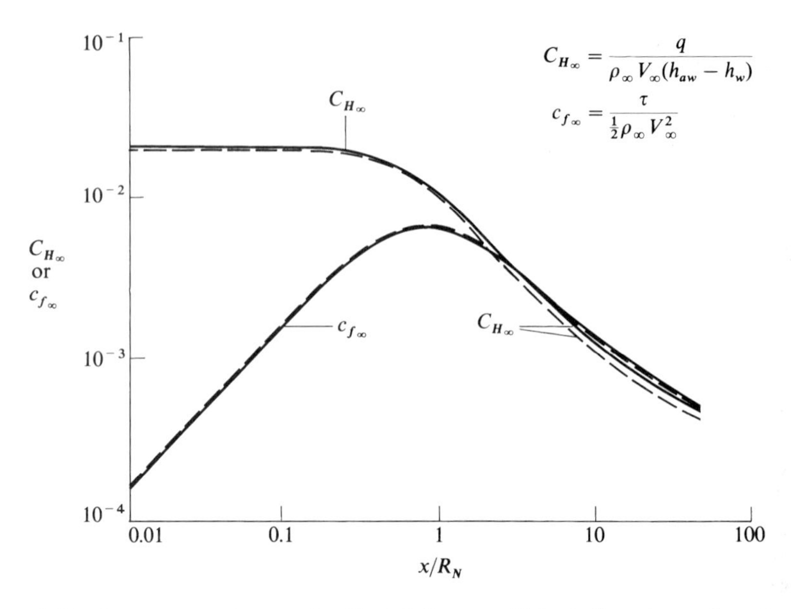

Fig. 6.23 Stanton number and skin-friction coefficient (based on freestream properties) along a hyperboloid (from [94]).

calculated for flow over an axisymmetric hyperboloid flying at 20,000 ft/s at an altitude of 100,000 ft, with a wall temperature of 1000 K. At these conditions, the boundary layer will involve dissociation, and such chemical reactions were included in the calculations of [94]. Chemically reacting boundary layers are the purview of Part 3; however, some results of [94] are presented here just to illustrate the finite difference method. For example, Fig. 6.22 gives the calculated velocity and temperature profiles at a station located at $x/R_N = 50$, where R_N is the nose radius. The local values of velocity and temperature at the boundary-layer edge are also quoted in Fig. 6.22. Considering the surface properties, the variations of C_H and c_f as functions of distance from the stagnation point are shown in Fig. 6.23. Note the following physical trends illustrated in Fig. 6.23:

1) The shear stress is zero at the stagnation point (as is always the case), then it increases around the nose, reaches a maximum, and decreases further downstream.

2) The values of C_H are relatively constant near the nose and then decrease further downstream.

3) Reynolds analogy can be written as

$$C_H = \frac{c_f}{2s} \tag{6.135}$$

where s is called the Reynolds analogy factor. For the flat-plate case, we see from Eq. (6.91) that $s = Pr^{2/3}$. However, clearly from the results of Fig. 6.23 we see that s is a variable in the nose region because C_H is relatively constant while c_f is rapidly increasing. In contrast, for the downstream region, c_f and

C_H are essentially equal, and we can state that Reynolds analogy becomes approximately $C_H/c_f = 1$. The point here is that Reynolds analogy is greatly affected by strong pressure gradients in the flow and hence loses its usefulness as an engineering tool in such cases, at least when C_H and c_f are based on freestream quantities as shown in Fig. 6.23.

6.7 Hypersonic Transition

To this point in our discussion, we have considered *laminar* hypersonic flows. Returning once again to the road map in Fig. 1.24, we have completed the first two items under the viscous-flow branch. In the present section, we will treat the next item, namely, transition from laminar to turbulent flow at hypersonic speeds. Also, in terms of our chapter road map in Fig. 6.4, we move back to the center row of four boxes arrayed horizontally, we are now at the second box from the left, labeled "boundary-layer transition."

There is a basic principle that applies universally in our world, in both physical science and in our daily human activities; simply stated, it is that nature, left to its own devices, always moves toward the state of maximum disorder. This is never more true than in the flow of a viscous fluid; such flows begin in the orderly, smooth manner that we define as laminar flow, but at some downstream region will transit into the disorderly, tortuous motion that we define as turbulent flow. Transition to turbulent flow has been a well-observed phenomena in fluid dynamics since the pioneering work of Osborne Reynolds in the 1880s (see Sec. 4.25 of [1] for an historical sketch of Reynolds, and Sec. 15.2 of [5] for a basic discussion of what is meant by transition from laminar to turbulent flow). On the other hand, although transition has been well observed, it certainly is not well understood, even to the present day. Turbulence, and transition to turbulence, is one of the unsolved problems in basic physics. Our only recourse in aerodynamics is to treat these problems in an approximate, engineering sense, depending always on as large a dose of empirical data as we can find and swallow. This situation is particularly severe at hypersonic speeds, where transition seems to exhibit some peculiar anomalies in comparison to our experience at lower speeds. All of the discussion in the present section is flavored by the preceding remarks.

First, let us address the matter of transition itself; the modeling of fully turbulent flows will be addressed in the next section. For simplicity, first consider the simple picture of transition, as sketched in Fig. 6.24 for flow over a flat surface. As discussed in any basic fluid-dynamics text (for example, see [5]), the flow starts out at the leading edge as laminar; this laminar flow is highly stable, and any disturbances are not amplified. However, at some location downstream the laminar flow becomes unstable, and any disturbances (say, from the freestream, or from the surface such as surface roughness) are now amplified. This point is labeled B in Fig. 6.24, for the beginning of transition. As the amplification of disturbances continues in this unstable flow, transition to turbulence takes place, finally becoming fully turbulent at point E in Fig. 6.24, where point E is the end of transition. The region between points B and E is called the transition region. (See [98] for a discussion of the basic theoretical aspects of

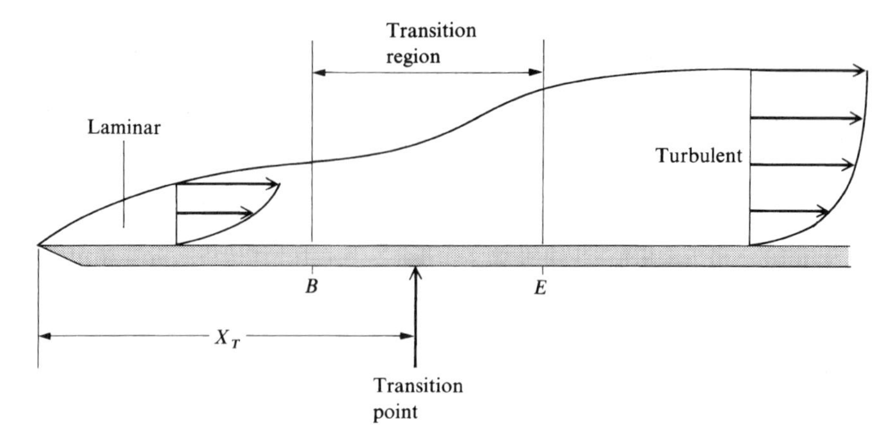

Fig. 6.24 Schematic of transition.

boundary-layer stability and transition to turbulent flow.) Because our knowledge of transition is so imprecise, including our knowledge of the extent of the transition region, engineering analyses frequently assume that transition takes place at a point, labeled the transition point in Fig. 6.24. For purposes of analysis, the flow is assumed laminar upstream of the transition point and fully turbulent downstream. The location of the transition point is given by x_T in Fig. 6.24, and we define a *transition Reynolds number* as

$$Re_T = \frac{\rho_e u_e x_T}{\mu_e} \tag{6.136}$$

For the accurate prediction of skin friction and aerodynamic heating to a body, knowledge of the transition Reynolds number is critical. To date, no theory exists for the accurate prediction of Re_T; any knowledge concerning its value for a given situation must be obtained from experimental data. If the desired application is outside the existing database, then an estimate of Re_T is essentially guesswork. For a state-of-the-art discussion of transition, see the definitive article by Reshotko [99].

Given this situation, in the present section we can only discuss some guidelines for transition at hypersonic speeds. Many of our remarks will be influenced by a recent survey by Stetson [100]. Indeed, Stetson begins by the flat statement that "there is no transition theory," although our database at hypersonic speeds is sufficient to establish some general trends based on experiment. The hypersonic transition Reynolds number can be expressed functionally as

$$Re_T = f\left(M_e,\, \theta_c,\, T_w,\, \dot{m},\, \alpha,\, k_R,\, E,\, \frac{\partial p}{\partial x},\, R_N,\, Re_\infty/\text{ft},\, \frac{x}{R_N},\, V,\, C,\, \frac{\partial w}{\partial z},\, T_0,\, d^*,\, \tau,\, Z\right)$$

where M_e is the Mach number at the edge of the boundary layer, θ_c is a characteristic defining the shape of the body (for a cone θ_c would be the cone angle), T_w

is the wall temperature, $\dot{m}$ is mass addition or removal at the surface, α is the angle of attack, k_R is a parameter expressing the roughness of the surface, E is a general term characterizing the environment (such as freestream turbulence, or acoustic disturbances propagating from the nozzle boundary layer in a wind tunnel), $\partial p/\partial x$ is the local pressure gradient, R_N is the radius of a blunt nose tip, Re_∞/ft is the Reynolds number per foot (to be discussed later), x/R_N is the location of the boundary layer while it is immersed in the entropy layer generated by the nose (effects of the entropy layer can be felt more than 180 nose radii downstream of the tip), V is an index of the vibration of the body, C is the body curvature, $\partial w/\partial z$ is the crossflow velocity gradient, T_0 is the stagnation temperature, d^* is a characteristic dimension of the body, τ is a chemical reaction time, and Z is an index of the magnitude of chemical reactions taking place in the boundary layer. One look at this list, and the reader is justified in becoming frustrated. Clearly, the transition Reynolds number is an elusive quantity, and it is no surprise that our knowledge of it is so imprecise. However, the situation is not hopeless; for any given situation Re_T will be dominated by only a few of the parameters just listed, and the others will be secondary. Let us examine those parameters that seem to be most important for hypersonic speeds.

6.7.1 Mach Number

The Mach number at the edge of the boundary layer M_e has a strong influence on the stability of the laminar boundary layer and through this on Re_T. Boundary-layer stability theory shows that stability of the laminar boundary layer is generally enhanced by an increasing Mach number, and hence Re_T is increased with increased M_e, especially above $M_e = 4$. This is dramatically shown in Fig. 6.25, obtained from [101]. Here we see a plot of Re_T vs M_e for sharp

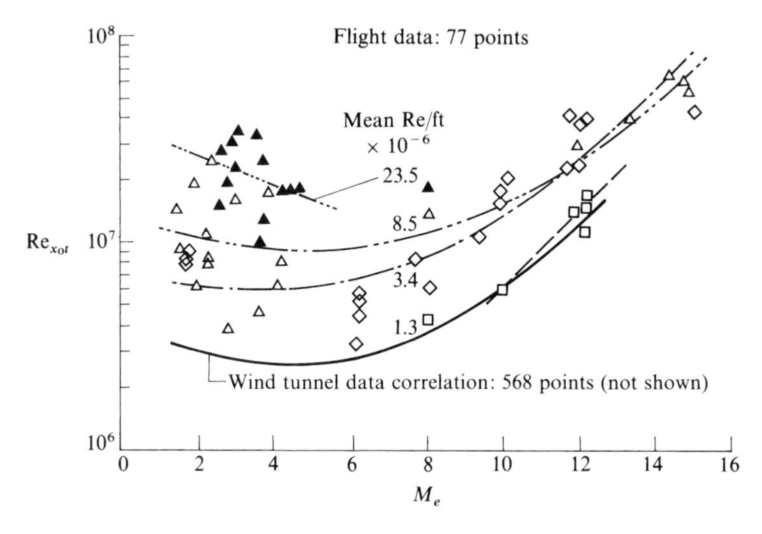

Fig. 6.25 Transition Reynolds-number data on sharp cones from wind tunnels and free flight (from Stetson [100]).

cones in both wind tunnels and free flight. Clearly, above Mach 4, Re_T increases rapidly with M_e. In basic fluid-dynamic courses, a virtual rule of thumb places the transition Reynolds number for incompressible flow over a flat plate near 5×10^5; in contrast, at high hypersonic Mach numbers Re_T can be on the order of 10^8. This effect of Mach number on transition is extremely beneficial. Because skin friction and aerodynamic heating are considerably smaller for laminar in comparison to turbulent flows, the relatively large region of laminar flow that can occur over a body at hypersonic speeds is a very advantageous design feature.

6.7.2 Environment

Transition is quite sensitive to disturbances that come from the environment, such as freestream turbulence, acoustic disturbances from sources either exterior or interior to the body, and disturbances that are introduced into wind-tunnel flows from the active turbulent boundary layer on the walls of the tunnel. These environmental phenomena can make dramatic changes in the transition behavior of a boundary layer. For this reason, wind-tunnel measurements of transition are always compromised by the environmental question; indeed, for hypersonic aerodynamics there is a prevailing feeling that the only meaningful transition data must be obtained from free-flight experiments. This feeling is reinforced by the data in Fig. 6.25 which, in addition to the effect of increasing M_e, also shows a marked difference in data obtained in wind tunnels compared to that obtained in free flight. Note that, as we might expect, the flight data are consistently higher than the correlation of wind-tunnel data.

6.7.3 Unit Reynolds Number

The unit Reynolds number is defined as the Reynolds number based on a unit length, for example, unit $Re = \rho_e u_e x/\mu_e$, where x is taken as 1 ft, or 1 m, yielding the unit Reynolds number per foot or per meter, respectively. There is no basic physical reason to expect the unit Reynolds number to influence transition; however, experimental data clearly show some correlation with unit Reynolds number. Considering again Fig. 6.25, we see that the flight data depend on unit Reynolds number, with Re_T increasing as unit Re increases. The role of unit Reynolds number in determining transition at hypersonic speeds has been the subject of much debate and even disbelief; however, the weight of experimental evidence clearly shows that unit Reynolds number plays a strong role in hypersonic transition. Let us accept this observation at face value here and wait for the future to explain its significance.

6.7.4 Angle of Attack

Three-dimensional flows can have a strong effect on boundary-layer transition. An example is given in Fig. 6.26, which shows the measured transition variation on sharp cones as a function of angle of attack (from [102]). Note that, as α is increased, transition moves rearward on the windward side and

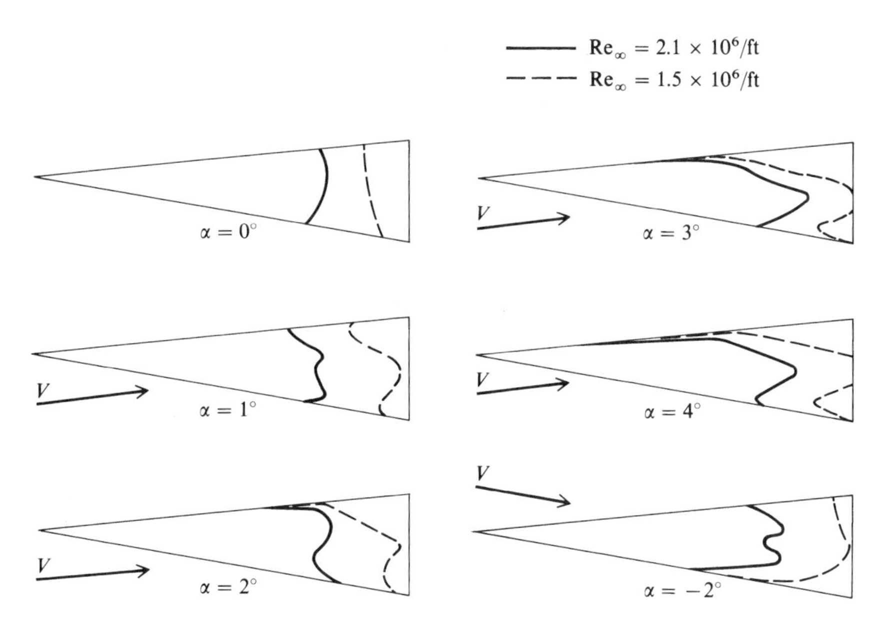

Fig. 6.26 Effect on angle of attack on boundary-layer transition on a sharp cone; θ_c = 8 deg (from DiCristina, [102]).

forward on the leeward side. This is exactly opposite to what might be expected intuitively from results at zero angle of attack. For example, consider the windward ray on the cone in Fig. 6.26. As the angle of attack increases, the local inviscid flow Mach number decreases, and the local Reynolds number increases. Based on experience at zero angle of attack, both of these changes should cause the transition point to move forward. However, Fig. 6.26 shows exactly the opposite. There is clearly an overriding three-dimensional effect. The trends shown in Fig. 6.26 have been observed by many investigators; they are well established in the literature. For example, additional angle-of-attack transition data are shown in Fig. 6.27, obtained from [103]. The case examined is a sharp cone with $\theta_c = 8$ deg at an angle of attack 2 deg in a Mach 6 airflow with $Re_\infty/\text{ft} = 9.7 \times 10^6$. Here, the radial distribution of the transition region is shown by the shaded region; the axial location of transition (measured along the surface from the tip) is plotted vs the radial angle ϕ. The windward ray is denoted by $\phi = 0$ deg, and the leeward ray by $\phi = 180$ deg. The bottom of the shaded region (labeled B) is the beginning of transition, and the top of the shaded region (labeled E) is the end of transition (corresponding to the sketch in Fig. 6.24). Note that the transition region moves upstream as we move around the cone from the windward to the leeward ray, consistent with the results shown in Fig. 6.26. Moreover, note that the length of the transition region decreases as we move around the cone. For comparison, the results for zero angle of attack (labeled $\alpha = 0$ deg) are also shown in Fig. 6.27. Clearly, there is a strong three-dimensional effect on transition. Superimposed on

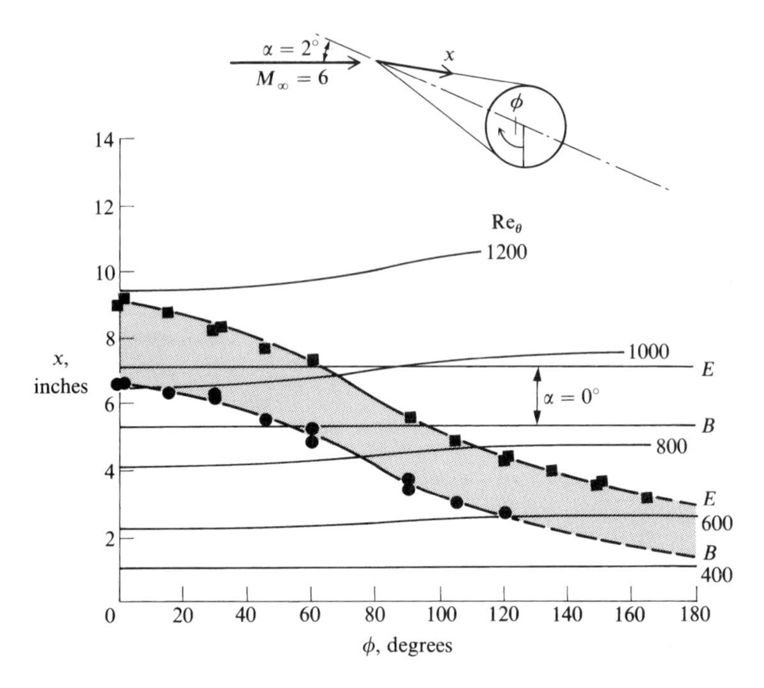

Fig. 6.27 Variation of transition region around a sharp cone at angle of attack: θ_c = 8 deg, $Re/\text{ft} = 9.7 \times 10^6$, M_∞ = 6, and x = 2 deg (from [100]).

Fig. 6.27 are lines of constant Reynolds number based on the boundary-layer momentum thickness Re_θ; the significance of Re_θ will be mentioned later.

6.7.5 Nose Bluntness

As stated in Part 1, the inviscid flow over a blunt-nosed slender body is characterized by the entropy layer created behind the highly curved bow shock wave and wetting the body downstream of the nose. Ramifications of this entropy layer are shown in Fig. 6.28, obtained from [100]. Here, inviscid flow calculations are shown for a blunted, 8-deg cone at zero angle of attack. The nose bluntness is small; the nose radius R_N is only 0.04 in., and the length of the cone is 14 in. The surface values of local Mach number, local static pressure (referenced to the pressure at the stagnation point p_{ST}), and local unit Reynolds number are plotted vs surface distance from the nose. The sharp cone values are given by the dashed lines at the right. In spite of the small nose bluntness, note the dramatic effect of the entropy layer; the local M and Re/ft vary strongly downstream of the nose and do not recover to the sharp cone values until the end of the 14-in. cone. In contrast, the pressure distribution recovers much earlier. This is characteristic of the entropy layer—the thermodynamic properties such as T (hence M through the speed of sound) and ρ are most influenced by the layer. Clearly, the transition behavior of a boundary layer should feel some

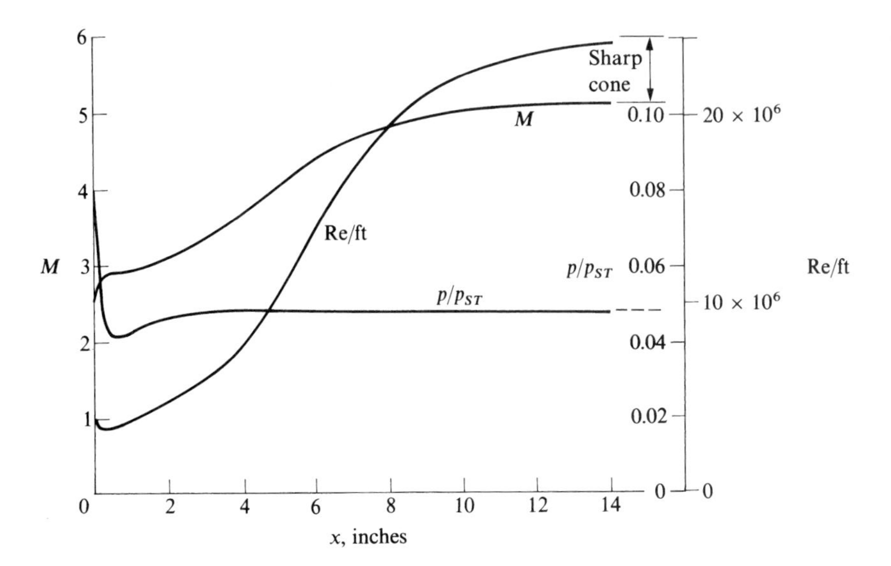

Fig. 6.28 Calculations of inviscid flow over a slender, blunted cone at $\alpha = 0$ deg, $M_\infty = 5.9$, and $\theta_c = 8$ deg. Nose-tip radius $R_N = 0.04$ in. (from [100]).

effect of this entropy layer (in comparison to a sharp cone). This is indeed the case, as shown in Fig. 6.29. This figure is very similar to Fig. 6.27 for a sharp cone, except now Fig. 6.29 includes the effect of nose bluntness, where $R_N = 0.2$ in. Compare Figs. 6.27 and 6.29 closely. Note that by adding a blunt nose to the cone, transition has been delayed to a distance further downstream of the nose tip. This is characteristic of small nose-tip bluntness; the transition Reynolds number is increased by such bluntness. In contrast, for large bluntness transition can occur prematurely on the nose itself, and hence the transition Reynolds number is greatly reduced. This nose-tip transition is often referred to as the "blunt nose paradox." This phenomena occurs in spite of the fact that a strong favorable pressure gradient is present on the nose, especially in the region around the sonic point. In general, favorable pressure gradients stabilize the laminar boundary, whereas adverse pressure gradients are destabilizing. The phenomena of nose-tip transition is contradictory to this general behavior—just another of nature's tricks associated with transition. In summary, we can clearly say that nose bluntness affects transition, but this effect can be different depending on the amount of nose bluntness.

6.7.6 Wall Temperature

Low-speed experiments have shown that wall temperature can have a major influence on transition; for boundary-layer cooling ($T_w < T_{aw}$) the laminar boundary layer is more stable, and transition is delayed, whereas for boundary-layer heating ($T_w > T_{aw}$) the laminar boundary layer is destabilized, and transition occurs earlier. At hypersonic speeds, however, the situation is not so

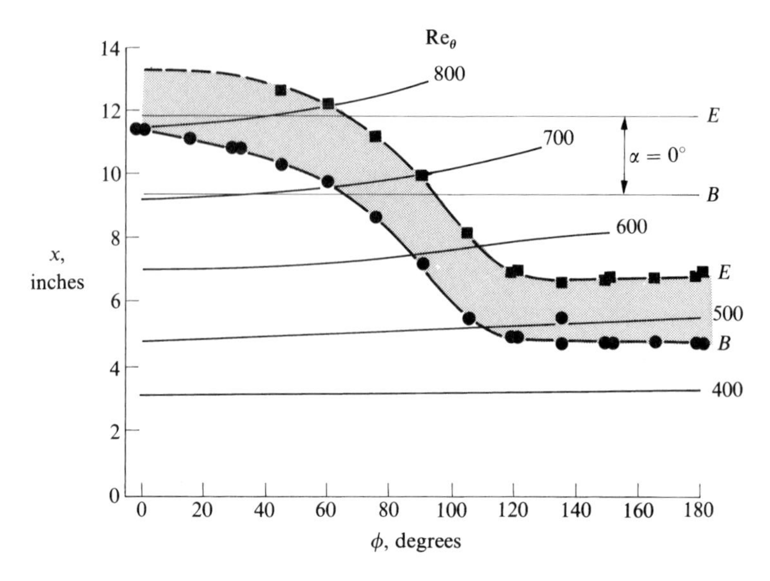

Fig. 6.29 Variation of transition region around a blunt cone at angle of attack: θ_c = 8 deg, $Re/\mathrm{ft} = 19.4 \times 10^6$, $M_{1f} = 6$, x = 2 deg, and R_N = 0.2 in. (from [100]).

clear. For moderate cooling, the hypersonic boundary layer is indeed stabilized, and the transition Reynolds number is increased (transition is delayed), just as observed in the low-speed case. However, for highly cool walls there is evidence of a reversal, where the transition Reynolds number actually decreases. As stated by Stetson in [100], "transition reversal, as a result of wall cooling, has remained a controversial subject." In the present book, we leave it at that, also.

This ends our discussion of the physical phenomena that affect transition at hypersonic speeds. We have highlighted only a few important trends—there are many others. The reader is encouraged to examine [100] for a more complete state-of-the-art discussion. We now proceed to examine a few methods, albeit very imprecise, for the prediction of transition.

6.7.7 *Prediction of Transition*

As unknown and tenuous as the phenomenon of transition is, in applied aerodynamics it is vital to have some engineering means of predicting the transition Reynolds number, even though it might be highly approximate. One prediction method that has been used for hypersonic transition is based on the transition Reynolds number referenced to the boundary-layer momentum thickness θ, where θ is defined as (for example, see [5])

$$\theta = \int_0^\delta \frac{\rho u}{\rho_e u_e}\left(1 - \frac{u}{u_e}\right) \mathrm{d}y \tag{6.137}$$

In turn, the transition Reynolds number can be referenced to the value of θ at transition θ_T:

$$Re_{\theta_T} = \frac{\rho_e u_e \theta_T}{\mu_e} \tag{6.138}$$

An empirical correlation for hypersonic transition that has found some use is

$$\frac{Re_{\theta_T}}{M_e} = 100 \tag{6.139}$$

where M_e is the local Mach number at the edge of the boundary layer. An expression similar to Eq. (6.139) was used for the preliminary design of the space shuttle.

Another prediction correlation, based on the cone data of [102], has been used recently by Bowcutt et al. [80] in a study of hypersonic waveriders, as follows:

$$\log_{10}(Re_T) = 6.421 \exp[1.209 \times 10^{-4} M_e^{2.641}] \tag{6.140}$$

Equation (6.140) is more convenient than Eq. (6.139) because it gives Re_T directly, rather than involving the momentum thickness. However, there is no reason to favor one correlation over the other. Furthermore, neither might be appropriate for new conditions outside the data on which they are based, and neither take into account many of the coupled physical phenomena discussed earlier. About all we can say in defense of Eqs. (6.139) and (6.140) (or others like them) is that they are better than nothing. In the design of hypersonic vehicles, it is usually necessary to make some estimate of where transition occurs, and this is where correlations such as Eqs. (6.139) or (6.140) are useful. However, the user must realize the uncertainty involved in such correlations—uncertainty that we cannot even quantize in most applications.

We end this discussion of transition with the following comments. The accurate prediction of transition at hypersonic speeds is currently one of the leading state-of-the-art questions. Its ultimate solution will most likely come when we obtain the ultimate understanding of the basic problem of turbulence itself. In the meantime, we must continue to make engineering estimates based on the most appropriate data available. Perhaps one of the most eye-opening aspects of the importance of transition are some unpublished design studies of hypersonic transatmospheric vehicles, where, depending on the criteria used for the transition Reynolds number, the weight of the vehicles varied by as much as 50%—truly a practical and driving motivation to improve our abilities in this area.

6.8 Hypersonic Turbulent Boundary Layer

At this point in our discussion, we now assume that the matter of where transition occurs has been reconciled, and we now ask the question: how do we analyze the turbulent boundary layer itself? There is no precise answer to this question; the analysis of turbulent boundary layers is in the same category as transition, that is, empirical data are required, and there is always an uncertainty (sometimes substantial) in the results. A huge amount of literature has been accumulated on turbulent boundary-layer analysis, covering the flight spectrum from incompressible to hypersonic. Whole books are devoted to this subject (for example, see [104] and [105]). Also, an extended discussion of hypersonic turbulent boundary layers is given in [106]. Therefore, an in-depth discussion of turbulence effects in hypersonic flow is beyond the scope of this book. Instead, our intent in the present section is to indicate trends and to discuss some pertinent results.

It is well known that, because of the large-scale turbulent motion, energy is transmitted more readily in turbulent boundary layers than in laminar. This is the reason for the fuller velocity profiles through a turbulent boundary layer, and hence the larger velocity gradients at the surface, as is emphasized in any first course in fluid mechanics. In turn, the skin friction and heat transfer are larger, sometimes markedly larger, for turbulent in comparison to laminar flows. These basic trends are no different at hypersonic conditions than they are for low-speed flow.

To include the effects of turbulence in any analysis or computation, it is first necessary to have a model for the turbulence itself. Turbulence modeling is a state-of-the-art subject, and a survey of such modeling as applied to computations is given in [107]. Again, it is beyond the scope of the present book to give a detailed presentation of various turbulence models; the reader is referred to the literature for such matters. Instead, we choose to discuss only one such model here, because 1) it is a typical example of an engineering-oriented turbulence model, 2) it is a model that has been used in many applications in turbulent supersonic and hypersonic flows, and 3) we will discuss several applications in subsequent chapters that use this model. The model is called the Baldwin–Lomax turbulence model, first proposed in [108]. It is in the class of what is called an eddy-viscosity model, where the effects of turbulence in the governing viscous flow equations (such as the boundary-layer equations or the Navier–Stokes equations) are included simply by adding an additional term to the transport coefficients. For example, in all of our previous viscous-flow equations, μ is replaced by $(\mu + \mu_T)$ and k by $(k + k_T)$, where μ_T and k_T are the eddy viscosity and eddy thermal conductivity respectively—both caused by turbulence. In these expressions, μ and k are denoted as the molecular viscosity and thermal conductivity, respectively. For example, the x-momentum boundary-layer equation for turbulent flow is written as

$$\rho u \frac{\partial u}{\partial x} + \rho v \frac{\partial u}{\partial y} = -\frac{\partial p}{\partial x} + \frac{\partial}{\partial y}\left[(\mu + \mu_T)\frac{\partial u}{\partial y}\right] \tag{6.141}$$

Moreover, the Baldwin–Lomax model is also in the class of algebraic, or zero-equation, models meaning that the formulation of the turbulence model utilizes just algebraic relations involving the flow properties. This is in contrast to one- and two-equation models that involve partial differential equations for the convection, creation, and dissipation of the turbulent kinetic energy and (frequently) the local vorticity. (See [105] for a concise description of such one- and two-equation turbulence models.)

The Baldwin–Lomax turbulence model is described next. We give just a "cookbook" prescription for the model; the motivation and justification for the model are described at length in [108]. This, like all other turbulence models, is highly empirical. The final justification for its use is that it yields reasonable results across a wide range of Mach numbers, from subsonic to hypersonic. The model assumes that the turbulent boundary layer is split into two layers, an inner and an outer layer, with different expressions for μ_T in each layer:

$$\mu_T = \begin{cases} (\mu_T)_{\text{inner}} y \leq y_{\text{crossover}} \\ (\mu_T)_{\text{outer}} y \geq y_{\text{crossover}} \end{cases} \tag{6.142}$$

where y is the local normal distance from the wall and the crossover point from the inner to the outer layer is denoted by $y_{\text{crossover}}$. By definition, $y_{\text{crossover}}$ is that point in the turbulent boundary layer where $(\mu_T)_{\text{outer}}$ becomes less than $(\mu_T)_{\text{inner}}$. For the inner region,

$$(\mu_T)_{\text{inner}} = \rho l^2 |\omega| \tag{6.143}$$

where

$$l = ky\left[1 - \exp\left(\frac{-y^+}{A^+}\right)\right] \tag{6.144}$$

$$y^+ = \frac{\sqrt{\rho_w \tau_w} y}{\mu_w} \tag{6.145}$$

and k and A^+ are two dimensionless constants, specified later. In Eq. (6.143), ω is the local vorticity, defined for a two-dimensional flow as

$$\omega = \frac{\partial u}{\partial y} - \frac{\partial v}{\partial x} \tag{6.146}$$

For the outer region,

$$(\mu_T)_{\text{outer}} = \rho K C_{\text{cp}} F_{\text{wake}} F_{\text{Kleb}} \tag{6.147}$$

where K and C_{cp} are two additional constants and F_{wake} and F_{Kleb} are related to the function

$$F(y) = y|\omega|\left[1 - \exp\left(\frac{-y^+}{A^+}\right)\right] \tag{6.148}$$

Equation (6.148) will have a maximum value along a given normal distance y; this maximum value and the location where it occurs are denoted by $F_{\max}$ and $y_{\max}$, respectively. In Eq. (6.147), F_{wake} is taken to be either $y_{\max} F_{\max}$ or C_{wk} $y_{\max} U_{\text{dif}}^2 / F_{\max}$, whichever is smaller, where C_{wk} is a constant, and

$$U_{\text{dif}} = \sqrt{u^2 + v^2} \tag{6.149}$$

Also, in Eq. (6.147), F_{Kleb} is the Klebanoff intermittency factor, given by

$$F_{\text{Kleb}}(y) = \left[1 + 5.5\left(C_{\text{Kleb}}\frac{y}{y_{\max}}\right)^6\right]^{-1} \tag{6.150}$$

The six dimensionless constants that appear in the preceding equations are $A^+ = 26.0$, $C_{\text{cp}} = 1.6$, $C_{\text{Kleb}} = 0.3$, $C_{\text{wk}} = 0.25$, $k = 0.4$, and $K = 0.0168$. These constants are taken directly from [108] with the understanding that, although they are not precisely the correct constants for most flows in general, they have been used successfully for a number of different applications. Note that, unlike many algebraic eddy-viscosity models that are based on a characteristic length, the Baldwin–Lomax model is based on the local vorticity ω. This is a distinct advantage for the analysis of flows without an obvious mixing length, such as separated flows. Note that, like all eddy-viscosity turbulent models, the value of μ just obtained is dependent on the flowfield properties themselves (for example, ω and ρ); this is in contrast to the molecular viscosity μ, which is solely property of the gas itself.

The molecular values of viscosity coefficient and thermal conductivity are related through the Prandtl number

$$k = \frac{\mu c_p}{Pr} \tag{6.151}$$

In lieu of developing a detailed turbulence model for the turbulent thermal conductivity k_T, the usual procedure is to define a turbulent Prandtl number as $Pr_T = \mu_T c_p / k_T$. Thus, analogous to Eq. (6.151), we have

$$k_T = \frac{\mu_T c_p}{Pr_T} \tag{6.152}$$

where the usual assumption is that $Pr_T = 1$. Therefore, μ_T is obtained from a given eddy-viscosity model (such as the Baldwin–Lomax model), and the corresponding k_T is obtained from Eq. (6.152).

The Baldwin–Lomax model just discussed is just one of many eddy-viscosity turbulence models that have been advanced over the years. For basic flows, such as flow over a flat plate, many of these models are quite accurate. Let us examine in more detail results obtained for hypersonic turbulent flow over a flat plate. Such solutions can be obtained by utilizing the boundary-layer equations (6.27–6.30) with $\partial p/\partial x = 0$ and with the transport properties μ and k directly replaced by the sums $(\mu + \mu_T)$ and $(k + k_T)$, respectively. Results for the variation of c_f with Mach number are given in Fig. 6.30, obtained from [107]. Here, calculations based on several turbulence models are made: an algebraic (zero-equation) model from [104], a two-equation model from [109]; and two different Reynolds-stress equations (which provide the turbulent stresses directly in the turbulent mean momentum equations) from [109] and [110]. The solid curve in Fig. 6.30 is a prediction by Van Driest [111], which is within 10% of available experimental data and which can be considered a standard for

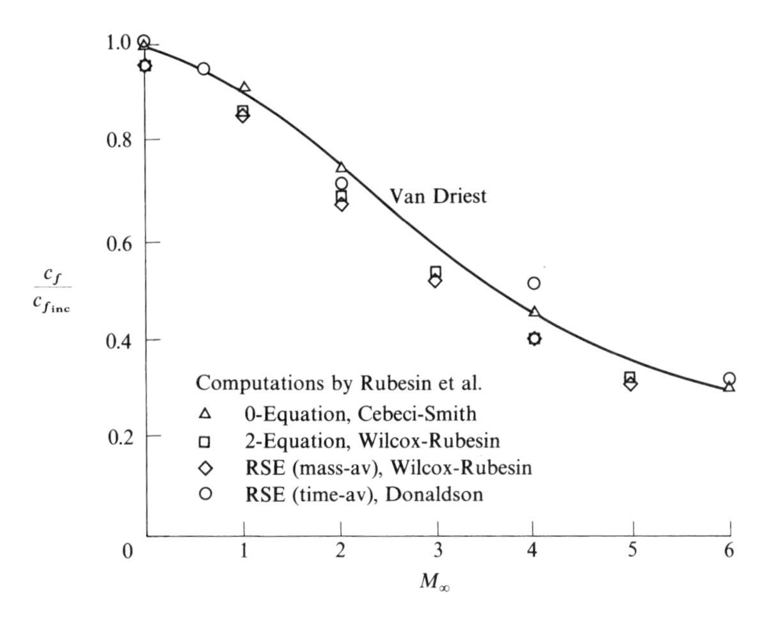

Fig. 6.30 Effects of compressibility on turbulent skin friction on a flat plate: adiabatic wall, where $Re_L = 10^7$ (from Marvin [107]).

comparison. Note that all of the models give essentially the same results. Also, note the important physical variation shown in Fig. 6.30, namely, that *the effect of increasing Mach number is to decrease c_f*. This is the same trend as shown for laminar flow in Fig. 6.11. However, comparing Figs. 6.11 and 6.30, we note that the Mach-number effect is stronger for turbulent flow; the turbulent c_f decreases faster with Mach number in comparison to the laminar results. This trend is further emphasized by the heat-transfer results shown in Fig. 6.31, obtained from [92]. Here, the Stanton number is plotted vs Re, with M_∞, as a parameter; lines for both laminar and turbulent flow are shown. Note in Fig. 6.31 that for a given Re the Mach-number effect is stronger on the turbulent results in comparison to the laminar results. Also note that for a given Me and Re the turbulent values of C_H are considerably larger than the laminar results, which demonstrates the importance of predicting hypersonic turbulent flows.

Some typical experimental heat-transfer data for hypersonic viscous flow over a sharp 8-deg cone at zero angle of attack are shown in Fig. 6.32, obtained from [102]. The freestream Mach number is 10, and the unit Reynold number of 2.1×10^6/ft. Here, $C_H\sqrt{Re_x}$ is plotted vs the running length along the surface of the cone, expressed in terms of the Reynolds number $Re_x = \rho_e u_e x/\mu_e$. At values of Re_x of 3×10^6 or less, the flow is laminar, and the measured Stanton number agrees very well with a theoretical laminar prediction (shown by the dashed line). Transition takes place above $Re_x = 3 \times 10^6$, with fully turbulent flow achieved about $Re_x = 7 \times 10^6$. This figure is shown for several reasons: 1) to illustrate some classical hypersonic results for heat transfer to a basic cone; 2) to further illustrate the phenomena of hypersonic transition; and 3) to

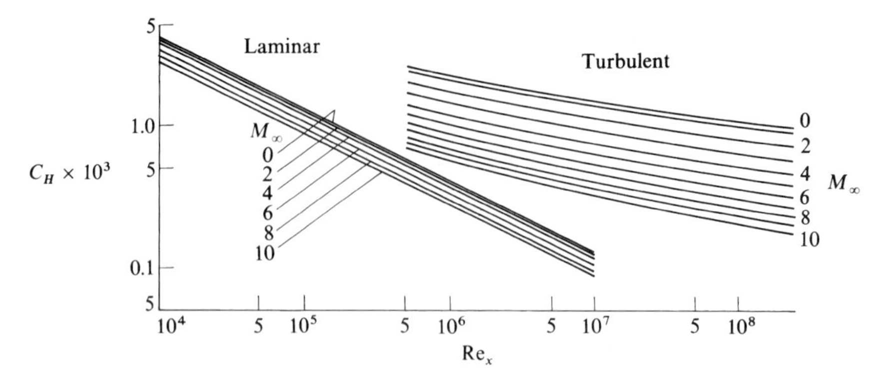

Fig. 6.31 Station number as a function of Reynolds and Mach numbers for an insulated flat plate (from Van Driest [92]).

demonstrate how much turbulent flow can increase the local heat-transfer rate—in the case shown here the increase is over a factor of three.

This concludes our discussion of hypersonic turbulent boundary layers. The subject is virtually inexhaustible, and our purpose here has been to give only its flavor. We have discussed a frequently used eddy-viscosity turbulence

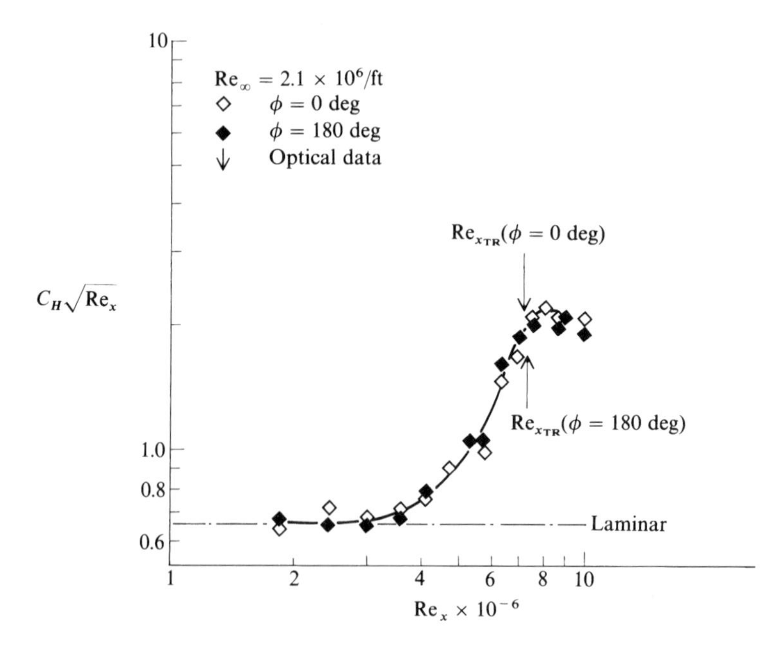

Fig. 6.32 Stanton number for a sharp cone: $\theta_c = 8$ deg, $M_\infty = 10$, and Re/ft = 2.1×10^6 (from [102]).

model, namely, the Baldwin–Lomax model, and we have shown some results for hypersonic turbulent flows over flat plates and cones. These basic flows were chosen to illustrate the trends associated with high-Mach-number effects on turbulent boundary-layer flows. For more detailed information on more complex flows, the reader is referred to the literature. In addition, for an in-depth study of the general aspects of hypersonic turbulent boundary layers, make certain to read the references given in this section.

6.9 Reference Temperature Method

In this section we discuss an approximate engineering method for predicting skin friction and heat transfer for both laminar and turbulent hypersonic flow. It is based on the simple idea of utilizing the formulas obtained from incompressible flow theory, wherein the thermodynamic and transport properties in these formulas are evaluated at some reference temperature indicative of the temperature somewhere inside the boundary layer. This idea was first advanced by Rubesin and Johnson in [112] and was modified by Eckert [113] to include a reference enthalpy. In this fashion, in some sense the classical incompressible formulas were "corrected" for compressibility effects. Reference temperature (or reference enthalpy) methods have enjoyed frequent application in engineering-oriented hypersonic analyses because of their simplicity. For this reason, we briefly describe the approach here.

Consider the incompressible laminar flow over a flat plate. The local skin-friction coefficient c_f, the overall skin-friction drag coefficient C_f, and the heat-transfer coefficient C_H, obtained from classical theory (for example, see [5] and [83], are respectively

$$c_f = \frac{0.664}{\sqrt{Re_x}} \tag{6.153}$$

$$C_f = \frac{1.328}{\sqrt{Re_c}} \tag{6.153a}$$

$$C_H = \frac{0.332}{\sqrt{Re_x}} Pr^{-2/3} \tag{6.154}$$

where Re and Pr are based on properties at the edge of the boundary layer, that is, $Re_x = \rho_e u_e x/\mu_e$, $Re_c = \rho_e u_e c/\mu_e$, and $Pr = \mu_e c_{p_e}/k_e$. In the preceding, x is the local distance downstream from the leading edge and c is the chord length of the plate.

Now consider the *compressible* laminar flow over a flat plate. In the reference temperature method, the compressible local skin-friction coefficient, the overall skin-friction drag coefficient, and the heat-transfer coefficient are given by expressions analogous to Eqs. (6.153), (6.153a), and (6.154):

$$c_f^* = \frac{0.664}{\sqrt{Re_c^*}} \tag{6.155}$$

$$C_f^* = \frac{1.328}{\sqrt{Re_c^*}} \tag{6.155a}$$

$$C_H^* = \frac{0.332}{\sqrt{Re_x^*}}(Pr^*)^{-2/3} \tag{6.156}$$

where c_f^*, C_f^*, C_H^*, Re_x^*, Re_c^*, and Pr^* are evaluated at a reference temperature T^*. That is,

$$c_f^* = \frac{\tau_w}{\frac{1}{2}\rho^* u_e^2} \tag{6.157a}$$

$$C_f^* = \frac{D_f}{\frac{1}{2}\rho^* u_e^2 S} \tag{6.157b}$$

$$C_H^* = \frac{q_w}{\rho^* u_e(h_{aw} - h_w)} \tag{6.157c}$$

$$Re_x^* = \frac{\rho^* u_e x}{\mu^*} \tag{6.158a}$$

$$Re_c^* = \frac{\rho^* u_e c}{\mu^*} \tag{6.158b}$$

$$Pr^* = \frac{\mu^* c_p^*}{k^*} \tag{6.158c}$$

where ρ^*, μ^*, c_p^* and k^* are evaluated for the reference temperature T^*. From Sec. 6.5 we know that, for compressible flow, c_f and C_H depend on M_e and T_w/T_e. Hence T^* must be a function of M_e and T_w/T_e. From [83] and [113] this function is

$$\frac{T^*}{T_e} = 1 + 0.032\, M_e^2 + 0.58\left(\frac{T_w}{T_e} - 1\right) \tag{6.159}$$

Return to Fig. 6.11, where the solid curves give the exact solutions for compressible laminar flow over a flat plate. The approximate results obtained from the reference temperature method using Eq. (6.155) where T^* is given by Eq. (6.159) are shown as dashed curves in Fig. 6.11. For most of the curves, the reference temperature method falls directly on the exact results, and hence no distinction can be made between the two sets of results; only for the insulated plate is there some discernible difference, and that is small.

To apply the preceding results to cones, simply multiply the right-hand sides of Eqs. (6.155) and (6.156) by the Mangler fraction $\sqrt{3}$. It makes sense that, everything else being equal, the skin friction and heat transfer to the cone should be higher than the flat plate. For the cone, there is a three-dimensional relieving effect that makes the boundary layer thinner. This in turn results in larger velocity and temperature gradients throughout the boundary layer including at the wall and hence yields a higher skin friction and heat transfer than in the two-dimensional boundary layer over a flat plate. Also, the idea of the reference temperature method has been carried over to general three-dimensional flows simply by defining Re_x^* as the running length Reynolds number along a *streamline* (where now x denotes distance along the streamline). This idea is discussed by Zoby et al. [114]. Moreover, in [114] a modified reference temperature approach using the Reynolds number based on momentum thickness is employed. See [114] for details.

For turbulent flow over a flat plate, a reasonable incompressible result is (see [83])

$$c_f = \frac{0.0592}{(Re_x)^{0.2}} \tag{6.160}$$

Carrying over the reference temperature concept to the turbulent case, the compressible turbulent flat-plate skin-friction coefficient can be approximated as

$$c_f^* = \frac{0.0592}{(Re_x^*)^{0.2}} \tag{6.161}$$

where Re_x^* is evaluated at the reference temperature given by Eq. (6.159). The turbulent flat-plate heat transfer can be estimated from a form of Reynolds analogy, written as

$$C_H = \frac{c_f}{2s} \tag{6.162}$$

where s is defined as the Reynolds analogy factor. For reasonable values of s for turbulent flow, see Van Driest [92].

We end this section with the following *caution*. The reference temperature method is *approximate*. Because of its simplicity along with (sometimes) reasonable accuracy, it is useful for preliminary design purposes. In Sec. 6.10, more will be said about its accuracy within the framework of approximate three-dimensional solutions. It is interesting, however, that Dorrance [106] has shown in the special case of the flat plate that the evaluation of the reference temperature is indeed an accurate representation, falling out of the detailed, exact laminar boundary-layer theory discussed in Sec. 6.5. In general, however, it must be realized that the best obtainable accuracy in predicting skin friction and heat transfer over general shapes can only be obtained by a detailed numerical solution of the governing boundary-layer equations (such as discussed in Sec. 6.6), or the complete Navier–Stokes equations, at the cost of considerable complexity and computer time.

Example 6.5

Repeat Example 6.1 except use the reference temperature method for its solution.

Solution: From Example 6.1, $M_\infty = 10$, $p_\infty = 1.01 \times 10^5\ \text{N/m}^2$, and $T_\infty = 288$ K. Also, because we are dealing with a flat plate with an adiabatic wall, from Example 6.3 we have

$$T_w = T_{\text{aw}} = 5155\ \text{K}$$

The reference temperature T^* is obtained from Eq. (6.159) repeated here:

$$\frac{T^*}{T_e} = 1 + 0.032\, M_e^2 + 0.58\left(\frac{T_w}{T_e} - 1\right)$$

where for the flat-plate case $T_e = T_\infty$ and $M_e = M_\infty$. Hence,

$$\frac{T^*}{T_\infty} = 1 + 0.032(10)^2 + 0.58\left(\frac{5155}{288} - 1\right) = 14$$

$$T^* = 14\, T_\infty = 14(288) = 4032\ \text{K}$$

The reference density ρ^* is obtained from the equation of state, recalling that the pressure is constant through the flat-plate boundary layer.

$$\rho^* = \frac{p_\infty}{\text{RT}^*} = \frac{1.01 \times 10^5}{(287)(4032)} = 0.0873\ \text{kg/m}^3$$

The coefficient of viscosity μ^* is obtained from Sutherland's law given by Eq. (6.84a):

$$\frac{\mu^*}{\mu_{\text{ref}}} = \left(\frac{T^*}{T_{\text{ref}}}\right)^{3/2} \frac{T_{\text{ref}} + S}{T^* + S}$$

where $\mu_{\text{ref}} = 1.789 \times 10^{-5}\ \text{kg/(m)(s)}$, $T_{\text{ref}} = 288$ K, and $S = 110$ K. This gives

$$\frac{\mu^*}{\mu_{\text{ref}}} = \left(\frac{4032}{288}\right)^{3/2} \frac{(288 + 110)}{(4032 + 110)} = 5.03$$

or

$$\mu^* = 5.03\mu_{\text{ref}} = 5.03(1.789 \times 10^{-5}) = 9 \times 10^{-5}\ \frac{\text{kg}}{\text{(m)(s)}}$$

Also, from Example 6.1, $u_e = 3402$ m/s, and $x = 0.5$ m. From Eq. (6.158a)

$$Re_x^* = \frac{\rho^* u_e^* x}{\mu^*} = \frac{(0.0873)(3402)(0.5)}{9 \times 10^{-5}} = 1.65 \times 10^6$$

From Eq. (6.155)

$$c_f^* = \frac{0.664}{\sqrt{Re_x^*}} = \frac{0.664}{\sqrt{1.65 \times 10^6}} = 5.17 \times 10^{-4}$$

From Eq. (6.157a)

$$\tau_w = \tfrac{1}{2}\rho^* u_e^2 c_f^* = \tfrac{1}{2}(0.0873)(3402)^2(5.17 \times 10^{-4})$$

$$\tau_w = \boxed{261 \text{ N/m}^2}$$

This compares favorably with the result of $\tau_w = 281.9 \text{ N/m}^2$ obtained in Example 6.1.

Note: In Example 6.1, we used Fig. 6.11 to obtain a value of $c_f\sqrt{Re_x}$. Although the curves given in Fig. 6.11 are obtained from the exact laminar compressible boundary-layer solution for a flat plate, for use in Example 6.1 we can only read the curves within some degree of graphical accuracy. Hence the preciseness of the answer obtained in Example 6.1 might be slightly compromised. Nevertheless, we can say with confidence that results obtained using the reference temperature method compare favorably with exact boundary-layer solutions.

Example 6.4 illustrates the real value of the reference temperature method. In just the few steps shown in this example, we obtained an answer for surface shear stress that otherwise requires a detailed solution of the boundary-layer equations as discussed in Secs. 6.4 and 6.5. The reference temperature method is a back-of-the-envelope solution that is ready made for engineering-oriented preliminary design studies. We will say more about this feature in the Design Examples at the end of this chapter.

6.9.1 Recent Advances: Meador–Smart Reference Temperature Method

The reference temperature method discussed in Sec. 6.9 is a concept that dates back to the late 1940s, but is still a work in progress. Very recently, Meador and Smart in [244] published improved formulas for the calculation of the reference temperature, one for laminar flow and another for turbulent flow. This result for a laminar flow is

$$\frac{T^*}{T_e} = 0.45 + 0.55\frac{T_w}{T_e} + 0.16\,r\left(\frac{\gamma - 1}{2}\right)M_e^2$$

where r is the recovery factor for laminar flow, $r = \sqrt{Pr^*}$.

This method gives a reference temperature equation for turbulent flow slightly different than that for laminar flow. For a turbulent flow, their equation is

$$\frac{T^*}{T_e} = 0.5\left(1 + \frac{T_w}{T_e}\right) + 0.16\,r\left(\frac{\gamma - 1}{2}\right)M_e^2$$

They also use a *local* turbulent skin-friction coefficient for incompressible flow as

$$c_f = \frac{\tau_w}{\frac{1}{2}\rho_\infty u_e^2} = \frac{0.02296}{(Re_x)^{0.139}}$$

When integrated over the entire plate of length c, this gives for the net skin-friction drag coefficient (prove it to yourself)

$$C_f = \frac{D_f}{\frac{1}{2}\rho_\infty V_\infty^2 S} = \frac{0.02667}{(Re_c)^{0.139}}$$

This leads to the analogous expressions for compressible flow with the properties evaluated, as usual, at the reference temperature.

$$c_f^* = \frac{\tau_w}{\frac{1}{2}\rho^* u_e^2} = \frac{0.02296}{(Re_x^*)^{0.139}}$$

and

$$C_f^* = \frac{D_f}{\frac{1}{2}\rho^* u_e^2 S} = \frac{0.02667}{(Re_c^*)^{0.139}}$$

The author has shown in [5] that the Meador–Smart reference temperature method gives slightly more accurate results than the older method described in Sec. 6.9, but for engineering purposes the differences are small. Of primary importance is the fact that the Meador–Smart method is a recent reinforcement of the viability of the reference temperature philosophy in general.

6.10 Hypersonic Aerodynamic Heating: Some Comments and Approximate Results Applied to Hypersonic Vehicles

The present chapter serves as an introduction to the basic physics of hypersonic viscous flow, with primary concentration on boundary-layer theory. We have discussed such diverse topics as exact solutions to hypersonic laminar boundary layers, the uncertainties and approximations associated with transition and turbulence, and an approximate "engineering" method of predicting local skin friction and heat transfer. In the process we have discussed many detailed fluid-dynamic aspects of hypersonic boundary layers. Therefore, it is appropriate at this stage in our discussion to recall the basic *practical* reasons for studying hypersonic viscous flows, as discussed in Sec. 6.1; namely, from the practical aspect of the design of hypersonic vehicles and facilities, we are vitally concerned with the prediction of *surface heat transfer* and *skin friction*. Moreover, of these two items, surface heat transfer is usually the dominant aspect that drives the design characteristics of conventional hypersonic vehicles, although skin friction is very important in tailoring the aerodynamic efficiency of slender vehicles. Because of the importance of aerodynamic heating at hypersonic speeds, the present section provides some elaboration on that topic.

Section 6.1 discussed some of the practical motivation for the concern about aerodynamic heating to hypersonic vehicles; at this stage, the reader should review Sec. 6.1 before progressing further. In particular, in Sec. 6.1 some estimates of the stagnation-point heating to a transatmospheric vehicle were given and compared to that for the space shuttle (see Fig. 6.1). We are now in a position to understand why the aerodynamic heating becomes so large at hypersonic speeds, as demonstrated by the following reasoning. The Stanton number was defined by Eq. (6.63) in terms of the local properties at the outer edge of the boundary layer. If we take the case of a flat plate parallel to the flow, these local properties are freestream values, and C_H can be written as

$$C_H = \frac{q_w}{\rho_\infty V_\infty (h_{\text{aw}} - h_w)}$$

or

$$q_w = \rho_\infty V_\infty (h_{\text{aw}} - h_w) C_H \tag{6.163}$$

Assuming an approximate recovery factor of unity, $h_{\text{aw}} = h_0$, where h_0 is the total enthalpy, defined as

$$h_0 = h_\infty + \frac{V_\infty^2}{2} \tag{6.164}$$

At hypersonic speeds, $V_\infty^2/2$ is much larger than h_∞, and from Eq. (6.164) h_0 is essentially given by

$$h_0 \approx \frac{V_\infty^2}{2} \tag{6.165}$$

Moreover, the surface temperature, although hot by normal standards, still must remain less than the melting or decomposition temperature of the surface material. Hence, the surface enthalpy h_{w} is usually much less than h_0 at hypersonic speeds.

$$h_0 \gg h_w \tag{6.166}$$

Combining Eqs. (6.163–6.166), we obtain the approximate relation that

$$\boxed{q_w \approx \tfrac{1}{2}\rho_\infty V_\infty^3 C_H} \tag{6.167}$$

The main purpose of Eq. (6.167) is to demonstrate that aerodynamic heating increases with the *cube* of the velocity and hence increases very rapidly in the hypersonic flight regime. By comparison, aerodynamic drag is given by

$$D = \tfrac{1}{2}\rho_\infty V_\infty^2 S C_D \tag{6.168}$$

which increases as the square of the velocity. Hence, at hypersonic speeds, aerodynamic heating increases much more rapidly with velocity than drag, and this is the primary reason why aerodynamic heating is a dominant aspect of hypersonic vehicle design. Moreover, from Eq. (6.167), we can understand why Fig. 6.1 indicates that the major aerodynamic heating for a transatmospheric vehicle is encountered during ascent rather than during entry. Some designs call for such a vehicle to accelerate to orbital velocity within the sensible atmosphere (using airbreathing propulsion); hence, high velocity will be combined with relatively high ρ_∞, which from Eq. (6.167) combine to yield very high heating values. In contrast, on atmospheric entry, the transatmospheric vehicle will follow a gliding flight path where deceleration to lower velocities will take place at higher altitudes, hence resulting in lower heating rates than are encountered during ascent. Please note that the preceding discussion is for general guidance only; Eq. (6.167) is approximate only, and moreover C_H and C_D in Eqs. (6.167) and (6.168) respectively both decrease as M_∞ increases (a general trend we have established frequently in our earlier discussions). However, the trends shown by these equations are correct, and they clearly demonstrate why aerodynamic heating progressively becomes more important, and indeed dominant, as the hypersonic flight regime is more deeply penetrated.

Now that we have established the importance of aerodynamic heating, it is instructional to examine various prediction methods for estimating the heat transfer to hypersonic vehicles. Within the context of the ideas presented in the present chapter, the most precise method would be as follows:

1) Calculate the inviscid three-dimensional flow over the vehicle by means of an appropriate numerical CFD technique, such as described in Secs. 5.3 and 5.5. The surface-flow properties from such a calculation will provide the outer-edge boundary conditions for a boundary layer calculation.

2) Using these outer-edge conditions, calculate the boundary-layer profiles by an exact finite difference method, such as described in Sec. 6.6. An important distinction must be made here, however. In Sec. 6.6, only two-dimensional boundary layers were discussed. These two-dimensional calculations could be employed in an approximate sense by following a surface streamline generated by the three-dimensional inviscid flow calculation and ignoring any crossflow gradients perpendicular to the streamline. However, in regions of large crossflow gradients, such a "locally two-dimensional" boundary-layer calculation is certainly not appropriate. The only true exact method would be to carry out a three-dimensional boundary-layer calculation. We have not discussed such three-dimensional boundary-layer calculations—they are beyond the scope of this book. Such calculations are a state-of-the-art research problem today. It is not just a simple matter of adding the third dimension to the boundary-layer equations and then routinely proceeding with a finite difference solution. Any numerical solution of the three-dimensional boundary-layer equations must pay close attention to various "regions of influence" somewhat analogous to those encountered in a method-of-characteristics analysis. However, three-dimensional boundary-layer solutions can, with some effort, be carried out (for example, see [83]). In any event, the locally two-dimensional or precise three-dimensional

boundary-layer solutions will provide detailed flowfield profiles through the boundary layer including of course the local temperature gradient at the surface.

3) Using this local temperature gradient at the surface, the local heat-transfer rate can be calculated: $q_w = k(\partial T/\partial y)_w$.

The application of this approach to calculating the aerodynamic heating distribution over a three-dimensional hypersonic vehicle, although feasible, is costly in terms of the large amount of computer time involved. Moreover, today—if such a detailed calculation is desired—a solution of the complete Navier–Stokes equations such as described in Chapter 8 might be the more appropriate choice. Such matters will be discussed in detail in Chapter 8.

Solutions for the aerodynamic heating distributions as just described are not yet practical for engineering analysis and design, where a large number of different cases are examined. For such applications simpler and, hence, more approximate methods are needed. In the remainder of this section, several such approximate methods are discussed.

In the extreme, perhaps the simplest method for estimating hypersonic aerodynamic heating is to use a generalized form of Eq. (6.167) as

$$q_w = \rho_\infty^N V_\infty^M C \tag{6.169}$$

Such a form was used in [79] for a preliminary analysis of aerodynamic heating to a transatmospheric vehicle and was the basis for the results shown in Fig. 6.1. For these calculations, the following values for N, M, and C were used, where the units for q_w, V_∞, and ρ_∞ were W/cm^2, m/s, and kg/m^3, respectively.

Stagnation point:

$$M = 3, \quad N = 0.5, \quad C = 1.83 \times 10^{-8} R^{-1/2}\left(1 - \frac{h_w}{h_0}\right)$$

where R is the nose radius in meters and h_w and h_0 are the wall and total enthalpies, respectively. With these values of M, N, and C, there is a direct similarity between the approximate Eq. (6.169) and the exact result given by Eq. (6.106). (The demonstration of this similarity is left as a homework problem.)

Laminar flat plate:

$$M = 3.2 \quad N = 0.5 \quad C = 2.53 \times 10^{-9}(\cos\phi)^{1/2}(\sin\phi)x^{-1/2}\left(1 - \frac{h_w}{h_0}\right)$$

where ϕ is the local body angle with respect to the freestream and x is the distance measured along the body surface in meters.

Turbulent flat plate:

$$N = 0.8$$

For $V_\infty \leqslant 3962$ m/s

$$M = 3.37$$

$$C = 3.89 \times 10^{-8}(\cos\phi)^{1.78}(\sin\phi)^{1.6}x_T^{-1/5}\left(\frac{T_w}{556}\right)^{-1/4}\left(1 - 1.11\frac{h_w}{h_0}\right)$$

For $V_\infty > 3962$ m/s

$$M = 3.7$$

$$C = 2.2 \times 10^{-9}(\cos\phi)^{2.08}(\sin\phi)^{1.6}x_T^{-1/5}\left(1 - 1.11\frac{h_w}{h_0}\right)$$

where x_T is the distance measured along the body surface in the turbulent boundary layer.

The preceding is an extreme example of an engineering method for estimating hypersonic aerodynamic heating, requiring the least amount of work and detail. The validity of these correlations is "reasonable" as long as the flight conditions are such that boundary-layer theory is valid. They are useful for preliminary analysis and are not recommended for more detailed work. They are presented here only as an example of the most approximate method for estimating hypersonic aerodynamic heating and for providing information on how the results shown earlier in Fig. 6.1 were obtained.

Note that the preceding method does not directly incorporate the variation of local inviscid flow properties along the surface. In contrast, the use of the reference enthalpy approach, described in Sec. 6.9, has this advantage. An example of an improved engineering method for predicting hypersonic aerodynamic heating, albeit still approximate, is the work of Zoby and Simmonds [115]. Here, the inviscid flow over a hypersonic vehicle is calculated using a version of the approximate thin shock-layer analysis of Maslen, the elements of which are given in Sec. 4.9. The local aerodynamic heating distributions are then obtained from standard incompressible formulas modified for compressible conditions by Eckert's reference enthalpy relation (see Sec. 6.9). Sample results are shown in Fig. 6.33, which gives the local laminar Stanton number (normalized by the stagnation-point value) for the windward centerline for a blunt 25-deg cone at various angles of attack. The freestream Mach number is 7.77. In this figure, s is the distance along the surface of the cone from the nose, and R is the nose radius. The open symbols are experimental data obtained from [116], and the curves are from the approximate calculations of [115]. Reasonable agreement is obtained between the calculations and experiment. Note the expected physical trends shown in Fig. 6.33, namely, 1) heat transfer decreases with distance from the nose and 2) heat transfer increases with increasing angle of attack along the windward centerline.

A more complex heat-transfer calculation applied to the space shuttle has been carried out by Hamilton et al. [117]. The exact three-dimensional inviscid flow is calculated by the time-dependent finite difference approach discussed in Sec. 5.3, yielding an inviscid streamline pattern over the windward surface of the space

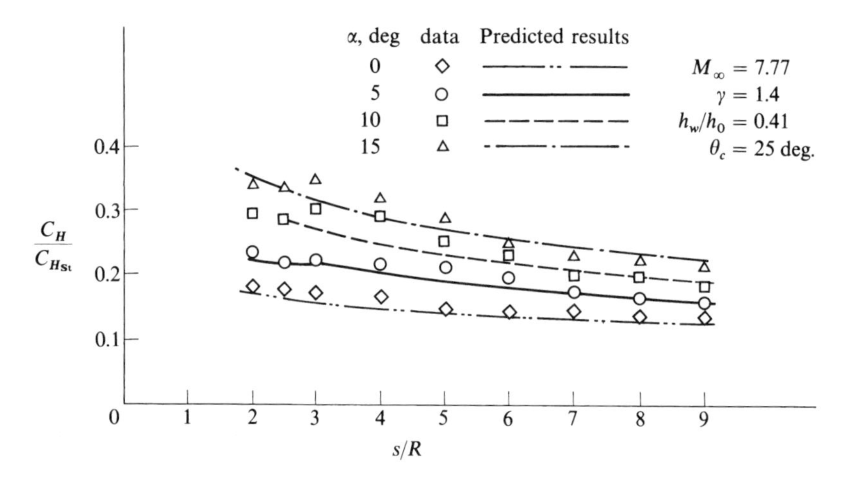

Fig. 6.33 Comparison of predicted [115] and measured [116] laminar heat-transfer rates on a blunt cone (from Zoby and Simmonds [115]).

shuttle as shown in Fig. 6.34 for $M_\infty = 9.15$ and $\alpha = 34.8$ deg. Then, following each streamline, the modified reference temperature method of [114] is used to calculate the aerodynamic heating distributions. The basic ideas of [114] have been discussed in Sec. 6.9; hence, no further elaboration will be given here.

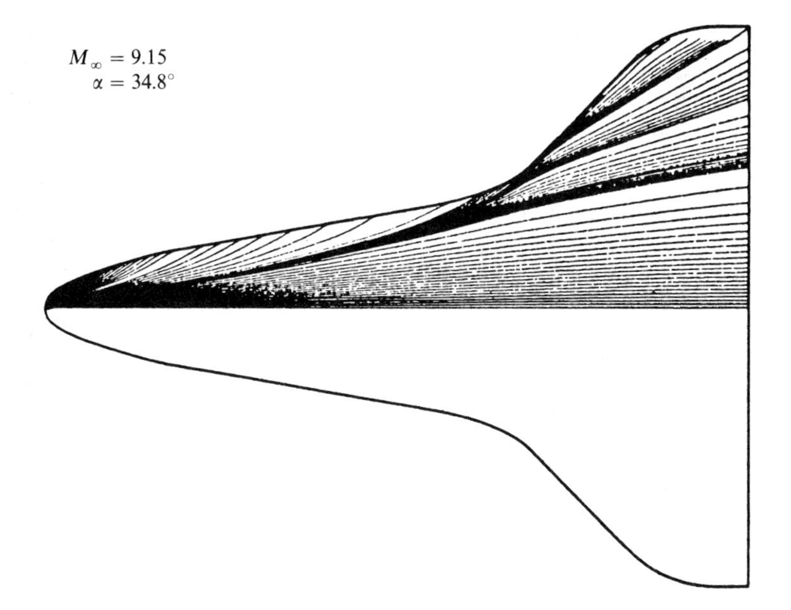

Fig. 6.34 Calculated streamline pattern on the space shuttle (from DeJarnette et al. [118]).

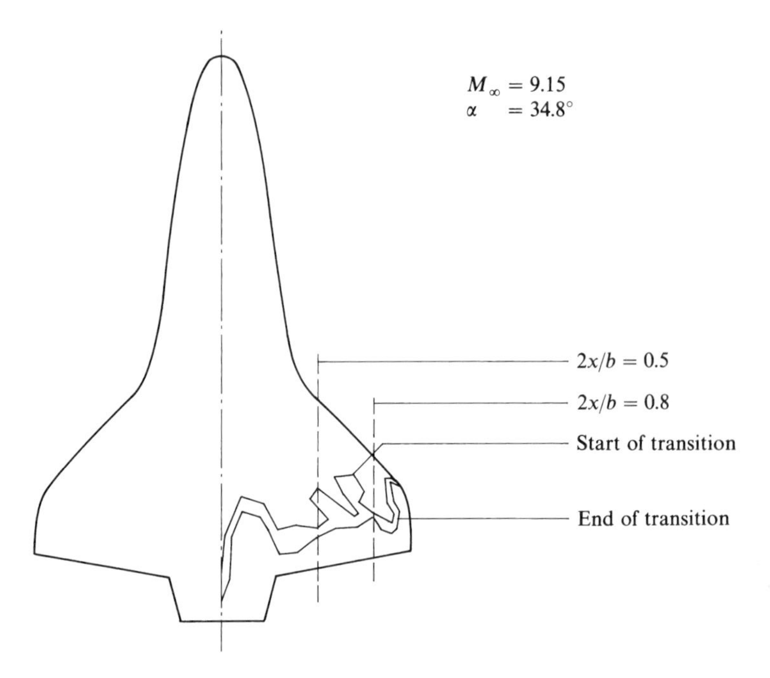

Fig. 6.35 Measured transition region on the space shuttle wing, from the STS-2 flight (from [118]).

Consider a "midwing" chord of the space shuttle located at $2x/b = 0.5$, as shown in Fig. 6.35. Also shown in Fig. 6.35 is the irregular pattern of transition observed from shuttle flight-test data. The calculated streamwise heat-transfer distributions along the chord at $2x/b = 0.5$ are shown in Fig. 6.36, obtained from the method of [114]. Both laminar and turbulent calculations are shown by the solid curves, as reported in [117]. The flight-test data are given by the open circles. These data are bracketed by the laminar and turbulent calculations. Near the leading edge, good agreement is obtained with the laminar calculations, and near the trailing edge good agreement is obtained with the turbulent calculations. This graphically demonstrates the accuracy that can be obtained with approximate heat-transfer calculations in complex flows. The behavior of the flight-test data in the transition region, which at first glance appears irregular (first laminar, then transitional, then laminar, then transitional, then laminar again, finally approaching fully turbulent flow at the trailing edge) is indeed totally consistent with the observed transition pattern shown in Fig. 6.35.

This is a good ending point for our discussion of approximate hypersonic heat-transfer calculations. There are other approximate methods that have been developed over the past years; this section has endeavored to indicate only a few recent approaches. An excellent and authoritative review of approximate aerodynamic heat-transfer methods was published by DeJarnette et al. [118], which the interested reader is encouraged to study carefully. Again, the purpose of this section has been to serve as a counterpoint to our previous discussions concerning exact

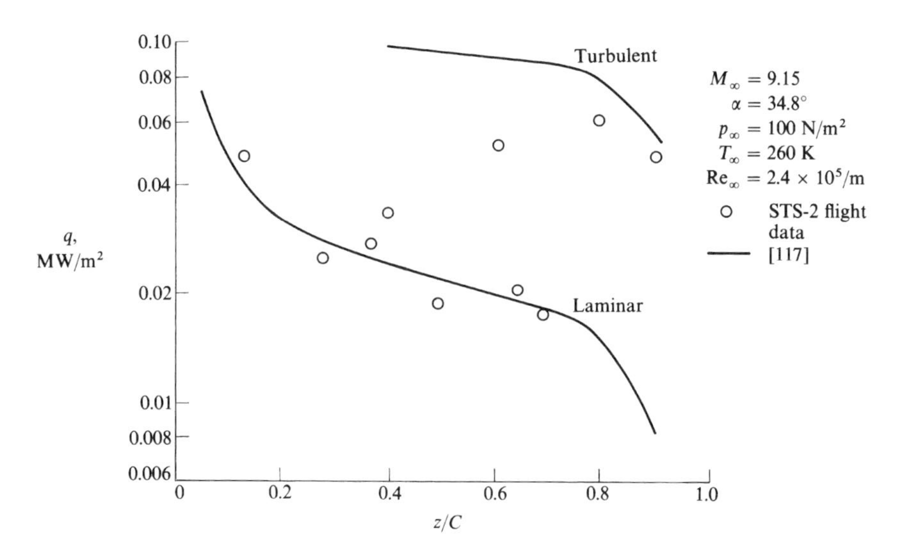

Fig. 6.36 Streamwise distribution of heating on the wing of the space shuttle at $2x/b = 0.5$, where $2x/b$ is the spanwise location shown in Fig. 6.35 (from [117]).

hypersonic boundary-layer calculations and to emphasize the usefulness of approximate heat-transfer analyses for engineering studies. The final choice of an exact or an approximate method for calculating hypersonic aerodynamic heating depends on the problem, the need for accuracy, and the resources available.

6.11 Entropy-Layer Effects on Aerodynamic Heating

Consider the inviscid hypersonic flow over a blunt-nosed body, such as sketched in Fig. 6.37. The surface streamline, which has passed through the normal portion of the bow shock wave, is indicated by the dashed line. Because the flow is inviscid and adiabatic, the entropy is constant along this streamline, and equal to the entropy behind a normal shock wave. According to the usual boundary-layer method, this streamline with its normal shock entropy would constitute the boundary condition at the outer edge of the boundary layer. On the other hand, return to Fig. 1.14 with the attendant discussion in Chapter 1 concerning the entropy layer. Recall that for some distance downstream of the blunt nose the thin boundary layer will be growing inside the entropy layer, and then the boundary layer will eventually "swallow" the entropy layer far enough downstream. In both cases, it is clear that the entropy at the outer edge of the boundary layer will *not* be the normal shock entropy. Therefore, the conventional boundary-layer assumption that the outer-edge boundary condition is given by the inviscid surface streamline as shown in Fig. 6.37 when dealing with blunt-nosed hypersonic bodies is not appropriate.

The interaction of the entropy layer and the boundary layer has been a challenging aerodynamic problem for years. Within the framework of boundary-layer

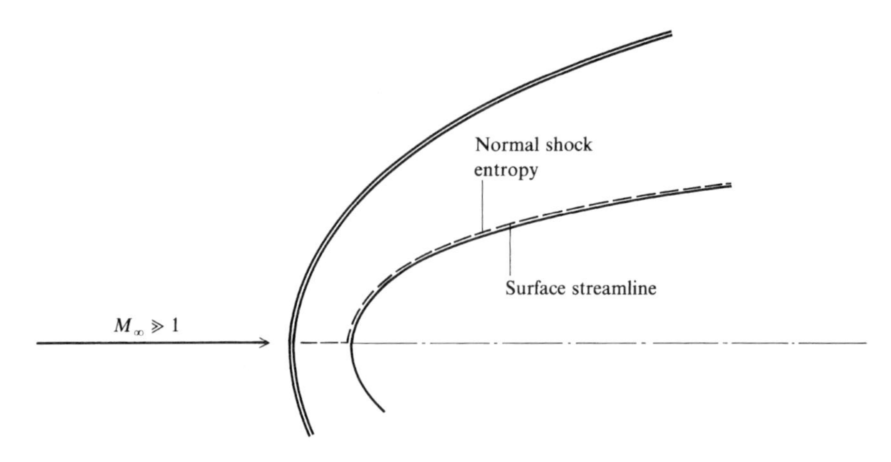

Fig. 6.37 Illustration of the surface streamline containing the normal shock entropy.

analysis, current practice is to estimate the boundary-layer thickness δ and then utilize the inviscid-flow properties located a distance δ from the wall as outer-boundary conditions for the boundary layer. This approximate approach has been used successfully by Zoby and colleagues in [114] and [115] and Hamilton et al. in [117]. The entropy layer can have an appreciable effect on the prediction of hypersonic aerodynamic heating. This is dramatically shown in Fig. 6.38, obtained from [118] and [119]. Here, the aerodynamic heat-transfer distribution along the space shuttle windward ray is shown at the velocity and altitude corresponding to maximum heating along the entry trajectory. The open circles are experimental data extrapolated from wind-tunnel data. The various curves are predictions of the heating distributions from [120] and [121] both making two sets of calculations, first assuming normal shock entropy at the outer edge of the boundary layer and then treating the variable entropy associated with the entropy-layer/boundary-layer interaction. The solid circles are from the calculations of [119], which also account for the entropy layer. Note two important aspects from Fig. 6.38: 1) the presence of the entropy layer *increases* the predicted values of q_w by at least 50%—a nontrivial amount and 2) the taking into account of the entropy layer by using boundary-layer outer-edge properties associated with the inviscid flow a distance δ from the wall gives good agreement with the experimental data.

Clearly, the presence of the entropy layer on a blunt-nosed hypersonic body has an important effect on aerodynamic heating predictions using boundary-layer techniques. However, the simple method just stated appears to be a reasonable approach to including the effect of the entropy layer. Indeed, the heat-transfer predictions shown earlier in Figs. 6.33 and 6.36 take into account the entropy layer as just described, and reasonable agreement with wind-tunnel and flight-test data is obtained.

Finally, the problems discussed in this section concerning the entropy layer are important for boundary-layer calculations. In contrast, when the entire

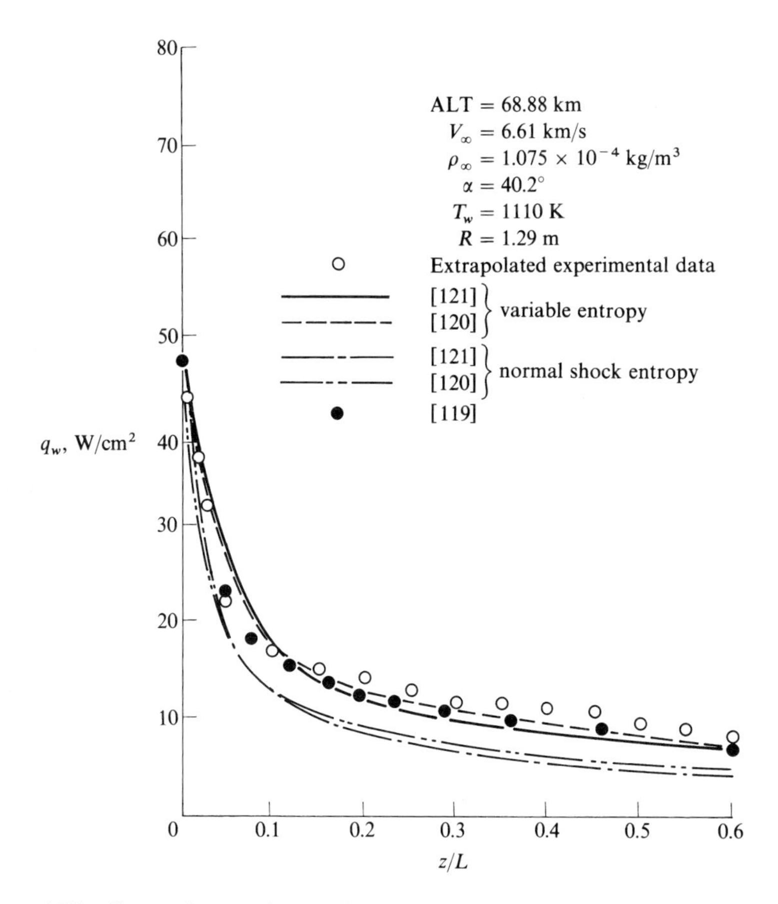

Fig. 6.38 Comparison of predicted shuttle windward-ray heat-transfer distributions; illustration of the entropy-layer effects (from [118]).

shock layer is treated as viscous from the body to the shock wave, the explicit treatment of the entropy layer is not needed. For such viscous shock layers, the interaction between the entropy layer and the shock layer "comes out in the wash"; no separate treatment is required because it is contained within the framework of a fully viscous calculation. Such fully viscous-flow calculations are discussed in Chapter 8.

6.12 Summary and Comments

In the present chapter, we have discussed some basic physical aspects of hypersonic viscous flow and have concentrated on the conventional boundary-layer concept with associated results at hypersonic conditions. Referring again to our road map in Fig. 1.24, we have covered the first five items listed under viscous flows, ranging from basic aspects to approximate engineering methods.

Examine these items in Fig. 1.24, and make certain that you feel comfortable with the associated material in this chapter before you progress further. In the next two chapters, we will treat hypersonic viscous flows by more general and modern (and hence more accurate) methods. However, the boundary-layer theory and results discussed in the present chapter constitute the "bread and butter" of many hypersonic viscous flow applications, and they provide a foundation on which the understanding of hypersonic viscous flow is built. Therefore, in the following paragraphs it is useful to highlight some of the material we have discussed.

The Navier–Stokes equations (6.1–6.5) are the fundamental governing equations for viscous flow. These are coupled nonlinear partial differential equations, difficult to solve by any approach other than detailed numerical solutions (to be discussed in Chapter 8). The boundary-layer equations, obtained from the Navier–Stokes equations by an order-of-magnitude reduction analysis, are simpler to solve and serve as a classical starting point for the analysis of viscous flows. For two-dimensional flow, the boundary-layer equations are as follows.

Continuity:

$$\frac{\partial(\rho u)}{\partial x} + \frac{\partial(\rho v)}{\partial y} = 0 \tag{6.27}$$

x Momentum:

$$\rho u \frac{\partial u}{\partial x} + \rho v \frac{\partial u}{\partial y} = -\frac{\mathrm{d}p_e}{\mathrm{d}x} + \frac{\partial}{\partial y}\left(\mu \frac{\partial u}{\partial y}\right) \tag{6.28}$$

y Momentum:

$$\frac{\partial p}{\partial y} = 0 \tag{6.29}$$

Energy:

$$\rho u \frac{\partial h}{\partial x} + \rho v \frac{\partial h}{\partial y} = \frac{\partial}{\partial y}\left(k \frac{\partial T}{\partial y}\right) + u \frac{\mathrm{d}p_e}{\mathrm{d}x} + \mu \left(\frac{\partial u}{\partial y}\right)^2 \tag{6.30}$$

For hypersonic flow, the constant pressure condition given by Eq. (6.29) can be relaxed; it is appropriate to allow a normal pressure gradient through a hypersonic boundary layer without invalidating the boundary-layer concept.

By transforming the boundary-layer equations through the Lees–Dorodnitsyn transformation,

$$\xi = \int_0^x \rho_e u_e \mu_e \,\mathrm{d}x \tag{6.33}$$

$$\eta = \frac{u_e}{\sqrt{2\xi}} \int_0^y \rho \,\mathrm{d}y \tag{6.34}$$

a form of the boundary-layer equations is obtained as displayed in Eqs. (6.55) and (6.58). In turn, these equations lead to self-similar solutions for the special cases of the flat plate and stagnation point. Defining the skin-friction coefficient as

$$c_f = \frac{\tau_w}{\frac{1}{2}\rho_e u_e^2} \tag{6.59}$$

the Nusselt number as

$$Nu = \frac{q_w x}{k_e(T_{\text{aw}} - T_w)} \tag{6.62}$$

and the Stanton number as

$$C_H = \frac{q_w}{\rho_e u_e (h_{\text{aw}-h_w)}} \tag{6.63}$$

where the Nusselt and Stanton numbers are alternative heat-transfer coefficients related by $Nu = C_H \, Re \, Pr$, we find that

$$C_f = \sqrt{2}\frac{\rho_w \mu_w}{\rho_e \mu_e}\frac{f''(0)}{\sqrt{Re_x}} \tag{6.71}$$

and

$$C_H = \frac{1}{\sqrt{2}}\frac{1}{\mu_e}\frac{k_w}{C_{pw}}\frac{\rho_w}{\rho_e}\frac{h_e}{(h_{\text{aw}} - h_w)}\frac{g'(0)}{\sqrt{Re_x}} \tag{6.77}$$

where $f' = u/u_e$ and $g = h/h_e$. The self-similar solutions for the transformed boundary-layer equations yield numbers for $f''(0)$ and $g'(0)$, giving the following laminar flow results.

Flat plate:

$$c_f = \frac{F(M_e, Pr, \gamma, T_w/T_e)}{\sqrt{Re_x}} \tag{6.76}$$

$$C_H = \frac{G(M_e, Pr, \gamma, T_w/T_e)}{\sqrt{Re_x}} \tag{6.81}$$

Stagnation point:

$$q_w = 0.57\, Pr^{-0.6}(\rho_e \mu_e)^{1/2}\sqrt{\frac{\mathrm{d}u_e}{\mathrm{d}x}}(h_{\text{aw}} - h_w) \qquad \text{(Cylinder)} \tag{6.106}$$

$$q_w = 0.763\, Pr^{-0.6}(\rho_e \mu_e)^{1/2}\sqrt{\frac{\mathrm{d}u_e}{\mathrm{d}x}}(h_{\text{aw}} - h_w) \qquad \text{(Sphere)} \tag{6.111}$$

The stagnation-point heat transfer for a sphere is larger than that for a cylinder because of the three-dimensional relieving effect. At a stagnation point, the skin friction is zero. For hypersonic flow, the velocity gradient $\mathrm{d}u_e/\mathrm{d}x$ is given from Newtonian flow as

$$\frac{\mathrm{d}u_e}{\mathrm{d}x} = \frac{1}{R}\sqrt{\frac{2(p_e - p_\infty)}{\rho_e}} \tag{6.121}$$

From this, we obtain the important result that

$$q_w \propto \frac{1}{\sqrt{R}} \tag{6.122}$$

In general, boundary layers encountering arbitrary streamwise gradients of velocity, pressure, and temperature at the outer edge are nonsimilar. For nonsimilar boundary layers, several methods of solution have been developed, including local similarity, the difference-differential method, and finite difference methods, the latter being the standard approach today.

Detailed boundary-layer solutions such as just mentioned yield the flowfield profiles through the boundary layer, as well as the velocity and temperature gradients at the surface, hence the surface skin friction and heat transfer. These detailed solutions frequently require extensive computer resources. For engineering preliminary analysis, simplified, more approximate methods are useful for rapid estimation of skin friction and aerodynamic heating. The reference temperature (or reference enthalpy) method is an excellent example of such an approximate approach. Calculations as elaborate as the estimation of space shuttle three-dimensional heat-transfer distributions have been made using the reference temperature concept.

Finally, the aspects of hypersonic transition and turbulent flow are extremely important for vehicle design and analysis. Sections 6.7 and 6.8 discuss these matters, emphasizing the basic aspects of transition and turbulence at hypersonic speeds and underscoring the great uncertainties that still exist in our predictions of such phenomena.

Design Example 6.1

The Configuration Based Aerodynamics (CBAERO) software package newly developed at the NASA Ames Research Laboratory is described in a Design Example at the end of Chapter 3. This is a modern design-oriented code for the engineering prediction of hypersonic vehicle aerodynamic characteristics. In Chapter 3, we examined the use of this code for the prediction of pressure distributions over hypersonic bodies of arbitrary shapes. In the present Design Example, we extend these considerations to aerodynamic heating. In CBAERO, the stagnation-point convective heating is predicted using the engineering correlations of Tauber, as discussed in Sec. 6.10 and given in detail in [79] and [245].

The aerodynamic heating over the remainder of the body is calculated by means of the reference enthalpy (reference temperature) method discussed in Sec. 6.9. See [228] and [229] for more details. We simply emphasize that the engineering methods described in this chapter for predicting aerodynamic heating and skin friction, especially the reference temperature method, are alive and well today in the world of modern hypersonic vehicle design.

Return to Fig. 3.29 showing the space shuttle covered by a distribution of panels over its surface used by Kinney and Garcia [229] to implement CBAERO. In [229], rather than plotting the heat-transfer rate itself, Kinney and Garcia plot the resulting surface temperature. They calculate the surface temperature on the basis of a simple energy balance at the surface, assuming that the surface is cooled only by heat radiated away from the surface (see Sec. 1.4 and Chapter 18 for discussions of surface radiative cooling). That is,

$$q_{\text{convective}} + q_{\text{radiative}} = \sigma \varepsilon T_w^4$$

where $q_{\text{convective}}$ is the convective heat transfer to the surface, $q_{\text{radiative}}$ is the radiative heat transfer to the surface stemming from thermal radiation from the hot gas in the shock layer (to be discussed in Chapter 18), and $\sigma \varepsilon T_w^4$ is the thermal energy radiated away from the hot wall. Here, σ is the Stefan–Boltzman constant, ε is the emissivity of the surface, and T_w is the wall temperature. For the atmospheric entry flight path of the space shuttle, $q_{\text{radiative}}$ is small and is usually neglected. The resulting calculations for the centerline temperature distribution over both the bottom and top surfaces of the space shuttle are shown in Fig. 6.39 for the case of $M_\infty = 24.87$ and an angle of attack of 40 deg. Two

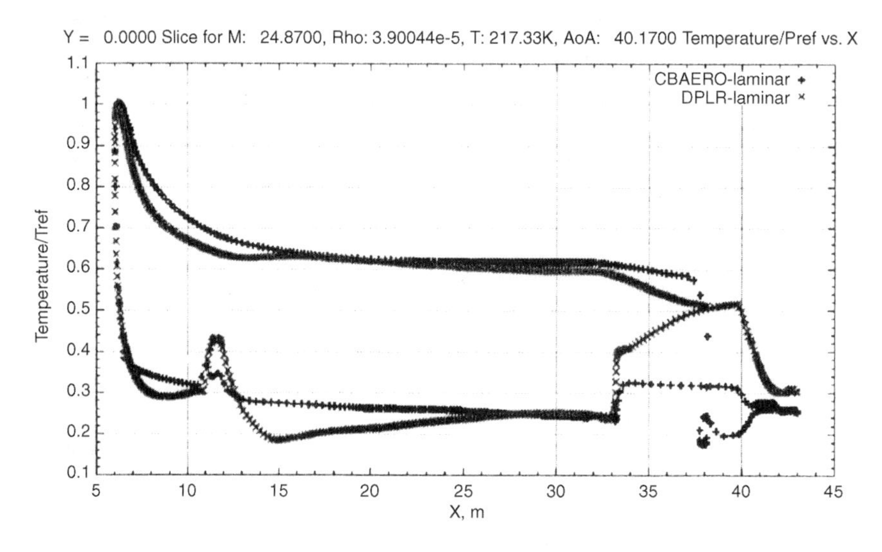

Fig. 6.39 Calculated centerline temperature distribution for the space shuttle, where M_∞ = 24.87 and angle of attack = 40 deg [229].

sets of results are shown, one from the engineering predictions of CBAERO and the other from a detailed calculation using a high-fidelity finite volume CFD solution of the Navier–Stokes equations developed at the NASA Ames Research Center and labeled DPLR. Laminar flow is assumed for both sets of calculations. In Fig. 6.39, the upper curves correspond to the bottom surface of the shuttle (the windward side), and the lower curves correspond to the top surface (the leeward side). Note that CBAERO does a good job of predicting the windward surface temperatures. On the leeward side, CBAERO captures the trend for increased heating on the canopy and the vertical tail, but quantitative agreement with the CFD solution from DPLR is not as good. Given the complexity of the leeward side flowfield, however, the engineering predictions from CBAERO are in the ballpark.

Return to Fig. 3.32 showing the Apollo command module as treated by Kinney and Garcia. Figure 6.40 shows the calculated surface temperature distribution along the centerline cut for the vehicle flying at Mach 28.6 at an 18.2 deg angle of attack. The upper curves pertain to the windward surface and the lower curves to the leeward surface. The overall temperature on the heat shield is reasonably predicted by CBAERO, including capturing the peak heating on the shoulder of the capsule. The off-shoulder distribution, however, deviates somewhat from the DPLR results.

Return to Fig. 3.34 showing the Project Fire II test vehicle as treated by Kinney and Garcia. Figure 6.41 shows the calculated aerodynamic heating rates in Watt/cm^2 along the centerline for the vehicle flying at Mach 35.75 at 0 deg angle of attack. For this case, the surface heating is caused both by convective and radiative heating, $q_{\text{convective}} + q_{\text{radiative}}$. (The engineering method used in

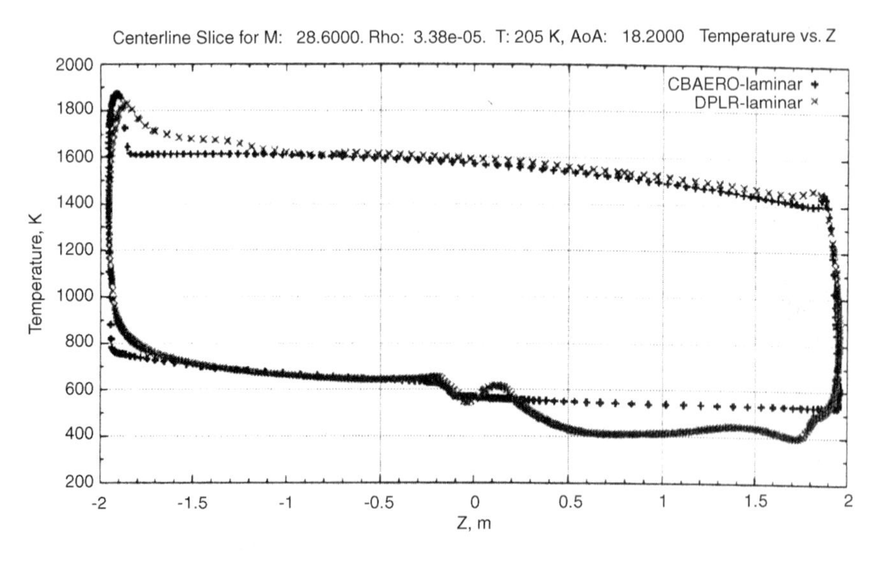

Fig. 6.40 Calculated temperature distribution along the centerline for the Apollo command module, where M_∞ = 28.6 and angle of attack = 18.2 deg [229].

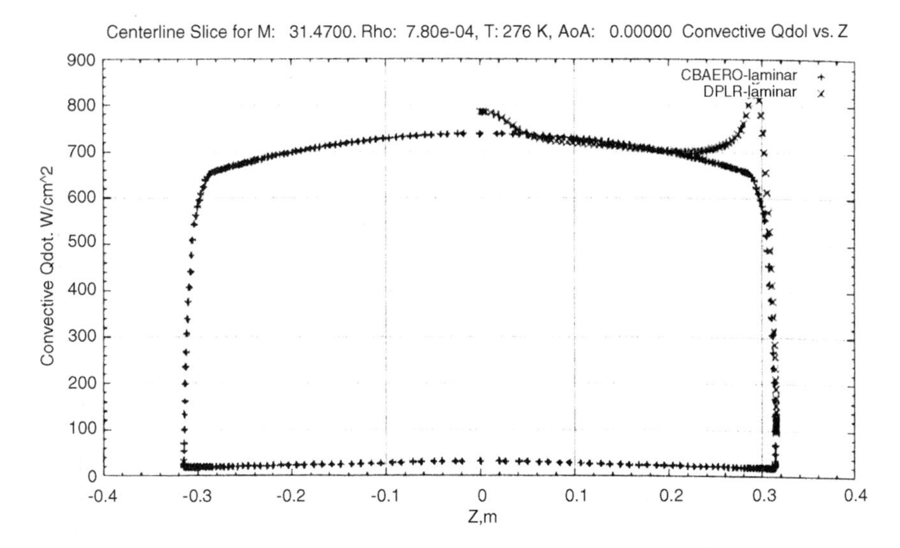

Fig. 6.41 Calculated convective heating distribution for the Fire II test vehicle, where M_∞ = 35.75 and angle of attack = zero [229].

CBAERO for predicting $q_{\text{radiative}}$ is discussed in Chapter 18.) Once again, the results from CBAERO compare favorably with the CFD results from DPLR except near the shoulder region. There, the Navier–Stokes solution predicts a local peak in heat transfer that is not simulated by the engineering predictions in CBAERO.

For more details on these engineering predictions, see [228] and [229].

Design Example 6.2: Hypersonic Waveriders—Part 2

This is continuation of the Design Example at the end of Chapter 5 entitled Hypersonic Waveriders—Part 1. There we identified what is a waverider and discussed the basic philosophy of their design, namely, that waverider shapes are carved out of inviscid streamsurfaces downstream of known supersonic and hypersonic shock waves. Moreover, we mentioned cases where hypersonic waverider shapes were optimized to obtain high values of $(L/D)_{\max}$, but that such optimizations did not include the skin-friction drag. As a result, the resulting waverider shapes when tested in wind tunnels did not achieve the expected values of $(L/D)_{\max}$. I suggest that you review the waverider Design Example at the end of Chapter 5 before proceeding further.

Appropriate to the present chapter on viscous flow, the present Design Example highlights the design of viscous optimized hypersonic waveriders, wherein skin-friction drag is included within the optimization process itself. Details of the design process can be found in [80] and [246] by Bowcutt et al., based on their work at the University of Maryland. This work, beginning in the late 1980s, led to a new class of waveriders where the optimization process is trying to reduce the wetted surface area, hence reducing skin-friction drag, while maximizing L/D. Because detailed viscous effects cannot be couched in simple analytical forms, the formal optimization methods based on the calculus of variations cannot be used. Instead, a *numerical* optimization technique was used based on the simplex method of Nelder and Mead [247]. By using a numerical optimization technique, other real configuration aspects could be included in the analysis in addition to viscous effects, such as blunted leading edges and an expansion upper surface (in contrast to the standard assumption of a freestream upper surface, i.e., an upper surface with all generators parallel to the freestream direction). The results of the study by Bowcutt et al. led to a new class of waveriders, namely, viscous optimized waveriders. Moreover, these waveriders produced relatively high values of (L/D), as will be discussed later.

For the viscous optimized waverider configurations, the following philosophy was followed:

1) The lower (compression) surface was generated by a streamsurface behind a conical shock wave. The inviscid conical flowfield was obtained from the numerical solution of the Taylor–Maccoll equation, derived for example in [4].

2) The upper surface was treated as an expansion surface, generated in a similar manner from the inviscid flow about a tapered, axisymmetric cylinder at zero angle of attack, and calculated by means of the axisymmetric method of characteristics.

3) The viscous effects were calculated by means of an integral boundary-layer analysis following surface streamlines, including transition from laminar to turbulent flow.

4) Blunt leading edges were included to the extent of determining the maximum leading-edge radius required to yield acceptable leading-edge surface temperatures, and then the leading-edge drag was estimated by modified Newtonian theory.

5) The final waverider configuration, optimized for maximum L/D at a given Mach number and Reynolds number with body fineness ratio as a constraint, was obtained from the numerical simplex method taking into account all of the effects itemized in 1–4 within the optimization process itself. For a highly detailed discussion of all of these items, see [248].

The following discussion provides some insight into the optimization process. First, assume a given conical shock wave in a flow at a given Mach number, say, a conical shock wave angle of $\theta_s = 11$ deg at Mach 6. As discussed at the end of Chapter 5, now trace a curve on the surface of the shock wave. The stream surface generated from this curve is a bottom surface of a waverider, and the curve itself forms the leading edge of the waverider. An infinite number of such curves can be traced on the conical shock wave, generating an infinite number of waverider shapes using the conical shock with $\theta_s = 11$ deg at $M_\infty = 6$. Indeed, some of these leading-edge curves are shown in Fig. 6.42. The optimization procedure progresses through a series of these leading-edge shapes, each one generating a new waverider with a certain lift-to-drag ratio, and finally settling on that particular leading-edge shape that yields the maximum value of L/D. This is the optimum waverider for the given generating conical shock wave angle of $\theta_s = 11$ deg. This resulting $(L/D)_{\max}$ is then plotted as a point in Fig. 6.43 for the conical shock wave angle $\theta_s = 11$ deg. Figure 6.43 also gives the

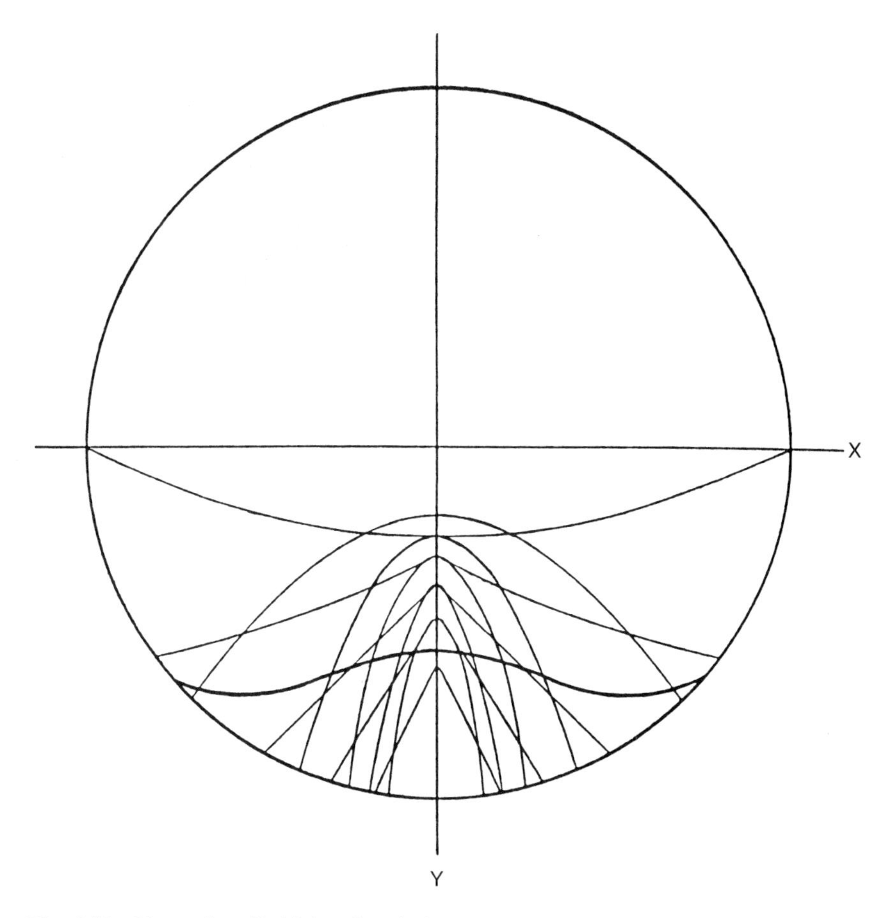

Fig. 6.42 Examples of initial and optimized waverider leading edge shapes (Bowcutt et al. [246]).

corresponding value of lift coefficient C_L and volumetric efficiency $\eta = V^{2/3}/S_p$, where V is the vehicle volume and S_p is the planform area. Now choose another conical shock angle for the generating flowfield, say, $\theta_s = 12$ deg, and repeat the preceding procedure, finding that leading-edge shape that yields the waverider shape that produces the highest (L/D). This result is now plotted in Fig. 6.43 for $\theta_s = 12$ deg. Then another conical shock wave angle, say, $\theta_s = 13$ deg, is chosen, and the process is repeated again, finding that particular waverider shape that produces the highest L/D. This point is now plotted in Fig. 6.43 for $\theta_s = 13$ deg. And so forth. The front views of these optimized waverider shapes are shown in Fig. 6.44, each one labeled according to its generating conical shock-wave angle. These same optimized waveriders are shown in perspective in Fig. 6.45. Returning to Fig. 6.43, note that the curve of L/D vs θ_s itself has a maximum value of (L/D), occurring in this case for $\theta_s = 12$ deg.

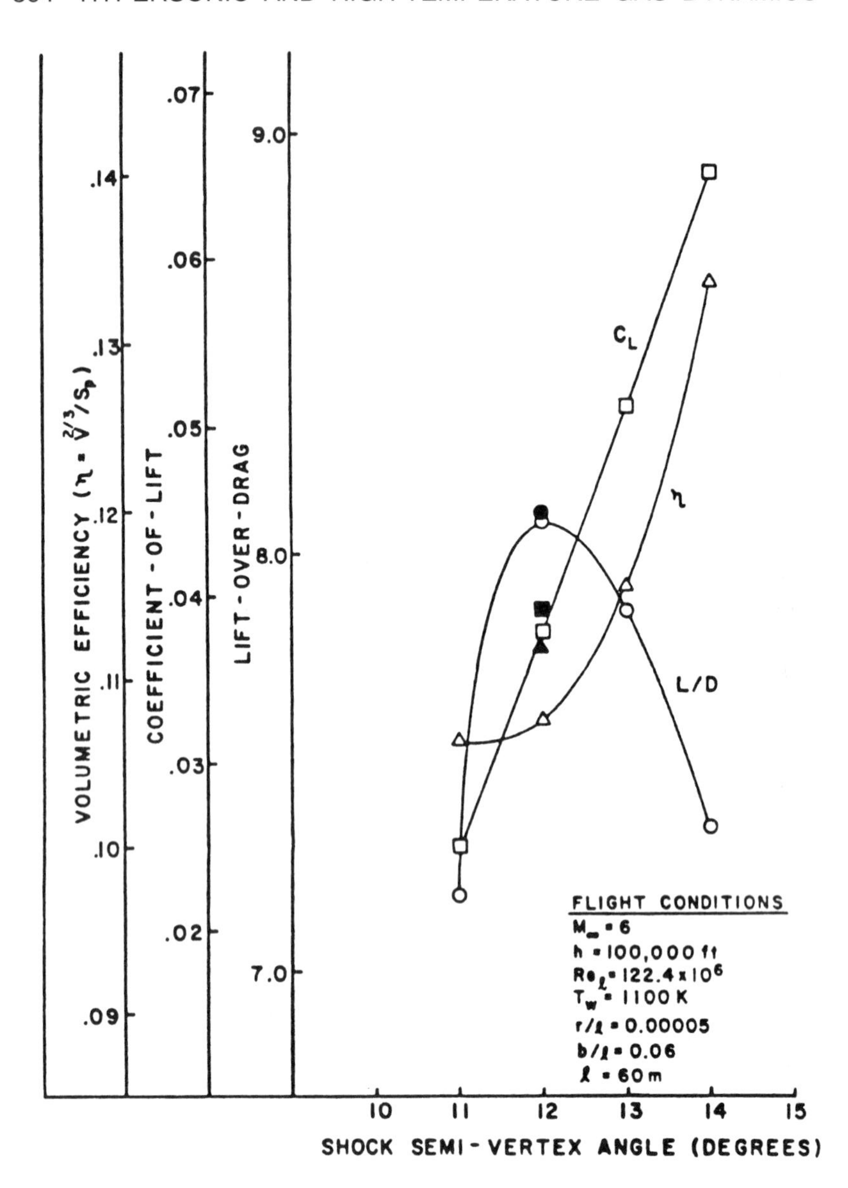

Fig. 6.43 Results for a series of optimized waveriders at Mach 6: l = length of waverider, b/l = body fineness ratio, and r = leading-edge radius [246].

This yields an "optimum of the optimums" and defines the final viscous optimized waverider at $M_\infty = 6$ for the flight conditions shown in Fig. 6.43. Finally, a summary three view of the best optimum (the optimum of the optimum) waverider, which here corresponds to $\theta_s = 12$ deg, is given in Fig. 6.46. Also, in Figs. 6.44–6.46 the lines on the upper and lower surfaces

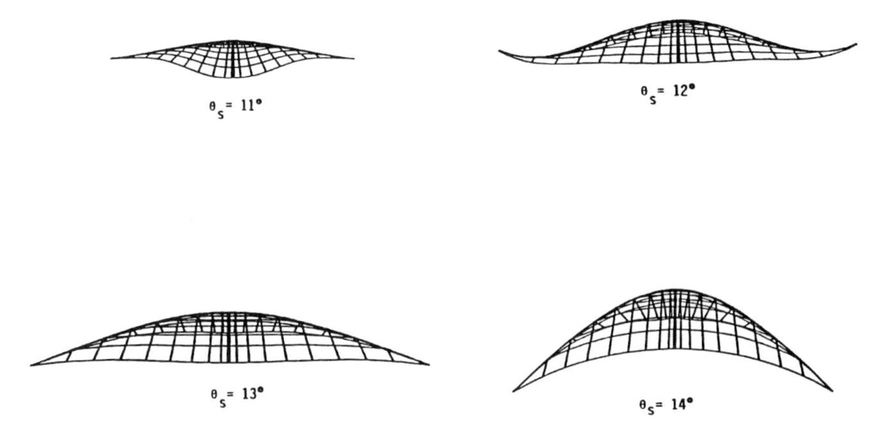

Fig. 6.44 Results for a series of optimized waveriders at Mach 6 [246].

of the waveriders are inviscid streamlines. Note in these figures that the shape of the optimum waverider changes considerably with θ_s. Moreover, examining (for example) Fig. 6.46, note the rather complex curvature of the leading edge in both the planform and front views; the optimization program is shaping the waverider to adjust both wave drag and skin-friction drag so that the overall L/D is a maximum. Indeed, it was observed that the best optimum shape at any given M_∞ results in the magnitudes of wave drag and skin-friction drag being approximately the same, never differing by more than a factor of two. For conical shock angles below the best optimum (for example, $\theta_s = 11$ deg in Figs. 6.44 and 6.45), skin-friction drag is greater than wave drag; in contrast, for conical shock angles

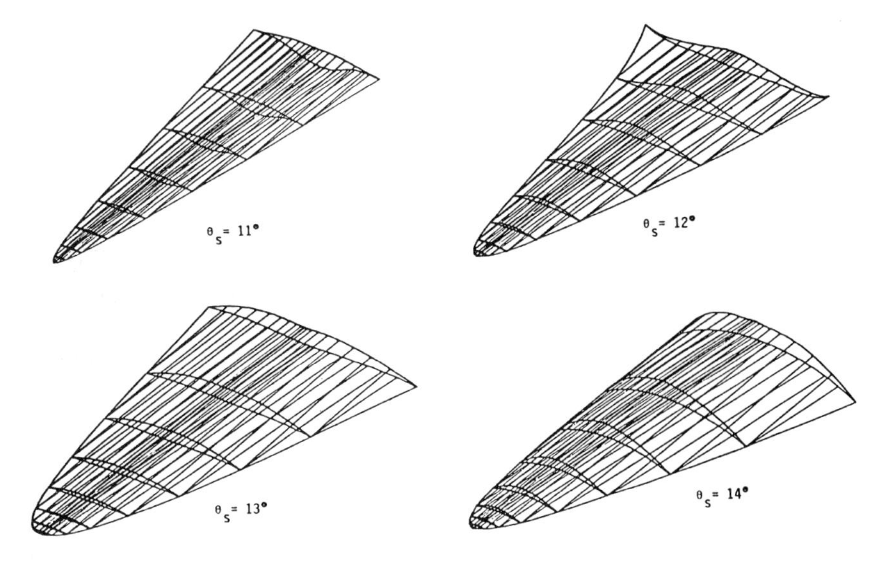

Fig. 6.45 Perspective views of a series of optimized waveriders at Mach 6 [246].

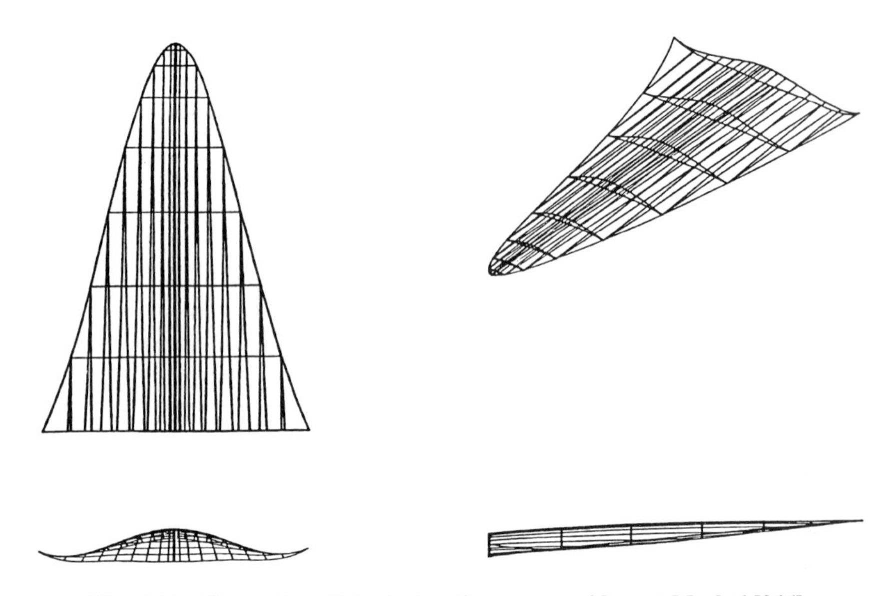

Fig. 6.46 Three view of the best optimum waveriders at Mach 6 [246].

above the best optimum (for example, $\theta_s = 13$ and 14 deg in Figs. 6.44 and 6.45), skin-friction drag is less than wave drag. (*Note*: For a hypersonic flat plate, using Newtonian theory and an average skin-friction coefficient, it can readily be shown that at maximum L/D, the wave drag is twice the friction drag.)

The results in Figs. 6.43–6.46 pertain to $M_\infty = 6$. An analogous set of results for the other extreme of the lifting hypersonic flight spectrum at $M_\infty = 25$ is given in Figs. 6.47–6.50. The aerodynamic characteristics of optimum waveriders for $\theta_s = 7$, 8, 9, and 10 deg are given as the open symbols Fig. 6.47. (The solid symbols will be discussed later.) The respective front views are shown in Fig. 6.48 and perspective views in Fig. 6.49. Finally, the best optimum Mach 25 waverider (which occurs at $\theta_s = 9$ deg) is summarized in Fig. 6.50. Comparing the optimum configuration at $M_\infty = 6$ (Fig. 6.46) with the optimum configuration at Mach 25 (Fig. 6.50), note that the Mach 25 shape has more wing sweep and pertains to a conical flowfield with a smaller wave angle, both of which are intuitively expected at higher Mach number. However, note from the flight conditions listed in Figs. 6.43 and 6.47 that the body slenderness ratio at $M_\infty = 6$ is constrained to be $b/\ell = 0.06$ (analogous to a supersonic transport such as the Concorde), but that $b/\ell = 0.09$ is the constraint chosen at $M_\infty = 25$ (analogous to a hydrogen-fueled hypersonic airplane). The two different slenderness ratios are chosen on the basis of reality for two different aircraft with two different missions at either extreme of the hypersonic flight spectrum. Also note in Figs. 6.48–6.50 that the optimization program has sculptured a best optimized configuration with a spline down the center of the upper surface—an interesting and curious result, caused principally by the competing effects of minimizing pressure and skin-friction drag, while meeting the slenderness ratio constraint.

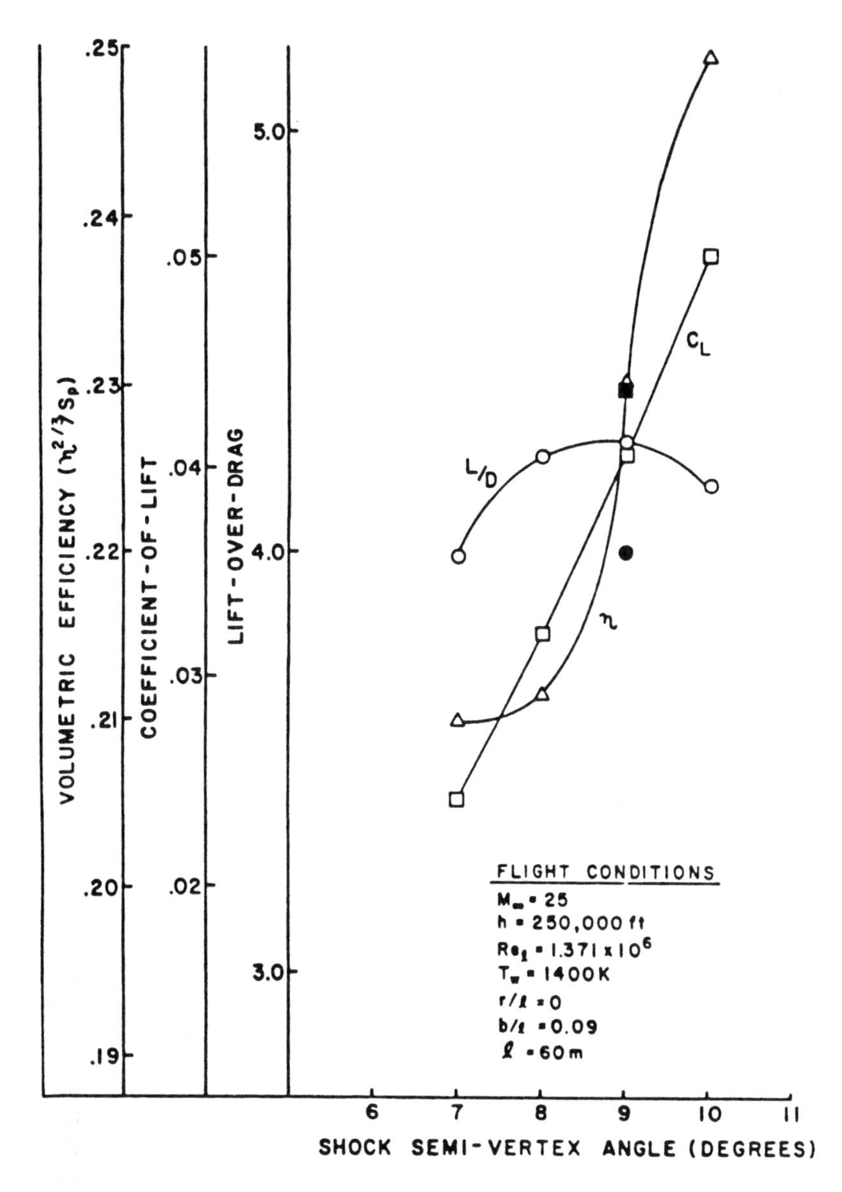

Fig. 6.47 Results for a series of optimized waveriders at Mach 25 [246].

Return to Fig. 6.47, and note the solid symbols. These pertain to the values of C_L and L/D obtained by setting the ratio of specific heats γ to 1.1 in order to assess possible effects of high-temperature chemically reacting flow. The solid symbols pertain to an optimized waverider at $\theta_s = 9$ deg with $\gamma = 1.1$. This is not necessarily the best optimum at Mach 25 with $\gamma = 1.1$; rather, it is just a

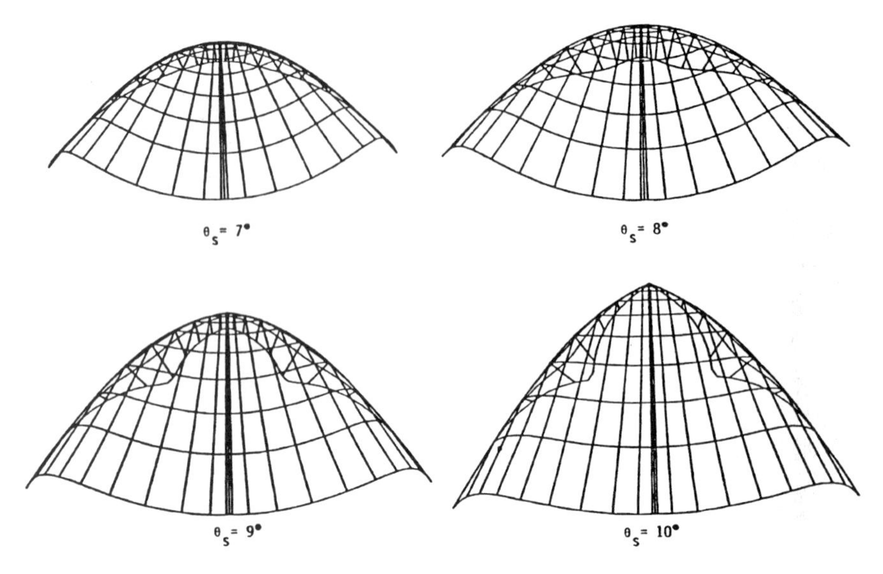

Fig. 6.48 Front views of series of optimized waveriders at Mach 25 [246].

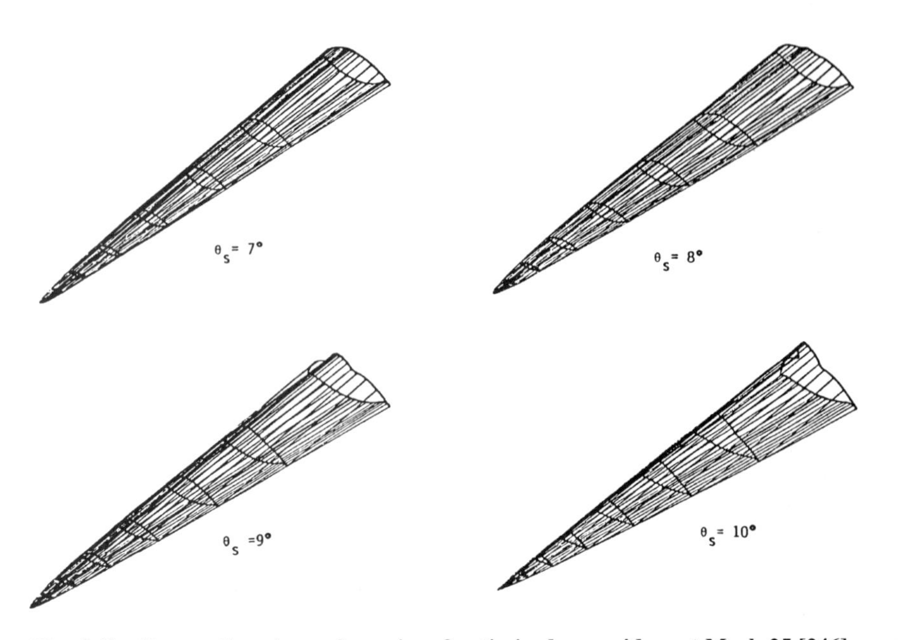

Fig. 6.49 Perspective views of a series of optimized waveriders at Mach 25 [246].

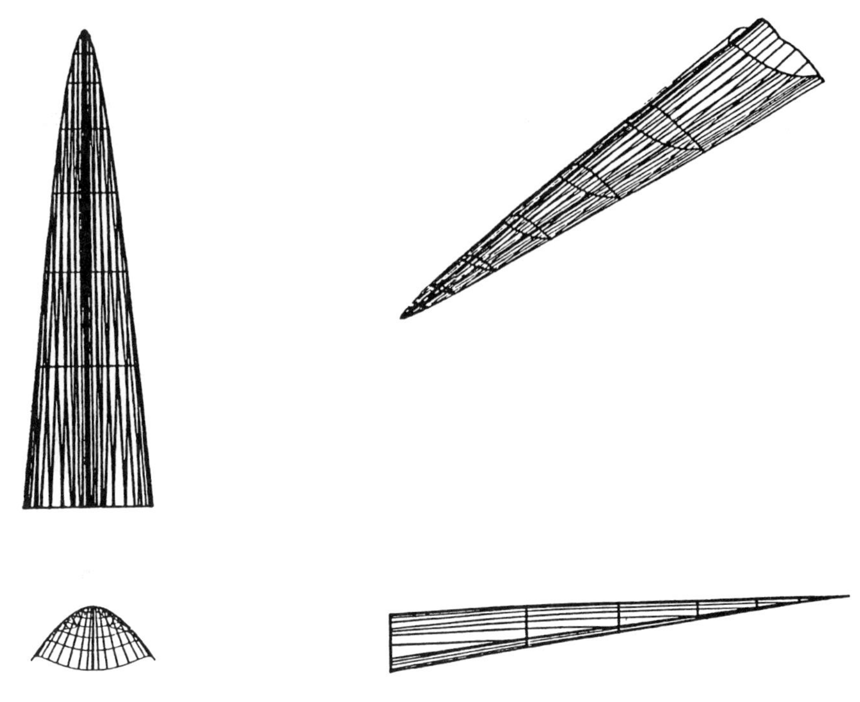

Fig. 6.50 Three view and perspective of the best optimized waverider at Mach 25 [246].

point of calculation to indicate that high-temperature effects have some impact on optimized waverider generation. This impact will be examined at the end of Chapter 14.

Recall that for supersonic and hypersonic vehicles, L/D markedly decreases as M_∞ increases. Indeed, Kuchemann [249] gives the following general empirical correlation for $(L/D)_{\max}$ based on actual flight-vehicle experience:

$$(L/D)_{\max} = \frac{4(M_\infty + 3)}{M_\infty}$$

This variation is shown as the solid curve in Fig. 6.51. This figure is important to our present discussion; it brings home the importance of the viscous optimized waveriders discussed in this design box. The Kuchemann curve (the solid curve) in Fig. 6.51 represents a type of "L/D barrier" for conventional vehicles, which is difficult to break. The open circles in Fig. 6.51, which form an almost "shotgun" scatter of points, are data for a variety of conventional vehicles representing various wind-tunnel and flight tests. (Precise identification of the sources for these points is given in [248].) The solid symbols pertain to the viscous optimized hypersonic waveriders discussed here. The solid squares are results for the waveriders based on conical generating flows discussed in this Design Example.

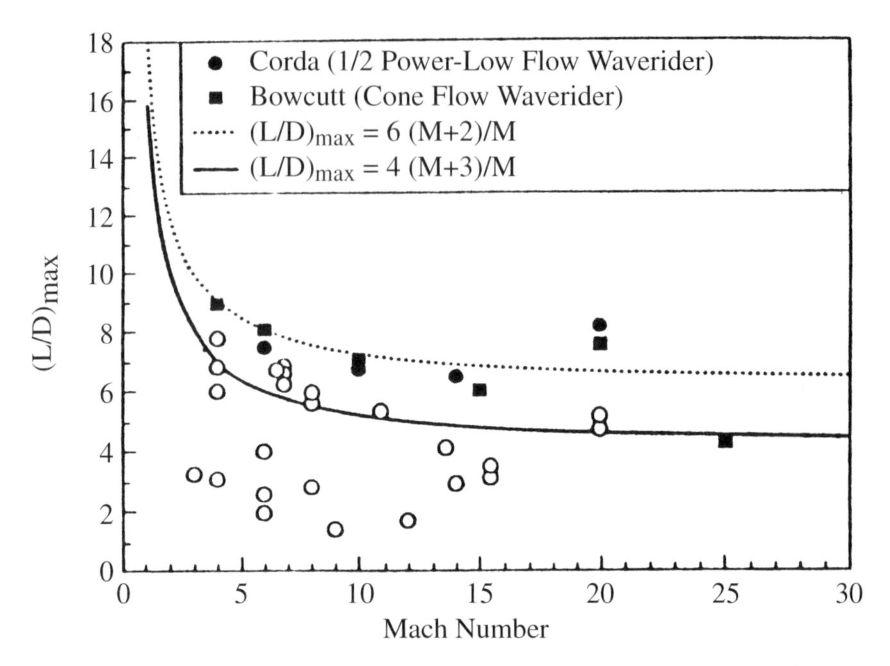

Fig. 6.51 Maximum lift-to-drag ratio comparison for various hypersonic configurations [246].

The solid circles are results for another family of viscous optimized waveriders based on the shock wave and downstream streamsurfaces generated by a one-half-power-law ogive-shaped body, obtained by Corda and Anderson [250]. From Fig. 6.51 we see that *the viscous optimized waveriders break the L/D barrier*, that is, they all give $(L/D)_{\max}$ values that lie above the Kuchemann curve. Indeed, the L/D variation of the viscous optimized waveriders is more closely given by

$$(L/D)_{\max} = \frac{6(M_\infty + 2)}{M_\infty}$$

This variation is shown as the dotted curve in Fig. 6.51. The importance of the viscous optimized waveriders is established by the results shown in Fig. 6.51. These results have been confirmed by various wind-tunnel tests. They are the reason for renewed interest in the waverider configuration as a hypersonic vehicle, particularly for sustained cruising in the atmosphere.

Hypersonic vehicle design is sensitive to the location of transition from laminar to turbulent flow, and the design of viscous optimized hypersonic waveriders is no exception. Figures 6.52–6.54 show results of a numerical experiment carried out at $M_\infty = 10$ wherein the transition location was varied over a wide latitude, ranging from all laminar flow on one hand, to almost all turbulent flow on the other hand, with various cases inbetween. Specific results at Mach 10 are given in Fig. 6.52; here values of (L/D) are given for optimized waveriders

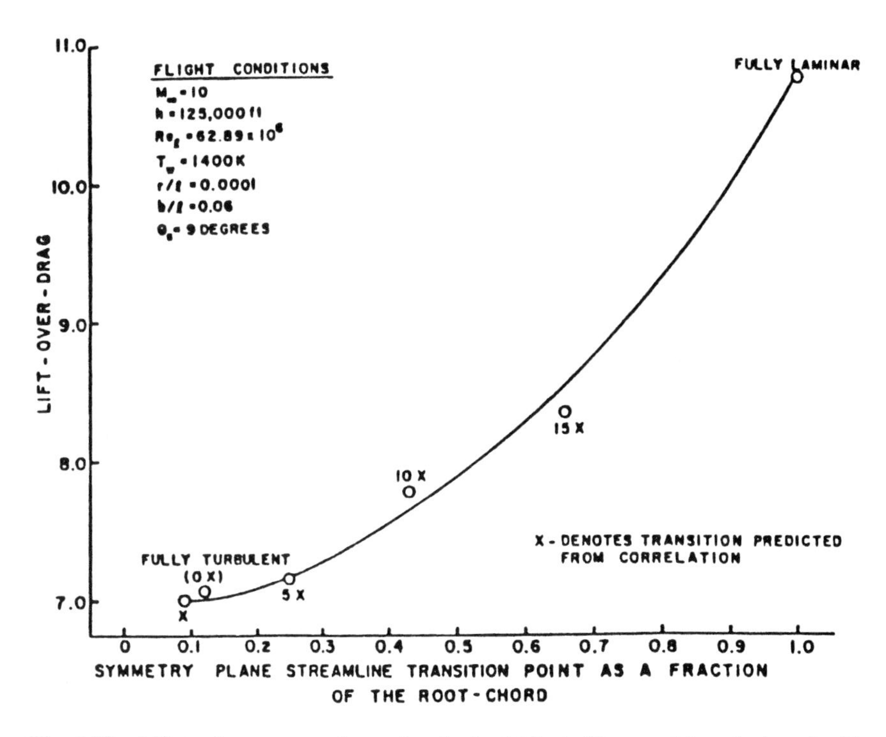

Fig. 6.52 Lift-to-drag comparison of optimized Mach 10 waveriders designed with various boundary-layer transition criteria [246].

as a function of assumed transition location. The point corresponding to the transition correlation described in [80] and [248] is denoted by x in Fig. 6.52. Other points in Fig. 6.52 labeled $5x$, $10x$, and $15x$ correspond to transition locations that are 5, 10, and 15 times the value predicted by the transition correlation. All of the data given in Fig. 6.52 pertain to optimized waveriders for $\theta_s = 9$ deg, which yields the best optimum at Mach 10 for the usual transition correlation. (Note, however, that $\theta_s = 9$ deg might not yield the best optimum for other transition locations; this effect was not investigated in [80] and [248].) The results in Fig. 6.52 demonstrate a major increase in (L/D) in going from almost all turbulent flow to all laminar flow. However, for the case where transition is changed by a factor of five, only a 2% change in L/D results. Even for the case where transition is changed by a factor of 10, a relatively small change in L/D of 11% results. On the other hand, the *shapes* of the resulting optimized waveriders are fairly sensitive to the transition location, as illustrated in Figs. 6.53 and 6.54. The conclusion to be made here is that waverider optimization is indeed relatively sensitive to transition location, and this underscores the need for reliable prediction of transition at hypersonic speeds.

Tieing up a loose end, the solid symbols in Fig. 6.43 correspond to calculations where an average overall skin-friction drag coefficient was calculated for the complete waverider configuration and subsequently used for the optimization

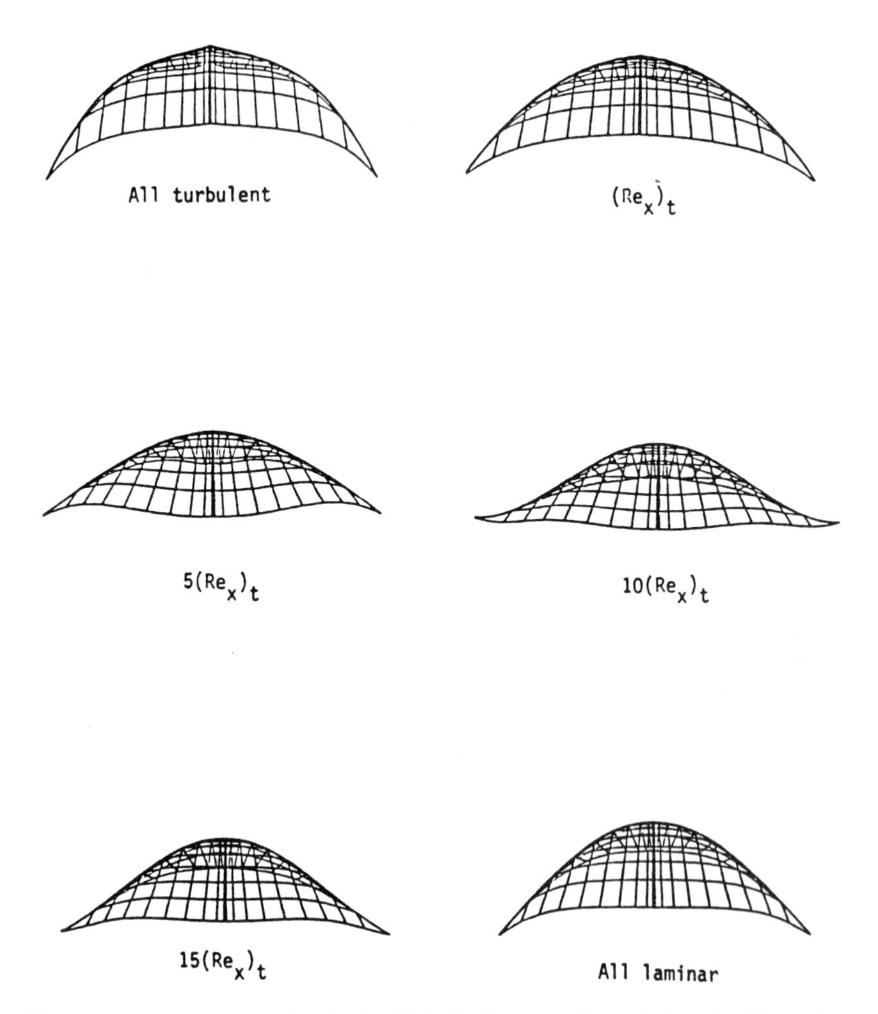

Fig. 6.53 Front views of optimized Mach 10 waveriders designed with various boundary-layer transition criteria [246].

process, rather than the use of the detailed shear-stress distributions employed in the standard calculations. Only a small difference exists between the two cases as seen in Fig. 6.43. Indeed, the resulting waverider shapes are virtually the same. The use of an average skin-friction drag coefficient greatly reduces the computer time for the optimization process.

The question of aerodynamic heating of viscous optimized waveriders was examined by Vanmol and Anderson in [251]. To minimize aerodynamic heating, the leading edges of waveriders must be blunt, as for all classes of hypersonic vehicles. Vanmol and Anderson used Tauber's engineering correlation for stagnation-point heat transfer [79] and [245] modified to apply to a swept leading

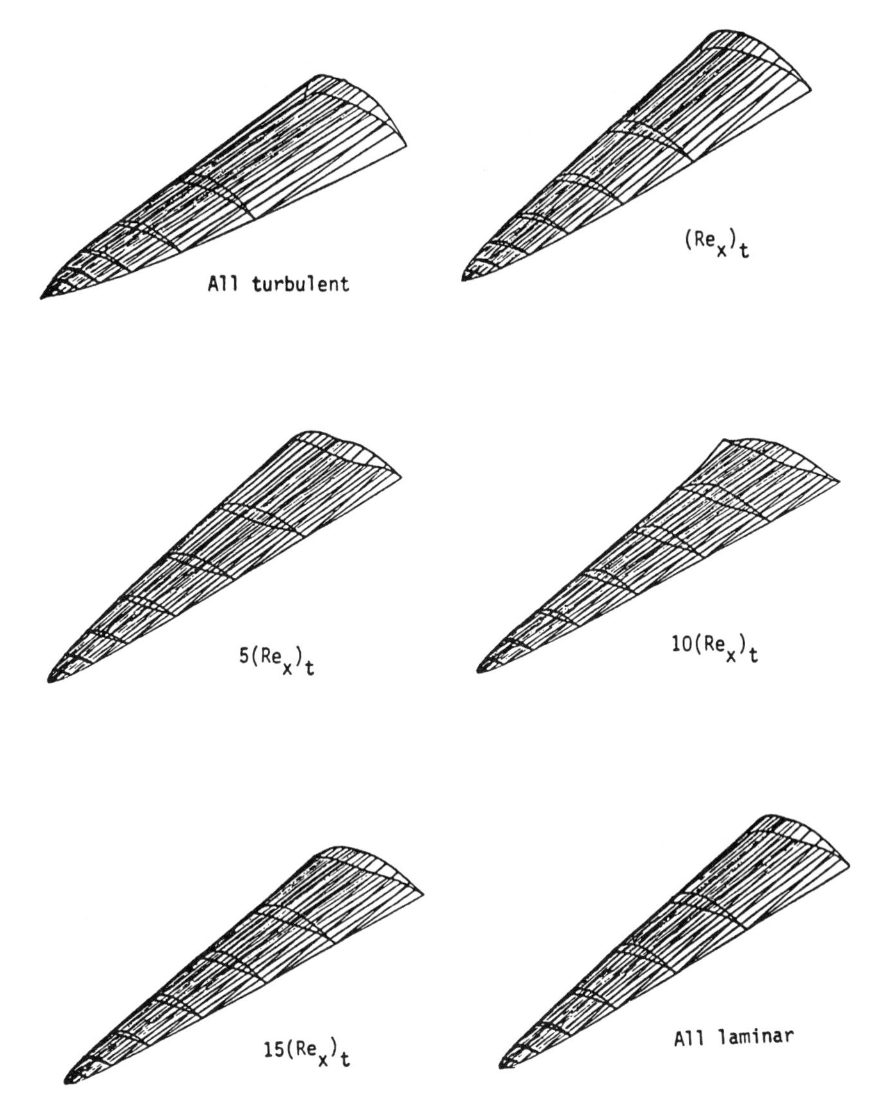

Fig. 6.54 Perspective views of optimized Mach 10 waveriders designed with various boundary-layer transition criteria [246].

edge using the correlation of Hamilton [252]. A swept blunt leading edge has no stagnation points, but rather an attachment line along the leading edge; the attachment line is simply the dividing line between the flow that wets the upper and lower surfaces. Moreover, there is a finite component of velocity along the attachment line. Thus, at the outer edge of the boundary layer on the leading edge there is a finite component of velocity in the direction of the attachment line, creating the possibility that transition from laminar to turbulent flow might occur along the attachment line. Where transition and turbulent flow do occur along the attachment line, the boundary layer over the top and bottom surfaces of the wing will be turbulent downstream of that portion of the attachment line. This attachment line transition was taken into account by Vanmol and Anderson; complete details are given in [253] by Vanmol. Transition along the attachment line was found to be important for some Mach- and Reynolds-number combinations. For the upper and lower surfaces of the waverider, the reference temperature method (see Sec. 6.9) was used.

This study used flight conditions associated with two different constant dynamic pressure flight trajectories through the atmosphere, one for $q_\infty = 0.2$ atm and the other for $q_\infty = 1.0$ atm. The results indicate that aerodynamic heating to hypersonic viscous optimized waveriders is manageable. For Mach numbers less than 10, passive radiative cooling of the surface is sufficient. For Mach numbers greater than 10, a combination of radiative cooling and state-of-the-art active cooling is sufficient. In addition, for a vehicle of 60 m length considered in this study, the degree of leading-edge bluntness demanded by aerodynamic heating constraints (a leading-edge radius distribution from 6 to 1.2 cm) does not materially degrade the aerodynamic performance of the waveriders. See [251] and [253] for more details.

In summary, viscous optimized hypersonic waveriders have been highlighted in this Design Example. They are a graphic example of the fundamental hypersonic flow aspects discussed so far in this book as applied to hypersonic vehicle design. Moreover, they offer a promising design option for future hypersonic vehicles. We will have more to say about these waveriders in Design Examples at the end of Chapters 7 and 14.

Problems

6.1 Starting with the Navier–Stokes equations in dimensional form, derive Eqs. (6.7–6.10).

6.2 Derive Eq. (6.58).

6.3 Derive Eqs. (6.109) and (6.110) for an *axisymmetric* stagnation point.

6.4 Consider the hypersonic laminar flow over a flat plate. When $Pr = 1$, show that enthalpy is a function of local velocity, that is, show that $h = h(u)$. Obtain this function.

6.5 Show similarities between the approximate Eq. (6.169) and exact results.

7
Hypersonic Viscous Interactions

It is yet too early to describe other promising methods of inquiry by which knowledge of the size and texture of the boundary layer may be obtained. It seems, however, that we are on the threshold of a new domain of great promise; research is needed, first for the advancement of our knowledge and then for its application.

Leonard Bairstow, English aerodynamicist, 1923

Interact—to act on each other.

From The American Heritage Dictionary of the English Language, 1976

Chapter Preview

Hypersonic flow—what is it? Section 1.3 served to answer this question. One of the physical characteristics defining hypersonic flow described in Sec. 1.3 is the rapid growth of the boundary-layer thickness with increasing Mach number, creating a substantial interaction with the outer inviscid flow and changing the flow properties of this inviscid flow. In turn, these changes feed back to affect the growth of the boundary layer. In Sec. 1.3, this phenomenon was labeled *viscous interaction*. Hypersonic viscous interaction is so important that it rates a chapter all by itself—*this chapter*. Hypersonic viscous interaction can have important effects on the surface-pressure distributions over vehicles, changing the lift and drag of such vehicles and markedly affecting their stability characteristics. Moreover, skin friction and heat transfer are increased by viscous interaction. So this is important stuff. How do you calculate the effects of viscous interaction? You will learn in this chapter. How can you tell under what flight conditions (say, what Mach numbers and Reynolds numbers) is viscous interaction important, and under what conditions is it not important? You will learn in this chapter.

Finally, we have another subject piggybacking on this chapter—hypersonic shock-wave/boundary-layer interactions. This is the interaction that occurs when and external shock wave impinges on a boundary layer. Such an interaction is not a part of the classic viscous interaction phenomena we just described, but is nevertheless an interaction between the external inviscid flow from whence the shock wave is coming and the boundary layer on which the shock wave impinges. Hence it is an appropriate subject under the title of this chapter. Moreover, it is an enormously important subject. Shock-wave impingement results in local peaks in heat transfer on the surface. At hypersonic speeds these peaks are so high that holes can be burned into the surface with sometimes disastrous consequences to hypersonic vehicles. Shock-wave impingement can also lead to separation of the boundary layer from the surface, and such flowfield separation can cause a large increase in drag and a substantial decrease in lift on a hypersonic vehicle. If the interaction occurs internally in an airbreathing scramjet engine, the engine performance can be severely reduced if not totally wiped out. This is important stuff. It is so important that decades of experimental study and numerical calculations have been focused on this problem. You will see some of this work described in this chapter.

The bottom line of this chapter is that viscous interactions, including shock-wave/boundary-layer interactions, are a vital part of the study of the fundamentals of hypersonic flow. Therefore, the material in this chapter is serious stuff—important stuff. Please treat it this way, and march on.

7.1 Introduction

In contrast to the preceding statement by the eminent British aerodynamicist, L. Bairstow, in 1923, it is no longer "too early to describe other promising methods" of studying viscous flows. Indeed, in the modern world of hypersonics, it is mandatory that we go beyond the original boundary-layer concept as introduced by Prandtl in 1904. The material in this chapter is one such example. Here, we will examine two important flow problems where the viscous boundary layer changes the nature of the outer inviscid flow, and in turn these inviscid changes feed back as changes in the boundary-layer structure. This gives rise to phenomena classified as *viscous interactions*. In hypersonic flow, there are two important viscous interactions: 1) pressure interaction, caused by the exceptionally thick boundary layers on surfaces under some hypersonic conditions, and 2) shock-wave/boundary-layer interaction, caused by the impingement of a strong shock wave on a boundary layer. The first item, pressure interaction, is frequently identified in the hypersonic literature as simply "viscous interaction." This is the physical effect described in Sec. 1.3.3 and sketched in Figs. 1.15 and 1.16. This material from Chapter 1 should be reviewed at this stage before progressing further. The viscous interaction described in Sec. 1.3.3 constitutes the subject for most of the present chapter.

The road map for this chapter is given in Fig. 7.1. We start out on the left side of the map and discuss the classic viscous interaction problem. This is divided into two categories, the strong interaction and the weak interaction. Then we move to the right side of the map and consider the shock-wave/boundary-layer interaction. This interaction is also divided into two categories, one where the boundary layer is laminar and the other where the boundary layer is turbulent.

The classic hypersonic interaction between the outer inviscid flow and the boundary layer is caused by the very large boundary-layer thicknesses, which can occur at hypersonic speeds. Indeed, it was stated in Sec. 1.3.3 that for a flat-plate laminar boundary layer δ grows as

$$\delta \propto \frac{M_e^2}{\sqrt{Re_x}} \tag{7.1}$$

Hence, for equal Reynolds number, δ grows as the square of the Mach number. We are now in a position to prove this, as follows. For a laminar boundary layer on a flat plate, the self-similar solution described in Sec. 6.5 leads to the familiar result that

$$\delta \propto \frac{x}{\sqrt{Re}} \tag{7.2}$$

Because of intense viscous dissipation in hypersonic boundary layers, the temperature can vary widely. In turn, ρ and μ can be strongly variable throughout the boundary layer. Let us choose to evaluate the Reynolds number in Eq. (7.2) using ρ_w and μ_w at the wall. Then,

$$\delta \propto \frac{x}{\sqrt{\rho_w u_e x/\mu_w}}$$

or

$$\delta \propto \frac{x}{\sqrt{\rho_e u_e x/\mu_e}}\sqrt{\frac{\rho_e}{\rho_w}}\sqrt{\frac{\mu_w}{\mu_e}} \tag{7.3}$$

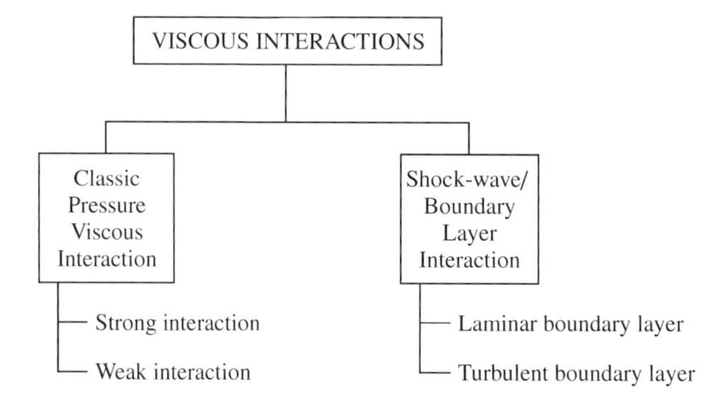

Fig. 7.1 Road map for Chapter 7.

From the equation of state, assuming $p_e = p_w = p =$ constant through the boundary layer,

$$\frac{\rho_e}{\rho_w} = \frac{p_e}{p_w}\frac{T_w}{T_e} = \frac{T_w}{T_e} \tag{7.4}$$

Also, assuming a linear dependency of μ on T,

$$\frac{\mu_w}{\mu_e} = \frac{T_w}{T_e} \tag{7.5}$$

Combining Eqs. (7.3–7.5), we have

$$\delta \propto \frac{x}{\sqrt{Re}}\left(\frac{T_w}{T_e}\right) \tag{7.6}$$

where Re is the conventional Reynolds number based on properties at the outer edge of the boundary layer, that is, $Re = \rho_e u_e x/\mu_e$. Assuming an adiabatic wall with recovery factor $r = 1$,

$$\frac{T_w}{T_e} = \frac{T_{aw}}{T_e} = \frac{T_0}{T_e} = 1 + \frac{\gamma - 1}{2} M_e^2 \tag{7.7}$$

For large M_e, Eq. (7.7) becomes

$$\frac{T_w}{T_e} = \frac{T_0}{T_e} = \frac{\gamma - 1}{2} M_e^2 \tag{7.8}$$

Substituting Eq. (7.8) into (7.6), we find

$$\boxed{\frac{\delta}{x} \propto \frac{M_e^2}{\sqrt{Re}}} \tag{7.9}$$

Clearly, *the thickness grows as the square of the Mach number*, and therefore *hypersonic boundary layers can be orders of magnitude thicker than low-speed boundary layers at the same Reynolds number.*

This thick hypersonic boundary layer displaces the outer inviscid flow, changing the nature of the inviscid flow. For example, inviscid flow over a flat plate is sketched in Fig. 7.2a; the streamlines are straight and parallel, and the pressure on the surface is constant (as sketched above the streamlines). In contrast, for hypersonic viscous flow with a thick boundary layer, the inviscid streamlines are displaced upward, creating a shock wave at the leading edge as sketched in Fig. 7.2b. Moreover, the pressure varies over the surface of the flat plate, as sketched above the flow picture in Fig. 7.2b. This is the source of the viscous interaction. The increased pressure (hence increased density) tends to make the boundary layer thinner than would be expected (although δ is still large on a relative scale), and hence the velocity and temperature gradients at the wall are increased. In turn, the skin friction and heat transfer are increased over their values that

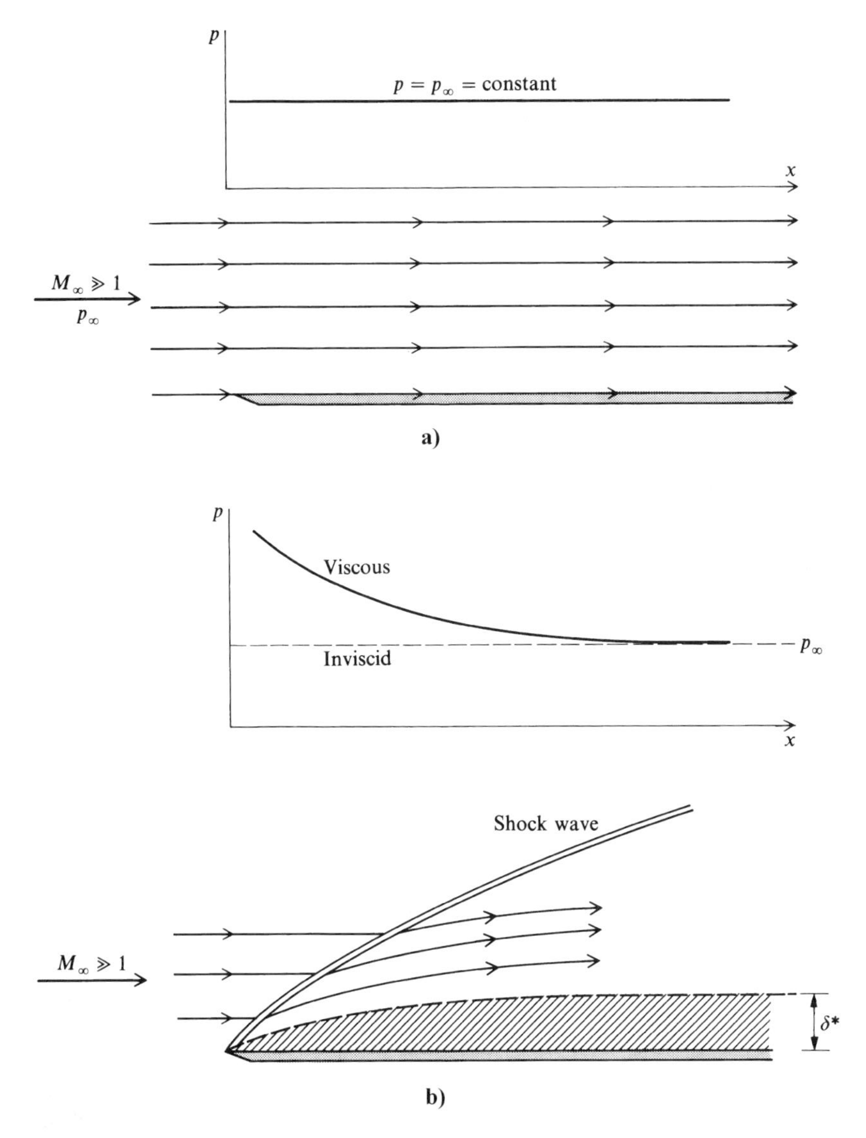

Fig. 7.2 Illustration of pressure distributions over a flat plate: a) inviscid flow and b) viscous flow.

would exist if a constant pressure equal to p_∞ were assumed. In the viscous interaction, the pressure increase, and the resulting c_f and C_H increases, become more severe closer to the leading edge. We will soon see that the magnitude of the viscous interaction increases as Mach number is increased and Reynolds number is decreased. Therefore, viscous interaction effects are important for slender hypersonic vehicles flying at high Mach numbers and high altitudes.

7.2 Strong and Weak Viscous Interactions: Definition and Description

Consider the sketch shown in Fig. 7.3, which illustrates the hypersonic viscous flow over a flat plate. Two regions of viscous interaction are illustrated here—the strong interaction region immediately downstream of the leading edge and the weak interaction region further downstream. By definition, the *strong interaction* region is one where the following physical effects occur:

1) In the leading-edge region, the rate of growth of the boundary-layer displacement thickness is large, that is, $d\delta^*/dx$ is large.

2) Hence, the incoming freestream "sees" an effective body with rapidly growing thickness; the inviscid streamlines are deflected upward, into the incoming flow, and a shock wave is consequently generated at the leading edge of the flat plate, that is, the inviscid flow is *strongly affected* by the rapid boundary-layer growth.

3) In turn, the substantial changes in the outer inviscid flow feedback to the boundary layer, affecting its growth and properties.

This mutual interaction process, where the boundary layer substantially affects the inviscid flow, which in turn substantially affects the boundary layer, is called a *strong viscous interaction*, as sketched in Fig. 7.3.

In contrast, further downstream a region of weak interaction is eventually encountered. By definition, the *weak interaction* region is one where the following physical effects occur:

1) The rate of growth of the boundary layer is moderate, that is, $d\delta^*/dx$ is reasonably small.

2) In turn, the outer inviscid flow is only weakly affected.

3) As a result, the changes in the inviscid flow result in a negligible feedback on the boundary layer, and this is ignored.

Therefore, as indicated in Fig. 7.3, the region of flow where the feedback effect is ignored is called a *weak viscous interaction.*

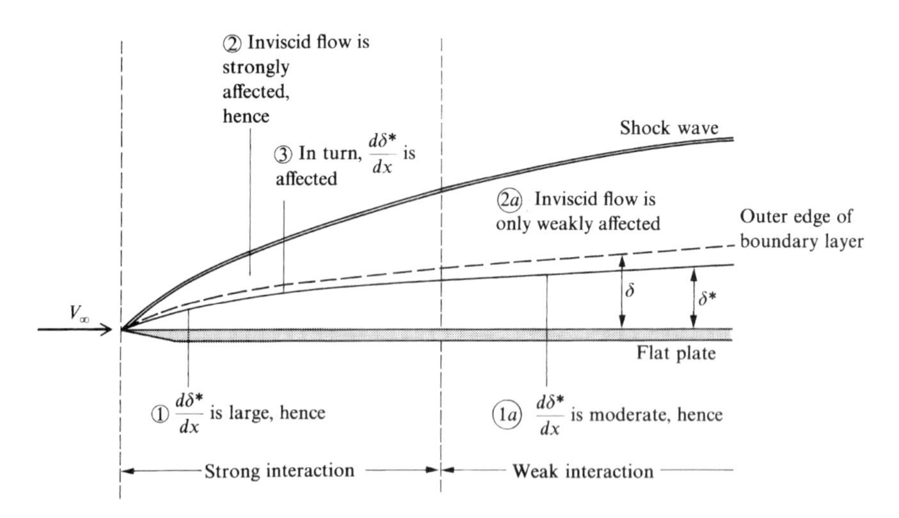

Fig. 7.3 Illustration of strong and weak viscous interactions.

The similarity parameter that governs laminar viscous interactions, both strong and weak, is "chi bar," defined as

$$\bar{\chi} = \frac{M_\infty^3}{\sqrt{Re}}\sqrt{C} \tag{7.10}$$

where

$$C = \frac{\rho_w \mu_w}{\rho_e \mu_e} \tag{7.11}$$

The value of $\bar{\chi}$ can be used to ascertain whether an interaction region is strong or weak; large values of $\bar{\chi}$ correspond to the strong interaction region, and small values of $\bar{\chi}$ denote a weak interaction region. The role of $\bar{\chi}$ as a similarity parameter is derived in the next section.

Finally, we emphasize again the major consequence of this viscous interaction, namely, the creation of an induced pressure change that can be substantial. This induced pressure change, sometimes called the induced pressure increment, is sketched in Fig. 7.4, where the actual pressure ratio p/p_∞ along the surface of the plate lies considerably above the inviscid flow value of unity. This type of

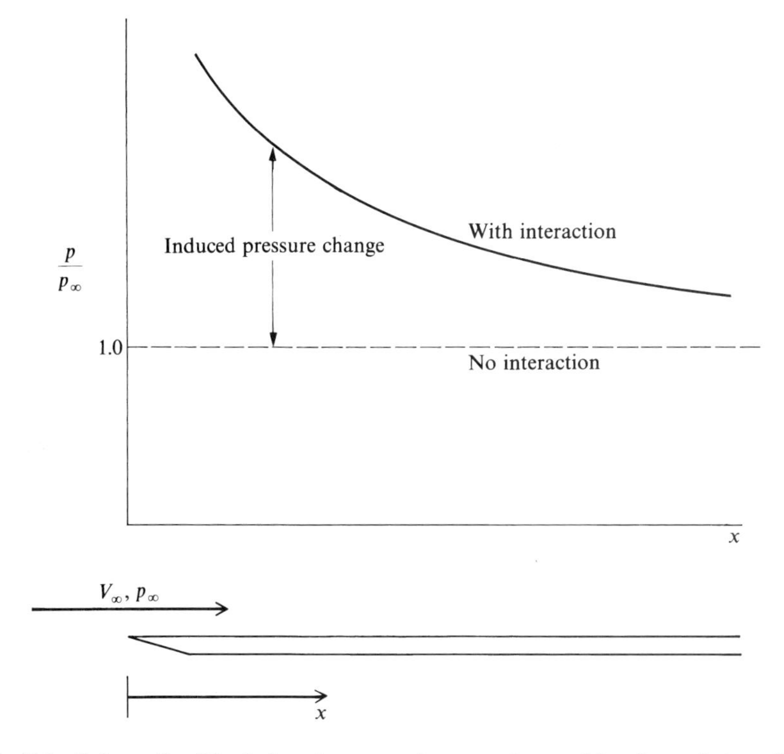

Fig. 7.4 Schematic of the induced pressure increment caused by viscous interaction.

effect was first reported by Becker [122] in 1950, who measured pressures near the leading edge of a wedge that were above the classical wedge pressure from oblique shock theory.

7.3 Role of $\bar{\chi}$ in Hypersonic Viscous Interaction

The induced pressure increment sketched in Fig. 7.4 is governed by the parameter $\bar{\chi}$, defined by Eqs. (7.10) and (7.11). The purpose of this section is to demonstrate this fact. The following analysis, patterned after that of Stollery in [123], is a physically based argument, with a minimum of mathematical detail, which illustrates the major role played by $\bar{\chi}$ in hypersonic viscous interactions.

The displacement thickness δ^*, shown in Fig. 7.3, can be expressed for a hypersonic laminar boundary layer on a flat plate as proportional to the familar result

$$\delta^* \propto \frac{x}{\sqrt{\overline{Re}}} \tag{7.12}$$

where, following the reference temperature method discussed in Sec. 6.9, $\overline{Re}$ is based on average properties within the boundary layer evaluated at the reference temperature given by Eq. (6.159). Equation (7.12) can then be written as

$$\delta^* \propto x\sqrt{\frac{\mu^*}{\rho^* V_\infty x}} = x\sqrt{\frac{\mu_\infty}{\rho_\infty V_\infty x}}\sqrt{\frac{\rho_\infty}{\rho^*}\frac{\mu^*}{\mu_\infty}} = \frac{x}{\sqrt{Re}}\sqrt{\frac{\rho_\infty}{\rho^*}\frac{\mu^*}{\mu_\infty}} \tag{7.13}$$

where Re is the usual Reynolds number based on freestream properties and ρ^* and μ^* are evaluated at the reference temperature T^*. From the equation of state,

$$\frac{\rho_\infty}{\rho^*} = \frac{T^*}{T_\infty}\frac{p_\infty}{p^*} \tag{7.14}$$

Assuming that pressure is constant through the boundary layer in the direction normal to the surface, we have $p^* = p_e$, where p_e is the pressure at the outer edge of the boundary layer. Keep in mind that, because of the viscous interaction effect, p_e is *not* equal to the freestream pressure p_∞. Thus, Eq. (7.14) can be written as

$$\frac{\rho_\infty}{\rho^*} = \frac{T^*}{T_\infty}\frac{p_\infty}{p_e} \tag{7.15}$$

Also, assume a variation of viscosity with temperature as

$$\frac{\mu^*}{\mu_\infty} = C\frac{T^*}{T_\infty} \tag{7.16}$$

where C is given by

$$\frac{\mu_w}{\mu_e} = C\frac{T_w}{T_e}$$

From the equation of state and recalling that pressure is constant through the boundary layer in the normal direction, $T_w/T_e = \rho_e/\rho_w$. Thus, the preceding relation becomes

$$\frac{\mu_w}{\mu_e} = C\frac{\rho_e}{\rho_w}$$

or

$$C = \frac{\rho_w \mu_w}{\rho_e \mu_e}$$

Therefore, C in Eq. (7.16) is the same as defined in Eq. (7.11) associated with the definition of $\bar{\chi}$. Substituting Eqs. (7.15) and (7.16) into (7.13), we have

$$\delta^* \propto \frac{x}{\sqrt{Re}}\sqrt{C\left(\frac{T^*}{T_\infty}\right)^2 \frac{p_\infty}{p_e}} \tag{7.17}$$

Examining Eq. (6.159) for the reference temperature, we see that T^*/T_e depends on M_e^2. Thus, with only a small approximation, we can assume that $T_e \approx T_\infty$ and $M_e \approx M_\infty$, and accept the following proportionality:

$$\frac{T^*}{T_\infty} \propto M_\infty^2 \tag{7.18}$$

Combining Eqs. (7.17) and (7.18), we have

$$\delta^* \propto \frac{x}{\sqrt{Re}} M_\infty^2 \sqrt{\frac{C}{p_e/p_\infty}} \tag{7.19}$$

Equation (7.19) is an intermediate result, to which we will return later. Note that it expresses the variation of δ^* in terms of the ratio of the boundary-layer edge pressure to freestream pressure. Since p_e is higher than p_∞ because of the rapid growth of the boundary layer (examine again Fig. 7.3), let us obtain an expression for p_e/p_∞ in terms of $d\delta^*/dx$.

In Sec. 2.3 we obtained from exact oblique shock theory an exact expression for p_2/p_1 in terms of the hypersonic similarity parameter. This result is given in Eq. (2.28), repeated here:

$$\frac{p_2}{p_1} = 1 + \frac{\gamma(\gamma+1)}{4}K^2 + \gamma K^2\sqrt{\left(\frac{\gamma+1}{4}\right)^2 + \frac{1}{K^2}} \tag{2.28}$$

Here, $K = M_1\theta$, where θ is the flow deflection angle across the oblique shock. Recall that the nomenclature in Chapter 2 used the usual shock conventions, with subscripts 1 and 2 denoting conditions upstream and downstream of the shock respectively. Let us now apply Eq. (2.28) to estimate the pressure at the outer edge of the boundary layer shown in Fig. 7.3. We will assume that the effective body thickness seen by the freestream is given by δ^*, with a slope equal to $\mathrm{d}\delta^*/\mathrm{d}x$. Using the tangent wedge method described in Sec. 3.6, Eq. (2.28) can be written as

$$\frac{p_e}{p_\infty} = 1 + \frac{\gamma(\gamma+1)}{4}K^2 + \gamma K^2\sqrt{\left(\frac{\gamma+1}{4}\right)^2 + \frac{1}{K^2}} \tag{7.20}$$

where $K = M_\infty(\mathrm{d}\delta^*/\mathrm{d}x)$.

Pause for a moment, and assess our progress so far. We have obtained an expression for δ^* in terms of p_e/p_∞ given by Eq. (7.19). In turn, we have developed an equation for p_e/p_∞ in terms of $\mathrm{d}\delta^*/\mathrm{d}x$, given by Eq. (7.20). These two equations provide the tools for analyzing the viscous interaction—the effect of the boundary layer on the outer inviscid flow [Eq. (7.20)] and the effect of the outer inviscid flow on the boundary layer [Eq. (7.19)]. However, the use of these two equations depends on whether we are dealing with the strong interaction or the weak interaction region as illustrated in Fig. 7.3. Let us consider each of these separately.

7.3.1 Strong Interaction

In the region of strong interaction, $\mathrm{d}\delta^*/\mathrm{d}x$ is large. Because $K = M_\infty(\mathrm{d}\delta^*/\mathrm{d}x)$, we therefore assume that $K^2 \gg 1$. With this, Eq. (7.20) becomes

$$\frac{p_e}{p_x} \approx \frac{\gamma(\gamma+1)}{2}K^2 = \frac{\gamma(\gamma+1)}{2}M_\infty^2\left(\frac{\mathrm{d}\delta^*}{\mathrm{d}x}\right)^2 \tag{7.21}$$

To couple the boundary layer with the outer inviscid flow, substitute Eq. (7.21) into (7.19), obtaining

$$\delta^* \propto \frac{x}{\sqrt{Re}}M_\infty^2\sqrt{C}\frac{1}{M_\infty(\mathrm{d}\delta^*/\mathrm{d}x)}$$

or

$$\delta^*\,\mathrm{d}\delta^* \propto \sqrt{\frac{C}{Re}}M_\infty\,x\,\mathrm{d}x \tag{7.22}$$

Recalling that $Re = \rho_\infty V_\infty\, x/\mu_\infty$, Eq. (7.22) can be written as

$$\delta^*\,\mathrm{d}\delta^* \propto \sqrt{\frac{C\mu_\infty}{\rho_\infty V_\infty}}M_\infty x^{1/2}\,\mathrm{d}x \tag{7.23}$$

Integrating Eq. (7.23), we obtain

$$(\delta^*)^2 \propto \sqrt{\frac{C\mu_\infty}{\rho_\infty V_\infty}} M_\infty x^{3/2}$$

or

$$\delta^* \propto \left(\frac{C\mu_\infty}{\rho_\infty V_\infty}\right)^{1/4} M_\infty^{1/2} x^{3/4} \tag{7.24}$$

Note an important physical result from Eq. (7.24). We are used to the conventional laminar boundary-layer result that δ^* grows parabolically, that is, as $x^{1/2}$. However, in the strong interaction region Eq. (7.24) demonstrates that

$$\boxed{\delta^* \propto x^{3/4}} \tag{7.25}$$

Differentiating Eq. (7.24), we obtain

$$\frac{d\delta^*}{dx} \propto \left(\frac{C\mu_\infty}{\rho_\infty V_\infty}\right)^{1/4} M_\infty^{1/2} x^{-1/4} \tag{7.26}$$

Hence, in the strong interaction region,

$$\boxed{\frac{d\delta^*}{dx} \propto x^{-1/4}} \tag{7.27}$$

Combining Eqs. (7.27) and (7.21), we also see in the strong interaction region that

$$\boxed{\frac{p_e}{p_\infty} \propto x^{-1/2}} \tag{7.28}$$

Hence, for strong viscous interaction, the variation of induced pressure with sketched in Fig. 7.4 is an inverse square-root variation. Finally, let us rewrite Eq. (7.26) as

$$\frac{d\delta^*}{dx} \propto \left(\frac{C}{Re}\right)^{1/4} M_\infty^{1/2} \tag{7.29}$$

Hence

$$K^2 = M_\infty^2 \left(\frac{d\delta^*}{dx}\right)^2 \propto \frac{M_\infty^3}{\sqrt{Re}} \sqrt{C} \equiv \bar{\chi} \tag{7.30}$$

Finally, substituting Eq. (7.30) into (7.20), we have (neglecting the $1/K^2$ term because $K \gg 1$)

$$\boxed{\frac{p_e}{p_\infty} = 1 + a_1\bar{\chi}} \tag{7.31}$$

where a_1 is a constant. Equation (7.31) is important. It demonstrates that, for strong viscous interaction, 1) p_e/p_∞ depends only on $\bar{\chi}$; *hence, $\bar{\chi}$ is the governing similarity parameter for induced pressure changes*, as sketched in Fig. 7.4; and 2) the induced pressure change varies *linearly* with $\bar{\chi}$.

Note: In examining Fig. 7.4, keep in mind that, for given freestream conditions, $\bar{\chi} \propto x^{-1/2}$. Hence, the abscissa of Fig. 7.4, which is the running length along the plate, can also be interpreted as a variation in $\bar{\chi}$, where $\bar{\chi}$ *decreases* with increasing x, that is, at the leading edge, $\bar{\chi} \to \infty$, and as x increases, $\bar{\chi}$ constantly decreases. For example, in a single wind-tunnel test at a given set of freestream conditions, one set of surface-pressure measurements gives data over a range of $\bar{\chi}$.

7.3.2 Weak Interaction

For weak viscous interactions, recall from Fig. 7.3 that $\mathrm{d}\delta^*/\mathrm{d}x$ is moderate. In fact, let us assume that $\mathrm{d}\delta^*/\mathrm{d}x$ is small enough that $K = M_\infty(\mathrm{d}\delta^*/\mathrm{d}x) < 1$, and hence $K^2 \ll 1$. With this Eq. (7.20) can be written as

$$\frac{p_e}{p_\infty} = 1 + \gamma K + \frac{\gamma(\gamma+1)}{4}K^2 \tag{7.32}$$

Because $K < 1$ and $K^2 \ll 1$, Eq. (7.32) given approximately $p_e/p_\infty \approx 1$; hence, from Eq. (7.19),

$$\delta^* \propto \frac{x}{\sqrt{Re}} M_\infty^2 \sqrt{C} \tag{7.33}$$

and

$$\frac{\mathrm{d}\delta^*}{\mathrm{d}x} \propto \sqrt{\frac{\mu_\infty}{\rho_\infty V_\infty}} M_\infty^2 \sqrt{C} x^{-1/2} = \frac{M_\infty^2}{\sqrt{Re}} \sqrt{C} \tag{7.34}$$

This is consistent with the definition of weak viscous interaction illustrated in Fig. 7.3; there is no feedback of the changes in the inviscid flow to the boundary layer. Consequently, from Eqs. (7.33) and (7.34) we obtain the familiar results that

$$\boxed{\delta^* \propto x^{1/2}} \tag{7.34a}$$

and

$$\boxed{\frac{\mathrm{d}\delta^*}{\mathrm{d}x} \propto x^{-1/2}} \tag{7.35}$$

Also,

$$K = M_\infty \frac{d\delta^*}{dx} \propto \frac{M_\infty^3}{\sqrt{Re}} \sqrt{C} = \bar{\chi} \tag{7.36}$$

Note from Eq. (7.36) that, in contrast to strong interaction theory where $K^2 \propto \bar{\chi}$, we find that for weak interaction theory, $K \propto \bar{\chi}$. Thus, from Eq. (7.32),

$$\boxed{\frac{p_e}{p_\infty} = 1 + b_1 \bar{\chi} + b_2 \bar{\chi}^2} \tag{7.37}$$

If $d\delta^*/dx$, hence K, is small enough, Eq. (7.37) can be further reduced to

$$\frac{p_e}{p_\infty} = 1 + b_1 \bar{\chi} \tag{7.38}$$

In summary, the analysis of this section has demonstrated that $\bar{\chi}$ is the governing parameter that dictates the induced pressure increment for hypersonic viscous interactions. Moreover, expressions for the induced pressures as a function of $\bar{\chi}$ have been obtained. In a more detailed analysis, Hayes and Probstein [46] have obtained the following results for air with $\gamma = 1.4$.

For an *insulated flat plate*, the *strong interaction* is

$$\frac{p}{p_\infty} = 0.514\bar{\chi} + 0.759 \tag{7.39}$$

and the *weak interaction* is

$$\frac{p}{p_\infty} = 1 + 0.31\bar{\chi} + 0.05\bar{\chi}^2 \tag{7.40}$$

For a *cold-wall case*, where $T_w \ll T_{aw}$, the *strong interaction* is

$$\frac{p}{p_\infty} = 1 + 0.15\bar{\chi} \tag{7.41}$$

and the *weak interaction* is

$$\frac{p}{p_\infty} = 1 + 0.078\bar{\chi} \tag{7.42}$$

Note that a cold wall mitigates to some extent the magnitude of the viscous interaction. This makes sense because for a cold wall the density in the boundary layer will be higher; hence, the boundary-layer thickness will be smaller, thus diminishing the root cause of the viscous interaction in the first place. Also note that the form of Eqs. (7.39–7.42) is consistent with that of Eqs. (7.31), (7.37), and (7.38).

Some classical results are shown in Fig. 7.5, obtained from [46]. Here, experimental data for p/p_∞ on an insulated flat plate (denoted by the circles and triangles) are compared with both strong and weak viscous interaction theory

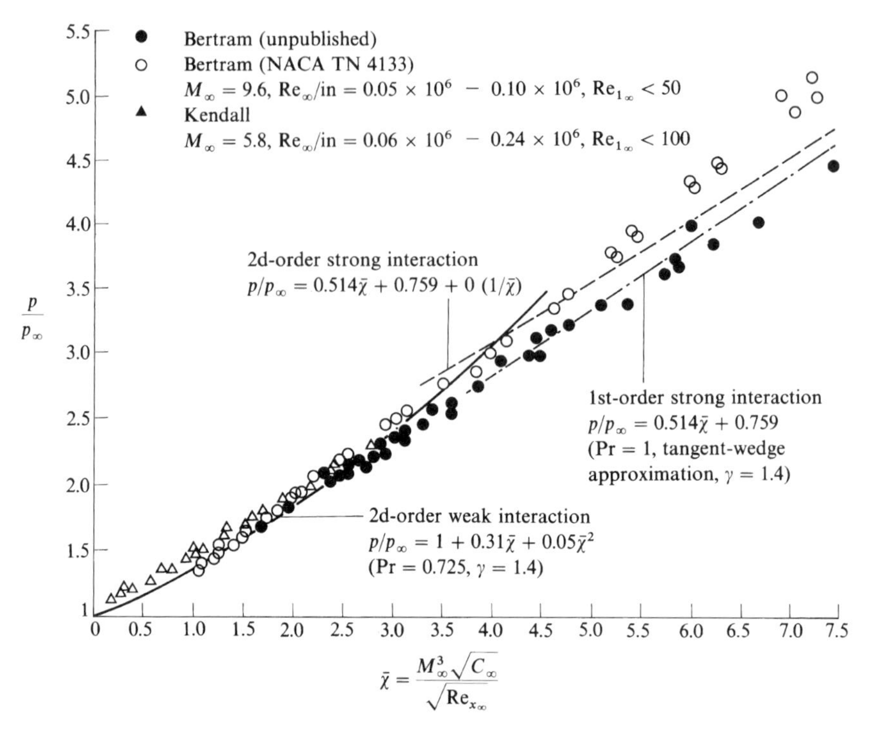

Fig. 7.5 Correlation of induced pressures (from Hayes and Probstein [46]).

(denoted by the curves.) Note that the data are reasonably correlated by $\bar{\chi}$, and that reasonable agreement is obtained between theory and experiment. Also note that, for all practical purposes, the strong and weak interaction regions appear to be described by

$$\bar{\chi} > 3$$

for strong interaction and

$$\bar{\chi} < 3$$

for weak interaction.

Additional experimental and theoretical data are given in Fig. 7.6, obtained from [81]. Here, the induced pressure increment is plotted vs $\bar{\chi}^{-1}$ for hypersonic flow over a flat plate. Measurements were made at Mach numbers of 5, 10, and 20. Looking at the right half of Fig. 7.6, we see again that the pressure data are correlated by $\bar{\chi}$ and agree well with weak and strong viscous interaction theory. Along the abscissa, $\bar{\chi}$ is increasing from right to left; hence, the left half of Fig. 7.6 corresponds to high values of $\bar{\chi}$, dictated by the low Reynolds numbers associated with x locations near the leading edge of the plate. As the

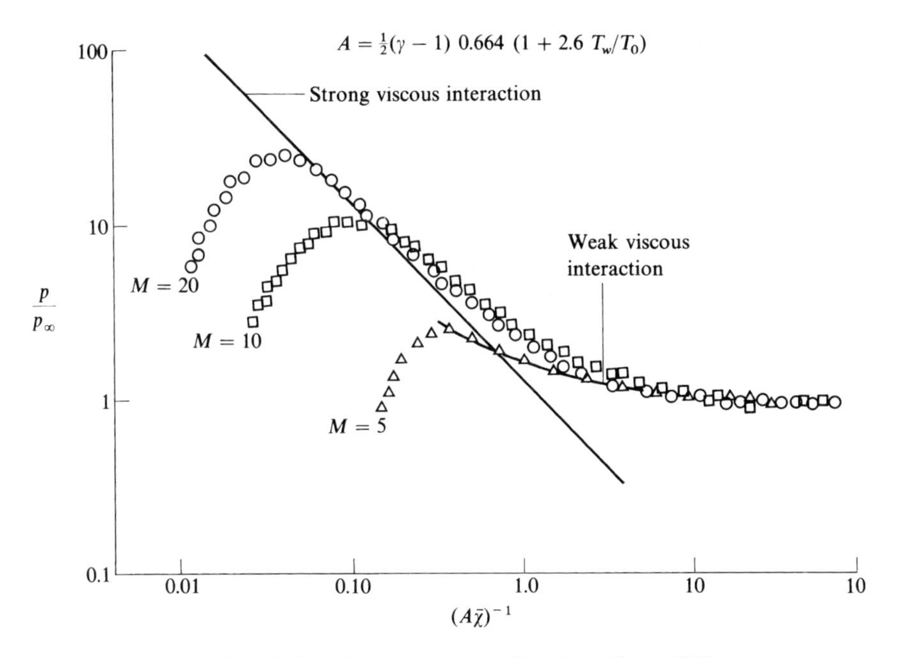

Fig. 7.6 Induced pressures on a flat plate (from [81]).

leading edge is approached more closely, low-density effects such as discussed in Sec. 1.3.5 are encountered, that is, the Knudsen number becomes large. Hence, in the immediate neighborhood of the leading edge, the continuum assumption breaks down, and the measured pressures decrease because of slip-flow effects. Of course, the continuum theory discussed in this chapter does not hold for such low-density conditions.

7.4 Other Viscous Interaction Results

In Sec. 7.3, $\bar{\chi}$ was demonstrated to be the proper viscous interaction correlation parameter for the induced pressure change p/p_∞. In contrast, a different correlation parameter governs pressure coefficient and force coefficients. This is easily seen by considering the pressure coefficient in the form given by Eq. (2.13) written as

$$C_p = \frac{2}{\gamma M_\infty^2}\left(\frac{p}{p_\infty} - 1\right)$$

Assuming that for hypersonic conditions $p/p_\infty \gg 1$, then the preceding becomes

$$C_p \approx \frac{2}{\gamma M_\infty^2}\frac{p}{p_\infty} \tag{7.43}$$

From the results of Sec. 7.3,

$$\frac{p}{p_\infty} \propto \bar{\chi} = \frac{M_\infty^3}{\sqrt{Re}}\sqrt{C}$$

Substituting this into Eq. (7.43), we have

$$C_p \propto \frac{M_\infty}{\sqrt{Re}}\sqrt{C} \equiv \overline{V} \tag{7.44}$$

Hence, we see that the proper viscous interaction correlation parameter for C_p is *not* $M_\infty^3\sqrt{C}/\sqrt{Re}$, but rather $M_\infty\sqrt{C}/\sqrt{Re}$, defined as $\overline{V}$ in Eq. (7.44). Moreover, because lift and wave drag coefficients are obtained by integrating C_p over a given body surface, then viscous interaction effects on both C_L and C_{Dw} are also correlated by $\overline{V}$, that is,

$$C_L = f_1(\overline{V})$$
$$C_{Dw} = f_2(\overline{V})$$

Viscous interaction effects on skin friction and heat-transfer coefficients are also correlated by $\overline{V}$. Both skin friction and heat transfer are *increased* by viscous interaction. Sample results are shown in Fig. 7.7, obtained from [81]. Here the skin-friction coefficient c_f is plotted vs Re for hypersonic laminar flow over a flat plate. Conventional boundary-layer theory shows that $c_f \propto 1/\sqrt{Re}$,

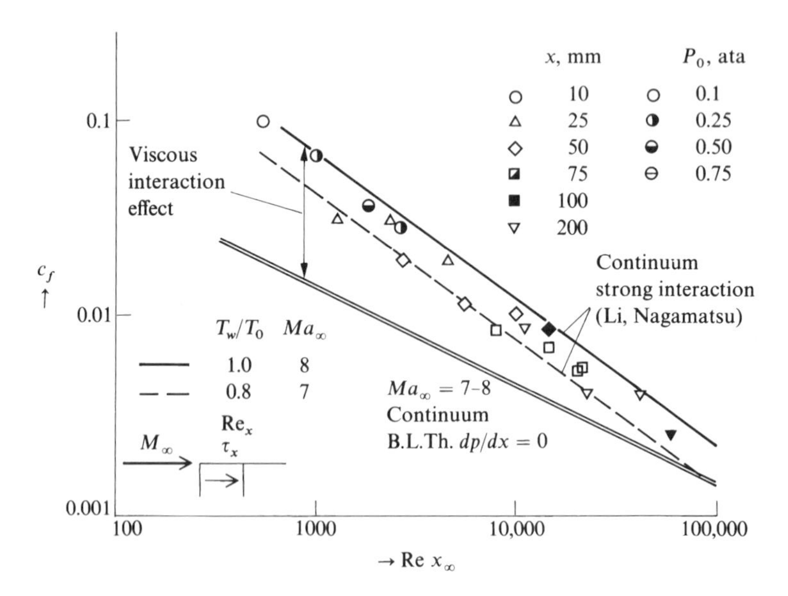

Fig. 7.7 Viscous interaction effect on skin friction (from [81]).

and this variation is given as the double line, which makes a slope of -0.5 on the log-log plot of Fig. 7.7. Experimental data are given by the symbols, and viscous interaction theory is given by the solid and dashed lines. Both the experimental data and the viscous interaction theory fall far above the conventional boundary-layer theory, thus demonstrating the important effect of viscous interactions on c_f. The fact that viscous interaction effects on c_f are correlated by $\overline{V}$ (rather than $\bar{\chi}$) is demonstrated in Fig. 7.8, obtained from [81]. Here, c_f is plotted vs $M_\infty/\sqrt{Re}$ (hence essentially $\overline{V}$). Note that experimental data obtained at different Mach and Reynolds numbers are correlated fairly well by this parameter.

The preceding discussion has centered on viscous interaction as it affects flow over a flat plate. This is because a flat plate is a simple configuration that allows us to highlight the physical aspects of viscous interaction. However, viscous interaction is a basic phenomenon that affects the hypersonic flow over any configuration. Another simple geometry is a sharp cone. Figure 7.9 gives experimental and theoretical data for the hypersonic flow over cones. Here, the induced pressure increment is plotted vs $\bar{\chi}_c$, where p is the actual cone surface pressure, p_c is the inviscid cone pressure, and $\bar{\chi}_c$ is defined as

$$\bar{\chi}_c = M_c^3 \sqrt{\frac{C}{Re_c}}$$

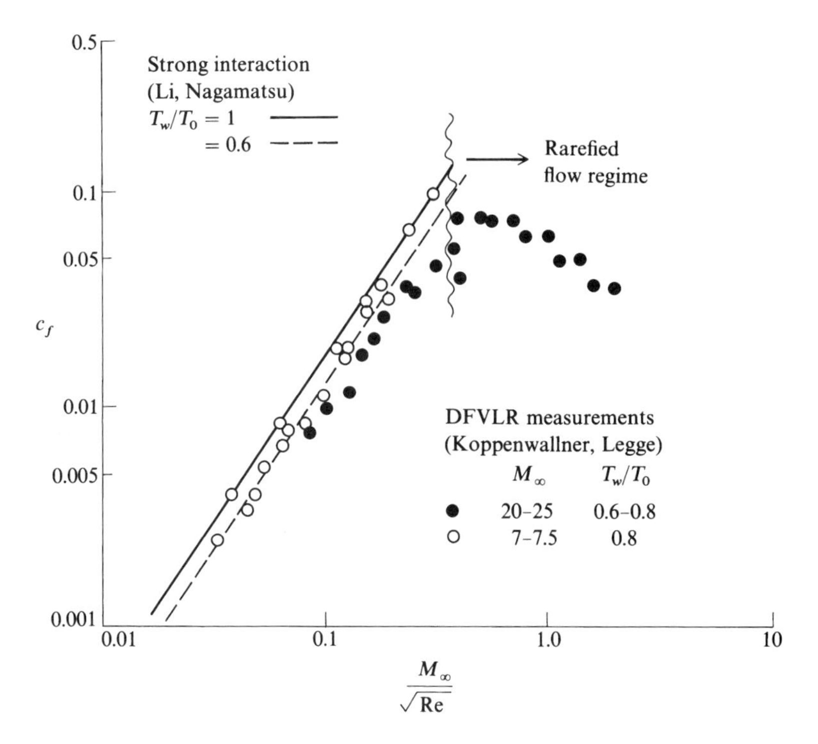

Fig. 7.8 Correlation of the viscous interaction effect on skin friction (from [81]).

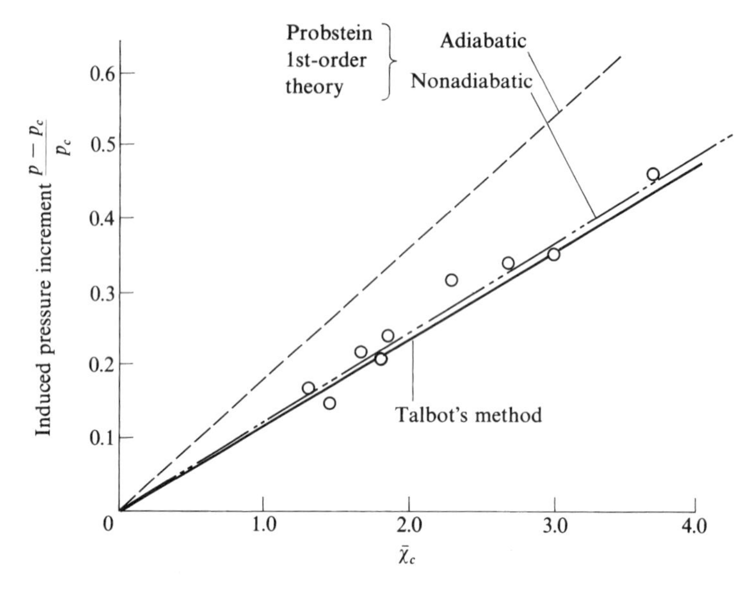

Fig. 7.9 Induced pressure increment vs the hypersonic interaction parameter (from [6]).

where the subscript c denotes inviscid cone surface properties. Also, here $C = (\rho\mu)_w/(\rho\mu)_c$. In Fig. 7.9, the circles denote experimental data obtained from [6], and the lines denote theoretical results from [124] and [125]. These theoretical analyses are approximate techniques for estimating the viscous interaction effect. Probstein [124] obtained analytic results using a Taylor-series expansion in powers of the slope of the boundary-layer displacement thickness. Talbot's method [125] is an approximate graphical approach coupling the displacement thickness slope with the inviscid flow over a cone. The major point to be noted from Fig. 7.9 is that $\bar{\chi}_c$ is a reasonable parameter for correlating the induced pressure increment on cones. As $\bar{\chi}_c$ increases (caused by either or both Mach number increasing and Reynolds number decreasing), the induced pressure increment increases. Moreover, this variation is linear, as seen in Fig. 7.9, and is consistent with the flat-plate results discussed earlier.

The overall effect of viscous interaction on a hypersonic flight vehicle is to *reduce* the lift-to-drag ratio L/D. This is illustrated in Fig. 7.10, obtained from [123] where maximum L/D is plotted vs $M_\infty/\sqrt{Re}$ (hence essentially $\overline{V}$) for a number of different generic vehicle shapes ranging from blunt to slender bodies. In all cases, $(L/D)_{\max}$ decreases as $\overline{V}$ increases. This is because viscous interaction effects increase pressure (hence wave drag) and skin friction (hence friction drag), both increasing the overall drag of the body. The viscous interaction effect on lift is minor because the increased pressure caused by viscous interaction acts on both the top and bottom of lifting surfaces and hence tends to cancel in the lift direction. Thus the degradation of $(L/D)_{\max}$ with increasing $\overline{V}$ shown in Fig. 7.10 is caused primarily by an increase in D.

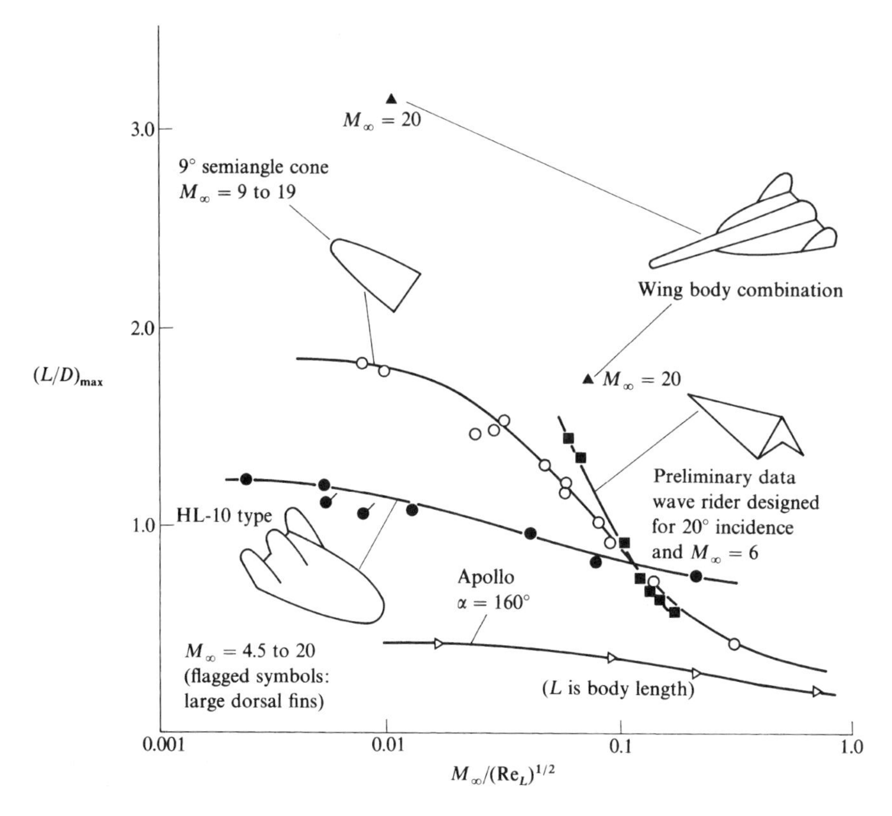

Fig. 7.10 Viscous effects on hypersonic maximum lift-to-drag ratio for five classes of vehicles correlated with the viscous interaction parameter (from Stollery [123]).

More recent work on viscous interaction correlations for force coefficients derived from the space shuttle program has identified a modified viscous interaction parameter as

$$\overline{V}' \equiv \frac{M_\infty}{\sqrt{Re}}\sqrt{C'}$$

where

$$C' = \frac{\rho'\mu'}{\rho_\infty \mu_\infty}$$

and where ρ' and μ' are evaluated at a reference temperature T' within the boundary layer

$$\frac{T'}{T_\infty} = 0.468 + 0.532\frac{T_w}{T_\infty} + 0.195\left(\frac{\gamma - 1}{2}\right)M_\infty^2 \tag{7.45}$$

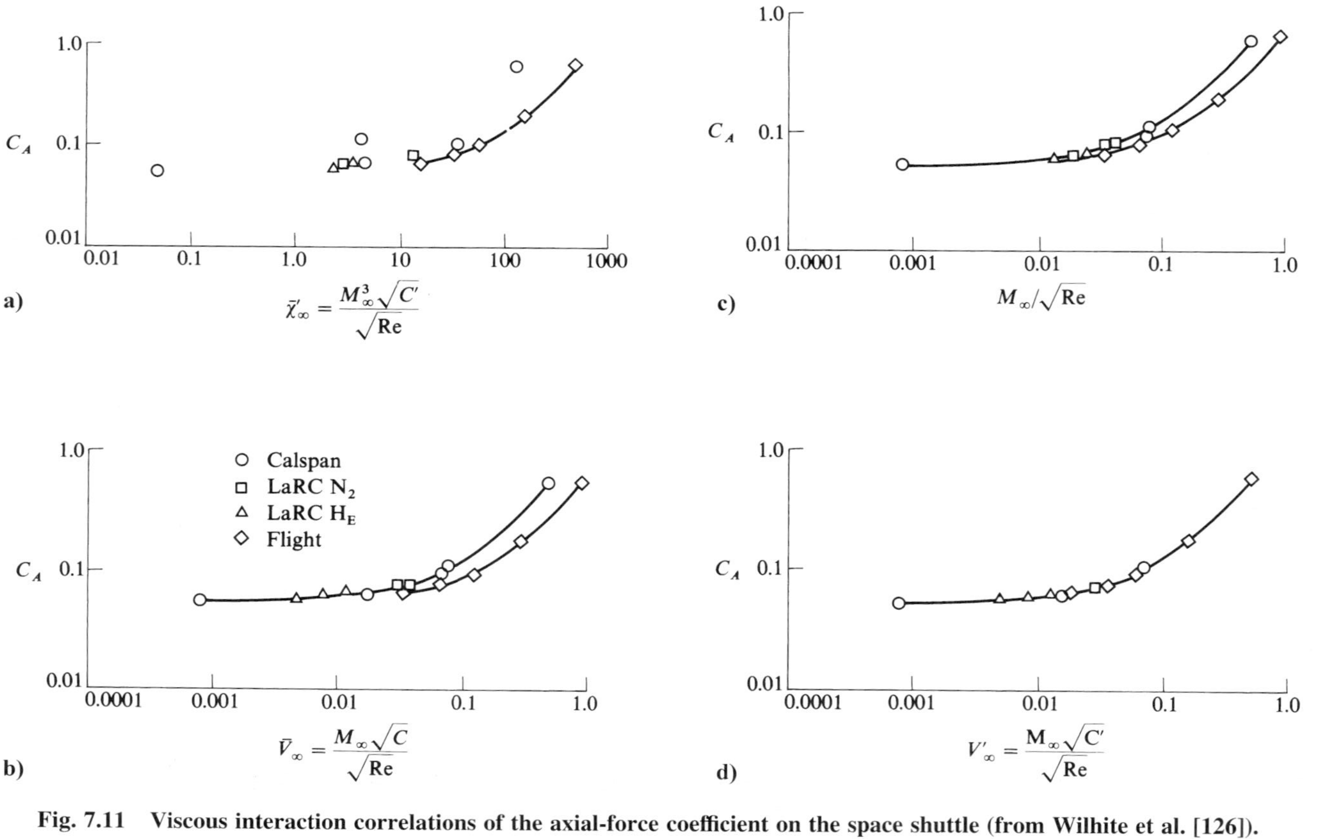

Fig. 7.11 Viscous interaction correlations of the axial-force coefficient on the space shuttle (from Wilhite et al. [126]).

The validity of $\overline{V}'$ as a viscous interaction parameter for force coefficients is demonstrated in Fig. 7.11, obtained from [126]. Here, experimental and flight data for the axial-force coefficient for the space shuttle are correlated by four different parameters. In Fig. 7.11a the data are plotted vs a modified form of $\bar{\chi}$, where the constant C' is evaluated at the references temperature given by Eq. (7.45). A poor correlation is obtained, as shown by the scattered data points. In Fig. 7.11b, the correlation parameter is $\overline{V}$; again, the data are scattered. In Fig. 7.11c, a simple $M_\infty/\sqrt{Re}$ correlation is attempted, but it also fails. Finally, in Fig. 7.11d, we see that the data collapse to the same curve when correlated vs $\overline{V}'$. This is the desired result, and it confirms the use of $\overline{V}'$ as a force coefficient correlation parameter.

Note that all of our discussion so far on viscous interaction has assumed a laminar boundary layer. This is usually the case that actually prevails; viscous interactions occur when $\bar{\chi}$, $\overline{V}$, or $\overline{V}'$ are large, and this corresponds to large M_∞ and/or small Re. In turn, in Sec. 6.7 we saw that large M_∞ and small Re promoted laminar flow. Hence, most viscous interaction theory is based on laminar flow. However, for the sake of completeness, we mention the work of Stollery (see [123]) on studies of viscous interactions associated with turbulent flow. His analysis identified the following viscous interaction parameters for turbulent flow.

Strong interaction:

$$\left(\frac{M_\infty^9 C}{Re}\right)^{2/7}$$

Weak interaction:

$$\left(\frac{M_\infty^9 C}{Re}\right)^{1/5}$$

This brings to an end our discussion of pressure-oriented viscous interactions. These viscous interactions are an important element of hypersonic viscous flow. By no means are all hypersonic flows dominated by viscous interactions. However, for those flow problems where $\bar{\chi}$ (or $\overline{V}$, or $\overline{V}'$) are large, viscous interactions will play an important role. Therefore, when analyzing any hypersonic viscous-flow problem, it is important to examine the associated values of the interaction parameters in order to ascertain whether or not the inclusion of viscous interaction effects is necessary.

7.5 Hypersonic Shock-Wave/Boundary-Layer Interactions

In this section we move to the right-hand side of our chapter road map in Fig. 7.1 and address a second type of viscous interaction, completely distinct from the pressure interaction discussed in Secs. 7.1–7.4, namely, the interaction that occurs when a shock wave impinges on a boundary layer. Such shock-wave/boundary-layer interactions are particularly important to hypersonic flow problems where aerodynamic heating is a major factor because there can be local peaks of heat transfer in the interaction region that can be extremely severe.

Fig. 7.12 X-15 hypersonic test aircraft (U.S. Air Force, Edwards Air Force Base).

A graphical practical example of this interaction heating is provided by one of the final flights of the X-15 hypersonic airplane vehicle in the 1960s (see Fig. 7.12). For this flight, which occurred on 3 October 1967, a dummy ramjet was hung below the fuselage of the X-15, with a pylon connecting the dummy ramjet and the lower surface of the fuselage. On that day, test pilot Pete Knight flew the X-15 at virtually maximum speed, reaching Mach 6.72 at slightly over 100,000 ft altitude. During the hypersonic flight, a shock wave from the ramjet nacelle impinged upon the pylon and burned a hole through the pylon surface. A photograph of this damage is shown in Fig. 7.13, obtained from [127]. The black bar that slashes across the bottom of the pylon is simply a graphical means of pointing out the burned interaction region. In addition, the bow shock wave from the pylon, impinging on the bottom surface of the X-15, also caused local heating damage, as seen at the top of Fig. 7.13. The ramjet model was burned completely off the pylon and punched a hole in the X-15 that allowed the extremely hot boundary-layer air to be rammed into the internal structure, thus weakening the aircraft. Fortunately, Knight was able to safely land the X-15; however, it was the worst case of damage caused by aerodynamic heating throughout the test history of the X-15. (A detailed description of this flight is presented by Richard Hallion in [128].) Clearly from this example, shock wave–boundary-layer interactions can have serious effects on hypersonic vehicles, and this only becomes more severe as the Mach number increases. (We note that the damage from the ramjet-pylon interaction might also have been contributed by a type-IV shock-shock interaction and the resulting supersonic jet impinging on the pylon. The type-IV shock-shock interaction is detailed in Sec. 5.5.)

The qualitative physical aspects of a two-dimensional shock-wave/boundary-layer interaction are sketched in Fig. 7.14. Here we see a boundary layer growing along a flat plate, where at some downstream location an incident shock wave impinges on the boundary layer. The large pressure rise across the shock wave acts as a severe adverse pressure gradient imposed on the boundary layer, thus causing the boundary layer to locally separate from the surface. Because the high pressure behind the shock feeds upstream through the subsonic portion of the boundary layer, the separation takes place ahead of the impingement point of the incident shock wave. In turn, the separated boundary layer induces a second shock wave, identified here as the induced separation shock. The

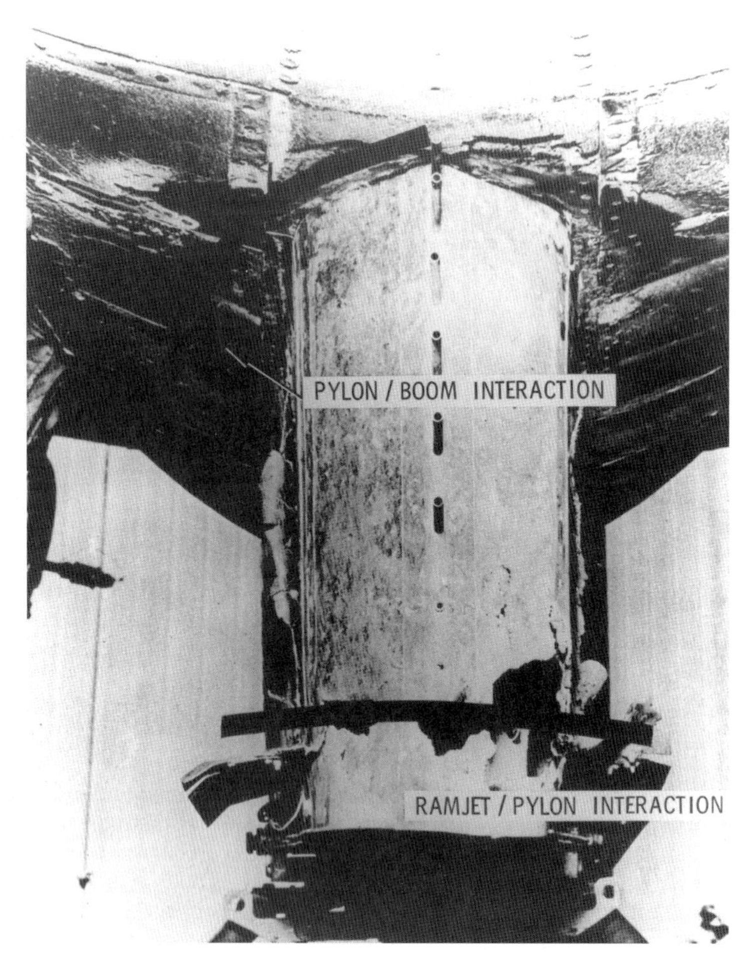

Fig. 7.13 Damage to the X-15 as a result of shock-wave impingement (from Neumann [127]).

separated boundary layer subsequently turns back toward the plate, reattaching to the surface at some downstream location, and causing a third shock wave called the *reattachment shock*. Between the separation and reattachment shocks, expansion waves are generated where the boundary layer is turning back toward the surface. At the point of reattachment, the boundary layer has become relatively thin, the pressure is high, and consequently this becomes a region of high local aerodynamic heating. Further away from the plate, the separation and reattachment shocks merge to form the conventional "reflected shock wave," which is expected from the classical inviscid picture (for example, see [4]). The scale and severity of the interaction picture shown in Fig. 7.14 depends on whether the boundary layer is laminar or turbulent. Because laminar boundary layers separate more readily than turbulent boundary layers (for example, see [1] and [5]),

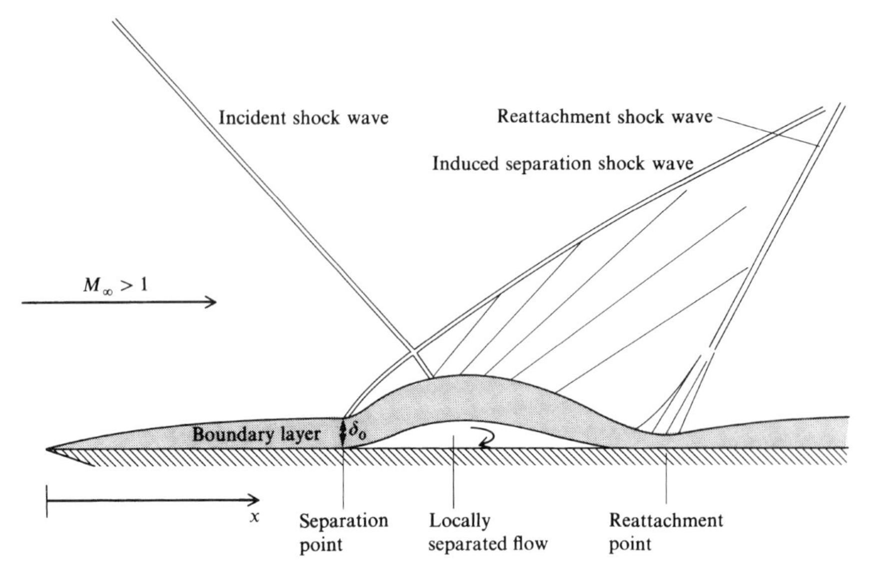

Fig. 7.14 Schematic of the shock-wave boundary-layer interaction.

the laminar interaction usually takes place more readily with more severe attendant consequences than the turbulent interaction. However, the general qualitative aspects of the interaction as sketched in Fig. 7.14 are the same.

The fluid-dynamic and mathematical details of the interaction region sketched in Fig. 7.14 are complex, and the full prediction of this flow is still a state-of-the-art research problem. However, great strides have been made in recent years with the application of computational fluid dynamics to this problem, and solutions of the full Navier–Stokes equations for the flow sketched in Fig. 7.14 have been obtained. Solutions of the full Navier–Stokes equations are described in Chapter 8. The purpose of the present section is simply to describe some basic physical aspects of the hypersonic shock-wave/boundary-layer interaction problem.

Experimental and computational data for the two-dimensional interaction of a shock wave impinging on a turbulent flat-plate boundary layer are given in Fig. 7.15, obtained from [108]. In Fig. 7.15a, the ratio of surface pressure to freestream total pressure is plotted vs distance along the surface (nondimensionalized by δ_0, the boundary-layer thickness ahead of the interaction). Here, x_0 is taken as the theoretical inviscid flow impingement point for the incident shock wave. The freestream Mach number is 3—not hypersonic, but certainly illustrative of the basic phenomena. The Reynolds number based on δ_0 is about 10^6. Note in Fig. 7.15a that the surface pressure first increases at the front of the interaction region (ahead of the theoretical incident shock impingement point), reaches a plateau through the center of the separated region, and then increases again as the reattachment point is approached. The pressure variation shown in Fig. 7.15a is typical of that for a two-dimensional shock-wave/boundary-layer

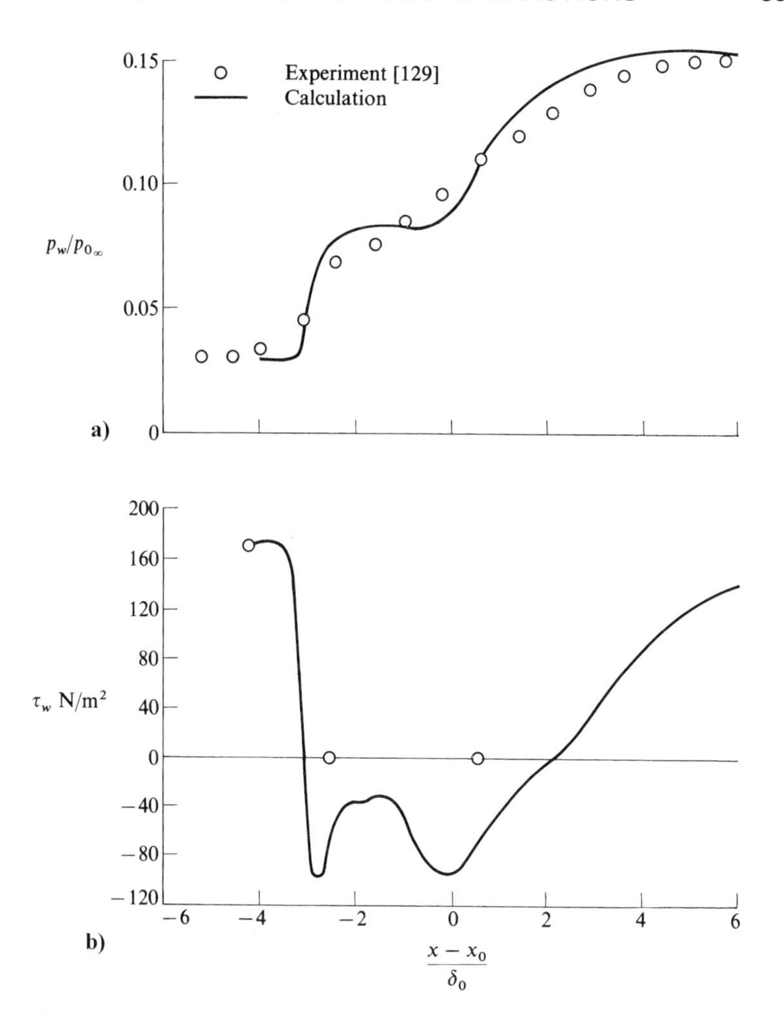

Fig. 7.15 Effects of shock-wave boundary-layer interaction on a) pressure distribution and b) shear stress, for Mach 3 flow over a flat plate. Turbulent flow (from [108]).

interaction. The open circles correspond to experimental measurements of Reda and Murphy [129]. The curve is obtained from a numerical solution of the thin-layer Navier–Stokes equations (see Chapter 8) as reported in [108], and using the Baldwin–Lomax turbulence model discussed in Sec. 6.8. In Fig. 7.15b the variation of surface shear stress is plotted vs distance along the wall. Note that in the separated region the shear stress plummets to zero, reverses its direction (negative values) in a rather complex variation, and then recovers to a positive value in the vicinity of the reattachment point. The two circles on the horizontal axis denote measured separation and reattachment points, and the curve is obtained from the calculations of [108].

An axisymmetric shock-wave/boundary-layer interaction is illustrated in Figs. 7.16–7.18, obtained from [130]. The experimental model and a sketch of the interaction region are shown in Fig. 7.16. Here, an ogive-cylinder is used as the test surface, and an annular shock-wave generator is mounted concentric with the cylinder axis. Shock waves of two different strengths are generated by different annular rings, one beveled at a deflection angle of $\alpha = 7.5$ deg and the other with $\alpha = 15$ deg. Test results obtained at $M_\infty = 7.2$ and a freestream

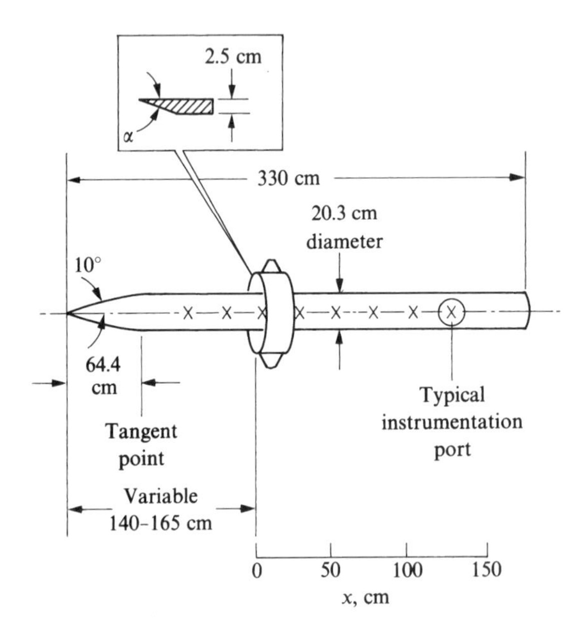

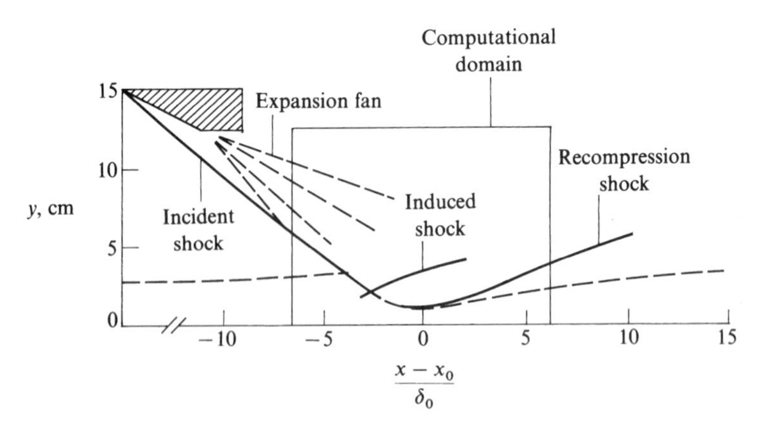

Fig. 7.16 Test model geometry and flowfield sketch for the shock-wave/boundary-layer interaction studied by Marvin et al. [130].

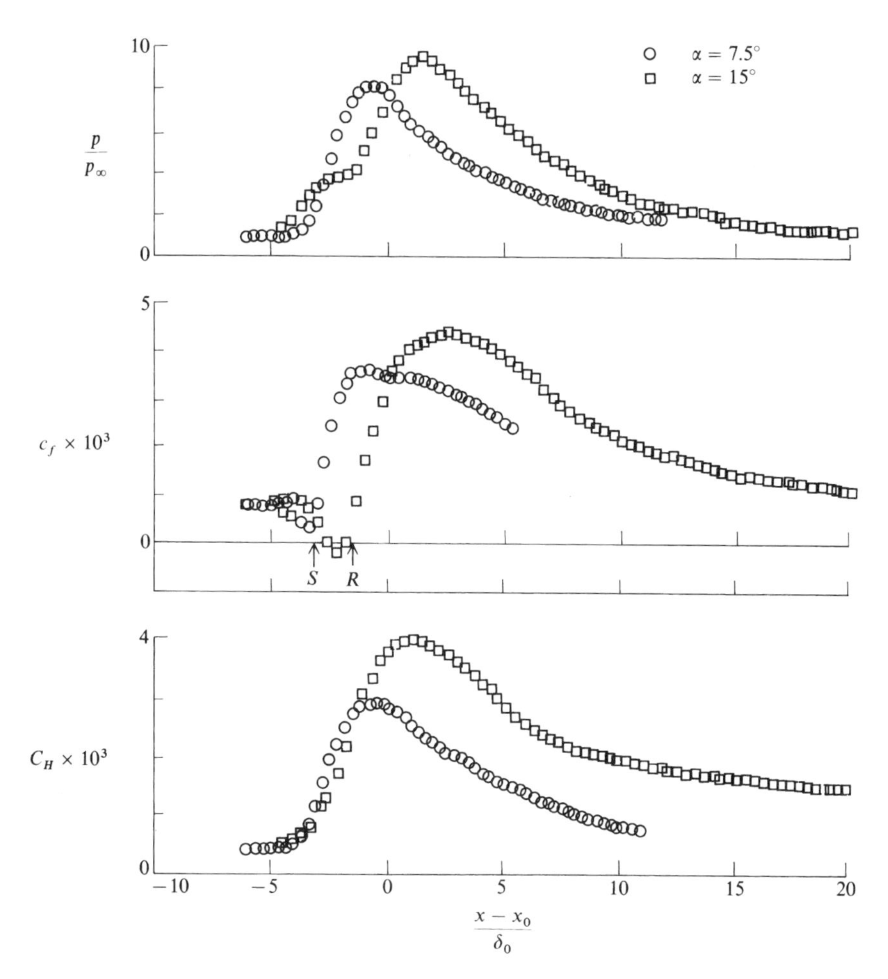

Fig. 7.17 Effects of shock-wave/boundary-layer interaction on pressure, skin friction, and heat-transfer distributions (from [130]).

unit Reynolds number of 10.9×10^6 per meter are shown in Fig. 7.17. One again, x_0 denotes the theoretical inviscid incident shock impingement point. Plotted in Fig. 7.17 are experimental results for p/p_∞, c_f, and C_H vs $(x - x_0)/\delta_0$. The boundary layer is turbulent. First, examine the results for the bevel angle $\alpha = 7.5$ deg, which produces a relatively weak shock wave. For this case, no flow separation can be seen. The pressure rises smoothly through the interaction region; the skin friction first decreases in the face of the adverse pressure gradient but then increases in the recompression region where the boundary layer becomes thinner. The heat transfer continually increases, following the same behavior as the pressure distribution. In contrast, the results for $\alpha = 15$ deg, which produces a stronger shock, show definite flow separation. The pressure distribution

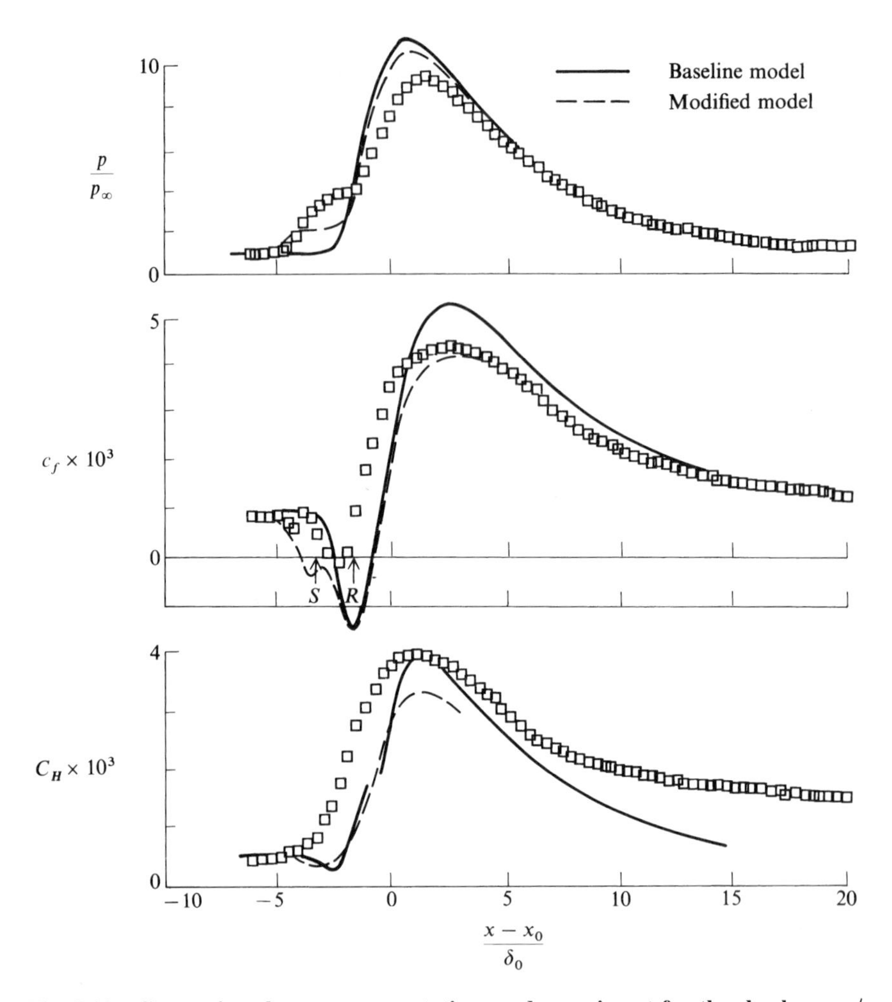

Fig. 7.18 Comparison between computations and experiment for the shock-wave/boundary-layer interaction on a flat plate (from [130]).

has a local plateau in the separation region, the skin friction goes negative in this region, and the heat transfer rises continually through the interaction region. The decay in p/p_∞, c_f, and C_H downstream of the interaction region is caused by the expansion wave from the annular ring (see Fig. 7.16). The experimental results for $\alpha = 15$ deg are repeated in Fig. 7.18, where they are compared with numerical calculations based on a solution of the Navier–Stokes equations. This solution uses MacCormack's time-marching procedure, which was described in Sec. 5.3 for inviscid flows, but here is applied to the Navier–Stokes equations. (Again, note that Navier–Stokes solutions are the subject of Chapter 8.) Two sets of calculations are shown, each made with a different algebraic eddy-viscosity model for the turbulent flow (for details on the models, see [130]). Neither calculation does a very adequate job in predicting the details of the turbulent shock-wave/boundary-layer interaction, thus demonstrating that improvements are needed in the state of the art for this problem.

Comparing the variations of p/p_∞ and C_H in Fig. 7.17, we have already noted that heat-transfer tends to follow the pressure distribution. This is somewhat to be expected on the basis of flat-plate results, as follows. From Eq. (6.81) for a laminar flow

$$C_H \propto \frac{1}{\sqrt{Re}} \propto \frac{1}{\sqrt{\rho_e}} \tag{7.46}$$

From the definition of C_H [Eq. (6.63)],

$$q_w = \rho_e u_e (h_{\mathrm{aw}} - h_w) C_H \tag{7.47}$$

Combining Eqs. (7.46) and (7.47), we have

$$q_w \propto \sqrt{\rho_e} \tag{7.48}$$

From the equation of state, $p = \rho RT$, Eq. (7.48) becomes

$$q_w \propto \sqrt{p_e} \tag{7.49}$$

Equation (7.49) holds for a laminar flow. In contrast, for a turbulent flow

$$C_H \propto \frac{1}{Re^{1/5}} \tag{7.50}$$

and hence, in combination with Eq. (7.47) and the equation of state, we have

$$q_w \propto p_e^{4/5} \tag{7.51}$$

From the results of Eqs. (7.49) and (7.51), it is no surprise that heat transfer and pressure tend to follow the same qualitative variations for a two-dimensional shock-wave/boundary-layer interaction. Indeed, Neumann [127] suggests the following relation between maximum pressure in the interaction $p_{\max}$, maximum heating $q_{\max}$, and the standard flat-plate values p_{fp} and q_{fp}:

$$\frac{q_{\max}}{q_{fp}} = \left(\frac{p_{\max}}{p_{fp}}\right)^n \tag{7.52}$$

where $n = 0.5$ for laminar flow and $n = 0.8$ for turbulent flow. To support this result, Neumann gives the correlation of turbulent shock-wave/boundary-layer interaction data shown as a log-log plot in Fig. 7.19. The data are obtained from various experiments ranging from Mach 6 to 10. The straight line in

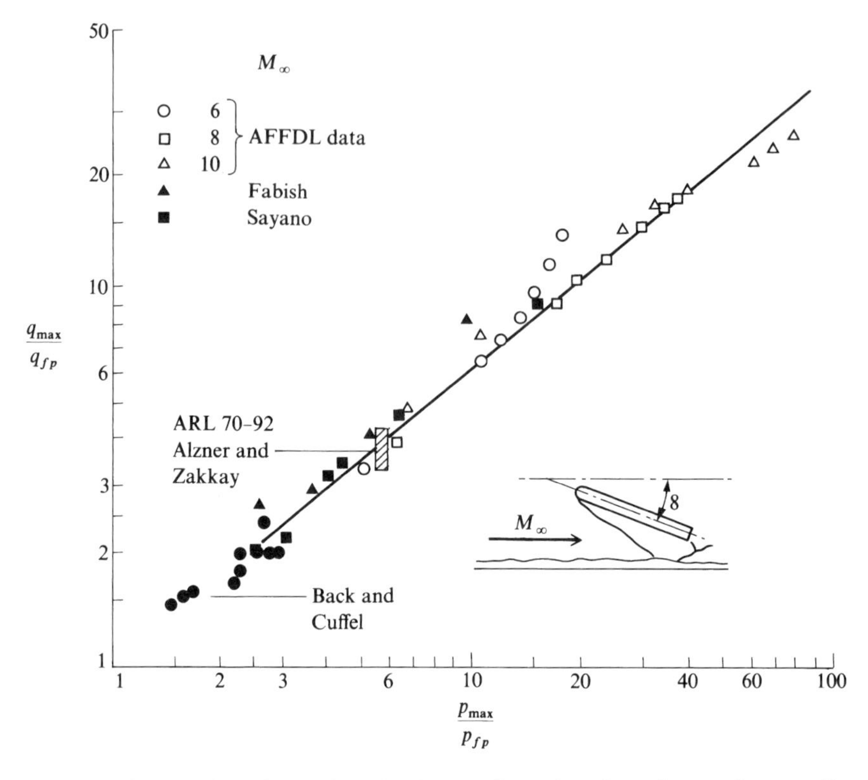

Fig. 7.19 Correlation of turbulent shock-wave/boundary-layer interaction on a flat plate, as given by Neumann [127].

Fig. 7.19 has a slope of 0.8, and the data are clustered around this line, thus confirming the variation given by Eq. (7.52).

An example of a three-dimensional shock-wave/boundary-layer interaction is the flow configuration shown in Fig. 7.20. Here we see a sharp wedge mounted on a flat plate. The interaction between the oblique shock wave from the leading edge of the wedge and the flat-plate boundary layer is a complex, three-dimensional problem. This flow problem has been studied experimentally Oskam et al. [131] and [132] and numerically by Knight [133]. In particular, using MacCormack's time-marching technique (see Sec. 5.3) to solve the complete Navier–Stokes equations and the Baldwin–Lomax turbulence model (see Sec. 6.8), Knight obtained the results shown in Fig. 7.21. Here, the pressure distribution is given as a function of z (the distance from the wedge surface) at a given axial location, $x/\delta_\infty = 14.1$, for a Mach 3 freestream and a wedge angle $\alpha = 9.72$ deg. In Fig. 7.21, z is nondimensionalized by δ_∞, the flat-plate boundary-layer thickness at $x = 0$ (the location of the wedge leading edge). The arrow in Fig. 7.21 denotes the theoretical z coordinate of the inviscid flow shock wave impinging on the flat-plate surface. The solid curve

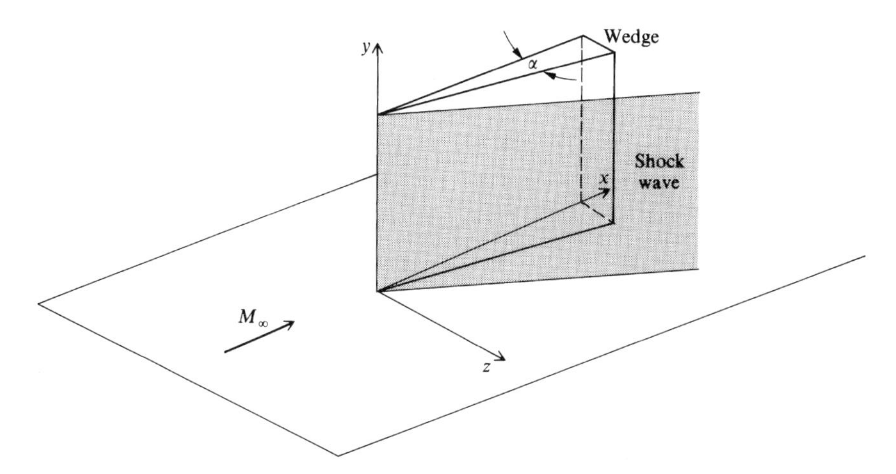

Fig. 7.20 Three-dimensional shock-wave/boundary-layer interaction geometry; wedge on a flat plate.

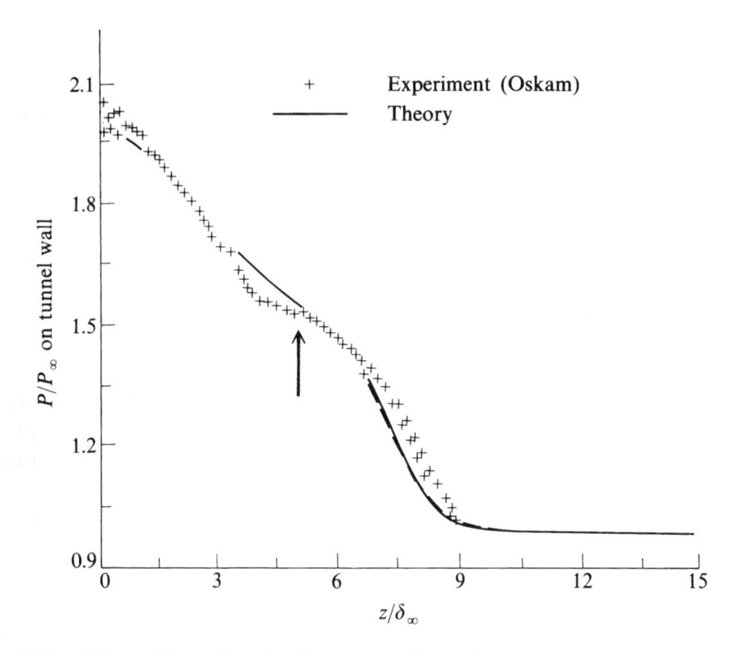

Fig. 7.21 Three-dimensional shock-wave/boundary-layer interaction results; comparison between computations and experiment for pressure distributions (from Knight [133]).

represents Knight's three-dimensional calculations, and the crosses are data from Oskam et al. [131] and [132]. Even in a three-dimensional flow, the pressure exhibits the familiar variation through the interaction region: 1) a rapid increase at the start of the interaction, 2) a plateau in the separated region, and 3) another rapid increase associated with reattachment and the near corner flow at the juncture of the wedge and the flat plate. The aerodynamic heating is shown in Fig. 7.22, where C_H/C_{H_∞} is plotted vs z/δ_∞; C_{H_∞} is the flat-plate value at $x = 0$. (Recall that $x = 0$ is the location of the wedge leading edge at a given distance downstream of the flat-plate leading edge.) As in the preceding figure, z is located at $x/\delta_\infty = 14.1$. Note the rapid increase in C_H through the interaction zone and the severe drop and subsequent recovery as the corner is approached. Again, the solid curve represents the calculations of Knight, and the crosses correspond to the data of Oskam et al. It is rather remarkable in both Figs. 7.21 and 7.22 that fairly reasonable agreement is obtained between the calculations and experiment, considering the complexity of the three-dimensional interaction.

With this, we end our discussion of the shock-wave/boundary-layer interaction. Our purpose has been to describe the basic physical nature of the interaction, without delving into the theoretical complexities of the problem. The literature should be consulted for more details.

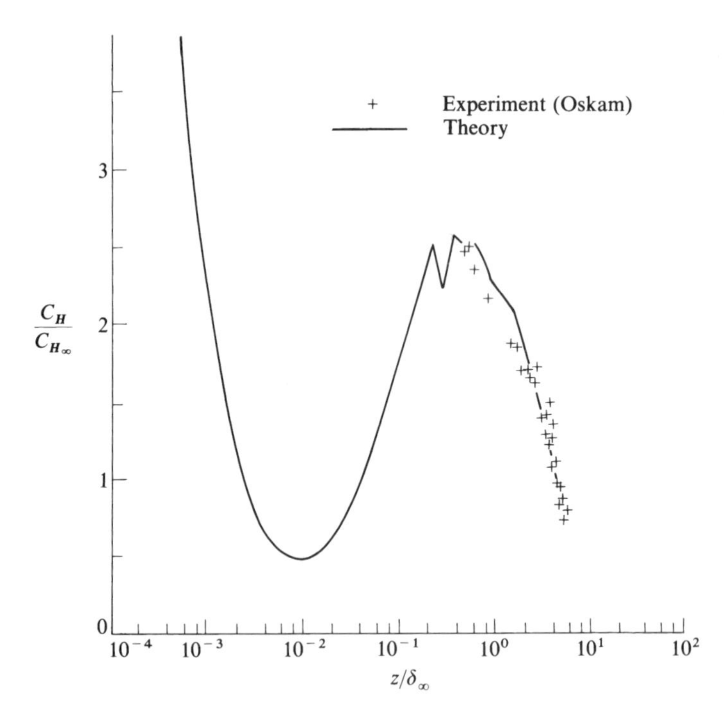

Fig. 7.22 Comparison between computations and experiment for heat-transfer distributions in a three-dimensional shock-wave/boundary-layer interaction (from [133]).

7.6 Summary and Comments

Return for a moment to the road map given in Fig. 1.24. In the present chapter we have discussed hypersonic viscous interactions of two types: the pressure interaction that occurs between a rapidly growing hypersonic boundary layer and the inviscid flow (usually identified simply as viscous interaction) and the interaction between an incident shock wave and a boundary layer. Both of these interactions are listed in Fig. 1.24 near the bottom of the branch dealing with hypersonic viscous flows. Clearly, we are nearing the completion of our discussion of such flows.

In the present chapter, we have shown that the laminar boundary-layer thickness grows as the Mach number squared:

$$\frac{\delta}{x} \propto \frac{M_e^2}{\sqrt{Re}} \tag{7.9}$$

Hence, at hypersonic speeds the boundary-layer thickness can be large. In turn, the rapidly growing boundary layer interacts with the outer inviscid flow, causing an increase in pressure (induced pressure), skin friction, and heat transfer. If the inviscid flow is strongly affected, these changes feed back to the boundary layer itself, causing a *strong* viscous interaction. If the inviscid flow is only weakly affected, it has only a negligible feedback effect on the boundary layer, causing a *weak* viscous interaction. The governing parameter for the induced pressure increment as a result of both strong and weak viscous interaction is

$$\bar{\chi} \equiv \frac{M_\infty^3}{\sqrt{Re}} \sqrt{C} \tag{7.10}$$

where

$$C = \frac{\rho_w}{\rho_e} \frac{\mu_w}{\mu_e} \tag{7.11}$$

For an insulated flat plate, the strong interaction is

$$\frac{p}{p_\infty} = 0.514\bar{\chi} + 0.759 \tag{7.39}$$

and the weak interaction is

$$\frac{p}{p_\infty} = 1 + 0.31\bar{\chi} + 0.05\bar{\chi}^2 \tag{7.40}$$

For a cold wall, where $T_w \ll T_{\mathrm{aw}}$, the strong interaction is

$$\frac{p}{p_\infty} = 1 + 0.15\bar{\chi} \tag{7.41}$$

and the weak interaction is

$$\frac{p}{p_\infty} = 1 + 0.07\bar{\chi} \tag{7.42}$$

From a comparison of experimental data with theory, the strong and weak interaction regions can be identified by

$$\bar{\chi} > 3$$

for strong interaction

$$\bar{\chi} < 3$$

for weak interaction.

The proper correlation parameter for viscous interaction effects on C_p is

$$\overline{V} \equiv \frac{M_\infty}{\sqrt{Re}}\sqrt{C} \tag{7.44}$$

$\overline{V}$ governs skin-friction coefficient. More recent work derived from work on the space shuttle has identified a modified viscous interaction parameter that correlates the axial-force coefficient:

$$\overline{V}' \equiv \frac{M_\infty}{\sqrt{Re}}\sqrt{C'}$$

where

$$C' \equiv \frac{\rho'\mu'}{\rho_\infty \mu_\infty}$$

and where ρ' and μ' are evaluated at a reference temperature given by

$$\frac{T'}{T_\infty} = 0.468 + 0.532\frac{T_w}{T_\infty} + 0.195\left(\frac{\gamma - 1}{2}\right)M_\infty^2$$

A second type of viscous interaction particularly important at hypersonic speeds is the shock-wave/boundary-layer interaction. Such an interaction is characterized by an incident shock, an induced separation shock, a reattachment shock, an embedded expansion wave, and a separated flow region. Shock-wave/boundary-layer interactions cause local peaks in aerodynamic heating that can have serious consequences on hypersonic vehicles.

Design Example 7.1: Hypersonic Waveriders—Part 3

In this chapter we have seen that the classic viscous interaction effect increases the pressure and skin-friction drag on a hypersonic body. What effect does this have on the design of the viscous optimized hypersonic waveriders discussed in the Design Example at the end of Chapter 6? This question was addressed by Chang in [254], with some results summarized in [255]. The local flat-plate viscous interaction analyses discussed in this chapter were applied locally along each streamline over the surface of the waverider. This is justified because the streamlines have very little transverse curvature; they are reasonably straight. Also, for the noninteraction case the pressure gradients along the surface streamlines are small, and therefore it is reasonable to apply the flat-plate viscous interaction analyses in this chapter locally at each point along a streamline. Using these viscous interaction analyses to obtain pressure and skin friction over the surface, the optimization approach described in the Design Example at the end of Chapter 6 was utilized to find the best viscous optimized waverider including the effect of viscous interaction.

These results showed that the shape of the viscous optimized waverider is greatly affected by viscous interaction. Figure 7.23 shows a viscous optimized hypersonic waverider for $M_\infty = 16$ at an altitude of 140,000 ft *not* including viscous interaction. In contrast, Fig. 7.24 shows a viscous optimized hypersonic waverider for the same flight conditions but *including* viscous interaction. Both waveriders are constrained to be 60 m long. The waverider optimized for viscous interaction (Fig. 7.24) is more slender with a smaller volume than the case not including viscous interaction (Fig. 7.23). The local values of $\bar{\chi}$ on the upper surface centerline ranged from 8 near the leading edge to 0.2 at the trailing edge, spanning both the strong and weak interaction regions. The resulting values of maximum L/D are compared here.

Optimized *without* viscous interaction (Fig. 7.23):

$$(L/D)_{\max} = 10.9$$

Optimized *with* viscous interaction (Fig. 7.24):

$$(L/D)_{\max} = 9.9$$

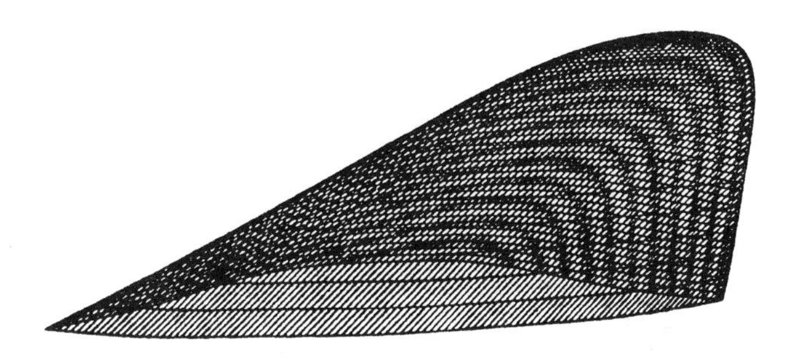

Fig. 7.23 Waverider shape *without* viscous interaction at Mach 16 (Anderson et al. [255]).

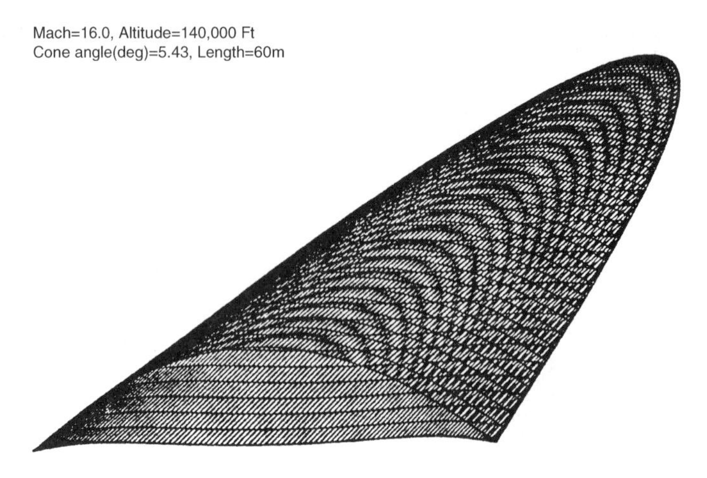

Fig. 7.24 Optimized waverider shape with viscous interaction at Mach 16 [255].

Clearly, the effects of viscous interaction causes a loss in $(L/D)_{\max}$; this loss is mainly because of an increase in drag.

Figures 7.25 and 7.26 show the corresponding results for $M_\infty = 25$ at an altitude of 230,000 ft. The shape of the viscous optimized waverider *not* including viscous interaction is shown in Fig. 7.25, and the shape *including* viscous interaction effects is shown in Fig. 7.26. The local values of $\bar{\chi}$ on the upper-surface centerline ranged from 110 near the leading edge to about 9 at the trailing edge—clearly well into the strong interaction region. The resulting values of maximum L/D are compared here.

Optimized *without* viscous interaction (Fig. 7.25):

$$(L/D)_{\max} = 5.26$$

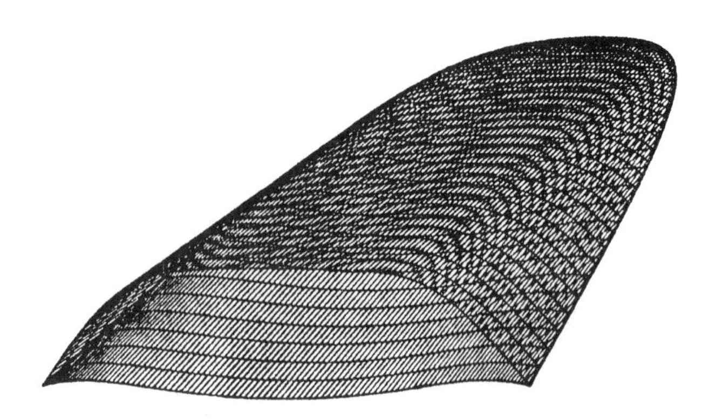

Fig. 7.25 Waverider shape *without* viscous interaction at Mach 25 [255].

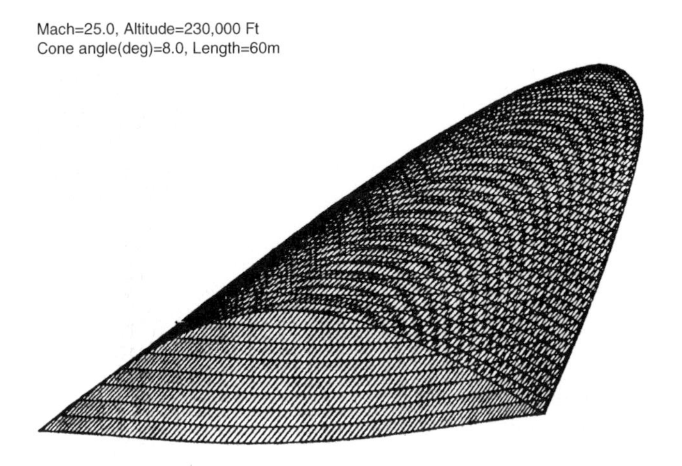

Fig. 7.26 Optimized waverider shape with viscous interaction at Mach 25 [255].

Optimized *with* viscous interaction (Fig. 7.26):

$$(L/D)_{\max} = 4.13$$

Clearly, the shape of viscous optimized waveriders and the resulting value of $(L/D)_{\max}$ are affected by viscous interactions. For the cases reported here, at Mach 16 and 140,000 ft viscous interaction reduces $(L/D)_{\max}$ by 9%, and at Mach 25 and 235,000 ft viscous interaction reduces $(L/D)_{\max}$ by 21%. Considerably more details on the analysis and many more results can be found in [254] and [255]. The large number of results reported in [254] have been used to lay out a region in the velocity-altitude map where viscous interaction is important for hypersonic waveriders. This is the shaded region in Fig. 7.27, defined as that region where viscous interaction effects cause more than a 5% decrease in $(L/D)_{\max}$ for waveriders of length equal to 60 m. The shaded region in Fig. 7.27 was mapped by running hundreds of different waverider optimizations for different Mach and Reynolds numbers. For waveriders larger than 60 m length, the boundary will move upward and to the right because viscous interaction effects are strongest near the leading edge and for longer waveriders the integrated effect of viscous interactions becomes proportionally smaller. In contrast, for a waverider smaller than 60 m length, the boundary will move down and to the left in Fig. 7.27. These results give some useful guidance for the design of waveriders insofar as the need to include viscous interaction effects is concerned.

We note that, to this author's knowledge, the results given in [254] and [255], summarized in this Design Example, are the first published data on the effects of viscous interaction on any type of waverider configurations.

Hypersonic waveriders have been chosen as the subject for three Design Examples in this book, not because this author is necessarily championing them as configurations for future hypersonic vehicles, but rather as design examples of the applications of some of the fundamental hypersonic aerodynamics discussed

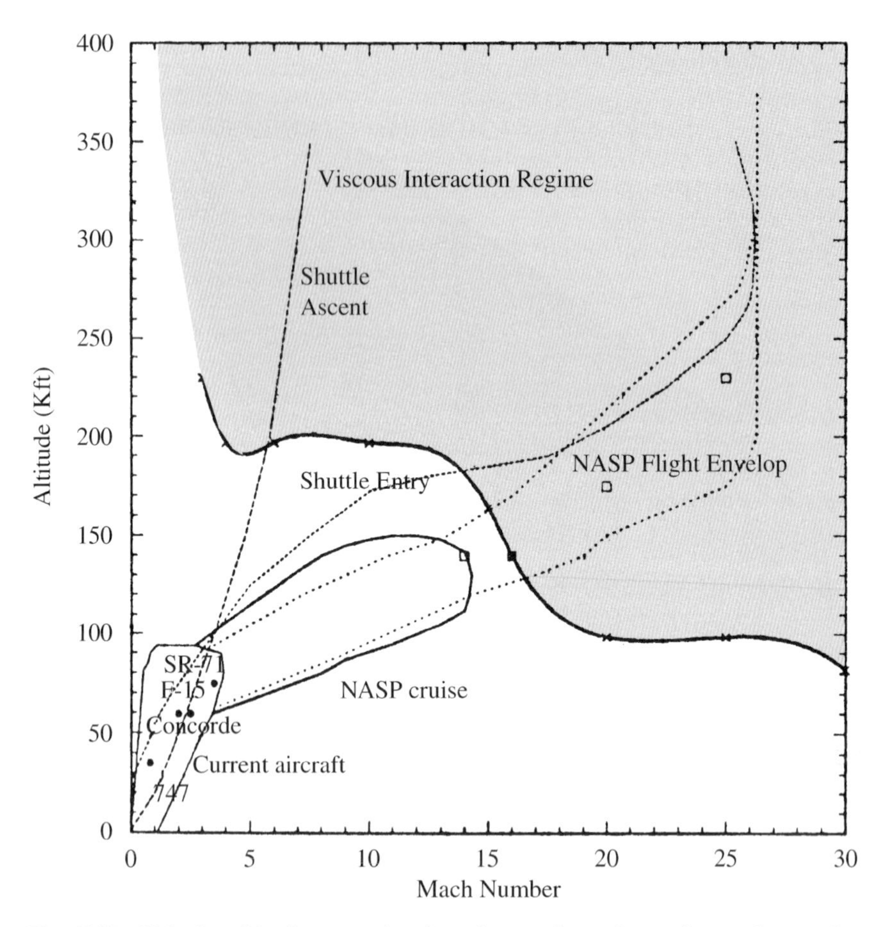

Fig. 7.27 Velocity-altitude map showing that region where viscous interaction effects cause more than a 5% decrease in maximum lift-to-drag ratio. Length of the waverider is 60 m [255].

in the main parts of the chapters. We have one more waverider Design Example to go, namely, the effect of chemically reacting flow on waverider design; this appears at the end of Chapter 14. High-altitude, low-density effects on viscous optimized waverider design are discussed in [256], but we will not highlight them here because low-density effects are not a fundamental subject covered in this book.

Hypersonic waveriders, however, *do* have promise as future hypersonic vehicles, for all of the reasons given in this and the previous hypersonic waverider design examples. There are many practical design problems to be overcome for this to happen. For example, propulsion must be integrated with the waverider aerodynamics. This matter has been studied at the University of Maryland by O'Neill and Lewis in [257]. There is the question of off-design performance of waveriders. The viscous optimized waveriders discussed in this book are point designed for a particular Mach number and altitude. What happens when these waveriders fly at different Mach numbers and altitudes? This subject has been studied, also at the University of Maryland, by Takashima and Lewis in [258], and as part of a broader propulsion-oriented investigation by O'Brien and Lewis in [259]. Because the present book is not a design text, and this work delves more deeply into design details, we will not discuss it further here. References [257–259] among others are readily available in the archive literature.

Problems

7.1 Consider a flat pate of length equal to 10 m. Assume this flat plate is flying the trajectory labeled as "high lift" ($m/C_L S = 50\ \text{kg/m}^2$) shown in Fig. 6.3. The wall temperature of the plate is held constant at 1000 K.

(a) Plot the variation of $\bar{\chi}$ at 0.5 m from the leading edge as a function of M_∞ as the flat plate files this trajectory.

(b) Repeat part a, except calculate $\bar{\chi}$ at the trailing edge of the plate. For simplicity, in the preceding assume that the angle of attack of the plate is essentially zero (although this violates the finite lift used in obtaining the trajectory in Fig. 6.3). The purpose of this problem is to obtain a "feel" for the values of $\bar{\chi}$ encountered by hypersonic vehicles during atmosphere flight.

7.2 Consider a flat plate with a 5-m length at zero angle of attack. The wall temperature is 1200 K. The freestream condition are $M_\infty = 25$ at a standard altitude of 280,000 ft. Calculate and plot the variation of pressure as a function of distance downstream of the leading edge. Compare this with the exact inviscid pressure. Comment on the impact of viscous interaction for this case.

8
Computational-Fluid-Dynamic Solutions of Hypersonic Viscous Flows

But as no two (theoreticians) agree on this (skin friction) or any other subject, some not agreeing today with what they wrote a year ago, I think we might put down all their results, add them together, and then divide by the number of mathematicians, and thus find the average coefficient of error.

Hiram Maxim, early aeronautical designer, 1908

The advent of the electronic computer completely altered the nature of the facilities available for numerical calculations. An electronic computer can perform all the functions of a desk calculator but at much higher speeds and, in addition, it can largely replace the operator as well!

K. N. Dodd, British mathematician, 1964

Chapter Preview

Enough already! It is time for us to treat a hypersonic aerodynamic flowfield in its full glory the way nature does—as a fully viscous flow at every point throughout the flowfield. In the previous chapters, as human beings we have intellectually idealized certain types of flow. We dealt with inviscid flows; they do not exist in real life, but many flowfields and practical engineering problems are closely approximated by the assumption of inviscid flow, and, as we have seen, a lot of good practical results for pressure distributions, lift, and wave drag can be obtained with this assumption. We also dealt with limited viscous effects—limited to the boundary layer adjacent to a surface. Again, a lot of good practical results for skin friction and heat transfer can be obtained from a boundary-layer analysis. In the modern hypersonic aerodynamics of today, these analyses still play a pivotal role, and they provide usually straightforward, rapid, and inexpensive methods for solving many engineering problems.

Flows involving flow separation, however, are a different story. None of the methods discussed in this book so far can handle separated flows. Local flow separation is frequently a consequence of shock-wave/boundary-layer interaction, as discussed in Sec. 7.5. The flow over the base of a hypersonic vehicle is separated flow; it is a major player in determining the base pressure, hence the base drag. A numerical solution of the full Navier–Stokes equations for viscous flow is usually required for the calculation of these types of flows.

The flows over hypersonic vehicles flying at very high Mach numbers at very high altitudes are also a different story. For these high Mach numbers and comparatively low Reynolds numbers, the boundary layers become so thick that the conventional boundary-layer equations and boundary-layer methods discussed in Chapter 6 are not valid. Indeed, the boundary layer can become so thick that it completely fills the thin shock layer between the shock wave and the body surface. When this happens, what we have is no longer a boundary layer in the classical sense but rather a *fully viscous shock layer*. Some type of numerical solution assuming that the complete shock layer is viscous is therefore required.

Numerical solutions of hypersonic fully viscous flows are the subject of this chapter. You might consider this to be the "crème de la crème" of hypersonic flowfield analysis, especially the full Navier–Stokes solutions discussed at the end of the chapter. In historical perspective, this chapter could not have been written 30 years ago. It is possible today only because of the rapid development of sophisticated numerical algorithms for the solution of the Navier–Stokes equations and the phenomenal increase in computer power that allows such algorithms to function. But nevertheless, here we are. This chapter is a suitable finish to our studies of nonreacting hypersonic flow in Parts 1 and 2 of this book. With this chapter, we go out in style. Enjoy.

8.1 Introduction

In the chapter's opening quotation, Hiram Maxim, inventor of the machine gun and the designer and builder of a large flying machine in the 1890s, is venting his frustration at the lack of applicability of mathematical theory to the practical problems of flight. In contrast, in the second quote we have, just 56 years later, K. N. Dodd remarking about the revolution precipitated by the high-speed digital computer. In the 56 years between these quotations, mathematical theory was indeed successfully applied to the practical problem of flight (see the historical notes in [1], [4], and [5]), and in the 43 years that have ensued from Dodd's quotation, we have indeed seen a most remarkable revolution in computing. These quotations are relevant to the present chapter because here we discuss the most "exact" analyses of hypersonic flows available, and these exact analyses are made possible only by the use of a high-speed computer. This chapter could not have been written 40 years ago; moreover, if he were alive today, Hiram Maxim would have to change his image of the theoretician.

To be more specific, this chapter deals with the application of computational fluid dynamics to hypersonic viscous flows. However, as in Chapter 5, our intent is not to elaborate on the details of CFD; the book by Anderson [52] serves this purpose. Instead, our objective here will be to present various approaches to the solution of hypersonic viscous flows that go beyond, and are more exact than, the boundary-layer analyses discussed in Chapter 6.

Thinking along another line, the weak and strong viscous interaction theories discussed in Chapter 7 are a product of the 1950s and 1960s, before the advent of computational fluid dynamics. They serve a useful purpose in providing convenient correlations and prediction expressions, albeit based on an approximate theory. The approximations involved separate calculations of the boundary layer and outer inviscid flow and then a coupling of these separate calculations to take into account the viscous interaction. Today, the viscous interaction effect can be calculated *exactly*, simply by treating the entire flowfield between the body and shock as *fully viscous*—no arbitrary division between a boundary layer and an inviscid flow needs to be made. Indeed, this is the natural and physically proper approach. The fully viscous-flow calculations are made with standard CFD techniques, to be discussed in the present chapter.

There is another reason to favor a fully viscous shock-layer analyses over the conventional boundary-layer approach. Recall from Sec. 6.4 that the derivation of the hypersonic boundary-layer equations by means of an order-of-magnitude reduction of the Navier–Stokes equations did not preclude a finite normal pressure gradient through the boundary layer, that is, it is compatible with boundary-layer hypothesis that $\partial p/\partial y \neq 0$ at hypersonic speeds. However, the classical first-order boundary-layer theory as discussed in Chapter 6 has no mechanism for computing $\partial p/\partial y$. Hence, for an analysis of a hypersonic viscous flow, it is inherently more appropriate to assume the flow is viscous throughout the entire flowfield and to compute this fully viscous flow by means of a system of equations more accurate than the boundary-layer equations. This is the purpose of the present chapter.

In the modern hypersonics of today, there are three approaches to the solution of a fully viscous flow, which have found widespread use. They are 1) viscous shock-layer solutions, 2) parabolized Navier–Stokes solutions, and 3) full Navier–Stokes solutions. Each of three approaches just listed utilizes systems of equations that are more accurate than the boundary-layer equations; going from items 1 to 3, the system of equations is progressively more accurate, finally ending with the complete Navier–Stokes equations with no basic simplifications whatsoever. Moreover, these CFD techniques go *far beyond* just the calculation of viscous interactions—they allow the detailed calculations of the complete flowfield over a body where the flow is assumed to be viscous at every point. Hence they provide *everything* about the flow, such as the shock shape, detailed flow variables between the shock and the body, skin friction, heat transfer, lift, drag, moments, etc. In the following sections, we will examine individually the approach taken by each of these methods, with the presentation of appropriate results. The intellectual path we take in this chapter is a very linear, very straightforward road, and therefore we have no need for a local chapter road map. Also, referring to our general road map in Fig. 1.24, we note that with this chapter we have come to the end of the line under viscous flows.

Finally, all of the techniques discussed here are derived in some form or another from the complete Navier–Stokes equations given by Eqs. (6.1–6.6). It is important to examine these equations again, and to review Sec. 6.2, before progressing further.

8.2 Viscous Shock-Layer Technique

Although all of the techniques discussed in this chapter deal with fully viscous flows, one technique has acquired the official label as the viscous shock-layer method. We will follow this terminology here. Specifically, in 1970, the late Tom Davis introduced a solution of a set of equations that approximate the Navier–Stokes equations and used them to solve for the fully viscous shock layer over a blunt body at hypersonic speeds (see [134]). His technique has subsequently become commonly known as the viscous shock-layer (VSL) technique. Davis' viscous shock-layer equations are obtained by writing the full Navier–Stokes equations (see Sec. 6.2) in boundary-layer coordinates (parallel and perpendicular respectively to the surface) and performing an order-of-magnitude analysis on the terms in the equations. Terms are kept up to second order in $1/\sqrt{Re}$. This leads to a system of equations, which is more powerful than the boundary-layer equations in that they hold across the entire shock layer, but which is far simpler than the full Navier–Stokes equations. Moreover, Davis' VSL equations are parabolic and therefore allow a *downstream-marching* finite difference solution, starting from some specified initial data plane. Of particular distinction is that the VSL equations take into account a pressure gradient in the normal direction to the surface, $\partial p/\partial n \neq 0$, in contrast to the familiar boundary-layer assumption. We have already seen from Sec. 6.4 that, for hypersonic flow, accounting for such a nonzero pressure gradient is quite appropriate.

The derivation and discussion of the basic equations ultimately used in the VSL technique can be found in [135]. The details are beyond the scope of the present book. However, the important ideas are as follows. The Navier–Stokes equations are first written in boundary-layer coordinates s and n parallel and perpendicular to the surface respectively, as shown in Fig. 8.1. The resulting equations are then nondimensionalized in two different ways: 1) one set of equations is obtained by forming nondimensional variables that are of order one near the body surface; and 2) a second set of equations is obtained by nondimensionalizing in terms of variables of order one in the nearly inviscid region far away from the surface. Terms in each of the two sets of equations that are third order or higher in terms of the inverse square root of the Reynolds number are dropped. Finally, after a comparison of the two sets of equations, one set is found from them that is valid to second order in both the inner and outer regions. These equations, as they appear in [134], are displayed next (see also [224]).

Continuity equation:

$$\frac{\partial}{\partial s^*}[(r^* + m^* \cos\phi)^m \rho^* u^*] + \frac{\partial}{\partial n^*}[(1 + \kappa^* n^*)(r^* + n^* \cos\phi)^m \rho^* v^*] = 0 \quad (8.1)$$

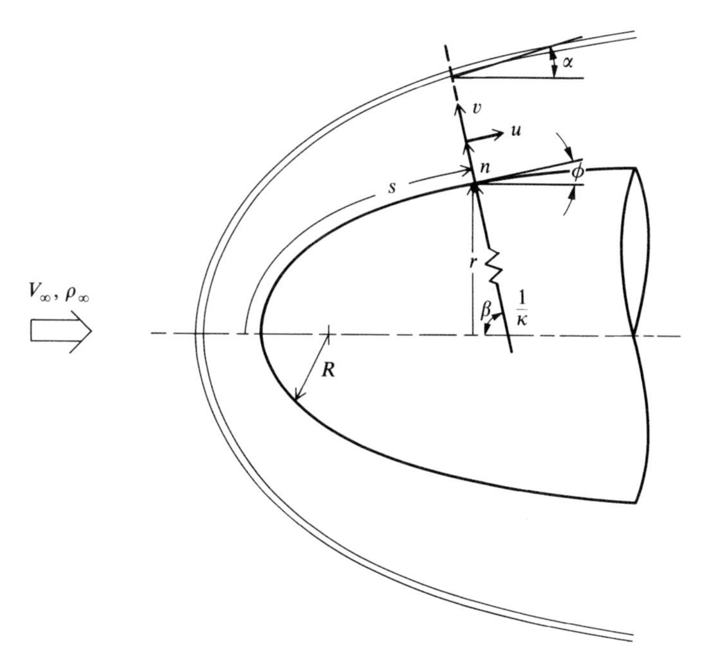

Fig. 8.1 Coordinate system for VSL equations.

s-Momentum equation:

$$\rho^*\left\{u^*\left[\frac{1}{(1+\kappa^*n^*)}\frac{\partial u^*}{\partial s^*}\right]+v^*\frac{\partial u^*}{\partial n^*}+\left[\frac{\kappa^*}{(1+\kappa^*n^*)}\right]u^*v^*\right\}+\frac{1}{(1+\kappa^*n^*)}\frac{\partial p^*}{\partial s^*}$$
$$=\left[\frac{\varepsilon^2}{(1+\kappa^*n^*)^2(r^*+n^*\cos\phi)^m}\right]\frac{\partial}{\partial n^*}[(1+\kappa^*n^*)^2(r^*+n^*\cos\phi)^m\tau] \tag{8.2}$$

where

$$\tau=\mu^*\left[\frac{\partial u^*}{\partial n^*}-\frac{\kappa^*u^*}{1+\kappa^*n^*}\right]$$

n-Momentum equation:

$$\rho^*\left\{u^*\left[\frac{1}{(1+\kappa^*n^*)}\frac{\partial v^*}{\partial s^*}\right]+v^*\frac{\partial v^*}{\partial n^*}-\left[\frac{\kappa^*}{(1+\kappa^*n^*)}\right]u^{*2}\right\}+\frac{\partial p^*}{\partial n^*}=0 \tag{8.3}$$

Energy equation:

$$\rho^*\left\{u^*\left[\frac{1}{(1+\kappa^* n^*)}\frac{\partial T^*}{\partial s^*}\right]+v^*\frac{\partial T^*}{\partial n^*}\right\}-\left\{u^*\left[\frac{1}{(1+\kappa^* n^*)}\frac{\partial p^*}{\partial s^*}\right]+v^*\frac{\partial p^*}{\partial n^*}\right\}$$
$$=\left[\frac{\varepsilon^2}{(1+\kappa^* n^*)(r^*+n^*\cos\phi)^m}\right]\frac{\partial}{\partial n^*}$$
$$\times\left[(1+\kappa^* n^*)(r^*+n^*\cos\phi)^m\left(\frac{\mu^*}{P_r}\right)\frac{\partial T^*}{\partial n^*}\right]+\left(\frac{\varepsilon^2}{\mu^*}\right)\tau^2 \tag{8.4}$$

In Eqs. (8.1–8.4), the nondimensional variables are defined as

$$s^*=\frac{s}{R}\qquad n^*=\frac{n}{R}\qquad r^*=\frac{r}{R}$$
$$\kappa^*=\frac{\kappa}{R}\qquad u^*=\frac{u}{V_\infty}\qquad v^*=\frac{v}{V_\infty}$$
$$T^*=\frac{T}{T_{\text{ref}}}\qquad p^*=\frac{p}{\rho_\infty V_\infty^2}$$
$$\rho^*=\frac{\rho}{\rho_\infty}\qquad \mu^*=\frac{\mu}{\mu_{\text{ref}}}$$

where, from Fig. 8.1, R is the nose radius, κ is the longitudinal body curvature, V_∞ and ρ_∞ are the freestream velocity and density, respectively, and T_{ref} and μ_{ref} are reference values; $T_{\text{ref}}=V_\infty^2, c_{p_\infty}$. Also, in Eqs. (8.2) and (8.4), ε is defined as

$$\varepsilon=\sqrt{\frac{\mu_{\text{ref}}}{\rho_\infty V_\infty R}}$$

Do not be intimidated by the form of the preceding equations; their seeming complexity is really caused by the curvilinear, boundary-oriented coordinate system. To recast them in the more familiar two-dimensional Cartesian coordinate system, set $m=0$, $\kappa^*=0$, $x^*=s^*$, and $y^*=n^*$, and express the variables in dimensional form, obtaining the following.

Continuity equation:

$$\frac{\partial(\rho u)}{\partial x}+\frac{\partial(\rho v)}{\partial y}=0 \tag{8.5}$$

x-Momentum equation:

$$\rho u\frac{\partial u}{\partial x}+\rho v\frac{\partial u}{\partial y}=-\frac{\partial p}{\partial x}+\frac{\partial}{\partial y}\left(\mu\frac{\partial u}{\partial y}\right) \tag{8.6}$$

y-Momentum equation:

$$\rho u \frac{\partial v}{\partial x} + \rho v \frac{\partial v}{\partial y} = -\frac{\partial p}{\partial y} \tag{8.7}$$

Energy equation:

$$\rho u \frac{\partial h}{\partial x} + \rho v \frac{\partial h}{\partial y} = \frac{\partial}{\partial y}\left(\kappa \frac{\partial T}{\partial y}\right) + u \frac{\partial p}{\partial x} + v \frac{\partial p}{\partial y} + \mu \left(\frac{\partial u}{\partial y}\right)^2 \tag{8.8}$$

where $h = c_p T$. Examine Eqs. (8.5–8.8) closely, and compare them with the boundary-layer equations given by Eqs. (6.27–6.30). We find that the *viscous shock-layer equations given by Eqs. (8.5–8.8) are essentially the boundary-layer equations with two notable exceptions*, as follows:

1) Equation (8.7) is a y-momentum equation that allows a finite value of $\partial p/\partial y$, unlike Eq. (6.29) for the classical boundary-layer case.

2) Equation (8.8) contains a normal pressure gradient term $v(\partial p/\partial y)$, which does not appear in the corresponding boundary-layer energy equation (6.30).

Therefore, in our hierarchy of solutions for a fully viscous flow, we can visualize the VSL equations as "one notch up" from the boundary-layer equations. However, in being so, the VSL equations have the distinct advantage of allowing a normal pressure gradient in the flow and hence can be integrated across the entire viscous flowfield. At the same time, the VSL equations retain the same convenience as the boundary-layer equations, namely, they can be solved by means of a downstream-marching finite difference procedure. An implicit method is employed, similar to that discussed in Sec. 6.6 for nonsimilar boundary layers. Because the flow conditions behind the shock wave are the outer boundary conditions on the viscous flowfield, and the shock shape is not known in advance, a global iteration is needed (using mass continuity) to obtain the shock shape and location. The shock wave is treated as a discontinuity, with either the exact oblique shock relations holding across the wave (see Chapter 2), or for very low-density cases a shock "slip" condition is used. The solution starts at the stagnation streamline, where the VSL equations become ordinary differential equations, and then marches downstream, solving the viscous flow across the shock layer at each streamwise station. See [134] for details on the numerical solution.

Some results obtained with the VSL equations are given by Davis in [134]. An analytical blunt-body shape was treated, namely a 45-deg hyperboloid. The flow conditions were $M_\infty = 10$, $\varepsilon = 0.1806$, $\gamma = 1.4$, $Pr = 0.7$, and $T_w/T_0 = 0.2$. Some results obtained from [134] are shown in Figs. 8.2–8.6. For example, in Fig. 8.2 we see the calculated variation of c_f vs distance along the surface, starting at the stagnation point. Here, unlike the usual convention where c_f is based on ρ_e and u_e at the edge of the boundary layer, the skin-friction coefficient in Fig. 8.2 is defined as $\tau/\frac{1}{2}\rho_\infty V_\infty$. This is because the shock layer is being treated as fully viscous, and a distinct boundary layer is therefore not an easily distinguished item. Note that the shear stress is zero at the stagnation point, increases rapidly over the blunt nose, reaches a maximum value about one nose radius downstream, and then progressively decreases further

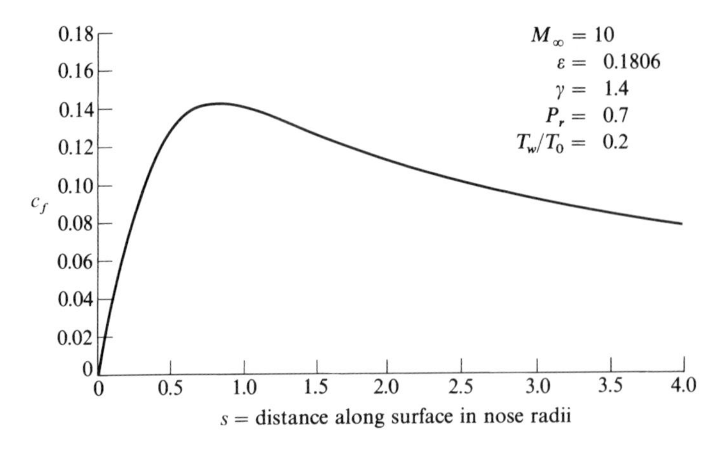

Fig. 8.2 Skin friction on a 45-deg hyperboloid: VSL calculations of Davis [134].

downstream. Tangential velocity profiles are shown in Fig. 8.3. Here, u_{sh} and n_{sh} are the velocity and shock-layer coordinate immediately behind the bow shock wave; both u_{sh} and n_{sh} are functions of location along the shock, that is, are functions of s. In Fig. 8.3, u/u_{sh} is plotted vs n/n_{sh} in the same manner as we

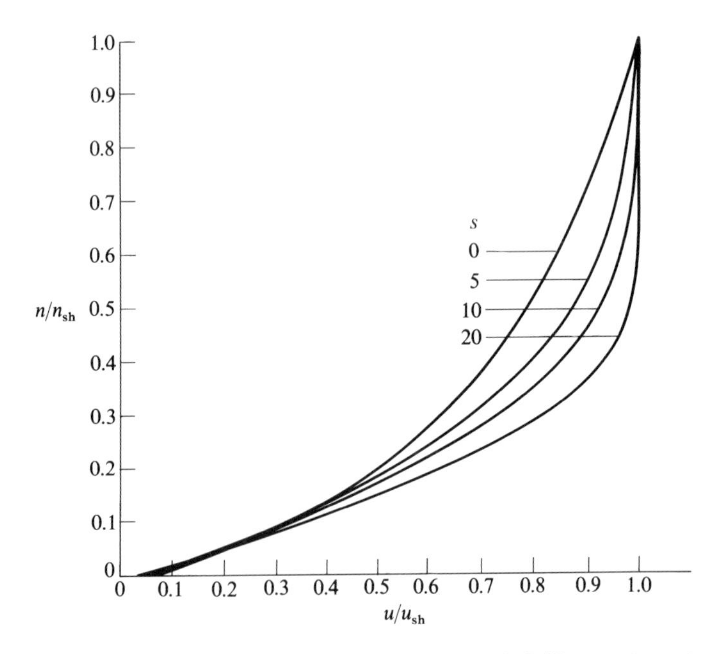

Fig. 8.3 Tangential velocity profiles on a 45-deg hyperboloid at various streamwise stations. Same conditions as Fig. 8.2 (from [134]).

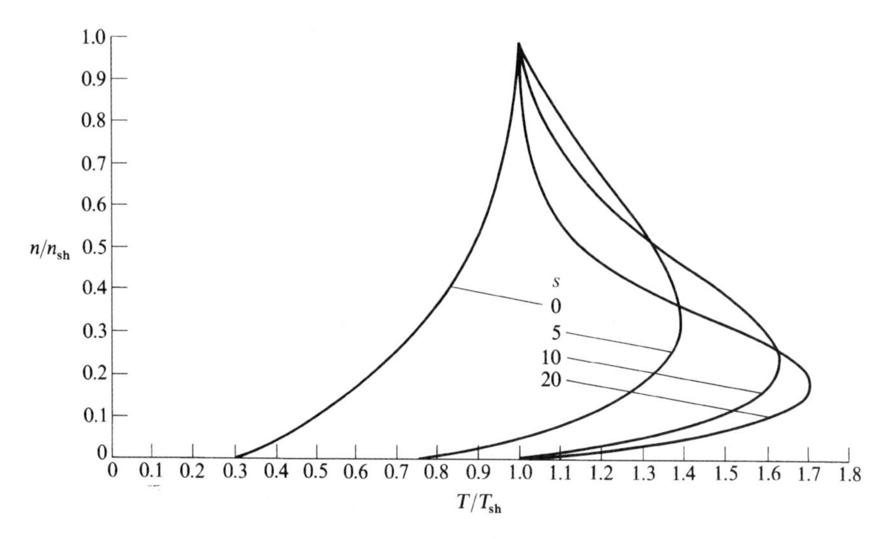

Fig. 8.4 Temperature profiles on a 45-deg hyperboloid at various streamwise stations. Same conditions as Fig. 8.2 (from [134]).

plotted boundary-layer profile in Chapter 6. Profiles are shown for different streamwise locations denoted by s. The stagnation streamline profile is given by $s = 0$. Note that, for the conditions shown, there are substantial velocity gradients all of the way across the shock layer. This is just the type of flow for which a

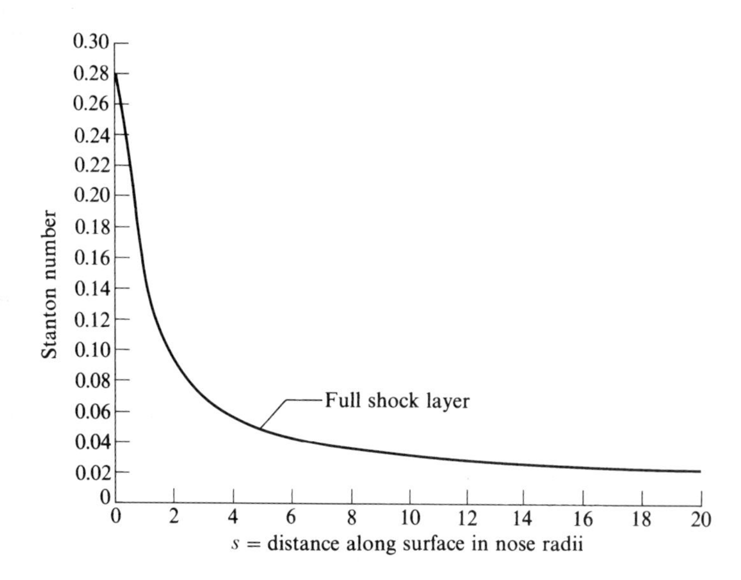

Fig. 8.5 Heat-transfer distribution over a 45-deg hyperboloid. Same conditions as Fig. 8.2 (from [134]).

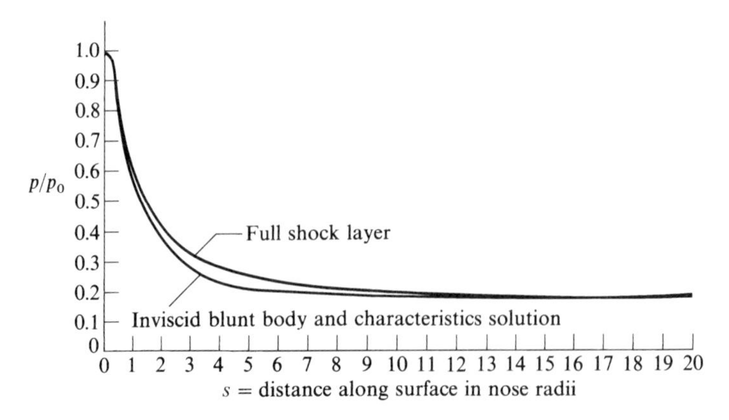

Fig. 8.6 Pressure distribution on a 45-deg hyperboloid. Same conditions as Fig. 8.2 (from [134]).

VSL solution is suited; a boundary-layer calculation would not be appropriate. In the same region, Fig. 8.4 shows temperature profiles across the shock layer, and the same comments can be made here. The heat-transfer distribution is given in Fig. 8.5; note that C_H monotonically decreases as a function of distance downstream of the stagnation point. The corresponding pressure distribution is shown in Fig. 8.6, which also shows the expected monotonic decrease with s. In Fig. 8.6, results are also shown for an inviscid flow calculation using a blunt-body solution in the nose region and continuing downstream with the method of characteristics. Note that the pressure distribution from the viscous shock-layer calculation is consistently higher than the inviscid pressure distribution. This is a clear demonstration of a mild viscous interaction effect occurring on the blunt body.

The VSL method has found wide application to chemically reacting viscous flows, as will be discussed in Part 3.

8.3 Parabolized Navier–Stokes Solutions

In this section we discuss a system of equations that contain more terms than the VSL equations and hence are theoretically more accurate, but which still are simpler than the full Navier–Stokes equations. This system is called the parabolized Navier–Stokes (PNS) equations. They are obtained from the full Navier–Stokes equations by dropping the viscous terms that involve derivatives in the streamwise direction. For example, the exact x component of the steady flow momentum equation is obtained from Eqs. (6.2) and (6.6) as

$$\rho u \frac{\partial u}{\partial x} + \rho v \frac{\partial u}{\partial y} + \rho w \frac{\partial u}{\partial z} = -\frac{\partial p}{\partial x} + \frac{\partial}{\partial x}\left(\lambda \nabla \cdot \mathbf{V} + 2\mu \frac{\partial u}{\partial x}\right) + \frac{\partial}{\partial y}\left[\mu\left(\frac{\partial v}{\partial x} + \frac{\partial u}{\partial y}\right)\right] + \frac{\partial}{\partial z}\left[\mu\left(\frac{\partial u}{\partial z} + \frac{\partial w}{\partial x}\right)\right] \tag{8.9}$$

The parabolized form of this equation is obtained by neglecting the viscous terms that involve the x derivatives, obtaining directly

$$\rho u \frac{\partial u}{\partial x} + \rho v \frac{\partial u}{\partial y} + \rho w \frac{\partial u}{\partial z} = -\frac{\partial p}{\partial x} + \frac{\partial}{\partial y}\left(\mu \frac{\partial u}{\partial y}\right) + \frac{\partial}{\partial z}\left(\mu \frac{\partial u}{\partial z}\right) \tag{8.10}$$

Of particular importance is the steady flow y-momentum equation, given in its exact form by Eqs. (6.3) and (6.6) as

$$\rho u \frac{\partial v}{\partial x} + \rho v \frac{\partial v}{\partial y} + \rho w \frac{\partial v}{\partial z} = -\frac{\partial p}{\partial y} + \frac{\partial}{\partial x}\left[\mu\left(\frac{\partial v}{\partial x} + \frac{\partial u}{\partial y}\right)\right] + \frac{\partial}{\partial y}\left(\lambda \nabla \cdot \mathbf{V} + 2\mu \frac{\partial v}{\partial y}\right) + \frac{\partial}{\partial z}\left[\mu\left(\frac{\partial w}{\partial y} + \frac{\partial v}{\partial z}\right)\right] \tag{8.11}$$

The parabolized form of this equation is also obtained by neglecting the viscous terms that involve the x derivatives, obtaining directly

$$\rho u \frac{\partial v}{\partial x} + \rho v \frac{\partial v}{\partial y} + \rho w \frac{\partial v}{\partial z} = -\frac{\partial p}{\partial y} + \frac{\partial}{\partial y}\left[(\lambda + 2\mu)\frac{\partial v}{\partial y} + \lambda \frac{\partial w}{\partial z}\right] + \frac{\partial}{\partial z}\left[\mu\left(\frac{\partial w}{\partial y} + \frac{\partial v}{\partial z}\right)\right] \tag{8.12}$$

Clearly, Eq. (8.12) takes into account a normal pressure gradient across the shock layer. Moreover, Eq. (8.12) is superior to the corresponding y-momentum equation contained in the VSL system. This can be seen by comparing the two-dimensional counterpart of Eq. (8.21), namely,

$$\rho u \frac{\partial v}{\partial x} + \rho v \frac{\partial v}{\partial y} = -\frac{\partial p}{\partial y} + \frac{\partial}{\partial y}\left[(\lambda + 2\mu)\frac{\partial v}{\partial y}\right] \tag{8.13}$$

with Eq. (8.7). Clearly, the PNS form given by Eq. (8.13) has a viscous term that is missing from the VSL form given by Eq. (8.7).

In summary, the parabolized Navier–Stokes equations are obtained from the full steady-flow Navier–Stokes equations [given by Eqs. (6.1–6.6) with all time derivatives set to zero] simply by neglecting all *viscous* terms that involve derivatives in the streamwise direction (in the x direction). For convenience, the resulting system of PNS equations is itemized here.

Continuity equation:

$$\frac{\partial(\rho u)}{\partial x} + \frac{\partial(\rho v)}{\partial y} + \frac{\partial(\rho w)}{\partial z} = 0 \tag{8.14}$$

x-Momentum equation:

$$\rho u \frac{\partial u}{\partial x} + \rho v \frac{\partial u}{\partial y} + \rho w \frac{\partial u}{\partial z} = -\frac{\partial p}{\partial x} + \frac{\partial}{\partial y}\left(\mu \frac{\partial u}{\partial y}\right) + \frac{\partial}{\partial z}\left(\mu \frac{\partial u}{\partial z}\right) \tag{8.15}$$

y-Momentum equation:

$$\begin{aligned} \rho u \frac{\partial v}{\partial x} + \rho v \frac{\partial v}{\partial y} + \rho w \frac{\partial v}{\partial z} &= -\frac{\partial p}{\partial y} + \frac{\partial}{\partial y}\left[(\lambda + 2\mu)\frac{\partial v}{\partial y} + \lambda \frac{\partial w}{\partial z}\right] \\ &\quad + \frac{\partial}{\partial z}\left[\mu\left(\frac{\partial w}{\partial y} + \frac{\partial v}{\partial z}\right)\right] \end{aligned} \tag{8.16}$$

z-Momentum equation:

$$\begin{aligned} \rho u \frac{\partial w}{\partial x} + \rho v \frac{\partial w}{\partial y} + \rho w \frac{\partial w}{\partial z} &= -\frac{\partial p}{\partial z} + \frac{\partial}{\partial y}\left[\mu\left(\frac{\partial w}{\partial y} + \frac{\partial v}{\partial z}\right)\right] \\ &\quad + \frac{\partial}{\partial z}\left[(\lambda + 2\mu)\frac{\partial w}{\partial z} + \lambda \frac{\partial v}{\partial z}\right] \end{aligned} \tag{8.17}$$

Energy equation:

$$\begin{aligned} &\rho u \frac{\partial}{\partial x}\left(e + \frac{V^2}{2}\right) + \rho v \frac{\partial}{\partial y}\left(e + \frac{V^2}{2}\right) + \rho w \frac{\partial}{\partial z}\left(e + \frac{V^2}{2}\right) \\ &= \rho \dot{q} + \frac{\partial}{\partial y}\left(k \frac{\partial T}{\partial y}\right) + \frac{\partial}{\partial z}\left(k \frac{\partial T}{\partial z}\right) - \left[\frac{\partial(pu)}{\partial x} + \frac{\partial(pv)}{\partial y} + \frac{\partial(pw)}{\partial z}\right] \\ &\quad + \frac{\partial}{\partial y}\left[u\mu\left(\frac{\partial u}{\partial y}\right)\right] + \frac{\partial}{\partial z}\left[u\mu\left(\frac{\partial u}{\partial z}\right)\right] + \frac{\partial}{\partial y}\left[v\lambda\left(\frac{\partial v}{\partial y} + \frac{\partial w}{\partial z}\right) + 2v\mu \frac{\partial v}{\partial y}\right] \\ &\quad + \frac{\partial}{\partial z}\left[v\mu\left(\frac{\partial w}{\partial y} + \frac{\partial v}{\partial z}\right)\right] + \frac{\partial}{\partial y}\left[w\mu\left(\frac{\partial w}{\partial y} + \frac{\partial v}{\partial z}\right)\right] \\ &\quad + \frac{\partial}{\partial z}\left[w\lambda\left(\frac{\partial v}{\partial y} + \frac{\partial w}{\partial z}\right) + 2w\mu \frac{\partial w}{\partial z}\right] \end{aligned} \tag{8.18}$$

Equations (8.14–8.18) constitute the parabolized Navier–Stokes equations; with two exceptions they are a mixed system of parabolic-hyperbolic partial differential equations and hence can be solved by a downstream-marching procedure starting from an initial data line across the flowfield. The two exceptions are as follows:

1) The pressure gradient terms $\partial p/\partial x$ allow the propagation of information upstream through the subsonic portion of the viscous flow near the body surface. Hence, the downstream-marching procedure is not well posed in this region. To preserve the parabolic nature of the PNS equations, the assumption is usually made that, in the subsonic region, the pressure is constant in the direction normal to the surface, equal to its value at the first grid point at which supersonic flow exists.

2) The volumetric heating term in Eq. (8.18), namely, $\rho \dot{q}$ can destroy the parabolic behavior of the system. For example, in a three-dimensional radiating flow, if $\dot{q}$ includes radiative energy absorbed at a point from all directions in the flow, the problem becomes elliptic in nature, and downstream marching is not valid. For flows where such volumetric heating does not occur, such a problem does not exist.

Numerical solutions of the PNS equations are usually carried out using an implicit finite difference method similar to that discussed in Sec. 6.6. Details concerning the numerical solution are nicely described in [224], hence, no further elaboration will be made here.

An excellent illustration of solutions obtained with the PNS equations is found in the work of McWherter et al. [136]. These results also have the advantage of illustrating a modern calculation of flowfields where viscous interaction effects are important. In [136], the flows over slender blunt-nosed cones at small angles of attack are calculated by two methods: 1) a classical inviscid flow/boundary-layer method, where the inviscid flow in the nose region is computed by means of the time-marching technique described in Sec. 5.3, the downstream inviscid flow is computed by means of the downstream-marching procedure described in Sec. 5.5, and the boundary-layer solution is an integral method following the inviscid, three-dimensional streamlines; and 2) a solution of the PNS equations as described earlier in this section. In the following figures, approach 1 will be labeled as 3DV, and approach 2 will be labeled PNS. Keep in mind that the 3DV method is a classical inviscid flow/boundary-layer approach that does not adequately account for strong viscous interactions; it does, however, contain an estimate of the induced pressure based on the displacement thickness variation. In contrast, the PNS method is a fully viscous shock-layer approach wherein strong viscous interactions are automatically accounted for, that is, they essentially "come out in the wash" during the course of such solutions. Emphasis is again made that, in the modern world of hypersonics, the proper conceptual treatment of viscous interactions is to assume the shock layer is fully viscous, such as in the PNS case.

Results for a relatively low-Mach-number hypersonic flow ($M_\infty = 5.95$) with high Reynolds number ($Re = 15.23 \times 10^6$) are shown in Fig. 8.7 for the flow over a blunt-nose 6-deg half-angle cone at an angle of attack $\alpha = 4$ deg. For this flow, the parameter $M_\infty^3/\sqrt{Re} = 0.054$; hence, viscous interaction effects should be negligible. In Fig. 8.7, the pressure distribution p/p_∞ is plotted vs the nondimensional distance downstream from the nose x/D_n, where D_n is the nose diameter. Four curves are shown, each corresponding to a circumferential angle ϕ around the cone measured from the windward ray, that is, $\phi = 0$ corresponds to the windward ray and $\phi = 180$ deg corresponds to the leeward ray. The 3DV calculations are given by the solid curves and the PNS calculations by the dotted curves. The solid symbols are experimental data obtained from [137]. Note the following information from Fig. 8.7:

1) On the windward side, the pressure rapidly expands over the blunt nose, overexpanding below the cone value, and then gradually recompressing further downstream. This overexpansion phenomenon is analogous to that shown in Fig. 5.16. Because of the very low value of $M_\infty^3/\sqrt{Re}$ for these data, the actual

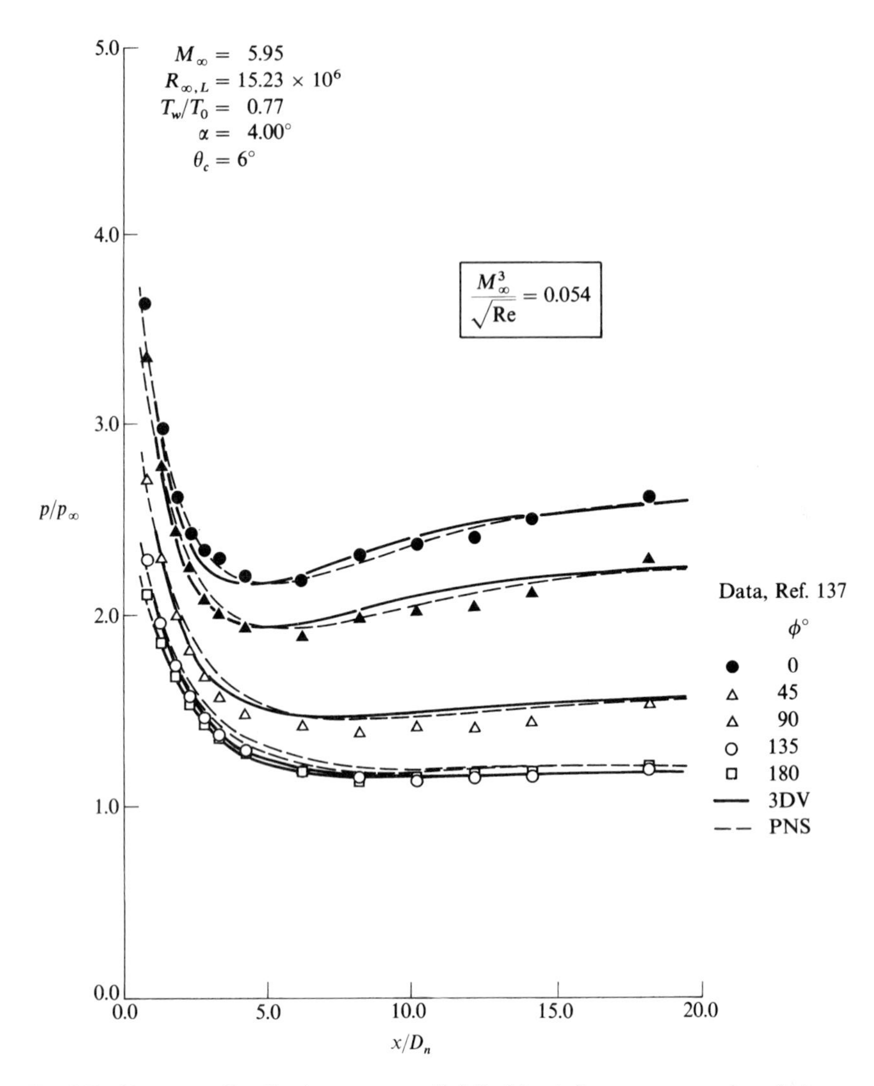

Fig. 8.7 Pressure distributions over a slightly blunted cone; comparison between experiment and computations (from McWherter et al. [136]).

pressure distribution over the blunted cone is mainly governed by inviscid-flow effects.

2) The 3DV and PNS calculations agree very closely with each other, another ramification of the negligible viscous interaction effects for the low-Mach-number and high-Reynolds-number conditions for Fig. 8.7.

3) The calculations agree well with experiment.

In contrast, results for a higher Mach number ($M_\infty = 9.82$) and lower Reynolds number ($Re = 0.459 \times 10^6$) are shown in Fig. 8.8. Here, a blunted

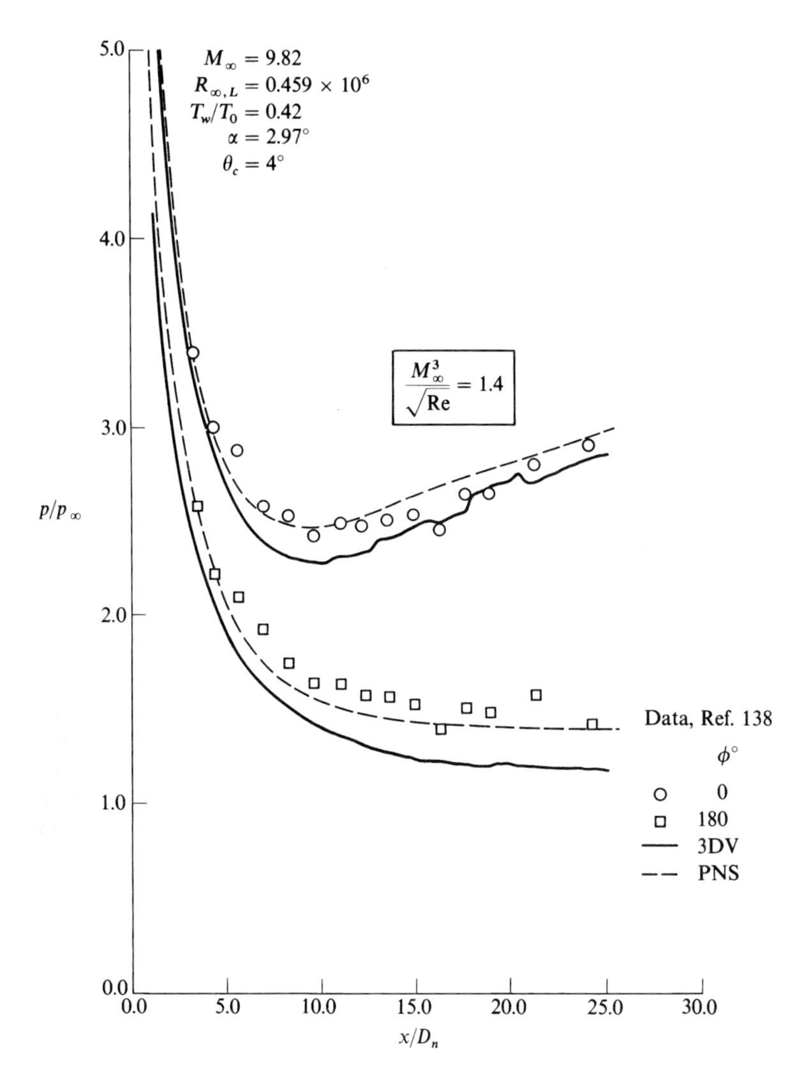

Fig. 8.8 Pressure distributions (as affected by viscous interaction) over a slightly blunted cone; comparison between experiment and computations (from [136]).

4-deg cone at an angle of attack of 2.97 deg is treated. For these conditions, $M_\infty^3/\sqrt{Re} = 1.4$, a high value which indicates that viscous interactions should be important. The experimental data shown in Fig. 8.8 are from [138]. Note the following information from Fig. 8.8:

1) The PNS calculations predict higher pressures than the 3DV calculations. This is because of the strong viscous interaction effect, which is automatically taken into account by the PNS method. The sizable difference between the PNS and 3DV curves is indeed the viscous interaction phenomena.

2) The PNS results agree favorably with the experimental data, especially on the leeward side ($\phi = 180$ deg), where the local Mach number is higher, the local Reynolds number is lower, and hence the viscous interaction effect is stronger.

Results for an almost identical case are shown in Fig. 8.9. Here, the axial-force coefficient C_A is plotted vs angle of attack. The experimental data are from [139]. Note that the PNS method predicts a much higher C_A than the 3DV method, again a graphic illustration of the viscous-interaction effect. Also note that the PNS results agree very well with experiment, thus demonstrating the superiority of a fully viscous shock-layer calculation in comparison to the classical inviscid flow/boundary-layer method for conditions where strong viscous interactions are important.

In summary, Figs. 8.7–8.9 illustrate an application of the PNS method to a basic hypersonic flow problem, as discussed in [136]. The reader is encouraged to study [136] closely for more details. Moreover, these figures demonstrate the value of a fully viscous shock-layer calculation for conditions where viscous

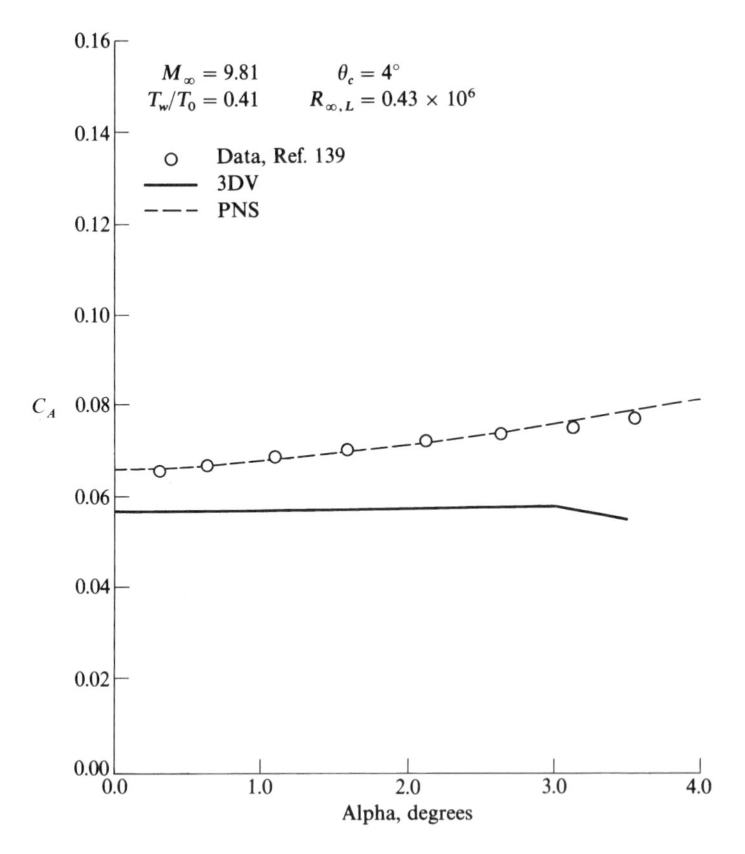

Fig. 8.9 Viscous interaction effects on axial-force coefficient on slightly blunted cones; comparison between experiment and computations (from [136]).

interactions are strong; the PNS method is a good example of a such a viscous shock-layer analysis. However, following the old adage that you "cannot get something for nothing." McWherter et al. in [136] point out the following, taken directly from their paper:

> The PNS solution generally requires a large amount of user interaction and the adjustment of various input parameters in order to obtain an accurate solution. The inviscid boundary-layer solution is straightforward to obtain and is, thus, very well suited for a design environment where rapid job turnaround and low user interaction requirements are significant considerations.

In other words, even though the PNS solutions are more accurate, it takes a lot more effort to obtain such solutions.

For a moment, let us consider the matter of flow separation. The classical boundary-layer equations discussed in Chapter 6 do not allow the calculation

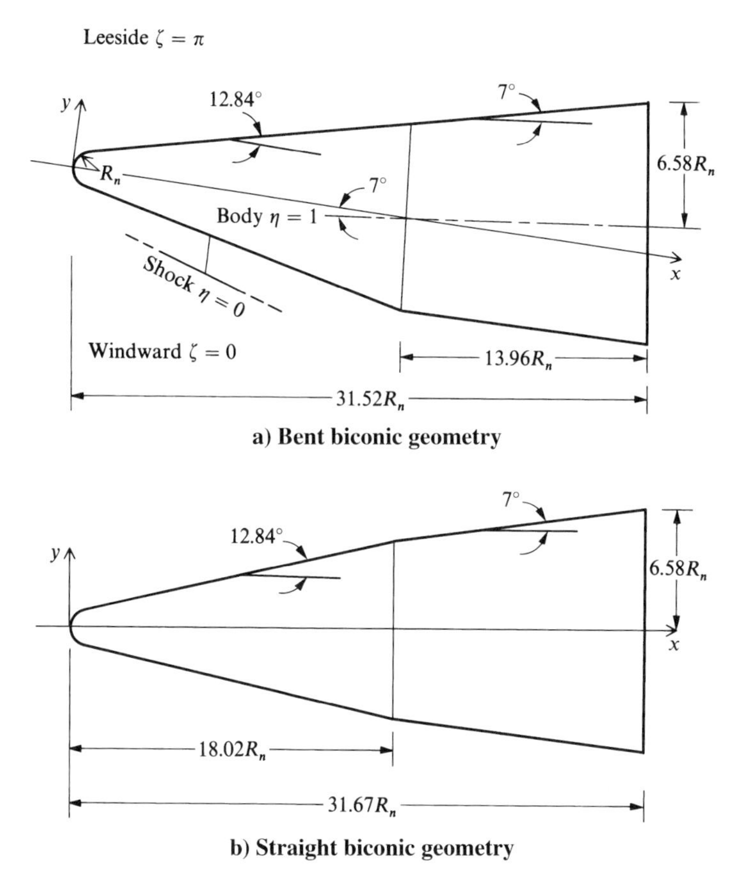

Fig. 8.10 Bent and straight biconic configurations for the calculations of Gnoffo [140].

of separated flows; such solutions "blow up" on the computer when zones of separated flow are encountered. Similarly, the VSL equations discussed in Sec. 8.2 do not allow the calculation of separated flow. The basic reason in both cases is that separated flow involves upstream of feeding of information in the flowfield, and downstream-marching methods, such as boundary layer and VSL calculations, do not allow or tolerate such upstream feeding. The same can almost be said for the PNS method, with one notable exception. Because of the nature of the z-momentum equation (8.17), the PNS method can predict flow separation in the *crossflow plane*; it cannot, however, handle separation in the streamwise direction. There are many problems where crossflow separation is the dominant mechanism, such as an axisymmetric body at angle of attack, and for these the PNS method does a reasonable job of handling the separated flow. For example, Fig. 8.10 shows a blunt-nose bent biconic body studied by Gnoffo in [140]. Solving the hypersonic flowfield over the body at $x = 20$ deg and $M_\infty = 6$ by means of a PNS solution, Gnoffo obtained the crossflow separation results shown in Fig. 8.11. Here, only a portion of the crossflow plane at $x/R_n = 7$ is shown; this portion is on the leeward side, near the top of

Fig. 8.11 Crossflow separation as predicted by the PNS calculations of Gnoffo [140].

the vehicle. We see the outer crossflow velocity vectors coming around the body from right to left and crossflow separation with reversed flow taking place near the surface of the body. This velocity field is the computed result from Gnoffo's PNS analysis. The separation lines agree well with experiment, as seen in Fig. 8.12. Here, we are looking at the top view of the bent biconic. The crosses represent the separation lines computed from the PNS method, and the dashed lines are experimental results obtained from surface oil flow visualization. Figure 8.12 also shows the computed and measured lines of local minimum pressure on the leeward surface. In all cases, agreement between experiment and the PNS calculations is very good. Hence, in Figs. 8.11 and 8.12 we see an important advantage of the PNS method over both the boundary layer and the VSL methods, namely, the ability to predict crossflow separation. However, we are reminded that none of these downstream-marching methods are capable of solving a separated flow in the streamwise direction.

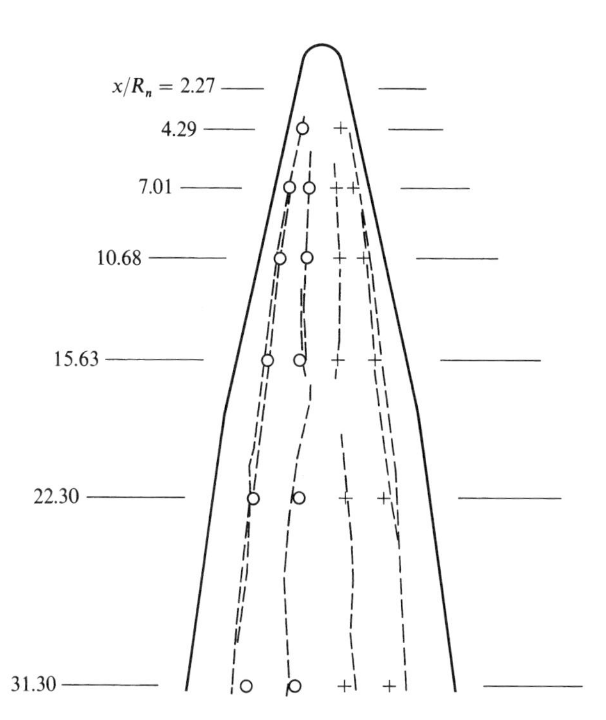

Fig. 8.12 Separation points and pressure minima for the bent biconic shown in Fig. 8.10; $\alpha = 20$ deg, $M_\infty = 6$, $Re_{\infty,L} = 8.2 \times 10^6$; comparison between experiment and the PNS calculations (from [140]).

8.4 Full Navier–Stokes Solutions

The *ultimate* in hypersonic viscous-flow calculations is the solution of complete Navier–Stokes equations, that is, the solution of Eqs. (6.1–6.6) with no reduction or simplification of any terms. Such full Navier–Stokes solutions were simply dreams in the minds of aerodynamicists as late as 1970. However, the modern techniques of computational fluid dynamics in combination with new supercomputers now allow the numerical solution of the Navier–Stokes equations; all that is needed for most practical problems is plenty of computer storage and running time. Such matters are the subject of this section.

Examine Eqs. (6.1–6.6) closely; they are a system of partial differential equations with a somewhat mixed hyperbolic, parabolic, and elliptic behavior. The elliptic behavior comes about as a result of the viscous terms in the x direction, which allow the upstream propagation of information via thermal conduction and viscosity. These are precisely the terms that are neglected in the PNS equations. Because of this elliptic nature, the full Navier–Stokes equations cannot be solved by a downstream-marching philosophy. However, recall that the problem with the inviscid blunt-body case as discussed in Sec. 5.3 was the mixed hyperbolic and elliptic behavior of the flowfield, and this problem was eventually solved by using the time-marching technique, also described in Sec. 5.3. The same holds true for numerical solutions of the full Navier–Stokes equations; such solutions must be time-marching solutions in order to take into account the elliptic behavior.

The time-marching solution of the Navier–Stokes equations is inherently straightforward and is patterned after the philosophy given in Sec. 5.3. Let us write Eqs. (6.1–6.5) such that the time derivatives are on the left side and all spatial derivatives are on the right side of the equations, as follows.

Continuity equation:

$$\frac{\partial \rho}{\partial t} = -\frac{\partial(\rho u)}{\partial x} - \frac{\partial(\rho v)}{\partial y} - \frac{\partial(\rho w)}{\partial z} \tag{8.19}$$

x-Momentum equation:

$$\frac{\partial u}{\partial t} = -u\frac{\partial u}{\partial x} - v\frac{\partial u}{\partial y} - w\frac{\partial u}{\partial z} + \frac{1}{\rho}\left(-\frac{\partial p}{\partial x} + \frac{\partial \tau_{xx}}{\partial x} + \frac{\partial \tau_{yx}}{\partial y} + \frac{\partial \tau_{zx}}{\partial z}\right) \tag{8.20}$$

y-Momentum equation:

$$\frac{\partial v}{\partial t} = -u\frac{\partial v}{\partial x} - v\frac{\partial v}{\partial y} - w\frac{\partial v}{\partial z} + \frac{1}{\rho}\left(-\frac{\partial p}{\partial y} + \frac{\partial \tau_{xy}}{\partial x} + \frac{\partial \tau_{yy}}{\partial y} + \frac{\partial \tau_{zy}}{\partial z}\right) \tag{8.21}$$

z-Momentum equation:

$$\frac{\partial w}{\partial t} = -u\frac{\partial w}{\partial x} - v\frac{\partial w}{\partial y} - w\frac{\partial w}{\partial z} + \frac{1}{\rho}\left(-\frac{\partial p}{\partial z} + \frac{\partial \tau_{xz}}{\partial x} + \frac{\partial \tau_{yz}}{\partial y} + \frac{\partial \tau_{zz}}{\partial z}\right) \tag{8.22}$$

Energy equation:

$$\frac{\partial}{\partial t}\left(e+\frac{V^2}{2}\right) = -u\frac{\partial}{\partial x}\left(e+\frac{V^2}{2}\right) - v\frac{\partial}{\partial y}\left(e+\frac{V^2}{2}\right) - w\frac{\partial}{\partial z}\left(e+\frac{V^2}{2}\right) + \dot{q}$$
$$+\frac{1}{\rho}\left[\frac{\partial}{\partial x}\left(k\frac{\partial T}{\partial x}\right) + \frac{\partial}{\partial y}\left(k\frac{\partial T}{\partial y}\right) + \frac{\partial}{\partial z}\left(k\frac{\partial T}{\partial z}\right) - \frac{\partial(pu)}{\partial x}\right.$$
$$-\frac{\partial(pv)}{\partial y} - \frac{\partial(pw)}{\partial z} + \frac{\partial(u\tau_{xx})}{\partial x} + \frac{\partial(u\tau_{yx})}{\partial y} + \frac{\partial(u\tau_{zx})}{\partial z} + \frac{\partial(v\tau_{xy})}{\partial x}$$
$$\left.+\frac{\partial(v\tau_{yy})}{\partial y} + \frac{\partial(v\tau_{zy})}{\partial z} + \frac{\partial(w\tau_{xz})}{\partial x} + \frac{\partial(w\tau_{yz})}{\partial y} + \frac{\partial(w\tau_{zz})}{\partial z}\right] \tag{8.23}$$

The time-marching solution of these equations is conceptually carried out as follows:

1) Cover the flowfield with grid points, and assume arbitrary values of all of the dependent variables at each grid point. This represents the assumed initial conditions at time $t = 0$.

2) Calculate the values of ρ, u, v, w, and $(e + V^2/2)$ from Eqs. (8.19–8.23) as function of time, using a time-marching finite difference method. One such method is the explicit predictor-corrector technique of MacCormack described in detail in Sec. 5.3. (Indeed, the reader should review this technique as described in Sec. 5.3 before progressing further.)

3) The final steady-state flow is obtained in the asymptotic limit of large times. In most cases, this is the desired result. However, the time-marching procedure can also be used to calculate the transient behavior of viscous flows, as well.

The numerical solution of the full Navier–Stokes equations for hypersonic flows is a state-of-the-art research problem at present. Many numerical approaches have been and are being developed and studied, both using explicit and implicit finite difference methods. See [224] for an organized presentation of such methods. Our purpose here is not to delve into any of these methods in detail, but rather to give the flavor of results obtained from such Navier–Stokes solutions.

At the beginning of this section, we stated that the "ultimate" in hypersonic viscous-flow calculations is the solution of the complete Navier–Stokes equations. Let us expand this statement by saying that the ultimate of the ultimate would be a full Navier–Stokes calculation of the flowfield over a complete, three-dimensional airplane configuration. Such a calculation has recently been made, for the first time in the history of aerodynamics, by Joe Shang at the Air Force Flight Dynamics Laboratory, and is described in [141]. Here, the viscous flow is calculated over the X-24C hypersonic research vehicle at $M_\infty = 5.95$. A three view of the X-24C is shown in Fig. 8.13. The calculation carried out by Shang has the following characteristics:

1) The complete Navier–Stokes equations were used in a conservation form derivable from Eqs. (8.19–8.23).

2) The Baldwin–Lomax turbulence model was employed (see Sec. 6.8).

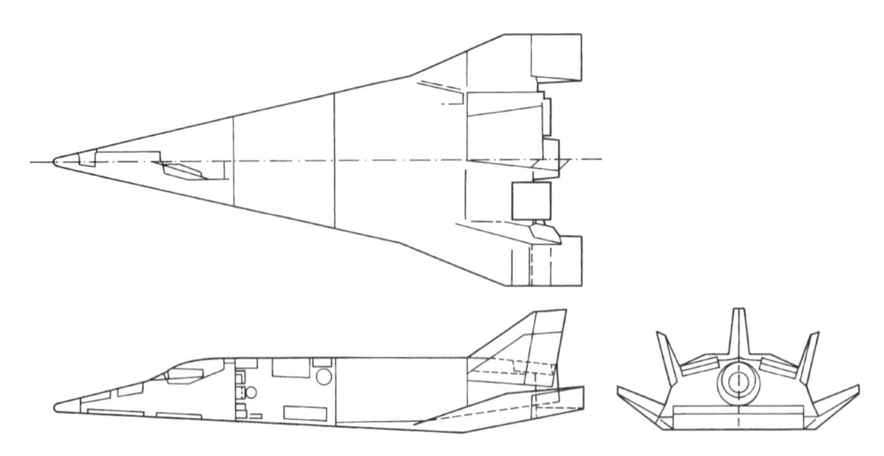

Fig. 8.13 Three view of the X-24C hypersonic test vehicle.

3) MacCormack's explicit predictor-corrector finite difference scheme in precisely the same form as described in Sec. 5.3 was used for the numerical solution of the Navier–Stokes equations.

4) The shock-capturing approach was taken (as defined in Sec. 5.5).

5) A mesh system consisting of 475,200 grid points was distributed over the flowfield.

Sample results from the calculation are shown in Figs. 8.14–8.17. In Figs. 8.14a and 8.14b, peripheral surface-pressure distributions are given as a function of normalized arc length at various streamwise stations denoted by x/R_n, where R_n is the nose radius. By peripheral distributions, what is meant is a distribution along a body surface generator that goes from the top to the bottom of the vehicle at a given streamwise station; these peripheral directions are clearly shown in Fig. 8.15, which is a perspective view of the X-24C. In Fig. 8.14, the normalized arc length is defined as the length measured from the top of the vehicle toward the bottom, divided by the total arc length of each individual cross section. For graphical clarity, each peripheral distribution at succeeding axial locations is displaced slightly to the right along the abscissa. Also in Fig. 8.14, the computed results are compared with the experimental data of [142].

Note that very good agreement is obtained between the calculations and experiment. The pressure distributions in Fig. 8.14a pertain to the front part of the vehicle, from the nose region to downstream of the canopy. For example, for $x/R_n = 15$ and 29.21, the pressures show a relatively constant value along the side of the vehicle and a compression at the lower corner of the essentially trapezoidal cross section (see Fig. 8.13). For $x/R_n = 43.25$, the initial compression is caused by the canopy, then a relatively constant pressure along the side and bottom, with the corner compression occurring again. In Fig. 8.14b, the pressure distributions pertain to the back part of the vehicle, and the various pressure spikes correspond to fins or a strake protruding into the oncoming flow. In Fig. 8.16, pitot-pressure contours are shown at $x/R_n = 108$. The experimental data are obtained from [143]. The outer contour corresponds to the shock

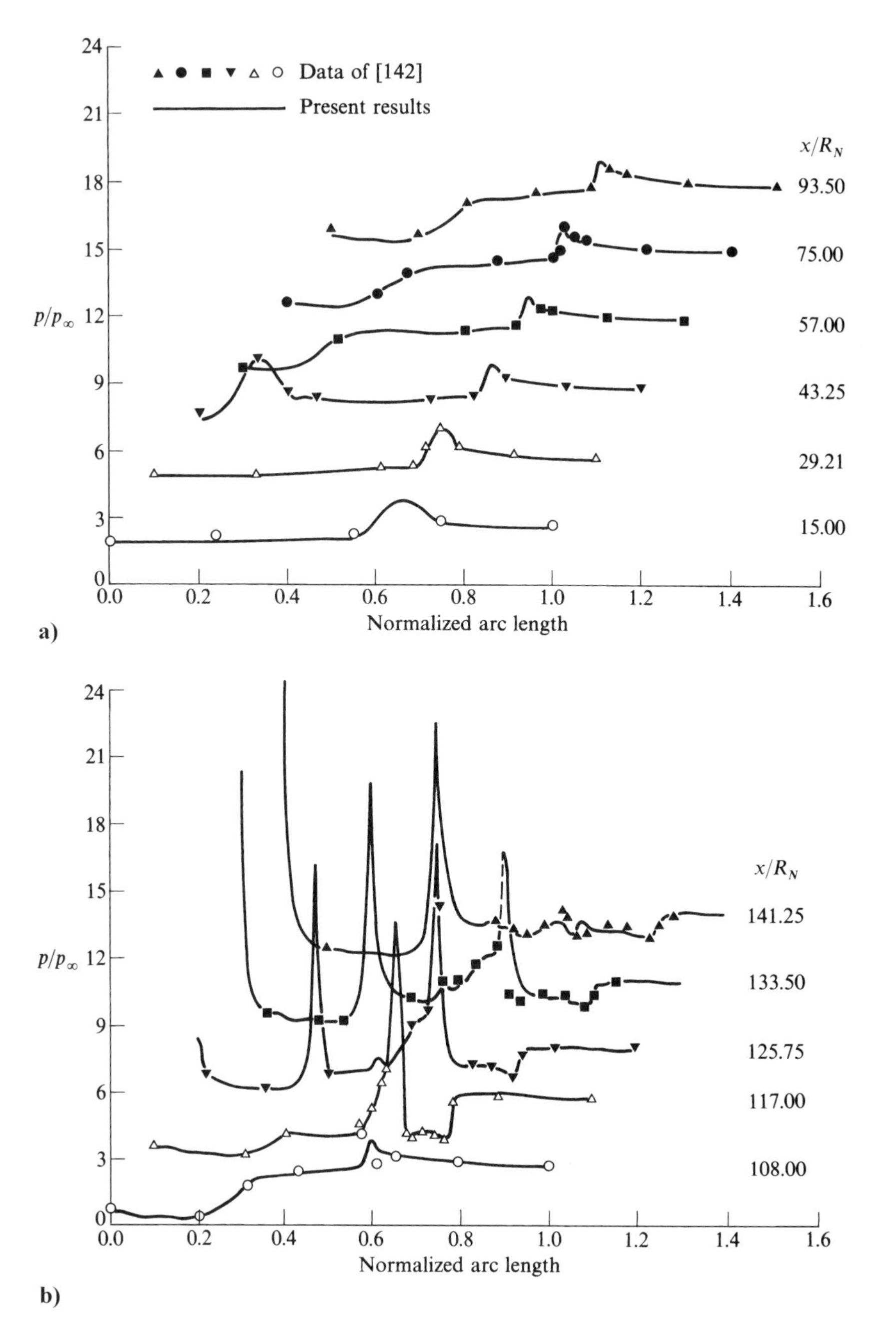

Fig. 8.14 Peripheral surface-pressure distributions around the X-24C, comparison between experiment [142] and the Navier–Stokes calculations of Shang [141].

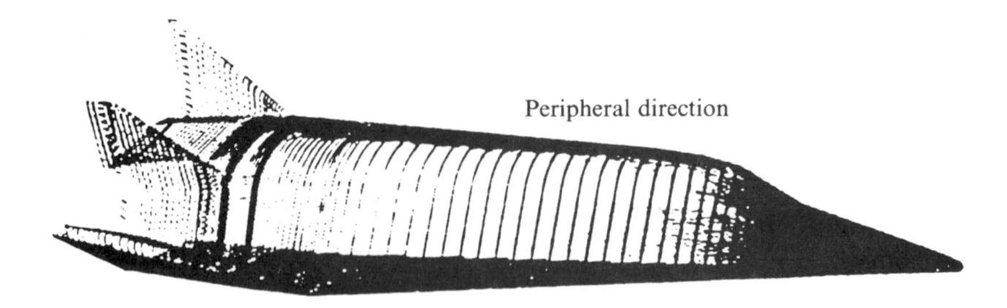

Fig. 8.15 Illustration of the peripheral direction around the X-24C for the data shown in Fig. 8.14.

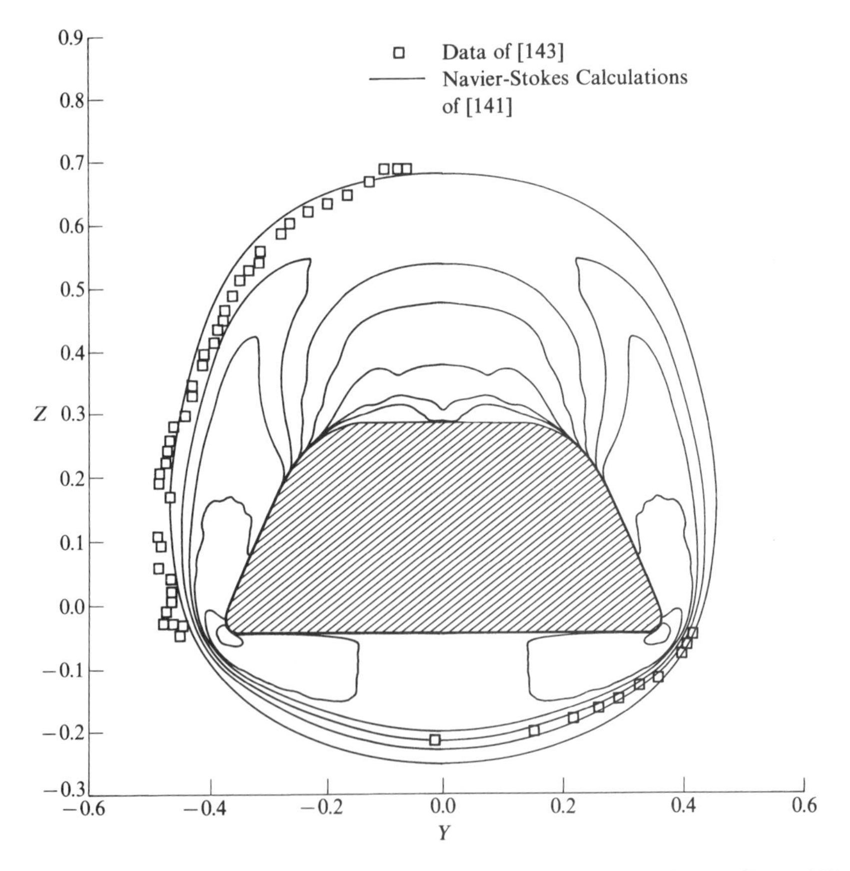

Fig. 8.16 Pitot-pressure contours at the longitudinal station $x/R_N = 108$; comparison between experiment and calculations (from [141]).

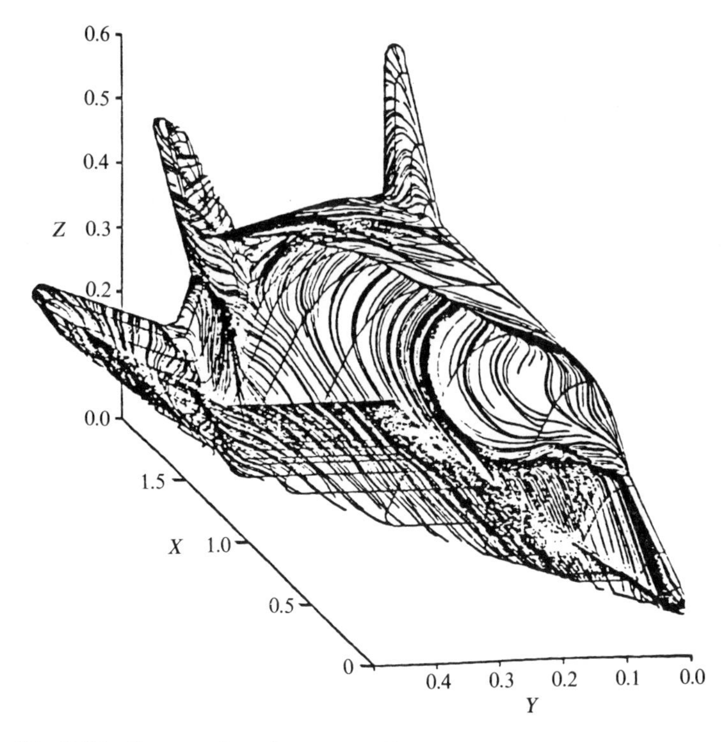

Fig. 8.17 Computed surface streamlines over the X-24C (from [141]).

shape wrapped around the vehicle; reasonable agreement between the measured and computer cross-sectional shock shape is obtained. Finally, the computed surface streamline pattern is shown in Fig. 8.17. The overall calculated aerodynamic lift and drag coefficients, obtained by integrating the calculated pressure and shear-stress distributions over the airplane surface, are compared with experimental measurements as shown in Table 8.1.

Note that the errors in C_L and C_D tend to cancel, giving a reasonably accurate estimate of lift-to-drag ratio, L/D.

The reader is strongly encouraged to study [141], not only because of its hallmark significance in hypersonic viscous flowfield calculations, but also because it contains some excellent color graphics presentations, which cannot be suitably reproduced in black and white in the present book.

Table 8.1 Lift and drag data

Type of data	C_L	C_D	L/D
Experimental data	3.676×10^{-2}	3.173×10^{-2}	1.158
Numerical results	3.503×10^{-2}	2.960×10^{-2}	1.183
Percent error	4.71	6.71	2.16

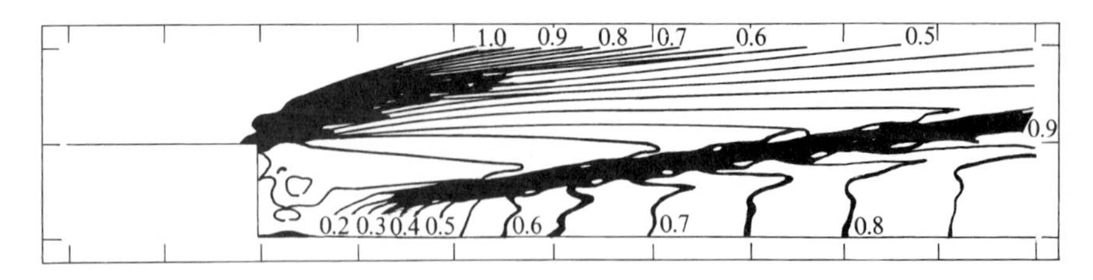

Fig. 8.18 Calculated pressure contours for the supersonic, separated flow over a rearward-facing step. Navier–Stokes calculations of Kuruvila and Anderson [144].

Finally, here is a word about flow separation. Time-marching solutions of the complete Navier–Stokes equations allow the calculation of fully separated flows in *any direction*, not just the crossflow direction as in the case of the PNS equations. This is a marked advantage of full Navier–Stokes solutions over the other methods presented in this chapter. A sample case is shown in Fig. 8.18, where the supersonic viscous flow over a rearward-facing step is calculated. The calculations involve a time-marching finite difference solution of the two-dimensional Navier–Stokes equations, as described in [144]. The freestream conditions above the step are $M_\infty = 4.08$, $T_\infty = 1046$ K, $\gamma = 1.31$ (to partially simulate dissociated air in a supersonic combustion ramjet environment), and $Re = 849$ based on step height. The wall temperature is given by $T_w/T_\infty = 0.2957$. In Fig. 8.18, the calculated pressure contours are given, which clearly show the expansion wave emanating from the top corner, the relatively constant pressure region in the recirculating separated flow behind the step, and the reattachment shock wave. Similar calculations can be found in [145] and [146]. In all of this work [144–146], the two-dimensional Navier–Stokes equations are solved using MacCormack's time-marching, predictor-corrector, finite difference scheme, as described in Sec. 5.3.

We conclude this section with discussions of two recent sets of calculations, both numerical solutions of the Navier–Stokes equations, both calculating the hypersonic viscous flow over a blunt nose, and both examining interesting but different physical mechanisms for reducing stagnation region heat transfer and nose wave drag. These calculations have direct relevance to our discussion in the present section, the aerodynamic heating discussions in Chapter 6, and the hypersonic blunt-body discussions in Chapter 5—a suitable conclusion to our presentation in Parts 1 and 2 of this book on hypersonic gas dynamics. (Part 3 focuses on high-temperature gas dynamics and is a related but self-contained entity.)

Meyer et al. [260] studied the hypersonic flow over a blunt-nosed two-dimensional body wherein a small diameter supersonic jet of air exhausted upstream from the centerline of the nose, penetrating into the shock layer and greatly altering the blunt-body flowfield. The flowfield was calculated using the time-marching Navier–Stokes code SPARK developed at the NASA Langley Research Center by Drummond et al. [261]. The flow was assumed to be laminar. Some results are shown in Figs. 8.19–8.22. Figure 8.19 shows pressure contours for the baseline, no-injection case for the nose region of a

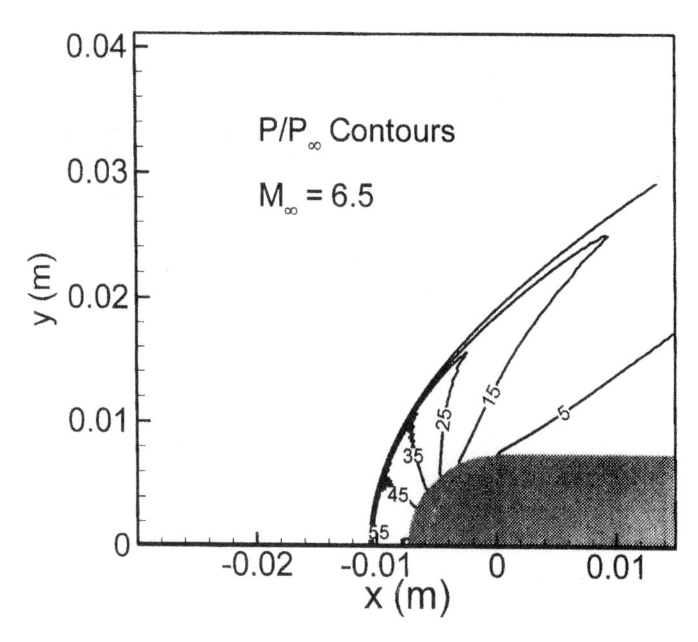

Fig. 8.19 **Nondimensional pressure distribution for the no-injection case (Meyer et al. [260]).**

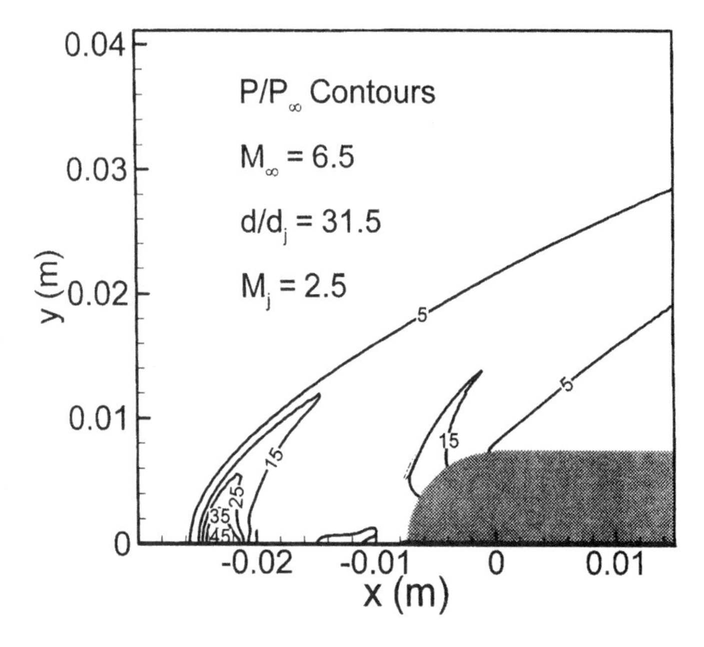

Fig. 8.20 **Nondimensional pressure distribution for the injection case [260].**

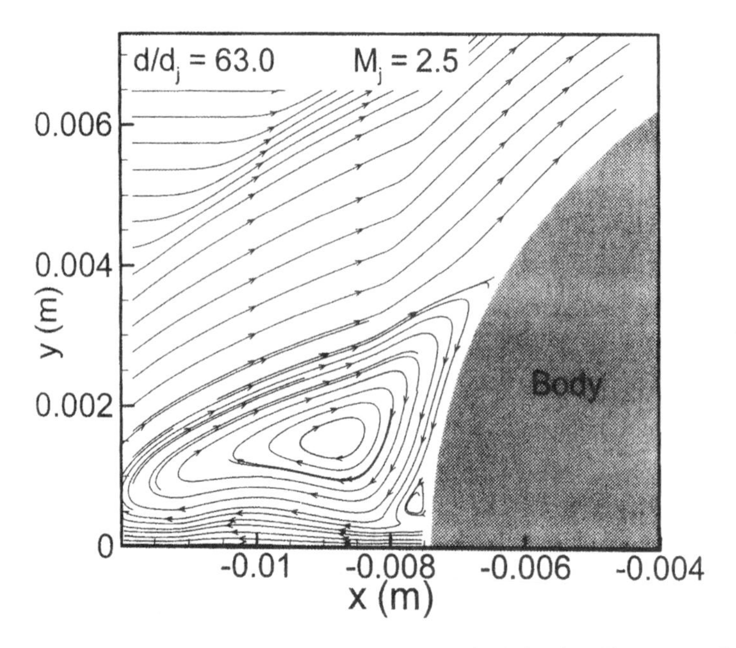

Fig. 8.21 Streamlines showing reattachment point for injection diameter ratio of 63 [260].

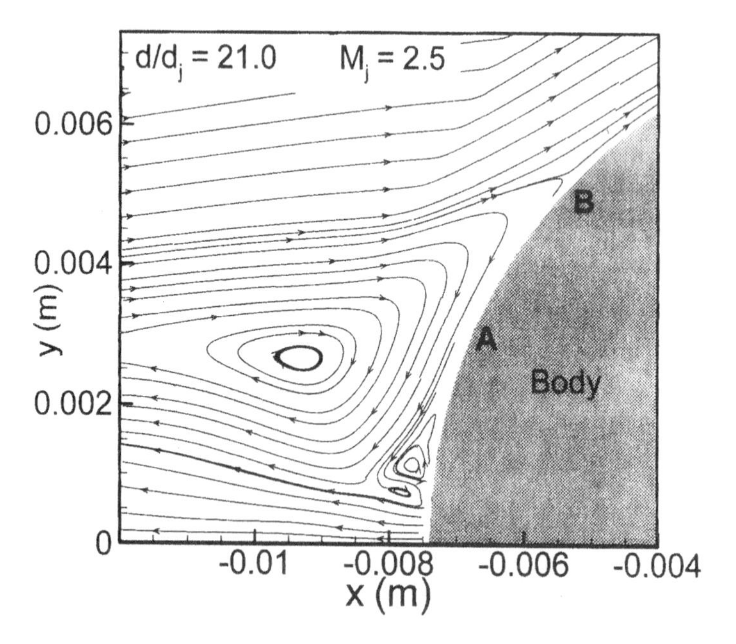

Fig. 8.22 Streamlines showing reattachment point for injection diameter ratio of 21 [260].

blunt slab at $M_\infty = 6.5$. Each contour is labeled according to its value of p/p_∞. The curved bow shock wave and shock detachment distance are nicely defined. Figure 8.20 shows the pressure contours when a supersonic jet exhausts into the flow from the centerline of the body; the jet Mach number M_j is 2.5, and the jet diameter d_j is 0.032 of that of the body nose diameter d. Comparing Figs. 8.19 and 8.20, we see that the jet has pushed the bow shock wave much further from the nose, increasing the shock detachment distance by about a factor of six, and greatly reducing the surface pressure on the nose. As a result, the nose wave drag is decreased. The jet itself, facing in the upstream direction, produces thrust in the downstream direction, thus adding to the force in the drag direction. But the reduction in wave drag more than compensates for this jet force. Letting D be the drag per unit span on the body including the added jet force in the drag direction and D_{ref} be the reference drag per unit span with no injection, for the case shown in Fig. 8.20, $D/D_{\text{ref}} = -0.452$, a marked reduction in the overall nose drag. Details of the calculated streamline shapes in the nose region are shown in Figs. 8.21 and 8.22. Figure 8.21 gives results for the smallest diameter jet; $d_j/d = 0.0159$, considered in the study, and Fig. 8.22 gives results for the largest diameter jet, $d_j/d = 0.0476$. For both cases shown, $M_\infty = 6.5$, and the jet Mach number is $M_j = 2.5$. Examining Fig. 8.21, we see a large region of recirculating separated flow induced by the jet in front of the nose; reattachment occurs at a point on the body at about $y = 0.0038$ m as measured in Fig. 8.21. For the case of the larger diameter jet shown in Fig. 8.22, the region of separated flow is even larger, and the reattachment point on the body is further downstream at about $y = 0.0051$ m, labeled as point B in Fig. 8.22. Reversed flow exists on the body for a substantial distance centered around point A, with an attendant reversed shear-stress direction. The low pressure in the separated flow region is responsible for the decrease in wave drag. It is also partly responsible for a dramatic reduction in local heat-transfer rates to the surface. The combined effects of lower pressure, increased shock detachment distance, and a cooler layer of air from the jet flowing over the body reduced the overall heat-transfer rate per meter to virtually zero; indeed, for some cases discussed by Meyer et al. the heat transfer reversed, with the flow actually cooling the body.

Shang [262] studied the effect of a plasma jet exhausting upstream into a hypersonic blunt-body flowfield. The study was both experimental and computational. The computations were made by a time-marching implicit Navier–Stokes solver. Calculated streamlines of the flow over an axisymmetric blunt body showing the interaction with a counterflow plasma jet are given in Fig. 8.23. The freestream Mach number was $M_\infty = 6$ and jet exit Mach number was $M_j = 3.28$. Note the massive region of separated flow ahead of the nose and the large increase in the bow shock detachment distance. Shang's calculations included not only the aerodynamic effects of the flowfield-jet interaction, but also nonequilibrium thermodynamic and chemical phenomena (of the type to be discussed in Part 3 of this book), and the electromagnetic-aerodynamic interaction. Like the results of Meyer et al. discussed earlier, Shang found a large drag reduction caused by the jet, and that most of this reduction is caused by the aerodynamic effect of the interacting flows, with negligible effects caused by nonequilibrium and electromagnetic phenomena.

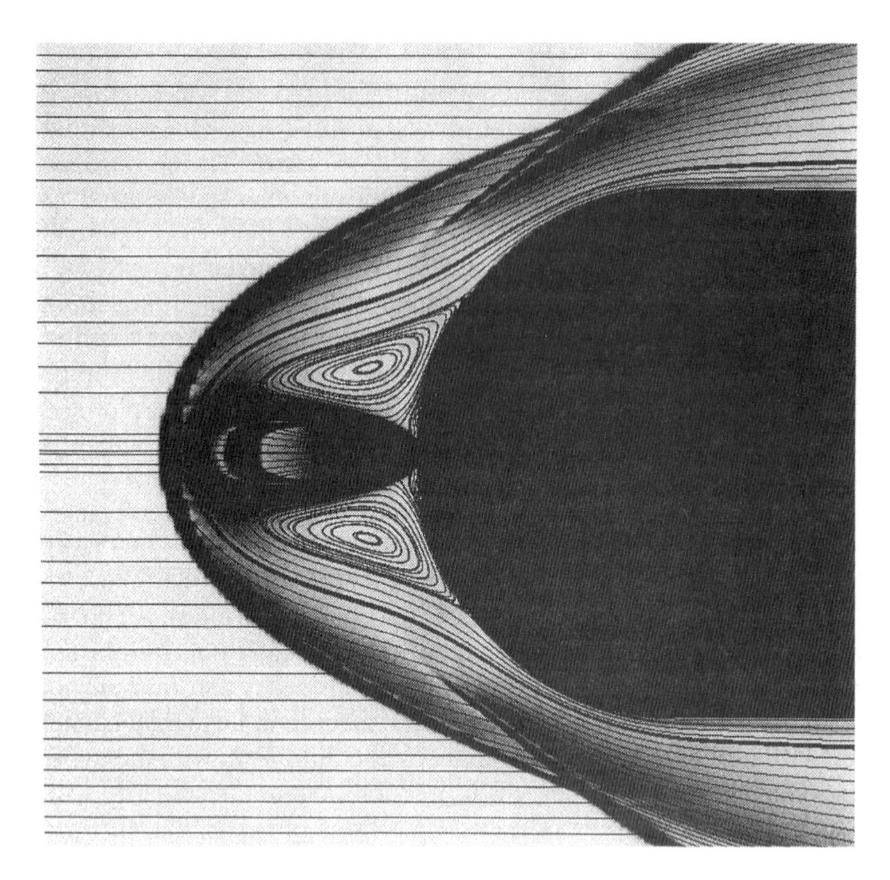

Fig. 8.23 Instantaneous streamlines of a counterflow jet (Shang [262]).

In summary, the results of Meyer et al., and of Shang, are excellent examples of recent hypersonic Navier–Stokes calculations as applied to an important and practical hypersonic aerodynamic problem, namely, the reduction of drag and hear transfer to a blunt-nosed body. In the modern hypersonic aerodynamics of today, Navier–Stokes calculations are an important tool for accurate flowfield analyses, and they take their place along with all of the other analytical and numerical methods discussed in Parts 1 and 2 of this book.

8.5 Summary and Comments

This brings to an end our discussion of various categories of fully viscous hypersonic flow calculations. In increasing order of accuracy and complexity, we have examined the following approaches.

1) *Viscous-shock-layer (VSL) method*: This approach uses a system of equations which is very much like the classical boundary-layer equations, but

which allows a finite normal pressure gradient via a more extensive y-momentum equation. The VSL method is a downstream-marching technique; it does not allow for any type of flow separation. The VSL method is in widespread use, and its relatively straightforward calculational procedure is appreciated by many engineers.

2) *Parabolized Navier–Stokes (PNS) method*: This approach uses a simplified version of the Navier–Stokes equations wherein the viscous terms involving streamwise derivatives are neglected. The PNS method allows a finite normal pressure gradient via a y-momentum equation, which, unlike the VSL method, retains some viscous terms. The PNS method is a downstream-marching technique: it allows for flow separation in the crossflow plane, but not in the streamwise direction. The PNS method is in very widespread use; indeed, it forms the basis of an industry-standard computer program, which is used by virtually all major aerodynamic laboratories and companies. This PNS code is sophisticated and requires much user effort to obtain accurate solutions; in this sense, it is a more demanding method than computer codes based on the VSL method.

3) *Full Navier–Stokes method*: Here, the complete Navier–Stokes equations are solved by means of a time-marching approach. This method is the ultimate in conceptual accuracy allowing for pressure gradients and flow separation to occur as would be the case in the natural flow problem. Such Navier–Stokes solutions, especially for three-dimensional flows, although carried out in practice today, are still state-of-the-art research calculations. Computer storage requirements and running times can be enormous for such calculations.

Also, this brings to an end our discussion of hypersonic viscous flows in general and hence an end to Part 2. Return again to our road map in Fig. 1.24. Looking down the column under the heading "viscous flows," we see that we have covered a number of important topics dealing with the combined effect of high Mach number and the transport phenomena of thermal conduction and viscosity. Recall that in Part 2 our intent has been to examine these effects without the extra complication of high-temperature effects. However, this is about as far as we should go along this route. In Part 3 to follow, we will examine such high-temperature effects, and we will revisit the problem of hypersonic viscous flows, this time including the chemical reactions and possible radiative-transfer effects that frequently dominate such flows in real life.

Part 3
High-Temperature Gas Dynamics

In Part 3, we discuss high-temperature effects in fluid flows. This is intimately related to hypersonic flow because any high-velocity flow will have regions where the temperature is high, and therefore physical–chemical processes can be strong enough to influence and even dominate the flow characteristics. In relation to our earlier discussions, recall that in Part 1 we examined the question: what happens to the fluid dynamics of an inviscid flow when the Mach number is made very large? In Part 2 we addressed the next logical question: what happens in a high-Mach-number flow when the transport, phenomena of viscosity and thermal conduction are included? Now, in Part 3, we consider the next logical question: what happens in a high-Mach-number flow when high temperatures are present? In this regard we will consider both inviscid and viscous high-temperature flows. However, the material in Part 3 goes beyond applications to just hypersonic flow: it is pertinent to any flow problem where high temperatures, hence physical-chemical processes, are important. Some examples are combustion phenomena, high-energy lasers, laser-matter interaction, flames, and rocket and jet-engine flowfields. Moreover, much of the basic material presented in Part 3 does not depend on our discussions in Parts 1 and 2; therefore, in this sense Part 3 stands as a self-contained presentation of high-temperature gas dynamics that can be studied in its own right. However, in the spirit of the present book, we will take many opportunities to relate the fundamentals of high-temperature gas dynamics to hypersonic flow.

9
High-Temperature Gas Dynamics: Some Introductory Considerations

In teaching, no doubt it is a good general principle 'to begin at the beginning', but to carry out the same it is necessary to know where that beginning is.

H. Middleton, British mathematician, 1883

Chapter Preview

The preceding quotation wisely states that in order to begin a new subject from the beginning it is necessary to know where that beginning is. The fundamentals of high-temperature effects on gas dynamic flows are generally not included in most introductory gas dynamics, compressible flow, and aerodynamics texts and classes. Most practicing engineers have not had the luxury of formal instruction in the basics of high-temperature gas dynamics. If you have never studied the physics, chemistry, and gas dynamics of high-temperature gases, then Part 3 of this book is for you. On the other hand, if you have studied and/or worked in the discipline, Part 3 is also for you because you might find new thoughts and new perspectives as you read through the material. In any event, with the present chapter we start at the very beginning of the principles of high-temperature gas dynamics. We will assume that you know nothing about the subject, and our purpose is, by the end of Part 3, to bring you to a certain degree of understanding and confidence that you will feel comfortable reading the literature, both classic and modern, in high-temperature gas dynamics, and that you can more effectively work in the discipline.

Let us start. We begin at the beginning.

9.1 Importance of High-Temperature Flows

On 24 July 1969, Apollo 11 successfully entered the atmosphere of the Earth, returning from the historic first manned flight to the moon. During its return to Earth, the Apollo vehicle acquired a velocity essentially equal to escape velocity from the Earth, approximately 11.2 km/s. At this entry velocity, the shock-layer temperature becomes very large. How large? Let us make an estimate based on the results of Chapter 2. The temperature ratio across a normal shock wave is given by Eq. (2.5). Let us assume that the temperature in the nose region of the Apollo lunar return vehicle is approximately that behind a normal shock wave, that is, as given by Eq. (2.5). Considering a given point on the entry trajectory, at an altitude of 53 km, the vehicle's Mach number is 32.5. At this altitude, the freestream temperature is $T_\infty = 283$ K. From Eq. (2.5), this yields a shock-layer temperature behind the shock of 58,128 K—*ungodly high*, but also *totally incorrect*. It is totally incorrect because Eq. (2.5), as many of the equations throughout all of the preceding chapters, is based on the assumption that the gas has constant specific heats. In our preceding calculation we have used for the ratio of specific heats $\gamma = 1.4$. In reality, at such high temperatures the gas becomes chemically reacting, and γ no longer equals 1.4 nor is it constant. A more realistic calculation, assuming the flow to be in local chemical equilibrium (a term to be defined later), yields a shock-layer temperature of 11,600 K—*also a very high temperature*, but considerably lower than the 58,128 K originally predicted. The major points here are as follows:

1) The temperature in the shock layer of a high-speed entry vehicle can be very high.

2) If this temperature is not calculated properly, huge errors will result. The assumption of constant $\gamma = 1.4$ does not even come close.

One of the functions of Part 3 of this book is to show how to make proper calculations of the temperature and indeed of all of the properties of a high temperature, chemically reacting flow. Some of the basic physical characteristics of high temperature hypersonic flows are discussed in Sec. 1.3.4; it is important for you to review Sec. 1.3.4 before progressing further.

The considerations just discussed are reinforced by the results shown in Fig. 1.18, taken from [4]. Here we see the temperature behind a normal shock wave in air plotted vs velocity at a standard altitude of 52 km. This temperature is indicative of the shock-layer temperature in the nose region of an atmospheric entry vehicle. Indeed, the entry velocities for various types of vehicles are noted on the abscissa, varying from the lower speeds of intermediate-range and intercontinental ballistic missiles (IRBMs and ICBMs), to the very high speed associated with the return of a space vehicle from Mars. Two curves are shown, one for a calorically perfect gas with constant $\gamma = 1.4$ and the other for an equilibrium chemically reacting gas. The upper curve for a constant $\gamma = 1.4$ shows an extremely rapid increase in temperature with velocity, leading to extraordinarily high predicted values of T_s at normal entry velocities. Of course, as described earlier, these predictions are totally incorrect. In contrast, the lower curve illustrates a calculation where the chemically reacting effects are properly taken into account. The temperatures here are still high, but considerably lower than those predicted on the basis of constant $\gamma = 1.4$. Note that the lower curve predicts, for the entry velocity of Apollo, a shock-layer temperature of about 11,600 K—the realistic temperature mentioned earlier. Figure 1.18 illustrates two important points that, for emphasis, we reiterate here: 1) at high velocities, the shock-layer temperatures are high, and 2) it is essential that this temperature be calculated properly.

The applications of the material to be discussed in Part 3 of this book are widespread. The following are listed as just a few examples.

9.1.1 Atmospheric Entry

We have already discussed this application to some extent. Here, we will just note the high-temperature regions in the flowfield around a blunt-nosed entry body, as sketched in Fig. 9.1. The massive amount of flow kinetic energy in a hypersonic freestream is converted to internal energy of the gas across the strong bow shock wave, hence creating very high temperatures in the shock layer near the nose. In addition, downstream of the nose region, where the shock-layer gas has expanded and cooled around the body, we have a boundary layer with an outer-edge Mach number that is still high; hence, the intense frictional dissipation within the hypersonic boundary layer creates high temperatures and can cause the boundary layer to become chemically reacting. Another aspect of entry-body flowfields occurs when ionization is present in the shock layer, hence providing large numbers of free electrons throughout the shock layer. This is illustrated in Fig. 9.2, where an entry body is sheathed in a flow with ions and free electrons. For air, the principle ionized species are NO^+, O^+, and N^+, along with the associated free electrons. The free electrons absorb

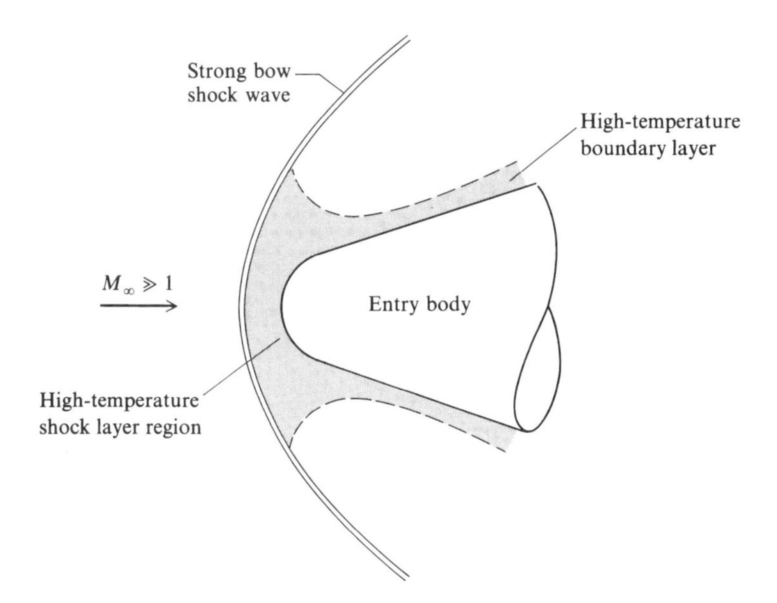

Fig. 9.1 Schematic of the high-temperature regions in an entry-body flowfield.

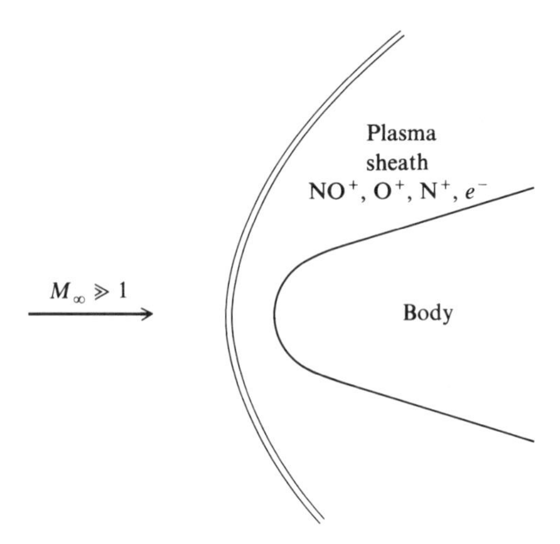

Fig. 9.2 Schematic of the plasma sheath around an entry body.

radio-frequency radiation and can cause a communications blackout to and from the vehicle during parts of the entry trajectory. In practice today, because of the availability of communications satellites in orbit around the Earth, radio transmissions from an entering space vehicle can be transmitted upward to a satellite

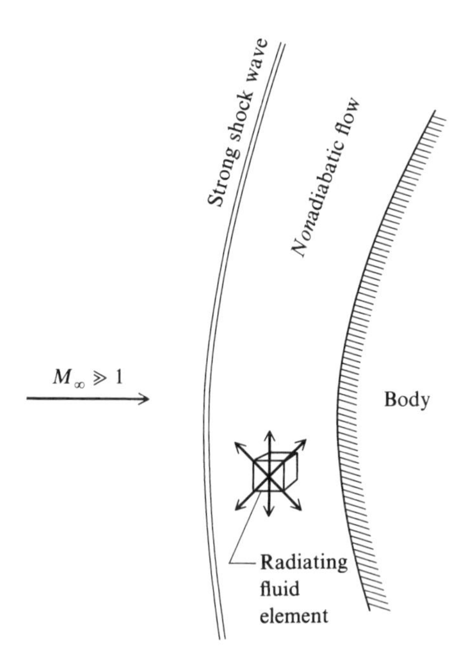

Fig. 9.3 Schematic of the nonadiabatic, radiating flowfield around a body.

through a less ionized part of the flowfield, and then relayed to the Earth. Therefore, the communications blackout is not quite the problem it used to be. Nevertheless, it is still a problem to be dealt with, and therefore the accurate prediction of the electron number density in the plasma sheath around the vehicle is frequently of high priority. Yet another aspect of entry-body flowfields is sketched in Fig. 9.3. If the shock-layer temperature is high enough, the fluid elements in the flow will emit and absorb radiation. This causes the flowfield to become *nonadiabatic*. Recall that throughout all of our inviscid flow considerations in Part 1, we assume the flow to be adiabatic. However, radiating shock layers will be nonadiabatic, and in such a case we lose some of the conceptual advantages we enjoyed in Part 1.

9.1.2 Rocket Engines

A schematic of a rocket engine is shown in Fig. 9.4. Here, a fuel and oxidizer are burned in a combustion chamber, and a chemically reacting gas subsequently expands through the nozzle of the engine. For the proper design of the engine, and the accurate prediction of rocket thrust and specific impulse, we need to know the properties of the products of combustion in the combustion chamber and the details of the chemically reacting flow through the nozzle. One question we can immediately ask is this: because the contours of supersonic nozzles are usually designed by the method of characteristics, what happens to the method of characteristics when the flow is chemically reacting? This question will be addressed in Chapters 14 and 15. The answers impact the proper design of rocket engine nozzles.

9.1.3 High-Enthalpy Wind Tunnels

For hypersonic wind-tunnel testing wherein the simulation of high-temperature flows over bodies is desired, a conventional hypersonic wind tunnel is not sufficient. Such conventional tunnels frequently use electrical resistance heaters to heat the reservoir air to just enough temperature (typically

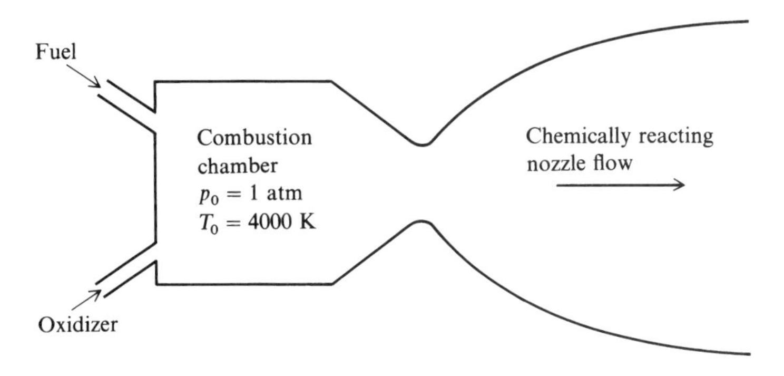

Fig. 9.4 Schematic of a rocket engine.

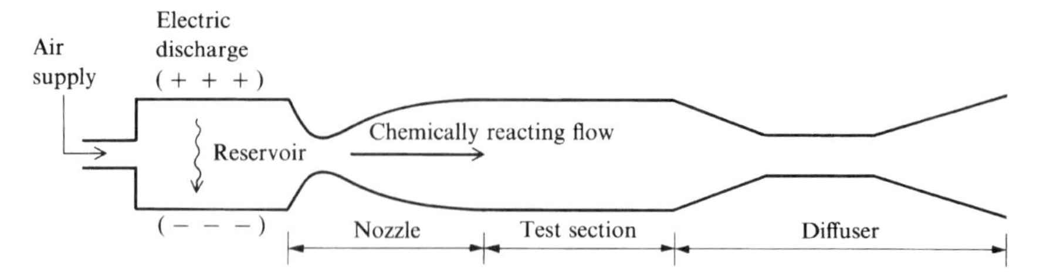

Fig. 9.5 Schematic of an arc tunnel.

1500 K) to avoid liquefaction of the air in the nozzle expansion and test section. To simulate shock-layer temperatures in the range of 5000 to 11,000 K, specialized high-enthalpy facilities are required. One such example is sketched in Fig. 9.5. Here we see an *arc tunnel*, wherein air is heated to high temperatures by an electric arc discharge in the reservoir and then the chemically reacting air expands through a hypersonic nozzle into the test section, exiting through a hypersonic diffuser. Another high-enthalpy device is the *shock tunnel* sketched in Fig. 9.6. Here, an incident shock moves from left to right in a shock tube, hitting the end wall and reflecting back from right to left. (See [4] for a basic description of shock tubes and their associated flow phenomena.) The reflected shock wave is shown in Fig. 9.6. Behind the reflected shock wave, the gas is at high pressure and temperature. A diaphragm mounted in the end wall is broken by the high pressure (or broken by some independent mechanical or electrical device), thus allowing the high-pressure, high-temperature chemically reacting gas to expand through the nozzle and pass through the test section and diffuser. Very high enthalpy and temperature levels (T as high as 11,000 K) can be produced in such shock tunnels; however, this is an impulse device with useful test times in the test section only on the order of a few milliseconds.

9.1.4 High-Power Lasers

We are familiar with small, desktop lasers that produce powers on the order of milliwatts. However, if we wish to have a laser that produces megawatts of power, we simply cannot scale such conventional lasers to large enough

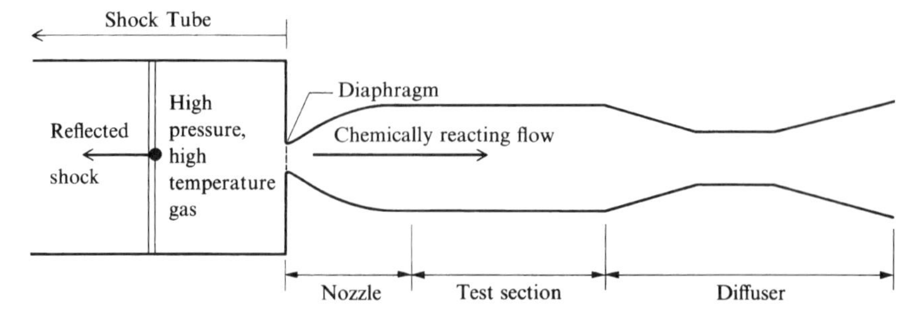

Fig. 9.6 Schematic of a shock tunnel.

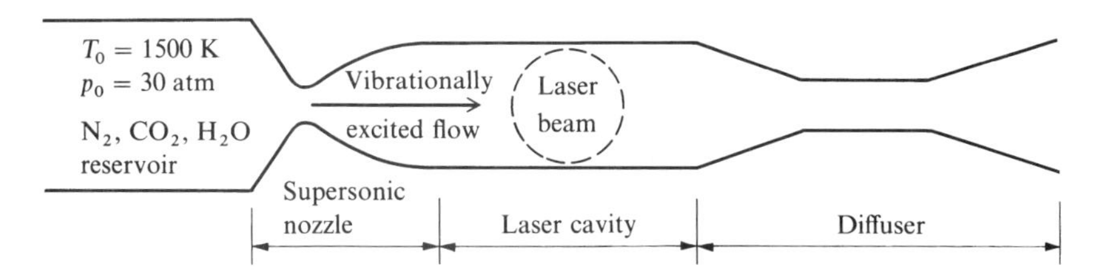

Fig. 9.7 Schematic of a gas dynamic laser.

sizes—the physics works against us. Therefore, since the late 1960s, several different classes of lasers, which can potentially produce megawatts of power, have been developed. Such high-power lasers have obvious applications in the commercial world (welding, for example) and in the military world as laser weapons (antiaircraft and antisatellite defense, for example). These high-power lasers are essentially high-temperature flow devices and hence are excellent applications of high-temperature gas dynamics. For example, Fig. 9.7 illustrates the concept of a *gas dynamic laser.* Here, a mixture of CO_2, N_2, and H_2O is heated to temperatures on the order of 1500 K in the reservoir. This temperature is high enough to vibrationally excite the molecules, but not high enough to cause chemical reactions. The vibrationally excited mixture expands rapidy through one or more supersonic nozzles; in this rapid expansion, the natural vibrational nonequilibrium processes turn the flow into a laser gas (with a population inversion). If reflecting mirrors are put on both sides of the test section (here called a laser cavity), an intense laser beam will be produced in a direction perpendicular to the page. Such a device is called a gas dynamic laser and it is a very interesting application of some of the principles of high-temperature gas dynamics. A gas dynamic laser is essentially a specialized supersonic wind tunnel that produces a high-power laser beam. For more details on gas dynamic lasers, see [147]. A second type of high-power laser, called the *electric discharge laser*, is sketched in Fig. 9.8. Here, a flow of CO_2, N_2, and He is passed through a duct, usually at subsonic speeds. An intense electric discharge is established across the flow, creating a vibrational nonequilibrium gas with laser properties. If mirrors are placed on both sides of the flow, a high-power laser beam will be produced in a direction perpendicular to the page. Such an electric discharge laser is

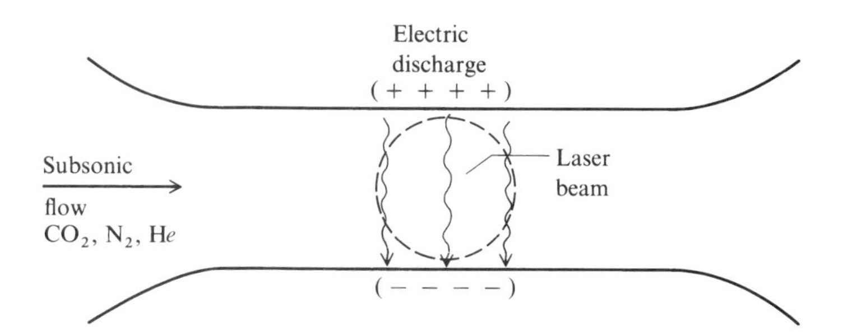

Fig. 9.8 Schematic of an electric discharge laser.

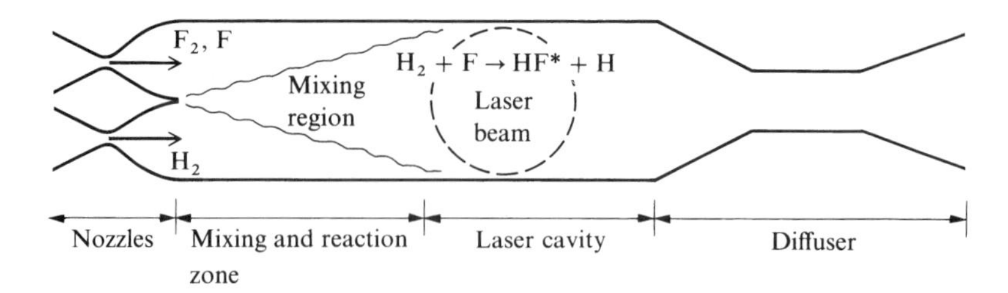

Fig. 9.9 Schematic of a chemical laser.

essentially a specialized subsonic wind tunnel that produces a laser beam. Finally, a third type of high-power laser, a *chemical laser*, is sketched in Fig. 9.9. Here, a supersonic stream containing atomic fluorine F is mixed with a supersonic flow of H_2. Downstream of the nozzles, in the chemically reacting zone, HF is produced in a vibrationally excited form via the chemical reaction $H_2 + F \rightarrow HF^* + H$, where the asterisk denotes a vibrationally excited species. The HF^* might have a population inversion and hence is a laser medium. If mirrors are placed on both sides of the flow, a high-power laser beam can be produced. Such a chemical laser is again a special type of "supersonic wind tunnel" that produces a laser beam. Other types of chemical lasers exist today using other gases as the laser medium. The physical and gas dynamic aspects of these high-power lasers are an excellent application of some of the fundamental material to be discussed in Part 3.

9.1.5 Ramjet and Scramjet Engines

A conventional ramjet engine is sketched in Fig. 9.10. Here, the engine is essentially an open duct wherein freestream air at high subsonic or supersonic

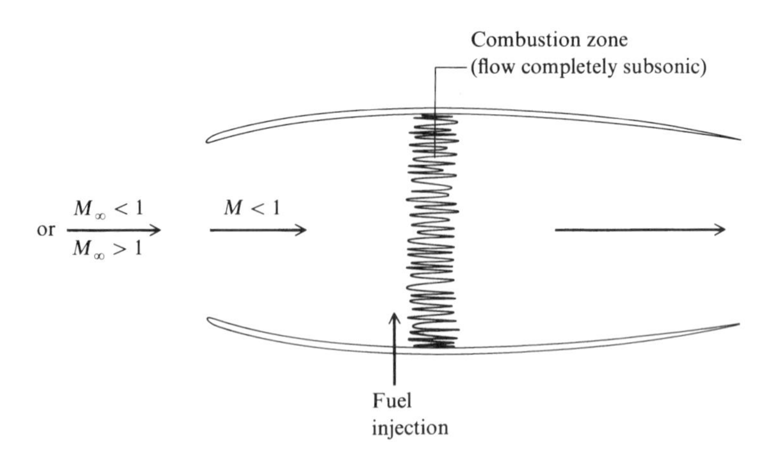

Fig. 9.10 Schematic of a conventional ramjet engine.

speeds is compressed and slowed to a low subsonic Mach number at the entrance to a combustor. Fuel is injected into the combustor, and burning takes place in a subsonic stream. Such conventional ramjets have propulsion advantages over the standard gas turbine engines in the Mach-number range from 2 to 5. However, they have a serious drawback at hypersonic speeds. To see this, consider a freestream at $M_\infty = 10$ and $T_\infty = 300\,\mathrm{K}$. Assume this stream is slowed adiabatically to a low subsonic velocity just in front of the combustor. We can make a crude estimate of the temperature of the air entering the combustor by using the adiabatic energy equation and assuming constant $\gamma = 1.4$, that is, $T_0/T_\infty = 1 + [(\gamma - 1)/2]M_\infty^2$ (for example, see [1], [4], and [5]). From this, we calculate an air temperature $T_0 = 6300\,\mathrm{K}$ entering the combustor. This is far above the adiabatic flame temperature of the fuel/air burning process in the combustor. Therefore, under these conditions, when the fuel is injected, it will simply decompose rather than burn, and the engine will be a drag device rather than a thrust device. To overcome this problem, the freestream air is not slowed to a low subsonic speed in the combustor: rather, it must be kept flowing at some supersonic speed where the temperature increase is not so great. Hence, the combustion process takes place in a supersonic stream. This is the essence of the supersonic combustion ramjet (scramjet), sketched in Fig. 9.11. Here, a hypersonic freestream is slowed to supersonic speeds by an inlet compression. Fuel (usually H_2) is injected into the supersonic stream, where it mixes and burns in a combustion region downstream of the fuel-injector strut. The mixture of burned gases subsequently expands through a supersonic nozzle at the back end of the engine, producing thrust. It is virtually certain that hypersonic cruise aircraft flying above Mach 5 or 6 will have to be powered by scramjet engines. The X-43 Hyper-X shown in

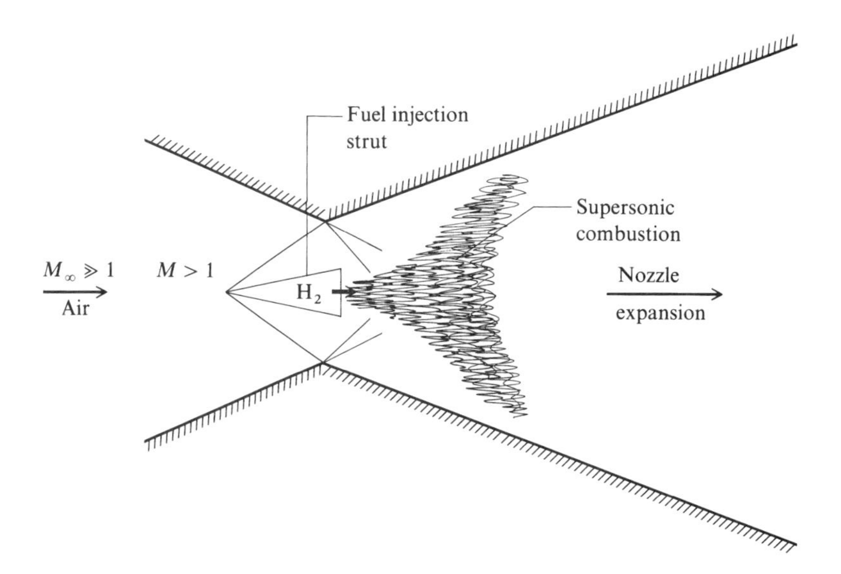

Fig. 9.11 Schematic of a supersonic combustion ramjet engine (scramjet).

Fig. 1.9 is such a vehicle. It is an example of a major, ongoing technology development program to create such vehicles.

Note: All of the sketches shown in Figs. 9.1–9.11 are conceptual only, completely devoid of detail. Their purpose is to convey the general principle being discussed, and actual machines and devices will in reality be more complex. Our purpose in this discussion has been to describe some applications of high-temperature gas dynamics. The intent is essentially motivational; as we proceed through Part 3, keep in mind that everything we discuss, no matter how obtuse and unrelated it might seem, is in reality absolutely necessary for the understanding of such applications. Also, the applications just discussed are just a small sample of the problems that demand an understanding of high-temperature gas dynamics. By the time you finish Part 3, you will have a much better understanding of the physical and gas dynamic processes that are the foundation of all of these applications—this, indeed, is our purpose here.

9.2 Nature of High-Temperature Flows

In the following chapters, we will delve into the details of high-temperature effects in gas dynamics. However, at this stage, let us address the question: what, in general, makes high-temperature flows any different to study than the flow of a gas with constant γ? The answer is as follows:

1) The thermodynamic properties (e, h, p, T, ρ, s, etc.) are completely different.

2) The transport properties (μ and k) are completely different. Moreover, the additional transport mechanism of diffusion becomes important, with the associated diffusion coefficients $D_{i,j}$.

3) High heat-transfer rates are usually a dominant aspect of any high-temperature application.

4) The ratio of specific heats, $\gamma = c_p/c_v$, is a variable. In fact, for the analysis of high-temperature flows, γ loses the importance it has for the classical constant γ flows, such as studied in Parts 1 and 2. From this point of view, all equations derived in Parts 1 and 2 under the assumption of a constant γ are not valid for a high-temperature gas. Such equations represent the vast majority of our results in Parts 1 and 2. In the process, we lose the ability for closed-form analyses using such equations.

5) In view of the preceding, virtually all analyses of high-temperature gas flows require some type of numerical, rather than closed-form, solutions.

6) If the temperature is high enough to cause ionization, the gas becomes a partially ionized plasma, which has a finite electrical conductivity. In turn, if the flow is in the presence of an exterior electric or magnetic field, then electromagnetic body forces act on the fluid elements. This is the purview of an area called magnetohydrodynamics (MHD).

7) If the gas temperature is high enough, there will be nonadiabatic effects caused by radiation to or from the gas.

For these reasons, a study of high-temperature flow is quite different from our previous considerations in Parts 1 and 2. A major purpose of Part 3 is to discuss how high-temperature effects are properly accounted for in gas dynamic analysis and to point out the differences in comparison to our previous work in Parts 1 and 2.

9.3 Chemical Effects in Air: The Velocity-Altitude Map

From the applications discussed in Sec. 9.1, it is clear that we are frequently concerned with air as the working gas. In future chapters, we will have frequent occasion to examine the chemical properties of high-temperature air in detail. However, in this section we simply ask the question: at what temperatures do chemically reacting effects become important in air? An answer is given in Fig. 9.12, which illustrates the ranges of dissociation and ionization in air at a pressure of 1 atm. Let us go through the following thought experiment. Imagine that we take the air in the room around us and progressively increase the temperature, holding the pressure constant at 1 atm. At about a temperature of 800 K, the vibrational energy of the molecules becomes significant (as noted on the right of Fig. 9.12). This is not a chemical reaction, but it does have some impact on the properties of the gas, as we will see in subsequent chapters. When the temperature reaches about 2000 K, the dissociation of O_2 begins. At 4000 K, the O_2 dissociation is essentially complete; most of the oxygen is in the form of atomic oxygen O. Moreover, by an interesting quirk of nature, 4000 K is the temperature at which N_2 begins to dissociate, as shown in Fig. 9.12. When

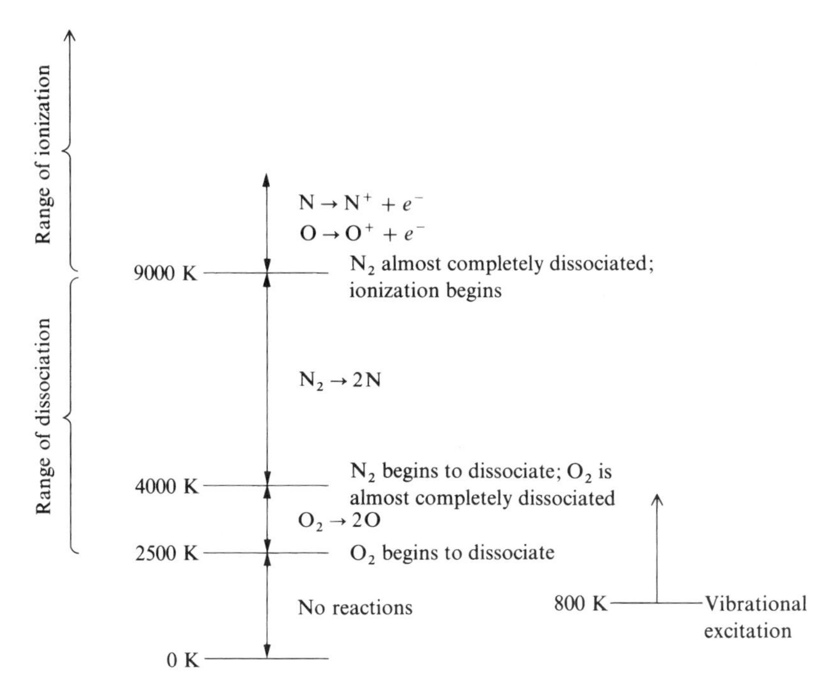

Fig. 9.12 Ranges of vibrational excitation, dissociation, and ionization for air at 1 atm pressure.

the temperature reaches 9000 K, most of the N_2 has dissociated. Coincidentally, this is the temperature at which both oxygen and nitrogen ionization occurs, and above 9000 K we have a partially ionized plasma consisting mainly of O, O^+, N, N^+, and electrons. Not shown in Fig. 9.12 (because it would become too cluttered) is a region of mild ionization that occurs around 4000 to 6000 K; here, small amounts of NO are formed, some of which ionize to form NO^+ and free electrons. In terms of the overall chemical composition of the gas, these are small concentrations; however, the electron number density as a result of NO ionization can be sufficient to cause the communications blackout discussed in Sec. 9.1. Reflecting upon Fig. 9.12, it is very useful to fix in your mind the "onset" temperatures: 800 K for vibrational excitation, 2500 K for O_2 dissociation, 4000 K for N_2 dissociation, and 9000 K for ionization. With the exception of vibrational excitation, which is not affected by pressure, if the air pressure is lowered, these onset temperatures decrease; conversely, if the air pressure is increased these onset temperatures are raised.

The information on Fig. 9.12 leads directly to the velocity-altitude map shown in Fig. 9.13. (Recall that we have led off Parts 1 and 2 with pertinent information on a velocity-altitude map; we do the same here.) In Fig. 9.13, we once again show the flight paths of lifting entry vehicles with different values of the lift parameters, m/C_LS. Superimposed on this velocity-altitude map are the flight regions associated with various chemical effects in air. The 10 and 90% labels at the top of Fig. 9.13 denote the effective beginning and end of various regions where these effects are important. Imagine that we start in the lower-left corner, and mentally "ride up" the flight path in reverse. As the velocity becomes

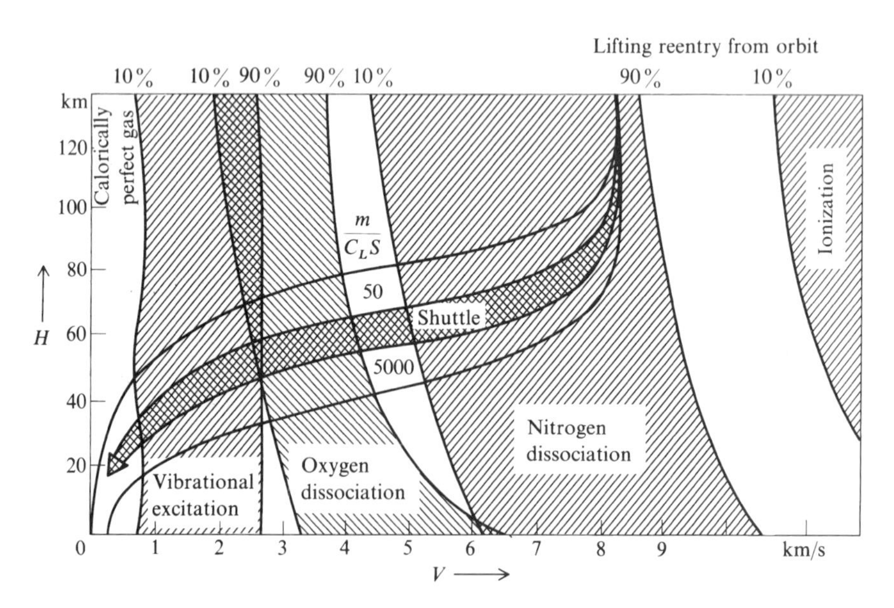

Fig. 9.13 Velocity-amplitude map with superimposed regions of vibrational excitation, dissociation, and ionization (from [79]).

larger, vibrational excitation is first encountered in the flowfield, at about $V = 1\ \text{km/s}$. At the higher velocity of about 2.5 km/s, the vibrational mode is essentially fully excited, and oxygen dissociation begins. This effect covers the shaded region labeled "oxygen dissociation." The O_2 dissociation is essentially complete at about 5 km/s, wherein N_2 dissociation commences. This effect covers the shaded region labeled "nitrogen dissociation." Finally, above 10 km/s, the N_2 dissociation is complete, and ionization begins. It is most interesting that regions of various dissociations and ionization are so separate on the velocity-altitude map, with very little overlap. This is, of course, consistent with the physical data shown in Fig. 9.12. In a sense, this is a situation when nature is helping to simplify things for us. Finally, we can make the following general observation from Fig. 9.13. The entry flight paths slash across major sections of the velocity-altitude map where chemical reactions and vibrational excitation are important. Indeed, the vast majority of any given flight path is in such regions. From this, we can clearly understand why high-temperature effects are so important to entry-body flows.

9.4 Summary and Comments

In this chapter we have discussed the importance of high-temperature gas dynamics by illustrating various practical engineering problems that are dominated by high-temperature effects. Moreover, we have examined in a very preliminary manner the basic nature of these effects. For air in particular, we have delineated various regions that are associated with different physical-chemical effects. The basic purpose of this chapter is simply to get the reader thinking about high-temperature flows and to give some appreciation for the nature of the problem.

At this stage in our discussion, it is worthwhile to return to our road map in Fig. 1.24, and chart our course. For the next few chapters, we will be dealing with the basic fundamentals of physical chemistry, statistical thermodynamics, and kinetic theory, all under the general heading of high-temperature flows in Fig. 1.24. However, these early discussions will not deal with flow problems at all; rather, they will lay the physical fundamentals that will be necessary to understand high-temperature effects in gases. Then, with these fundamentals in hand, we will tackle the analysis of high-temperature flows, both inviscid and viscous, that is, be prepared to study some chemistry and physics in the next few chapters, because it is absolutely necessary for our later applications to high-temperature flow problems. However, as we delve into this chemistry and physics, never lose sight of the fact that our ultimate purpose is to apply such chemical and physical aspects to flows associated with practical problems such as described in this chapter.

10
Some Aspects of the Thermodynamics of Chemically Reacting Gases (Classical Physical Chemistry)

Thermodynamics provides laws that govern the transfer of energy from one system to another, the transformation of energy from one form to another, the utilization of energy for useful work, and the transformation of matter from one molecular, atomic, or nuclear species to another.

Frederick D. Rossini, physical chemist, 1995

Chapter Preview

The thermodynamics and chemistry of high-temperature gases is the chassis on which high-temperature gas dynamics moves. So we start here at the beginning. Historically, this is not a new subject. It is rooted in the development of classical physical chemistry that has taken place over the past one-and-a-half centuries. But it is just as applicable today as it was then. In this chapter we discuss only those aspects of classical physical chemistry that we need to apply for our study of high-temperature gas dynamics. In subsequent chapters these classical results will be embellished by concepts from modern 20th century physics.

A warning—parts of this chapter are not necessarily easygoing. But expecting this, if you slow down, reread, and think the thoughts, you will master these parts and forge ahead. In this chapter you will begin to learn how to handle the thermodynamics of a high-temperature gas, to begin to appreciate what is meant by equilibrium and nonequilibrium conditions, and to calculate the chemical composition of an equilibrium chemically reacting mixture. Then we will expand on these concepts in subsequent chapters, where your understanding will gradually mature. So this chapter constitutes an essential part of our beginning. Read on, and visit the world of classical thermodynamics and chemistry.

10.1 Introduction: Definition of Real Gases and Perfect Gases

In this chapter, we will deal with chemical thermodynamics from a classical point of view, that is, we will deal, for the most part, with *macroscopic* properties of a system without appealing to the individual molecular and atomic particles that make up the system. This will be in contrast to Chapters 11 and 12, where we will consider the *microscopic* picture, dealing with the system made up of individual particles, with the macroscopic properties of the system being given by suitable averages over the particles.

To begin our chemical thermodynamic discussion, we have to distinguish between a *real gas* and *perfect gas*. These are defined as follows. Consider the air around you as made up of molecules that are in random motion, frequently colliding with neighboring molecules. Imagine that you pluck one of these molecules out of the air around you. Examine it closely. You will find that a force field surrounds this molecule, as a result of the electromagnetic action of the electrons and nuclei of the molecule. In general, this force field will reach out from the given molecule and will be felt by neighboring molecules, and vice versa. Thus, the force field is called an *intermolecular force*. A schematic of a typical intermolecular force field caused by a single particle is shown in Fig. 10.1. Here, the intermolecular force is sketched as a function of distance away from the particle. Note that at small distances, the force is strongly repulsive, tending to push the two molecules away from each other. However, as we move further away from the molecule, the intermolecular force rapidly decreases and becomes a weak attractive force, tending to attract molecules

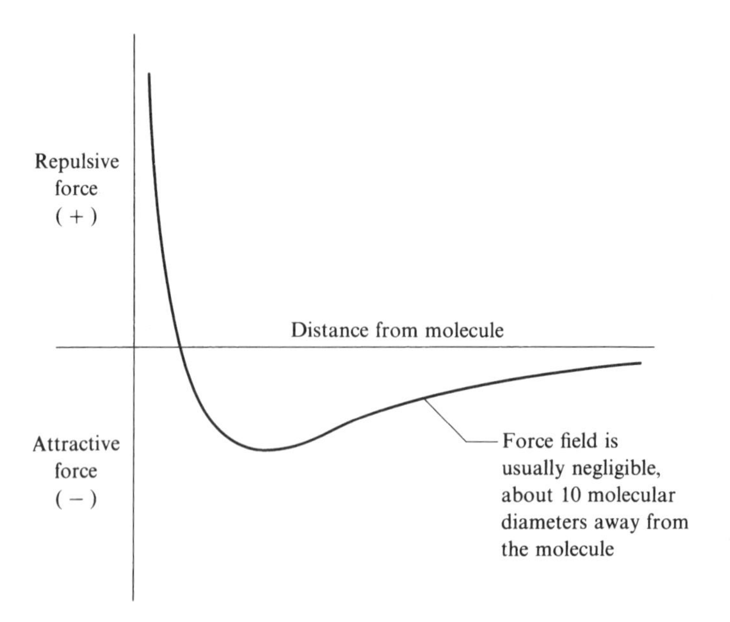

Fig. 10.1 Sketch of the intermolecular force variation.

together. At distances approximately 10 molecular diameters away from the molecule, the magnitude of the intermolecular force is negligible. Because the molecules are in constant motion, and this motion is what generates the macroscopic thermodynamic properties of the system, then the intermolecular force should affect these macroscopic properties. This leads to the following definition:

> *Real gas* is a gas where intermolecular forces are important and must be accounted for.

On the other hand, if the molecules are spaced, on the average, more than 10 moleculear diameters apart, the magnitude of the intermolecular force is very small (see Fig. 10.1) and can be neglected. This, for example, is the case for air at standard conditions. This leads to the next definition:

> *Perfect gas* is a gas where intermolecular forces are negligible.

For most problems in aerodynamics, the assumption of a perfect gas is very reasonable. We have made this assumption throughout Parts 1 and 2 of this book. (The quantitative ramifications of a perfect gas are discussed in the next section.) Conditions that require the assumption of a real gas are very high pressures ($p \approx 1000\,\text{atm}$) and/or low temperatures ($T \approx 30\,\text{K}$). Under these conditions the molecules in the system will be packed closely together and will be moving slowly with consequent low inertia. Thus, the intermolecular force has every opportunity to act on the molecules in the system and in turn to modify the macroscopic properties of the system. In contrast, at lower pressures ($p \approx 10\,\text{atm}$, for example) and higher temperatures ($T = 300\,\text{K}$, for example), the molecules are widely spaced apart and are moving more rapidly with consequent higher inertia. Thus, on the average, the intermolecular force has little effect on the particle motion and therefore on the macroscopic properties of the system. Repeating again, we can assume such a gas to be a *perfect gas*, where the intermolecular force can be ignored. Deviations from perfect-gas behavior tend to be proportional to p/T^3, which makes qualitative sense based on the preceding discussion. Unless otherwise stated, in the present book, we will always deal with a perfect gas as defined herein; this is compatible with about 99% of all practical aerodynamic problems.

The road map for this chapter is given in Fig. 10.2. We will start with the left-hand column and discuss the equation of state, identify various types (classifications) of gases, and discuss their ramifications. Then we move to the center column and discuss some classical thermodynamics of chemically reacting gases, concentrating on the first and second laws of thermodynamics. Finally, we move to the right-hand column where we break the ice on how you can approach the calculation of the composition of a chemically reacting mixture in equilibrium. We begin to address the question: if you have a mixture of chemically reacting gases at a given pressure and temperature, how do you calculate the amount of each individual chemical species in the mixture when the mixture is in chemical equilibrium? In the process, we will begin to understand just what is meant by the term equilibrium in this context.

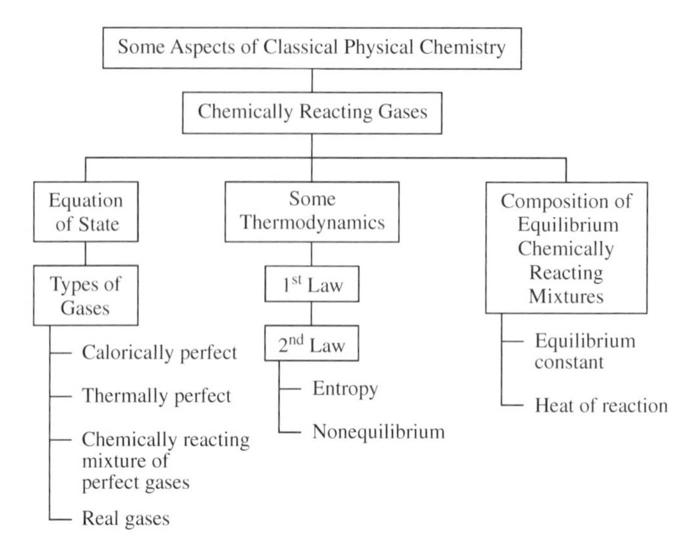

Fig. 10.2 Road map for Chapter 10.

10.2 Various Forms of the Perfect-Gas Equation of State

Consider a thermodynamic system of a given volume V, as sketched in Fig. 10.3. Denote the total mass of the system by M, the total number of gas particles by N, and the total number of moles by $\mathcal{N}$. (Recall from chemistry that one mole of a substance is an amount of mass of that substance equal to its molecular weight, i.e., if we were dealing with O_2 which has a molecular weight of 32, then 32 kg of O_2 would constitute one kg · mole, 32 g of O_2 would constitute one gm · mole, and 32 slugs of O_2 would constitute one slug · mole. Note that the kg, g, and slug just used as identifiers of what type of mole is being considered are simple *adjectives* in front of the word "mole"; they are *not* separate units and in a numerical calculation cannot be separately canceled, i.e., a kg · mole is one

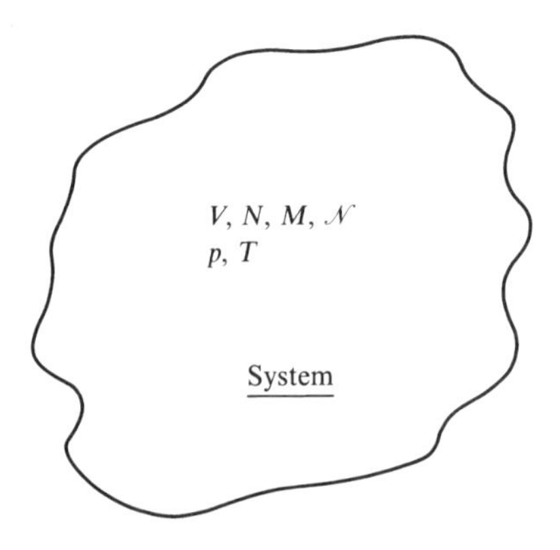

Fig. 10.3 Thermodynamic system.

single entity, *not* a kg multiplying a mole.) The pressure and temperature of the system in Fig. 10.3 are p and T, respectively.

The equation of state relates the quantities shown in Fig. 10.3. This equation cannot be derived from the principles of classical macroscopic thermodynamics; it must be considered as a given result—a postulate. Historically, it was first synthesized from laboratory measurements by Robert Boyle in the 17th century, Jacques Charles in the 18th century, and Joseph Gay-Lussac and John Dalton around 1800. The empirical result that unfolded from these observations was

$$\boxed{pV = MRT} \tag{10.1}$$

where R is the *specific gas constant*, which has different values for different gases. In macroscopic thermodynamics, we take Eq. (10.1) as a given result. Moreover, because the empirical data on which Eq. (10.1) is based were obtained with gases at near standard conditions, where intermolecular forces are negligible, then Eq. (10.1) is called the *perfect-gas equation of state*. In Chapters 11 and 12, we will see the perfect-gas equation of state can be *derived* from first principles, using the concepts of either statistical mechanics or kinetic theory, both disciplines being developed in the late 19th century and early 20th century. These theories take a microscopic approach to the gas, in contrast to the macroscopic approach of classical thermodynamics, which we are discussing in this chapter.

Several forms of Eq. (10.1) can be obtained as follows. Divide Eq. (10.1) by M, yielding

$$\boxed{pv = RT} \tag{10.2}$$

where v is the *specific volume*, defined as the volume per unit mass. Divide Eq. (10.1) by V, obtaining

$$\boxed{p = \rho RT} \tag{10.3}$$

where ρ is the *density*. Now consider the molecular weight of the gas denoted by $\mathcal{M}$. From Eq. (10.1) we have

$$pV = \frac{M}{\mathcal{M}}(\mathcal{M}R)T \tag{10.4}$$

Keeping in mind that $\mathcal{M}$ is the mass per mole, then $M/\mathcal{M}$ is $\mathcal{N}$, the total number of moles in the system. Also, $\mathcal{M}R$ is the *universal gas constant* $\mathcal{R}$. Note that

$$R = \frac{\mathcal{R}}{\mathcal{M}} \tag{10.5}$$

Hence, Eq. (10.4) becomes

$$\boxed{pV = \mathcal{N}\mathcal{R}T} \tag{10.6}$$

Divide Eq. (10.6) by $\mathscr{N}$. We obtain

$$\boxed{p\mathscr{V} = \mathscr{R}T} \tag{10.6a}$$

where $\mathscr{V}$ is the *molar volume*, or volume per mole. Dividing Eq. (10.6) by V, we have

$$\boxed{p = C\mathscr{R}T} \tag{10.7}$$

where C is the *concentration*, or moles per unit volume. If we divide Eq. (10.6) by the total mass M, we have

$$\boxed{pv = \eta\mathscr{R}T} \tag{10.8}$$

where η is the *mole-mass ratio*, or number of moles per unit mass. Finally, let N_A denote Avagadro's number, which is the number of particles per mole; for a kg · mole, $N_A = 6.02 \times 10^{26}$ particles per kg · mole. Using Eq. (10.6), we have

$$pV = (\mathscr{N}N_A)\frac{\mathscr{R}}{N_A}T \tag{10.8a}$$

In Eq. (10.8a), $\mathscr{N}N_A$ is physically the number of particles in the system N. Also, $\mathscr{R}/N_A$ is the gas constant per particle, which is defined as the Boltzmann constant k. Thus, Eq. (10.8a) becomes

$$\boxed{pV = NkT} \tag{10.9}$$

Dividing Eq. (10.9) by V, we have

$$\boxed{p = nkT} \tag{10.10}$$

where n is the *number density*, or number of particles per unit volume.

Starting with Eq. (10.1), review all of the preceding equations that are in boxes. They represent nine different forms of the perfect-gas equation of state. They all mean the same thing; the different forms are just expressed in terms of different quantities. Make certain that you feel comfortable with these equations and with all of the defined terms (such as mole-mass ratio, concentration, etc.) before progressing further. These equations are good for a perfect gas consisting of a single chemical species, and they are also valid for a chemically reacting mixture of perfect gases. Equations (10.1–10.10) hold whether or not the gas is chemically reacting. For a nonreacting gas $\mathscr{M}$, hence R from Eq. (10.5), is constant. In contrast, for a reacting gas $\mathscr{M}$ is a variable; hence,

$$R = \frac{\mathscr{R}}{\mathscr{M}} \tag{10.11}$$

is also a variable. However, even though R is a variable, Eqs. (10.1), (10.2), and (10.3) involving R are still valid as long as we are dealing with a chemically reacting mixture of perfect gases, that is, a mixture where intermolecular forces are negligible.

Do not be confused by the variety of gas constants that are associated with the preceding equations. They are easily sorted out as follows.

1) When the equation deals with moles, use the universal gas constant $\mathscr{R}$, which is the gas constant per mole. It is the same for all gases and equal to the following in the SI and English engineering systems of units respectively:

$$\mathscr{R} = 8314\ \text{J}/(\text{kg} \cdot \text{mol K})$$

$$\mathscr{R} = 4.97 \times 10^4\ (\text{ft lb})/(\text{slug} \cdot \text{mol}\ ^\circ\text{R})$$

2) When the equation deals with mass, use the specific gas constant R, which is the gas constant per unit mass. It is different for different gases and is related to the universal gas constant through Eq. (10.11). Because air is of special importance to many high-temperature applications, we give the following data for air at standard conditions:

$$R = 287\ \text{J}/(\text{kg K})$$

$$R = 1716\ (\text{ft lb})/(\text{slug}\ ^\circ\text{R})$$

3) When the equation deals with particles, use the Boltzmann constant k, which is the gas constant per particle. Values for k in the two systems of units are

$$k = 1.38 \times 10^{-23}\ \text{J/K}$$

$$k = 0.565 \times 10^{-23}\ (\text{ft lb})/^\circ\text{R}$$

In all of the preceding relations, p is the pressure of the gas mixture, which can consist of a number of different chemical species. Let us now introduce the concept of *partial pressure* by means of the following thought experiment. Imagine that you go to a store and buy a special vacuum cleaner that selectively "vacuums up" oxygen in the air. You come home, seal all of the doors and windows in a given room (which is at an air pressure of 1 atm), and turn on the special vacuum. After all of the O_2 has been vacuumed up, only N_2 remains in the room (assuming air to consist of 20% O_2 and 80% N_2).

Question: What is the gas pressure in the room?

Answer: Because 20% of the gas molecules have been taken out, the resulting pressure is 0.8 atm.

Moreover, because the gas now consists of only N_2, this 0.8 atm is defined as the partial pressure of N_2, designated by p_{N_2}. Now, assume that you go to a different store, and buy a special vacuum cleaner that selectively vacuums up only N_2. Assume that you go back to a sealed room containing air at 1 atm, and turn on this vacuum cleaner. After all of the N_2 has been removed, only O_2 remains in the room, and the pressure will be 0.2 atm. This is defined as the partial pressure

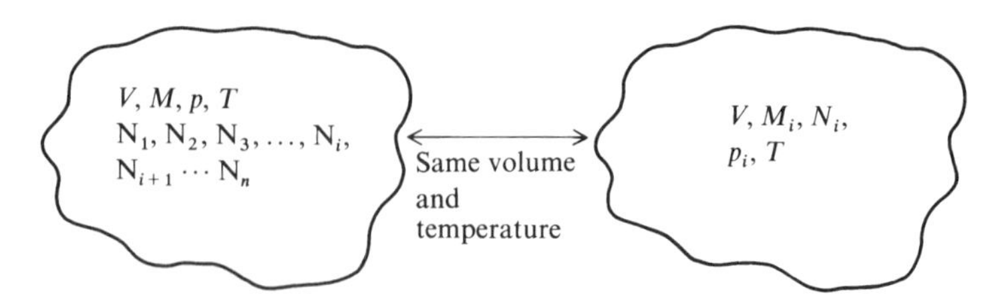

Fig. 10.4 Systems for the definition of partial pressure.

of O_2, designated by p_{O_2}. This very simplistic thought experiment, which, of course, could never be carried out in real life, is given only to introduce the general concept of partial pressure, which can be stated as follows. Consider a system of volume V consisting of a mixture of n different chemical species, each with a different number of particles, $N_1, N_2, \ldots, N_n$, as sketched on the left of Fig. 10.4. Now consider a single species in the mixture, say, the ith species. By definition, the partial pressure of species i, p_i is the pressure that *would* exist in the system if all of the other species were removed, and the N_i particles were the only ones occupying the whole system at the volume V and temperature T. This is illustrated on the right of Fig. 10.4. Returning to the left side of Fig. 10.4, we now construe p to be the "total" pressure of the gas mixture, made up of the individual partial pressures of the n species. Indeed, for a perfect gas (no intermolecular forces), we have

$$\boxed{p = \sum_i p_i} \tag{10.12}$$

where the summation is carried out over all of the chemical species in the mixture. Equation (10.12) is called Dalton's law of partial pressures.

For a perfect gas, p_i also obeys various equations of state, analogous to those discussed earlier for a gas mixture. For example, analogous to Eq. (10.1), we have

$$\boxed{p_i V = M_i R_i T} \tag{10.13}$$

where R_i is the specific gas constant for species i, M_i is the mass of species i in the system, and V is the volume of the system, as usual. Dividing Eq. (10.13) by M_i, we have

$$\boxed{p_i v_i = R_i T} \tag{10.14}$$

where v_i is the specific volume based on species i, that is, the volume per unit mass of species i; $v_i = V/M_i$. Dividing Eq. (10.13) by V, we obtain

$$\boxed{p_i = \rho_i R_i T} \tag{10.15}$$

where ρ_i is the density of species i, that is, the mass of species i per unit volume of mixture. Let $\mathcal{M}_i$ be the molecular weight of species i, that is, the mass of i per mole of i. Then, from Eq. (10.13),

$$p_i V = \frac{M_i}{\mathcal{M}_i}(\mathcal{M}_i R_i)T$$

or

$$\boxed{p_i V = \mathcal{N}_i \mathcal{R} T} \tag{10.16}$$

where

$$R_i = \frac{\mathcal{R}}{\mathcal{M}_i} \tag{10.17}$$

and $\mathcal{N}_i$ is the number of moles of species i in the mixture. Dividing Eq. (10.16) by $\mathcal{N}_i$, we have

$$\boxed{p_i \mathcal{V}_i = \mathcal{R} T} \tag{10.18}$$

where $\mathcal{V}_i$ is the volume per mole of species i. Dividing Eq. (10.16) by V, we obtain

$$\boxed{p_i = C_i \mathcal{R} T} \tag{10.19}$$

where C_i is the concentration of species i, that is, the number of moles of species i per unit volume of mixture. Dividing Eq. (10.16) by the total mass of the system M, we have

$$\boxed{p_i v = \eta_i \mathcal{R} T} \tag{10.20}$$

where η_i is the mole-mass ratio of species i, that is, the number of moles of i per unit mass of mixture. Recalling that N_A is the number of particles per mole (Avogadro's number), Eq. (10.16) can be written as

$$p_i V = (\mathcal{N}_i N_A)\left(\frac{\mathcal{R}}{N_A}\right)T$$

or

$$\boxed{p_i V = N_i k T} \tag{10.21}$$

Finally, dividing Eq. (10.21) by V, we have

$$\boxed{p_i = n_i k T} \tag{10.22}$$

where n_i is the number density of species i, that is, the number of i particles per unit volume of the mixture.

As tedious as it might seem, it is necessary for you to feel comfortable with all of the different forms of the equation of state as already obtained. Also it is important to have a clear understanding of the defined terms such as C_i, $\mathcal{M}_i$, η_i, etc. These equations and terms are used throughout the literature in high-temperature reacting gases. They are presented in the detail just shown for your convenience and ease of understanding.

10.3 Various Descriptions of the Composition of a Gas Mixture

How do we describe the *composition* of a chemical reacting gas mixture? In other words, what terms are used to describe how much of each species i is present in the mixture? The most useful and frequently used terms are itemized here.

1) The partial pressures p_i: If we know all of the partial pressures of the mixture, the chemical composition is uniquely defined.

2) The concentrations C_i: If we know all of the values of C_i, the chemical composition is uniquely defined.

3) The mole-mass ratio η_i: If we know all the values of η_i, the chemical composition is uniquely defined.

4) The mole fraction X_i, defined as the number of moles of species i per mole of mixture. If we know all of the values of X_i, the chemical composition is uniquely defined.

5) The mass fraction c_i, defined as the mass of i per unit mass of mixture: From the definition, we can write

$$c_i = \frac{\rho_i}{\rho}$$

If we know all of the values of c_i, the chemical composition is uniquely defined. All of the quantities just listed are *intensive* variables, that is, they do not depend on the extent of the system. Also, for gas dynamic problems, variables based on per unit mass are particularly useful. Thus, in our considerations to follow, we will be particularly interested in mass fraction c_i and mole-mass ratio η_i. However, if the chemical composition is given in terms of *any* of the preceding variables, we can always calculate directly the values of the other variables. For example, assume that the composition is described in terms of p_i. The mole fractions X_i are obtained directly by dividing Eq. (10.16) by Eq. (10.6):

$$\frac{p_i V}{p V} = \frac{\mathcal{N}_i \mathcal{R} T}{\mathcal{N} \mathcal{R} T}$$

or

$$\boxed{\frac{p_i}{p} = \frac{\mathcal{N}_i}{\mathcal{N}} \equiv X_i} \tag{10.23}$$

where $p = \sum_i p_i$. The mass fraction can be obtained from X_i as

$$\boxed{c_i = X_i \left(\frac{\mathcal{M}_i}{\mathcal{M}} \right)} \tag{10.24}$$

Equation (10.24) is based just on the physical meanings of c_i, X_i, $\mathcal{M}_i$, and $\mathcal{M}$. (Work this through in your mind.) Equation (10.24) can also be formally obtained by dividing Eq. (10.15) by Eq. (10.3), and using the relations given in Eqs. (10.11) and (10.17). Other relations are the subject of homework problem 10.1.

Note that, based on the definitions of c_i and X_i, the following relations hold:

$$\sum_i c_i = 1$$

and

$$\sum_i X_i = 1$$

Before leaving this section, let us examine two related questions. First, how are R and R_i related; the former is the specific gas constant for the mixture, and the latter is the specific gas constant for the species i. From Eqs. (10.12), (10.3), and (10.15), we have

$$p = \sum_i p_i$$

$$\rho RT = \sum_i \rho_i R_i T$$

$$R = \sum_i \frac{\rho_i}{\rho} R_i$$

$$\boxed{R = \sum_i c_i R_i} \tag{10.25}$$

Hence, for a chemically reacting mixture, the value of R for the mixture can be obtained from a simple summation of the R_i multiplied by their respective mass fractions. Secondly, how do we obtain the mixture molecular weight $\mathcal{M}$. For example, this is needed in Eq. (10.24), and it also represents an alternate calculation of R through the relation $R = \mathcal{R}/\mathcal{M}$. An answer is obtained from Eq. (10.25) written as

$$\frac{\mathcal{R}}{\mathcal{M}} = \sum_i c_i \frac{\mathcal{R}}{\mathcal{M}_i}$$

Solving for $\mathcal{M}$, we have

$$\boxed{\mathcal{M} = \frac{1}{\sum_i c_i/\mathcal{M}_i}} \tag{10.26}$$

Equation (10.26) allows the calculation of $\mathcal{M}$ from the composition of the gas mixture given in terms of mass fraction c_i. An alternative relation in terms of the mole fraction is

$$\boxed{\mathcal{M} = \sum_i X_i \mathcal{M}_i} \tag{10.27}$$

Equation (10.27) is based simply on the physical meaning of the terms involved; work it through yourself.

10.4 Classification of Gases

For the analysis of gas dynamic problems, we can identify four categories of gases, as follows.

10.4.1 Calorically Perfect Gas

By definition, a calorically perfect gas is one with constant specific heats c_p and c_v. In turn, the ratio of specific heats $\gamma = c_p/c_v$ is constant. For this gas, the enthalpy and internal energy are functions of temperature, given explicity by

$$h = c_p T$$

and

$$e = c_v T$$

The perfect-gas equation of state holds, for example,

$$pv = RT$$

where R is a constant. In the introductory study of compressible flow, the assumption of a calorically perfect gas is almost always made (for example, see [4] and [5]); hence, the thermodynamics of a calorically perfect gas is probably quite familiar to you. Indeed, for the hypersonic analyses contained in Parts 1 and 2 of this book, the assumption of a calorically perfect gas was almost universally made, and many of the detailed formulas and results were obtained under the assumption of constant γ.

10.4.2 Thermally Perfect Gas

By definition, a thermally perfect gas is one where c_p and c_v are *variables* and specifically are functions of temperature only.

$$c_p = f_1(T)$$
$$c_v = f_2(T)$$

Differential changes in the h and e are related to differential changes in T via

$$\mathrm{d}h = c_p \, \mathrm{d}T$$
$$\mathrm{d}e = c_v \, \mathrm{d}T$$

Hence, h and e are functions of T only, that is,

$$h = h(T)$$
$$e = e(T)$$

The perfect-gas equation of state holds, for example,

$$pv = RT$$

where R is a constant. In Chapter 11 we will demonstrate that the temperature variation of specific heats, hence the whole nature of a thermally perfect gas, is caused by the excitation of vibrational energy within the molecules of the gas and to the electronic energy associated with electron motion within the atoms and molecules.

10.4.3 Chemically Reacting Mixture of Perfect Gases

Here we are dealing with a multispecies, chemically reacting gas where intermolecular forces are neglected; hence, each individual species obeys the perfect-gas equation of state in such forms as given by Eqs. (10.13–10.22). At this stage, we need to make a distinction between equilibrium and nonequilibrium chemically reacting gases. Our discussion here will be preliminary; a more fundamental understanding of the meaning of equilibrium and nonequilibrium systems will evolve in subsequent chapters. For the time being, imagine that you take the air in the room around you, and instantly increase the temperature to 5000 K, holding the pressure constant at 1 atm. We know from Fig. 9.12 that dissociation will occur. Indeed, let us allow some time (maybe several hundred milliseconds) for the gas properties to "settle out," and come to some steady state at 5000 K and 1 atm. The chemical composition that finally evolves in the limit of "large" times (milliseconds) is the *equilibrium* composition at 5000 K and 1 atm. In contrast, during the first few milliseconds immediately after we instantly increase the temperature to 5000 K, the dissociation reactions are just beginning to take place, and the variation of the amount of O_2, O, N_2, N, etc. in the gas is changing as a function of time. This is a *nonequilibrium* system. After the lapse of a sufficient time, the amounts of O_2, O, N_2, etc. will approach some steady values, and these steady values are the equilibrium values. It is inferred from the preceding that, once the system is in equilibrium, then the equilibrium values of c_{O_2}, c_{N_2}, c_O, c_N, etc. will depend only on the pressure and temperature, that is, at 5000 K and 1 atm, the equilibrium chemical composition is uniquely defined. We will prove this later in the present chapter. In contrast, for the nonequilibrium system, c_{O_2}, c_{N_2}, c_O, etc. depend not only on p and T, but also on *time*. If the nonequilibrium system were a fluid element rapidly expanding through a shock-tunnel nozzle, another way of stating this effect is to say that c_{O_2}, c_{N_2}, etc. depend on the "history" of the flow.

With these thoughts in mind, we can define a chemically reacting mixture of perfect gases as follows. Consider a system at pressure p and temperature T. For convenience, assume a unit mass for the system. The number of particles of each different chemical species per unit mass of mixture are given by $N_1, N_2, \ldots, N_n$. For each *individual* chemical species present in the mixture (assuming a perfect gas), the enthalpy and internal energy per unit mass of i, h_i, and e_i, respectively, will be functions of T (i.e., each individual species, by itself, behaves as a thermally perfect gas). However, h and e for the chemically reacting mixture depend not only on h_i and e_i, but also on how much of each species is present. Therefore, for a chemically reacting mixture of perfect gases, in the general nonequilibrium case, we write

$$h = h(T, N_1, N_2, N_3, \ldots, N_n)$$
$$e = e(T, N_1, N_2, N_3, \ldots, N_n)$$
$$c_p = f_1(T, N_1, N_2, N_3, \ldots, N_n)$$
$$r = f_2(T, N_1, N_2, N_3, \ldots, N_n)$$

where, in general, $N_1, N_2, N_3, \ldots, N_n$ depend on p, T, and the "history of the gas flow." The perfect-gas equation of state still holds:

$$pv = RT$$

However, here R is a variable because in a chemically reacting gas the molecular weight of the mixture $\mathcal{M}$ is a variable, and $R = \mathcal{R}/\mathcal{M}$.

For the special case of an *equilibrium* gas, the chemical composition is a unique function of p and T; hence, $N_1 = f_1(p, T)$, $N_2 = f_2(p, T)$, etc. Therefore, the preceding results for h, e, c_p, and c_v become

$$h = h(T, p)$$
$$e = e_1(T, p) = e_2(T, v)$$
$$c_p = f_1(T, p)$$
$$c_v = f_2(T, p) = f_3(T, v)$$

In the preceding, it is frequently convenient to think of e and c_v as functions of T and v rather than T and p. It does not make any difference, however, because for a themodynamic system in equilibrium (including an equilibrium chemically reacting system) the state of the system is uniquely defined by *any two state variables.* The choice of T and p, or T and v, in the preceding, is somewhat arbitrary in this sense.

10.4.4 Real Gas

Here, we must take into account the effect of intermolecular forces. We could formally consider a chemically reacting gas as well as a nonreacting real gas. However, in practice, a gas behaves as a real gas under conditions of very high pressure and low temperature—conditions that accentuate the influence of intermolecular forces on the gas. For these conditions, the gas is rarely chemically reacting. Therefore, for simplicity, we will consider a nonreacting gas here. Recall that for both the cases of a calorically perfect gas and a thermally perfect gas, h and e were functions of T only. For a real gas, with intermolecular forces, h and e depend on p (or v) as well:

$$h = h(T, p)$$
$$e = e(T, v)$$
$$c_p = f_1(T, p)$$
$$c_v = f_2(T, v)$$

Moreover, the perfect-gas equation of state is no longer valid here. Instead, we must use a real-gas equation of state, of which there are many versions. Perhaps the most familiar is the Van der Waals equation, given by

$$\left(p + \frac{a}{v^2}\right)(v - b) = RT \tag{10.28}$$

where a and b are constants that depend on the type of gas. Note that Eq. (10.28) reduces to the perfect-gas equation of state when $a = b = 0$. In Eq. (10.28) the terms a/v^2 take into account the intermolecular force effects, and b takes into account the actual volume of the system occupied by the volume of the gas particles themselves.

In summary, the preceding discussion has presented four different categories of gases. Any existing analyses of thermodynamic and gas dynamic problems will fall in one of these categories; they are presented here so that you can establish an inventory of such gases in your mind. It is extremely helpful to keep these categories in mind while performing any study of gas dynamics.

Also, to equate these different categories to a practical situation, let us once again take the case of air. Imagine that you take the air in the room around you and begin to increase its temperature. At room temperature, air is essentially a calorically perfect gas, and it continues to act as a calorically perfect gas until the temperature reaches approximately 800 K. Then, as the temperature increases further, we see from Fig. 9.12 that vibrational excitation becomes important. When this happens, air acts as a thermally perfect gas. Finally, above 2500 K, chemical reactions occur, and air becomes a chemically reacting mixture of perfect gases. If we were to go in the opposite direction, that is, reduce the air temperature considerably below room temperature, and/or increase the pressure to a very high value, say, 1000 atm, then the air would behave as a real gas.

Finally, it is important to note a matter of nomenclature. We have followed classical physical chemistry in defining a gas where intermolecular forces are important as a real gas. Unfortunately, an ambiguous term has evolved in the aerodynamic literature that means something quite different. In the 1950s, aerodynamicists were suddenly confronted with hypersonic entry vehicles at velocities as high as 26,000 ft/s (8 km/s). As discussed in Chapter 9, the shock layers around such vehicles were hot enough to cause vibrational excitation, dissociation, and even ionization (see Fig. 9.13). These were "real" effects that happened in air in "real life." Hence, it became fashionable in the aerodynamic literature to denote such conditions as real-gas effects. For example, the categories just itemized as a thermally perfect gas, and as a chemically reacting mixture of perfect gases, would come under the classification of real-gas effects in some of the aerodynamic literature. But in light of classical physical chemistry, this is truly a *misnomer*. A real gas is truly one in which intermolecular forces are important, and this has nothing to do with vibrational excitation or chemical reactions. Therefore, in this book we will talk about *high-temperature effects* and will discourage the use of the incorrect term real-gas effects.

With this, we have completely navigated the left column of our chapter road map in Fig. 10.2. We now move on to the center column and concentrate on some thermodynamics.

10.5 First Law of Thermodynamics

Classical thermodynamics is based on two fundamental laws, the first of which is described here. Consider a system of gas as sketched in Fig. 10.5. The system is separated from its surroundings by a boundary. The gas in the system is composed of molecules, each of which has a particular energy (to be discussed in Chapter 11). The internal energy E of the system is equal to this molecular energy summed over all of the molecules in the system. A change in this internal energy $\mathrm{d}E$ can be brought about by 1) adding heat δQ across the boundary of the system and 2) doing work on this system δW. This is sketched in Fig. 10.5. Here $\mathrm{d}E$ is an infinitesimally small change in internal energy, and δQ and δW are small increments in heat and work. As you may recall from thermodynamics, $\mathrm{d}E$ is an exact differential, related just to the change in state of the system, whereas δQ and δW are not exact differentials because they depend on the *process* by which heat is added or work is done. With these items in mind, the *first law of thermodynamics* is written as

$$\boxed{\delta Q + \delta W = \mathrm{d}E} \tag{10.29}$$

Equation (10.29) holds for any type of gas, nonreacting or reacting, perfect or real.

If we have no shaft work done on the system (no mechanical device sticking through the boundary of the system and performing work, such as a paddle wheel or turbine), then the work done on or by the system is caused by the compression ($\mathrm{d}V$ negative) or expansion ($\mathrm{d}V$ positive) of the volume of the system. For example, imagine the system is the air inside an inflated balloon. Grab hold of the balloon, and squeeze it with your hands. You are doing work on the air in the balloon, and in the process you are decreasing its volume. The amount of work associated with an infinitesimally small volume change $\mathrm{d}V$ is

$$\delta W = -p\,\mathrm{d}V \tag{10.30}$$

The derivation of Eq. (10.30) can be found in any good thermodynamics text and is also given in Chapter 4 of [1]. Equation (10.30) is based on purely

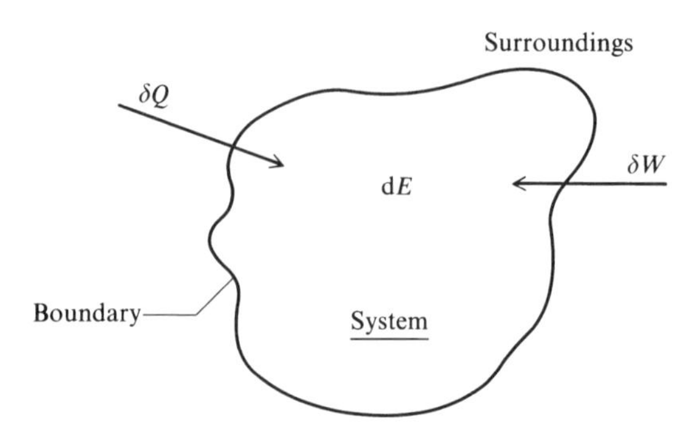

Fig. 10.5 System to illustrate the first law of thermodynamics.

mechanical reasoning; it is good for any type of gas, reacting or nonreacting, perfect or real.

Combining Eqs. (10.29) and (10.30), we have

$$\delta Q = \mathrm{d}E + p\,\mathrm{d}V \tag{10.31}$$

In gas dynamics, we like to deal in terms of per unit mass, hence, considering Eq. (10.31) in terms of per unit mass, we write

$$\boxed{\delta q = \mathrm{d}e + p\,\mathrm{d}v} \tag{10.32}$$

where δq is the heat added per unit mass. Also, the definition of enthalpy is

$$h = e + pv$$

Hence,

$$\mathrm{d}h = \mathrm{d}e + p\,\mathrm{d}v + v\,\mathrm{d}p$$

or

$$\mathrm{d}e = \mathrm{d}h - p\,\mathrm{d}v - v\,\mathrm{d}p \tag{10.33}$$

Substituting Eq. (10.33) into (10.32), we obtain

$$\boxed{\delta q = \mathrm{d}h - v\,\mathrm{d}p} \tag{10.34}$$

Equations (10.32) and (10.34) are very useful alternate forms of the first law of thermodynamics. We will have many occasions to refer to these forms. Also, in the spirit of distinguishing the fundamentals of chemical thermodynamics from the reader's prior experience in classical thermodynamics, we will be constantly reminding the reader about what does, and what does not, apply to a chemically reacting gas. For example, the entire development in this section is based on a general thermodynamic system; hence, everything in this section applies to any type of gas.

A large number of basic thermodynamic relations can be obtained from the first law. We will obtain one of them here because it bears on the difference between reacting and nonreacting gases. Specifically, we consider the relation

$$c_p - c_v = R \tag{10.35}$$

Equation (10.35) is a familiar relation from elementary thermodynamics dealing with calorically perfect gases.

Question: Does Eq. (10.35) hold for a thermally perfect gas? A chemically reacting gas? A real gas?

To answer this question, let us obtain a general expression for the difference in specific heats, $c_p - c_v$, using the first law. Assuming an equilibrium system, we can write

$$e = e(T,\ v)$$

or, from the definition of a perfect differential

$$de = \left(\frac{\partial e}{\partial T}\right) dT + \left(\frac{\partial e}{\partial v}\right)_T dv \tag{10.36}$$

Recall from thermodynamics that c_v is defined as

$$c_v = \left(\frac{\partial e}{\partial T}\right)_v \tag{10.37}$$

Substituting Eq. (10.37) into (10.36), we have

$$de = c_v \, dT + \left(\frac{\partial e}{\partial v}\right)_T dv \tag{10.38}$$

Substituting Eq. (10.38) into the first law in the form given by Eq. (10.32), we have

$$\delta q = c_v \, dT + \left[\left(\frac{\partial e}{\partial v}\right)_T + p\right] dv \tag{10.39}$$

Now recall the definition of c_p. In analogy with Eq. (10.37), we usually write

$$c_p \equiv \left(\frac{\partial h}{\partial T}\right)_p \tag{10.40}$$

However, more fundamentally, we recall the basic definition of specific heat as the heat added per unit change in temperature, that is, $\delta q / dT$. Because ∂q can be added in an infinite number of different ways (different processes), the $\delta q / dT$ is not unique. However, if we stipulate that δq is added at constant pressure, $(\delta q / dT)_p$, then we have a unique quantity, and it is defined as c_p. Thus,

$$c_p \equiv \left(\frac{\delta q}{dT}\right)_p \tag{10.41}$$

Considering a constant-pressure process, Eq. (10.39) can be written as

$$\left(\frac{\delta q}{dT}\right)_p = c_v + \left[\left(\frac{\partial e}{\partial v}\right)_T + p\right]\left(\frac{\partial v}{\partial T}\right)_p \tag{10.42}$$

From the relation given by Eq. (10.41), we can then express Eq. (10.42) as

$$\boxed{c_p - c_v = \left[\left(\frac{\partial e}{\partial v}\right)_T + p\right]\left(\frac{\partial v}{\partial T}\right)_p} \tag{10.43}$$

Equation (10.43) is a general relation for the difference $c_p - c_v$ for any type of gas, real or perfect, reacting or nonreacting. Specializing Eq. (10.43) we have, from the perfect-gas equation of state,

$$v = \frac{RT}{p}$$

For a nonreacting gas, where R is constant, this yields

$$\left(\frac{\partial v}{\partial T}\right)_p = \frac{R}{p} \tag{10.44}$$

Substitute Eq. (10.44) into (10.43), obtaining

$$c_p - c_v = R + \frac{R}{p}\left(\frac{\partial e}{\partial v}\right)_T \tag{10.45}$$

For both cases of a calorically perfect gas and a thermally perfect gas, where e is a function of T only, $(\partial e/\partial v)_T = 0$, and Eq. (10.45) yields

$$c_p - c_v = R$$

which is Eq. (10.35). However, for a chemically reacting gas, e is a function of both T and $v, e = e(T, v)$, and therefore $(\partial e/\partial v)_T$ has a finite value. Thus, Eq. (10.43) must be used to obtain $c_p - c_v$ for a chemically reacting mixture. For a real gas, because Eq. (10.44) does not hold, we also have to use Eq. (10.43) to obtain the difference in specific heats. Thus, we have answered our question, namely, that the familiar relation given in Eq. (10.35) holds for a calorically perfect or a thermally perfect gas, but it does not hold for a chemically reacting gas or a real gas.

10.6 Second Law of Thermodynamics

The second law of thermodynamics is involved with the concept of entropy s, conventionally defined as

$$\mathrm{d}s \equiv \frac{\delta q_{\text{rev}}}{T} \tag{10.46}$$

Here, we are considering a system originally in state 1 where the entropy is s_1, undergoing an infinitesimal change to state 2, where the entropy is $s_2 = s_1 + \mathrm{d}s$. The process by which this change is taking place can be reversible or irreversible. (An irreversible process is one that involves the dissipative effects of viscosity, thermal conduction, or mass diffusion, and/or where the system is in nonequilibrium. A reversible process is one that involves none of the above.) This change from state 1 to state 2 is illustrated in Fig. 10.6. Here we see two possible paths by which the system can change from state 1 to state 2. State 1 is the same in both cases, and state 2 is the same in both cases. Because s is a state variable, then $\mathrm{d}s$ is the same in both cases. At the top of Fig. 10.6, the change in

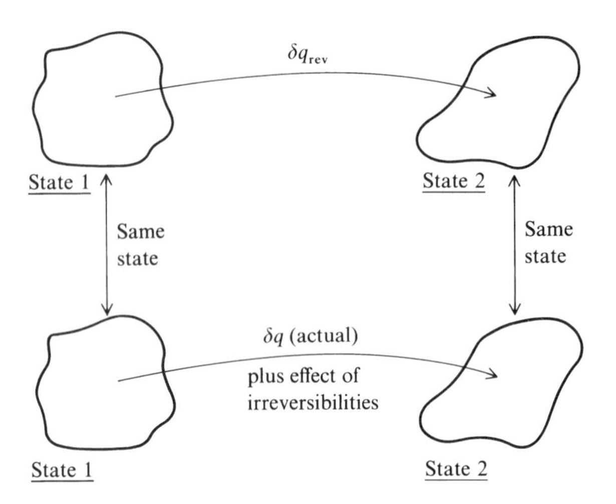

Fig. 10.6 Systems to illustrate the second law of thermodynamics.

state of the system is imagined as taking place as a result of the reversible addition of heat δq_{rev}, and ds is given by Eq. (10.46). If the process were truly reversible, then δq_{rev} would be the actual heat added. However, if the process is irreversible, the δq_{rev} is *not* the actual heat added; rather, δq_{rev} is an "artificial" number contrived to satisfy Eq. (10.46). Thus, if the process shown at the top of Fig. 10.6 is irreversible, then δq_{rev} is a number that is chosen to satisfy Eq. (10.46) for the given ds. A more satisfactory way of looking at this situation is sketched at the bottom of Fig. 10.6. Here we see the same two states, with the same entropy change ds as the upper sketch. However, now δq denotes the *actual* heat added, no matter whether the process is reversible or irreversible. If the process were reversible, the change in entropy ds would be caused totally by δq, which would be given by $\delta q = T ds$ from Eq. (10.46). However if the process is irreversible, then ds is caused only *in part* by δq and is also caused by the effect of any irreversibilities taking place in the system (because of friction, thermal conduction, diffusion, and/or nonequilibrium). For an irreversible process, the value of δq is less than it would be for a reversible process, the difference being supplied by the irreversibilities. Thus, in contrast to Eq. (10.46), it is more satisfying to write the change of entropy as follows:

$$\boxed{ds = \frac{\delta q}{T} + ds_{irrev}} \tag{10.46a}$$

In Eq. (10.46a), δq is the *actual* amount of heat added during the process, and ds_{irrev} is the generation of entropy caused by the dissipative phenomena itemized earlier. These dissipative phenomena always *increase* the entropy; hence,

$$ds_{irrev} > 0$$

Therefore, from Eq. (10.46a),

$$\boxed{ds > \frac{\delta q}{T}} \tag{10.47}$$

If the process is *adiabatic*, where by definition $\delta q = 0$, then

$$\boxed{ds > 0} \tag{10.48}$$

Equations (10.47) and (10.48) are statements of the second law of thermodynamics. Moreover, all of the statements and equations in the section are very fundamental; they apply in general for nonreacting or reacting gases and for perfect or real gases. The second law dictates that, no matter what process is acting on the system, the inequalities given in Eqs. (10.47) or (10.48) must be satisfied.

10.7 Calculation of Entropy

Because entropy is a state variable, it must be uniquely related to other state variables. The purpose of this section is to obtain such a relationship.

For a system in equilibrium, entropy can be expressed as a function of T and p. This is true for all of the categories of gases discussed in Sec. 10.4. For example, even for calorically or thermally perfect gases, where enthalpy and internal energy are functions of temperature only, the entropy still depends on both T and p. Indeed, a familiar expression for the entropy change between two states, $s_2 - s_1$, for a calorically perfect gas is

$$s_2 - s_1 = c_p \ln \frac{T_2}{T_1} - R \ln \frac{p_2}{p_1} \tag{10.49}$$

[The reader should check his or her memory bank or standard texts such as [1] and [5] for Eq. (10.49).] If s_2 represents the entropy s at a given temperature $T = T_2$ and $p = p_2$, and if s_1 represents a reference value of entropy s_{ref} at temperature $T_1 = T_{\text{ref}}$ and pressure $p_1 = p_{\text{ref}}$, then Eq. (10.49) can be written as

$$s = c_p \ln \frac{T}{T_{\text{ref}}} - R \ln \frac{p}{p_{\text{ref}}} + s_{\text{ref}} \tag{10.50}$$

Keep in mind that Eq. (10.50) holds for a calorically perfect gas only.

Let us obtain an expression somewhat analogous to Eq. (10.50) that holds for an equilibrium chemically reacting mixture. In the following, we will denote quantities per mole by capital letters, that is, S, H, E, etc. will hereafter denote the entropy, enthalpy, internal energy, etc. per mole. Let S_i be the entropy of species i per mole of species i. Then we can write for the mixture,

$$S = \sum_i X_i S_i \tag{10.51}$$

where S is the entropy per mole of mixture, S_i is the entropy of species i per mole of i, and X_i is the mole fraction. Let us obtain an expression for S_i and insert it into Eq. (10.51) to obtain the entropy of the reacting mixture.

To begin with, consider a system with just the pure species i at the temperature T and pressure p_i. Assume there is one mole of species i in the system, so that the entropy is S_i and the enthalpy is H_i. From the first law in the form of Eq. (10.34), but written per mole,

$$\delta Q = \mathrm{d}H_i - \mathscr{V}_i \, \mathrm{d}p_i \tag{10.52}$$

From Eq. (10.46) written per mole,

$$T \, \mathrm{d}S_i = \delta Q_{\mathrm{rev}} \tag{10.53}$$

We can assume a reversible process, and combine Eqs. (10.52) and (10.53), obtaining

$$\mathrm{d}S_i = \frac{\mathrm{d}H_i}{T} - \frac{\mathscr{V}_i}{T} \, \mathrm{d}p_i \tag{10.54}$$

From Eq. (10.18), we have

$$\frac{\mathscr{V}_i}{T} = \frac{\mathscr{R}}{p_i} \tag{10.55}$$

Substituting Eq. (10.55) into (10.54), we have

$$\mathrm{d}S_i = \frac{\mathrm{d}H_i}{T} - \mathscr{R} \frac{\mathrm{d}p_i}{p_i} \tag{10.56}$$

Species i by itself is a thermally perfect gas, where by definition $C_{pi} = f(T)$. From Sec. 10.4, we can write

$$\mathrm{d}H_i = C_{pi} \, \mathrm{d}T \tag{10.57}$$

where C_{pi} is the specific heat at constant pressure for one mole of species i. Substituting Eq. (10.57) into (10.56), we have

$$\mathrm{d}S_i = C_{pi} \frac{\mathrm{d}T}{T} - \mathscr{R} \frac{\mathrm{d}p_i}{p_i} \tag{10.58}$$

Integrating Eq. (10.58) between a reference condition with entropy $S_{i,\mathrm{ref}}$ at reference temperature T_{ref} and reference pressure p_{ref}, and a state where the entropy is S_i at a temperature and pressure of T and p_i, respectively, we have

$$\int_{S_{i,\mathrm{ref}}}^{S_i} \mathrm{d}S_i = \int_{T_{\mathrm{ref}}}^{T} C_{pi} \frac{\mathrm{d}T}{T} - \mathscr{R} \int_{p_{\mathrm{ref}}}^{p_i} \frac{\mathrm{d}p_i}{p_i}$$

or

$$S_i = \int_{T_{\text{ref}}}^{T} C_{pi} \frac{\mathrm{d}T}{T} - \mathscr{R}\, \ell n \frac{p_i}{p_{\text{ref}}} + S_{i,\text{ref}} \tag{10.59}$$

Equation (10.59) is an expression for the entropy per mole of the pure species i.

The entropy of the mixture is obtained by combining Eqs. (10.51) and (10.59)

$$S = \sum_i X_i \left[\int_{T_{\text{ref}}}^{T} C_{pi} \frac{\mathrm{d}T}{T} - \mathscr{R}\, \ell n \frac{p_i}{p_{\text{ref}}} \right] + \sum_i X_i S_{i,\text{ref}}$$

or

$$\boxed{S = \sum_i X_i \left[\int_{T_{\text{ref}}}^{T} C_{pi} \frac{\mathrm{d}T}{T} - \mathscr{R}\, \ell n \frac{p_i}{p_{\text{ref}}} \right] + S_{\text{ref}}} \tag{10.60}$$

where S_{ref} is the reference entropy level for the mixture, given by $S_{\text{ref}} = \sum_i X_i S_{i,\text{ref}}$. Equation (10.60) gives the entropy per mole of mixture for an equilibrium chemically reacting mixture, as a function of the temperature T, and the individual partial pressures p_i. In turn, as we will see in Sec. 10.9, the individual p_i are functions of the temperature T and the total mixture pressure p. Hence Eq. (10.60) for an equilibrium chemically reacting gas gives $S = S(p,\ T)$ and is the direct analog of Eq. (10.50) for a catorically perfect gas.

10.8 Gibbs Free Energy and the Entropy Produced by Chemical Nonequilibrium

In Sec. 10.7, we obtained an expression for the entropy of an equilibrium chemically reacting mixture. Now consider two different equilibrium states of this mixture, where the temperatures and pressures are $T_2,\ p_2$, and $T_1,\ p_1$, respectively. The corresponding change in entropy $S_2 - S_1$ can be found by using Eq. (10.60). Because entropy is a state variable, $S_2 = S(T_2,\ p_2)$ and $S_1 = S(T_1,\ p_1)$. Once two states are specified, $S_2 - S_1$ is totally independent of any possible process by which state 1 is changed into state 2. However, we do know from Eq. (10.46a) that, no matter what the process might be, in general part of the entropy change is caused by heat exchange and the other part caused by irreversibilities:

$$\mathrm{d}S = \frac{\delta Q}{T} + \mathrm{d}S_{\text{irrev}} \tag{10.61}$$

Although $S_2 - S_1$ is independent of the process between the two specified equilibrium states, it is sometimes very useful to know, for a specific process connecting states 1 and 2, how much of the $S_2 - S_1$ is caused by irreversibilities during the process. This is particularly true when the irreversibilities are caused by chemical *nonequilibrium* during the process. Therefore, the purpose of this section is to calculate the change in entropy as a result of a chemical nonequilibrium process.

To accomplish this, let us introduce a new, defined thermodynamic variable, namely, the Gibbs free energy per mole of mixture, denoted by G. By definition

$$G \equiv H - TS \tag{10.62}$$

Hence,

$$\mathrm{d}G = \mathrm{d}H - T\,\mathrm{d}S - S\,\mathrm{d}T$$

or

$$\mathrm{d}H = \mathrm{d}G + T\,\mathrm{d}S + S\,\mathrm{d}T \tag{10.63}$$

Putting Eq. (10.63) "on hold" for a moment, return to Eq. (10.61). For δQ in this equation, we have from the first law [Eq. (10.34)]

$$\delta Q = \mathrm{d}H - \mathcal{V}\,\mathrm{d}p$$

Hence, Eq. (10.61) is written as

$$\mathrm{d}S = \frac{\mathrm{d}H}{T} - \frac{\mathcal{V}}{T}\mathrm{d}p + \mathrm{d}S_{\mathrm{irrev}}$$

or

$$T\,\mathrm{d}S = \mathrm{d}H - \mathcal{V}\,\mathrm{d}p + T\,\mathrm{d}S_{\mathrm{irrev}} \tag{10.64}$$

Now, combine Eqs. (10.63) and (10.64):

$$T\,\mathrm{d}S = \mathrm{d}G + T\,\mathrm{d}S + S\,\mathrm{d}T - \mathcal{V}\,\mathrm{d}p + T\,\mathrm{d}S_{\mathrm{irrev}}$$

The terms involving $T\,\mathrm{d}S$ cancel, leaving

$$\mathrm{d}G = -S\,\mathrm{d}T + \mathcal{V}\,\mathrm{d}p - T\,\mathrm{d}S_{\mathrm{irrev}} \tag{10.65}$$

Before examining Eq. (10.65), further note that, if we had an equilibrium mixture, because G is a state variable we could write $G = G(p, T)$. However, because we are treating a chemical nonequilibrium process here, where the number of particles of species i per mole of mixture N_i is a function of not only T and p but also the timewise history of the process, we must write

$$G = G(T, p, N_1, N_2, \ldots, N_i, \ldots, N_n) \tag{10.66}$$

The total differential of Eq. (10.66) is

$$\mathrm{d}G = \left(\frac{\partial G}{\partial T}\right)_p \mathrm{d}T + \left(\frac{\partial G}{\partial p}\right)_T \mathrm{d}p + \sum_i \frac{\partial G}{\partial N_i}\mathrm{d}N_i \tag{10.67}$$

Comparing Eqs. (10.65) and (10.67) term by term, we see that

$$\boxed{S = -\left(\frac{\partial G}{\partial T}\right)_p \qquad \mathcal{V} = \left(\frac{\partial G}{\partial p}\right)_T} \tag{10.67a}$$

and

$$-T\,\mathrm{d}S_{\text{irrev}} = \sum_i \frac{\partial G}{\partial N_i}\,\mathrm{d}N_i \tag{10.68}$$

Equation (10.67a) expresses two basic thermodynamic relations, and Eq. (10.68) gives the change in entropy caused by the irreversible effect of nonequilibrium chemical reactions. Let us cast Eq. (10.68) in slightly different form. For a chemically reacting mixture, the value of G per mole of mixture is related to g_i', which is defined as the Gibbs free energy of species i per particle through

$$G = \sum_i N_i g_i' \tag{10.69}$$

where, again, N_i is the number of i particles per mole of mixture. Equation (10.69) is based on simple physical definitions, that is, g_i' is the Gibbs free energy of species i per particle of $i, N_i g_i'$ is then the Gibbs free energy caused by species i per mole of mixture, and the summation over all of the species gives the Gibbs free energy of the mixture per mole of mixture. Noting that Eq. (10.69) can be expanded as

$$G = N_1 g_1' + N_2 g_2' + \cdots + N_i g_i' + \cdots + N_n g_n'$$

then,

$$\frac{\partial G}{\partial N_i} = g_i' \tag{10.70}$$

Substituting Eq. (10.70) into (10.68), we have

$$\boxed{\mathrm{d}S_{\text{irrev}} = -\frac{1}{T}\sum_i g_i'\,\mathrm{d}N_i} \tag{10.71}$$

Letting N_A denote Avogadro's number (number of particles per mole), Eq. (10.71) becomes

$$\mathrm{d}S_{\text{irrev}} = -\frac{1}{T}\sum_i (N_A g_i')\left(\frac{\mathrm{d}N_i}{N_A}\right) \tag{10.71a}$$

However,

$$N_A g_i' = G_i \qquad \text{(Gibbs free energy of species } i \text{ per mole of } i\text{)}$$

$$\frac{\mathrm{d}N_i}{N_A} = \mathrm{d}\mathcal{N}_i \qquad \text{(change in the number of moles of } i\text{)}$$

Then, Eq. (10.71a) becomes

$$\boxed{\mathrm{a}S_{\text{irrev}} = -\frac{1}{T}\sum_i G_i\,\mathrm{d}\mathcal{N}_i} \tag{10.72}$$

Equation (10.72) is the final result expressing the infinitesimal increase in entropy caused by a nonequilibrium chemically reacting process to the corresponding infinitesimal nonequilibrium changes in the number of moles $d\mathcal{N}_i$. From the second law (see Sec. 10.6)

$$dS_{irrev} > 0$$

On the other hand, if the change in thermodynamic state is taking place through an infinite number of local equilibrium states, the process is reversible, and we have

$$dS_{irrev} = 0 \tag{10.73}$$

From Eq. (10.72) this implies an equilibrium relation between the $\mathcal{N}_i$ such that $\sum_i G_i d\mathcal{N}_i$ is zero for an equilibrium mixture. This provides a mechanism for calculating the equilibrium chemical composition of the mixture, as discussed in the next section.

With this, we have completely navigated the center column of our chapter road map in Fig. 10.2. We now move on to the right-hand column.

10.9 Composition of Equilibrium Chemically Reacting Mixtures: The Equilibrium Constant

Let us review what we have stated about an equilibrium chemically reacting mixture:

1) For any equilibrium thermodynamic system, any thermodynamic state variable is a function of any other *two* state variables, for example,

$$H = H(T, p)$$
$$E = E(T, v)$$
$$\text{etc.}$$

2) For an equilibrium chemically reacting mixture, the *chemical composition* is also a unique function of any two state variables, for example,

$$N_i = f_1(T, p) = f_2(T, v) = f_3(T, S)$$
$$X_i = f_3(T, p) = f_4(T, v) = f_5(T, S)$$
$$\text{etc.}$$

The second statement is certainly inferred by our discussion surrounding Fig. 9.12, where the ranges of dissociation and ionization are given in terms of temperature for a fixed pressure of 1 atm. In chemically reacting flows, it is particularly convenient to think in terms of the equilibrium chemical composition as given by the local T and p. Hence, the purpose of the present section is to answer the following question: for an equilibrium chemically reacting mixture at a given

p and T, what is the equilibrium chemical composition? The answer is derived from Eq. (10.72) applied to an equilibrium system where by definition $dS_{irrev} = 0$. Hence, for an equilibrium chemically reacting mixture,

$$\boxed{\sum_i G_i d\mathcal{N}i = 0} \tag{10.74}$$

To solve for the equilibrium composition, we will use Eq. (10.74) in a slightly different form, as developed next. A general statement of a chemical reaction is given by the following chemical equation:

$$\nu_i' A_1 + \nu_2' A_2 + \cdots + \nu_k' A_k \longrightarrow \nu_{k+1}' A_{k+1} + \nu_{k+2}' A_{k+2} + \cdots + \nu_j' A_j \tag{10.75}$$

where A_1, A_2, etc. denote different chemical species and ν_1', ν_2', etc. are the *stoichiometric mole numbers* for each chemical species. By definition, the chemical species, $A_1, A_2, \ldots, A_k$ on the left side of Eq. (10.75) are called the *reactants*, and $A_{k+1}, A_{k+2}, \ldots, A_j$ on the right side are called the *products*. Another convention is to write Eq. (10.75) in the form

$$0 = \sum_{i=1}^{j} \nu_i A_i \tag{10.76}$$

where ν_i is the stoichiometric mole number associated with species A_i and where ν_i is *negative for the reactants* and *positive for the products*. The symbol A_i is simply a symbol for the chemical species i. For example, consider the following chemical equation:

$$H_2 + O \longrightarrow OH + H \tag{10.77}$$

In the terms of the convention set by Eq. (10.76), for this chemical equation we have

$$\begin{aligned} A_1 &= H_2 & \nu_1 &= -1 \\ A_2 &= O & \nu_2 &= -1 \\ A_3 &= OH & \nu_3 &= 1 \\ A_4 &= H & \nu_4 &= 1 \end{aligned}$$

Another example is

$$\tfrac{1}{2} O_2 \longrightarrow O \tag{10.78}$$

where

$$\begin{aligned} A_1 &= O_2 & \nu_1 &= -\tfrac{1}{2} \\ A_2 &= O & \nu_2 &= 1 \end{aligned}$$

In this section, we will adopt the convention of writing a general chemical equation in the form of Eq. (10.76). Examine Eq. (10.76) more closely. Note that any change in the number of moles of species i, denoted by $d\mathcal{N}_i$, as a result of a specific chemical equation as represented by Eq. (10.76), must be proportional to ν_i. For example, if we had a chemically reacting mixture where Eq. (10.78) was the only reaction, then a 0.1 mole increase in O is accompanied by a 0.05 mole decrease in O_2; the changes in molar composition are in the ratio of $\nu_2 : \nu_1 = 1 : -\frac{1}{2}$. In general, for a system governed by Eq. (10.76), we can write

$$d\mathcal{N}_1 : d\mathcal{N}_2 : \cdots d\mathcal{N}_j = \nu_1 : \nu_2 \cdots \nu_j$$

Let the common proportionality constant be denoted by $d\xi$, such that

$$\frac{d\mathcal{N}_1}{\nu_1} = \frac{d\mathcal{N}_2}{\nu_2} = \cdots \frac{d\mathcal{N}_j}{\nu_j} = d\xi$$

or

$$\begin{aligned} d\mathcal{N}_1 &= \nu_1 \, d\xi \\ d\mathcal{N}_2 &= \nu_2 \, d\xi \\ &\vdots \\ d\mathcal{N}_j &= \nu_j \, d\xi \end{aligned} \tag{10.79}$$

A physical interpretation of ξ can be obtained as follows. Integrate Eq. (10.79) from a reference condition where $\xi = 0$ and $\mathcal{N}_i = \mathcal{N}_{i,\mathrm{ref}}$, to the condition where the corresponding quantities are ξ and $\mathcal{N}_i$,

$$\int_{\mathcal{N}_{i,\mathrm{ref}}}^{\mathcal{N}_i} d\mathcal{N}_i = \int_0^{\xi} \nu_i \, d\xi$$

or

$$\mathcal{N}_i - \mathcal{N}_{i,\mathrm{ref}} = \nu_i \xi \tag{10.80}$$

From Eq. (10.80), ξ can be seen as an index that describes the degree to which the chemical reaction has advanced from the reference condition. For example, when $\xi = 0$ the chemical composition is at the reference condition. When ξ is greater than zero, the reaction has taken place to the extent that $\mathcal{N}_i$ has changed from $\mathcal{N}_{i,\mathrm{ref}}$ to its current value $\mathcal{N}_i$. Hence, we can properly define ξ as the *degree of advancement* because it is an index of the advancement of the mole fractions caused by the chemical reaction given by Eq. (10.76) and where the quantitative amount of advancement is given by Eq. (10.80). In turn, the differential $d\xi$ is simply an infinitesimal change in the degree of advancement. Returning to Eq. (10.79), we can write for any species i,

$$d\mathcal{N}_i = \nu_i \, d\xi \tag{10.81}$$

where for a given reaction $d\xi$ is the same common factor for all species appearing in the reaction. Also, please note that our discussion from Eq. (10.75) to here does not depend in any way on the system being in equilibrium; the preceding relations and concepts apply in general to both equilibrium and nonequilibrium systems.

Let us now return to our discussion of equilibrium chemically reacting systems and to Eq. (10.74), which describes such systems. Substituting Eq. (10.81) into (10.74), we have

$$\sum_i G_i \nu_i \, \mathrm{d}\xi = 0 \tag{10.82}$$

Because $\mathrm{d}\xi$ is the same for all species in a given chemical reaction, it is a constant value in Eq. (10.82). Hence, from Eq. (10.82),

$$\mathrm{d}\xi \left(\sum_i G_i \nu_i \right) = 0$$

or

$$\boxed{\sum_i \nu_i G_i = 0} \tag{10.83}$$

Equation (10.83) is an alternate form of Eq. (10.74) and is therefore also a *condition for equilibrium.* Moreover, ξ has dropped out of Eq. (10.83); the concept of the degree of advancement was useful for the derivation of Eq. (10.83), but we will have no need for it in our future discussions.

The purpose of this section is to obtain a procedure for calculating the composition of an equilibrium chemically reacting mixture. Equation (10.83) leads to that procedure, after the following development. From the definition given in Eq. (10.62) we write, for species i,

$$G_i = H_i - TS_i \tag{10.84}$$

where G_i, H_i, and S_i are the Gibbs free energy, enthalpy, and entropy of species i per mole of i. Substituting Eq. (10.59) for S_i into Eq. (10.84), we have

$$G_i = H_i - T\left[\int_{T_{\mathrm{ref}}}^{T} C_{pi} \frac{\mathrm{d}T}{T} - \mathscr{R} \, \ln \frac{p_i}{p_{\mathrm{ref}}} + S_{i,\mathrm{ref}} \right] \tag{10.85}$$

For a moment, assume that p_i in Eq. (10.85) is one atmosphere, and let $G_i^{p_i=1}$ be the value of G_i evaluated at $p_i = 1$ atm. Then, from Eq. (10.85),

$$G_i^{p_i=1} = H_i - T\left[\int_{T_{\mathrm{ref}}}^{T} C_{p_i} \frac{\mathrm{d}T}{T} + \mathscr{R} \, \ln p_{\mathrm{ref}} + S_{i,\mathrm{ref}} \right] \tag{10.86}$$

Combining Eqs. (10.85) and (10.86), we have

$$G_i = G_i^{p_i=1} + \mathscr{R}T \, \ln p_i \tag{10.87}$$

Substituting Eq. (10.87) into the condition for equilibrium given by Eq. (10.83), we have

$$\sum_i \nu_i G_i = \sum_i \nu_i (G_i^{p_i=1} + \mathscr{R}T \ \ell n \ p_i) = 0$$

$$\sum_i \nu_i G_i^{p_i=1} + \mathscr{R}T \sum_i \nu_i \ \ell n \ p_i = \sum_i \nu_i G_i^{p_i=1} + \mathscr{R}T \sum_i \ \ell n \ p_i^{\nu_i} = 0$$

or

$$\sum_i \ \ell n \ p_i^{\nu_i} = -\sum_i \nu_i \frac{G_i^{p_i=1}}{\mathscr{R}T}$$

or

$$\prod_i p_i^{\nu_i} = \exp\left(-\sum_i \nu_i \frac{G_i^{p_i=1}}{\mathscr{R}T}\right) \tag{10.88}$$

Consider the physical meaning of $\sum_i \nu_i G_i^{p_i=1}$, with Eq. (10.76) in mind. With ν_i negative for the reactants and positive for the products, $\sum_i \nu_i G_i^{p_i=1}$ is simply the Gibbs free energy of the products minus the Gibbs free energy of the reactants for the given chemical equation, with all species evaluated at 1 atm pressure. Denote this difference by $\Delta G^{p=1}$ where, by definition,

$$\Delta G^{p=1} \equiv \sum_i \nu_i G_i^{p_i=1} = (G^{p=1} \text{ for products}) - (G^{p=1} \text{ for reactants}) \tag{10.89}$$

For example, for the reaction

$$OH + H_2 \longrightarrow H_3O + H$$

the change in Gibbs free energy at 1 atm is given by

$$\Delta G^{p=1} = G_{H_2O}^{p=1} + G_H^{p=1} - G_{OH}^{p=1} - G_{H_2}^{p=1}$$

and for the reaction

$$\tfrac{1}{2}O_2 \longrightarrow O$$

we have

$$\Delta G^{p=1} = G_O^{p=1} - \tfrac{1}{2} G_{O_2}^{p=1}$$

Inserting the definition given by Eq. (10.89) into (10.88), we have

$$\boxed{\prod_i p_i^{\nu_i} = e^{-\Delta G^{p=1}/\mathscr{R}T}} \tag{10.90}$$

In Eq. (10.90), $\Delta G^{p=1}$ depends only on T. Hence,

$$e^{-\Delta G^{p=1}/\mathscr{R}T} = f(T) \equiv K_p(T) \tag{10.91}$$

where $K_p(T)$ is defined as the equilibrium constant for the given chemical reaction. Hence, Eq. (10.90) becomes

$$\boxed{\prod_i p_i^{\nu_i} = K_p(T)} \tag{10.91a}$$

Emphasis is made that the equilibrium constant in Eq. (10.91) is strictly a function of temperature only. For a given chemical reaction, $K_p(T)$ can sometimes be obtained from experiment and can always be calculated from statistical thermodynamics, as discussed in Chapter 11.

Equation (10.91a) is the crux of the present section. It is a form of a general principle called the *law of mass action*, which essentially ensures the preservation of total mass during a chemical reaction. With it, we can establish a method for the calculation of the equilibrium composition of a chemically reacting mixture, as follows.

Assume that we have a system of chemically reacting gases in equilibrium at a given p and T. For clarity, it is best to consider a specific case. Let us consider the combustion chamber of a rocket engine, where H_2 is injected as the fuel and O_2 is injected as the oxidizer. The H_2 and O_2 are injected at a given ratio to each other, that is, at a given fuel/oxidizer ratio. In the combustion chamber, the H_2 and O_2 will chemically react through numerous different chemical equations. Assume that the products of combustion finally form an equilibrium chemically reacting system at a specific p and T. Assume that we know p and T by some independent means. Furthermore, assume that the products of combustion form an equilibrium chemically reacting mixture containing the following species: H_2, H, O_2, O, OH, and H_2O.

Question: What amounts of these species are present in the system in equilibrium at the given p and T?

To answer this question, consider the following chemical equations involving the mixture species, along with the definitions of the respective equilibrium constants from Eq. (10.91).

$$\tfrac{1}{2}H_2 \longrightarrow H: \qquad \frac{p_H}{\sqrt{p_{H_2}}} = K_{p,1} \tag{10.92}$$

$$\tfrac{1}{2}O_2 \longrightarrow O: \qquad \frac{p_O}{\sqrt{p_{O_2}}} = K_{p,2} \tag{10.93}$$

$$H_2 + O \longrightarrow OH + H \qquad \frac{p_{OH}\,p_H}{p_{H_2}\,p_O} = K_{p,3} \tag{10.94}$$

$$OH + H \longrightarrow H_2O \qquad \frac{p_{H_2O}}{p_{OH}\,p_H} = K_{p,4} \tag{10.95}$$

Equations (10.92–10.95) are relations for the partial pressures that must be satisfied for this mixture. These equations constitute a set of four equations with six unknowns. To close the system, it is initially tempting to add two more chemical equations, with their associated equilibrium constant expressions. However, this would be inappropriate because the chemically reacting system, in addition to obeying Eqs. (10.92–10.95), must satisfy two constraints. The first of these is Dalton's law of partial pressures, expressed by Eq. (10.12), that is, $p = \sum_i p_i$. For the given system, this is

$$p_{H_2} + p_H + p_{O_2} + p_O + p_{OH} + p_{H_2O} = p \tag{10.96}$$

where p is the given pressure of the mixture, that is, the total pressure of the mixture. The second constraint is associated with the number of hydrogen and oxygen nuclei in the mixture, denoted by N_H and N_O, respectively. Because we are not dealing with nuclear reactions, N_H and N_O remain constant in the mixture, and their ratio N_H/N_O is a *known* quantity, fixed by the given fuel/oxidizer ratio at which the hydrogen and oxygen are being pumped into the combustion chamber. For convenience, let us assume a unit mass for the equilibrium chemically reacting mixture; hence, N_H and N_O represent the number of nuclei per unit mass of mixture. Also, note that each H_2 molecule contributes two hydrogen nuclei to the mixture, each H_2O molecule contributes two hydrogen nuclei, and the OH molecule and H atom contribute one hydrogen nucleus each. Therefore, to count the total number of hydrogen nuclei per unit mass of mixture, we write

$$N_H = N_A(2\eta_{H_2} + \eta_H + 2\eta_{H_2O} + \eta_{OH})$$

where N_A is Avogadro's number (number of particles per mole) and η_i is the familiar mole-mass ratio defined in Eq. (10.20), that is, the number of moles of species i per unit mass of mixture. Similarly, to count the total number of oxygen nuclei per unit mass of mixture, we have

$$N_O = N_A(2\eta_{O_2} + \eta_O + \eta_{H_2O} + \eta_{OH})$$

Hence, the ratio N_H/N_O is given by

$$\frac{N_H}{N_O} = \frac{N_A(2\eta_{H_2} + \eta_H + 2\eta_{H_2O} + \eta_{OH})}{N_A(2\eta_{O_2} + \eta_O + \eta_{H_2O} + \eta_{OH})} \tag{10.97}$$

However, when Eq. (10.20) in the form of

$$\eta_i = \frac{p_i v}{\mathcal{R}T}$$

is substituted into Eq. (10.97), we obtain

$$\frac{N_H}{N_O} = \frac{N_A(2p_{H_2} + p_H + 2p_{H_2O} + p_{OH})}{N_A(2p_{O_2} + p_O + 2p_{H_2O} + p_{OH})} \tag{10.98}$$

Important: Equations (10.92–10.96) and (10.98) constitute six algebraic equations for the six unknown partial pressures. This determines the chemical composition of the equilibrium chemically reacting mixture at the given T and p, which was the objective of this section.

In Eqs. (10.92–10.95), the values of $K_p(T)$ for the given temperature can be obtained from the literature; [148] and [149] are excellent sources for such data. Moreover, $K_p(T)$ can be calculated directly from the results of statistical thermodynamics, to be discussed in Chapter 11. As a reminder, in Eq. (10.96), p is the given pressure of the mixture, and in Eq. (10.98) N_H/N_O is a known ratio determined from the fuel/oxidizer ratio. Also, the chemical equations chosen in Eqs. (10.92–10.95) must be independent, that is, one cannot be obtained by adding or subtracting the others; this is to ensure that the system of the algebraic equations is an independent system. Outside of this consideration, the choice of the chemical reactions used for Eqs. (10.92–10.95) is rather arbitrary, just as long as they involve the relevant species, and the associated equilibrium constants can be obtained. Implicit in the preceding calculation is the proper choice of the relevant chemical species present in the mixture. In the preceding example, we *assumed* that the products of combustion were primarily H_2, H, O_2, O, OH, and H_2O. In such an equilibrium calculation, we must be certain to assume all of the relevant species that might be present; for example, if we had not included OH in the preceding calculation, the results would be different, and they would be deficient. There is no routine method that allows you to choose all of the relevant species automatically. You have to make a good educated guess, based on prior knowledge of the system. A safe approach is to assume all species that are made up of all possible combinations of the various elements present; if many of these assumed species are negligible, then they will show up in the calculation as trace species only.

In summary, in the present section, we have done the following:

1) We have developed a method for calculating the chemical composition of an equilibrium chemically reacting mixture. (Note that we have obtained the equilibrium composition in terms of the partial pressures p_i; however, once the p_i are known, we can obtain the composition in whatever other terms we wish, as explained in Sec. 10.3.)

2) We have demonstrated that this equilibrium composition is a unique function of p and T because a) Eqs. (10.92–10.95) require a knowledge of the K_p, which are functions of T only, and b) Eq. (10.96) requires a knowledge of p.

We will defer further considerations of the calculation of equilibrium chemical compositions until Chapter 11.

10.10 Heat of Reaction

An important term in chemical thermodynamics is the heat of reaction, to be defined in the present section. To introduce this concept, let us consider a specific example, as follows. Consider the chemical reaction given by

$$H_2 + \tfrac{1}{2}O_2 \longrightarrow OH + H \tag{10.99}$$

Assume that we have a system made up of one mole of H_2 and a half-mole of O_2, at a reference temperature T_{ref}. These are the *reactants*. The enthalpy of the reactants is

$$(H_{H_2} + \tfrac{1}{2}H_{O_2})^{T_{ref}} = \text{enthalpy of the reactants at } T_{ref}$$

Now allow the reactants to form the products OH and H as shown in Eq. (10.99); these represent the *products* of the reaction. Furthermore, carry out the reaction at constant pressure. Finally, extract or add enough heat from or to the system so that the products are also at the reference temperature T_{ref}. Then, the enthalpy of the products is given as

$$(H_{OH} + H_H)^{T_{ref}} = \text{enthalpy of the products at } T_{ref}$$

By definition, the heat that was added to or subtracted from the preceding system is called the *heat of reaction* for the chemical reaction at the reference temperature T_{ref}. In turn, for the assumed constant pressure, from the first law in the form of Eq. (10.34), we know that the heat added or subtracted is equal to the change in enthalpy. Therefore, we will consider the heat of reaction at a given reference temperature T_{ref} for a given chemical reaction to be denoted by $\Delta H_R^{T_{ref}}$ and to be defined by

$$\begin{aligned}\Delta H_R^{T_{ref}} &= (\text{enthalpy of the products at } T_{ref}) \\ &\quad - (\text{enthalpy of the reactants at } T_{ref})\end{aligned}$$

For the chemical reaction given by Eq. (10.99),

$$\Delta H_R^{T_{ref}} = (H_{OH} + H_H - H_{H_2} - \tfrac{1}{2}H_{O_2})^{T_{ref}} \tag{10.100}$$

(*Note*: Keep in mind in the preceding discussion that H_i denotes the enthalpy of species i per mole of i.)

In general, consider the generic chemical reaction given by Eq. (10.76), repeated here:

$$0 = \sum_{i=1}^{j} \nu_i A_i$$

The heat of reaction for this chemical equation at the reference temperature is, by definition,

$$\boxed{\Delta H_R^{T_{ref}} = \sum_i \nu_i H_{A_i}^{T_{ref}}} \tag{10.101}$$

The values of $\Delta H_R^{T_{\text{ref}}}$ for various chemical reactions can be constructed from the data given in [148] and [149] and can also be calculated from the results of statistical thermodynamics discussed in Chapter 11. The concept of the heat of reaction is very important in evaluating the chemical energy changes that take place in chemically reacting flowfields, as we will see in subsequent chapters.

10.11 Summary and Comments

Referring to our road map in Fig. 1.24, the present chapter represents just a start to our introduction to some basic effects from physical chemistry. In particular, we have discussed the macroscopic picture painted by classical thermodynamics. We have examined the different categories of gases: 1) calorically perfect gases, 2) thermally perfect gases, 3) chemically reacting mixtures of perfect gases, and 4) real gases. We have presented a number of different forms for the perfect-gas equation of state, applicable to chemically reacting mixtures as well as to individual species. Make certain to review carefully the earlier sections of this chapter so that you have these details well in mind.

The two basic laws from classical thermodynamics are as follows.

1) First law of thermodynamics, with alternative forms:

$$\delta q + \delta w = \mathrm{d}e \tag{10.29}$$

$$\delta q = \mathrm{d}e + p\,\mathrm{d}v \tag{10.32}$$

$$\delta q = \mathrm{d}h + v\,\mathrm{d}p \tag{10.34}$$

2) Second law of thermodynamics:

$$\mathrm{d}s = \frac{\delta q}{T} + \mathrm{d}s_{\text{irrev}} \tag{10.46}$$

where $\mathrm{d}s_{\text{irrev}} > 0$. Hence,

$$\mathrm{d}s > \frac{\delta q}{T} \tag{10.47}$$

or, for an adiabatic process,

$$\mathrm{d}s > 0 \tag{10.48}$$

All of the preceding results apply to any general gas, real or perfect, reacting or nonreacting.

For a chemically reacting mixture of perfect gases, we have the following results:

$$S = \sum_i X_i \left[\int_{T_{\text{ref}}}^{T} C_{p_i} \frac{\mathrm{d}T}{T} - \mathscr{R}\,\ell n \frac{p_i}{p_{\text{ref}}} \right] + S_{\text{ref}} \tag{10.60}$$

and

$$dS_{\text{irrev}} = -\frac{1}{T}\sum_i G_i \, d\mathcal{N}_i \tag{10.72}$$

where $G_i = H_i - TS_i$. Using the fact that $dS_{\text{irrev}} = 0$ for an equilibrium system, we obtained

$$\prod_i p_i^{\nu_i} = K_p(T) \tag{10.91}$$

where $K_p(T)$ is the equilibrium constant, a function of T only, where K_p is given by

$$K_p(T) = e^{-\Delta G^{p=1}/\mathcal{R}T} \tag{10.90}$$

Along with other relations, the equilibrium constant allows the calculation of the equilibrium chemical composition. We have seen that the equilibrium composition is a unique function of T and p for the mixture.

The heat of formation at a reference temperature T_{ref} is defined for a given chemical reaction as

$$\Delta H_R^{T_{\text{ref}}} = (\text{enthalpy of products at } T_{\text{ref}}) - (\text{enthalpy reactants at } T_{\text{ref}})$$

For the generic chemical reactions given by

$$0 = \sum_i \nu_i A_i$$

we have

$$\Delta H_R^{T_{\text{ref}}} \equiv \sum_i \nu_i H_{A_i}^{T_{\text{ref}}} \tag{10.101}$$

Finally, we note that classical thermodynamics must treat the equation of state as an empirically defined relation, or as a postulate; it cannot be derived from first principles in classical thermodynamics. Moreover, classical thermodynamics cannot provide values of K_p from first principles; they must be obtained from measurement. In contrast, statistical thermodynamics *can* provide both the equation of state and values of K_p from first principles; this is the subject of the next chapter.

Note also that in the present chapter we have derived an expression for entropy of a chemically reacting mixture, given by Eq. (10.60). However, we have not obtained analogous expressions for enthalpy H or internal energy E for a chemically reacting mixture. This is intentional because for a proper interpretation of H and E we need to examine the principles of statistical thermodynamics. Hence, the calculation of H and E are deferred until Chapter 11.

Problems

10.1 At a given T and p, the composition of high-temperature air is given by $p_O = 0.163$ atm, $p_{O_2} = 0.002$ atm, $p_{N_2} = 0.33$ atm, and $p_N = 0.005$ atm. Calculate for each species: (a) mole fraction, (b) mass fraction, and (c) mole-mass ratio. Also, obtain the molecular weight and the specific gas constant for the mixture.

10.2 Consider the combustion chamber of a rocket engine using liquid H_2 and liquid O_2 as the fuel and oxidizer. For a fuel/oxidizer ratio of 0.1 by mass and $p = 10$ atm and $T = 3500$ K, calculate the equilibrium chemical composition of the gas in the combustion chamber. Assume the following species are present: H_2, H, O_2, O, OH, and H_2O. (The equilibrium constants are intentionally not given here in order to give you the opportunity to look them up in any of the standard references and thus become familiar with such references.)

10.3 Derive the following relation between the equilibrium constant and the heat of reaction:

$$\frac{\mathrm{d}\, \ell n K_p}{\mathrm{d}T} = \frac{\Delta H_R}{\mathcal{R} T^2}$$

This equation is called Van't Hoff's equation.

10.4 Consider a chemically reacting mixture. Let N_i and N denote the number of particles of species i and the total number of particles, respectively. The definition of mole fraction is then $\chi_i = N_i/N$. Now let the mixture be perturbed slightly (say, by a slight change in p and/or T). There will be corresponding changes in N_i and N, given by $\mathrm{d}N_i$ and $\mathrm{d}N$, respectively. Prove that, although $N_i/N = \chi_i$,

$$\frac{\mathrm{d}N_i}{N} \neq \mathrm{d}\chi_i$$

Hence, there must be some mathematical caution in using relations involving $\mathrm{d}N_i$ and $\mathrm{d}\chi_i$ for a chemically reacting gas.

10.5 There is a standard NASA computer program that uses a Gibbs free energy minimization technique for the calculation of equilibrium chemical compositions. Just on the basis of this label, how would you think such a technique is related to our calculation described in the present chapter?

11
Elements of Statistical Thermodynamics

$S = k \log W$

Inscription on the tombstone for Ludwig Boltzmann,
Vienna, Austria

Chapter Preview

Once again we use the phrase: This is where the rubber meets the road. This time it is for the calculation of the thermodynamic properties of an equilibrium chemically reacting mixture from first principles. Using some of the basic fundamentals from the previous chapter, we now heavily overlay some important results from quantum mechanics and quantum statistical mechanics, and we obtain equations from which we can directly calculate the internal energy, enthalpy, entropy, and any other thermodynamic variable for an equilibrium high-temperature gas. This is the essence of the discipline of statistical thermodynamics. These equations are the basis of modern computer subroutines that generate from scratch the equilibrium high-temperature thermodynamic properties and chemical composition used for numerical solutions of high-temperature chemically reacting flow. So this is important stuff. It is also intellectually beautiful stuff, put together by powerful minds. One of those powerful minds was that of Ludwig Boltzmann, a famous physicist and mathematician from the late 19th century. The simple inscription on Boltzmann's tombstone, given in the preceding quotation, is an equation that is the kingpin of statistical thermodynamics. You will find this equation in the middle of this chapter. The fact that it is the only inscription on the tombstone of such an important human being attests to its importance and to the importance of statistical thermodynamics in the physics of the modern world. It also attests to the reason why you should give this chapter your closest attention. In the process, allow yourself to enjoy the intellectual beauty of the flow of ideas and ingenious mental constructions in this chapter. Then you will be ready to take the next steps into high-temperature gas dynamics in the following chapters.

11.1 Introduction

In Chapter 10 we discussed some aspects of the thermodynamics of chemically reacting gases from a classical point of view. Note that in Chapter 10 we obtained relations *between* various thermodynamic properties; we did not explicitly calculate *values* of these properties from first principles. For example, Eq. (10.43) is a general relationship between c_p, c_v, and R for a chemically reacting mixture of perfect gases; it does not enable us to calculate a value for either c_p or c_v. Similarly, Eq. (10.60) is an expression for S as a function of T and the p_i. However, to obtain an actual number for S, we need the values of C_{p_i}, which cannot be obtained from classical thermodynamic theory. For such properties, classical thermodynamics must rely on experimental data.

In contrast, the results of statistical thermodynamics do allow the calculation of thermodynamic properties from first principles, as long as we are dealing with equilibrium systems. The purpose of this chapter is to develop such results. In turn, we will see that these results are very accurate and extremely practical in the analysis of high-temperature flows.

To elaborate, an essential ingredient of any high-temperature flowfield analysis is the knowledge of the thermodynamic properties of the gas. For example, consider again the flowfield over the X-24 shown in Fig. 8.17. Assume that the gas is in local thermodynamic and chemical equilibrium (concepts to be more fully examined later). The unknown flowfield variables, and how they can be obtained, are itemized as follows:

$$\left.\begin{array}{l}\rho = \text{density}\\ V = \text{velocity}\\ h = \text{enthalpy}\end{array}\right\} \quad \text{Obtained from a simultaneous solution of the continuity, momentum, and energy equations}$$

$$\left.\begin{array}{l}T = T(\rho, h)\\ p = p(\rho, h)\end{array}\right\} \quad \text{Obtained from the equilibrium thermodynamic properties of high-temperature air}$$

In the preceding, conceptually see that two thermodynamic variables ρ and h are obtained from the flowfield conservation equations and that the remaining thermodynamic variables T, p, e, s, etc., can be obtained from a knowledge of ρ and h. In general, for a gas in equilibrium, any two thermodynamic state variables uniquely define the complete thermodynamic state of the gas. The question posed here is that, given two thermodynamic state variables in an equilibrium high-temperature gas, *how* do we obtain values of the remaining state variables? There are two answers. One is to *measure* these properties from experiment. However, it is very difficult to carry out accurate experiments on gases at temperatures above a few thousand degrees; such temperatures are usually achieved in the laboratory for only short periods of time in devices such as shock tubes or by pulsed laser radiation absorption. The other answer is to *calculate* these properties. Fortunately, the powerful discipline of statistical mechanics developed over the last century, along with the advent of quantum mechanics in the early 20th century, gives us a relatively quick and extremely accurate method of calculating

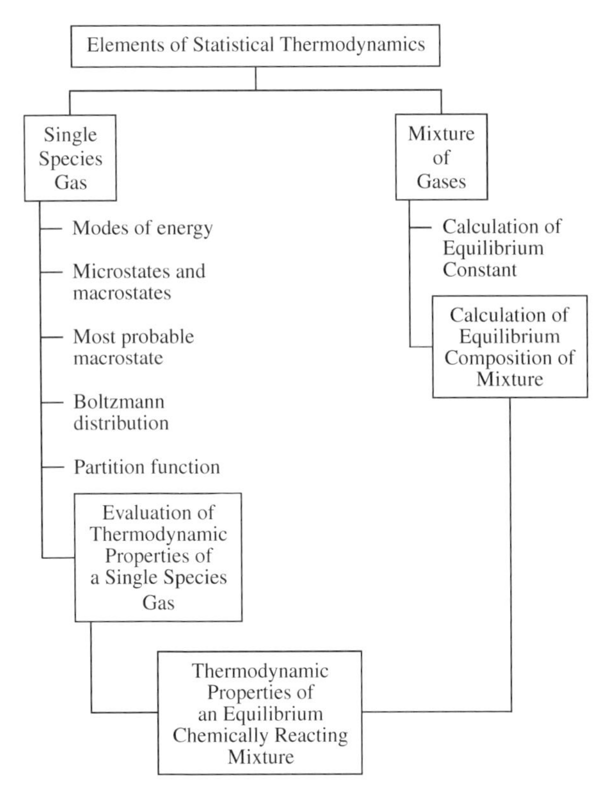

Fig. 11.1 Road map for Chapter 11.

equilibrium thermodynamic properties of high-temperature gases. These concepts form the basis of *statistical thermodynamics*, the elements of which will be developed and used in the following sections.

The road map for this chapter is given in Fig. 11.1. Our journey consists of two roads that ultimately come together at the end. First, we proceed down the left-hand column, developing the statistical thermodynamics of a single-species gas, culminating in practical equations for the thermodynamic properties of a single-species gas. Then we move to the right-hand column and develop the statistical thermodynamic expressions for equilibrium constants that allow the calculation of the chemical composition of an equilibrium reacting mixture. Finally, we combine the results of both columns to obtain the thermodynamic properties an equilibrium chemically reacting mixture.

11.2 Microscopic Description of Gases

In the development of statistical thermodynamics, we concentrate on the microscopic picture of a gas, that is, we assume the gas consists of a large number of individual molecules, and we examine the nature of these molecules. For example, a molecule is a collection of atoms bound together by a rather

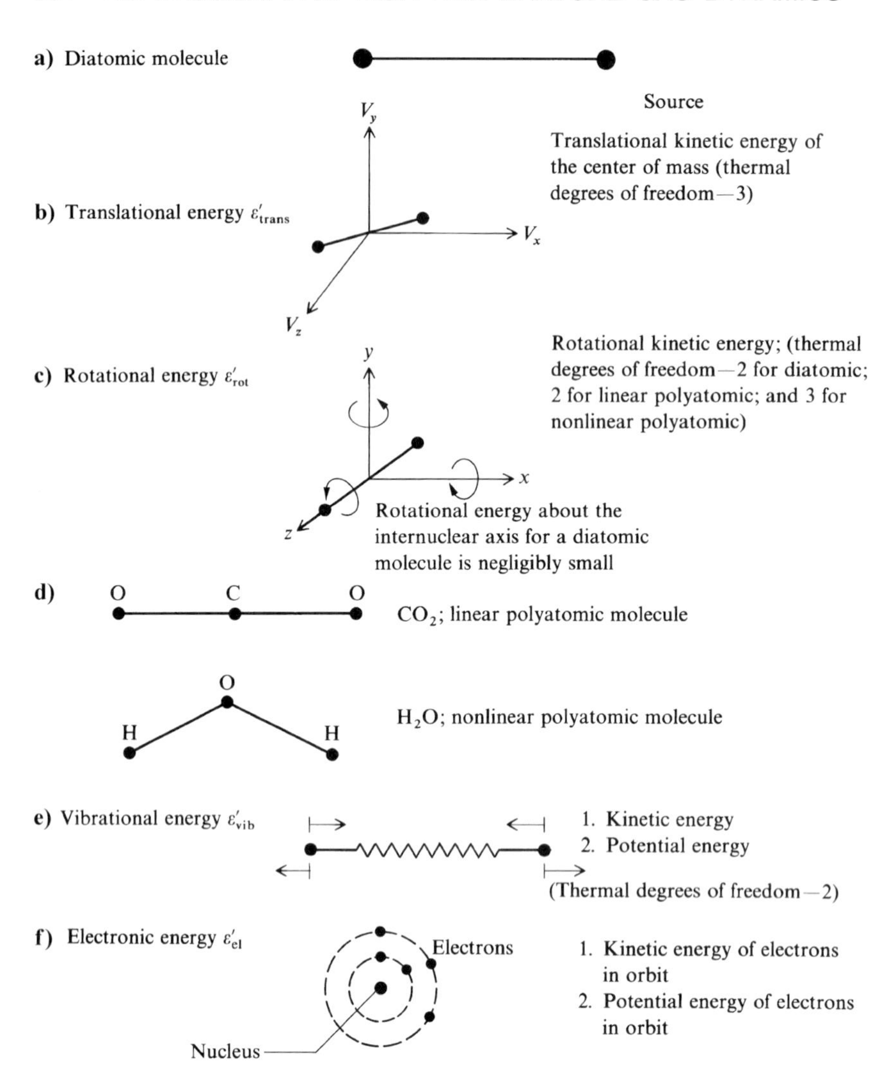

Fig. 11.2 Modes of molecular energy.

complex intramolecular force. A simple concept of a diatomic molecule (two atoms) is the "dumbbell" model sketched in Fig. 11.2a. The molecule has several modes (forms) of energy, as follows:

1) It is moving through space, and hence it has translational energy ε'_{trans}, as sketched in Fig. 11.2b. The source of this energy is the translational kinetic energy of the center of mass of the molecule. Because molecular translational velocity can be resolved into three components (such as V_x, V_y, and V_z in the xyz Cartesian space shown in Fig. 11.2b), the molecule is said to have three geometric degrees of freedom in translation. Because motion along each coordinate direction contributes to the total kinetic energy, the molecule is also said to have three thermal degrees of freedom.

2) It is rotating about the three orthogonal axes in space, and hence it has *rotational energy* $\varepsilon'_{\mathrm{rot}}$, as sketched in Fig. 11.2c. The source of this energy is the rotational kinetic energy associated with the molecule's rotational velocity and its moment of inertia. However, for the diatomic molecule shown in Fig. 11.2c, the moment of inertia about the internuclear axis (the z axis) is very small, and therefore the rotational kinetic energy about the z axis is negligible in comparison to rotation about the x and y axes. Therefore, the diatomic molecule is said to have only two geometric as well as two thermal degrees of freedom. The same is true for a linear polyatomic molecule such as CO_2 shown in Fig. 11.2d. However, for nonlinear molecules, such as H_2O, also shown in Fig. 11.2d, the number of geometric (and thermal) degrees of freedom in rotation are three.

3) The atoms of the molecule are vibrating with respect to an equilibrium location within the molecule. For a diatomic molecule, this vibration is modeled by a spring connecting the two atoms, as illustrated in Fig. 11.2e. Hence the molecule has *vibrational energy* $\varepsilon'_{\mathrm{vib}}$. There are two sources of this vibrational energy: the kinetic energy of the linear motion of the atoms as they vibrate back and forth, and the potential energy associated with the intramolecular force (symbolized by the spring). Hence, although the diatomic molecule has only one geometric degree of freedom (it vibrates along one direction only, namely, that of the internuclear axis), it has *two* thermal degrees of freedom because of the contribution of both kinetic and potential energy. For polyatomic molecules, the vibrational motion is more complex, and numerous fundamental vibrational modes can occur, with a consequent large number of degrees of freedom.

4) The electrons are in motion about the nucleus of each atom constituting the molecule, as sketched in Fig. 11.2f. Hence, the molecule has electronic energy $\varepsilon'_{\mathrm{el}}$. There are two sources of electronic energy associated with each electron: kinetic energy because of its translational motion throughout its orbit about the nucleus, and potential energy because of its location in the electromagnetic force field established principally by the nucleus. Because the overall electron motion is rather complex, the concepts of geometric and thermal degrees of freedom are usually not useful for describing electronic energy.

Therefore, we see that the total energy of a molecule ε' is the sum of its translational, rotational, vibrational, and electronic energies.

For molecules:

$$\varepsilon' = \varepsilon'_{\mathrm{trans}} + \varepsilon'_{\mathrm{rot}} + \varepsilon'_{\mathrm{vib}} + \varepsilon'_{\mathrm{el}}$$

For a single atom, only the translational and electronic energies exist.

For atoms:

$$\varepsilon' = \varepsilon'_{\mathrm{trans}} + \varepsilon'_{\mathrm{el}}$$

The results of quantum mechanics have shown that each of the preceding energies is *quantized*, that is, they can exist only at certain discrete values, as schematically shown in Fig. 11.3. This is a dramatic result. Intuition, based on our personal observations of nature, would tell us that at least the translational and rotational energies could be any value chosen from a continuous range of values (i.e., the complete real number system). However, our daily experience

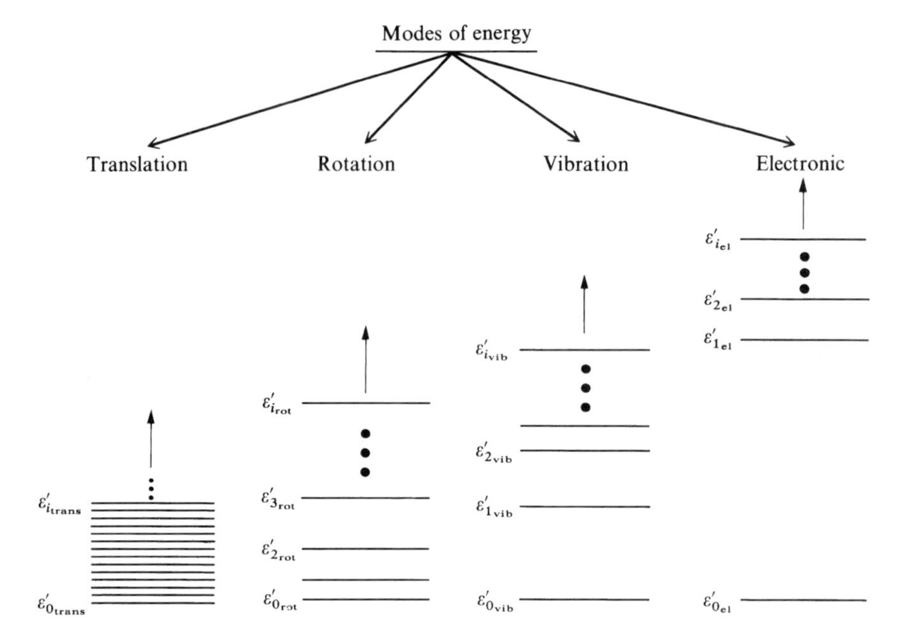

Fig. 11.3 Schematic of energy levels for the different molecular energy modes.

deals with the macroscopic, not the microscopic world, and we should not always trust our intuition when extrapolated to the microscopic scale of molecules. A major benefit of quantum mechanics is that it correctly describes microscopic properties, some of which are contrary to intuition. In the case of molecular energy, *all* modes are quantized, even the translational mode. These quantized energy levels are symbolized by the ladder-type diagram shown in Fig. 11.3, with the vertical height of each level as a measure of its energy. Taking the vibrational mode for example, the lowest possible vibrational energy is symbolized by $\varepsilon'_{0_{\text{vib}}}$. The next allowed quantized value is $\varepsilon'_{1_{\text{vib}}}$, then $\varepsilon'_{2_{\text{vib}}}, \ldots, \varepsilon'_{i_{\text{vib}}}, \ldots$. The energy of the ith vibrational energy level is $\varepsilon'_{i_{\text{vib}}}$, and so forth. Note that, as illustrated in Fig. 11.3, the spacing between the translational energy levels is very small, and if we were to look at this translational energy level diagram from across the room, it would look almost continuous. The spacings between rotational energy levels are much larger than between the translational energies; moreover, the spacing between two adjacent rotational levels increases as the energy increases (as we go up the ladder in Fig. 11.3). The spacings between vibrational levels are much larger than between rotational levels; also, contrary to rotation, adjacent vibrational energy levels become more closely spaced as the energy increases. Finally, the spacings between electronic levels are considerably larger than between vibrational levels and the difference between adjacent electronic levels decreases at higher electronic energies. The quantitative calculation of all of these energies will be given in Sec. 11.7.

Again, examining Fig. 11.3, note that the lowest allowable energies are denoted by $\varepsilon'_{0_{\text{trans}}}$, $\varepsilon'_{0_{\text{rot}}}$, $\varepsilon'_{0_{\text{vib}}}$, and $\varepsilon'_{0_{\text{el}}}$. These levels are defined as the *ground state* for the molecule. They correspond to the energy that the molecule would have if the gas were theoretically at a temperature of absolute zero; hence, the values are also called the *zero-point energies* for the translational, rotational, vibrational, and electronic modes, respectively. It will be shown in Sec. 11.7 that the rotational zero-point energy is precisely zero, whereas the zero-point energies for translation, vibration, and electronic motion are not. This says that, if the gas were theoretically at absolute zero, the molecules would still have some finite translational motion (albeit very small) as well as some finite vibrational motion. Moreover, it only makes common sense that some electronic motion should theoretically exist at absolute zero, or otherwise the electrons would fall into the nucleus and the atom would collapse. Therefore, the total zero-point energy for a molecule is denoted by ε_0', where

$$\varepsilon_0' = \varepsilon_{0_{\text{trans}}}' + \varepsilon_{0_{\text{vib}}}' + \varepsilon_{0_{\text{el}}}'$$

recalling that $\varepsilon_{0_{\text{rot}}}' = 0$.

It is common to consider the energy of a molecule as measured above its zero-point energy. That is, we can define the translational, rotational, vibrational, and electronic energies all *measured above the zero-point energy as* $\varepsilon_{j_{\text{trans}}}$, $\varepsilon_{k_{\text{rot}}}$, $\varepsilon_{l_{\text{vib}}}$, and $\varepsilon_{m_{\text{el}}}$, respectively, where

$$\begin{aligned}
\varepsilon_{j_{\text{trans}}} &= \varepsilon_{j_{\text{trans}}}' - \varepsilon_{0_{\text{trans}}}' \\
\varepsilon_{k_{\text{rot}}} &= \varepsilon_{k_{\text{rot}}}' \\
\varepsilon_{l_{\text{vib}}} &= \varepsilon_{l_{\text{vib}}}' - \varepsilon_{0_{\text{vib}}}' \\
\varepsilon_{m_{\text{el}}} &= \varepsilon_{m_{\text{el}}}' - \varepsilon_{0_{\text{el}}}'
\end{aligned}$$

(Note that the *unprimed* values denote energy measured *above* the zero-point value.) In light of the preceding, we can write the *total* energy of a molecule as ε_i', where

$$\varepsilon_i' = \underbrace{\varepsilon_{j_{\text{trans}}} + \varepsilon_{k_{\text{rot}}} + \varepsilon_{l_{\text{vib}}} + \varepsilon_{m_{\text{el}}}}_{\substack{\text{All are measured above the} \\ \text{zero-point energy; thus, all} \\ \text{are equal to zero at } T = 0 \text{ K.}}} + \underbrace{\varepsilon_0'}_{\substack{\text{This represents zero-point} \\ \text{energy, a fixed quantity for} \\ \text{a given molecular species that} \\ \text{is equal to the energy of the} \\ \text{molecule at absolute zero.}}}$$

For an atom, the total energy can be written as

$$\varepsilon_i' = \varepsilon_{j_{\text{trans}}} + \varepsilon_{m_{\text{el}}} + \varepsilon_0'$$

If we examine a single molecule at some given instant in time, we would see that it simultaneously has a zero-point energy ε_0' (a fixed value for a given molecular species), a quantized electronic energy measured above the zero-point $\varepsilon_{m_{\text{el}}}$, a quantized vibrational energy measured above the zero point $\varepsilon_{l_{\text{vib}}}$, and so forth for rotation and translation. The total energy of the molecule at this given instant is

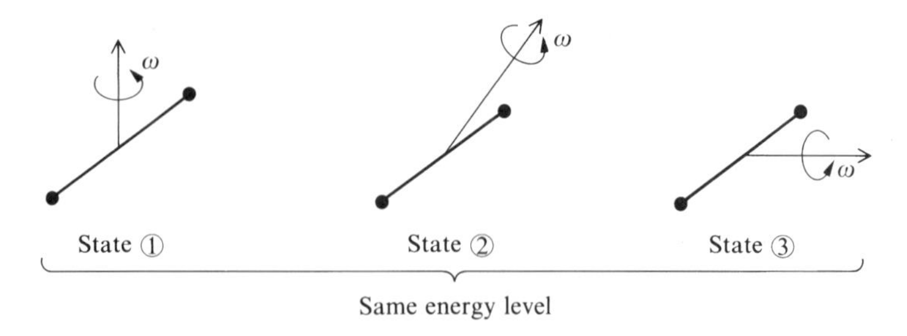

Fig. 11.4 Illustration of different energy states for the same energy level.

ε_i'. Because ε_i' is the sum of individually quantized energy levels, then ε_i' itself is quantized. Hence, the allowable *total* energies can be given on a single energy level diagram, where ε_0', ε_1', $\varepsilon_2', \ldots, \varepsilon_i' \ldots$ are the quantized values of the total energy of the molecule.

In the preceding paragraphs, we have gone to some length to define and explain the significance of molecular energy *levels*. In addition to the concept of an energy level, we now introduce the idea of an energy *state*. For example, quantum mechanics identifies molecules not only with regard to their energies, but also with regard to angular momentum. Angular momentum is a vector quantity and therefore has an associated direction. For example, consider the rotating molecule shown in Fig. 11.4. Three different orientations of the angular momentum vector are shown; in each orientation, assume the energy of the molecule is the same. Quantum mechanics shows that molecular orientation is also quantized, that is, it can point only in certain directions. In all three cases shown in Fig. 11.4, the rotational energy is the same, but the rotational momentum has different *directions*. Quantum mechanics sees these cases as different and distinguishable *states*. Different states associated with the same energy can also be defined for electron angular momentum, electron, and nuclear spin, and the rather arbitrary lumping together of a number of closely spaced translational levels into one approximate level with many states.

In summary we see that, for any given energy level ε_i', there can be a number of different states that all have the same energy. This number of states is called the degeneracy or statistical weight of the given level ε_i' and is denoted by g_i. This concept is exemplified in Fig. 11.5, which shows energy levels in the vertical

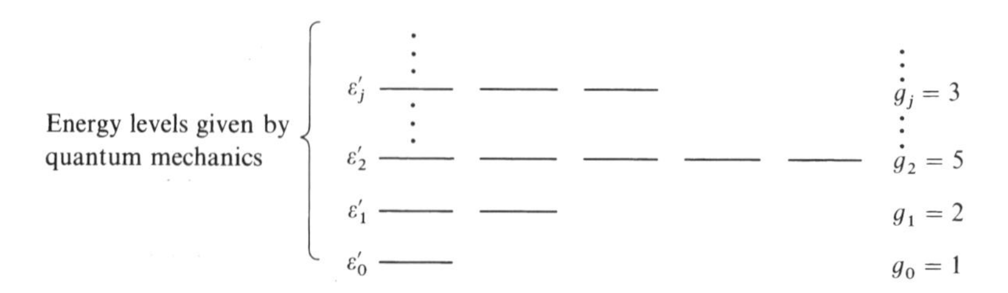

Fig. 11.5 Illustration of statistical weights.

direction, with the corresponding states as individual horizontal lines arrayed to the right at the proper energy value. For example, the second energy level is shown with five states, all with an energy value equal to ε_2'; hence, $g_2 = 5$. The values of g_i for a given molecule are obtained from quantum theory and/or spectroscopic measurements.

Now consider a system consisting of a fixed number of molecules N. Let N_j be the number of molecules in a given energy level ε_j'. The value N_j is defined as the population of the energy level. Obviously,

$$N = \sum_j N_j \tag{11.1}$$

where the summation is taken over all energy levels. The different values of N_j associated with the different energy levels ε_j' form a set of numbers, which is defined as the *population distribution*. If we look at our system of molecules at one instant in time, we will see a given set of N_j, that is, a certain population distribution over the energy levels. Another term for this set of numbers, synonymous with population distribution, is *macrostate*. Because of molecular collisions, some molecules will change from one energy level to another. Hence, when we look at our system at some later instant in time, there might be a different set of N_j and hence a different population distribution or macrostate. Finally, let us denote the total energy of the system as E, where

$$E = \sum_j \varepsilon_j' N_j \tag{11.2}$$

The schematic in Fig. 11.6 reinforces the preceding definitions. For a system of N molecules and energy E, we have a series of quantized energy levels, ε_0', $\varepsilon_1', \ldots, \varepsilon_j', \ldots$, with corresponding statistical weights, $g_0, g_1, \ldots, g_j, \ldots$. At some given instant, the molecules are distributed over the energy levels in a distinct way $N_0, N_1, \ldots, N_j, \ldots$, constituting a distinct macrostate. In the next instant, because of molecular collisions, the populations of some levels can change, creating a different set of N_j and hence a different macrostate.

Over a period of time, one particular macrostate, that is, one specific set of N_j, will occur much more frequently than any other. This particular macrostate is

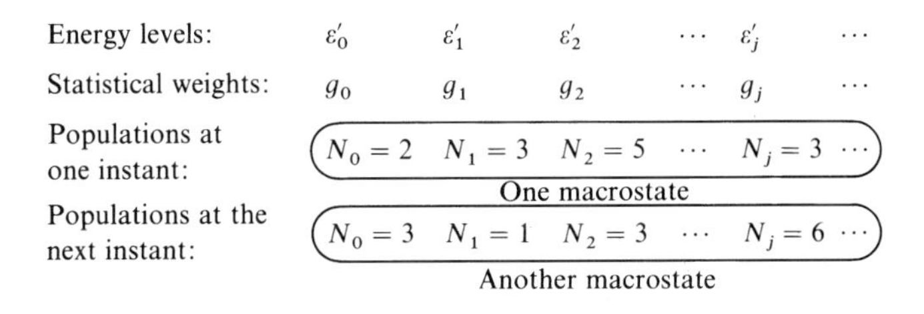

Fig. 11.6 Illustration of macrostates.

called the *most probable macrostate* (or *most probable distribution*). It is the macrostate that occurs when the system is in *thermodynamic equilibrium.* In fact, this is the *definition* of thermodynamic equilibrium within the framework of statistical mechanics. The central problem of statistical thermodynamics, and the one to which we will now address ourselves, is as follows.

> Given a system with a fixed number of identical particles, $N = \sum_j N_j$, and a fixed energy $E = \sum_j \varepsilon_j' N_j$, find the most probable macrostate

To solve the preceding problem, we need one additional definition, namely that of a *microstate*. Consider the schematic shown in Fig. 11.7, which illustrates a given macrostate. (For purposes of illustration, we choose $N_0 = 2$, $N_1 = 5$, $N_2 = 3$, etc.) Here, we display each statistical weight for each energy level as a vertical array of boxes. For example, under ε_1', we have $g_1 = 6$, and hence six boxes, one for each different energy state with the same energy ε_1'. In the energy level ε_1', we have five molecules ($N_1 = 5$). At some instant in time, these five molecules individually occupy the top three and lower two boxes under g_1, with the fourth box left vacant (i.e., no molecules at that instant have the energy state represented by the fourth box). The way that the molecules are distributed over the available boxes defines a microstate of the system, say, microstate I as shown in Fig. 11.7. At some later instant, the $N_1 = 5$ molecules can be distributed differently over the g_1 = six states, say leaving

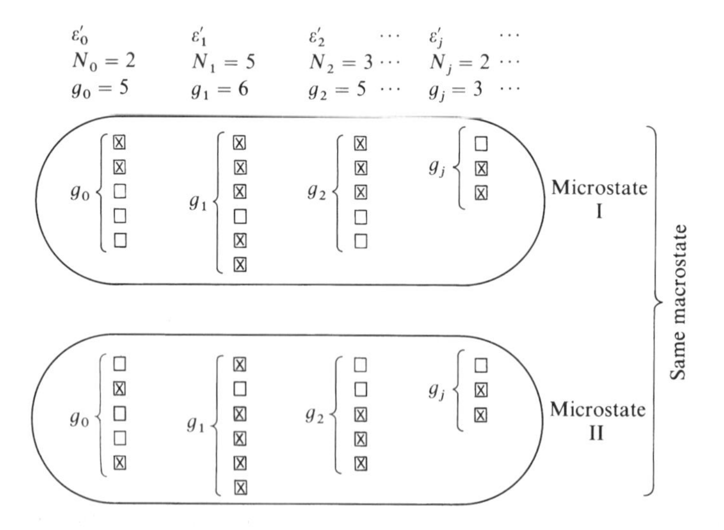

Fig. 11.7 Illustration of microstates.

the second box vacant. This represents another, different microstate, labeled microstate II in Fig. 11.7. Shifts over the other vertical arrays of boxes between microstates I and II are shown in Fig. 11.7. However, in both cases, N_0 still equals 2, N_1 still equals 5, etc.—that is, the macrostate is still the same. Thus, any one macrostate can have a number of different microstates, depending on which of the degenerate states (the boxes in Fig. 11.7) are occupied by the molecules. In any given system of molecules, the microstates are constantly changing because of molecular collisions. Indeed, it is a central assumption of statistical thermodynamics that each microstate of a system occurs with equal probability. Therefore, it is easy to reason that *the most probable macrostate is that macrostate which has the maximum number of microstates*. If each microstate appears in the system with equal probability, and there is one particular macrostate that has considerably more microstates than any other, then that is the macrostate we will see in the system most of the time. This is indeed the situation in most real thermodynamic systems. Figure 11.8 is a schematic that plots the number of microstates in different macrostates. Note there is one particular macrostate, namely, macrostate D, that stands out as having by far the largest number of microstates. This is the *most probable macrostate*; this is the macrostate that is usually seen and constitutes the situation of thermodynamic equilibrium in the system. Therefore, if we can count the number of microstates in any given macrostate, we

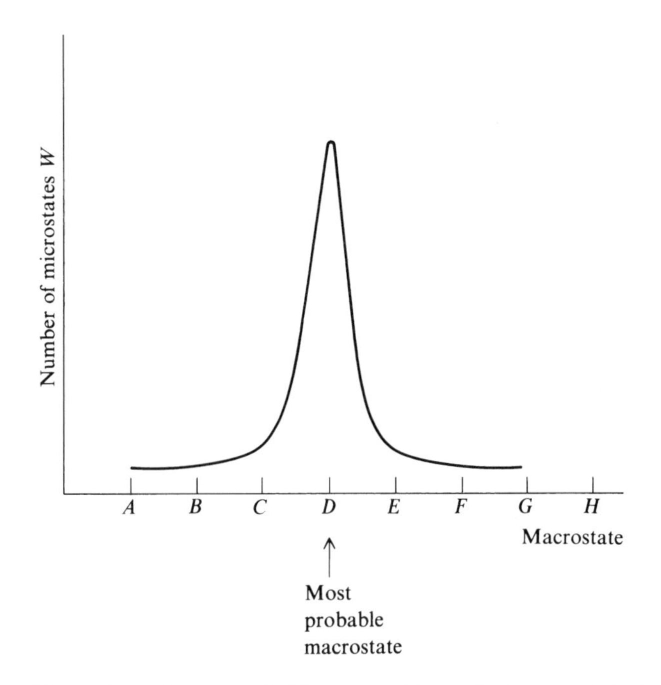

Fig. 11.8 Illustration of most probable macrostate as that macrostate that has the maximum number of microstates.

can easily identify the most probable macrostate. This counting of microstates is the subject of Sec. 11.3. In turn, after the most probable macrostate is identified, the equilibrium thermodynamic properties of the system can be computed. Such thermodynamic computations will be discussed in subsequent sections.

11.3 Counting the Number of Microstates for a Given Macrostate

Molecules and atoms are constituted from elementary particles—electrons, protons, and neutrons. Quantum mechanics makes a distinction between two different classes of molecules and atoms, depending on their number of elementary particles, as follows:

1) Molecules and atoms with an *even* number of elementary particles obey a certain statistical distribution called *Bose–Einstein statistics*. Let us call such molecules or atoms *bosons*.

2) Molecules and atoms with an *odd* number of elementary particles obey a different statistical distribution called *Fermi–Dirac* statistics. Let us call such molecules or atoms *fermions*.

There is an important distinction between the preceding two classes, as follows:

1) For bosons the number of molecules that can be in any one degenerate state (in any one of the boxes in Fig. 11.7) is *unlimited* (except, of course, that it must be less than or equal to N_j).

2) For fermions only one molecule can be in any given degenerate state at any instant.

This distinction has a major impact on the counting of microstates in a gas.

First, let us consider Bose–Einstein statistics. For the time being, consider one energy level by itself, say, ε_j'. This energy level has g_j degenerate states and N_j molecules. Consider the g_j states as the g_j containers diagrammed here:

$$\overbrace{x\ x\ x}^{1} \mid \overbrace{x\ x}^{2} \mid \overbrace{\quad\quad}^{3} \mid \overbrace{x}^{4} \mid \ \cdots \ \mid \overbrace{x\ x}^{g_j}$$

Distribute the N_j molecules among the containers, such as three molecules in the first container, two molecules in the second, etc., where the molecules are denoted by x in the preceding diagram. The vertical bars are partitions that separate one container from another. The distribution of molecules over these containers represents a distinct microstate. If a molecule is moved from container 1 to container 2, a different microstate is formed. To count the total number of different microstates possible, first note that the number of permutations between the symbols x and | is

$$[N_j + (g_j - 1)]!$$

This is the number of distinct ways that the N_j molecules and the $g_j - 1$ partitions can be arranged. However, the partitions are indistinguishable; we have counted them too many times. The $g_j - 1$ partitions can be permuted $(g_j - 1)!$ different ways. The molecules are also indistinguishable. They can be permuted $N_j!$ different ways without changing the picture just drawn. Therefore, there are $(g_i - 1)!$ $N_j!$ different permutations that yield the identical picture drawn above, that is, the same microstate. Thus, the number of different ways that N_j indistinguishable molecules can be distributed over g_j states is

$$\frac{(N_j + g_j - 1)!}{(g_j - 1)!\,N_j!}$$

This expression applies to one energy level ε_j' with population N_j and gives the number of different microstates just caused by the different arrangements within ε_j'. Consider now the whole set of N_j distributed over the complete set of energy levels. (Keep in mind that the given set of N_j defines a particular macrostate.) Letting W denote the total number of microstates for a given macrostate, the preceding expression, multiplied over all of the energy levels, yields

$$W = \prod_j \frac{(N_j + g_j - 1)!}{(g_j - 1)!\,N_j!} \tag{11.3}$$

Note that W is a function of all the N_j values, $W = W(N_1, N_2, \ldots, N_j, \ldots)$. The quantity W is called the *thermodynamic probability* and is a measure of the "disorder" of the system (as will be discussed later). In summary, Eq. (11.3) is the way to count the number of microstates in a given macrostate as long as the molecules are bosons.

Next, let us consider Fermi–Dirac statistics. Recall that, for fermions only one molecule can be in any given degenerate state at any instant, that is, there can be no more than one molecule per container. This implicitly requires that $g_j \geq N_j$. Consider the g_j containers. Take one of the molecules, and put it in one of the containers. There will be g_j choices, or ways of doing this. Take the next particle, and put it in one of the remaining containers. However, there are now only $g_i - 1$ choices because one of the containers is already occupied. Finally, placing the remaining molecules over the remaining containers, we find that the number of ways N_j particles can be distributed over g_j containers, with only one particle (or less) per container, is

$$g_j(g_j - 1)(g_j - 2)\cdots[g_j - (N_j - 1)] \equiv \frac{g_j!}{(g_j - N_j)!}$$

However, the N_j molecules are indistinguishable; they can be permuted $N_j!$ different ways without changing the preceding picture. Therefore, the number of different microstates just caused by the different arrangements with ε_j' is

$$\frac{g_j!}{(g_j - N_j)!\,N_j!}$$

Considering all energy levels, the total number of microstates for a given macrostate for fermions is

$$W = \prod_j \frac{g_j!}{(g_j - N_j)!\,N_j!} \tag{11.4}$$

In summary, if we are given a specific population distribution over the energy levels of a gas, that is, a specific set of N_j that is, a specific macrostate, Eq. (11.3) or (11.4) allows us to calculate the number of microstates for that given macrostate for bosons or fermions, respectively. It is again emphasized that W is a function of the N_j and hence is a different number for different macrostates. Moreover, as sketched in Fig. 11.8, there will in general be a certain macrostate, that is, a certain distribution for N_j, for which W will be considerably larger than for any other macrostate. This, by definition, will be the *most probable macrostate*. The precise solution for these N_j associated with the most probable macrostate is the subject of Sec. 11.4.

11.4 Most Probable Macrostate

The most probable macrostate is defined as that macrostate which contains the maximum number of microstates, that is, which has $W_{\max}$. Let us solve for the most probable macrostate, that is, let us find the specific set of N_j, which allows the maximum W.

First consider the case for bosons. From Eq. (11.3) we can write

$$\ln W = \sum_j [\ln (N_j + g_j - 1)! - \ln (g_j - 1)! - \ln N_j!] \tag{11.5}$$

Recall that we are dealing with the combined translational, rotational, vibrational, and electronic energies of a molecule, and that the closely spaced translational levels can be grouped into a number of degenerate states with essentially the same energy. Therefore, in Eq. (11.5), we can assume that $N_j \gg 1$ and $g_j \gg 1$, and hence that $N_j + g_j - 1 \approx N_j + g_j$ and $g_j - 1 \approx g_j$. Moreover, we can employ Sterling's formula

$$\ln a! = a \ln a - a \tag{11.6}$$

for the factorial terms in Eq. (11.5). Consequently, Eq. (11.5) becomes

$$\ln W = \sum_j [(N_j + g_j) \ln (N_j + g_j) - (N_j + g_j) - g_j \ln g_j + g_j - N_j \ln N_j + N_j]$$

Combining terms, this becomes

$$\ln W = \sum_j \left[N_j \ln \left(1 + \frac{g_j}{N_j}\right) + g_j \ln \left(\frac{N_j}{g_j} + 1\right) \right] \tag{11.7}$$

Recall that $\ell n\, W = f(N_j) = f(N_0, N_1, N_2, \ldots, N_j, \ldots)$. Also, to find the maximum value of W,

$$\mathrm{d}(\ell n W) = 0 \tag{11.8}$$

From the chain rule of differentiation,

$$\mathrm{d}(\ell n\, W) = \frac{\partial(\ell n\, W)}{\partial N_0}\mathrm{d}\, N_0 + \frac{\partial(\ell n\, W)}{\partial N_1}\mathrm{d}\, N_1 + \cdots + \frac{\partial(\ell n\, W)}{\partial N_j}\mathrm{d}\, N_j + \cdots \tag{11.9}$$

Combining Eqs. (11.8) and (11.9),

$$\mathrm{d}(\ell n\, W) = \sum_j \frac{\partial(\ell n\, W)}{\partial N_j}\mathrm{d}\, N_j = 0 \tag{11.10}$$

From Eq. (11.7)

$$\frac{\partial(\ell n\, W)}{\partial N_j} = \ell n\left(1 + \frac{g_j}{N_j}\right) \tag{11.11}$$

Substituting Eq. (11.11) into (11.10)

$$\mathrm{d}(\ell n\, W) = \sum_j \left[\ell n\left(1 + \frac{g_j}{N_j}\right)\right]\mathrm{d}\, N_j = 0 \tag{11.12}$$

In Eq. (11.12), the variation of N_j is not totally independent; $\mathrm{d}N_j$ is subject to two physical constraints, namely, the following:

1) $N = \sum_j N_j = \text{const}$, and hence

$$\sum_j \mathrm{d}\, N_j = 0 \tag{11.13}$$

2) $E = \sum_j \varepsilon_j' N_j = \text{const}$, and hence

$$\sum_j \varepsilon_j' \mathrm{d}N_j = 0 \tag{11.14}$$

Letting α and β be two Lagrange multipliers (two constants to be determined later), Eqs. (11.13) and (11.14) can be written as

$$-\sum_j \alpha\, \mathrm{d}\, N_j = 0 \tag{11.15}$$

$$-\sum_j \beta \varepsilon_j' \mathrm{d}\, N_j = 0 \tag{11.16}$$

Adding Eqs. (11.12), (11.15), and (11.16), we have

$$\sum_j \left[\ell n \left(1 + \frac{g_j}{N_j} \right) - \alpha - \beta \varepsilon'_j \right] \mathrm{d} N_j = 0 \tag{11.17}$$

From the standard method of Lagrange multipliers, α and β are defined such that each term in brackets in Eq. (11.17) is zero, that is,

$$\ell n \left(1 + \frac{g_j}{N_j} \right) - \alpha - \beta \varepsilon'_j = 0$$

or

$$1 + \frac{g_j}{N_j} = e^{\alpha} e^{\beta \varepsilon'_j}$$

or

$$\boxed{N_j^* = \frac{g_j}{e^{\alpha} e^{\beta \varepsilon'_j} - 1}} \tag{11.18}$$

The asterisk has been added to emphasize that N_j^* corresponds to the maximum value of W via Eq. (11.8), that is, N_j^* corresponds to the most probable distribution of particles over the energy levels ε'_j. Equation (11.18) gives the *most probable macrostate* for bosons. That is, the set of values obtained from Eq. (11.18) for all energy levels

$$N_0^*, N_1^*, N_2^*, \ldots, N_j^*, \ldots$$

is the most probable macrostate.

An analogous derivation for fermions, starting from Eq. (11.4), yields for the most probable distribution

$$\boxed{N_j^* = \frac{g_j}{e^{\alpha} e^{\beta \varepsilon'_j} + 1}} \tag{11.19}$$

which differs from the result for bosons [Eq. (11.18)] only by the sign in the denominator. The details of that derivation are left to the reader.

11.5 Limiting Case: Boltzmann Distribution

At very low temperature, say, less than 5 K, the molecules of the system are jammed together at or near the ground energy levels, and therefore the degenerate states of these low-lying levels are highly populated. As a result, the differences between Bose–Einstein statistics [Eq. (11.18)] and Fermi–Dirac statistics [Eq. (11.19)] are important. In contrast, at higher temperatures the molecules are distributed over many energy levels, and therefore the states are generally sparsely populated, that is, $N_j \ll g_j$. For this case, the denominators of

Eqs. (11.18) and (11.19) must be very large

$$e^{\alpha}e^{\beta\varepsilon'_j} - 1 \gg 1$$

and

$$e^{\alpha}e^{\beta\varepsilon'_j} + 1 \gg 1$$

Hence, in the high-temperature limit the unity term in these denominators can be neglected, and both Eqs. (13.18) and (13.19) reduce to

$$\boxed{N_j^* = g_j e^{-\alpha}e^{\beta\varepsilon'_j}} \tag{11.20}$$

This limiting case is called the Boltzmann limit, and Eq. (11.20) is termed the *Boltzmann distribution*, named after the famous 19th-century physicist, Ludwig Boltzmann (1844–1906). Because all gas dynamics problems generally deal with temperatures far above 5 K, the Boltzmann distribution is appropriate for all of our future considerations. That is, in our future discussions we will deal with Eq. (11.20) rather than Eqs. (11.18) or (11.19).

We still have two items of unfinished business with regard to the Boltzmann distribution, namely, α and β in Eq. (11.20). The link between classical and statistical thermodynamics is β. It can readily be shown (for example, see. p. 118 of [150] that

$$\beta = \frac{1}{kT}$$

where k is the Boltzmann constant [see Eq. (10.9)] and T is the temperature of the system. We will prove this relation in Sec. 11.6. Hence, Eq. (11.20) can be written as

$$N_j^* = g_j e^{-\alpha}e^{-\varepsilon'_j/kT} \tag{11.21}$$

To obtain an expression for α, recall that $N = \sum_j N_j^*$. Hence, from Eq. (11.21),

$$N = \sum_j g_j e^{-\alpha}e^{-\varepsilon'_j/kT} = e^{-\alpha}\sum_j g_j e^{-\varepsilon'_j/kT}$$

Here

$$e^{-\alpha} = \frac{N}{\sum_j g_j e^{-\varepsilon'_j/kT}} \tag{11.22}$$

Substituting Eq. (11.22) into (11.21), we obtain

$$\boxed{N_j^* = N\frac{g_j e^{-\varepsilon'_j/kT}}{\sum_j g_j e^{-\varepsilon'_j/kT}}} \tag{11.23}$$

The Boltzmann distribution, given by Eq. (11.23), is important. It is the most probable distribution of the molecules over all of the energy levels ε_j' of the system. Also, recall from Sec. 11.2 that ε_j' is the total energy, including the zero-point energy. However, Eq. (11.23) can also be written in terms of ε_j, the energy measured above the zero point, as follows. Because $\varepsilon_j' = \varepsilon_j + \varepsilon_0$, then

$$\frac{e^{-\varepsilon_j'/kT}}{\sum_j g_j e^{-\varepsilon_j'/kT}} = \frac{e^{-(\varepsilon_j+\varepsilon_0)/kT}}{\sum_j g_j e^{-(\varepsilon_j+\varepsilon_0)/kT}} = \frac{e^{-\varepsilon_0/kT} e^{-\varepsilon_j/kT}}{e^{-\varepsilon_0/kT} \sum_j g_j e^{-\varepsilon_j/kT}} = \frac{e^{-\varepsilon_j/kT}}{\sum_j g_j e^{-\varepsilon_j/kT}}$$

Hence, Eq. (11.23) becomes

$$\boxed{N_j^* = N \frac{g_j e^{-\varepsilon_j/kT}}{\sum_j g_j e^{-\varepsilon_j/kT}}} \tag{11.24}$$

where the energies are measured above the zero point. Finally, the *partition function Q* (or sometimes called the "state sum") is defined as

$$Q \equiv \sum_j g_j e^{-\varepsilon_j/kT}$$

and the Boltzmann distribution, from Eq. (11.24), can be written as

$$\boxed{N_j^* = N \frac{g_j e^{-\varepsilon_j/kT}}{Q}} \tag{11.25}$$

The partition function is a very useful quantity in statistical thermodynamics, as we will soon appreciate. Moreover, it is a function of the volume as well as the temperature of the system, as will be demonstrated later:

$$Q = f(T, V)$$

In summary, the Boltzmann distribution given, for example, by Eq. (11.25), is extremely important. Equation (11.25) should be interpreted as follows. *For molecules or atoms of a given species, quantum mechanics says that a set of well-defined energy levels ε_j exists, over which the molecules or atoms can be distributed at any given instant, and that each energy level has a certain number of degenerate states g_j. For a system of N molecules or atoms at a given T and V, Eq. (11.25) tells us how many such molecules or atoms N_j^* are in each energy level ε_j when the system is in thermodynamic equilibrium.*

11.6 Evaluation of Thermodynamic Properties in Terms of the Partition Function

The preceding formalism will now be cast in a form to yield practical thermodynamic properties for a high-temperature gas. In this section, properties such as internal energy will be expressed in terms of the partition function. In turn, in Sec. 11.7 the partition function will be developed in terms of T and V. Finally, in Sec. 11.8, the results will be combined to give practical expressions for the thermodynamic properties.

First consider the internal energy E, which is one of the most fundamental and important thermodynamic variables. From the microscopic viewpoint, for a system in equilibrium

$$E = \sum_j \varepsilon_j N_j^* \tag{11.26}$$

Note that in Eq. (11.26) E is measured above the zero-point energy. Combining Eq. (11.26) with the Boltzmann distribution given by Eq. (11.25), we have

$$E = \sum_j \varepsilon_j N \frac{g_j e^{-\varepsilon_j/kT}}{Q} = \frac{N}{Q} \sum_j g_j \varepsilon_j e^{-\varepsilon_j/kT} \tag{11.27}$$

Recall from the preceding section that

$$Q \equiv \sum_j g_j e^{-\varepsilon_j/kT} = f(V, T)$$

Hence

$$\left(\frac{\partial Q}{\partial T}\right)_v = \frac{1}{kT^2} \sum_j g_j \varepsilon_j e^{-\varepsilon_j/kT}$$

or

$$\sum_j g_j \varepsilon_j e^{-\varepsilon_j/kT} = kT^2 \left(\frac{\partial Q}{\partial T}\right)_v \tag{11.28}$$

Substituting Eq. (11.28) into (11.27),

$$E = \frac{N}{Q} kT^2 \left(\frac{\partial Q}{\partial T}\right)_v$$

or

$$\boxed{E = NkT^2 \left(\frac{\partial \ln Q}{\partial T}\right)_v} \tag{11.29}$$

This is the internal energy for a system of N molecules or atoms.

If we have 1 mol of atoms or molecules, then $N = N_A$, Avogadro's number. Also, $N_A k = \mathcal{R}$, the universal gas constant. Consequently, for the internal energy per mole, Eq. (11.29) becomes

$$E = \mathcal{R} T^2 \left(\frac{\partial \ln Q}{\partial T}\right)_v \tag{11.30}$$

In gas dynamics, a unit mass is a more fundamental quantity than a unit mole. Let M be the mass of the system of N molecules and m be the mass of an individual molecule. Then $M = Nm$. From Eq. (11.29), the internal energy *per unit mass* e is

$$e = \frac{E}{M} = \frac{NkT^2}{Nm}\left(\frac{\partial \ell n\, Q}{\partial T}\right)_v \tag{11.31}$$

However, $k/m = R$, the specific gas constant, and therefore Eq. (11.31) becomes

$$\boxed{e = RT^2\left(\frac{\partial \ell n\, Q}{\partial T}\right)_v} \tag{11.32}$$

The specific enthalpy is defined as

$$h = e + pv = e + RT$$

Hence, from Eq. (11.32)

$$\boxed{h = RT + RT^2\left(\frac{\partial \ell n\, Q}{\partial T}\right)_v} \tag{11.33}$$

Note that Eqs. (11.32) and (11.33) are "hybrid" equations, that is, they contain a mixture of thermodynamic variables such as e, h, and T, and a statistical variable Q.

Similar expressions for other thermodynamic variables can be obtained. Indeed, at this point we introduce the major link between classical thermodynamics and statistical thermodynamics, as follows. In Chapter 10 we introduced entropy in the classical sense, defined by Eqs. (10.46) or (10.46a). Now, we broaden the concept of entropy by considering S as a measure of the *disorder* of a system. The word "disorder" is used in a somewhat qualitative sense. We know that nature, when left to her own desires, always tends toward a state of maximum disorder. (Parents know that childrens' bedrooms, when left to themselves, tend to a state of maximum disorder, and it takes work to put the rooms back in order. The average child's room is a "high-entropy" system.) What do we mean by disorder in a thermodynamic system? Consider the following examples. First, examine the air around you. At atmospheric pressure, the mean distance between molecules is about 10 molecular diameters. If you are searching for some specific molecules, say those with a velocity near 300 m/s, you have to search a certain volume of the system before finding them. Now, increase the pressure to 10 atm. Because of the higher pressure, the molecules are packed more closely together; the average spacing between the molecules in this case is reduced to about five molecular diameters. Now, when you search for a certain number of molecules with a velocity near 300 m/s, you only have to search a smaller volume before finding them—an easier task than before. In this sense, the higher-pressure system with its molecules packed more closely together has a higher state of order or a lower state of disorder. Connecting entropy with disorder, this implies that, as p increases, S decreases. This is indeed confirmed by Eqs. (10.50) and (10.60). As a second example, imagine that we take the air around us and simply increase the temperature. The molecules will move faster. Once again, if we are searching for some specific molecules, they become harder to find as they move faster. Hence, the higher-temperature system has more "disorder" to it. Once again, associating S with disorder, this implies that S increases with an increase in temperature. This is confirmed by Eqs. (10.50) and (10.60). So we have a case for associating the classically defined entropy with the amount of disorder in the system. With this in mind, we ask the question: what represents an index of disorder in terms of statistical thermodynamics. The answer lies in the concept of *microstates*; a system with a certain number of microstates has a certain amount of disorder. The larger the number of microstates, the more is the disorder. Therefore, it makes sense to postulate a functional relationship between S and the maximum number of microstates, as given by the thermodynamic probability W_{max}, defined in Sec. 11.3

$$S = S(W_{max}) \tag{11.34}$$

Moreover, if we have two systems with S_1, W_1, and S_2, W_2, respectively, and we add these systems, the entropy is additive, $S_1 + S_2$, but the thermodynamic probability of the combined systems is the product of the two individual systems, $W_1 W_2$ (because each microstate of the first system can exist in the combined system with *each one* of the microstates of the second system). This suggests that Eq. (11.34) should be of the form

$$S = (\text{const}) \ln W_{max} \tag{11.35}$$

Equation (11.35) was first postulated by Ludwig Boltzmann, and the constant is named in his honor, namely,

$$\boxed{S = k \ln W_{max}} \tag{11.36}$$

where k is the familiar Boltzmann constant. *Equation (11.36) is the bridge between classical thermodynamics (represented by S) and statistical thermodynamics (represented by W).* It is so important to the modern world of physics that it is inscribed on Boltzmann's tombstone in Vienna, as indicated by the quotation given just below the title of this chapter. (Return to p. 501 and take note.)

Let us insert into Eq. (11.36) an expression for the thermodynamic probability obtained in the Boltzmann limit, defined in Sec. 11.5 as the case where $N_j \ll g_j$. Using the approximate result that $\ln(1 + x) \approx x$ for $x \ll 1$, Eq. (11.7) becomes, in the Boltzmann limit,

$$\ln W = \sum_j \left[N_j \ln \frac{g_j}{N_j} + N_j \right] = \sum_j N_j \left(\ln \frac{g_j}{N_j} + 1 \right) \tag{11.37}$$

Considering the maximum value of W, namely, $W_{\max}$, N_j is given by N_j^* obtained from the Boltzmann distribution derived in Sec. 11.5. In particular, from Eq. (11.25), and reverting to the use of β as given in Eq. (11.20), we have

$$N_j^* = \frac{N}{Q} g_j e^{-\beta\varepsilon_j}$$

or

$$\frac{g_j}{N_j^*} = \frac{Q}{N} e^{\beta\varepsilon_j} \tag{11.38}$$

Substituting Eq. (11.38) into (11.37), we have

$$\ell n\, W_{\max} = \sum_j N_j^* \ell n \frac{Q}{N} + \sum_j N_j^* + \sum_j N_j \beta\varepsilon_j$$

or

$$\ell n\, W_{\max} = \sum_j N_j^* \ell n \frac{Q}{N} + N + \beta E$$

or

$$\ell n\, W_{\max} = N\left(\ell n \frac{Q}{N} + 1\right) + \beta E \tag{11.39}$$

Substituting Eq. (11.39) into (11.36), we have

$$S = kN\left(\ell n \frac{Q}{N} + 1\right) + k\beta E \tag{11.40}$$

Note that, in our reversion to β, we are treating β as an unknown. We are now at a point where we can prove that $\beta = 1/kT$, which was just stated in Sec. 11.5. Consider the classical relation given in Eq. (10.31), combined with Eq, (10.46):

$$T\,\mathrm{d}S = \mathrm{d}E + p\,\mathrm{d}V \tag{11.41}$$

From Eq. (11.41), we can form the partial derivative

$$\left(\frac{\partial S}{\partial E}\right)_v = \frac{1}{T} \tag{11.42}$$

Similarly, from Eq. (11.40), we have

$$\left(\frac{\partial S}{\partial E}\right)_v = k\beta \tag{11.43}$$

Equation (11.42) is from classical thermodynamics; Eq. (11.43) is from statistical thermodynamics. Equating the right-hand sides of Eqs. (11.42) and (11.43), we find

$$\beta = \frac{1}{kT} \tag{11.44}$$

which was stated without proof in Sec. 11.5.

With this result, Eq. (11.40) can be written as

$$S = kN\left(\ell n\,\frac{Q}{N} + 1\right) + \frac{E}{T} \tag{11.45}$$

Combining Eqs. (11.45) and (11.29), we have

$$\boxed{S = Nk\left(\ell n\,\frac{Q}{N} + 1\right) + NkT\left(\frac{\partial\,\ell n\,Q}{\partial T}\right)_v} \tag{11.46}$$

Equation (11.46) is the statistical thermodynamic result for entropy in terms of Q.

Returning to Eq. (11.41), we form the partial derivatives

$$T\left(\frac{\partial S}{\partial V}\right)_T = \left(\frac{\partial E}{\partial V}\right)_T + p \tag{11.47}$$

Note that we are dealing with a single chemical species and that the gas is thermally perfect. Thus $(\partial E/\partial V)_T = 0$, and from Eq. (11.47)

$$p = T\left(\frac{\partial S}{\partial V}\right)_T \tag{11.48}$$

From Eq. (11.45),

$$\left(\frac{\partial S}{\partial V}\right)_T = Nk\left(\frac{\partial\,\ell n\,Q}{\partial V}\right)_T + \frac{1}{T}\left(\frac{\partial E}{\partial V}\right)_T = Nk\left(\frac{\partial\,\ell n\,Q}{\partial V}\right)_T \tag{11.49}$$

Combining Eqs. (11.48) and (11.49), we have

$$\boxed{p = NkT\left(\frac{\partial\,\ell n\,Q}{\partial V}\right)_T} \tag{11.50}$$

Equation (11.50) is the statistical thermodynamic result for pressure in terms of Q.

In all of the preceding equations, Q is the key factor. If Q can be evaluated as a function of V and T, the thermodynamic state variables can then be calculated. This is the subject of Sec. 11.7.

11.7 Evaluation of the Partition Function in Terms of *T* and *V*

Because the partition function is defined as

$$Q \equiv \sum_j g_j e^{-\varepsilon_j/kT}$$

we need expressions for the energy levels ε_j in order to further evaluate Q. The quantized levels for translational, rotational vibrational, and electronic energies are given by quantum mechanics. We state these results without proof here; see the classic books by Herzberg [151] and [152] for details.

Recall that the total energy of a molecule is

$$\varepsilon' = \varepsilon'_{\text{trans}} + \varepsilon'_{\text{rot}} + \varepsilon'_{\text{vib}} + \varepsilon'_{\text{el}}$$

In the preceding, from quantum mechanics,

$$\varepsilon'_{\text{trans}} = \frac{h^2}{8m}\left(\frac{n_1^2}{a_1^2} + \frac{n_2^2}{a_2^2} + \frac{n_3^2}{a_3^2}\right)$$

where n_1, n_2, n_3 are quantum numbers that can take the integral values 1, 2, 3, etc., and a_1, a_2, and a_3 are linear dimensions that describe the size of the system. The values of a_1, a_2, and a_3 can be thought of as the lengths of three sides of a rectangular box. (Also note in the preceding that h denotes Planck's constant, not enthalpy as before. To preserve standard nomenclature in both gas dynamics and quantum mechanics, we will live with this duplication. It will be clear which quantity is being used in our future expressions.) Also,

$$\varepsilon'_{\text{rot}} = \frac{h^2}{8\pi^2 I} J(J+1)$$

where J is the rotational quantum number, $J = 0, 1, 2$, etc., and I is the moment of inertia of the molecule. For vibration,

$$\varepsilon'_{\text{vib}} = h\nu(n + \tfrac{1}{2})$$

where n is the vibrational quantum number, $n = 0, 1, 2$, etc., and ν is the fundamental vibrational frequency of the molecule. For the electronic energy, no simple expression can be written, and hence it will continue to be expressed simply as ε'_{el}.

In the preceding, I and ν for a given molecule are usually obtained from spectroscopic measurements; values for numerous different molecules are tabulated in [152] among other sources. Also note that $\varepsilon'_{\text{trans}}$ depends on the size of the system through a_1, a_2, and a_3, whereas $\varepsilon'_{\text{rot}}$, $\varepsilon'_{\text{vib}}$, and ε'_{el} do not. Because of this spatial dependence of $\varepsilon'_{\text{trans}}$, Q depends on V as well as T. Finally, note that the lowest quantum number defines the zero-point energy for each mode, and from the preceding expressions, the zero-point energy for rotation is precisely zero, whereas it is a finite value for the other modes. For example,

$$\varepsilon'_{\text{trans}_0} = \frac{h^2}{8m}\left(\frac{1}{a_1^2} + \frac{1}{a_2^2} + \frac{1}{a_3^2}\right)$$

$$\varepsilon'_{\text{rot}_0} = 0$$

$$\varepsilon'_{\text{vib}_0} = \frac{1}{2}h\nu$$

In the preceding, $\varepsilon'_{\text{trans}_0}$ is very small, but it is finite. In contrast, $\varepsilon'_{\text{vib}_0}$ is a larger finite value, and $\varepsilon'_{\text{el}_0}$, although we do not have an expression for it, is larger yet.

Let us now consider the energy measured above the zero point:

$$\varepsilon_{\text{trans}} = \varepsilon'_{\text{trans}} - \varepsilon_{\text{trans}_0} \approx \frac{h^2}{8m}\left(\frac{n_1^2}{a_1^2} + \frac{n_2^2}{a_2^2} + \frac{n_3^2}{a_3^2}\right)$$

(Here, we are neglecting the small but finite value of $\varepsilon_{\text{trans}_0}$.)

$$\varepsilon_{\text{rot}} = \varepsilon'_{\text{rot}} - \varepsilon_{\text{rot}_0} = \frac{h^2}{8\pi^2 I} J(J+1)$$

$$\varepsilon_{\text{vib}} = \varepsilon'_{\text{vib}} - \varepsilon_{\text{vib}_0} = nh\nu$$

$$\varepsilon_{\text{el}} = \varepsilon'_{\text{el}} - \varepsilon_{\text{el}_0}$$

Therefore, the total energy is

$$\varepsilon' = \varepsilon_{\text{trans}} + \varepsilon_{\text{rot}} + \varepsilon_{\text{vib}} + \varepsilon_{\text{el}} + \varepsilon_0$$

Now, let us consider the total energy measured above the zero point ε, where

$$\varepsilon = \underbrace{\varepsilon' - \varepsilon_0}_{\substack{\text{Sensible energy, that is,}\\ \text{energy measured above}\\ \text{zero-point energy}}} = \underbrace{\varepsilon_{\text{trans}} + \varepsilon_{\text{rot}} + \varepsilon_{\text{vib}} + \varepsilon_{\text{el}}}_{\substack{\text{All measured above the zero-point}\\ \text{energy. Thus, all are equal to zero}\\ \text{at } T = 0\,\text{K.}}}$$

Recall from Eqs. (11.24) and (11.25) that Q is defined in terms of the sensible energy, that is, the energy measured above the zero point:

$$Q \equiv \sum_{l} g_j e^{-\varepsilon_j/kT}$$

where

$$\varepsilon_j = \varepsilon_{i_{\text{trans}}} + \varepsilon_{j_{\text{rot}}} + \varepsilon_{n_{\text{vib}}} + \varepsilon_{l_{\text{el}}}$$

Hence,

$$Q = \sum_i \sum_j \sum_n \sum_l g_i g_j g_n g_l \exp\left[-\frac{1}{kT}(\varepsilon_{i_{\text{trans}}} + \varepsilon_{j_{\text{rot}}} + \varepsilon_{n_{\text{vib}}} + \varepsilon_{l_{\text{el}}})\right]$$

or

$$Q = \left[\sum_i g_i \exp\left(-\frac{\varepsilon_{i_{\text{trans}}}}{kT}\right)\right]\left[\sum_i g_i \exp\left(-\frac{\varepsilon_{j_{\text{rot}}}}{kT}\right)\right] \times \left[\sum_n g_n \exp\left(-\frac{\varepsilon_{n_{\text{vib}}}}{kT}\right)\right]\left[\sum_l g_l \exp\left(-\frac{\varepsilon_{l_{\text{el}}}}{kT}\right)\right] \tag{11.51}$$

Note that the sums in each of the parentheses in Eq. (11.51) are partition functions of *each mode* of energy. Thus, Eq. (11.51) can be written as

$$Q = Q_{\text{trans}} Q_{\text{rot}} Q_{\text{vib}} Q_{\text{el}}$$

The evaluation of Q now becomes a matter of evaluating individually Q_{trans}, Q_{rot}, Q_{vib}, and Q_{el}.

First, consider Q_{trans}:

$$Q_{\text{trans}} = \sum_i g_{i_{\text{trans}}} \exp\left(-\frac{\varepsilon_{i_{\text{trans}}}}{kT}\right)$$

In the preceding, the summation is over all energy levels, each with g_i states. Therefore, the sum can just as well be taken over all energy states and written as

$$\begin{aligned} Q_{\text{trans}} &= \sum_j \exp\left(-\frac{\varepsilon_{j_{\text{trans}}}}{kT}\right) \\ &= \sum_{n_1=1}^{\infty}\sum_{n_2=1}^{\infty}\sum_{n_3=1}^{\infty} \exp\left[-\frac{h^2}{8mkT}\left(\frac{n_1^2}{a_1^2}+\frac{n_2^2}{a_2^2}+\frac{n_3^2}{a_3^2}\right)\right] \\ &= \left[\sum_{n_1=1}^{\infty} \exp\left(-\frac{h^2}{8mkT}\frac{n_1^2}{a_1^2}\right)\right]\left[\sum_{n_2=1}^{\infty} \exp\left(-\frac{h^2}{8mkT}\frac{n_2^2}{a_2^2}\right)\right] \\ &\quad \times \left[\sum_{n_3=1}^{\infty} \exp\left(-\frac{h^2}{8mkT}\frac{n_3^2}{a_3^2}\right)\right] \end{aligned} \tag{11.52}$$

If each of the terms in each preceding summation were plotted vs n, an almost continuous curve would be obtained because of the close spacings between the translational energies. As a result, each summation can be replaced by an integral, resulting in

$$Q_{\text{trans}} = a_1 \frac{\sqrt{2\pi mkT}}{h} a_2 \frac{\sqrt{2\pi mkT}}{h} a_3 \frac{\sqrt{2\pi mkT}}{h}$$

or

$$\boxed{Q_{\text{trans}} = \left(\frac{2\pi mkT}{h^2}\right)^{3/2} V} \tag{11.53}$$

where $V = a_1 a_2 a_3 =$ volume of the system.

To evaluate the rotational partition function, we use the quantum mechanical results $g_J = 2J + 1$. Therefore,

$$Q_{\text{rot}} = \sum_J g_J \exp\left(-\frac{\varepsilon_J}{kT}\right) = \sum_{J=0}^{\infty} (2J+1) \exp\left[-\frac{h^2}{8\pi^2 IkT} J(J+1)\right]$$

Again, if the summation is replaced by an integral,

$$\boxed{Q_{\text{rot}} = \frac{8\pi^2 IkT}{h^2}} \tag{11.54}$$

To evaluate the vibrational partition function, results from quantum mechanics give $g_n = 1$ for all energy levels of a diatomic molecule. Hence,

$$Q_{\text{vib}} = \sum_n g_n e^{-\varepsilon_n/kT} = \sum_{n=0}^{\infty} e^{-nh\nu/kT}$$

This is a simple geometric series, with a closed-form expression for the sum:

$$\boxed{Q_{\text{vib}} = \frac{1}{1 - e^{-h\nu/kT}}} \tag{11.55}$$

To evaluate the electronic partition function, no closed-form expression analogous to the preceding results is possible. Rather, the definition is used, namely,

$$Q_{\text{el}} \equiv \sum_{l=0}^{\infty} g_l e^{-\varepsilon_l/kT} = g_0 + g_1 e^{-\varepsilon_1/kT} + g_2 e^{-\varepsilon_2/kT} + \cdots \tag{11.56}$$

where spectroscopic data for the electronic energy levels, ε_1, ε_2, etc., are inserted directly in the preceding terms. Usually ε_l for the higher electronic energy levels is so large that terms beyond the first three shown in Eq. (11.56) can be neglected for $T \leq 15{,}000$ K.

Many results have been packed into this section, and the reader without previous exposure to quantum mechanics might feel somewhat uncomfortable. However, the purpose of this section has been to establish results for the partition function in terms of T and V: Eqs. (11.53–11.56) are those results. The discussion surrounding these equations removes, we hope, some of the mystery about their origin.

11.8 Practical Evaluation of Thermodynamic Properties for a Single Chemical Species

We now arrive at the focus of all of the preceding discussion in the chapter, namely, the evaluation of the high-temperature thermodynamic properties of a single-species gas. We will emphasize the specific internal energy e; other properties are obtained in an analogous manner.

First, consider the translational energy. From Eq. (11.53),

$$\ell n\, Q_{\text{trans}} = \frac{3}{2}\ell n\, T + \frac{3}{2}\ell n\, \frac{2\pi m k}{h^2} + \ell n\, V$$

Therefore,

$$\left(\frac{\partial\, \ell n\, Q_{\text{trans}}}{\partial T}\right)_V = \frac{3}{2}\frac{1}{T} \tag{11.57}$$

Substituting Eq. (11.57) into (11.32), we have

$$e_{\text{trans}} = RT^2 \frac{3}{2}\frac{1}{T}$$

$$\boxed{e_{\text{trans}} = \frac{3}{2} RT} \tag{11.57a}$$

Considering the rotational energy, we have from Eq. (11.54)

$$\ell n\, Q_{\text{rot}} = \ell n\, T + \ell n\, \frac{8\pi^2 I k}{h^2}$$

Thus,

$$\frac{\partial\, \ell n\, Q_{\text{rot}}}{\partial T} = \frac{1}{T} \tag{11.58}$$

Substituting Eq. (11.58) into (11.32), we obtain

$$\boxed{e_{\text{rot}} = RT} \tag{11.59}$$

$$\ell n\, Q_{\text{vib}} = -\ell n\,(1 - e^{-h\nu/kT})$$

Thus,

$$\frac{\partial \ell n Q_{\text{vib}}}{\partial \text{T}} = \frac{h\nu/kT^2}{e^{h\nu/kT} - 1} \tag{11.60}$$

Substituting Eq. (11.60) into (11.32), we obtain

$$\boxed{e_{\text{vib}} = \frac{h\nu/kT}{e^{h\nu/kT} - 1} RT} \tag{11.61}$$

Let us examine the preceding results in light of a classical theorem from kinetic theory, the theorem of equipartition of energy. Established before the turn of the century, this theorem states that each thermal degree of freedom of the molecule contributes $\frac{1}{2}kT$ to the energy of each molecule or $\frac{1}{2}RT$ to the energy per unit mass of gas. For example, in Sec. 11.2, we demonstrated that the translational motion of a molecule or atom contributes three thermal degrees of freedom; hence, because of equipartition of energy, the translational energy per unit mass should be $3(\frac{1}{2}RT) = \frac{3}{2}RT$. This is precisely the result obtained in Eq. (11.57a) from the modern principles of statistical thermodynamics. Similarly, for a diatomic molecule the rotational motion contributes two thermal degrees of freedom; therefore, classically $e_{\text{rot}} = 2(\frac{1}{2}RT) = RT$, which is in precise agreement with Eq. (11.59).

At this stage, you might be wondering why we have gone to all of the trouble of the preceding section if the principle of equipartition of energy will give us the results so simply. Indeed, extending this idea of the vibrational motion of a diatomic molecule, we recognize that the two vibrational thermal degrees of freedom should result in $e_{\text{vib}} = 2(\frac{1}{2}RT) = RT$. However, this is not confirmed by Eq. (11.61). Indeed the factor $(h\nu/kT)(e^{h\nu/kT} - 1)$ is less than unity except when $T \to \infty$ when it approaches unity; thus, in general, $e_{\text{vib}} < RT$, in conflict with classical theory. This conflict was recognized by scientists at the turn of the century, but it required the development of quantum mechanics in the 1920s to resolve the problem. Classical results are based on our macroscopic observations of the physical world, and they do not necessarily describe phenomena in the microscopic world of molecules. This is a major distinction between classical and quantum mechanics. As a result, the equipartition of energy principle is misleading. Instead, Eq. (11.61), obtained from quantum considerations, is the proper expression for vibrational energy.

In summary we have, for atoms,

$$\underbrace{e}_{\substack{\text{Internal energy per unit}\\ \text{mass measured above}\\ \text{zero-point energy}\\ \text{(sensible energy)}}} = \underbrace{\tfrac{3}{2}RT}_{\substack{\text{Translational}\\ \text{energy}}} + \underbrace{e_{\text{el}}}_{\substack{\text{Electronic energy}\\ \text{obtained directly}\\ \text{from spectroscopic}\\ \text{measurement}}} \tag{11.62}$$

and for molecules

$$\underbrace{e}_{\substack{\text{Sensible}\\\text{energy}}} = \underbrace{\tfrac{3}{2}RT}_{\substack{\text{Translational}\\\text{energy}}} + \underbrace{RT}_{\substack{\text{Rotational}\\\text{energy}}} + \underbrace{\frac{hv/kT}{e^{hv/kT}-1}RT}_{\substack{\text{Vibrational}\\\text{energy}}} + \underbrace{e_{\text{el}}}_{\substack{\text{Electronic}\\\text{energy}}} \tag{11.63}$$

In addition, recalling the specific heat at constant volume, $c_v \equiv (\partial e/\partial T)_v$, Eq. (11.62) yields for atoms

$$\boxed{c_v = \tfrac{3}{2}R + \frac{\partial e_{\text{el}}}{\partial T}} \tag{11.64}$$

and Eq. (11.63) yields for molecules

$$\boxed{c_v = \tfrac{3}{2}R + R + \frac{(hv/kT)^2 e^{hv/kT}}{(e^{hv/kT}-1)^2}R + \frac{\partial e_{\text{el}}}{\partial T}} \tag{11.65}$$

In light of the preceding results, we are led to the following important conclusions:

1) From Eqs. (11.62–11.65), we note that both e and c_v are functions of T only. This is the case for a thermally perfect, nonreacting gas, as defined in Sec. 10.4, that is,

$$e = f_1(T) \qquad \text{and} \qquad c_v = f(T)$$

This result, obtained from statistical thermodynamics, is a consequence of our assumption that the molecules are independent (no intermolecular forces) during the counting of microstates and that each microstate occurs with equal probability. If we included intermolecular forces, such would not be the case.

2) For a gas with only translational and rotational energy, we have the following.

For atoms:

$$c_v = \tfrac{3}{2}R$$

For diatomic molecules:

$$c_v = \tfrac{5}{2}R$$

That is, c_v is constant. This is the case of calorically perfect gas, as also defined in Sec. 10.4. For air at or around room temperature, $c_v = \frac{5}{2}R$, $c_p = c_v + R = \frac{7}{2}R$, and hence $\gamma = c_p/c_v = \frac{7}{5} = 1.4 = \text{const}$. So we see that air under normal conditions has translational and rotational energy, but no significant vibrational energy, and that the results of statistical thermodynamics predict $\gamma = 1.4 = \text{const}$—which we have assumed in all of the preceding chapters.

However, when the air temperature reaches 600 K or higher, vibrational energy is no longer negligible. Under these conditions, we say that "vibration is excited"; consequently, $c_v = f(T)$ from Eq. (11.65) and γ is no longer constant. For air at such temperatures, the constant γ results from the preceding chapters are no longer strictly valid. Instead, we have to redevelop our gas dynamics using results for a thermally perfect gas such as Eq. (11.65). This will be the subject of subsequent chapters.

3) In the theoretical limit of $T \to \infty$, Eq. (11.65) predicts $c_v \to \frac{7}{2}R$, and again we would expect c_v to be a constant. However, long before this would occur, the gas would dissociate and ionize as a result of the high temperature, and c_v would vary as a result of chemical reactions. This case will be addressed in subsequent sections.

4) Note that Eqs. (11.62) and (11.63) give the internal energy measured above the zero point. Indeed, statistical thermodynamics can only calculate the *sensible* energy or enthalpy; an absolute calculation of the total energy is not possible because we cannot in general calculate values for the zero-point energy. The zero-point energy remains a useful theoretical concept especially for chemically reacting gases, but not one for which we can obtain an absolute numerical value. This will also be elaborated upon in subsequent sections.

5) The theoretical variation of c_v for air as a function of temperature is sketched in Fig. 11.9. This sketch is qualitative only and is intended to show that, at very low temperatures (below 1 K), only translation is fully excited, and hence $c_v = \frac{3}{2}R$. (We are assuming here that the gas does not liquify at low temperatures.) Between 1 and 3 K, rotation comes into play, and above 3 K rotation and translation are fully excited, where $c_v = \frac{5}{2}R$. Then, above 600 K, vibration comes into play, and c_v is a variable until approximately 2000 K.

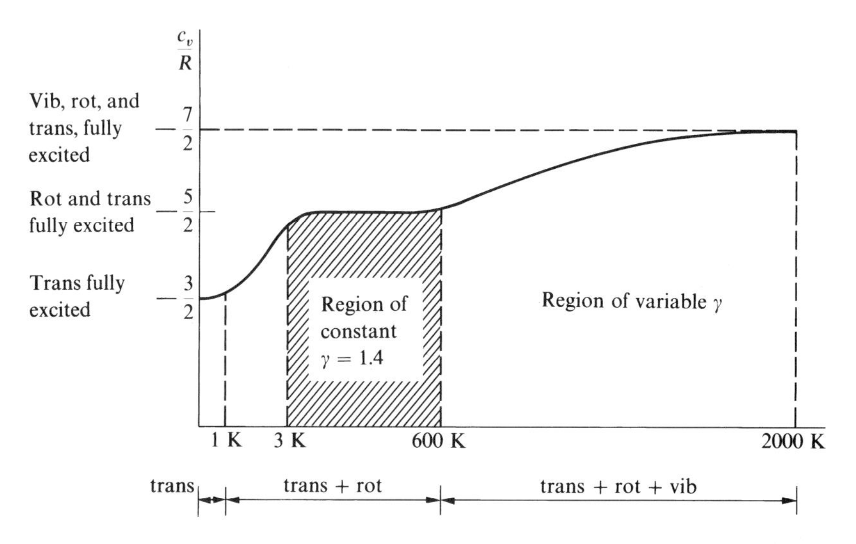

Fig. 11.9 Schematic of the temperature variation of the specific heat for a diatomic gas.

Above that temperature, chemical reactions begin to occur, and c_v experiences large variations, as will be discussed later. The shaded region in Fig. 11.9 illustrates the regime where all of our previous gas dynamic results assuming a calorically perfect gas are valid. The purpose of Part 3 of this book is to explore the high-temperature regime where γ is no longer constant and where vibrational and chemical reactions effects become important.

Consider again the perfect-gas equation of state, discussed in Sec. 10.2. In Chapter 10, we emphasized that, within the framework of classical thermodynamics, the equation of state had to be postulated—it could not be obtained from first principles. However, within the framework of statistical thermodynamics, the equation of state can be obtained from first principles, as follows. Consider Eq. (11.50) repeated here:

$$p = NkT\left(\frac{\partial \ln Q}{\partial V}\right)_T \tag{11.50}$$

Examining the partition functions in Sec. 11.7, the only one that depends on V is Q_{trans}. Hence, from Eq. (11.53)

$$\left(\frac{\partial \ln Q}{\partial V}\right)_T = \left(\frac{\partial \ln Q_{\text{trans}}}{\partial V}\right)_T = \frac{1}{V}$$

Substituting this result into Eq. (11.35), we have

$$p = NkT\left(\frac{1}{V}\right)$$

or

$$pV = NkT$$

However, this is precisely the perfect-gas equation of state given by Eq. (10.9). Hence, the formalism of statistical thermodynamics leads directly to a *derivation* of the perfect-gas equation of state.

Question: Because the perfect-gas equation of state holds for a gas where intermolecular forces are negligible, where have we made such an assumption within our development of statistical thermodynamics?

The answer is in the implicit assumption that the particles in our statistical thermodynamic system are independent and indistinguishable. If there were intermolecular forces acting on the particles, they could not be treated as independent, and, for example, the quantum mechanical expressions for energy in Sec. 11.7 would not be valid. Indeed, our assumption of specific energy states for each particle, unperturbed by outside influences, is analogous to ignoring intermolecular forces.

11.9 Calculation of the Equilibrium Constant

The concept of the equilibrium constant was introduced in Sec. 10.9 from a classical thermodynamic point of view. However, classical thermodynamics does not provide a theoretical calculation of values for K_p from first principles; statistical thermodynamics, on the other hand, does. The purpose of this section is to develop such a calculation. Also, with this section we move to the right-hand column of our chapter road map in Fig. 11.1, and we begin our consideration of mixtures of chemically reacting gases.

Note that the theory and results obtained in the preceding sections apply to a single chemical species. However, most high-temperature gases of interest are mixtures of several species. Let us now consider the statistical thermodynamics of a *mixture* of gases; the results obtained in this section represent an important ingredient for our subsequent discussions on equilibrium chemically reacting gases.

First, consider a gas mixture composed of three arbitrary chemical species A, B, and AB. The chemical equation governing a reaction between these species is

$$AB \rightleftharpoons A + B$$

Assume that the mixture is confined in a given volume at a given constant *pressure* and *temperature*. (We have already seen from Chapter 10 that p and T are important variables in dealing with chemically reacting mixtures.) We assume that the system has existed long enough for the composition to become fixed, that is, the preceding reaction is taking place an equal number of times to both the right and left. (The forward and reverse reactions are balanced.) This is the case of *chemical equilibrium*. Therefore, let N^{AB}, N^A, and N^B be the number of AB, A, and B particles, respectively, in the mixture at chemical equilibrium. Moreover, the A, B, and AB particles each have their own set of energy levels, populations, and degeneracies:

$$\begin{array}{ll}
\varepsilon'_0, \varepsilon'^A_1, \varepsilon'^A_2, \ldots, \varepsilon'^A_j, \ldots & \varepsilon'_0, \varepsilon'^B_1, \varepsilon'^B_2, \ldots, \varepsilon'^B_j, \ldots \\
N^A_0, N^A_1, N^A_2, \ldots, N^A_j, \ldots & N^B_0, N^B_1, N^B_2, \ldots, N^B_j, \ldots \\
g^A_0, g^A_1, g^A_2, \ldots, g^A_j, \ldots & g^B_0, g^B_1, g^B_2, \ldots, g^B_j, \ldots
\end{array}$$

$$\begin{array}{c}
\varepsilon'^{AB}_0, \varepsilon'^{AB}_1, \varepsilon'^{AB}_2, \ldots, \varepsilon'^{AB}_j, \ldots \\
N^{AB}_0, N^{AB}_1, N^{AB}_2, \ldots, N^{AB}_j, \ldots \\
g^{AB}_0, g^{AB}_1, g^{AB}_2, \ldots, g^{AB}_j, \ldots
\end{array}$$

A schematic of the energy levels is given in Fig. 11.10. Recall that, in most cases, we do not know the absolute values of the zero-point energies, but, in general, we know that $\varepsilon'^A_0 \neq \varepsilon'^B_0 \neq \varepsilon'^{AB}_0$. Therefore, the three energy level ladders shown in Fig. 11.10 are at different heights. However, it is possible to find the change in zero-point energy for the reaction

$$\underbrace{AB}_{\text{Reactant}} \longrightarrow \underbrace{A + B}_{\text{Products}}$$

$$\begin{array}{ccccc}
\begin{bmatrix}\text{Change in zero} \\ \text{point energy}\end{bmatrix} & \equiv & \begin{bmatrix}\text{zero-point energy} \\ \text{of products}\end{bmatrix} & - & \begin{bmatrix}\text{zero-point energy} \\ \text{of reactants}\end{bmatrix} \\
\Delta\varepsilon_0 & = & (\varepsilon'^A_0 + \varepsilon'^B_0) & - & \varepsilon'^{AB}_0
\end{array}$$

This relationship is illustrated in Fig. 11.11.

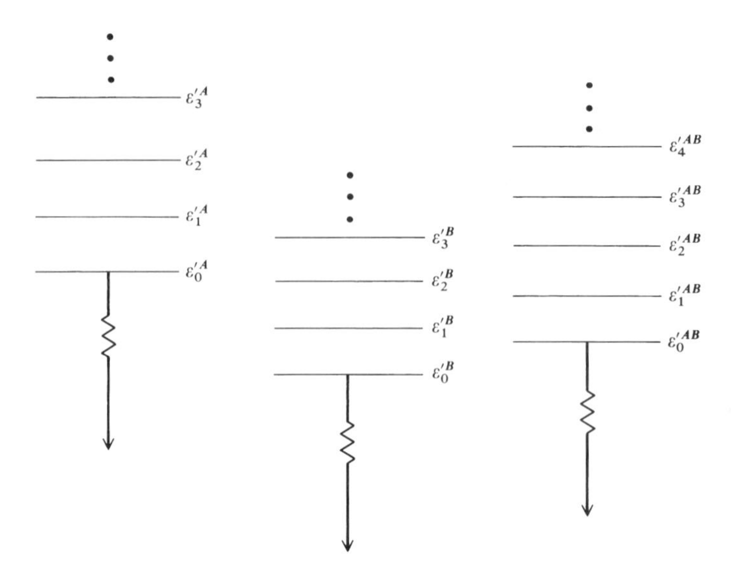

Fig. 11.10 Schematic of energy levels for three different chemical species.

The equilibrium mixture of A, B, and AB particles has two constraints.

1) The total energy E is constant:

$$E^A = \sum_j N_j^A \varepsilon_j'^A = \sum_j N_j^A (\varepsilon_j^A + \varepsilon_0^A)$$

$$E^B = \sum_j N_j^B \varepsilon_j'^B = \sum_j N_j^B (\varepsilon_j^B + \varepsilon_0^B)$$

$$E^{AB} = \sum_j N_j^{AB} \varepsilon_j'^{AB} = \sum_j N_j^{AB} (\varepsilon_j^{AB} + \varepsilon_0^{AB})$$

$$\boxed{E = E^A + E^B + E^{AB} = \text{const}} \tag{11.66}$$

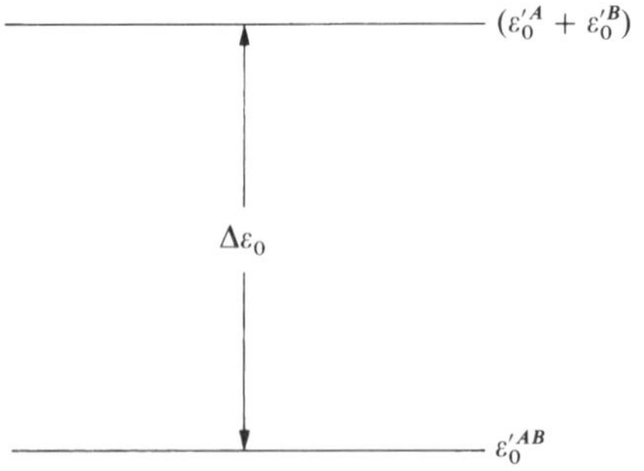

$(\varepsilon_0'^A + \varepsilon_0'^B)$

$\Delta\varepsilon_0$

$\varepsilon_0'^{AB}$

Fig. 11.11 Illustration of the meaning of change in zero-point energy.

2) Total number of A particles N_A, both free and combined (such as in AB), must be constant. This is essentially the same as saying that the total number of A nuclei stays the same, whether it is in the form of pure A or combined in AB. We are not considering nuclear reactions here—only chemical reactions that rearrange the electron structure. Similarly, the total number of B particles N_B, both free and combined, must also be constant:

$$\sum_j N_j^A + \sum_j N_j^{AB} = N_A = \text{const}$$

$$\sum_j N_j^B + \sum_j N_j^{AB} = N_B = \text{const} \tag{11.67}$$

To obtain the properties of the system in chemical equilibrium, we must find the most probable macrostate of the system, much the same way as we proceeded in Secs. 11.3 and 11.4 for a single species. The theme is the same; only the details are different. Consult [150] and [153] for those details. From this statistical thermodynamic treatment of the mixture, we find

$$\boxed{N_j^A = N^A \frac{g_j^A e^{-\varepsilon_j A/kT}}{Q^A}} \tag{11.68a}$$

$$\boxed{N_j^B = N^B \frac{g_j^B e^{-\varepsilon_j B/kT}}{Q^B}} \tag{11.68b}$$

$$\boxed{N_j^{AB} = N^{AB} \frac{g_j^{AB} e^{-\varepsilon_j AB/kT}}{Q^{AB}}} \tag{11.68c}$$

and

$$\boxed{\frac{N^A N^B}{N^{AB}} = e^{-\Delta\varepsilon_0/kT} \frac{Q^A Q^B}{Q^{AB}}} \tag{11.69}$$

Recall that N^A, N^B, and N^{AB} are the actual number of A, B, and AB particles present in the mixture; do not confuse these with N_A and N_B, which were defined as the number of A and B nuclei.

Equations (11.68a–11.68c) demonstrate that a Boltzmann distribution exists independently for each one of the three chemical species. More important, however, Eq. (11.69) gives some information on the relative amounts of A, B, and AB in the mixture. Equation (11.69) is called the law of mass action, and it relates the amounts of different species to the change in zero-point energy $\Delta\varepsilon_0$ and to the ratio of partition functions for each species.

For gas dynamic calculations, there is a more useful form of Eq. (11.69) as follows: From Sec. 10.2, we can write the perfect-gas equation of state for the mixture as

$$pV = NkT \tag{11.70}$$

For each species i, the partial pressure p_i can be written as

$$p_iV = N_ikT \tag{11.71}$$

The partial pressure is discussed at length in Sec. 10.2; it is the pressure that would exist if N_i particles of species i were the only matter filling the volume V. Letting N_i equal N^A, N^B, and N^{AB}, respectively, and defining the corresponding partial pressures, p_A, p_B, and p_{AB}, Eq. (11.71) yields

$$\frac{N^AN^B}{N^{AB}} = \frac{p_Ap_B}{p_{AB}}\frac{V}{kT} \tag{11.72}$$

Combining Eqs. (11.72) and (11.69), we have

$$\frac{p_Ap_B}{p_{AB}} = \frac{kT}{V}e^{-\Delta\varepsilon_0/kT}\frac{Q^AQ^B}{Q^{AB}} \tag{11.73}$$

Recall from Eqs. (11.52) and (11.53) that Q is proportional to the volume V. Therefore, in Eq. (11.73) the V cancel, and we obtain

$$\frac{p_Ap_B}{p_{AB}} = f(T)$$

This function of temperature is the equilibrium constant for the reaction $AB \leftrightharpoons A + B$, $K_p(T)$, as defined in Eqs. (10.90) and (10.91).

$$\boxed{\frac{p_Ap_B}{p_{AB}} = K_p(T)} \tag{11.74}$$

In Eq. (10.90), K_p is given in terms of $\Delta G^{p=1}$, which from classical thermodynamics must be treated as a measured quantity. In contrast, in Eq. (11.73), K_p is given in terms of the partition functions and the change in zero-point energy $\Delta\varepsilon_0$. Theoretical expressions for the partition functions are given in Sec. 11.7. The treatment of $\Delta\varepsilon_0$ will be discussed in Sec. 11.12.

Generalizing the preceding results, consider the chemical equation

$$0 = \sum_i v_iA_i \tag{11.75}$$

as first discussed in Sec. 10.9. Recall that v_i is the stoichiometric mole number for species i and A_i is the chemical symbol for species i. In Eq. (11.75) v_i is positive for products and negative for reactants. Then the equilibrium constant is obtained from Eqs. (11.73) and (11.74) as

$$K_p(T) \equiv \prod_i p_i^{\nu_i} = \left(\frac{kT}{V}\right)^{\Sigma\nu_i} e^{-\Delta\varepsilon_0/kT} \prod_i Q_i^{\nu_i} \tag{11.76}$$

Equation (11.76) is another form of the law of mass action, and it is extremely useful in the calculation of the composition of an equilibrium chemically reacting mixture. Some typical reactions, with their associated equilibrium constants, are

$$N_2 \rightleftharpoons 2N: \qquad K_{p,N_2} = \frac{(p_N)^2}{p_{N_2}}$$

$$H_2O_2 \rightleftharpoons 2H + 2O: \quad K_{p,H_2O_2} = \frac{(p_H)^2(p_O)^2}{p_{H_2O_2}}$$

In summary, we have made three important accomplishments in this section:

1) We have obtained the equilibrium constant, Eqs. (11.74) or (11.76), from the formal approach of statistical thermodynamics.

2) We have shown it to be a function of temperature only, Eq. (11.74).

3) We have demonstrated a formula from which it can be calculated based on a knowledge of the partition functions, Eq. (11.76). Indeed, tables of equilibrium constants for many basic chemical reactions have been calculated and are given in [148] and [149].

In perspective, the first part of this chapter has developed the high-temperature properties of a single species. Now, in order to focus on the properties of a chemically reacting mixture (such as high-temperature air), we must know *what* chemical species are present in the mixture and in what *quantity*. After these questions are answered, we can sum over all of the species and find the thermodynamics properties of the mixture. These matters are the subject of the next few sections.

11.10 Chemical Equilibrium—Some Further Comments

Consider air at normal room temperature and pressure. The chemical composition under these conditions is approximately 79% N_2, 20% O_2, and 1% trace species such as Ar, He, CO_2, H_2O, etc., by volume. Ignoring these trace species, we can consider that normal air consists of two species, N_2, and O_2. However, if we heat this air to a high temperature, where 2500 K $< T <$ 9000 K, chemical reactions will occur among the nitrogen and oxygen. Some of the important reactions in this temperature range are

$$O_2 \rightleftharpoons 2O \tag{11.77a}$$

$$N_2 \rightleftharpoons 2N \tag{11.77b}$$

$$N + O \rightleftharpoons NO \tag{11.77c}$$

$$N + O \rightleftharpoons NO^+ + e^- \tag{11.77d}$$

That is, at high temperatures, we have present in the air mixture not only O_2 and N_2, but O, N, NO, NO^+, and e^- as well. Moreover, if the air is brought to a given T and p, and then left for a period of time until the preceding reactions are occurring an equal amount in both the forward and reverse directions, we approach the condition of chemical equilibrium. For air in chemical equilibrium at a given p and T, the species O_2, O, N_2, N, NO, NO^+, and e^- are present in specific, fixed amounts, which are unique functions of p and T. Indeed, for any equilibrium chemically reacting gas, the chemical composition (the types and amounts of each species) is determined uniquely by p and T, as we discussed in Sec. 10.9.

From a statistical thermodynamic point of view, a system is in chemical equilibrium when it is characterized by the maximum number of microstates, that is, when the thermodynamic probability is maximum, W_{max}. This helps to broaden our concept of chemical equilibrium; namely, in an equilibrium chemically reacting mixture, the particles of each chemical species are distributed over their respective energy levels according to a local Boltzmann distribution for each species [Eqs. (11.68a–11.68c)].

11.11 Calculation of the Equilibrium Composition for High-Temperature Air

In Sec. 10.9, we established a procedure for the calculation of the chemical composition for an equilibrium chemically reacting gas; the material in that section was illustrated by considering a system of hydrogen and oxygen. In Sec. 11.9, the concept of chemical equilibrium and the equilibrium constant were developed from a statistical thermodynamic point of view; these concepts lead to results and methods that are identical to those from the classical viewpoint discussed in Sec. 10.9.

In the present section, we will review the calculational procedure for obtaining the equilibrium composition of a chemically reacting gas as discussed in Sec. 10.9. Moreover, because of the importance of high-temperature air in many practical applications, we will utilize a N_2-O_2 system in our example here.

Consider a system of high-temperature air at a given T and p, and assume that the following species are present: N_2, O_2, N, O, NO, NO^+, e^-. We want to solve for p_{O_2}, p_O, p_{N_2}, p_N, p_{NO}, p_{NO^+}, and p_{e^-} at the given mixture temperature and pressure. We have seven unknowns; hence, we need seven independent equations. The first equation is Dalton's law of partial pressures, which states that the total pressure of the mixture is the sum of the partial pressures (recall that Dalton's law holds only for perfect gases, i.e., gases wherein intermolecular forces are negligible):

I. $$p = p_{O_2} + p_O + p_{N_2} + p_N + p_{NO} + p_{NO^+} + p_{e^-} \tag{11.78}$$

In addition, using Eq. (11.76) we can define the equilibrium constants for the chemical reactions (11.77a–11.77d) as

II. $$\frac{(p_O)^2}{p_{O_2}} = K_{p,O_2}(T) \tag{11.79}$$

III.

$$\frac{(p_{\rm N})^2}{p_{\rm N_2}} = K_{p,{\rm N_2}}(T) \tag{11.80}$$

IV.

$$\frac{p_{\rm NO}}{p_{\rm N}p_{\rm O}} = K_{p,{\rm NO}}(T) \tag{11.81}$$

V.

$$\frac{p_{\rm NO}p_{e^-}}{p_{\rm N}p_{\rm O}} = K_{p,{\rm NO^+}}(T) \tag{11.82}$$

In Eqs. (11.79–11.82), the equilibrium constants K_p are known values, calculated from statistical mechanics as described earlier or obtained from thermodynamic measurements. They can be found in established tables, such as the JANAF tables [149]. However, Eqs. (11.78–11.82) constitute only five equations—we still need two more. The other equations come from the indestructibility of matter, as follows:

Fact: The number of O nuclei, both in the free and combined state, must remain constant. Let $N_{\rm O}$ denote the number of oxygen nuclei per unit mass of mixture.

Fact: The number of N nuclei, both in the free and combined state, must remain constant. Let $N_{\rm N}$ denote the number of nitrogen nuclei per unit mass of mixture.

Then, from the definition of Avogadro's number N_A and the mole-mass ratios η_i

$$N_{\rm A}(2\eta_{\rm O_2} + \eta_{\rm O} + \eta_{\rm NO} + \eta_{\rm NO^+}) = N_{\rm O} \tag{11.83}$$

$$N_{\rm A}(2\eta_{\rm N_2} + \eta_{\rm N} + \eta_{\rm NO} + \eta_{\rm NO^+}) = N_{\rm N} \tag{11.84}$$

However, from Eq. (10.20),

$$\eta_i = p_i \frac{v}{\mathcal{R}T} \tag{11.85}$$

Dividing Eqs. (11.83) and (11.84) and substituting Eq. (11.85) into the result, we have

VI.

$$\frac{2p_{\rm O_2} + p_{\rm O} + p_{\rm NO} + p_{\rm NO^+}}{2p_{\rm N_2} + p_{\rm N} + p_{\rm NO} + p_{\rm NO^+}} = \frac{N_{\rm O}}{N_{\rm N}} \tag{11.86}$$

Equation (11.86) is called the mass-balance equation. Here, the ratio $N_{\rm O}/N_{\rm N}$ is known from the original mixture at low temperature. For example, assuming at normal conditions that air consists of 80% N_2 and 20% O_2,

$$\frac{N_{\rm O}}{N_{\rm N}} = \frac{0.2}{0.8} = 0.25$$

Finally, to obtain our last remaining equation, we state the fact that electric charge must be conserved, and hence

$$\eta_{NO^+} = \eta_{e^-} \tag{11.87}$$

Substituting Eq. (11.85) into (11 87), we have

VII.

$$p_{NO^+} = p_{e^-} \tag{11.88}$$

In summary, Eqs. (11.78–11.82), (11.86), and (11.88) are seven nonlinear, simultaneous, algebraic equations that can be solved for the seven unknown partial pressures. Furthermore. Eq. (11.78) requires pressure p as input, and Eqs. (11.79–11.82) require the temperature T in order to evaluate the equilibrium constants. Hence, these equations clearly demonstrate that, for a given chemically reacting mixture, the equilibrium composition is a function of T and p, as was discussed at length in Chapter 10.

The preceding procedure, carried out for high-temperature air, is an example of a general procedure that applies to any chemically reacting mixture in chemical equilibrium. In general, if the mixture has $\sum$ species and ϕ elements, then we need $\sum - \phi$ independent chemical equations [such as Eqs. (11.77a–11.77d)] with the appropriate equilibrium constants. The remaining equations are obtained from the mass-balance equations and Dalton's law of partial pressures. In the preceding example for air, $\sum = 7$, and $\phi = 3$. (The elements are O, N, and e^-.) Therefore, we needed $\sum - \phi = 4$ independent chemical equations with four different equilibrium constants. These four equations were Eqs. (11.77a–11.77d).

The calculation of a chemical equilibrium composition is conceptually straightforward, as indicated in this section. However, the solution of a system of many nonlinear, simultaneous algebraic equations is not a trivial undertaking by hand, and today such calculations are almost always performed on a high-speed digital computer using custom-designed algorithms.

Also, let us emphasize a point made in Sec. 10.9, namely, that the specific chemical species to be solved are chosen at the beginning of the problem. This choice is important; if a major species is not considered (e.g., if N had been left out of our preceding calculations), the final results for chemical equilibrium will not be accurate. The proper choice of the type of species in the mixture is a matter of experience and common sense. If there is any doubt, it is always safe to assume all possible combinations of the atoms and molecules as potential species, then, if many of the choices turn out to be trace species, the results of the calculation will state so. At least in this manner, the possibility of overlooking a major species is minimized.

An example of results obtained from the preceding analysis is given in Fig. 11.12. Here, the equilibrium composition of high-temperature air (in terms of mole fraction) is given as a function of T at $p = 1$ atm. Note the following trends from Fig. 11.12—trends that we have mentioned in Chapters 9 and 10.

1) The O_2 begins to dissociate above 2000 K and is virtually completely dissociated above 4000 K.

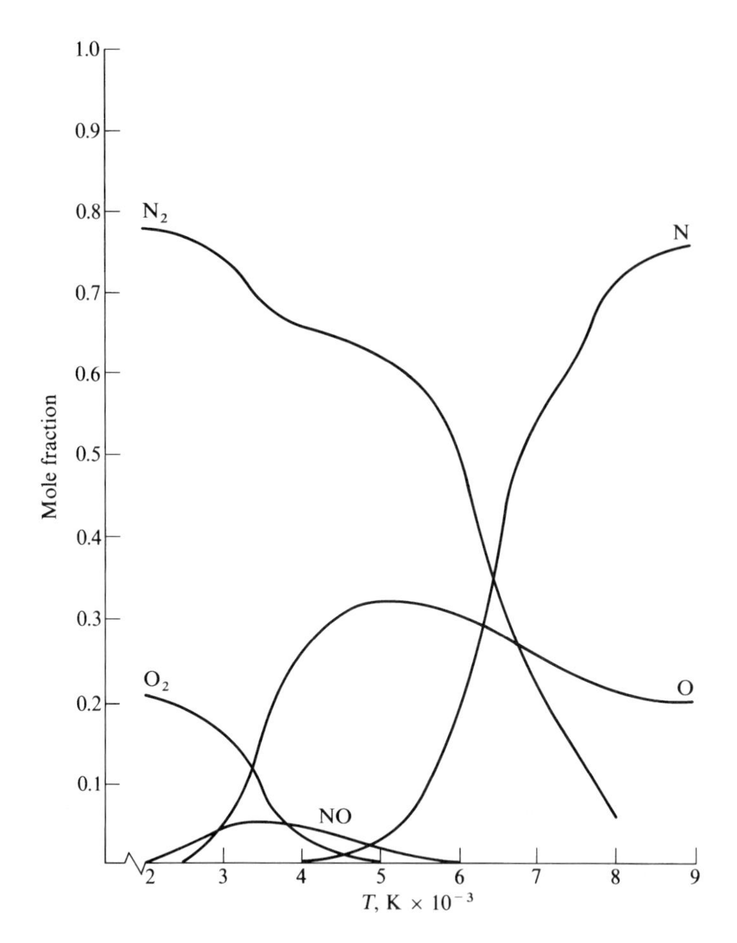

Fig. 11.12 Composition of equilibrium air vs temperature at 1 atm.

2) The N_2 begins to dissociate above 4000 K and is virtually completely dissociated above 9000 K.

3) The NO is present between 2000 and 6000 K, with a peak mole fraction occurring about 3500 K.

A point should be made about the variation of X_O as shown in Fig. 11.12. The curve for X_O has a local maximum around 5000 K and then decreases at higher temperatures in the range from 5000 to 9000 K. This does not mean, however, that the total amount of O atoms is decreasing in this range. Rather, it is a consequence of the definition of mole fraction; because $X_O \equiv \mathcal{N}_O/\mathcal{N}$, where $\mathcal{N}_O$ is the number moles of O and $\mathcal{N}$ is the total number of moles, then X_O decreases between 5000 and 9000 K simply because the total number of moles of the mixture $\mathcal{N}$ is increasing (as a result of the dissociation of N_2, for example).

It is important to keep in mind the effect of pressure on these results. If the pressure were increased to, say, 10 atm then all of the curves in Fig. 11.12

would qualitatively shift to the right, that is, the various dissociation processes would be delayed to higher temperatures. On the other hand, if p were decreased to, say, 0.1 atm, then all of the curves in Fig. 11.12 would qualitatively shift to the left, that is, dissociation would occur at lower temperature. Hence, raising the pressure decreases the amount of dissociation, and lowering the pressure increases the amount of dissociation. In a very qualitative sense, it is convenient to keep in mind that, in an equilibrium chemically reacting mixture, an increase in pressure "squeezes out" some of the amount of dissociation.

Extensive calculations of the equilibrium properties of high-temperature air, including the equilibrium composition, can be found in [154]. Indeed, the data plotted in Fig. 11.12 were obtained from the detailed tabulations found in [154].

11.12 Thermodynamic Properties of an Equilibrium Chemically Reacting Gas

In perspective, to this point in our discussion of the properties of high-temperature gases we have accomplished two major goals:

1) From Secs. 11.2–11.8, we have obtained formulas for calculating the thermodynamic properties of a given single species. In terms of our chapter road map in Fig. 11.1, we traveled down all of the left-hand column, arriving at the box at the bottom of that column.

2) From Secs. 11.9–11.11 and Sec. 10.9, we have seen how to calculate the *amount* of each species in an equilibrium chemically reacting mixture. In terms of our chapter road map, we traveled down all of the right-hand column, arriving at the box at the bottom of that column.

In this section, we now combine the preceding knowledge to obtain the thermodynamic properties of an equilibrium chemically reacting mixture, that is, we go to the intersection of both columns of the road map in Fig. 11.1 and deal with the final box at the bottom. Because of its importance to gas dynamics, we will concentrate on the enthalpy of the mixture.

For a chemically reacting mixture, we have seen that the enthalpy of the mixture per unit mass of mixture is given by

$$h = \sum_i c_i h_i = \sum_i \eta_i H_i \tag{11.89}$$

where c_i is the mass fraction of species i, h_i is the enthalpy of species i per unit mass of i, η_i is the mole-mass ratio, and H_i is the enthalpy of species i per mole of i. We can also write for the enthalpy of the mixture per mole of mixture

$$H = \sum_i X_i H_i \tag{11.90}$$

where X_i is the mole fraction of species i.

Let us now examine the meaning of H_i more closely:

$$\underbrace{H_i}_{\substack{\text{Absolute enthalpy}\\ \text{of species } i \text{ per}\\ \text{mole of } i}} = \underbrace{(H - E_0)_i}_{\substack{\text{Sensible enthalpy}\\ \text{of species } i \text{ per}\\ \text{mole of } i}} + \underbrace{E_{0_i}}_{\substack{\text{Zero-point energy}\\ \text{of species } i \text{ per}\\ \text{mole of } i}} \tag{11.91}$$

The sensible enthalpy is obtained from statistical mechanics, as we have already seen from Sec. 11.8.

$$(H - E_0)_i = (E - E_0)_i + \mathcal{R}T$$

$$(H - E_0)_i = \underbrace{\frac{3}{2}\mathcal{R}T}_{\text{Translation}} + \underbrace{\mathcal{R}T}_{\text{Rotation}} + \underbrace{\frac{h\nu/kT}{e^{h\nu/kT} - 1}\mathcal{R}T + \mathcal{R}T}_{\text{Vibration}} + \text{electronic energy} \tag{11.92}$$

Note that, $(H - E_0)_i$ is a function of T only. Also, E_{0_i} is the zero-point energy of species i, that is, the energy of the species at $T = 0$ K; it is a constant for a given chemical species. The relationship is schematically shown in Fig. 11.13. As discussed in Secs. 11.2 and 11.7, the absolute value of E_{0_i} usually cannot be calculated or measured; nevertheless, it is an important theoretical quantity. For example, in a complex chemically reacting mixture, we should establish some reference level from which all of the energies of the given species can be measured. Many times there is some difficulty and confusion in establishing what this level should be. However, by carrying through our concept of the absolute zero-point energy E_{0_i}, the choice of a proper reference level will soon become apparent.

Because the absolute value of E_{0_i} generally cannot be obtained, how can we calculate a number for h from Eq. (11.89) or H from Eq. (11.90)? The answer lies in the fact that we never need an absolute number for h. In all thermodynamic and gas dynamic problems, we deal with changes in enthalpy and internal energy. For example, in Chapter 2 dealing with shock waves, we were always interested in the change $h_2 - h_1$ across the shock. In the general conservation equations from Parts 1 and 2, we dealt with the derivatives $\partial h/\partial x$, $\partial h/\partial y$, $\partial h/\partial z$, $\partial h/\partial t$, which are changes in enthalpy. Letting points 1 and 2 denote two different

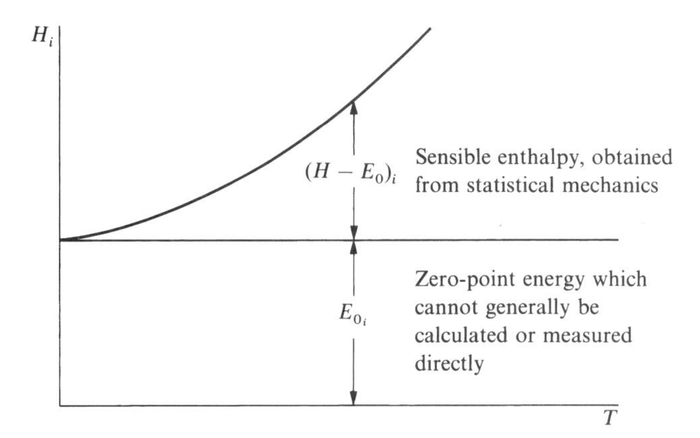

Fig. 11.13 Schematic showing the contrast between sensible enthalpy and zero-point energy.

locations in a flowfield, we have, from Eq. (11.89),

$$h_1 = \sum_i (\eta_i H_i)_l = \sum [\eta_i (H - E_0)_i]_1 + \sum (\eta_i E_{0i})_1$$

or

$$h_1 = h_{\text{sens}_1} + e_{0_1} \tag{11.93}$$

where h_{sens_1} and e_{0_i} are the sensible enthalpy and zero-point energy, respectively, per unit mass of mixture at point 1. Similarly, at point 2,

$$h_2 = \sum_i (\eta_i H_i)_2 = \sum [\eta_i (H - E_0)_i]_2 + \sum (\eta_i E_{0_i})_2$$

or

$$h_2 = h_{\text{sens}_2} + e_{0_2} \tag{11.94}$$

Subtracting Eq. (11.93) from (11.94) we have

$$\underbrace{h_2 - h_1}_{\text{Change in enthalpy}} = \underbrace{(h_{\text{sens}_2} - h_{\text{sens}_1})}_{\text{Change in sensible enthalpy}} + \underbrace{(e_{0_2} - e_{0_1})}_{\text{Change in zero-point energy}}$$

or

$$\boxed{\Delta h = \Delta h_{\text{sens}} + \Delta e_0} \tag{11.95}$$

In Eq. (11.95) we have circumvented the need to know the absolute value of the zero-point energy; rather, what we need now is a value for the *change* in zero-point energy Δe_0. The value of Δe_0 *can* be obtained from measurements, as discussed next.

The change in zero-point energy is related to the concept of the *heat of formation* for a given species. When a chemical reaction represents the formation of a single chemical species from its "elements" at standard conditions, the heat of reaction (as discussed in Sec. 10.10) is called the *standard heat of formation.* The standard conditions are those of the stable elements at the standard temperature, $T_s = 298.16$ K. (The quotation marks around the word "elements" reflects that some elements at the standard conditions are really diatomic molecules, not atoms. For example, nitrogen and oxygen are always found at standard conditions in the form N_2 and O_2, not N and O.) To illustrate, consider the formation of H_2O from its "elements" at standard conditions:

$$\underbrace{H_2 + \tfrac{1}{2}O_2}_{\text{At, } T_s} T_s \longrightarrow \underbrace{H_2O}_{\text{At } T_s}$$

Then, by definition

$$\underbrace{(\Delta H_f)^{T_s}_{\mathrm{H_2O}}}_{\substack{\text{Standard heat of}\\ \text{formation of } \mathrm{H_2O}}} \equiv \underbrace{\mathrm{H}^{T_s}_{\mathrm{H_2O}} - H^{T_s}_{\mathrm{H_2}} - \tfrac{1}{2}H^{T_s}_{\mathrm{O_2}}}_{\substack{\text{Enthalpy of the product minus}\\ \text{the enthalpy of reactants,}\\ \text{all at } T_s}}$$

In an analogous fashion, let us define the *heat of formation at absolute zero*. Here, both the product and reactants are assumed to be at absolute zero. For example,

$$\underbrace{\mathrm{H_2} + \tfrac{1}{2}\mathrm{O_2}}_{\text{At } T = 0\,\mathrm{K}} \longrightarrow \underbrace{\mathrm{H_2O}}_{\text{At } T = 0\,\mathrm{K}}$$

Letting $(\Delta H_f)^{\circ}_{\mathrm{H_2O}}$ denote the heat of formation of H_2O at absolute zero, we have

$$(\Delta H_f)^{\circ}_{\mathrm{H_2O}} \equiv \overset{\circ}{H}_{\mathrm{H_2O}} - \overset{\circ}{H}_{\mathrm{H_2}} - \tfrac{1}{2}\overset{\circ}{H}_{\mathrm{O_2}} \tag{11.96}$$

However, the enthalpy of any species at absolute zero is, by definition, its zero-point energy. Hence, Eq. (11.96) becomes

$$(\Delta H_f)^{\circ}_{\mathrm{H_2O}} \equiv (E_0)_{\mathrm{H_2O}} - (E_0)_{\mathrm{H_2}} - \tfrac{1}{2}(E_0)_{\mathrm{O_2}} \tag{11.97}$$

Note that the preceding expressions are couched in terms of energy per mole. However, the heat of formation of species i per unit mass $(\Delta h_f)_i$ is easily obtained as

$$(\Delta h_f)_i = \frac{(\Delta H_f)_i}{\mathcal{M}_i}$$

Also, the heats of formation for many species have been measured, and are tabulated in such references as the JANAF tables and NASA SP-3001 (see [149] and [148], respectively).

We now state the following theorem: In a chemical reaction, the change in zero-point energy (zero-point energy of the products minus the zero-point energy of the reactants) is equal to the difference between the heats of formation of the products at $T = 0$ K and the heats of formation of the reactants at $T = 0$ K.

Proof of the preceding theorem is obtained by induction from examples. For example, consider the water-gas reaction:

$$\mathrm{CO_2} + \mathrm{H_2} \longrightarrow \mathrm{H_2O} + \mathrm{CO}$$

By definition of the change in zero-point energy,

$$\Delta E_0 = (E_0)_{\mathrm{H_2O}} + (E_0)_{\mathrm{CO}} - (E_0)_{\mathrm{CO_2}} - (E_0)_{\mathrm{H_2}} \tag{11.98}$$

By definition of the heat of formation at absolute zero, we have

$$H_2 + \tfrac{1}{2}O_2 \longrightarrow H_2O: \quad (\Delta H_f)^{\circ}_{H_2O} = (E_0)_{H_2O} - (E_0)_{H_2} - \tfrac{1}{2}(E_0)_{O_2} \tag{11.99}$$

$$C + \tfrac{1}{2}O_2 \longrightarrow CO: \quad (\Delta H_f)^{\circ}_{CO} = (E_0)_{CO} - (E_0)_{C} - \tfrac{1}{2}(E_0)_{O_2} \tag{11.100}$$

$$C + O_2 \longrightarrow CO_2: \quad (\Delta H_f)^{\circ}_{CO_2} = (E_0)_{CO_2} - (E_0)_{C} - \tfrac{1}{2}(E_0)_{O_2} \tag{11.101}$$

$$H_2 \longrightarrow H_2: \quad (\Delta H_f)^{\circ}_{H_2} = 0 \tag{11.102}$$

Adding Eqs. (11.99) and (11.100) and subtracting Eqs. (11.101) and (11.102), we have

$$\begin{aligned}(\Delta H_f)^{\circ}_{H_2O} + (\Delta H_f)^{\circ}_{CO} - (\Delta H_f)^{\circ}_{CO_2} - (\Delta H_f)_{H_2} &= (E_0)_{H_2O} \\ + (E_0)_{CO} - (E_0)_{CO_2} - (E_0)_{H_2} &\equiv \Delta E_0\end{aligned}$$

Thus, for the water-gas reaction, we have just shown that

$$\Delta E_0 = (\Delta H_f)^{\circ}_{H_2O} + (\Delta H_f)^{\circ}_{CO} - (\Delta H_f)^{\circ}_{CO_2} - (\Delta H_f)^{\circ}_{H_2} \tag{11.103}$$

This is precisely the statement of the preceding theorem!

Compare Eqs. (11.98) and (11.103). It appears that the terms $(E_0)_{H_2O}$, $(E_0)_{CO}$, $(E_0)_{CO_2}$, and $(E_0)_{H_2}$ can be replaced in a one-to-one correspondence by $(\Delta H_f)^{\circ}_{H_2O}$ $(\Delta H_f)^{\circ}_{CO}$, $(\Delta H_f)^{\circ}_{CO_2}$, and $(\Delta H_f)^{\circ}_{H_2}$. Therefore. let us reorient our thinking about the enthalpy of a gas mixture. We have been writing

$$h = \sum_i \eta_i H_i = \underbrace{\sum_i \eta_i (H - E_0)_i}_{\text{Sensible enthalpy of the mixture}} + \underbrace{\sum_i \eta_i E_{0_i}}_{\text{Zero-point energy of the mixture}} \tag{11.104}$$

Let us replace the preceding with

$$h = \underbrace{\sum_i \eta_i (H - E_0)_i}_{\substack{\text{Sensible enthalpy,}\\ \text{obtained for example}\\ \text{from statistical}\\ \text{mechanics}}} + \underbrace{\sum_i \eta_i (\Delta H_f)^{\circ}_i}_{\substack{\text{“Effective” zero-}\\ \text{point energy,}\\ \text{obtained from}\\ \text{tables}}} \tag{11.105}$$

Equations (11.104) and (11.105) yield different absolute numbers for h; however, from the preceding theorem the values for *changes in enthalpy* Δh will be the same whether Eq. (11.104) or (11.105) is used. *Therefore, we are led to an important change in our interpretation of enthalpy, namely, from now on we will think of enthalpy as given by Eq. (11.105) with the term involving the heat of formation at absolute zero as an "effective" zero-point energy*. In terms of enthalpy per unit mass, we write

$$h = \sum_i c_i h_i$$

where

$$h_i = (h - e_0)_i + (\Delta h_f)_i^\circ$$

Thus

$$h = \sum_i c_i (h - e_0)_i + \sum_i c_i (\Delta h_f)_i^\circ \qquad (11.106)$$

[Note that in Eqs. (11.105) and (11.106), the effective zero-point energy $\sum_i \eta_i (\Delta H)_i^\circ = \sum_i c_i (\Delta h_f)_i^\circ$ is sometimes called the chemical enthalpy in the literature.]

With the preceding, we end our discussion on the calculation of the thermodynamic properties of an equilibrium chemically reacting mixture. In summary, we have shown the following:

1) The sensible enthalpy of a mixture can be obtained from the following: a) the sensible enthalpy for each species as given by the formulas of statistical mechanics for example, Eqs. (11.62), (11.63), and (11.92); and b) knowledge of the equilibrium composition described in terms of p_i, X_i, η_i, or c_i.

2) The zero-point energy can be treated as an effective value by using the heats of formation at absolute zero in its place. Therefore, Eq. (11.105) or (11.106) can be construed as the enthalpy of a gas mixture.

11.13 Equilibrium Properties of High-Temperature Air

As discussed in Chapter 9, many applications in high-temperature gas dynamics involve high-temperature air. Therefore, in this section we will highlight the equilibrium thermodynamic properties of high-temperature air.

For a moment, return to the discussion in Sec. 11.1 concerning the calculation of a flowfield in local thermodynamic and chemical equilibrium (terms to be made more precise in subsequent chapters). Note that in order to solve such an equilibrium flowfield we need to express two thermodynamic state variables (such as T and p) in terms of two other state variables (such as ρ and h). The choice of convenient dependent and independent variables is somewhat determined by the way that the flowfield solution is set up and generally varies from one application to another. In any event, the high-temperature thermodynamic properties of any equilibrium chemically reacting mixture, air or other mixtures, are obtained from statistical thermodynamic calculations as discussed in the preceding sections. However, in what *manner* are these high-temperature thermodynamic properties actually entered into a flow calculation? Or, another way to ask the same question is: in what *form* can you actually find these properties in the literature so that you can use them for a flow calculation?

For high-temperature air, the answer to the preceding questions is as follows. There are several options:

1) The equations from statistical thermodynamics, given in this chapter, can be entered directly in the flow calculations, and the thermodynamic properties can be generated "from scratch" internally in the calculation.

2) Tables of thermodynamic properties of high-temperature air exist; an excellent source of such tabulated data is the work of Hilsenrath and Klein [154]. These tables were calculated from the methods discussed in this chapter. In turn, the tabular data can be fed into a computer and can be used in numerical flowfield calculations via a "table look-up" procedure that interpolates between discrete entries from the tables. Also, the tables in [154] are useful for approximate hand calculations of simple problems.

3) Also useful for hand calculations are graphical plots of high-temperature air properties. Indeed, a large Mollier diagram is helpful in such cases. A small section of the Mollier diagram for air is given in Fig. 11.14.

4) The tabulated data discussed in item 2 can be cast in the form of polynomial correlations that are easy and convenient to apply within the framework of a flowfield calculation. An excellent and frequently used set of correlations for high-temperature air was obtained by Tannehill and Mugge as, given in [155]. Because of their convenience, these correlations are given in detail later in this section.

All of the four options just listed have been used in calculations of equilibrium air chemically reacting flowfields; the choice of any particular option is a function of the particular problem and the inclination of the user. However, for gas mixtures other than air, such as hydrocarbon mixtures associated with combustion or ablation problems, there are usually no tabulations available. (Because

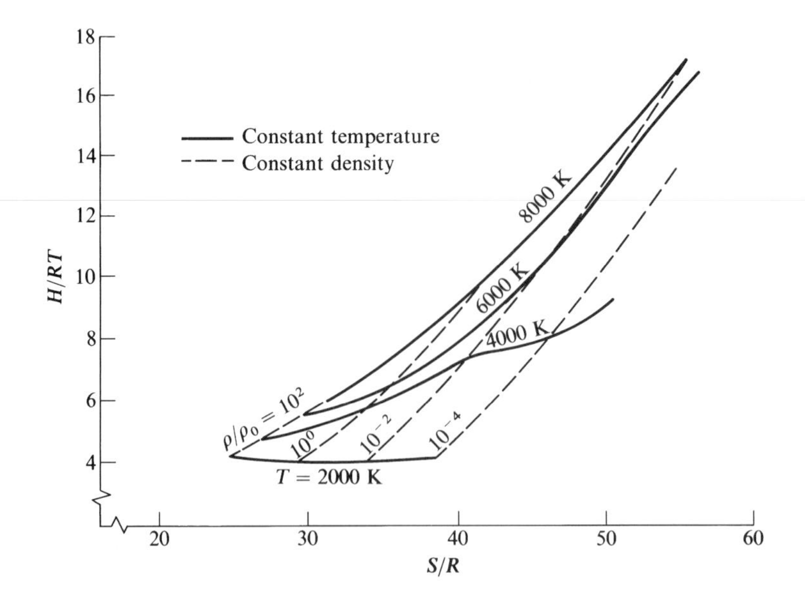

Fig. 11.14 Mollier diagram for high-temperature equilibrium air.

an infinite number of different mixtures can exist, it makes no sense to construct an infinite number of different tables.) Hence, there are no Mollier diagrams or polynomial correlations for the mixture properties. In this case, option 1 is the only recourse.

Let us make some important comments about the physical variations of high-temperature properties as reflected in the Mollier diagram in Fig. 11.14. A Mollier diagram is a plot of enthalpy vs entropy, where the various curves on the diagram correspond to constant temperature, constant density, or some other constant state variable. In Fig. 11.14, note that, at low temperatures (below 2000 K), the constant-temperature line (isothermal line) is essentially horizontal, indicating that H depends only on T. In light of our definitions in Sec. 10.4, this demonstrates that air is thermally perfect below about 2000 K. In contrast, at higher temperature (say 8000 K), the isothermal line is clearly not horizontal; we see in Fig. 11.14 that H increases rapidly with S, even though T is constant. This is a characteristic of a chemically reacting gas.

Question: Why does H increase, even though T is constant?

The answer can be constructed by following one of the high-temperature isothermal lines in Fig. 11.14 say, the line for $T = 4000$ K. Note that, as S increases, the density (and hence the pressure) decreases along this line. In turn, as the pressure decreases, the relative amount of dissociation increases along this isothermal line. (Recall our discussion about the effects of pressure on dissociation in Sec. 11.11.) This means that more atoms are present in the mixture. As noted earlier in Fig. 11.11, two atoms have a combined zero-point energy that is higher than the zero-point energy of the original single molecule. Hence, as pressure decreases at constant T, the effective zero-point energy of the mixture *increases*, and because H in Fig. 11.14 *includes* this effective zero-point energy (obtained from the heats of formation as discussed in Sec. 11.12), then H increases at constant T.

Finally, let us repeat that a particularly convenient method of entering high-temperature equilibrium air properties into a flowfield calculation is by way of polynomial correlations of the calculated and tabulated data. Because the correlations of Tannehill and Mugge [155] are widely used, they are, in part, itemized next. In terms of pressure as a function of internal energy and density

$$p = p(e, \rho)$$

we have

$$p = \rho e(\tilde{\gamma} - 1) \tag{11.107}$$

where $\tilde{\gamma}$ is given by

$$\begin{aligned}\tilde{\gamma} = a_1 + a_2 Y + a_3 Z + a_4 YZ + a_5 Y^2 + a_6 Z^2 + a_7 YZ^2 + a_8 Z^3 \\ + \frac{a_9 + a_{10} Y + a_{11} Z + a_{12} YZ}{1 + \exp[(a_{13} + a_{14} Y)(Z + a_{15} Y + a_{16})]}\end{aligned} \tag{11.108}$$

and where $Y = \log(\rho/1.292)$ and $Z = \log(e/78408.4)$. The units for p are N/m^2, the units for ρ are kg/m^3, and the units for e are m^2/s^2. The coefficients $a_1, a_2, \ldots, a_{16}$

Table 11.1 Coefficients for Eqs. (11.107) and (11.108): $p = p(e, \rho)$ (from [155])

Density range	Curve range	a_1	a_2	a_3	a_4	a_5	a_6	a_7	a_8	a_9	a_{10}
$Y > -0.50$	$Z \leqslant 0.65$	1.40000	0	0	0	0	0	0	0	0	0
	$0.65 < Z \leqslant 1.68$	1.45510	−0.000102	−0.081537	0.000166	0	0	0	0	0.128647	−0.049454
	$1.68 < Z \leqslant 2.46$	1.59608	−0.042426	−0.192840	0.029353	0	0	0	0	−0.019430	0.005954
	$Z > 2.46$	1.54363	−0.049071	−0.153562	0.029209	0	0	0	0	−0.324907	−0.077599
$-4.5 < Y \leqslant -0.50$	$Z \leqslant 0.65$	1.40000	0	0	0	0	0	0	0	0	0
	$0.65 < Z \leqslant 1.54$	1.44813	0.001292	−0.073510	−0.001948	0	0	0	0	0.054745	−0.013705
	$1.54 < Z \leqslant 2.22$	1.73158	0.003902	−0.272846	0.006237	0	0	0	0	0.041419	0.037475
	$2.22 < Z \leqslant 2.90$	1.59350	0.075324	−0.176186	−0.026072	0	0	0	0	−0.200838	−0.058536
	$Z > 2.90$	1.12688	−0.025957	0.013602	0.013772	0	0	0	0	−0.127737	−0.087942
$-7 \leqslant Y \leqslant -4.5$	$Z \leqslant 0.65$	1.40000	0	0	0	0	0	0	0	0	0
	$0.65 < Z \leqslant 1.50$	1.46543	0.007625	−0.254500	−0.017244	0.000292	0.355907	0.015422	−0.163235	0	0
	$1.50 < Z \leqslant 2.20$	2.02636	0.058493	−0.454886	−0.027433	0	0	0	0	−0.165265	−0.014275
	$2.20 < Z \leqslant 3.05$	1.60804	0.034791	−0.188906	−0.010927	0	0	0	0	−0.124117	−0.007277
	$3.05 < Z \leqslant 3.38$	1.25672	0.007073	−0.039228	0.000491	0	0	0	0	0.721798	0.073753
	$Z > 3.38$	−84.0327	−0.331761	72.2066	0.491914	0.001153	−20.3559	−0.070617	1.90979	0	0

Density range	Curve range	a_{11}	a_{12}	a_{13}	a_{14}	a_{15}	a_{16}	K_1	K_2	K_3
$Y > -0.50$	$Z \leqslant 0.65$	0	0	0	0	0	0	0	0	0
	$0.65 < Z \leqslant 1.68$	−0.101036	0.035518	−15.0	0	0	−1.420	0.000450	0.203892	0.101797
	$1.68 < Z \leqslant 2.46$	0.026097	−0.006164	−15.0	0	0	−2.050	−0.006609	0.127637	0.297037
	$Z > 2.46$	0.142408	0.022071	−10.0	0	0	−2.708	−0.000081	0.226601	0.170922
$-4.5 < Y \leqslant -0.50$	$Z \leqslant 0.65$	0	0	0	0	0	0	0	0	0
	$0.65 < Z \leqslant 1.54$	−0.055473	0.021874	−10.0	0	0	−1.420	−0.001973	0.183233	−0.059952
	$1.54 < Z \leqslant 2.22$	0.016984	−0.018038	−10.0	3.0	−0.023	−2.025	−0.013027	0.074270	0.012839
	$2.22 < Z \leqslant 2.90$	0.099687	0.025287	−10.0	5.0	0	−2.700	0.004342	0.212192	−0.001293
	$Z > 2.90$	0.043104	0.023547	−20.0	4.0	0	−3.30	0.006348	0.209716	−0.006001
$-7 \leqslant Y - 4.5$	$Z \leqslant 0.65$	0	0	0	0	0	0	0	0	0
	$0.65 < Z \leqslant 1.50$	0	0	0	0	0	0	−0.000954	0.171187	0.004567
	$1.50 < Z \leqslant 2.20$	0.136685	0.010071	−30.0	0	−0.0095	−1.947	0.008737	0.184842	−0.302441
	$2.20 < Z \leqslant 3.05$	0.069839	0.003985	−30.0	0	−0.007	−2.691	0.017884	0.153672	−0.930224
	$3.05 < Z \leqslant 3.38$	−0.198942	−0.021539	−50.0	0	−0.0085	−3.334	0.002379	0.217959	0.005943
	$Z > 3.38$	0	0	0	0	0	0	0.006572	0.183396	−0.135960

are given in Table 11.1. (The coefficients K in Table 11.1 pertain to the speed of sound, to be discussed later.) In terms of temperature as a function of internal energy and density

$$T = T(e, \rho)$$

we have

$$\log\left(\frac{T}{151.78}\right) = b_1 + b_2 Y + b_3 Z + b_4 YZ + b_5 Y^2 + b_6 Z^2 + b_7 Y^2 Z + b_8 YZ^2 + \frac{b_9 + b_{10} Y + b_{11} Z + b_{12} YZ + b_{13} Z^2}{1 + \exp[(b_{14} Y + b_{15})(Z + b_{16})]} \tag{11.109}$$

where $Y = \log(\rho/1.225)$, $X = \log(p/1.0314 \times 10^5)$, $Z = X - Y$, and the pressure p is first found from Eq. (11.107). The units for p are $\mathrm{N/m^2}$, and the units for T are K. The coefficients, $b_2, b_2, \ldots, b_{16}$ are given in Table 11.2. In terms of specific enthalpy as a function of pressure and density

$$h = h(p, \rho)$$

we have

$$h = \frac{p}{\rho}\left[\frac{\tilde{\gamma}}{\tilde{\gamma} - 1}\right] \tag{11.110}$$

where

$$\tilde{\gamma} = c_1 + c_2 Y + c_3 Z + c_4 YZ + \frac{c_5 + c_6 Y + c_7 Z + c_8 YZ}{1 + \exp[c_9(X + c_{10} Y + c_{11})]} \tag{11.111}$$

and where $Y = \log(\rho/1.292)$, $X = \log(p/1.013 \times 10^5)$, and $Z = X - Y$. The coefficients $c_1, c_2, \ldots, c_{11}$ are tabulated in Table 11.3. In terms of temperature as a function of pressure and density, we have

$$\log\left(\frac{T}{T_0}\right) = d_1 + d_2 Y + d_3 Z + d_4 YZ + d_5 Z^2 + \frac{d_6 + d_7 Y + d_8 Z + d_9 YZ + d_{10} Z^2}{1 + \exp[d_{11}(Z + d_{12})]} \tag{11.112}$$

where $T_0 = 288.16$ K, $Y = \log(\rho/1.225)$, $X = \log(p/1.0134 \times 10^5)$, and $Z = X - Y$. The coefficients $d_1, d_2, \ldots, d_{12}$ are given in Table 11.4.

The accuracy of the Tannehill and Mugge correlations is demonstrated in Fig. 11.15, where T is plotted vs p for constant ρ. The solid curves are from the correlations just given, and the points are from tabulated data calculated by means of statistical thermodynamics as described in this chapter. Note that the correlations are an excellent representation of the tabulated results. For more details about the correlations, see [155].

Table 11.2 Coefficients for Eq. (11.109): $T = T(e, \rho)$ (from [155])

Density range	Curve range	b_1	b_2	b_3	b_4	b_5	b_6	b_7	b_8
$Y > -0.50$	$Z \leqslant 0.48$	$\log(\rho/R_\rho T_0)$	0	0	0	0	0	0	0
	$0.48 < Z \leqslant 1.07$	0.279268	0	0.992172	0	0	0	0	0
	$Z > 1.07$	0.233261	−0.056383	1.19783	0.063121	−0.165985	0	0	0
$-4.5 < Y \leqslant -0.5$	$Z \leqslant 0.48$	$\log(\rho/R_\rho T_0)$	0	0	0	0	0	0	0
	$0.48 < Z \leqslant 0.9165$	0.284312	0.001644	0.987912	0	0	0	0	0
	$0.9165 < Z \leqslant 1.478$	0.502071	−0.012990	0.774818	0.025397	0	0	0	0
	$1.478 < Z \leqslant 2.176$	1.02294	0.021535	0.427212	0.006900	0	0	0	0
	$Z > 2.176$	1.47540	0.129620	0.254154	−0.046411	0	0	0	0
$-7.0 < Y \leqslant -4.5$	$Z < 0.30$	$\log(\rho/R_\rho T_0)$	0	0	0	0	0	0	0
	$0.30 < Z \leqslant 1.00$	0.271800	0.000740	0.990136	−0.004947	0	0	0	0
	$1.00 < Z \leqslant 1.35$	1.39925	0.167780	−0.143168	−0.159234	0	0	0	0
	$1.35 < Z \leqslant 1.179$	1.11401	0.002221	0.351875	0.017246	0	0	0	0
	$1.79 < Z \leqslant 2.47$	1.01722	−0.017918	0.473523	0.025456	0	0	0	0
	$Z > 2.47$	−45.0871	−9.00504	35.8685	6.79222	−6.77699	−0.064705	0.025325	−1.27370

		b_9	b_{10}	b_{11}	b_{12}	b_{13}	b_{14}	b_{15}	b_{16}
$Y > -0.50$	$Z \leqslant 0.48$	0	0	0	0	0	0	0	0
	$0.48 < Z \leqslant 1.07$	0	0	0	0	0	0	0	0
	$Z > 1.07$	−0.814535	0.099233	0.602385	−0.067428	−0.093991	5.0	−20.0	−1.78
$-4.5 < Y \leqslant -0.5$	$Z \leqslant 0.48$	0	0	0	0	0	0	0	0
	$0.48 < Z \leqslant 0.9165$	0	0	0	0	0	0	0	0
	$0.9165 < Z \leqslant 1.478$	0.009912	−0.150527	−0.000385	0.105734	0	0	−15.0	−1.28
	$1.478 < Z \leqslant 2.176$	−0.427823	−0.211991	0.257096	0.101192	0	0	−12.0	−1.778
	$Z > 2.176$	−0.221229	−0.057077	0.158116	0.030430	0	5.0	0	−2.40
$-7.0 < Y \leqslant -4.5$	$Z < 0.30$	0	0	0	0	0	0	0	0
	$0.30 < Z \leqslant 1.00$	0.990717	0.175194	−0.982407	−0.159232	0	0	−20.0	−0.88
	$1.00 < Z \leqslant 1.35$	−0.027614	−0.090761	0.307036	0.121621	0	0	−20.0	−1.17
	$1.35 < Z \leqslant 1.179$	−1.15099	−0.173555	0.673342	0.088399	0	0	−20.0	−1.56
	$1.79 < Z \leqslant 2.47$	−2.17978	−0.334716	0.898619	0.127386	0	0	−20.0	−2.22
	$Z > 2.47$	0	0	0	0	0	0	0	0

Table 11.3 Coefficients for Eqs. (11.110) and (11.111): $h = h(p, \rho)$ (from [155])

Density range	Curve range	c_1	c_2	c_3	c_4	c_5	c_6
$Y > -0.50$	$Z \leqslant 0.30$	1.40000	0	0	0	0	0
	$0.30 < Z \leqslant 1.15$	1.42598	0.000918	−0.092209	−0.002226	0.019772	−0.036600
	$1.15 < Z \leqslant 1.60$	1.64689	−0.062133	−0.334994	0.063612	−0.038332	−0.014468
	$Z > 1.60$	1.48558	−0.453562	0.152096	0.303350	−0.459282	0.448395
$-4.50 < Y \leqslant -0.5$	$Z \leqslant 0.30$	1.40000	0	0	0	0	0
	$0.30 < Z \leqslant 0.98$	1.42176	−0.000366	−0.083614	0.000675	0.005272	−0.115853
	$0.98 < Z \leqslant 1.38$	1.74436	−0.035354	−0.415045	0.061921	0.018536	0.043582
	$1.38 < Z \leqslant 2.04$	1.49674	−0.021583	−0.197008	0.030886	−0.157738	−0.009158
	$Z > 2.04$	1.10421	−0.033664	0.031768	0.024335	−0.178802	−0.017456
$-7 \leqslant Y \leqslant -4.5$	$Z < 0.398$	1.40000	0	0	0	0	0
	$0.398 < Z \leqslant 0.87$	1.47003	0.007939	−0.244205	−0.025607	0.872248	0.049452
	$0.87 < Z \leqslant 1.27$	3.18652	0.137930	−1.89529	−0.103490	−2.14572	−0.272717
	$1.27 < Z \leqslant 1.863$	1.63963	−0.001004	−0.303549	0.016464	−0.852169	−0.101237
	$Z > 1.863$	1.55889	0.055932	−0.211764	−0.023548	−0.549041	−0.101758

		c_7	c_8	c_9	c_{10}	c_{11}
$Y > -0.50$	$Z \leqslant 0.30$	0	0	0	0	0
	$0.30 < Z \leqslant 1.15$	−0.077469	0.043878	−15.0	−1.0	−1.040
	$1.15 < Z \leqslant 1.60$	0.073421	−0.002442	−15.0	−1.0	−1.360
	$Z > 1.60$	0.220546	−0.292293	−10.0	−1.0	−1.600
$-4.50 < Y \leqslant -0.5$	$Z \leqslant 0.30$	0	0	0	0	0
	$0.30 < Z \leqslant 0.98$	−0.007363	0.146179	−20.0	−1.0	−0.860
	$0.98 < Z \leqslant 1.38$	0.044353	−0.049750	−20.0	−1.04	−1.336
	$1.38 < Z \leqslant 2.04$	0.123213	−0.006553	−10.0	−1.05	−1.895
	$Z > 2.04$	0.080373	0.002511	−15.0	−1.08	−2.650
$-7.0 \leqslant Y \leqslant -4.5$	$Z < 0.398$	0	0	0	0	0
	$0.398 < Z \leqslant 0.87$	−0.764158	0.000147	−20.0	−1.0	−0.742
	$0.87 < Z \leqslant 1.27$	2.06586	−0.223046	−15.0	−1.0	−1.041
	$1.27 < Z \leqslant 1.863$	0.503123	0.043580	−10.0	−1.0	−1.544
	$Z > 1.863$	0.276732	0.046031	−15.0	−1.0	−2.250

Table 11.4 Coefficients for Eq. (11.112): $T = T(p, \rho)$ (from [155])

Density range	Curve range	d_1	d_2	d_3	d_4	d_5	d_6
$Y > -0.50$	$0.48 < Z \leqslant 0.90$	0.27407	0	1.00082	0	0	0
	$Z > 0.90$	0.235869	−0.043304	1.17619	0.046498	−0.143721	−1.37670
$-4.5 < Y \leqslant -0.5$	$0.48 < Z \leqslant 0.9165$	0.281611	0.001267	0.990406	0	0	0
	$0.9165 < Z \leqslant 1.478$	0.457643	−0.034272	0.819119	0.046471	0	−0.073233
	$1.478 < Z \leqslant 2.176$	1.04172	0.041961	0.412752	−0.009329	0	−0.434074
	$Z > 2.176$	0.418298	−0.252100	0.784048	0.144576	0	−2.00015
$-7 \leqslant Y \leqslant -4.5$	$0.30 < Z \leqslant 1.07$	2.72964	0.003725	0.938851	−0.011920	0	0.682406
	$1.07 < Z \leqslant 1.57$	2.50246	−0.042827	1.12924	0.041517	0	1.72067
	$1.57 < Z \leqslant 2.24$	2.44531	−0.047722	1.00488	0.034349	0	1.95893
	$Z > 2.24$	2.50342	0.026825	0.838860	−0.009819	0	3.58284
		d_7	d_8	d_9	d_{10}	d_{11}	d_{12}
$Y > -0.50$	$0.48 < Z \leqslant 0.90$	0	0	0	0	0	0
	$Z > 0.90$	0.160465	1.08988	−0.083489	−0.217748	−10.0	−1.78
$-4.5 < Y \leqslant -0.5$	$0.48 < Z \leqslant 0.9165$	0	0	0	0	0	0
	$0.9165 < Z \leqslant 1.478$	−0.169816	0.043264	0.111854	0	−15.0	−1.28
	$1.478 < Z \leqslant 2.176$	−0.196914	0.264883	0.100599	0	−15.0	−1.778
	$Z > 2.176$	−0.639022	0.716053	0.206457	0	−10.0	−2.40
$-7 \leqslant Y \leqslant -4.5$	$0.30 < Z \leqslant 1.07$	0.089153	−0.646541	−0.070769	0	−20.0	−0.82
	$1.07 < Z \leqslant 1.57$	0.268008	−1.25038	−0.179711	0	−20.0	−1.33
	$1.57 < Z \leqslant 2.24$	0.316244	−1.01200	−0.151561	0	−20.0	−1.88
	$Z > 2.24$	0.533853	−1.36147	−0.195436	0	−20.0	−2.47

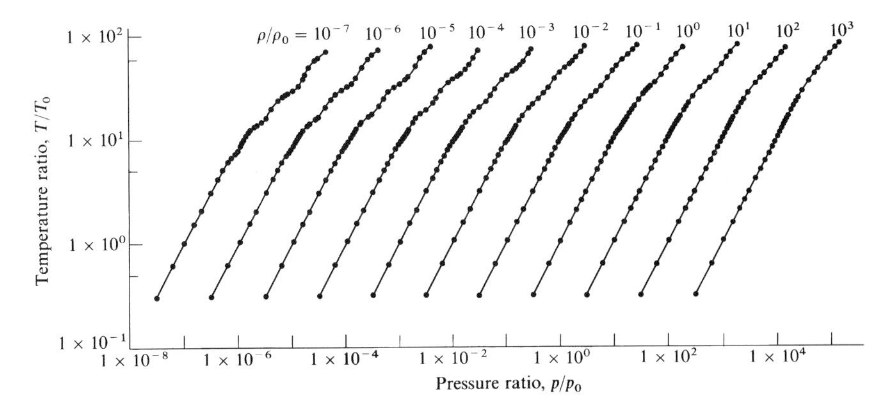

Fig. 11.15 Comparison of curve fits with tabulated data for high temperature air (from [155]).

11.14 Summary and Comments

In this chapter, we have seen how to calculate from first principles the thermodynamic properties of equilibrium chemically reacting mixtures. The calculations are obtained from the powerful concepts of statistical thermodynamics. The bridge between the classical thermodynamics discussed in Chapter 10 and the statistical thermodynamics discussed in the present chapter is

$$S = k \ln W_{\max} \tag{11.36}$$

where S is the entropy and $W_{\max}$ (the thermodynamic probability) is the total number of microstates in the most probable macrostate. Moreover, the population distribution associated with the most probable macrostate, which in terms of statistical thermodynamics is the state of equilibrium, is

$$N_j^* = N \frac{g_j e^{-\varepsilon_j/kT}}{Q} \tag{11.25}$$

where Q is the partition function given by

$$Q = \sum_j g_j e^{-\varepsilon_j/kT}$$

When the partition functions are evaluated for the various modes of molecular energy, that is, translational, rotational, and vibrational, we have, for the internal energy per unit mass for a given pure chemical species i,

$$e_i = \underbrace{\frac{3}{2} R_i T}_{\text{Translation}} + \underbrace{R_i T}_{\text{Rotation}} + \underbrace{\frac{h\nu_i/kT}{e^{h\nu_i/kT} - 1} R_i T}_{\text{Vibration}} + \underbrace{(e_{\text{el}})_i}_{\substack{\text{Sensible}\\ \text{electronic}\\ \text{energy}}} + \underbrace{(\Delta h_f)_i^{\circ}}_{\substack{\text{Effective}\\ \text{zero point}\\ \text{energy}}}$$

where $(\Delta h)_i^\circ$ is the heat of formation at absolute zero of species i per unit mass of i. In turn, the equilibrium chemical composition at a specified T and p can be obtained by using the equilibrium constant, where

$$\prod_i p_i^{\nu_i} = K_p(T) = \left(\frac{kT}{V}\right)^{\sum_i \nu_i} e^{-\Delta\varepsilon_0/kT} \prod_i Q_i \tag{11.76}$$

With the equilibrium composition expressed in terms of, say, mass fraction c_i, we have, for the internal energy of the chemically reacting mixture per unit mass of mixture,

$$e = \sum_i c_i e_i$$

where e includes the effective zero-point energy by way of the heats of formation.

Other thermodynamic properties can be obtained in like fashion. For high-temperature air, the results of calculations from statistical thermodynamics have been presented in the form of graphs, tables, and correlations.

This brings to an end our discussion of equilibrium thermodynamic properties of high-temperature chemically reacting gases. We have acquired the necessary tools to make calculations of equilibrium, chemically reacting flowfields. However, such calculations will be deferred until Chapter 14. In the meantime, we will continue our discussion of basic physical chemistry effects in Chapters 12 and 13, branching out to more extensive considerations of nonequilibrium processes.

Problems

11.1 Consider pure O_2. Calculate the equilibrium sensible enthalpy per kg at $T = 3000$ K. For O_2, $\nu = 4.73 \times 10^{13}$ s^{-1}. Ignore the electronic energy. *Note*: Boltzmann constant is $k = 1.38 \times 10^{-23}$ J/K, and Planck's constant is $h = 6.625 \times 10^{-34}$ (J)(s).

11.2 In problem 11.1, calculate the equilibrium sensible electronic energy per kg by *two* different methods, and compare the results. How does the electronic energy compare with the combined translational, rotational, and vibrational energies that were calculated in problem 11.1? For the electronic levels of O_2: $g_0 = 3$, $g_1 = 2$, and $\varepsilon_1/k = 11{,}390$ K. Ignore all higher electronic levels.

11.3 Consider air in chemical equilibrium at 0.1 atm and $T = 4500$ K. Assume the chemical species present are O_2, O, N_2, and N. (Ignore NO.) Calculate the enthalpy in joules per kilogram. Note the Following physical data: $K_{p,O_2} = 12.19$ atm, $K_{p,N_2} = 0.7899 \times 10^{-4}$ atm; for N_2, $\Delta H_f^\circ = 0$ and $\nu = 7.06 \times 10^{13}$ (s^{-1}); for N, $\Delta H_f^\circ = 4.714 \times 10^8$J/(kg · mol); for O_2, $\Delta H_f^\circ = 0$, and $\nu = 4.73 \times 10^{13}$ (s^{-1}); for O, $\Delta H_f^\circ = 2.47 \times 10^8$ J/(kg · mol). Ignore the electronic levels.

12
Elements of Kinetic Theory

So many of the properties of matter, especially when in the gaseous form, can be deduced from the hypothesis that their minute parts are in rapid motion, the velocity increasing with the temperature, that the precise nature of this motion becomes a subject of rational curiosity.

James Clark Maxwell, 1860

Chapter Preview

This chapter has to do with the rattling about of individual particles—molecules, atoms, electrons—and their collisions with each other. This is the level at which nature really works in a gas and is the essence of kinetic theory. How frequently does a given particle collide with its neighboring particles, that is, how many collisions per second does a particle in a gas experience? On the average, how far does a particle move in between collisions? Finally, what is the velocity of a given particle, on the average? Answers are given in this chapter. The reason that we are interested in such answers is that energy transfer between different modes of energy (translational, rotational, vibrational, and electronic) takes place by way of particle collisions. Chemical reactions also occur as a result of particle collisions. Collisions take time to occur, especially a sufficient number of collisions to bring about the energy transfer and chemical reactions. If the gas on the whole is simply sitting around, going nowhere, we can simply wait for the required number of particle collisions to occur. After that, the gas will be at equilibrium, with the equilibrium thermodynamic and chemical properties discussed in the preceding chapters. On the other hand, if, for example, the gas is zipping through a rocket engine or rapidly flowing over a hypersonic vehicle, the fluid elements of the gas might be far downstream of the exit of the rocket engine or way behind the trailing edge of the vehicle before the necessary number of particle collisions can occur within the fluid element. This creates a highly nonequilibrium situation. To deal with such a nonequilibrium flow, we have to deal with the frequency of the particle collisions in the flow. This is one of the reasons why you are about to study this chapter. Another reason is that the transport properties of a gas, the viscosity coefficient, thermal conductivity, and diffusion coefficient, depend on the mean distance a particle moves between one collision and the next. To obtain values of these transport coefficients for a high-temperature chemically reacting gas, we have to look at this mean distance between collisions, called the mean free path. So this chapter, as short as it is, is important to the remainder of our considerations in this book. Do not shortchange it.

12.1 Introduction

Return for a moment to our road map in Fig. 1.24. We are still working with the first item under high-temperature flows, namely, a discussion of basic physical chemistry effects. Keep in mind that the purpose of this item is to present basic concepts and develop essential equations for the understanding and analysis of high-temperature gas flows. In this sense, we are building a storehouse of "tools" to be used in Chapters 14 through 18. In Chapters 10 and 11 we established some tools that will be useful in the study of equilibrium chemically reacting flows. In Chapter 13 we will develop tools for the analysis of nonequilibrium flows. The function of the present chapter is to introduce some elementary concepts from kinetic theory that are necessary for understanding the tools to be developed in Chapter 13.

To set the perspective, in the classical thermodynamics of Chapter 10 we dealt with the system as composed of a continuous substance that interacted (by way of work and heat addition) with its surroundings. In Chapter 11 we took a more microscopic point of view and were concerned with the system as being made up of individual particles with translational, rotational, vibrational, and electronic energies. The macroscopic properties of the system are simply reflections of suitable statistical averages over all of the particles. In the present chapter, we continue with the microscopic point of view and narrow our attention to just the rapid translational motion of such particles. We will see that some important characteristics of gases are dominated by this translational motion. A study of such matters is the purview of the science of *kinetic theory*. In the present chapter we will introduce only those aspects of kinetic theory necessary for our future work with high-temperature flows. Hence, the present chapter does not constitute a rigorous and thorough presentation of kinetic theory. You are encouraged to study [150], [156], and [157] for definitive presentations. Also, this chapter is so short and straightforward that no chapter road map is required to help you navigate the flow of ideas.

12.2 Perfect-Gas Equation of State (Revisited)

In Sec. 10.2 we introduced the perfect-gas equation of state as an *empirical* result. In Sec. 11.8 we *derived* the equation of state from the principles of statistical thermodynamics. In the present section, we will again *derive* a form of the equation of state, this time using a simplified picture of molecular motion. The purpose for revisiting the equation of state here is that the derivation provides some useful insight to the molecular properties of gases.

Consider a gas contained within the cubical box sketched in Fig. 12.1. Single out a given gas particle at some instant in time and at some location P_1. This particle has a translational velocity denoted by $\boldsymbol{C}$, with x, y, and z components of velocity denoted by C_x, C_y, and C_z, respectively. Here, we treat the gas particle as a structureless "billiard ball," translating in space and frequently colliding with neighboring particles. Indeed, it is such molecular collisions that, given enough time, establish a state of *equilibrium* in the system. Assume that the gas in the box is in equilibrium. This implies that at any given point P_1, if a given particle with velocity $\boldsymbol{C}_1$ collides with another particle, causing a change in velocity, then there is another collision between other particles in the same neighborhood, which causes one of those other particles to have the velocity $\boldsymbol{C}_1$ at point P_1. The net result is as if the original particle simply continued at the velocity $\boldsymbol{C}_1$. With this picture, we can visualize a particle traversing the box with a constant velocity in the x direction, given by C_x. When the particle reaches the right face of the box in Fig. 12.1, it is assumed to specularly reflect from the surface at point P_2. That is, if $\boldsymbol{C}_1$ is the velocity just before impacting the surface at point P_2, and $\boldsymbol{C}_2$ is the velocity immediate after impact, then $|\boldsymbol{C}_1| = |\boldsymbol{C}_2|$, $C_{x_2} = -C_{x_1}$, $C_{y_2} = C_{y_1}$, and $C_{z_2} = C_{z_1}$. During the impact, the particle experiences a change in momentum in the x direction given by $2mC_x$, where m is the mass of the particle. Over a unit time (say, 1 s), the particle makes a number of traverses back and forth

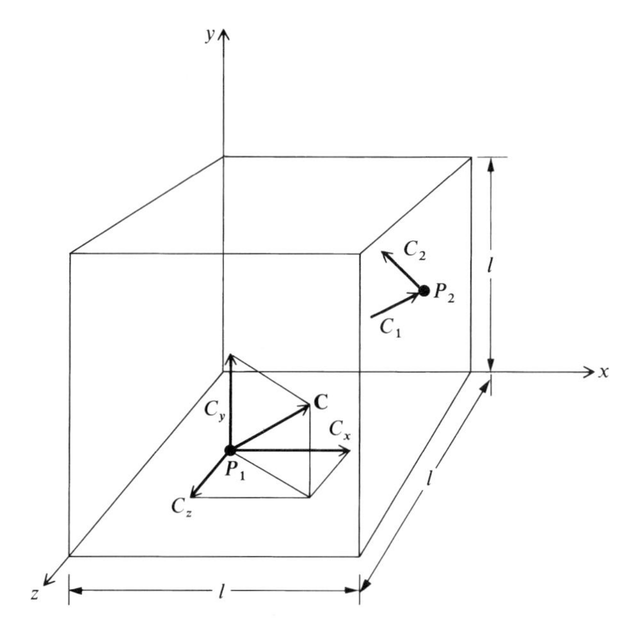

Fig. 12.1 Particle moving in a box; illustration of particle velocity components.

across the box in the x direction. Counting a complete traverse as going and coming to and from the right-hand face, the number of complete traverses per unit time is $C_x/2l$, where l is the length of the box along the x axis. Hence, the time rate of change of momentum experienced by the particle when impacting the right-hand face is given by $(2mC_x)(C_x/2l) = mC_x^2/l$. From Newton's second law, the time rate of change of momentum is equal to force. Hence, the force exerted by the particle on the right-hand face is also mC_x^2/l. Because pressure is force per unit area and the area of the face is l^2, then the pressure exerted by the particle on the right-hand face is given by $mC_x^2/l^3 = mC_x^2/V$, where V is the volume of the system. Now assume that we have a large number of particles in the box, each with a different mass m_i and different velocity $\boldsymbol{C}_i$. Then, the pressure exerted on the right-hand face by the particles in the system is

$$p = \frac{1}{V}\sum_i m_i C_{i,x}^2 \tag{12.1}$$

where the summation is taken over all of the particles. If we construct an expression for the pressure exerted on the upper face (perpendicular to the y axis) using identical reasoning, a similar result is obtained as

$$p = \frac{1}{V}\sum_i m_i C_{i,y}^2 \tag{12.2}$$

Similarly, for the pressure exerted on the face perpendicular to the z axis, we have

$$p = \frac{1}{V}\sum_i m_i C_{i,z}^2 \tag{12.3}$$

Adding Eqs. (12.1–12.3), we have

$$p = \frac{1}{3V}\sum_i m_i (C_{i,x}^2 + C_{i,y}^2 + C_{i,z}^2) = \frac{1}{3V}\sum_i m_i C_i^2 \tag{12.4}$$

where C_i is the magnitude of velocity for the ith particle. However, the total *kinetic energy* for the system E'_{trans} is given by

$$E'_{\text{trans}} = \frac{1}{2}\sum_i m_i C_i^2 \tag{12.5}$$

Combining Eqs. (12.4) and (12.5), we have

$$\boxed{pV = \tfrac{2}{3}E'_{\text{trans}}} \tag{12.6}$$

Equation (12.6) is the kinetic theory equivalent of the perfect-gas equation of state. It can only be related to temperature through classical thermodynamics because T is a variable that originated with classical thermodynamics. For example, assume that we have a mole of particles in the system. Then V

in Eq. (12.6) becomes the molar volume $\mathcal{V}$, and E_{trans} is the kinetic energy per mole

$$\boxed{p\mathcal{V} = \tfrac{2}{3} E_{\text{trans}}} \tag{12.7}$$

From Eq. (10.6a), we also have

$$p\mathcal{V} = \mathcal{R}T \tag{12.8}$$

Comparing Eqs. (12.7) and (12.8), we have

$$\boxed{E_{\text{trans}} = \tfrac{3}{2}\mathcal{R}T} \tag{12.9}$$

a result that we already know from Eq. (11.57). Hence, our simple kinetic theory model leads to the same result as obtained by statistical mechanics for the translation energy. If we divide Eq. (12.9) by Avogadro's number N_A, then

$$\frac{E_{\text{trans}}}{N_A} = \frac{3}{2}\frac{\mathcal{R}}{N_A}T$$

or

$$\boxed{\varepsilon_{\text{trans}} = \tfrac{3}{2}kT} \tag{12.9a}$$

Equation (12.9a) establishes the physical link between the thermodynamic variable T and the molecular picture, that is, temperature is a direct index for the mean kinetic energy of a particle in the system. The higher the temperature, the higher is the mean molecular kinetic energy in direct proportion.

Equations (12.6) and (12.7) are interesting in their own right. They establish a relation between the product of pressure and volume and the molecular kinetic energy of the system. Therefore, the pV product can be interpreted as a measure of energy in the system.

Finally, return to Eq. (12.4), and divide both sides by the total mass of the system M, where $m = \sum_i m_i$.

$$\frac{pV}{M} = \frac{1}{3}\frac{\sum_i m_i C_i^2}{\sum_i m_i} \tag{12.10}$$

Note that $M/V = \rho$, and define a mean square velocity $\bar{C}^2$ as

$$\bar{C}^2 \equiv \frac{\sum_i m_i C_i^2}{\sum_i m_i} \tag{12.11}$$

Then, Eq. (12.10) becomes

$$\boxed{\frac{p}{\rho} = \frac{1}{3}\bar{C}^2} \tag{12.12}$$

Equation (12.12) is another form of the kinetic theory equivalent of the perfect-gas equation of state. Using Eq. (10.3), we find from Eq. (12.11) that

$$\bar{C^2} = 3RT \tag{12.12a}$$

or, the rms molecular velocity is given by

$$\sqrt{\bar{C^2}} = \sqrt{3RT} \tag{12.13}$$

Return to Eq. (12.9a) for a moment. The translational kinetic energy for a particle $\varepsilon_{\text{trans}}$ is given by $\frac{1}{2}m_i C_i^2$, where m_i is the mass of the particle. On the other hand, from Eq. (12.9a), $\varepsilon_{\text{trans}}$ is also given by $\frac{3}{2}kT$, independent of the mass of the particle. Hence, for a gas mixture at temperature T the heavy particles will be moving more slowly, on the average, than the light particles. This is an interesting physical characteristic to keep in mind.

12.3 Collision Frequency and Mean Free Path

Consider a particle of molecular diameter d moving at the mean molecular velocity $\bar{C}$. [Note that $\bar{C}$ and $(\bar{C^2})^{1/2}$ are slightly different values, to be explained later.] Continuing with the billiard ball model of Sec. 12.2, whenever this molecule comes into contact with a like molecule, the separation of the centers of the two molecules is also d, as sketched in Fig. 12.2. This separation can be viewed as a radius of influence, in that any colliding molecule whose center comes within a distance d of the given molecule is going to cause a collision. Therefore, as our given molecule moves through space, its radius of influence will sweep out a cylindrical volume per unit time equal to $\pi d^2 \bar{C}$, as sketched in Fig. 12.3. If n is the number density, that is, the number of particles per unit volume, then our given particle will experience $n\pi d^2 \bar{C}$ collisions per second. This is defined as the *single particle collision frequency*, denoted by Z'. Hence

$$Z' = n\pi d^2 \bar{C} \tag{12.14}$$

We define the *mean free path*, denoted by λ, as the mean distance traveled by a particle between collisions. Because in unit time the particle travels a distance $\bar{C}$, and it experiences Z' collisions during this time, then

$$\lambda = \frac{\bar{C}}{Z'} = \frac{1}{n\pi d^2} \tag{12.15}$$

The preceding analysis is very simplified, and more accurate results for collision frequency and mean free path for a gas are slightly different than given by Eqs. (12.14) and (12.15). In the preceding analysis, we have imagined a particle with radius of influence d sweeping out a volume in space, and we have implicitly assumed that other molecules are simply present inside this volume. In reality, the other molecules are moving, and for more accuracy we should

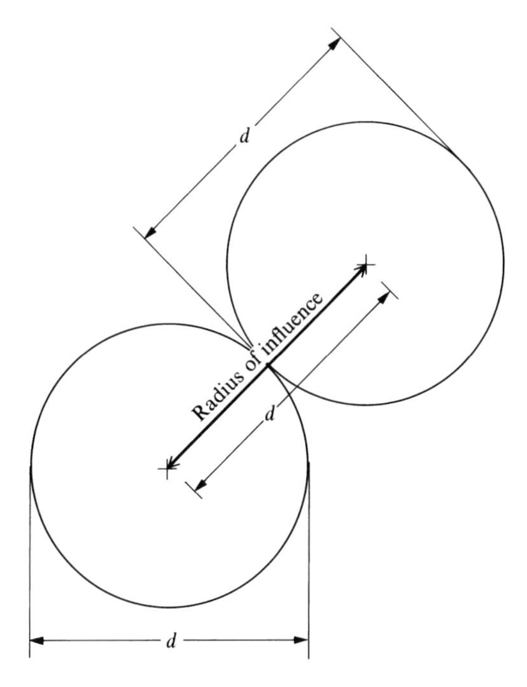

Fig. 12.2 Illustration of radius of influence.

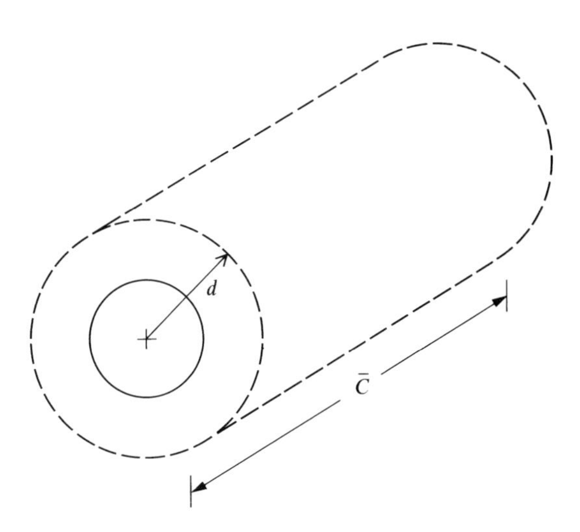

Fig. 12.3 Cylindrical volume swept out in 1 s by a particle with a radius of influence *d*, moving at a mean speed $\bar{C}$.

take into account the *relative* velocity between the molecules rather than the mean velocity of just one molecule. This requires a more sophisticated analysis beyond the scope of this book. However, the results are given next; they can be found derived in detail in Chapter 2 of [150]. The single particle collision frequency between a single molecule of chemical species A and the molecules of chemical species B is given by

$$Z_{AB} = n_B \pi d^2 \bar{C}_{AB} \tag{12.16}$$

where $\bar{C}_{AB}$ is a mean relative velocity between A and B molecules given by

$$\bar{C}_{AB} = \sqrt{\frac{8kT}{\pi m^*_{AB}}} \tag{12.17}$$

Hence, we can write

$$\boxed{Z_{AB} = n_B \pi d^2_{AB} \sqrt{\frac{8kT}{\pi m^*_{AB}}}} \tag{12.18}$$

In Eqs. (12.17) and (12.18), m^*_{AB} is the *reduced mass*, defined as

$$m^*_{AB} \equiv \frac{m_A m_B}{m_A + m_B} \tag{12.19}$$

where m_A and m_B are the masses of the A and B particles, respectively. For single-species gas, the single particle collision frequency is given by

$$\boxed{Z = \frac{n}{\sqrt{2}} \pi d^2 \bar{C} = \frac{n}{\sqrt{2}} \pi d^2 \sqrt{\frac{8kT}{\pi m}}} \tag{12.20}$$

Note that the difference between the simple result given by Eq. (12.14) and the more accurate result given by Eq. (12.20) is the factor $\sqrt{2}$, which takes into account the relative velocities between particles. Also note that Eq. (12.20) for a single-species gas does *not* fall out directly by simply inserting $m_A = m_B$ in Eqs. (12.18) and (12.19). To specialize Eq. (12.18) for a single-species gas, it must be divided by an additional factor of 2 because of the collision counting procedure used to derive Eq. (12.18). See [150] for the details.

Finally, for the mean free path of a single-species gas, taking into account the relative velocities of the molecules, it can be shown (see [150]) that

$$\boxed{\lambda = \frac{1}{\sqrt{2}\pi d^2 n}} \tag{12.21}$$

Note: In all of the preceding equations, the quantity πd^2 is frequently called the *collision cross section*, denoted by σ. For an accurate evaluation of collision frequency and mean free path, we need appropriate values of σ. These are obtained in various ways from experiment. For our purposes, we will assume that σ is a known quantity that can be obtained from the literature. Also note that, although we did not derive Eqs. (12.18), (12.20), and (12.21), they are certainly plausible based on our simple derivations of the similar but less exact equations given by Eqs. (12.14) and (12.15).

The principal reason for displaying the results shown in Eqs. (12.20) and (12.21) is to indicate how collision frequency and mean free path vary with the pressure and temperature of the gas. For example, because from the equation of state,

$$n = \frac{p}{kT}$$

we see from Eq. (12.20) that

$$\boxed{Z \propto \frac{p}{\sqrt{T}}} \tag{12.22}$$

and from Eq. (12.21) we see that

$$\boxed{\lambda \propto \frac{T}{p}} \tag{12.23}$$

Note that gases at high temperatures and low pressures are characterized by low collision frequencies and high mean free paths. These trends will be important in our discussions of nonequilibrium phenomena in subsequent chapters.

12.4 Velocity and Speed Distribution Functions: Mean Velocities

In this section we introduce the concept of a velocity distribution function as follows. Consider a system of N particles distributed in some manner (not necessarily uniformly) throughout physical space, as sketched in Fig. 12.4a. The instantaneous location of a particle is given by the location vector $\boldsymbol{r}$. For each particle, there is a corresponding point in the x-y-z physical space. The system of N particles is then represented by a cloud of N points in Fig. 12.4a. Also, at the same instant a given particle has a velocity $\boldsymbol{C}$, as represented in the velocity space shown in Fig. 12.4b. For each particle, there is a corresponding point in the C_x-C_y-C_z velocity space. Therefore, the system of N particles is also represented by a cloud of N points in Fig. 12.4b. Now consider a point in Fig. 12.4a denoted by $\boldsymbol{r}$, and a unit volume in physical space centered around that point. Simultaneously, consider a point in Fig. 12.4b denoted by $\boldsymbol{C}$, and a unit volume in velocity space centered around that point. Then, by definition, the distribution

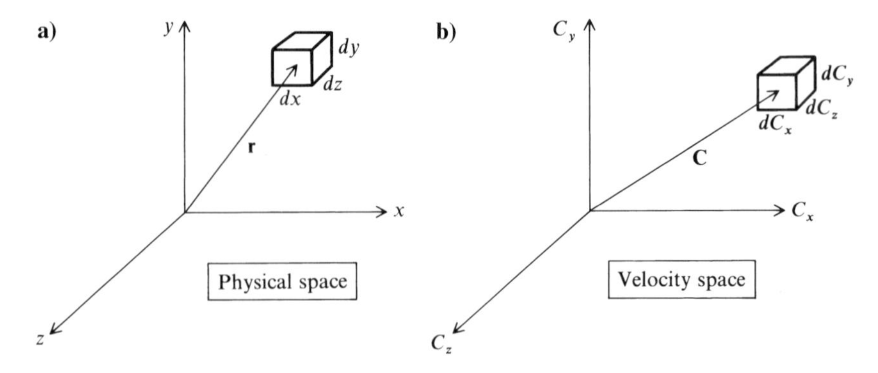

Fig. 12.4 Illustration of volume elements in physical and velocity spaces.

function $f(\boldsymbol{r}, \boldsymbol{C})$ is defined as the number of particles per unit volume of physical space at $\boldsymbol{r}$ with velocities per unit volume of velocity space at $\boldsymbol{C}$. In other words, let $\mathrm{d}x\,\mathrm{d}y\,\mathrm{d}z$ be an elemental volume in physical space, and $\mathrm{d}C_x\,\mathrm{d}C_y\,\mathrm{d}C_z$ be an elemental volume in velocity space, then

$$f(x, y, z, C_x, C_y, C_z)\ \mathrm{d}x\ \mathrm{d}y\ \mathrm{d}z\ \mathrm{d}C_x\ \mathrm{d}C_y\ \mathrm{d}C_z$$

represents the number of particles located between x and $x + \mathrm{d}x$, y and $y + \mathrm{d}y$, and z and $z + \mathrm{d}z$ with velocities that range from C_x to $C_x + \mathrm{d}C_x$, C_y to $C_y + \mathrm{d}C_y$, and C_z to $C_z + \mathrm{d}C_z$. Keep in mind that the gaseous system is composed of particles in constant motion in space, and that they collide with neighboring molecules, thus changing their velocities (in both magnitude and direction). Therefore, in the most general case of a nonequilibrium gas, the particles will be distributed nonuniformly throughout space and time, that is, the number of points within the element $\mathrm{d}x\ \mathrm{d}y\ \mathrm{d}z$ in Fig. 12.4a will be a function of $\boldsymbol{r}$ and t, and the number of points at any instant within the element $\mathrm{d}C_x\,\mathrm{d}C_y\,\mathrm{d}C_z$ in Fig. 12.4b might be changing with time.

The concept of the distribution function is a fundamental tool in classical kinetic theory. If we integrate f over all space and all velocities, we have

$$\int_{-\infty}^{\infty}\int_{-\infty}^{\infty}\int_{-\infty}^{\infty}\int_{-\infty}^{\infty}\int_{-\infty}^{\infty}\int_{-\infty}^{\infty} f(x, y, z, C_x, C_y, C_z) \times\ \mathrm{d}x\ \mathrm{d}y\ \mathrm{d}z\ \mathrm{d}C_x\ \mathrm{d}C_y\ \mathrm{d}C_z = N \tag{12.24}$$

One of the intrinsic values of the distribution function f is that the average value of any physical quantity Q, which is a function of space and/or velocity, $Q = Q(x, y, z, C_x, C_y, C_z)$ can be obtained from

$$\bar{Q} = \frac{1}{N}\int_{-\infty}^{\infty}\int_{-\infty}^{\infty}\int_{-\infty}^{\infty}\int_{-\infty}^{\infty}\int_{-\infty}^{\infty}\int_{-\infty}^{\infty} Qf\ \mathrm{d}x\ \mathrm{d}y\ \mathrm{d}z\ \mathrm{d}C_x\ \mathrm{d}C_y\ \mathrm{d}C_z \tag{12.25}$$

where $\bar{Q}$ is the average value of the property Q.

Now consider the special case of a gas in translational equilibrium. In terms of kinetic theory, a gas in equilibrium has the particles distributed uniformly throughout space (i.e., the number density n is a constant, independent of x, y, and z), and the number of molecular collisions that tend to decrease the number of points in the volume $\mathrm{d}C_x\,\mathrm{d}C_y\,\mathrm{d}C_z$ in velocity space (see Fig. 12.4b) is exactly balanced by other molecular collisions that increase the number of points in this elemental volume. For this case, f becomes essentially a *velocity distribution function*, $f = f(C_x, C_y, C_z)$. For a gas in translational equilibrium, f takes on a specific form that can be rigorously derived by examining the detailed collision processes within the gas. It is beyond the scope of this book to take the time and space for such a derivation; however, an excellent discussion is given in Chapter 2 of Vincenti and Kruger [150], which should be consulted for details. The result for the equilibrium velocity distribution function is

$$\boxed{f(C_x, C_y, C_z) = N\left(\frac{m}{2\pi kT}\right)^{3/2} \exp\left[-\frac{m}{2kT}(C_x^2 + C_y^2 + C_z^2)\right]} \tag{12.26}$$

Equation (12.26) is called the *Maxwellian distribution*; it gives the number of particles per unit volume of velocity space located by the velocity vector $\boldsymbol{C}$ in Fig. 12.4b. Keep in mind that Eq. (12.26) is a *velocity* distribution function, denoting both magnitude and direction.

In a system in equilibrium, f is a symmetric function, that is, $f(C_x, C_y, C_z) = f(-C_x, C_y, C_z) = f(C_x, -C_y, C_z)$, etc. Thus, for an equilibrium system, the velocity direction is not germane—we are concerned only with the *magnitudes* of the particle velocities, that is, the *speed* of the particles. Hence, we can introduce a *speed distribution function* $X(C)$ as follows. Consider the velocity space shown in Fig. 12.5. All particles on the surface of the sphere of radius C have the same speed. Now consider the space between the sphere of radius C and another sphere of radius $C + \mathrm{d}C$, where $\mathrm{d}C$ is an incremental change in speed. The volume of this space is $4\pi C^2 \mathrm{d}C$. Because the number of particles per unit volume of velocity space is given by Eq. (12.26), we then have for the number of particles in the space between the two spheres

$$4\pi N\left(\frac{m}{2\pi kT}\right)^{3/2} C^2 \exp\left(-\frac{mC^2}{2kT}\right) \mathrm{d}C$$

This gives the number of particles in the system with speeds between C and $C + \mathrm{d}C$. In turn, the number of particles with speed C per unit velocity change, which is defined as the *speed distribution* function χ, is given by

$$\boxed{\chi = 4\pi N\left(\frac{m}{2\pi kT}\right)^{3/2} C^2 e^{-(mC^2/2kT)}} \tag{12.27}$$

Equation (12.27) is plotted in Fig. 12.6. Clearly we see that, for a system in equilibrium at a given temperature, all of the particles do not move at the same speed; quite the contrary, some of the particles are moving slowly, others are moving

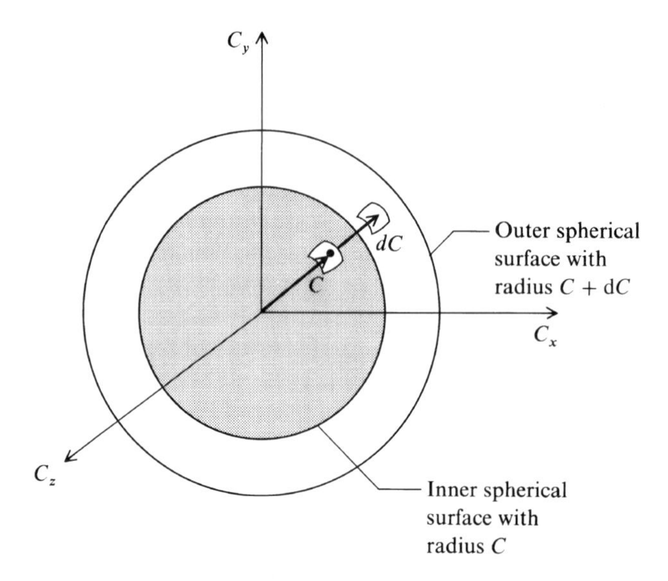

Fig. 12.5 Concentric spherical surfaces with radii C and $C + dC$.

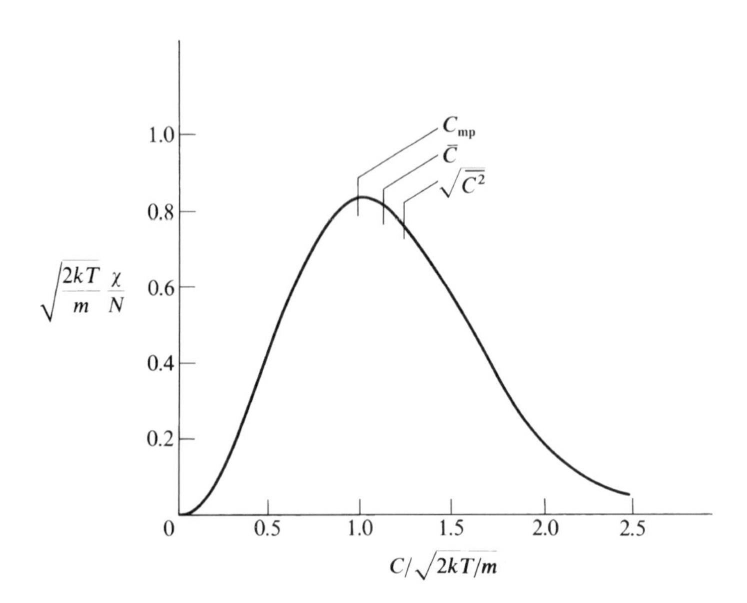

Fig. 12.6 Speed distribution function and the values of most probable speed C_{mp}, mean speed $\bar{C}$, and rms speed $(\bar{C}^2)^{1/2}$.

more rapidly, and Eq. (12.27) gives the distribution of these speeds over all the particles in the system.

Also noted in Fig. 12.6 are three speeds, defined as follows.

1) *Most probable speed*: This is the speed corresponding to the maximum value of χ, and it can be obtained by differentiating Eq. (12.27). The result is

$$\boxed{C_{\text{mp}} = \sqrt{2RT}} \tag{12.28}$$

where, as you recall, $R = k/m$.

2) *Average speed*: This is obtained from Eq. (12.25) by inserting $Q = C$. The result is

$$\boxed{\bar{C} = \sqrt{\frac{8RT}{\pi}}} \tag{12.29}$$

This is the speed that was used in Eqs. (12.14–12.16) and Eq. (12.20) dealing with collision frequency and the mean free path.

3) *Root-mean-square speed*: This is obtained from Eq. (12.25) by inserting $Q = C^2$. The result is

$$\boxed{\sqrt{\overline{C^2}} = \sqrt{3RT}} \tag{12.30}$$

This is the same result as obtained in Eq. (12.13) during our discussion of the perfect-gas equation of state.

The derivation of Eqs. (12.28–12.30) is left as a homework problem.

The molecular speeds given by Eqs. (12.28–12.30) are nearly equal to the speed of sound in a gas. We know that, for a perfect gas, the speed of sound is given by

$$a = \sqrt{\gamma RT}$$

which, for $\gamma = 1.4$, yields $a \approx 0.91 C_{\text{mp}}$. This makes sense because the energy of sound is transmitted through a gas by molecular collisions, and therefore the speed of this transmission should be somewhat related to the molecular speeds.

12.5 Summary and Comments

This brings to an end our elementary discussion of kinetic theory. We will revisit the discipline of kinetic theory in Chapter 16, when we discuss transport properties of high-temperature gases. However, for the time being we have shown the following:

1) The pV product is a form of energy in a gas, given by

$$pV = \tfrac{2}{3} E_{\text{trans}} \tag{12.7}$$

2) The equation of state for a perfect gas can be derived from kinetic theory as

$$\frac{p}{\rho} = \frac{1}{3}\bar{C}^2 \tag{12.12}$$

3) The single-particle collision frequency between a particle of A species with those of B species is

$$Z_{AB} = n_B \pi d_{AB}^2 \sqrt{\frac{8kT}{\pi m_{AB}^*}} \tag{12.18}$$

where m_{AB}^* is the reduced mass given by

$$m_{AB}^* = \frac{m_A m_B}{m_A + m_B} \tag{12.19}$$

For a gas made up of pure chemical species, the single-particle collision frequency is given by

$$Z = \frac{n}{\sqrt{2}} \pi d^2 \sqrt{\frac{8kT}{\pi m}} \tag{12.20}$$

4) The mean free path is given by

$$\lambda = \frac{1}{\sqrt{2}\pi d^2 n} \tag{12.21}$$

5) The variations of Z and λ with p and T are given by

$$Z \propto \frac{p}{\sqrt{T}} \tag{12.22}$$

and

$$\lambda \propto \frac{T}{p} \tag{12.23}$$

6) The Maxwellian distribution function for velocities in an equilibrium gas is

$$f = N\left(\frac{m}{2\pi kT}\right)^{3/2} \exp\left[-\frac{m}{2kT}(C_x^2 + C_y^2 + C_z^2)\right] \tag{12.26}$$

7) The corresponding speed distribution function is

$$\chi = 4\pi N\left(\frac{m}{2\pi kT}\right)^{3/2} C^2 e^{-(mC^2/2kT)} \tag{12.27}$$

8) The following speeds are obtained from Eq. (12.27).
Most probable speed:

$$C_{\text{mp}} = \sqrt{2RT} \tag{12.28}$$

Average speed:

$$\bar{C} = \sqrt{\frac{8RT}{\pi}} \tag{12.29}$$

Root-mean-square speed:

$$\sqrt{C^2} = \sqrt{3RT} \tag{12.30}$$

Problems

12.1 The single-particle collision frequency in a given gas is 2×10^{16} collisions per second. When the pressure and temperature are both increased by a factor of 4, what is the collision frequency?

12.2 Derive Eqs. (12.28), (12.29), and (12.30).

13
Chemical and Vibrational Nonequilibrium

Remember, then, that science is the guide of action; that the truth it arrives at is not that which we can ideally contemplate without error, but that which we may act upon without fear; and you cannot fail to see that scientific thought is not an accomplishment or condition of human progress, but human progress itself.

William Kingdon Clifford, 1872

Chapter Preview

This chapter is the antithesis of Chapter 11. In Chapter 11, we studied how to calculate the *equilibrium* thermodynamic properties and chemical composition of a chemically reacting mixture. In the present chapter, we learn how to calculate the *nonequilibrium* thermodynamic properties and chemical composition of a chemically reacting mixture. Take a chunk of the air around you as you are reading this book, and imagine that (somehow) you instantaneously increase the temperature of this chunk of air to 1500 K. The vibrational energy of the air will increase, but it will not be instantaneous. Because molecules have their vibrational energy changed by collisions with other molecules, and these collisions take time (as discussed in Chapter 12), your chunk of air will experience a *time rate of increase* of vibrational energy. Finally, if you wait around long enough, the vibrational energy of your chunk of air will come up to its equilibrium value (the subject of Chapter 11). But in the meantime, what has been the time rate of increase of vibrational energy? The answer is the stuff of the present chapter.

Now, take your chunk of air and (somehow) instantaneously increase its temperature to 5000 K. Chemical reactions will occur; the oxygen and nitrogen molecules will dissociate, forming oxygen and nitrogen atoms. Nitric oxide will be produced. But these chemical reactions do not go to completion instantaneously. Chemical reactions take place through collisions of molecules, and these collisions take time. Your chunk of air will experience a time rate of increase of oxygen atoms, a time rate of decrease of oxygen molecules, a time rate of formation of nitric oxide, and so forth. Finally, if you wait around long enough, the chemical composition of your chunk of air will arrive at its equilibrium value (the subject of Chapter 11). But in the meantime, what has been the time rate of increase of oxygen atoms? Of nitrogen atoms? And so forth. The answer is the stuff of this chapter.

Why do we need to know these answers? Imagine that you now sling your chunk of air at high velocity over a hypersonic vehicle at the same time that the vibrational energy and chemical composition of the chunk of air are changing, in such a fashion that the equilibrium vibrational energy and chemical composition is not achieved in the chunk of air until it is 100 m behind the vehicle. Then the flowfield over the vehicle will be a *nonequilibrium flowfield*, and to analyze the thermodynamic and aerodynamic characteristics of this nonequilibrium flow, the answers to be developed in this chapter are absolutely essential.

So read on with interest. The analyses of nonequilibrium processes have a certain intellectual beauty to them. As you read on with interest, also sit back and enjoy.

13.1 Introduction

All vibrational and chemical processes take place by molecular collisions and/or radiative interactions. Considering just molecular collisions, visualize, for example, an O_2 molecule colliding with other molecules in the system. If the O_2 vibrational energy is in the ground level before collision, it might or might not be vibrationally excited after the collision. Indeed, in general the O_2 molecule must experience a large number of collisions, typically on the order of 20,000, before it will become vibrationally excited. The actual number of collisions required depends on the type of molecule and the relative kinetic energy between the two colliding particles—the higher the kinetic energy (hence the higher the gas temperature), the fewer collisions are required for vibrational energy exchange. Moreover, as the temperature of the gas is increased, and hence the molecular collisions become more violent, it is probable that the O_2 molecule will be torn apart (dissociated) by collisions with other particles. However, this requires a large number of collisions, on the order of 200,000. The important point to note here is that vibrational and chemical changes take place as a result of collisions. In turn, collisions take time to occur. Hence, vibrational and chemical changes in a gas take time to occur. The precise amount of time depends on the molecular collision frequency Z, defined in Sec. 12.3. The results given by Eq. (12.22) show that $Z \propto p/\sqrt{T}$; hence, the collision frequency is low for low pressures and very high temperatures.

The equilibrium systems considered in Chapters 10 and 11 assumed that the gas has had enough time for the necessary collisions to occur and that the properties of the system at a fixed p and T are constant, independent of time.

However, there are many problems in high-speed gas dynamics where the gas is not given the luxury of the necessary time to come to equilibrium. A typical example is the flow across a shock wave, where the pressure and temperature are rapidly increased within the shock front. Consider a fluid element passing through this shock front. When its p and T are suddenly increased, its equilibrium vibrational and chemical properties will change. The fluid element will start to seek these new equilibrium properties, but this requires molecular collisions, and hence time. By the time enough collisions have occurred and equilibrium properties have been approached, the fluid element has moved a certain distance downstream of the shock front. Hence, there will be a certain region immediately behind the shock wave where equilibrium conditions do not prevail—there will be a nonequilibrium region. To study the nonequilibrium region, additional techniques must be developed that take into account the time required for molecular collisions. Such techniques are the subject of this chapter. The detailed study of both equilibrium and nonequilibrium flows through shock waves, as well as many other types of flows, will be made in Chapters 15–17.

Because of the straightforward linear arrangement of this short chapter, a local chapter road map is not needed. In terms of the road map for the book in Fig. 1.24, this chapter wraps up the first item on the right-hand column under high-temperature flows.

13.2 Vibrational Nonequilibrium: The Vibrational Rate Equation

In this section we derive an equation for the time rate of change of vibrational energy of a gas as a result of molecular collisions—the vibrational rate equation. In turn, this equation will be coupled with the continuity, momentum, and energy equations in subsequent chapters for the study of certain types of nonequilibrium flows.

Consider a diatomic molecule with a vibrational energy level diagram as illustrated in Fig. 13.1. Focus on the ith level. The population of this level N_i is increased by particles jumping up from the $i - 1$ level (transition a shown in Fig. 13.1) and by particles dropping down from the $i + 1$ level (transition b). The population N_i is decreased by particles jumping up to the $i + 1$ level (transition c) and dropping down to the $i - 1$ level (transition d). For the time being, consider just transition c. Let $P_{i,i+1}$ be the probability that a molecule in the ith level, upon collision with another molecule, will jump up to the $i + 1$ level. $P_{i,i+1}$ is called the transition probability and can be interpreted on a dimensional basis as the number of transitions per collision per particle (of course keeping in mind that a single transition requires many collisions). The value of $P_{i,i+1}$ is always *less* than unity. Also, let Z be the collision frequency as given by Eq. (12.20), where Z is the number of collisions per particle per second. Hence, the product $P_{i,i+1}Z$ is physically the number of transitions per particle per second. If there are N_i particles in level i, then $P_{i,i+1}ZN_i$ is the total number of transitions per second for the gas from the ith to the $i + 1$ energy level. Similar definitions can be made for transitions a, b, and d in Fig. 13.1. Therefore, on purely physical grounds, using the preceding, definitions, we can

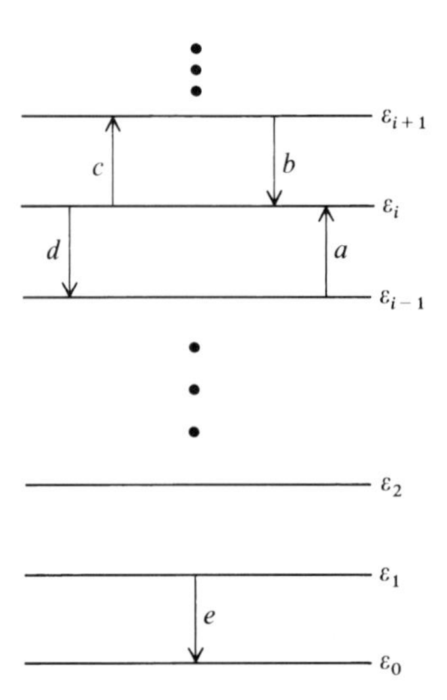

Fig. 13.1 Single quantum transition for vibrational energy exchange.

write the net rate of change of the population of the ith level as

$$\frac{dN_i}{dt} = \underbrace{P_{i+1,i}ZN_{i+1} + P_{i-1,i}ZN_{i-1}}_{\text{Rate of increase of } N_i} - \underbrace{P_{i,i+1}ZN_i - P_{i,i-1}ZN_i}_{\text{Rate of decrease of } N_i}$$

To simplify the preceding equation, define a vibrational rate constant $k_{i+1,i}$ such that $P_{i+1,i}Z \equiv k_{i+1,i}$, similarly for the other transitions. Then the preceding equation becomes

$$\boxed{\frac{dN_i}{dt} = k_{i+1,i}N_{i+1} + k_{i-1,i}N_{i-1} - k_{i,i+1}N_i - k_{i,i-1}N_i} \tag{13.1}$$

Equation (13.1) is called the master equation for vibrational relaxation.

For a moment, consider that the gas is in equilibrium. Hence, from the Boltzmann distribution, Eq. (11.25), and the quantum mechanical expression for vibrational energy, $h\nu(n + \frac{1}{2})$, given in Sec. 11.7

$$\frac{N_i^*}{N_{i-1}^*} = \frac{e^{-\varepsilon_i/kT}}{e^{-\varepsilon_{i-1}/kT}} = e^{-h\nu/kT} \tag{13.2}$$

Moreover, in equilibrium, each transition in a given direction is exactly balanced by its counterpart in the opposite direction—this is called the principle of detailed balancing. That is, the number of transitions a per second must exactly equal the number of transitions d per second:

$$k_{i-1,i}N_{i-1}^* = k_{i,i-1}N_i^*$$

or

$$k_{i-1,i} = k_{i,i-1} \frac{N_i^*}{N_{i-1}^*} \tag{13.3}$$

Combining Eqs. (13.2) and (13.3), we have

$$\boxed{k_{i-1,i} = k_{i,i-1} e^{-h\nu/kT}} \tag{13.4}$$

Equation (13.4) is simply a relation between reciprocal rate constants; hence, it holds for nonequilibrium as well as equilibrium conditions. Taking a result from quantum mechanics, it can also be shown that all of the rate constants for higher-lying energy levels can be expressed in terms of the rate constant for transition e in Fig. 13.1, that is, the transition from $i = 1$ to $i = 0$:

$$\boxed{k_{i,i-1} = ik_{1,0}} \tag{13.5}$$

From Eq. (13.5), we can also write

$$k_{i+1,i} = (i+1)k_{1,0} \tag{13.6}$$

Combining Eqs. (13.4) and (13.5), we have

$$k_{i-1,i} = ik_{1,0} e^{-h\nu/kT} \tag{13.7}$$

and from Eqs. (13.4), (13.5), and (13.6), we have

$$k_{i,i+1} = k_{i+1,i} e^{-h\nu/kT} = (i+1)k_{1,0} e^{-h\nu/kT} \tag{13.8}$$

Substituting Eqs. (13.5–13.8) into Eq. (13.1), we have

$$\frac{\mathrm{d}N_i}{\mathrm{d}t} = (i+1)k_{1,0}N_{i+1} + ik_{1,0}e^{-h\nu/kT}N_{i-1} - (i+1)k_{1,0}e^{-h\nu/kT}N_i - ik_{1,0}N_i$$

or

$$\frac{\mathrm{d}N_i}{\mathrm{d}t} = k_{1,0}\{-iN_i + (i+1)N_{i+1} + e^{-h\nu/kT}[-(i+1)N_i + iN_{i-1}]\} \tag{13.9}$$

In many gas dynamics problems, we are more interested in energies than populations. Let us convert Eq. (13.9) into a rate equation for e_{vib}. Assume that we are dealing with a unit mass of gas. From Secs. 11.2 and 11.7,

$$e_{\text{vib}} = \sum_{i=1}^{\infty} \varepsilon_i N_i = \sum_{i=1}^{\infty} (ih\nu)N_i = h\nu \sum_{i=0}^{\infty} iN_i$$

Hence

$$\frac{de_{\text{vib}}}{dt} = h\nu \sum_{i=1}^{\infty} i \frac{dN_i}{dt} \tag{13.10}$$

Substitute Eq. (13.9) into (13.10):

$$\frac{de_{\text{vib}}}{dt} = h\nu k_{1,0} \sum_{i=1}^{\infty} \{-i^2 N_i + i(i+1)N_{i+1} + e^{-h\nu/kT}[-i(i+1)N_i + i^2 N_{i-1}]\} \tag{13.11}$$

Considering the first two terms in Eq. (13.11) and letting $s = i + 1$,

$$\begin{aligned}
\sum_{i=1}^{\infty}[-i^2 N_i + i(i+1)N_{i+1}] &= -\sum_{i=1}^{\infty} i^2 N_i + \sum_{s=2}^{\infty}(s-1)sN_s \\
&= -\sum_{i=1}^{\infty} i^2 N_i + \sum_{s=2}^{\infty} s^2 N_s - \sum_{s=2}^{\infty} sN_s \\
&= -N_1 - \sum_{i=2}^{\infty} i^2 N_i + \sum_{s=2}^{\infty} s^2 N_s - \sum_{s=2}^{\infty} sN_s \\
&= -N_1 - \sum_{s=2}^{\infty} sN_s = -\sum_{i=1}^{\infty} iN_i = -\sum_{i=0}^{\infty} iN_i
\end{aligned}$$

Also, a similar reduction for the last two terms in Eq. (13.11) leads to

$$\sum_{i=1}^{\infty}[-i(i+1)N_i + i^2 N_{i-1}] = \sum_{i=0}^{\infty}(i+1)N_i$$

Thus, Eq. (13.11) becomes

$$\begin{aligned}
\frac{de_{\text{vib}}}{dt} &= h\nu k_{1,0} \sum_{i=0}^{\infty}[-iN_i + e^{-h\nu/kT}(i+1)N_i] \\
&= h\nu k_{1,0}\left[e^{-h\nu/kT} \sum_{i=0}^{\infty} N_i - (1 - e^{-h\nu/kT}) \sum_{i=0}^{\infty} iN_i\right]
\end{aligned} \tag{13.12}$$

However

$$\sum_{i=0}^{\infty} N_i = N$$

and

$$e_{\text{vib}} = \sum_{i=0}^{\infty} \varepsilon_i N_i = h\nu \sum_{i=0}^{\infty} iN_i$$

Therefore

$$\sum_{i=0}^{\infty} iN_i = \frac{e_{\text{vib}}}{h\nu}$$

Thus, Eq. (13.12) can be written as

$$\frac{\mathrm{d}e_{\text{vib}}}{\mathrm{d}t} = h\nu k_{1,0}\left[e^{-h\nu/kT}N - (1 - e^{-h\nu/kT})\frac{e_{\text{vib}}}{h\nu}\right]$$

or

$$\frac{\mathrm{d}e_{\text{vib}}}{\mathrm{d}t} = k_{1,0}(1 - e^{-h\nu/kT})\left[\frac{h\nu N}{e^{h\nu/kT} - 1} - e_{\text{vib}}\right] \tag{13.13}$$

However, recalling that we are dealing with a unit mass and hence N is the number of particles per unit mass, we have, from Sec. 10.2, that $Nk = R$, the specific gas constant. Then, considering one of the expressions in Eq. (13.13),

$$\frac{h\nu N}{e^{h\nu/kT} - 1} = \frac{h\nu/kT}{e^{h\nu/kT} - 1}(NkT) = \frac{h\nu/kT}{e^{h\nu/kT} - 1}RT \tag{13.14}$$

The right-hand side of Eq. (13.14) is simply the equilibrium vibrational energy from Eq. (11.61); we denote it by e_{vib}^{eq}. Hence, from Eq. (13.14)

$$\frac{h\nu N}{e^{h\nu/kT} - 1} = e_{\text{vib}}^{eq} \tag{13.15}$$

Substituting Eq. (13.15) into Eq. (13.13),

$$\frac{\mathrm{d}e_{\text{vib}}}{\mathrm{d}t} = k_{1,0}(1 - e^{-h\nu/kT})(e_{\text{vib}}^{eq} - e_{\text{vib}}) \tag{13.16}$$

In Eq. (13.16), the factor $k_{1,0}(1 - e^{-h\nu/kT})$ has units of s^{-1}. Therefore, we define a vibrational relaxation time τ as

$$\tau \equiv \frac{1}{k_{1,0}(1 - e^{-h\nu/kT})}$$

Thus, Eq. (13.16) becomes

$$\boxed{\frac{\mathrm{d}e_{\text{vib}}}{\mathrm{d}t} = \frac{1}{\tau}(e_{\text{vib}}^{eq} - e_{\text{vib}})} \tag{13.17}$$

Equation (13.17) is called the vibrational rate equation, and it is the main result of this section. Equation (13.17) is a simple differential equation that relates the time rate of change of e_{vib} to the difference between the equilibrium value it is seeking and its local instantaneous nonequilibrium value.

The physical implications of Eq. (13.17) can be seen as follows. Consider a unit mass of gas in equilibrium at a given temperature T. Hence,

$$e_{\text{vib}} = e_{\text{vib}}^{eq} = \frac{h\nu/kT}{e^{h\nu/kT} - 1} RT \tag{13.18}$$

Now let us instantaneously excite the vibrational mode above its equilibrium value (say, by the absorption of radiation of the proper wavelength, e.g., we "zap" the gas with a laser). Let e_{vib_o} denote the instantaneous value of e_{vib} immediately after the excitation at time $t = 0$. This is illustrated in Fig. 13.2. Note that $e_{\text{vib}_o} > e_{\text{vib}}^{eq}$. Because of molecular collisions, the excited particles will exchange this "excess" vibrational energy with the translational and rotational energy of the gas, and after a period of time e_{vib} will decrease and approach its equilibrium value. This is illustrated by the solid curve in Fig. 13.2. However, note that, as the vibrational energy drains away, it reappears in part as an increase in translational energy. Because the temperature of the gas is proportional to the translational energy [see Eq. (11.57)], T increases. In turn, the equilibrium value of vibrational energy, from Eq. (13.18), will also increase. This is shown by the dashed line in Fig. (13.2). At large times, e_{vib} and e_{vib}^{eq} will asymptotically approach the same value.

The relaxation time τ in Eq. (13.17) is a function of both local pressure and temperature. This is easily recognized because τ is a combination of the transition probability P and the collision frequency Z, both defined earlier. In turn, P depends on T (on the relative kinetic energy between colliding particles), and

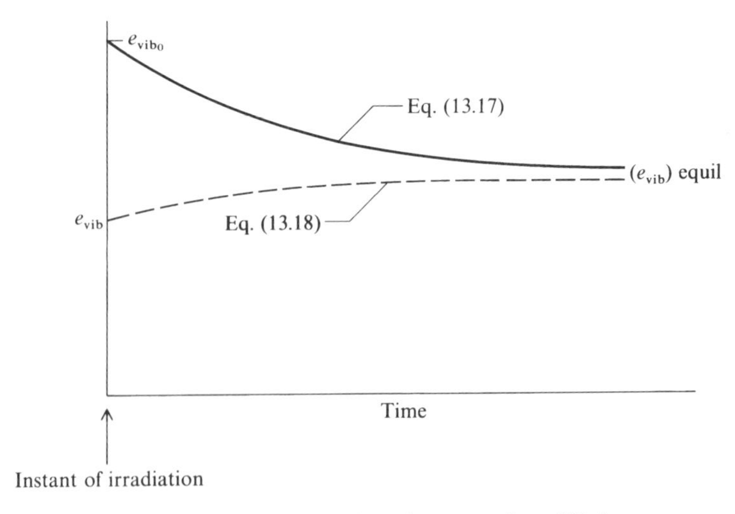

Fig. 13.2 Vibrational relaxation toward equilibrium.

$Z \propto p/\sqrt{T}$. For most diatomic gases, the variation of τ is given by the form

$$\tau p = C_1 e^{(C_2/T)^{1/3}}$$

or

$$\ell n \tau p = \ell n C_1 + \left(\frac{C_2}{T}\right)^{1/3} \tag{13.19}$$

The value of C_1 and C_2 must be obtained from experimental measurements. At this stage in our discussion, we raise a problem that has always plagued the analysis of nonequilibrium systems, namely, the uncertainties that exist in the rate data, such as in the measured values of C_1 and C_2. Such measurements must be made in high-temperature gases, and the experimental facility for generating such high-temperature gases is a shock tube, with testing times on the order of tens of microseconds. (See, for example, [4] for a discussion of shock tubes.) It is no wonder that a large scatter invariably occurs in the data, with associated uncertainties in the values of C_1 and C_2. Nevertheless, reasonable values of C_1 and C_2 that reflect the literature are given in [150] and are summarized in Table 13.1 for the pure gases O_2, N_2, and NO.

The temperature range of the data listed in Table 13.1 is approximately from 800 to 6000 K. Also, the preceding values of C_1 are C_2 are different if the O_2, N_2, or NO are in a mixture of different gases. For example, if O_2 is in a bath of N_2 molecules, then the vibrational relaxation time for O_2 caused by collisions with N_2 is quoted in [158] as given by $C_1 = 1.36 \times 10^{-4}$ atm-μs and $C_2 = 2.95 \times 10^6$ K.

In summary, the nonequilibrium variation of vibrational energy is given by the vibrational rate equation expressed as Eq. (13.17). Note that in Eq. (13.17) both τ and e_{vib}^{eq} are variables, with $\tau = (p, T)$ from Eq. (13.19) and $e_{\text{vib}}^{eq} = e(T)$ from Eq. (13.18). However, a word of caution is given. Equation (13.17) has certain limitations that have not been stressed during the preceding derivation, namely, it holds only for diatomic molecules that are harmonic oscillators. The use of $e_{\text{vib}} = h\nu n$, obtained from Sec. 11.7, is valid only if the molecule is a harmonic oscillator. Moreover, from Fig. 13.1 we have considered only single quantum jumps between energy levels, that is, we did not consider transitions, say, from the ith directly to the $i+2$ level. Such multiple quantum jumps can occur for anharmonic molecules, but their transition probabilities are very small. In spite of these restrictions, experience has proven that Eq. (13.17) is

Table 13.1 Vibrational rate data for Eq. (13.19)

Species	C_1 atm-μs	C_2, K
O_2	5.42×10^{-5}	2.95×10^6
N_2	7.12×10^{-3}	1.91×10^6
NO	4.86×10^{-3}	1.37×10^5

reasonably valid for real problems dealing with diatomic gases, and it is employed in almost all nonequilibrium analyses of such gases.

Recent developments in the study of vibrational nonequilibrium flows have highlighted a further limitation of Eq. (13.17), as follows. The energy level transitions included in the master equation (13.1) are so-called translation-vibration (T-V) transfers. Here, a molecule upon collision with another will gain or lose vibrational energy, which then reappears as a decrease or increase in translational kinetic energy of the molecules. For example, a T-V transfer in CO can be given as

$$\mathrm{CO}(n) + \mathrm{CO}(n) \rightleftharpoons \mathrm{CO}(n-1) + \mathrm{CO}(n) + \mathrm{KE}$$

where a CO molecule in the nth vibrational level drops to the $(n-1)$ level after collision, with the consequent release of kinetic energy (KE). However, vibration-vibration (V-V) transfers also occur, where the vibrational quantum lost by one molecule is gained by its collision partner. For example, a V-V transfer in CO can be given as

$$\mathrm{CO}(n) + \mathrm{CO}(n) \rightleftharpoons \mathrm{CO}(n+1) + \mathrm{CO}(n-1)$$

The preceding equation assumes a harmonic oscillator, where the spacings between all energy levels are the same. However, all molecules are in reality anharmonic oscillators, which results in unequal spacings between vibrational energy levels. Thus, in a V-V transfer involving anharmonic molecules, there is a small amount of translational energy exchanged in the process, as follows:

$$\mathrm{CO}(n) + \mathrm{CO}(n) \rightleftharpoons \mathrm{CO}(n+1) + \mathrm{CO}(n-1) + \mathrm{KE}$$

During an expansion process (decreasing temperature), the V-V transfers among anharmonic molecules result in an overpopulation of some of the higher energy levels than would be the case of a harmonic oscillator. This is called anharmonic pumping and is particularly important in several types of gas dynamic and chemical lasers. The reverse effect occurs in a compression process (increasing temperature). In cases where anharmonic pumping is important, Eq. (13.17) is not valid, and the analysis must start from a master rate equation [such as Eq. (13.1)] expanded to include V-V transfers. For a fundamental discussion of the anharmonic pumping effect at an introductory level, see pp. 112–120 of [147].

Vibrational nonequilibrium effects are particularly important in hypersonic wind-tunnel nozzle expansions, in the expanding high-temperature flow over blunt-nosed bodies, and in the region immediately behind a strong shock wave. These matters will be discussed in Chapter 15.

13.3 Chemical Nonequilibrium: The Chemical Rate Equation

Consider a system of oxygen in chemical equilibrium at $p = 1$ atm and $T = 3000$ K. Although Fig. 11.11 is for air, it clearly demonstrates that the oxygen under these conditions should be partially dissociated. Thus, in our system, both O_2 and O will be present in their proper equilibrium amounts. Now, assume that somehow T is instantaneously increased to, say, 4000 K. Equilibrium conditions at this higher temperature demand that the amount of O_2 decrease and the amount of O increase. However, as explained in Sec. 13.1, this change in composition takes place via molecular collisions, and hence it takes time to adjust to the new equilibrium conditions. During this non-equilibrium adjustment period, chemical reactions are taking place at a definite net rate. The purpose of this section is to establish relations for the finite time rate of change of each chemical species present in the mixture—the chemical rate equations.

Continuing with our example of a system of oxygen, the only chemical reaction taking place is

$$O_2 + M \longrightarrow 2O + M \tag{13.20}$$

where M is a collision partner; it can be either O_2 or O. In terms of notation, in Eq. (10.19) the symbol C_i denoted the concentration of species i (moles of i per unit volume). Here we introduce an alternative notation for concentration, where $[O_2]$, $[N_2]$, etc. denote the concentrations of O_2, N_2, etc. Such a bracket notation was not used in Chapter 10 because it would make the equation of state look "funny." However, in equations dealing with chemical nonequilibrium, the use of $[O_2]$ to denote the concentration of O_2 is convenient. Using the bracket notation for concentration, we denote the number of moles of O_2 and O per unit volume of the mixture by $[O_2]$ and $[O]$, respectively. Empirical results have shown that the time rate of formation of O atoms via Eq. (13.20) is given by

$$\frac{\mathrm{d}[O]}{\mathrm{d}t} = 2k[O_2][M] \tag{13.21}$$

where $\mathrm{d}[O]/\mathrm{d}t$ is the reaction rate, k is the reaction rate constant, and Eq. (13.21) is called a reaction rate equation. The reaction rate constant k is a function of T only. Equation (13.21) gives the rate at which the reaction given in Eq. (13.20) goes from left to right; this is called the forward rate, and k is really the forward rate constant k_f:

$$O_2 + M \xrightarrow{k_f} 2O + M$$

Hence, Eq. (13.21) is more precisely written as follows.

Forward rate:

$$\frac{\mathrm{d}[O]}{\mathrm{d}t} = 2k_f[O_2][M] \tag{13.22}$$

The reaction in Eq. (13.20) that would proceed from right to left is called the *reverse reaction*, or *backward reaction*,

$$O_2 + M \xleftarrow[k_b]{} 2O + M$$

with an associated *reverse* or *backward rate constant* k_b, and a *reverse* or *backward rate* given by the following.

Reverse rate:

$$\frac{d[O]}{dt} = -2k_b[O]^2[M] \tag{13.23}$$

Note that in both Eqs. (13.22) and (13.23), the right-hand side is the product of the concentrations of those particular colliding molecules that produce the chemical change, raised to the power equal to their stoichiometric mole number in the chemical equation. Equation (13.22) gives the time rate *of increase* of O atoms because of the forward rate, and Eq. (13.23) gives the time rate of *decrease* of O atoms because of the reverse rate. However, what we would actually observe in the laboratory is the *net* time rate of change of O atoms caused by the *combined* forward and reverse reactions

$$O_2 + M \underset{k_b}{\overset{k_f}{\rightleftharpoons}} 2O + M$$

and the net reaction rate is given by the following.

Net rate:

$$\boxed{\frac{d[O]}{dt} = 2k_f[O_2][M] - 2k_b[O]^2[M]} \tag{13.24}$$

Now consider our system to again be in chemical equilibrium; hence, the composition is fixed with time. Then $d[O]/dt \equiv 0$, $[O_2] \equiv [O_2]^*$, and $[O] \equiv [O]^*$ where the asterisk denotes equilibrium conditions. In this case, Eq. (13.24) becomes

$$0 = 2k_f[O_2]^*[M]^* - 2k_b[O]^{*2}[M]^*$$

or

$$k_f = k_b \frac{[O]^{*2}}{[O_2]^*} \tag{13.25}$$

Examining the chemical equation just given, we can define the ratio $[O]^{*2}/[O_2]^*$ in Eq. (13.25) as an equilibrium constant *based on concentrations* K_c. This is related to the equilibrium constant based on partial pressures K_p, defined in Sec. 10.9. From Eq. (10.19) it directly follows for the preceding oxygen reaction that

$$K_c = \frac{1}{\mathscr{R}T} K_p$$

In general, we have

$$K_c = \left(\frac{1}{\mathscr{R}T}\right)^{\Sigma_i \nu_i} K_p$$

Hence, Eq. (13.25) can be written as

$$\boxed{\frac{k_f}{k_b} = K_c} \tag{13.26}$$

Equation (13.26), although derived by assuming equilibrium, is simply a relation between the forward and reverse rate constants, and therefore it holds in general for nonequilibrium conditions. Therefore, the net rate, Eq. (13.24), can be expressed as

$$\boxed{\frac{\mathrm{d}[\mathrm{O}]}{\mathrm{d}t} = 2k_f[M]\left\{[\mathrm{O}_2] - \frac{1}{K_c}[\mathrm{O}]^2\right\}} \tag{13.27}$$

In practice, values for k_f are found from experiment, and then k_b can be directly obtained from Eq. (13.26). Keep in mind that k_f, k_b, K_c, and K_p for a given reaction are all functions of temperature only. Also, k_f in Eq. (13.27) is generally different depending on whether the collision partner M is chosen to be O_2 or O.

The preceding example has been a special application of the more general case of a reacting mixture of n different species. Consider the general chemical reaction (but it must be an elementary reaction, as defined later)

$$\sum_{i=1}^{n} \nu_i' X_i \underset{k_b}{\overset{k_f}{\rightleftharpoons}} \sum_{i=1}^{n} \nu_i'' X_i \tag{13.28}$$

where ν_i' and ν_i'' represent the stoichiometric mole numbers of the reactants and products, respectively. (Note that in our preceding example for oxygen where the chemical reaction was $O_2 + M \rightleftharpoons 2O + M$, $\nu'_{O_2} = 1$, $\nu'_{O} = 0$, $\nu''_{O_2} = 0$, $\nu'_M = 1$, $\nu''_M = 1$, and $\nu''_O = 2$.) For the preceding general reaction, Eq. (13.28), we write the following.

Forward rate:

$$\boxed{\frac{\mathrm{d}[X_i]}{\mathrm{d}t} = (\nu_i'' - \nu_i')k_f \prod_i [X_i]^{\nu_i'}} \tag{13.29}$$

Reverse rate:

$$\boxed{\frac{\mathrm{d}[X_i]}{\mathrm{d}t} = -(\nu_i'' - \nu_i')k_b \prod_i [X_i]^{\nu_i''}} \tag{13.30}$$

Net rate:

$$\boxed{\frac{\mathrm{d}[X_i]}{\mathrm{d}t} = (\nu_i'' - \nu_i')\left\{k_f \prod_i [X_i]^{\nu_i'} - k_b \prod_i [X_i]^{\nu_i''}\right\}} \tag{13.31}$$

Equation (13.31) is a generalized net rate equation; it is a general form of the law of mass action first introduced in Sec. 10.9. In addition, the relation between k_f and k_b given by Eq. (13.26) holds for the general reaction given in Eq. (13.28).

The chemical rate constants are generally measured experimentally. Although methods from kinetic theory exist for their theoretical estimation, such results are sometimes uncertain by orders of magnitude. The empirical results for many reactions can be correlated in the form

$$\boxed{k = Ce^{-\varepsilon_a/kT}} \tag{13.32}$$

where ε_a is defined as the *activation energy* and C is a constant. Equation (13.32) is called the *Arrhenius equation.* An improved formula includes a preexponential temperature factor

$$\boxed{k = c_1 T^{\alpha} e^{-\varepsilon_0/kT}} \tag{13.33}$$

where c_1, α, and ε_0 are all found from experimental data.

Returning to the special case of a dissociation reaction such as for diatomic nitrogen

$$N_2 + M \xrightarrow{k_f} 2N + M$$

the dissociation energy ε_d is denned as the difference between the zero-point energies,

$$\varepsilon_d \equiv \Delta\varepsilon_0 = 2(\varepsilon_0)_N - (\varepsilon_0)_{N_2}$$

For this reaction, the rate constant is expressed as

$$k_f = c_f T^{\alpha} e^{-\varepsilon_d/kT} \tag{13.34}$$

where the activation energy $\varepsilon_a = \varepsilon_d$. Physically, the dissociation energy is the energy required to dissociate the molecule at $T = 0$ K. It is obviously a finite number: it takes energy—sometimes a considerable amount of energy—to tear a molecule apart. In contrast, consider the recombination reaction

$$N_2 + M \underset{k_b}{\longleftarrow} 2N + M$$

Here, no relative kinetic energy between the two colliding N atoms is necessary to bring about a change; indeed, the role of the third body M is to carry away some of the energy that must be given up by the two colliding N atoms before they can recombine. Hence, for recombination, there is no activation energy; $\varepsilon_a = 0$. Thus, the recombination rate constant is written as

$$k_b = c_b T^{\eta b} \tag{13.35}$$

with no exponential factor.

Finally, it is important to note that all of the preceding formalism applies only to elementary reactions. An elementary chemical reaction is one that takes place in a single step. For example, a dissociation reaction such as

$$O_2 + M \longrightarrow 2O + M$$

is an elementary reaction because it literally takes place by a collision of an O_2 molecule with another collision partner, yielding directly two oxygen atoms. On the other hand, the reaction

$$2H_2 + O_2 \longrightarrow 2H_2O \tag{13.36}$$

is not an elementary reaction. Two hydrogen molecules do not come together with one oxygen molecule to directly yield two water molecules, even though if we mixed the hydrogen and oxygen together in the laboratory our naked eye would observe what would appear to be the direct formation of water. Reaction (13.36) does not take place in a single step. Instead, Eq. (13.36) is a statement of an overall reaction that actually takes place through a series of elementary steps:

$$H_2 \longrightarrow 2H \tag{13.37a}$$

$$O_2 \longrightarrow 2O \tag{13.37b}$$

$$H + O_2 \longrightarrow OH + O \tag{13.37c}$$

$$O + H_2 \longrightarrow OH + H \tag{13.37d}$$

$$OH + H_2 \longrightarrow H_2O + H \tag{13.37e}$$

Equations (13.37a–13.37e) constitute the reaction mechanism for the overall reaction (13.36). Each of Eqs. (13.37a–13.37e) is an elementary reaction.

We again emphasize that Eqs. (13.21) through (13.35) apply only for elementary reactions. In particular, the law of mass action given by Eq. (13.31) is valid for elementary reactions only. We *cannot* write Eq. (13.31) for reaction (13.36), but we can apply Eq. (13.31) to *each one* of the elementary reactions that constitute the reaction mechanism (13.37a–13.37e).

13.4 Chemical Nonequilibrium in High-Temperature Air

We again highlight the importance of air in high-speed compressible flow problems. For the analysis of chemical nonequilibrium effects in high-temperature air, the following reaction mechanism occurs, valid below 9000 K:

$$O_2 + M \underset{k_{b_1}}{\overset{k_{f_1}}{\rightleftharpoons}} 2O + M \tag{13.38}$$

$$N_2 + M \underset{k_{b_2}}{\overset{k_{f_2}}{\rightleftharpoons}} 2N + M \tag{13.39}$$

$$NO + M \underset{k_{b_3}}{\overset{k_{f_3}}{\rightleftharpoons}} N + O + M \tag{13.40}$$

$$O_2 + N \underset{k_{b_4}}{\overset{k_{f_4}}{\rightleftharpoons}} NO + O \tag{13.41}$$

$$N_2 + O \underset{k_{b_5}}{\overset{k_{f_5}}{\rightleftharpoons}} NO + N \tag{13.42}$$

$$N_2 + O_2 \underset{k_{b_6}}{\overset{k_{f_6}}{\rightleftharpoons}} 2NO \tag{13.43}$$

$$N + O \underset{k_{b_7}}{\overset{k_{f_7}}{\rightleftharpoons}} NO^+ + e^- \tag{13.44}$$

Equations (13.38–13.40) are dissociation reactions. Equations (13.41) and (13.42) are bimolecular exchange reactions (sometimes called the "shuffle" reactions); they are the two most important reactions for the formation of nitric oxide (NO) in air. Equation (13.44) is called a dissociative-recombination reaction because the recombination of the NO^+ ion with an electron produces not NO but rather a dissociated product $N + O$. Note that the preceding reactions are not all independent; for example, Eq. (13.43) can be obtained by adding Eqs. (13.41) and (13.42). However, in contrast to the calculation of an equilibrium composition as discussed in Secs. 10.9 and 11.11, for a nonequilibrium reaction mechanism the chemical equations do *not* have to be independent. In such a nonequilibrium case, the kinetic reaction mechanism can contain a large number of elementary chemical reactions, many of which are not independent. What is important is that all pertinent reactions that can affect the rate process must be included. This is quite different from the reactions used to calculate an equilibrium composition. For such equilibrium calculations, all we need are $\Sigma - \phi$ independent chemical reaction, where Σ is the number of species and ϕ is the number of elements, as discussed in Sec. 11.11. The actual reactions used are somewhat arbitrary, as long as they involve the various species in the equilibrium mixture, and as long as they are independent (i.e., as long as one chemical equation cannot be obtained by adding and/or subtracting any of the other chemical equations). This is in direct contrast to a nonequilibrium system, where the specification of a detailed kinetic mechanism with many participating reactions is necessary.

From the preceding reaction mechanism for air [Eqs. (13.38–13.44)], let us construct the rate equation for NO. Reactions (13.40–13.43) involve the production and extinction of NO. Moreover, in reaction (13.40) the collision partner M can be any of the different species, each requiring a different rate constant. That is, Eq. (13.40) is really the following equations:

$$NO + O_2 \underset{k_{b_{3a}}}{\overset{k_{f_{3a}}}{\rightleftharpoons}} N + O + O_2 \tag{13.40a}$$

$$NO + N_2 \underset{k_{b_{3b}}}{\overset{k_{f_{3b}}}{\rightleftharpoons}} N + O + N_2 \tag{13.40b}$$

$$NO + NO \underset{k_{b_{3c}}}{\overset{k_{f_{3c}}}{\rightleftharpoons}} N + O + NO \tag{13.40c}$$

$$NO + O \underset{k_{b_{3d}}}{\overset{k_{f_{3d}}}{\rightleftharpoons}} N + O + O \tag{13.40d}$$

$$NO + N \underset{k_{b_{3e}}}{\overset{k_{f_{3e}}}{\rightleftharpoons}} N + O + N \tag{13.40e}$$

$$NO + NO \underset{k_{b_{3f}}}{\overset{k_{f_{3f}}}{\rightleftharpoons}} N + O + NO \tag{13.40f}$$

$$NO + e^- \underset{k_{b_{3g}}}{\overset{k_{f_{3g}}}{\rightleftharpoons}} N + O + e^- \tag{13.40g}$$

Thus, the chemical rate equation for NO is

$$\begin{aligned}
\frac{d[NO]}{dt} = {} & -k_{f_{3a}}[NO][O_2] + k_{b_{3a}}[N][O][O_2] \\
& - k_{f_{3b}}[NO][N_2] + k_{b_{3b}}[N][O][N_2] \\
& - k_{f_{3c}}[NO]^2 + k_{b_{3c}}[N][O][NO] \\
& - k_{f_{3d}}[NO][O] + k_{b_{3d}}[N][O^2] \\
& - k_{f_{3e}}[NO][N] + k_{b_{3e}}[N]^2[O] \\
& - k_{f_{3f}}[NO][NO]^+ + k_{b_{3f}}[N][O][NO^-] \\
& - k_{f_{3g}}[NO][e^-] + k_{b_{3g}}[N][O][e^-] \\
& + k_{f_4}[O_2][N] - k_{b_4}[NO][O] \\
& + k_{f_5}[N_2][O] - k_{b_5}[NO][N] \\
& + 2k_{f_6}[N_2][O_2] - 2k_{b_6}[NO]^2
\end{aligned} \tag{13.45}$$

There are rate equations similar to Eq. (13.45) for O_2, N_2, O, N, NO^+, and e^-. Clearly, you can see that a major aspect of such a nonequlibrium analysis is simply bookkeeping, making certain to keep track of all of the terms in the equations.

Values of the rate constants for high-temperature air are readily available in the literature. See, for example, [158–162]. Again, keep in mind that there is always some uncertainty in the published rate constants; they are difficult to measure experimentally and very difficult to calculate accurately. Hence, any nonequilibrium analysis is a slave to the existing rate data.

For temperatures above 9000 K, where ionization of the atoms takes place, a kinetic mechanism more complex than that given by Eqs. (13.40a–13.40g) is needed. Such a mechanism, along with the appropriate rate constants, is given in Table 13.2. These data were compiled by Dunn and Kang in [162], and the table is readily found in [158]. In this table, a form of the rate constant similar to that given in Eq. (13.33) is used, specifically

$$k_f = C_f T^{\eta_f} e^{-K_f/\mathscr{R}T} \tag{13.46}$$

where $\mathscr{R} = 1.986\,\text{cal}/(\text{g}\cdot\text{mol})$ K. The units of k_f are expressed in cm^3, $\text{g}\cdot\text{mol}$, and seconds, in the combination appropriate for the given chemical equation. A typical temperature variation for one of the reactions, namely, the $O_2 + O_2 \rightleftharpoons 2O + O_2$ reaction, is shown in Fig. 13.3. Clearly, the value of k_f changes rapidly with temperature.

Table 13.2 and Fig. 13.3 are given here, not to say that the rate data are precise, because they are not. Uncertainties exist in all of these data. Rather, Table 13.2 and Fig. 13.3 are representative of the rate data in modern use and are given here simply to serve as an example. If you wish to carry out a serious nonequilibrium analysis, the suggestion is made to always canvas the existing literature for the most accurate rate data, even to the extent of talking with the physical chemistry community, before embarking on any extensive calculations.

13.4.1 Two-Temperature Kinetic Model

The temperature T that appears in all preceding sections is labeled the translational temperature; it is a measure of the collective translational energy of the particles of the gas through Eq. (12.9a). It is the temperature that enters into the molecular collision frequency through Eqs. (12.18–12.20). Therefore, it is natural that the vibrational relaxation time τ is a function of T through Eq. (13.19), and the chemical rate constants are functions of T through Eqs. (13.32) and (13.33).

In the special case that both vibrational and chemical *nonequilibrium* simultaneously exist in a mixture of gases, there is a coupling between the chemical reaction rates and the vibrational relaxation rates that affects the values of each. Molecules that are highly excited vibrationally, that is, in the higher-lying vibrational energy levels, are more readily dissociated by molecular collisions. They simply require less energy exchange during collisions to dissociate. Therefore, the chemical kinetic rates for dissociation are going to be faster if

Table 13.2 Kinetic mechanism for high-temperature air for the Dunn/Kang model

Reaction	C_f	η_f	K_f
$O_2 + N = 2O + N$	3.6000E18	−1.00000	118,800
$O_2 + NO = 2O + NO$	3.6000E18	−1.00000	118,800
$N_2 + O = 2N + O$	1.9000E17	−0.50000	226,000
$N_2 + NO = 2N + NO$	1.9000E17	−0.50000	226,000
$N_2 + O_2 = 2N + O_2$	1.9000E17	−0.50000	226,000
$NO + O_2 = N + O + O_2$	3.9000E20	−1.5	151,000
$NO + N_2 = N + O + N_2$	3.9000E20	−1.5	151,000
$O + NO = N + O_2$	3.2000E9	1	39,400
$O + N_2 = N + NO$	7.0000E13	0	76,000
$N + N_2 = 2N + N$	4.0850E22	−1.5	226,000
$O + N = NO^+ + e^-$	1.4000E06	1.50000	63,800
$O + e^- = O^+ + 2e^-$	3.6000E31	−2.91	316,000
$N + e^- = N^+ + 2e^-$	1.1000E32	−3.14	338,000
$O + O = O_2^+ + e^-$	1.6000E17	−0.98000	161,600
$O + O_2^+ = O_2 + O^+$	2.9200E18	−1.11000	56,000
$N_2 + N^+ = N + N_2^+$	2.0200E11	0.81000	26,000
$N + N = N_2^+ + e^-$	1.4000E13	0	135,600
$O + NO^+ = NO + O^+$	3.6300E15	−0.6	101,600
$N_2 + O^+ = O + N_2^+$	3.4000E19	−2.00000	46,000
$N + NO^+ = NO + N^+$	1.0000E19	−0.93	122,000
$O_2 + NO^+ = NO + O_2^+$	1.8000E15	0.17000	66,000
$O + NO^+ = O_2 + N^+$	1.3400E13	0.31	154,540
$O_2 + O = 2O + O$	9.0000E19	−1	119,000
$O_2 + O_2 = 2O + O_2$	3.2400E19	−1	119,000
$O_2 + N_2 = 2O + N_2$	7.2000E18	−1	119,000
$N_2 + N_2 = 2N + N_2$	4.7000E17	−0.5	226,000
$NO + O = N + 2O$	7.8000E20	−1.5	151,000
$NO + N = O + 2N$	7.8000E20	−1.5	151,000
$NO + NO = N + O + NO$	7.8000E20	−1.5	151,000
$O_2 + N_2 = NO + NO^+ + e^-$	1.3800E20	−1.84	282,000
$NO + N_2 = NO^- + e^- + N_2$	2.2000E15	−0.35	216,000

the gas is already highly excited vibrationally. Similarly, because dissociation more readily occurs from the higher-lying vibrational levels, there is a preferential depopulation of the higher-lying vibrational energy levels when dissociation is going on, thus affecting the instantaneous vibrational energy of the gas and the vibrational relaxation time.

The precise accounting of this mutual coupling phenomena on both the vibrational and chemical rates is complex. An attempt to deal approximately with this coupling involves the definition of a vibrational temperature T_{vib} as follows. Consider a given molecular species, say, O_2. If the instantaneous nonequilibrium value of vibrational energy per unit mass of this species is e_{vib} as dictated by the vibrational rate equation (13.17), then we define an

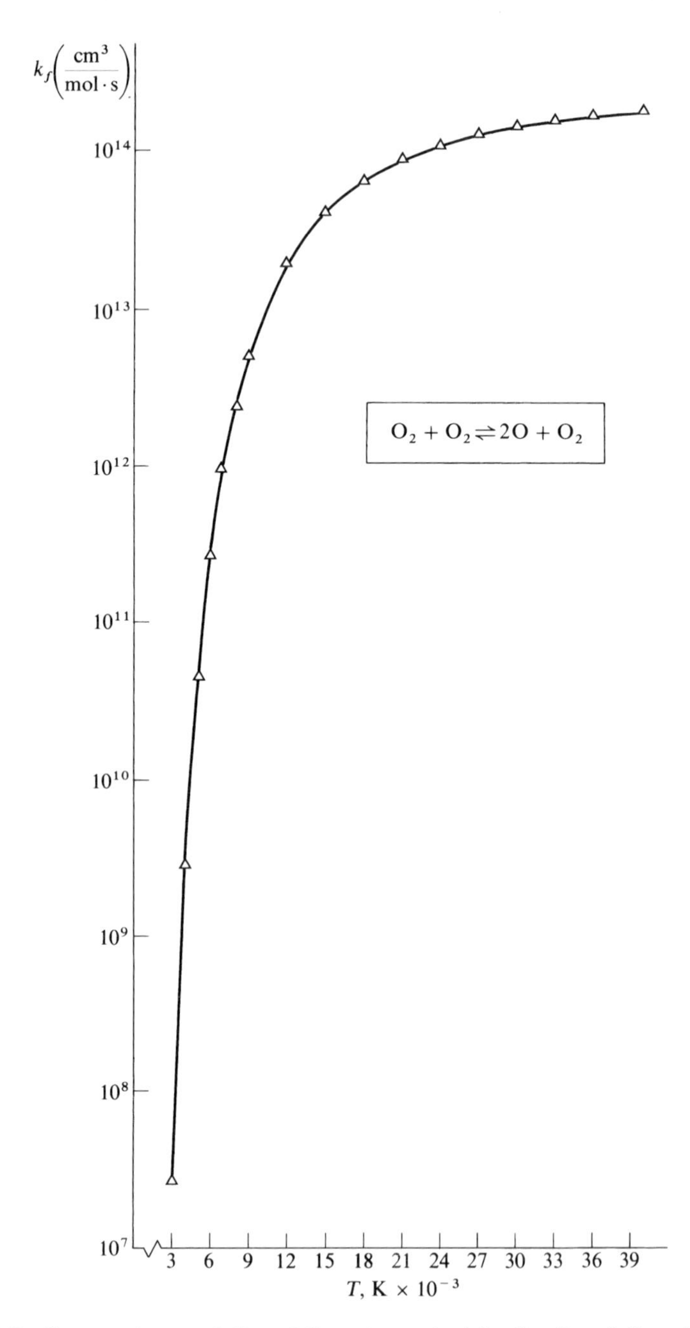

Fig. 13.3 Temperature variation of the rate constant for the dissociation of oxygen.

instantaneous vibrational temperature for that species T_{vib} as that governing a *local* Boltzmann distribution of the particles over the vibrational energy levels at the temperature T_{vib}, with the value of e_{vib} given by

$$e_{\text{vib}} = \left[\frac{h\nu/kT_{\text{vib}}}{e^{h\nu/kT_{\text{vib}}} - 1}\right] R\,T_{\text{vib}}$$

The higher is T_{vib}, the more particles there are in the higher-lying vibrational energy levels, and the easier (hence faster) is the dissociation. To account for this effect on the chemical rate constant, Park [263] suggests that the instantaneous vibrational and translational temperatures be combined to form an average temperature T_a, where

$$T_a = \sqrt{T_{\text{vib}} T}$$

and that T_a rather than T be used to calculate the chemical rate constants. When the gas is in vibrational equilibrium, $T_{\text{vib}} = T$, and hence $T_a = T$. It is only when vibrational nonequilibrium exits that T_a is different from T. If the local nonequilibrium vibrational energy is higher than the local equilibrium value ($T_{\text{vib}} > T$), then $T_a > T$, and the chemical reaction rate for that species will be faster than if vibrational equilibrium existed. Similarly, if the local nonequilibrium vibrational energy is lower than the local equilibrium value ($T_{\text{vib}} < T$), then $T_a < T$, and the chemical reaction rate will be slower than if vibrational equilibrium existed.

Park gives a kinetic mechanism for high-temperature air and associated rate constants in [263]. Table 13.3 lists Park's selected rate constants in units of $\text{cm}^3\,\text{mol}^{-1}\,\text{s}^{-1}$ as a function of T_a, T, and T_e, where T_e is the electron temperature, an index of the translational energy of the electrons in the gas. For some cases, Park generalizes the average temperature as

$$T_a = T_{\text{vib}}^{q} T^{1-q}$$

where q is between 0.3 and 0.5. The purpose of Table 13.3 in this book is simply to illustrate a set of rate constants that are based on the two-temperature kinetic model discussed in this section. Once again, if you are embarking on a serious nonequilibrium analysis, you are encouraged to first visit your local physical chemist to obtain the latest rate data and kinetic mechanism.

13.5 Chemical Nonequilibrium in H_2-Air Mixtures

As described in Chapters 1 and 9, future airbreathing hypersonic vehicles will be powered by supersonic combustion ramjet engines (scramjets). The fuel for the scramjets will most likely be hydrogen, and therefore the chemical kinetics of H_2-air mixtures is of vital importance to the supersonic combustion process. With this in mind, Tables 13.4 and 13.5 give a typical kinetic mechanism for H_2-air chemically reacting mixtures, along with the reaction rate constant data wherein k_f is in the same form as given by Eq. (13.46).

Table 13.3 Selected rate coefficients for reactions in air ($cm^3\ mol^{-1}\ s^{-1}$)[a]

No.	Reactions	Rate expression	Remark
1	$O_2 + O_2 \rightarrow O + O + O_2$	$2 \times 10^{21}\ T_a^{-1.5} \exp(-59{,}500/T_a)$	——
2	$O_2 + NO \rightarrow O + O + NO$	$2 \times 10^{21}\ T_a^{-1.5} \exp(-59{,}500/T_a)$	Estimated
3	$O_2 + N_2 \rightarrow O + O + N_2$	$2 \times 10^{21}\ T_a^{-1.5} \exp(-59{,}500/T_a)$	——
4	$O_2 + O \rightarrow O + O + O$	$10^{22}\ T_a^{-1.5} \exp(-59{,}500/T_a)$	——
5	$O_2 + N \rightarrow O + O + N$	$10^{22}\ T_a^{-1.5} \exp(-59{,}500/T_a)$	Estimated
6	$NO + O_2 \rightarrow N + O + NO$	$5 \times 10^{15} \exp(-75{,}500/T_a)$	Estimated
7	$NO + NO \rightarrow N + O + NO$	$1.1 \times 10^{17} \exp(-75{,}500/T_a)$	——
8	$NO + N_2 \rightarrow N + O + N_2$	$5 \times 10^{15} \exp(-75{,}500/T_a)$	——
9	$NO + O \rightarrow N + O + O$	$1.1 \times 10^{17} \exp(-75{,}500/T_a)$	Estimated
10	$NO + N \rightarrow N + O + N$	$1.1 \times 10^{17} \exp(-75{,}500/T_a)$	Estimated
11	$N_2 + O_2 \rightarrow N + N + O_2$	$7 \times 10^{21}\ T_a^{-1.6} \exp(-113{,}200/T_a)$	Estimated
12	$N_2 + NO \rightarrow N + N + NO$	$7 \times 10^{21}\ T_a^{-1.6} \exp(-113{,}200/T_a)$	Estimated
13	$N_2 + N_2 \rightarrow N + N + N_2$	$7 \times 10^{21}\ T_a^{-1.6} \exp(-113{,}200/T_a)$	——
14	$N_2 + O \rightarrow N + N + O$	$3 \times 10^{22}\ T_a^{-1.6} \exp(-113{,}200/T_a)$	Estimated
15	$N_2 + N \rightarrow N + N + N$	$3 \times 10^{22}\ T_a^{-1.6} \exp(-113{,}200/T_a)$	——
16	$N_2 + e \rightarrow N + N + e$	$3 \times 10^{24}\ T_e^{-1.6} \exp(-113{,}200/T_e)$	Estimated
17	$N_2 + O \rightarrow NO + N$	$6.4 \times 10^{17}\ T_a^{-1} \exp(-38{,}200/T_a)$	——
18	$NO + O \rightarrow O_2 + N$	$8.4 \times 10^{12} \exp(-19{,}400/T_a)$	——
19	$N + O \rightarrow NO^+ + e$	$5.3 \times 10^{12} \exp(-31{,}900/T_a)$	——
20	$N + N \rightarrow N_2^+ + e$	$2 \times 10^{13} \exp(-67{,}500/T_a)$	——
21	$O + O \rightarrow O_2^+ + e$	$1.1 \times 10^{13} \exp(-80{,}600/T_a)$	——
22	$O + e \rightarrow O^+ + e + e$	$3.9 \times 10^{33}\ T_e^{-3.78} \exp(-158{,}500/T_e)$	Estimated
23	$N + e \rightarrow N^+ + e + e$	$2.5 \times 10^{33}\ T_e^{-3.82} \exp(-168{,}200/T_e)$	——
24	$NO^+ + O \rightarrow N^+ + O_2$	$10^{12}\ T^{0.5} \exp(-77{,}200/T)$	——
25	$O_2^+ + N \rightarrow N^+ + O_2$	$8.7 \times 10^{13}\ T^{0.14} \exp(-28{,}600/T)$	——
26	$NO + O^+ \rightarrow N^+ + O_2$	$1.4 \times 10^{5}\ T^{1.9} \exp(-15{,}300/T)$	——
27	$O_2^+ + N_2 \rightarrow N_2^+ + O_2$	$9.9 \times 10^{12} \exp(-40{,}700/T)$	——
28	$O_2^+ + O_2 \rightarrow Ok + O_2$	$4 \times 10^{12}\ T^{-0.09} \exp(-18{,}000/T_a)$	——
29	$NO^+ + N \rightarrow O^+ + N_2$	$3.4 \times 10^{13}\ T_a^{-1.08} \exp(-12{,}800/T)$	——
30	$NO^+ + O_2 \rightarrow O_2^+ + NO$	$2.4 \times 10^{13}\ T^{0.41} \exp(-32{,}600/T)$	——
31	$NO^+ + O \rightarrow O_2^+ + N$	$7.2 \times 10^{12}\ T^{0.29} \exp(-48{,}600/T)$	——
32	$O^+ + N_2 \rightarrow N_2^+ + O$	$9 \times 10^{11}\ T^{0.36} \exp(-22{,}800/T)$	——
33	$NO^+ + N \rightarrow N_2^+ + O$	$7.2 \times 10^{13} \exp(-35{,}500/T)$	——

[a]T is the translational–rotational temperature, T_e is the election temperature, and T_a is the geometrical average between T and the vibrational temperature T_v, $T_a = T_v^q\ T^{1=q}$, where q is between 0.3 and 0.5.

Tables 13.4 and 13.5 are obtained from [158], based on data supplied by C. Jachimowski frorn the NASA Langley Research Center. In Table 13.5 are some third-body efficiencies for several reactions where H_2 and H_2O are the third bodies, that is, the collision partner denoted by M in some of the chemical equations.

Table 13.4 Kinetic mechanism for H_2 air from Jachimowski

Reaction	C_f	η_f	K_f
$H_2 + O_2 = OH + OH$	1.7E13	0	48,000
$OH + H_2 = H_2O + H$	2.2E13	0	5,150
$H + O_2 = OH + O$	2.20E14	0	16,800
$O + H_2 = OH + H$	1.80E10	1	8,900
$OH + OH = H_2O + O$	6.3E12	0	1,090
$H + OH = H_2O + M$	2.20E22	−2	0
$H + O = OH + M$	6.00E16	−0.6	0
$H + H = H_2 + M$	6.40E17	−1	0
$H + O_2 = HO_2 + M$	1.70E15	0	−1,000
$HO_2 + H = H_2 + O_2$	1.30E13	0	0
$HO_2 + H = OH + OH$	1.40E14	0	1,080
$HO_2 + O = OH + O_2$	1.50E13	0	950
$HO_2 + OH = H_2O + O_2$	8.00E12	0	0
$HO_2 + HO_2 = H_2O_2 + O_2$	2.00E12	0	0
$H + H_2O_2 = H_2 + HO_2$	1.40E12	0	3,600
$O + H_2O_2 = OH + HO_2$	1.40E13	0	6,400
$OH + H_2O_2 = H_2O + HO_2$	6.10E12	0	1,430
$M + H_2O_2 = 2OH + M$	1.20E17	0	45,500
$O + O = O_2 + M$	6.00E13	0	−1,000
$N + N = N_2 + M$	2.80E17	−0.75	0
$N + O_2 = NO + O$	6.40E9	1.0	6,300
$N + NO = N_2 + O$	1.60E13	0	0
$N + OH = NO + H$	6.30E11	0.5	0
$H + NO = HNO + M$	5.40E15	0	− 600
$H + HNO = NO + H_2$	4.80E12	0	0
$O + HNO = NO + OH$	5.00E11	0.5	0
$OH + HNO = NO + H_2O$	3.60E13	0	0
$HO_2 + HNO = NO + H_2O_2$	2.00E12	0	0
$HO_2 + NO = NO_2 + OH$	3.43E12	0	−260
$H + NO_2 = NO + OH$	3.50E14	0	1,500
$O + NO_2 = NO + O_2$	1.00E13	0	600
$M + NO_2 = NO + O$	1.16E16	0	66,000

13.6 Summary and Comments

In this chapter we have considered some of the elementary characteristics of gases in both vibrational and chemical nonequilibrium.

To analyze and compute the time rate of change of vibrational energy in a gas, the vibrational rate equation can be used:

$$\frac{\mathrm{d}e_{\mathrm{vib}}}{\mathrm{d}t} = \frac{1}{\tau}(e_{\mathrm{vib}}^{eq} - e_{\mathrm{vib}}) \tag{13.17}$$

Table 13.5 Third body efficiencies for the kinetic mechanism for H_2 air from Jachimowski

	Third-body efficiencies[a]			
Reaction	Third body	Efficiency	Third body	Efficiency
$H + OH + M = H_2O + M$	H_2	1.0	H_2O	6.0
$H + O + M = OH + M$	H_2	1.0	H_2O	5.0
$H + H + M = H_2 + M$	H_2	2.0	H_2O	6.0
$H + O_2 + M = HO_2 + M$	H_2	2.0	H_2O	16.0
$M + H_2O_2 = 2OH + M$	H_2	1.0	H_2O	15.0

[a]All other third bodies have efficiency of 1.0.

where the relaxation time τ is given by

$$\tau p = C_1 e^{(C_2/T)^{1/3}}$$

Frequently, the preceding equations are called the Landau–Teller rate model.

To analyze and compute the finite-rate chemical kinetic processes in any gas mixture, it is necessary to do the following:

1) Define the reaction mechanism [such as reactions (13.38–13.44)].

2) Obtain the rate constants from the literature, usually in the form of Eq. (13.33).

3) Write all of the appropriate rate equations, such as Eq. (13.45).

4) Solve the rate equations simultaneously to obtain the time variation of the species concentrations, that is, $[O_2] = f_1(t)$, $[O] = f_2(t)$, etc. This is a job for a high-speed digital computer. Indeed, most modern analyses of chemical nonequilibrium systems would not be practicably possible without computers.

Finally, we will see how these considerations are used in the analyses of nonequilibrium high-temperature flowfields in Chapters 15 and 17.

14
Inviscid High-Temperature Equilibrium Flows

Equilibrium: Any condition in which all acting influences are cancelled by others resulting in a stable, balanced, or unchanging system.

The American Heritage Dictionary of the English Language, 1969

Chapter Preview

Finally, after four chapters of basic physics and chemistry, we are going to look at some high-temperature gas dynamic flows. But we are not going off the deep end and dealing with all kinds of complexities; rather, in this chapter we are going to examine some basic high-speed flows such as shock waves, nozzle flows, flows over cones, and flows over blunt-nosed bodies. You can hardly get more basic. These flows are the bread and butter of classical compressible flow, except here we examine how these classic flows are changed by high-temperature effects. In fact, the underlying question addressed throughout the remainder of this book is: how do the high-temperature physics and chemistry discussed in the preceding four chapters affect and change some otherwise familiar and classic flowfields? The answers will be graphic, sometimes unexpected, and always fascinating.

As with all major subjects, we cannot do the whole thing at once. So in this chapter we take our first step into the study of high-temperature gas dynamics by assuming that both vibrational and chemical equilibrium exist throughout the flowfield—*equilibrium flows*. In the whole panoply of high-temperature gas dynamics, equilibrium flows are usually the most straightforward to calculate and understand. So we start here. Ready ... set ... go!

14.1 Introduction

It is worthwhile at this stage to return to our road map in Fig. 1.24. All of our discussion in Part 3 of this book has, so far, been centered in the first item under high-temperature flows in Fig. 1.24, namely, a presentation of basic physical chemistry effects. The material covered under this item has been in the spirit of "tool building," that is, acquiring the necessary tools (concepts, definitions, equations, etc.) from physical chemistry to allow us to properly analyze and understand high-temperature flows. We now have enough of these tools to study inviscid high-temperature flows—the next item on the road map in Fig. 1.24. In particular, the subject of the present chapter is *equilibrium* inviscid flows; the matter of nonequilibrium inviscid flows is the subject of Chapter 15.

Before progressing further, it is important to examine more closely what is meant by high-temperature *equilibrium* flow.

> *Definition*: Flow in *local thermodynamic equilibrium*—a local Boltzmann distribution [Eq. (11.25)] exists at each point in the flow at the local temperature T. Hence, at each point in the flow, the energy of each species is given by Eq. (11.62) or (11.63).

> *Definition*: Flow in *local chemical equilibrium*—the local chemical composition at each point in the flow is the same as that determined by the chemical equilibrium calculations described in Secs. 10.9, 11.9, and 11.11 (using the equilibrium constants) at the local values of T and p.

How close an actual high-temperature flow comes to these ideal conditions of local thermodynamic equilibrium and local chemical equilibrium depends

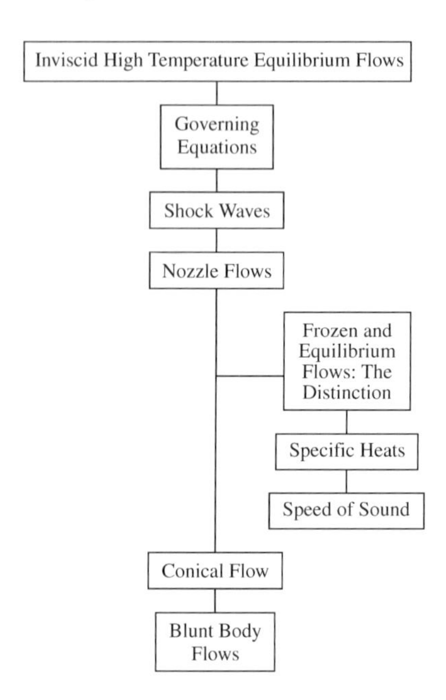

Fig. 14.1 Road map for Chapter 14.

on the collision frequency and the flow time, as will be explained in Chapter 15. In the present chapter we will simply assume that local equilibrium conditions hold at each point in the flowfield, and we will examine the nature of high-temperature flows under such equilibrium conditions.

The road map for this chapter is given in Fig. 14.1. We start with the governing flow equations for inviscid equilibrium flow. We then apply these equations to study shock wave and nozzle flows. Then we take a side trip to contrast two extremes—frozen flow and equilibrium flow—and to look at the thermodynamic properties of specific heat and the speed of sound in these two extremes. Then we come back to the main route and finish our journey with a study of equilibrium conical flows and blunt-body flows.

14.2 Governing Equations for Inviscid High-Temperature Equilibrium Flow

Consider again Eqs. (4.1–4.5). Examine these equations carefully; they are frequently called the *Euler* equations, and they are the governing equations for inviscid flow used throughout Part 1 of this book. These equations are derived in most basic fluid-dynamics texts: for example, see [4] and [5]. Think over the nature of these derivations, and if necessary review them in these references. We will proceed from here assuming that you are familiar with the derivations of Eqs. (4.1–4.5).

Question: Do these equations hold for high-temperature, chemically reacting equilibrium flows?

The answer lies in their derivation. For example, the continuity equation (4.1) is simply a statement of global mass conservation, which holds whether or not the flow is chemically reacting. Similarly, Eqs. (4.2–4.4) are basically Newton's second law, which is also independent of chemically reacting effects. Finally, Eq. (4.5) is the energy equation (or more precisely the entropy equation) for adiabatic flow; it stems from the combined first and second laws of thermodynamics presented in Chapter 10. These laws hold for any type of gas and hence are applicable to high-temperature chemically reacting flows. Thus, the answer to the question is *yes*; Eqs. (4.1–4.5) hold for a high-temperature, chemically reacting, inviscid, equilibrium flow. [On the other hand, Eq. (4.6), which was used frequently throughout Part 1, does not hold for such a flow; it is a specialized form assuming constant γ, and hence applies only to a calorically perfect gas.]

Note that Eq. (4.5) is a statement that the entropy of a moving fluid element is constant in an adiabatic, inviscid flow. For a high-temperature gas, this remains true as long as the flow is in local equilibrium—the case treated in this chapter. However, for nonequilibrium flow (to be discussed in Chapter 15), we know from results such as Eq. (10.72) that there is an entropy increase caused by the irreversible effect of the nonequilibrium process. Hence, Eq. (4.5) does not hold for a nonequilibrium inviscid flow. For such a case, and essentially for all general high-temperature flows, it is recommended that we deal with another variable rather than entropy in the energy equation. There are a number of alternate forms of the energy equation for adiabatic, inviscid flows—for example, see Chapter 6 of [4]. Let us choose the total enthalpy as our dependent variable, and write the energy equation for an adiabatic inviscid flow as

$$\rho \frac{\mathrm{D}h_0}{\mathrm{D}t} = \rho \frac{\partial h_0}{\partial t} + \rho u \frac{\partial h_0}{\partial x} + \rho v \frac{\partial h_0}{\partial y} + \rho w \frac{\partial h_0}{\partial z} = \frac{\partial p}{\partial t} \tag{14.1}$$

which holds for both equilibrium and nonequilibrium flows.

In summary, the governing equations for an inviscid, high-temperature, equilibrium flow are, from Eqs. (4.1–4.4) and (14.1), written in vector and substantial derivative notation.

Continuity:

$$\frac{\partial \rho}{\partial t} + \nabla \cdot (\rho \mathbf{V}) = 0 \tag{14.2}$$

Momentum:

$$\rho \frac{\mathrm{D}\mathbf{V}}{\mathrm{D}t} = -\nabla p \tag{14.3}$$

Energy:

$$\rho \frac{\mathrm{D}h_0}{\mathrm{D}t} = \frac{\partial p}{\partial t} \tag{14.4}$$

where

$$h_0 = h + \frac{V^2}{2} \tag{14.5}$$

So we see that high-temperature effects do not change the basic form of these equations; they are the same as used in many of our earlier analyses in Part 1.

Question: Why does the energy equation *not* have an extra term that deals with the energy changes caused by chemical reactions (exothermic or endothermic) in the flow?

The answer is that h in Eqs. (14.4) and (14.5) *contains* the effective zero-point energies, namely, the heats of formation, as explained in Sec. 11.12. In this fashion, the local energy exchanges caused by chemical reactions are automatically accounted for when h is treated as the absolute enthalpy in the form given by Eq. (11.105). When this is done, no explicit heat-addition term appears in Eq. (14.4) to account for chemical reactions. (In some literature, a chemical heat-addition term is included in the energy equation; in such cases the enthalpy is the sensible rather than absolute value. Review Sec. 11.12 for the difference between sensible and absolute enthalpy.)

Recall our discussion in Sec. 11.1 concerning the unknown variables in an equilibrium chemically reacting flowfield and how they are obtained in principle. We now see in detail how they are obtained. The flow equations (14.2–14.4) constitute three equations for four unknowns: ρ, $\mathbf{V}$, p, and h. This system of equations must be completed by the addition of the equilibrium thermodynamic properties for the gas. Conceptually, we can write these properties in the form

$$T = T(\rho, h) \tag{14.6}$$

$$p = p(\rho, h) \tag{14.7}$$

Therefore, Eqs. (14.2–14.4), (14.6), and (14.7) constitute five equations for the five unknowns: ρ, $\mathbf{V}$, p, h, and T. Note that T is not only an important flowfield variable, but it is absolutely necessary for the evaluation of the equilibrium constants and the internal energy and enthalpy from the expressions given by statistical thermodynamics (see Chapter 11). Recall from Sec. 11.13 that, in a given calculation, Eqs. (14.6) and (14.7) can take the form of any of the following:

1) The first is a direct calculation of the equilibrium thermodynamic properties from the equations of statistical thermodynamics (Chapter 11) carried out in parallel with the solutions of the flow equations. In terms of a computer calculation, this can be viewed as a computer subroutine that generates the properties directly from the statistical mechanical equations.

2) Next is tabulation of the equilibrium thermodynamic properties (if one is available for the particular gas you are dealing with). For high-temperature air, [154] is a good example of such tabulations.

3) They can also be correlations of the equilibrium thermodynamic properties (again, if they are available). For air, [155] is a good example of such correlations.

4) Graphical plots of the equilibrium thermodynamic properties (again, if they are available) are the final form. For example, a large Mollier diagram for high-temperature air is available in many laboratories and companies for such purposes.

Finally, we note that analytical, closed-formed solutions of Eqs. (14.2–14.4), and (14.6) and (14.7) have not yet been obtained in the literature, even for the simplest type of high-temperature flow problem. It is almost axiomatic that high-temperature effects, even in the most straightforward case of inviscid one-dimensional flow, force the solutions of such problems to be numerical. This aspect of the analysis of high-temperature flows will be amply demonstrated in the subsequent sections and chapters.

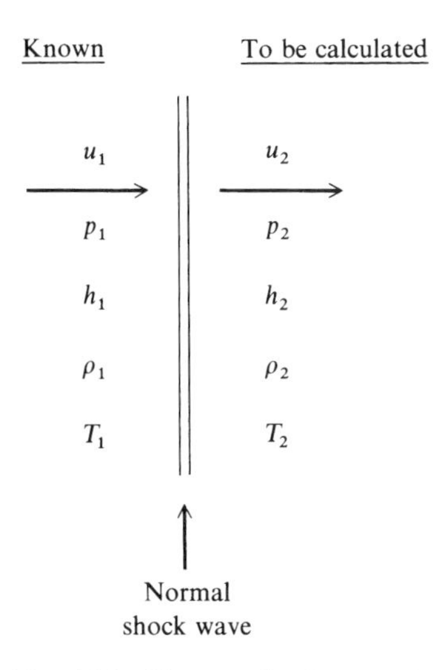

Fig. 14.2 Normal shock geometry.

14.3 Equilibrium Normal and Oblique Shock-Wave Flows

Consider a stationary normal shock wave as sketched in Fig. 14.2. Assume that the shock is strong enough; hence, T_2 is high enough, such that vibrational excitation and chemical reactions occur behind the shock front. Moreover, assume that local thermodynamic and chemical equilibrium hold behind the shock. All conditions ahead of the shock wave (region 1) are known. Our objective is to calculate properties behind the shock.

The governing equations for the flow across a normal shock are derived in many basic texts; for example, see [4]. They can be obtained by writing Eqs. (14.2–14.4) for steady, one-dimensional flow, and integrating between points in front of and behind the shock. They are as follows.

Continuity:

$$\rho_1 u_1 = \rho_2 u_2 \tag{14.8}$$

Momentum:

$$p_1 + \rho_1 u_1^2 = p_2 + \rho_2 u_2^2 \tag{14.9}$$

Energy:

$$h_1 + \frac{u_1^2}{2} = h_2 + \frac{u_2^2}{2} \tag{14.10}$$

Equations (14.8–14.10) are the familiar basic normal shock equations; consistent with the discussion in Sec. 14.2, these equations are general; they hold for both reacting and nonreacting gases.

In addition, the equilibrium thermodynamic properties for the high-temperature gas are assumed known from the techniques discussed in Chapter 11. These can take the form of tables or graphs, or can be calculated directly from the equations developed in Chapter 11. In any event, we can consider these properties in terms of the following functional relations ("equations of state," if you will):

$$\rho_2 = \rho(p_2, h_2) \tag{14.11}$$

$$T_2 = T(p_2, h_2) \tag{14.12}$$

Recall that, for a calorically perfect gas, Eqs. (14.8–14.12) yield a series of closed-form algebraic relations for p_2/p_1, T_2/T_1, M_2, etc., as functions of M_1 (for example, see [4] and [5]). Unfortunately, no simple formulas can be obtained when the gas is vibrationally excited and/or chemically reacting. For such high-temperature cases, Eqs. (14.8–14.12) must be solved numerically. To set up such a numerical solution, let us first rearrange Eqs. (14.8–14.10). From Eq. (14.8)

$$u_2 = \frac{\rho_1 u_1}{\rho_2} \tag{14.13}$$

Substitute Eq. (14.13) into (14.9):

$$p_1 + \rho_1 u_1^2 = p_2 + \rho_2 \left(\frac{\rho_1 u_1}{\rho_2}\right)^2 \tag{14.14}$$

Solving Eq. (14.14) for p_2, we have

$$\boxed{p_2 = p_1 + \rho_1 u_1^2 \left(1 - \frac{\rho_1}{\rho_2}\right)} \tag{14.15}$$

In addition, substituting Eq. (14.13) into (14.10), we have

$$h_1 + \frac{u_1^2}{2} = h_2 + \frac{(\rho_1 u_1/\rho_2)^2}{2} \tag{14.16}$$

Solving Eq. (14.16) for h_2,

$$\boxed{h_2 = h_1 + \frac{u_1^2}{2}\left[1 - \left(\frac{\rho_1}{\rho_2}\right)^2\right]} \tag{14.17}$$

Because all of the upstream conditions, ρ_1, u_1, p_1, h_1, etc., are known, Eqs. (14.15) and (14.17) express p_2 and h_2, respectively, in terms of only one unknown, namely, ρ_1/ρ_2. This establishes the basis for an iterative numerical solution, as follows:

1) Assume a value for ρ_1/ρ_2. (A value of 0.1 is usually good for a starter.)

2) Calculate p_2 from Eq. (14.15) and h_2 from Eq. (14.17).

3) With the values of p_2 and h_2 just obtained, calculate ρ_2 from Eq. (14.11).

4) Form a new value of ρ_1/ρ_2 using the value of ρ_2 obtained from step 3.

5) Use this new value of ρ_1/ρ_2 in Eqs. (14.15) and (14.17) to obtain new values of p_2 and h_2, respectively. Then repeat steps 3–5 until convergence is obtained, that is, until there is only a negligible change in ρ_1/ρ_2 from one iteration to the next. (This convergence is usually very fast, typically requiring less than five iterations.)

6) At this stage, we now have the correct values of p_2, h_2, and ρ_2. Obtain the correct value of T_2 from Eq. (14.12).

7) Obtain the correct value of u_2 from Eq. (14.13).

By means of steps 1–7, we can obtain all properties behind the shock wave for given properties in front of the wave.

There is a basic practical difference between the shock results for a calorically perfect gas and those for a chemically reacting gas. For a calorically perfect gas (see [4] and [5])

$$\frac{p_2}{p_1} = f_1(M_1)$$

$$\frac{\rho_2}{\rho_1} = f_2(M_1)$$

$$\frac{h_2}{h_1} = f_3(M_1)$$

Note that in this case only M_1 is required to obtain the ratios of properties across a normal shock wave. In contrast, for an *equilibrium chemically reacting gas*, we have seen that

$$\frac{p_2}{p_1} = g_1(u_1, p_1, T_1)$$

$$\frac{\rho_2}{\rho_1} = g_2(u_1, p_1, T_1)$$

$$\frac{h_2}{h_1} = g_3(u_1, p_1, T_1)$$

Note that in this case *three* freestream parameters are necessary to obtain the ratios of properties across a normal shock wave. This makes plenty of sense—the equilibrium composition behind the shock depends on p_2 and T_2, which in turn are governed in part by p_1 and T_1. Hence, in addition to the upstream velocity u_1, the normal shock properties must depend also on p_1 and T_1. By this same reasoning, if no chemical reactions take place, but the vibrational and electronic energies are excited (a thermally perfect gas), then the downstream normal shock properties depend on two upstream conditions, namely, u_1 and T_1.

Also note that, in contrast to a calorically perfect gas, the Mach number no longer plays a pivotal role in the results for normal shock waves in a high-temperature gas. In fact, for most high-temperature flows in general, the Mach number is not a particularly useful quantity. The flow of a chemically reacting gas is mainly governed by the primitive variables of velocity, temperature, and pressure. For an equilibrium gas, the Mach number is still uniquely defined as V/a, and it can be used along with other determining variables—it just does not hold a dominant position as in the case of a calorically perfect gas. For a non-equilibrium gas, however, there is some ambiguity even in the definition of Mach number (to be discussed in Chapter 15), and hence the Mach number further loses significance for such cases.

For high-temperature air, a comparison between calorically perfect gas and equilibrium chemically reacting gas results was shown in Fig. 1.18. Here, the temperature behind a normal shock wave is plotted vs upstream velocity for conditions at a standard altitude of 52 km. The equilibrium results are plotted directly from normal shock tables prepared by the Cornell Aeronautical Laboratory (now CALSPAN Corporation) and published in [163] and [164]. These reports should

Table 14.1 Properties across a normal shock wave in air at a velocity of 36,000 ft/s and an altitude of 170,000 ft

Flow property	For calorically perfect air, $\gamma = 1.4$	For equilibrium chemically reacting air (CAL Report AG-1729-A-2)
p_2/p_1	1233	1387
ρ_2/ρ_1	5.972	15.19
h_2/h_1	206.35	212.8
T_2/T_1	206.35	41.64

be consulted for equilibrium normal shock properties associated with air in the standard atmosphere. From Fig. 1.18, the calorically perfect results considerably overpredict the temperature, and for obvious reasons. For a calorically perfect gas, the directed kinetic energy of a flow ahead of the shock is mostly converted to translational and rotational molecular energy behind the shock. On the other hand, for a thermally perfect and/or chemically reacting gas, the directed kinetic energy of the flow, when converted across the shock wave, is shared across all molecular modes of energy, and/or goes into zero-point energy of the products of chemical reaction. Hence, the temperature (which is a measure of translational energy only) is less for such a case.

For further comparison, consider a reentry vehicle at 170,000-ft standard altitude with a velocity of 36,000 ft/s. The properties across a normal shock wave for this case are tabulated in Table 14.1. Note from that tabulation that chemical reactions have the strongest effect on temperature, for the reasons given earlier. This is generally true for all types of chemically reacting flows—the temperature is by far the most sensitive variable. In contrast, the pressure ratio is affected only by a small amount. Pressure is a "mechanically" oriented variable; it is governed mainly by the fluid mechanics of the flow and not so much by the thermodynamics. This is substantiated by examining the momentum equation, namely, Eq. (14.9). For high-speed flow, $u_2 \ll u_1$, and $p_2 \gg p_1$. Hence from Eq. (14.9),

$$p_2 \approx \rho_1 u_1^2$$

This is a common hypersonic approximation; note that p_2 is mainly governed by the freestream velocity, and that thermodynamic effects are secondary.

In an equilibrium dissociating and ionizing gas, increasing the pressure at constant temperature tends to decrease the atom and ion mass fractions, that is, increasing the pressure tends to inhibit dissociation and ionization. The consequences of this effect on equilibrium normal shock properties are shown in Fig. 14.3, where the temperature ratio across the shock is plotted vs upstream velocity for three different values of upstream pressure. Note that T_2/T_1 is higher at higher pressures; the gas is less dissociated and ionized at higher pressure, and hence more energy goes into translational molecular motion behind the shock rather than into the zero-point energy of the products of dissociation.

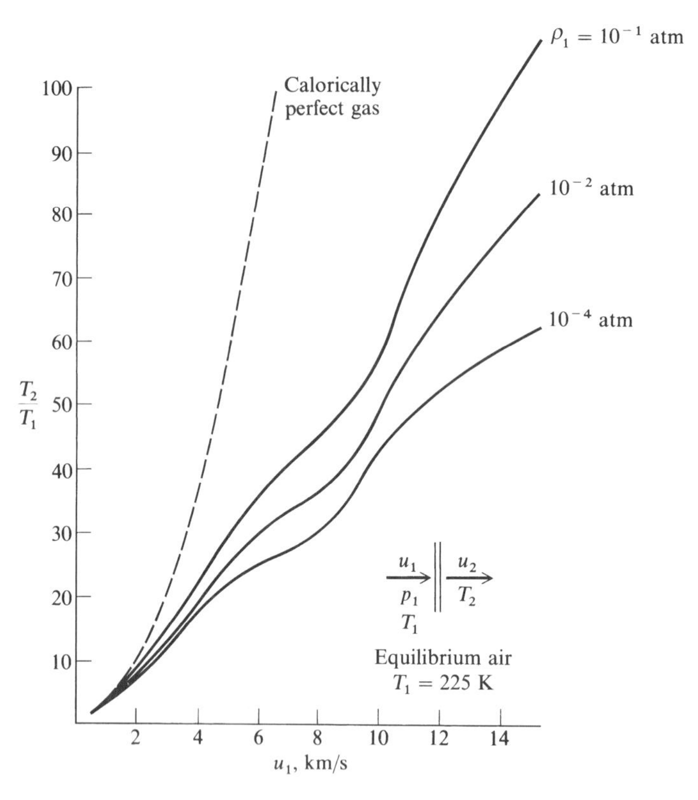

Fig. 14.3 Influence of pressure on the normal shock temperature in equilibrium air.

Because of the importance of high-temperature effects in high-speed atmospheric flight, more detailed results for equilibrium normal shock properties in air are given in Figs. 14.4 and 14.5, obtained from [165]. In Figs. 14.4a and 14.4b, the temperature behind a normal shock wave is plotted as a function of velocity in front of the wave, with altitude as a parameter. The velocity range in Fig. 14.4a is below orbital velocity, and hence the results are affected primarily by dissociation. In contrast, the velocities in Fig. 14.4b cover the superorbital range (above 26,000 ft/s) and therefore reflect the effects of substantial ionization. Note again the effect of pressure in these results; at a given velocity, T_2 increases with decreasing altitude (i.e., increasing pressure) because the amount of dissociation and ionization in an equilibrium gas is decreased at higher pressure. Also note the general magnitude of the temperatures encountered. At $u_1 = 10{,}000$ ft/s (typical of a hypersonic cruise transport), $T_2 \approx 3000$ K. At $u_2 = 26{,}000$ ft/s (orbital velocity typical of a space shuttle or trans-atmospheric vehicle), $T_2 \approx 7000$ K. For atmospheric entry at escape velocity, $u_1 = 36{,}000$ ft/s (typical of Apollo-type vehicles and aeroassisted orbital transfer vehicles), $T_2 \approx 11{,}000$ K. Moreover, Fig. 14.4a illustrates that chemically reacting effects begin to impact the normal shock properties at velocities above 6000 ft/s (approximately Mach 6). The density ratio across a normal shock

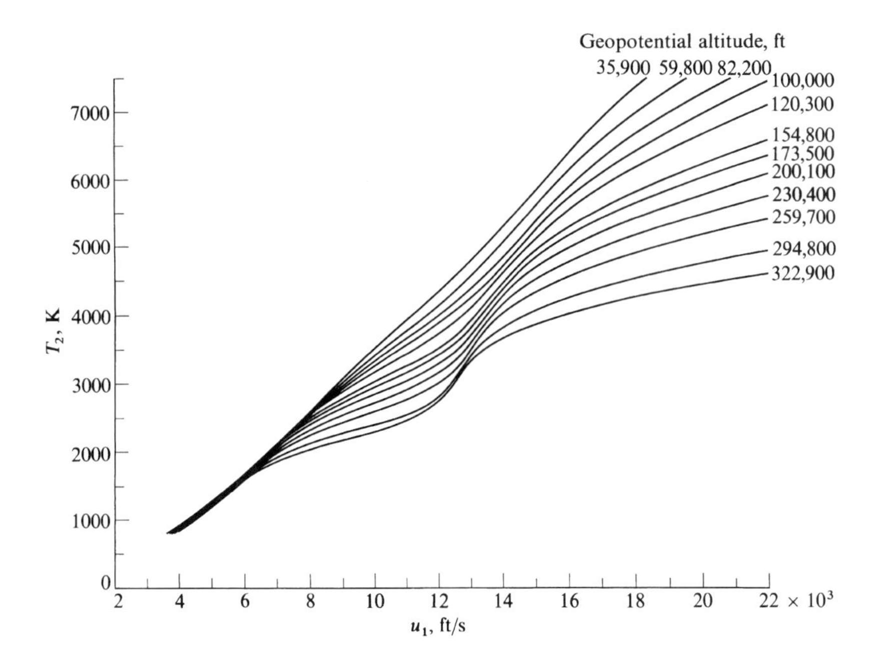

Fig. 14.4a Variation of normal shock temperature with velocity and altitude; velocity range below orbital velocity (from Huber [165]).

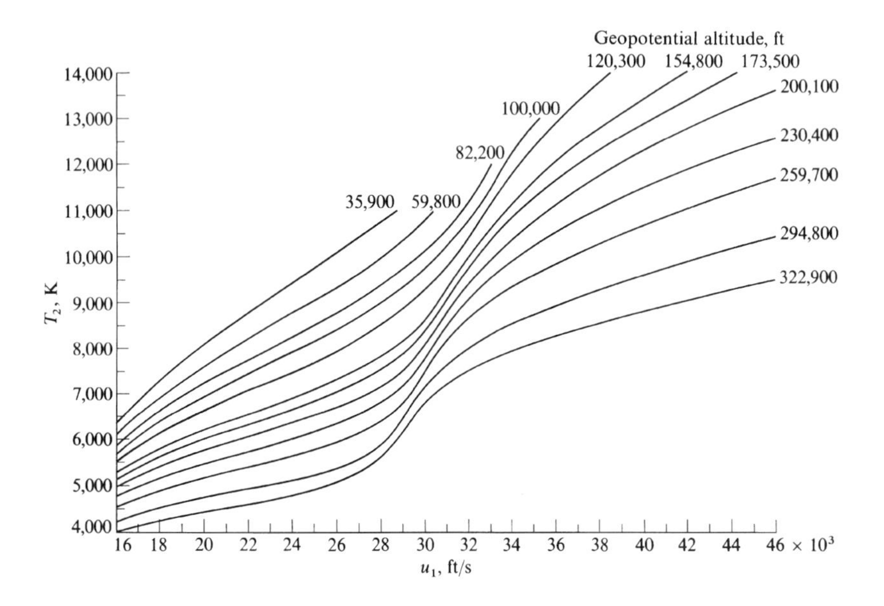

Fig. 14.4b Variation of normal shock temperature with velocity and altitude; velocity range near and above orbital velocity (from [165]).

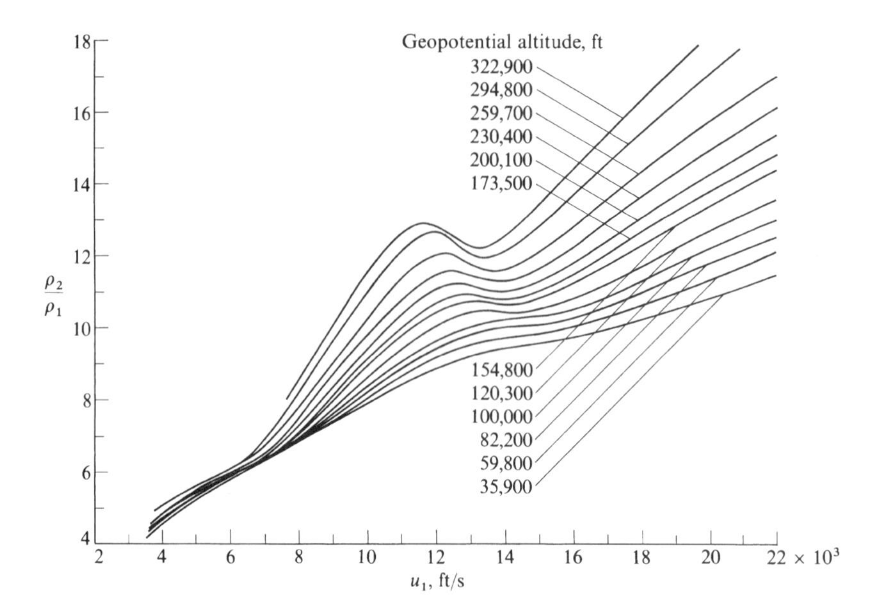

Fig. 14.5a Variation of normal shock density with velocity and altitude; velocity range below orbital velocity (from [165]).

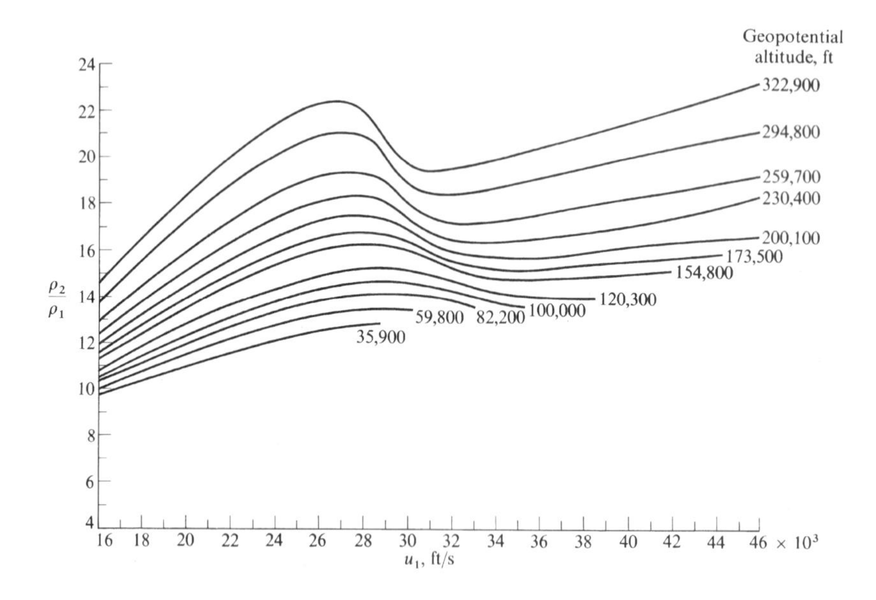

Fig. 14.5b Variation of normal shock density with velocity and altitude; velocity range near and above orbital velocity (from [165]).

wave ρ_2/ρ_1 is shown in Figs. 14.5a and 14.5b, plotted vs velocity with altitude as a parameter. Recall from Eq. (2.4) that, for a calorically perfect gas, ρ_2/ρ_1 approaches the limiting value of $(\gamma + 1)/(\gamma - 1)$ as $M_\infty \to \infty$. For air with $\gamma = 1.4$, this limiting ratio is 6. Note from Figs. 14.5a and 14.5b that ρ_2/ρ_1 is strongly affected by chemical reactions and that its values range far above 6—reaching as high as 22.

The value of ρ_2/ρ_1 has an important effect on the shock detachment distance in front of a hypersonic blunt body. The flow of a calorically perfect gas over a hypersonic blunt body is discussed in Secs. 5.3 and 5.4. An approximate expression for the shock detachment distance δ on a sphere of radius R is (see [46])

$$\frac{\delta}{R} = \frac{\rho_1/\rho_2}{1 + \sqrt{2(\rho_1/\rho_2)}} \tag{14.18}$$

In the limit of high velocities, ρ_1/ρ_2 becomes small compared to unity, and Eq. (14.18) is approximated by

$$\frac{\delta}{R} \approx \frac{\rho_1}{\rho_2} = \frac{1}{(\rho_2/\rho_1)} \tag{14.19}$$

Therefore, the value of the density ratio across a normal shock wave has a major impact on shock detachment distance; the higher the density ratio ρ_2/ρ_1, the smaller is δ. From Figs. 14.5a and 14.5b, we see that the effect of chemical reactions is to increase ρ_2/ρ_1, which, in turn, decreases the shock detachment distance. Therefore, in comparison to the calorically perfect-gas blunt-body results discussed in Secs. 5.3 and 5.4, the shock wave for a chemically reacting gas (at the same velocity and altitude conditions) will lie closer to the body. This is emphasized schematically in Fig. 14.6, where δ_{cp} and δ_{cR} are the shock detachment distances for a calorically perfect gas and a chemically reacting gas, respectively.

Let us now turn our attention to oblique shock waves in an equilibrium gas. The flow across an oblique shock is sketched in Fig. 14.7. It is readily shown (see [4] and [5]) that the component of flow velocity tangential to a straight oblique shock wave is preserved across the shock, that is, $V_{t,1} = V_{t,2}$ in Fig. 14.7. This is a basic mechanical result obtained from the momentum equation, and hence it is *not* influenced by high-temperature effects. In turn, the thermodynamic changes across the oblique shock are dictated only by the component of the upstream velocity perpendicular to the shock $V_{n,1}$. Therefore, we have for the high-temperature equilibrium flow across an oblique shock wave the same basic, familiar results from classic shock-wave theory, namely, that the properties behind the oblique shock are the same as the properties across a normal shock with upstream velocity $u_1 = V_{n,1}$. (The exception to this is, of course, the flow velocity behind the oblique shock V_2, where V_2 must be obtained by the vector addition of the tangential component $V_{t,2}$ and the normal component $V_{n,2}$, with $V_{n,2}$ satisfying the normal shock results.) Consequently, the normal shock analysis involving Eqs. (14.13–14.17) also applies to the equilibrium flow across an oblique shock wave.

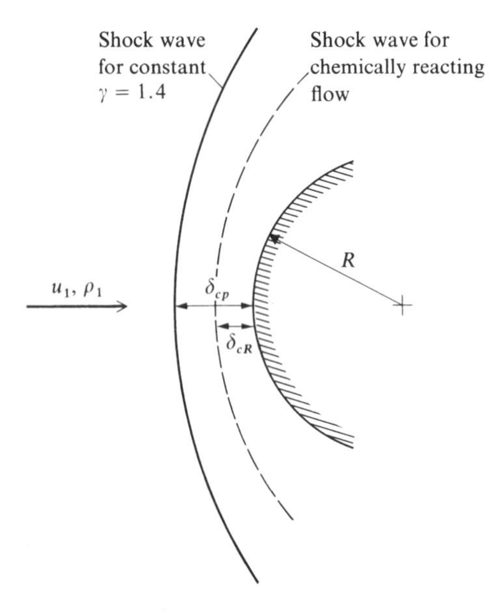

Fig. 14.6 Relative locations of blunt-body bow shock waves for calorically perfect and chemically reacting gases.

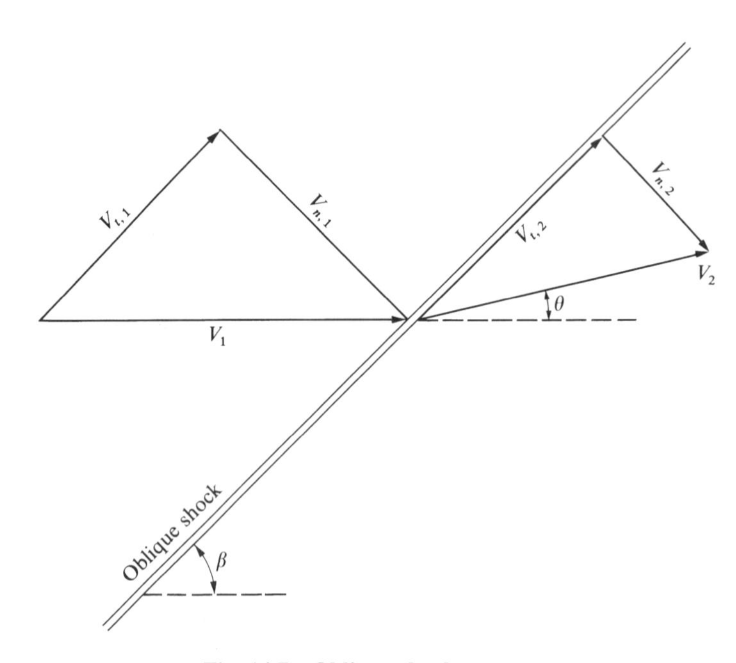

Fig. 14.7 Oblique shock geometry.

From, the oblique shock picture in Fig. 14.7, we have

$$\tan(\beta - \theta) = \frac{V_{n,2}}{V_{t,2}} \tag{14.20}$$

where, as in Chapter 2, θ is the deflection angle and β is the wave angle. Because $V_{t,2} = V_{t,1}$, Eq. (14.20) can be written as

$$\tan(\beta - \theta) = \frac{V_{n,2}}{V_{t,1}} = \frac{V_{n,2}}{V_{n,1}} \frac{V_{n,1}}{V_{t,1}}$$

or

$$\tan(\beta - \theta) = \frac{V_{n,2}}{V_{n,1}} \tan\beta \tag{14.21}$$

Equation (14.21) is, for the equilibrium high-temperature case, the analog of Eq. (2.16) for the calorically perfect-gas case; Eq. (14.21) relates the wave angle β, the deflection angle θ, and the upstream velocity V_1 (via its components $V_{n,1}$ and $V_{n,2}$). The solution of Eqs. (14.21) combined with the normal shock numerical solution described by Eqs. (14.13–14.17) yields a θ-β-V diagram for equilibrium flow across oblique shocks, which is a direct analog to the familiar θ-β-M diagram obtained for a calorically perfect gas shown in Fig. 2.3. An equilibrium θ-β-V diagram for high-temperature air is given in Fig. 14.8a, obtained from [166]. The results shown in Fig. 14.8a are for an altitude of 100,000 ft, that is, for a fixed p_1 and T_1. The equilibrium chemically reacting results are given by the solid curves for different values of V_1. These are compared with the calorically perfect-gas results with $\gamma = 1.4$, given by the dashed curves. From these results, note the following aspects:

1) Figure 14.8a for equilibrium chemically reacting air is qualitatively similar to Fig. 2.3 for calorically perfect air.

2) For the equilibrium chemically reacting results, Mach number M_1 is *not* an important parameter, as discussed earlier for the normal shock case. Rather, the oblique shock results depend on *velocity* V_1 as well as p_1 and T_1 (or equivalently, as in the case of Fig. 14.8a, on V_1 and altitude).

3) The density ratio effect is strongly evident in Fig. 14.8a. Consider the "weak-shock solutions" given by the lower portion of the θ-β-V curves. (See [4] and [5] for a discussion of weak-shock and strong-shock cases for oblique shock waves.) In Fig. 14.8a, for a given deflection angle θ the equilibrium results for the wave angle β (solid curves) are less than those for a calorically perfect gas with constant $\gamma = 1.4$ (dashed curves). This implies that the oblique shock wave will lie closer to the surface for the chemically reacting equilibrium case, as sketched in Fig. 14.8b. In this sense, Fig. 14.8b is, for the flow over a wedge, the analog to Fig. 14.6 for the flow over a blunt body. Of course, the reason why the chemically reacting oblique shock lies closer to the surface is because of the increased density ratio ρ_2/ρ_1 across the wave, just as in the case of the normal shock. The reverse is true for the strong-shock solutions given by the upper portions of the curves in Fig. 14.8a. Here, the wave angle is

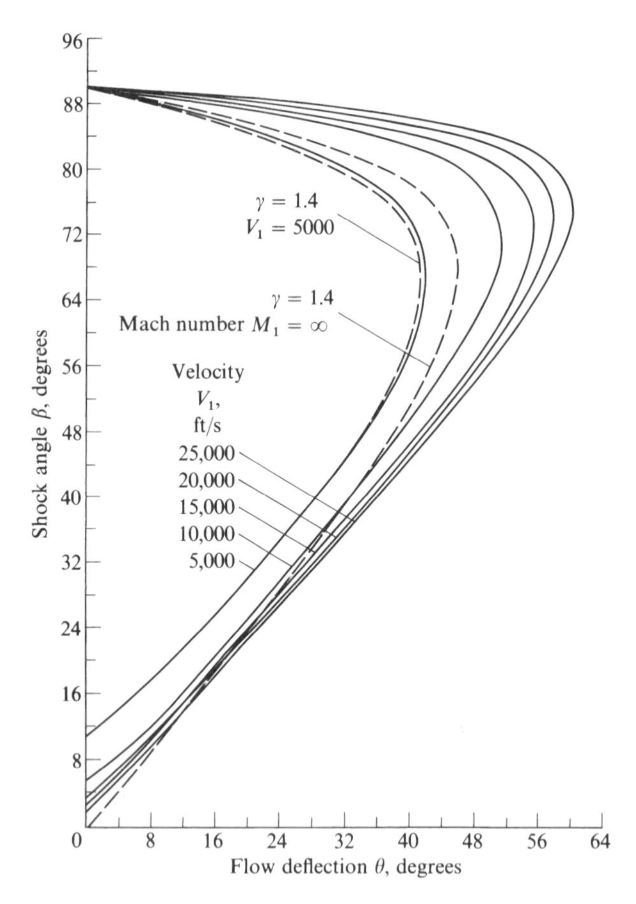

Fig. 14.8a Deflection angle-wave angle-velocity diagram for oblique shocks in high-temperature air at 100,000 ft altitude (from Moeckel [166]).

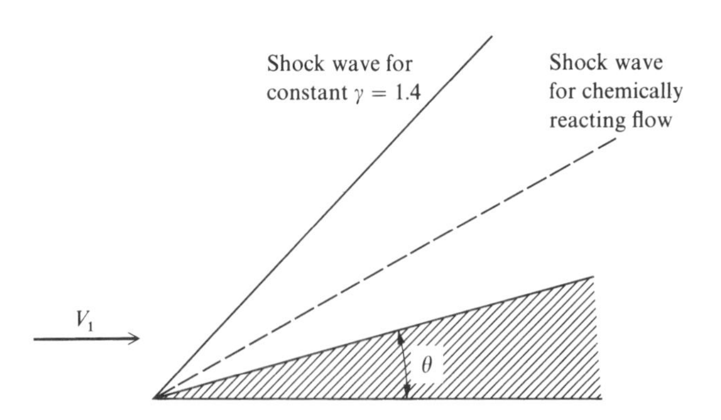

Fig. 14.8b Comparison of oblique shock waves for a calorically perfect gas vs an equilibrium chemically reacting gas.

greater for the chemically reacting case. (However, keep in mind that in the vast majority of actual applications it is the weak-shock solution that prevails.)

4) The maximum deflection angle θ allowed for the solution of a straight oblique shock wave is increased by chemically reacting effects.

An interesting study of equilibrium properties behind normal and oblique shock waves for velocity-altitude points following trajectories for a transatmospheric vehicle was recently given by Bussing and Eberhardt in [158]. The trajectories are shown on the velocity-altitude map in Fig. 14.9. The two lower curves correspond to two possible ascents of a hypothetical transatmospheric vehicle. The other curves show the ascent and entry flight paths of the space shuttle, for comparison. The equilibrium chemical composition behind a normal shock is shown in Fig. 14.10, and that for an oblique shock with $\beta = 30$ deg is shown in Fig. 14.11. In both figures, the values of p_1 and T_1 that correspond to the various M_1 values on the abscissa are those that pertain to the standard altitudes as dictated by the upper and lower trajectories in Fig. 14.9. In Fig. 14.10 for the normal shock case, note the progressive increase in dissociation as M_1 increases above a value of 6. Also note that the mole fraction of ions is exclusively because of NO^+, and that this mole fraction is small. In contrast, for the case with $\beta = 30$ deg shown in Fig. 14.11, dissociation does not become important until M_1 is well above 12, and that ionization is virtually nonexistent.

This concludes our discussion of normal and oblique shock waves for flow in local thermodynamic and chemical equilibrium. These are very basic flows, and they clearly exhibit the type of high-temperature effects associated with compression-type flows. Make certain that you feel comfortable with these results, both quantitative and qualitative, before progressing further.

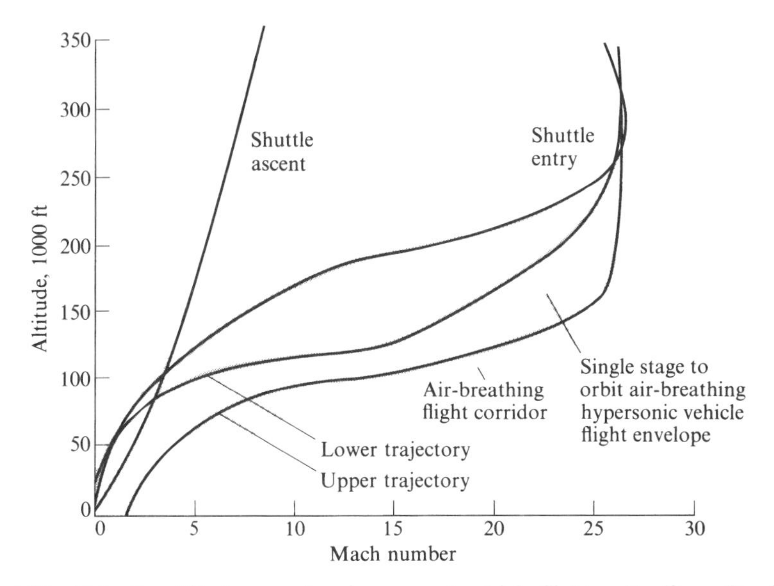

Fig. 14.9 Velocity-altitude map showing several vehicle flight paths (from Bussing and Eberhardt [158]).

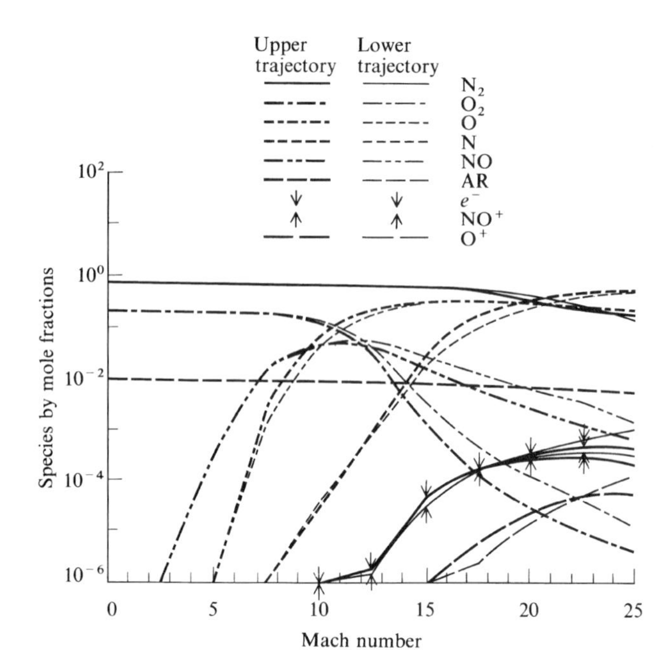

Fig. 14.10 Equilibrium chemical species variations behind a normal shock, following the trajectories shown in Fig. 14.9 (from [158]).

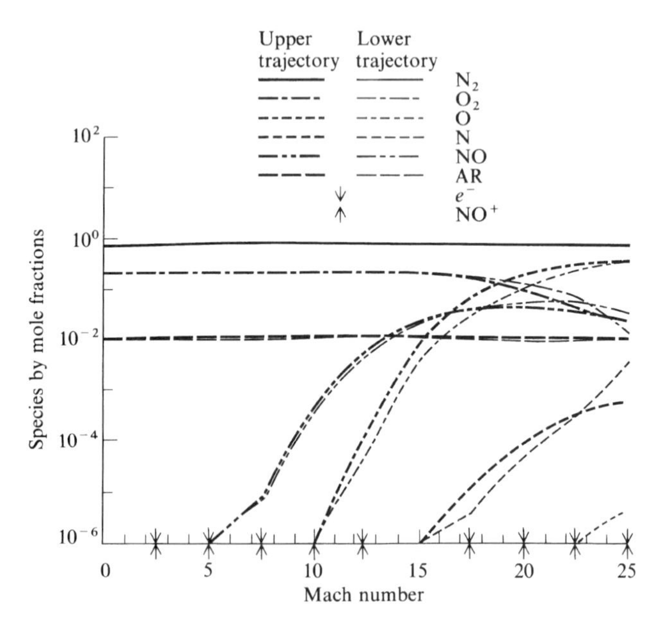

Fig. 14.11 Equilibrium chemical species variations behind a 30-deg oblique shock, following the trajectories shown in Fig. 14.9 (from [158]).

14.4 Equilibrium Quasi-One-Dimensional Nozzle Flows

Consider the inviscid, adiabatic high-temperature flow through a convergent-divergent Laval nozzle, as sketched at the top of Fig. 14.12. As usual, the reservoir pressure and temperature are denoted by p_0 and T_0, respectively. The throat conditions are denoted by an asterisk and exit conditions by a subscript e. This nozzle could be a high-temperature wind tunnel, where air is heated in the reservoir, for example, by an electric arc (an arc tunnel) or by shock waves (a shock tunnel). In a shock tunnel, the nozzle is placed at the end of a shock tube, and the reservoir is essentially the hot, high-pressure gas behind a reflected shock wave (see Sec. 9.1). The nozzle in Fig. 14.12 could also be a rocket engine, where the reservoir conditions are determined by the burning of fuel and oxidizer in the combustion chamber. In either case—the high-temperature wind tunnel or the rocket engine—the flow through the nozzle is chemically reacting. Assuming local chemical equilibrium throughout the flow, let us examine the properties of the nozzle expansion.

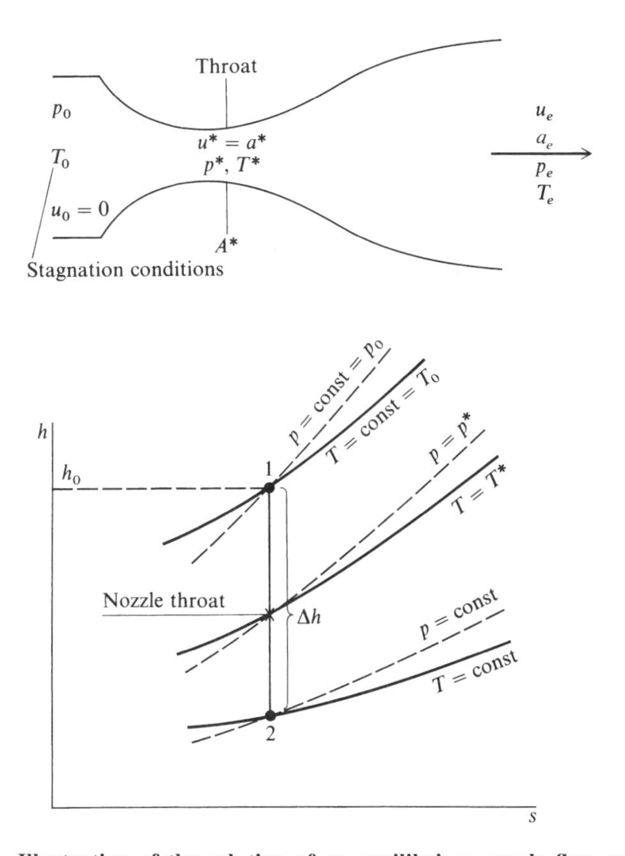

Fig. 14.12 Illustration of the solution of an equilibrium nozzle flow on a Mollier diagram.

First, let us pose the question: is the chemically reacting flow isentropic? On a physical basis, the flow is both inviscid and adiabatic. However, this does not guarantee in general that the chemically reacting flow is irreversible. If we deal with an equilibrium chemically reacting flow, we can write the combined first and second laws of thermodynamics from Eqs. (10.34) and (10.46) as

$$T\,\mathrm{d}s = \mathrm{d}h - v\,\mathrm{d}p \tag{14.22}$$

From Eq. (14.5) we have, for an adiabatic steady flow, $h_0 =$ constant or, in differential form,

$$\mathrm{d}h + V\,\mathrm{d}V = 0 \tag{14.23}$$

From Eq. (14.3), written along a streamline, we obtain a familiar form of Euler's equation as

$$\mathrm{d}p = -\rho V\,\mathrm{d}V \tag{14.24}$$

This can be rearranged as

$$V\,\mathrm{d}V = -\frac{\mathrm{d}p}{\rho} = -v\,\mathrm{d}p \tag{14.25}$$

As explained in Sec. 14.2, all of these equations hold for chemically reacting flow. Combining Eqs. (14.23) and (14.25), we have

$$\mathrm{d}h - \frac{\mathrm{d}p}{\rho} = \mathrm{d}h - v\,\mathrm{d}p = 0 \tag{14.26}$$

Substituting Eq. (14.26) into (14.22), we obtain

$$T\,\mathrm{d}s = 0 \tag{14.27}$$

Hence, the equilibrium chemically reacting nozzle flow is isentropic. Moreover, because Eq. (14.27) was obtained by combining the energy and momentum equations, the assumption of isentropic flow can be used in place of either the momentum or energy equations in the analysis of the flow.

It is a general result that equilibrium chemical reactions do not introduce irreversibilities into the system; if an equilibrium reacting system starts at some conditions p_1 and T_1, deviates from these conditions for some reason, but then returns to the original p_1 and T_1, the chemical composition at the end returns to what it was at the beginning. Equilibrium chemical reactions are reversible. Hence, any shockless, inviscid, adiabatic, equilibrium chemically reacting flow is isentropic. This is *not* true if the flow is nonequilibrium, as will be discussed in Chapter 15.

All of the preceding results and statements hold for any general shockless flow in local thermodynamic and chemical equilibrium. Let us now address the specific aspects of quasi-one-dimensional flow. As defined and discussed in [4] and [5], quasi-one-dimensional flow is a flow where the cross-sectional area is a variable $A = A(x)$, but where all of the flow properties across any given cross section are assumed to be uniform. Hence, a flow that is in reality two or three dimensional is assumed to have properties that vary only in the x direction; $p = p(x)$, $V = u = u(x)$, $T = T(x)$, etc. Various aspects of quasi-one-dimensional flow for a calorically perfect gas are discussed in Chapter 5 of [4], such matters should be reviewed by the reader before progressing further.

Let us pose another question: for an equilibrium, chemically reacting, quasi-one-dimensional nozzle flow, does sonic flow exist at the throat? We have already established that the flow is isentropic. This is the only necessary condition for the derivation of the area-velocity relation, derived in most compressible flow texts (see [4] and [5]). This relation is given by

$$\frac{\mathrm{d}A}{A} = (M^2 - 1)\frac{\mathrm{d}u}{u} \tag{14.28}$$

which holds for a general gas. In turn, when $M = 1$, $\mathrm{d}A/A = 0$, and therefore sonic flow does exist at the throat of an equilibrium chemically reacting nozzle flow. The same is not true for a nonequilibrium flow, as will be discussed in Chapter 15.

We are now in a position to solve the equilibrium chemically reacting nozzle flow. A graphical solution is the easiest to visualize. Consider that we have the equilibrium gas properties on a Mollier diagram, as sketched in Fig. 14.12. Recall from Fig. 11.14 that a Mollier diagram is a plot of h vs s, and lines of constant p and constant T can be traced on the diagram. Hence, referring to Fig. 14.12, a given point on the Mollier diagram gives not only h and s, but p and T at that point as well (and any other equilibrium thermodynamic property, because the state of an equilibrium system is completely specified by any two-state variables). Let point 1 in Fig. 14.12 denote the known reservoir conditions in the nozzle. Because the flow is isentropic, conditions at all other locations throughout the nozzle must fall somewhere on the vertical line through point 1 in Fig. 14.12. In particular, choose a value of $u = u_2 \neq 0$. The point in Fig. 14.12 that corresponds to this velocity (point 2) can be found from Eqs. (14.4) and (14.5) as follows:

$$h_0 = \text{const}$$

Hence,

$$h_1 + \frac{u_1^2}{2} = h_2 + \frac{u_2^2}{2} = h_0 \tag{14.29}$$

Thus,

$$\Delta h = h_0 - h_2 = \frac{u_2^2}{2} \tag{14.30}$$

Therefore, for a given velocity u_2, Eq. (14.30) locates the appropriate point on the Mollier diagram. In turn, the constant-pressure and -temperature lines that run through point 2 define the pressure p_2 and temperature T_2 associated with the chosen velocity u_2. In this fashion, the variation of the thermodynamic properties through the nozzle expansion can be calculated as a function of velocity u for given reservoir conditions.

For an equilibrium gas, the speed of sound, $a^2 \equiv (\partial p/\partial \rho)_s$, is also a unique function of the thermodynamic state. This will be discussed in more detail in Sec. 14.6. For example,

$$a = a(h, s) \tag{14.31}$$

Thus, at each point on the Mollier diagram in Fig. 14.12, there exists a definite value of a. Moreover, at some point along the vertical line through point 1, the speed of sound a will equal the velocity u at that point. Such a point is marked by an asterisk in Fig. 14.12. At this point, $u = a = u^* = a^*$. Because we demonstrated earlier that sonic flow corresponds to the throat in an equilibrium nozzle flow, then this point in Fig. 14.12 must correspond to the throat. The pressure, temperature, and density at this point are p^*, T^*, and ρ^*, respectively. Thus, from the continuity equation for quasi-one-dimensional flow, we have

$$\rho u A = \rho^* u^* A^* \tag{14.32}$$

or

$$\frac{A}{A^*} = \frac{\rho^* u^*}{\rho u} \tag{14.33}$$

Therefore, Eq. (14.33) allows the calculation of the nozzle area ratio as a function of velocity through the nozzle.

In summary, using the Mollier diagram in Fig. 14.12, we can compute the appropriate values of u, p, T, and A/A^* through an equilibrium nozzle flow for given reservoir conditions. An alternative to this graphical approach is a straightforward numerical integration of Eqs. (14.23), (14.24), and (14.32) along with tabulated values of the equilibrium thermodynamic properties. The integration starts from known conditions in the reservoir and marches downstream. Such a numerical integration solution is left for the reader to construct.

In either case, numerical or graphical, it is clear that the familiar closed-form algebraic relations that can be obtained for a calorically perfect gas (see [4] and [5]) are not obtainable for chemically reacting nozzle flows. This is analogous to the case of chemically reacting flow through a shock wave discussed in Sec. 14.2. In fact, by now the reader should suspect, and correctly so, that closed-form algebraic relations cannot be obtained for any high-temperature chemically reacting flow of interest. Numerical or graphical solutions are necessary for such cases.

Recall that, for a calorically perfect gas, the nozzle flow characteristics were governed by the local Mach number only. For example, from [4] and [5], for a calorically perfect gas,

$$\frac{A}{A^*} = f_1(M)$$

$$\frac{T}{T_0} = f_2(M)$$

$$\frac{p}{p_0} = f_3(M)$$

In contrast, for an equilibrium chemically reacting gas,

$$\frac{A}{A^*} = g_1(p_0, T_0, u)$$

$$\frac{T}{T_0} = g_2(p_0, T_0, u)$$

$$\frac{p}{p_0} = g_3(p_0, T_0, u)$$

Note, as in the case of a normal shock, that the nozzle flow properties depend on three parameters. Also, once again we see that Mach number is not the pivotal parameter for a chemically reacting flow.

Some results for the equilibrium supersonic expansion of high-temperature air are shown in Fig. 14.13. Here the mole-mass ratios for N_2, O_2, N, O, and NO are given as a function of area ratio for $T_0 = 8000$ K and $p_0 = 100$ atm. At these conditions, the air is highly dissociated in the reservoir. However, as the gas expands through the nozzle, the temperature decreases, and as a result the oxygen and nitrogen recombine. This is reflected in Fig. 14.13, which shows η_O and η_N decreasing and η_{O_2} and η_{N_2} increasing as the gas expands supersonically from $A/A^* = 1$ to 1000.

A typical result from equilibrium chemically reacting flow through a rocket nozzle is shown in Fig. 14.14. Here, the equilibrium temperature distribution is compared with that for a calorically perfect gas as a function of area ratio. The reservoir conditions are produced by the equilibrium combustion of an oxidizer (N_2O_2) with a fuel (half N_2H_4 and half unsymmetrical dimethyl hydrazine) at an oxidizer-to-fuel ratio of 2.25 and a chamber pressure of 4 atm. The calorically perfect gas is assumed to have a constant $\gamma = 1.20$. It is important to note from Fig. 14.14 that the equilibrium temperature is higher than that for the calorically perfect gas. This is because, as the gas expands and becomes cooler, the chemical composition changes from a high percentage of atomic species (O and H) in the reservoir with an attendant high zero-point energy to a high percentage of molecular products (H_2O), CO, etc. in the nozzle expansion with an

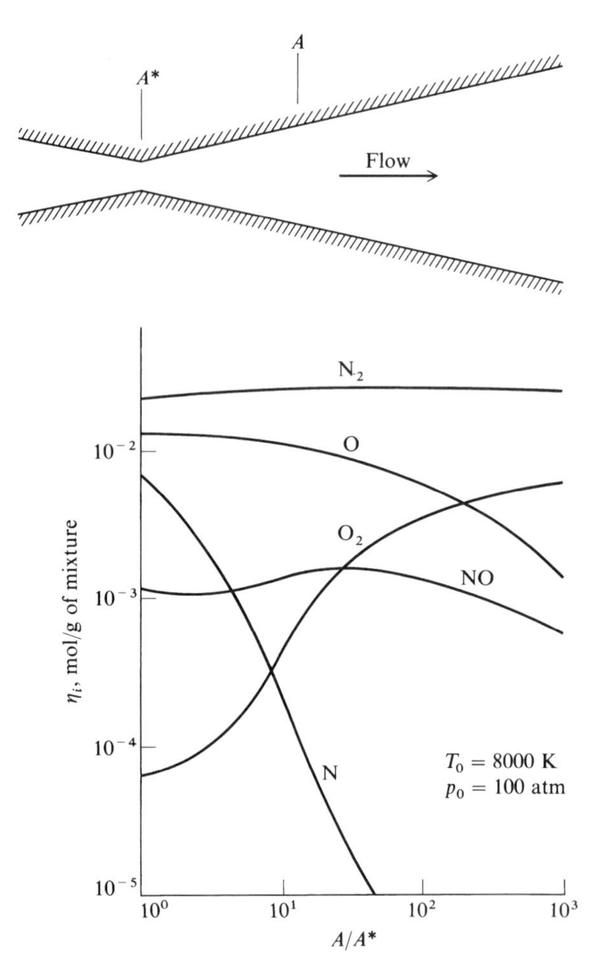

Fig. 14.13 Chemical composition for the equilibrium nozzle expansion of high-temperature air (from [264]).

attendant lower zero-point energy. That is, the gas recombines, giving up chemical energy, which serves to increase the translational energy to the molecules, hence resulting in a higher static temperature that would exist in the nonreacting case. Note that the trend shown in Fig. 14.14 for nozzle flow is exactly the opposite of that shown in Figs. 1.18 and 14.3 for shock waves. For nozzle flow, the equilibrium temperature is always higher than that for a calorically perfect gas; for flow behind a shock wave, the equilibrium temperature is always lower than that for a calorically perfect gas. In the former case, the reactions are exothermic, and energy is dumped into the translational molecular motion; in the latter, the reactions are endothermic, and energy is taken from the translational mode.

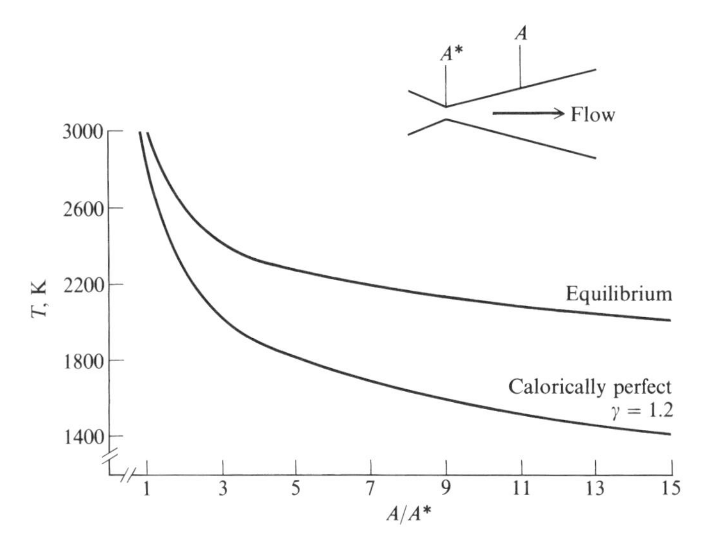

Fig. 14.14 Comparison between equilibrium and calorically perfect results for the flow through a rocket engine.

Without going into the details, the two- or three-dimensional nature of nozzle flows in local thermodynamic and chemical equilibrium can be calculated by means of the method of characteristics. The philosophy and execution of the method of characteristics for such a case is no different than discussed in Sec. 5.2; the compatibility equations and characteristic lines are exactly the same—only the high-temperature thermodynamic properties given in concept by Eqs. (14.6) and (14.7) have to be included. See [53] for more details.

14.5 Frozen and Equilibrium Flows: The Distinction

Referring to our chapter road map in Fig. 14.1, we now make a temporary detour to the right-hand column. To this point in the present chapter, we have discussed flows in local thermodynamic and chemical equilibrium, as defined in Sec. 14.1. In reality, such flows never occur precisely in nature. This is because all chemical reaction and vibrational energy exchanges require a certain number of molecular collisions to occur; because the gas particles experience a finite collision frequency (see Chapter 12), such reactions and energy exchanges require a finite time to occur, as described in Chapter 13. Therefore, in the hypothetical case of local equilibrium flow discussed in the present chapter where the equilibrium properties of a moving fluid element demand instantaneous adjustments to the local T and p as the element moves through the field, the reaction rates have to be infinitely large. *Therefore, equilibrium flow implies infinite chemical and vibrational rates.*

The opposite of this situation is a flow where the reaction rates are precisely zero—so-called frozen flow. As a result, the chemical composition of frozen flow remains constant throughout space and time. (This is true for an inviscid flow; for a viscous flow the composition of a given fluid element can change via diffusion, even though the flow is chemically frozen. Diffusion effects are discussed in Chapters 16 and 17.)

To reinforce the distinction between equilibrium and frozen flows, the qualitative difference between chemical equilibrium and frozen nozzle flows is sketched in Fig. 14.15 for a case of fully dissociated oxygen in the reservoir. Examining Fig. 14.15c, the flow starts out with oxygen atoms in the reservoir ($c_O = 1$, $c_{O_2} = 0$). If we have equilibrium flow, as the temperature decreases throughout the expansion the oxygen atoms will recombine; hence, c_O decreases, and c_{O_2} increases as a function of distance through the nozzle. If the expansion

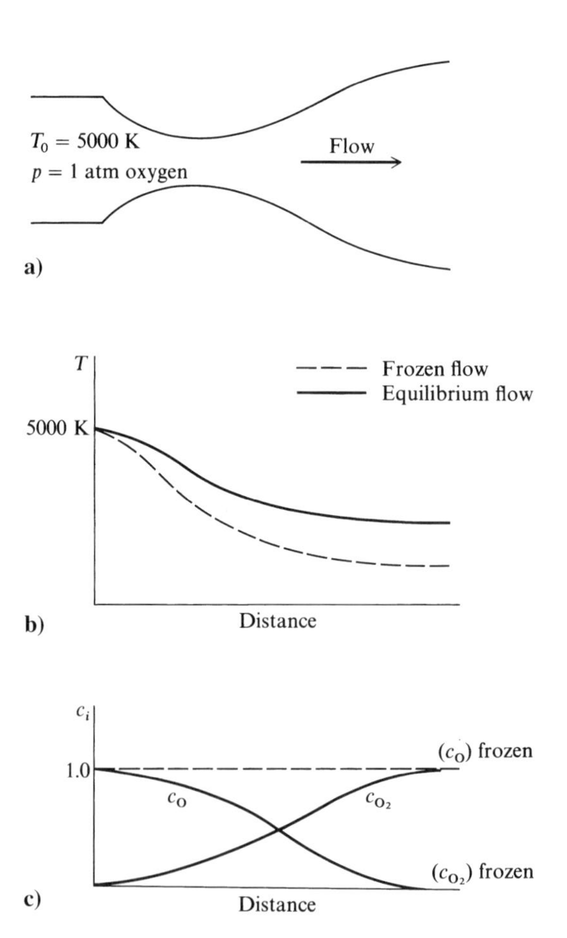

Fig. 14.15 Schematic comparing equilibrium and frozen chemically reacting flows through a nozzle.

(area ratio) is large enough such that the exit temperature is near room temperature, equilibrium conditions demand that virtually all of the oxygen atoms recombine, and, for all practical purposes, $c_{O_2} = 1$ and $c_O = 0$ at the exit. These equilibrium distributions are shown by the solid curves in Fig. 14.15. In contrast, if the flow is chemically frozen, then by definition the mass fractions are constant as a function of distance through the nozzle (the dashed lines in Fig. 14.15c). Recombination is an exothermic reaction; hence, the equilibrium expansion results in the chemical zero-point energy of the atomic species being transferred into the translational, rotational, and vibrational modes of molecular energy. (The zero-point energy of two O atoms is much higher than the zero-point energy of one O_2 molecule. When two O atoms recombine into one O_2 molecule, the decrease in zero-point energy results in an increase in the internal molecular energy modes.) As a result, the temperature distribution for equilibrium flow is higher than that for frozen flow, as sketched in Fig. 14.15.

For vibrationally frozen flow, the vibrational energy remains constant throughout the flow. Consider a nonreacting vibrationally excited nozzle expansion as sketched in Fig. 14.16. Assume that we have diatomic oxygen in the reservoir at a temperature high enough to excite the vibrational energy, but low enough such that dissociation does not occur. If the flow is in local thermodynamic equilibrium, the translational, rotational, and vibrational energies are given by Eqs. (11.57), (11.59), and (11.61), respectively. The energies decrease through the nozzle, as shown by the solid curves in Fig. 14.16c. However, if the flow is vibrationally frozen, then e_{vib} is constant throughout the nozzle and is equal to its reservoir value. This is shown by the horizontal dashed line in Fig. 14.16c. In turn, because energy is permanently sealed in the frozen vibrational mode, less energy is available for the translational and rotational modes. Thus, because T is proportional to the translational energy, the frozen flow temperature distribution is less than that for equilibrium flow, as shown in Fig. 14.16b. In turn, the distributions of e_{trans} and e_{rot} will be lower for vibrationally frozen flow, as shown in Fig. 14.16c.

It is left as an exercise for the reader to compare the equilibrium and frozen flows across a normal shock wave.

Note that a flow which is both chemically and vibrationally frozen has constant specific heats. This is nothing more than the flow of a calorically perfect gas; we have treated the topic in Chapters 1 through 8.

As a final note in this section, although precisely equilibrium or precisely frozen flows never occur in nature, there are a large number of flow applications that come very close to such limiting situations and can be analyzed using one or the other of these assumptions. (This is in the same category as saying that precisely isentropic flow never exists in real life, but there are many practical problems that can be very accurately analyzed by making the assumption of isentropic flow.) The judgment as to whether a given flow in real life is close enough to either equilibrium or frozen flow depends on the comparison between reaction times and flow times, to be described in Sec. 15.1. Suffice it to say here that the study of flows in local thermodynamic and chemical equilibrium (the subject of this chapter) is very practical and is applicable to many real flow problems.

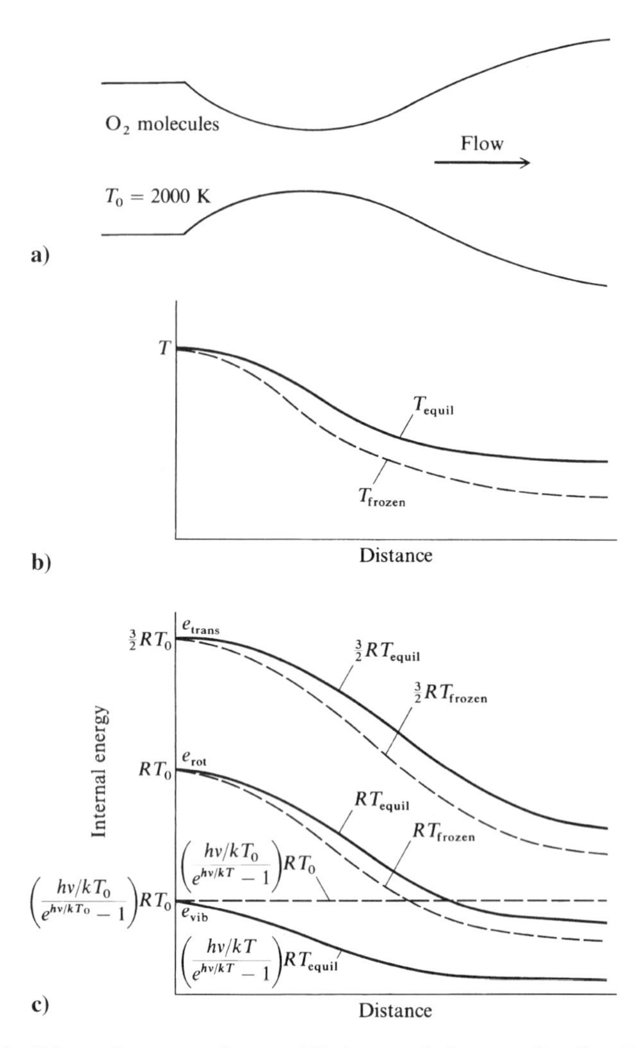

Fig. 14.16 Schematic comparing equilibrium and frozen vibrationally relaxing flows through a nozzle.

14.6 Equilibrium and Frozen Specific Heats

In this section, we temporarily deviate from our discussion of flow problems and reexamine a thermodynamic property of chemically reacting gases, namely, the specific heat. To understand the significance of this section, look back over the flow problems described in Secs. 14.2–14.4. Note that the governing equations do not involve the specific heats c_p or c_v; rather, the energy equation deals with the more fundamental variables of enthalpy or internal energy. The reason for this is developed in the present section.

Let us derive an expression for the specific heat of an equilibrium chemically reacting gas, as follows.

The enthalpy of a chemically reacting mixture can be obtained from Eq. (11.89), repeated here:

$$h = \sum_i c_i h_i \tag{11.89}$$

By definition, the specific heat at constant pressure c_p is

$$c_p = \left(\frac{\partial h}{\partial T}\right)_p \tag{14.34}$$

Thus, for a chemically reacting mixture, Eqs. (11.89) and (14.34) give

$$c_p = \left[\frac{\partial}{\partial T}\left(\sum_i c_i h_i\right)\right]_p$$

$$c_p = \sum_i c_i \left(\frac{\partial h_i}{\partial T}\right)_p + \sum_i h_i \left(\frac{\partial c_i}{\partial T}\right)_p \tag{14.35}$$

In Eq. (14.35), $(\partial h_i/\partial T)_p$ is the specific heat per unit mass for the pure species i, c_{p_i}. Hence, Eq. (14.35) becomes

$$\boxed{c_p = \sum_i c_i c_{p_i} + \sum_i h_i \left(\frac{\partial c_i}{\partial T}\right)_\rho} \tag{14.36}$$

Equation (14.36) is an expression for the specific heat of a chemically reacting mixture. If the flow is frozen, by definition there are no chemical reactions, and therefore in Eq. (14.36) the term $(\partial c_i/\partial T)_p = 0$. Thus, for a frozen flow, the specific heat becomes, from Eq. (14.36),

$$c_p = c_{p_f} = \sum_i c_i c_{p_i} \tag{14.37}$$

In turn, the frozen flow specific heat, denoted in Eq. (14.37) by c_{p_f}, can be inserted in Eq. (14.36), yielding, for a chemically reacting gas,

$$\boxed{\underbrace{c_p}_{\substack{\text{Specific heat at}\\ \text{constant pressure for}\\ \text{the reacting mixture}}} = \underbrace{c_{p_f}}_{\substack{\text{Frozen}\\ \text{specific}\\ \text{heat}}} + \underbrace{\sum_i h_i \left(\frac{\partial c_i}{\partial T}\right)_p}_{\substack{\text{Contribution caused by}\\ \text{chemical reaction}}}} \tag{14.38}$$

Considering the internal energy of the chemically reacting gas given by

$$e = \sum_i c_i e_i$$

and using the definition of specific heat at constant volume

$$c_v = \left(\frac{\partial e}{\partial T}\right)_v$$

we obtain in a similar fashion

$$\boxed{c_v = c_{v_f} + \sum_i e_i \left(\frac{\partial c_i}{\partial T}\right)_v} \tag{14.39}$$

where

$$c_{v_f} = \sum_i c_i c_{v_i} \tag{14.40}$$

Equations (14.38) and (14.39) are conceptually important. Throughout our calorically perfect-gas discussions in Chapters 1 through 8, we were employing c_p and c_v as expressed by Eqs. (14.37) and (14.40). Now, for the case of a chemically reacting gas, we see from Eqs. (14.38) and (14.39) that an extra contribution, namely,

$$\sum_i h_i \left(\frac{\partial c_i}{\partial T}\right)_p \qquad \text{or} \qquad \sum_i e_i \left(\frac{\partial c_i}{\partial T}\right)_v$$

is made to the specific heats purely because of the reactions themselves. The magnitude of this extra contribution can be very large and usually dominates the value of c_p or c_v.

For practical cases, it is not possible to find analytic expressions for $(\partial c_i/\partial T)_p$ or $(\partial c_i/\partial T)_v$. For an equilibrium mixture, they can be evaluated numerically by differentiating the data from an equilibrium calculation, such as was described in Secs. 11.12 and 11.13. Such evaluations have been made, for example, by Frederick Hansen in NASA TR-50 (see [167]). Figure 14.17 is taken directly from Hansen's work and shows the variation of c_v for air with temperature at several different pressures. The humps in each curve reflect the reaction term in Eq. (14.39),

$$\sum_i e_i \left(\frac{\partial c_i}{\partial T}\right)_v$$

and are caused consecutively by dissociation of oxygen, dissociation of nitrogen, and then, at very high temperatures, the ionization of both O and N. (Note that the ordinate of Fig. 14.17 is a nondimensionalized specific heat, where $\mathscr{R}$ is the universal gas constant, $\mathscr{M}_0$ is the initial molecular weight of undissociated air, $\mathscr{M}$ is molecular weight at a given T and p, and C_v is the molar specific heat.)

Because c_p and c_v for a chemically reacting mixture are functions of both T and p (or T and v) and because they exhibit such wild variations as seen in Fig. 14.17, they are not usually employed directly in calculations of inviscid

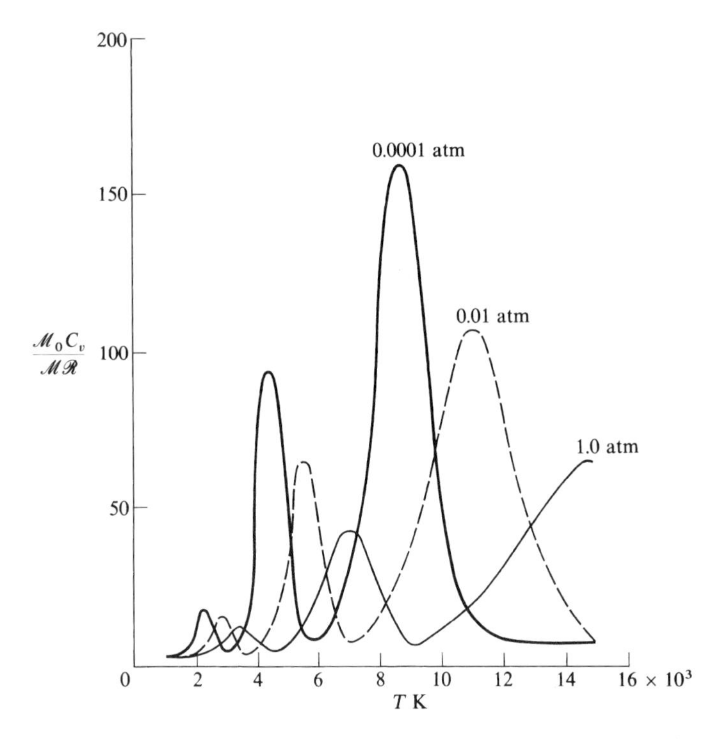

Fig. 14.17 Specific heat of equilibrium air at constant pressure as a function of temperature (from Hansen [167]).

high-temperature flows. Note that, in our preceding discussions on shock waves (Sec. 14.3) and nozzle flows (Sec. 14.4), h or e were used for a solution rather than c_p or c_v. However, it is important for an overall understanding of high-temperature flows to know how and why the specific heats vary. This has been the purpose of the preceding discussion. Moreover, in Chapter 17 dealing with chemically reacting viscous flows, the Prandtl and Lewis numbers are identified as important similarity parameters, both of which involve c_p. Thus, we need values for c_p for chemically reacting viscous flows in order to evaluate local values of the Prandtl and Lewis numbers.

14.7 Equilibrium Speed of Sound

In general, the speed of sound in a gas is given by (see [4] and [5])

$$a = \sqrt{\left(\frac{\partial p}{\partial \rho}\right)_s}$$

This is a physical fact and is not changed by the presence of chemical reactions. Furthermore, for a calorically perfect gas, $a = \sqrt{\gamma RT}$. But what is the value of speed of sound in an equilibrium reacting mixture? How do we calculate it? Is it equal to $\sqrt{\gamma RT}$? The purpose of this section is to address these questions.

Consider an equilibrium chemically reacting mixture at a fixed p and T. Therefore, the chemical composition is uniquely fixed by p and T. Imagine a sound wave passing through this equilibrium mixture. Inside the wave, p and T will change slightly. If the gas remains in local chemical equilibrium through the internal structure of the sound wave, the gas composition is changed locally within the wave according to the local variations of p and T. For this situation, the speed of the sound wave is called the equilibrium speed of sound, denoted by a_e. In turn, if the gas is in motion at the velocity V, then V/a_e is defined as the equilibrium Mach number M_e.

To obtain a quantitative relation for the equilibrium speed of sound, consider the first and second laws of thermodynamics from Eqs. (10.32), (10.34), and (10.46):

$$T\,\mathrm{d}s = \mathrm{d}e + p\,\mathrm{d}v \tag{14.41}$$

$$T\,\mathrm{d}s = \mathrm{d}h - v\,\mathrm{d}p \tag{14.42}$$

The process through a sound wave is isentropic; hence, Eqs. (14.41) and (14.42) become

$$\mathrm{d}e + p\,\mathrm{d}v = 0 \tag{14.43}$$

and

$$\mathrm{d}h - v\,\mathrm{d}p = 0 \tag{14.44}$$

For an equilibrium chemically reacting gas

$$e = e(v, T)$$

Thus, the total differential is

$$\mathrm{d}e = \left(\frac{\partial e}{\partial v}\right)_T \mathrm{d}v + \left(\frac{\partial e}{\partial T}\right)_v \mathrm{d}T$$

$$\mathrm{d}e = \left(\frac{\partial e}{\partial v}\right)_T \mathrm{d}v + c_v\,\mathrm{d}T \tag{14.45}$$

Similarly,

$$h = h(p, T)$$

$$\mathrm{d}h = \left(\frac{\partial h}{\partial p}\right)_T \mathrm{d}p + \left(\frac{\partial h}{\partial T}\right)_p \mathrm{d}T$$

$$\mathrm{d}h = \left(\frac{\partial h}{\partial p}\right)_T \mathrm{d}p + c_p\,\mathrm{d}T \tag{14.46}$$

Note that, in Eqs. (14.45) and (14.46), c_v and c_p are given by Eqs. (14.39) and (14.36), respectively. Substituting Eqs. (14.45) into (14.43),

$$\left(\frac{\partial e}{\partial v}\right)_T \mathrm{d}v + c_v\,\mathrm{d}T + p\,\mathrm{d}v = 0$$

$$c_v\,\mathrm{d}T + \left[p + \left(\frac{\partial e}{\partial v}\right)_T\right]\mathrm{d}v = 0 \tag{14.47}$$

Substituting Eq. (14.46) into (14.44),

$$\left(\frac{\partial h}{\partial p}\right)_T \mathrm{d}p + c_p\,\mathrm{d}T - v\,\mathrm{d}p = 0$$

$$c_p\,\mathrm{d}T + \left[\left(\frac{\partial h}{\partial p}\right)_T - v\right]\mathrm{d}p = 0 \tag{14.48}$$

Dividing Eq. (14.48) by (14.47),

$$\frac{c_p}{c_v} = \frac{[(\partial h/\partial p)_T - v]\;\mathrm{d}p}{[(\partial e/\partial v)_T + p]\;\mathrm{d}v} \tag{14.49}$$

However, $v = 1/\rho$; hence, $\mathrm{d}v = -\mathrm{d}\rho/\rho^2$. Thus, Eq. (14.49) becomes

$$\frac{c_p}{c_v} = \frac{[(\partial h/\partial p)_T - v]}{[(\partial e/\partial v)_T + p]}(-\rho^2)\frac{\mathrm{d}p}{\mathrm{d}\rho} \tag{14.50}$$

Because we are dealing with isentropic conditions within the sound wave, any changes $\mathrm{d}p$ and $\mathrm{d}\rho$ within the wave must take place isentropically. Thus, $\mathrm{d}p/\mathrm{d}\rho \equiv (\partial p/\partial \rho)_s \equiv a_e^2$. Hence, Eq. (14.50) becomes

$$\left(\frac{\partial p}{\partial \rho}\right)_s = \frac{c_p}{c_v}\frac{1}{\rho^2}\frac{[(\partial e/\partial v)_T + p]}{[1/\rho - (\partial h/\partial p)_T]}$$

or

$$a_e^2 = \frac{c_p}{c_v}\frac{p}{\rho}\frac{[1 + (1/p)(\partial e/\partial v)_T]}{[1 - \rho(\partial h/\partial p)_T]} \tag{14.51}$$

As usual, let $\gamma \equiv c_p/c_v$. Also, note from the equation of state that $p/\rho = RT$. Thus, Eq. (14.51) becomes

$$\boxed{a_e^2 = \gamma RT\frac{[1 + (1/p)(\partial e/\partial v)_T]}{[1 - \rho(\partial h/\partial p)_T]}} \tag{14.52}$$

Equation (14.52) gives the equilibrium speed of sound in a chemically reacting mixture.

Equation (14.52) gives an immediate answer to one of the questions asked at the beginning of this section. The speed of sound in an equilibrium reacting mixture is not equal to the simple result $\sqrt{\gamma RT}$ obtained for a calorically perfect gas. However, if the gas is calorically perfect, then $h = c_p T$ and $e = c_v T$. In turn, $(\partial h/\partial p)_T = 0$ and $(\partial e/\partial v)_T = 0$, and Eq. (14.52) reduces to the familiar result

$$a_f = \sqrt{\gamma RT} \tag{14.53}$$

The symbol a_f is used in Eq. (14.53) to denote the frozen speed of sound because a calorically perfect gas assumes no reactions. Equation (14.53) is the speed at which a sound wave will propagate when no chemical reactions take place internally within the wave, that is, when the flow inside the wave is frozen.

For a thermally perfect gas, $h = h(T)$ and $e(T)$. Hence, again Eq. (14.52) reduces to Eq. (14.53).

Clearly, the full Eq. (14.51) must be used whenever $(\partial e/\partial v)_T$ and $(\partial h/\partial p)_T$ are finite. This occurs for two cases: 1) when the gas is chemically reacting and 2) when intermolecular forces are important, that is, when we are dealing with a real gas (see Sec. 10.4). In both of the preceding cases, $h = h(T, p)$ and $e = e(T, v)$, and hence Eq. (14.52) must be used.

Note from Eq. (14.52) that the equilibrium speed of sound is a function of both T and p, unlike the case for a calorically or thermally perfect gas where it depends on T only. This is emphasized in Fig. 14.18, which gives the equilibrium speed of sound for high-temperature air as a function of both T and p. In addition, note in Fig. 14.18 that the frozen speed of sound is given by a constant horizontal line at $a^2\rho/p = 1.4$, and that the difference between the frozen and equilibrium speed of

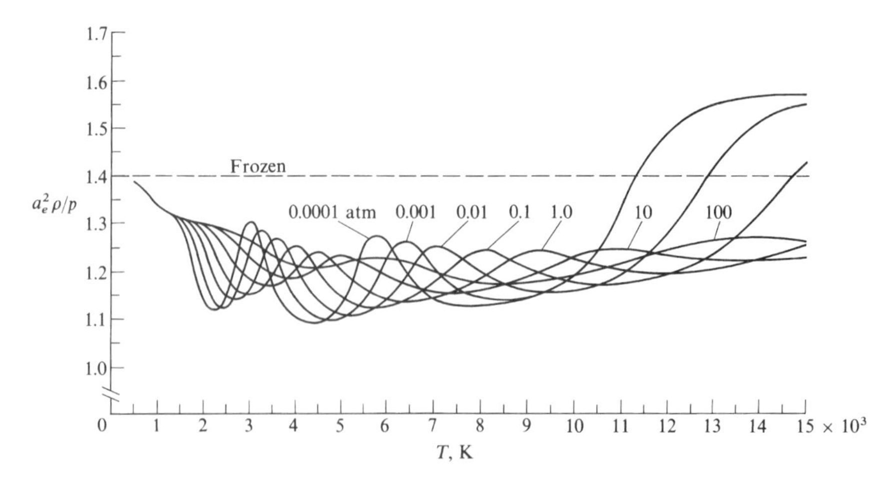

Fig. 14.18 Equilibrium speed of sound for air as a function of temperature (from [167]).

sound in air can be as large as 20% under practical conditions. In turn, this once again underscores the ambiguity in the definition of Mach number for high-temperature flows. The frozen Mach number $M_f = V/a_f$ and the equilibrium Mach number $M_e = V/a_e$ can differ by a substantial amount. Hence, Mach number is not particularly useful in this context.

Finally, note that the derivatives of e and h in Eq. (14.52) must be obtained numerically from the high-temperature equilibrium properties of the mixture. Although Eq. (14.52) is in a useful form to illustrate the physical aspects of the equilibrium speed of sound, it does not constitute a closed-form formula from which, given the local p and T, a value of a_e can be immediately obtained. Rather, the derivatives must be evaluated numerically, as has been carried out by Hansen [167] and others. Indeed, a correlation for the equilibrium speed of sound in high-temperature air is given by Tannehill and Mugge in [155] as

$$a = \left[e\left\{ K_1 + (\tilde{\gamma} - 1)\left[\tilde{\gamma} + K_2\left(\frac{\partial \tilde{\gamma}}{\partial \ln e}\right)_p \right] + K_3\left(\frac{\partial \tilde{\gamma}}{\partial \ln \rho}\right)_e \right\} \right]^{1/2} \tag{14.54}$$

where K_1, K_2, and K_3 are given in Table 11.1 found in Sec. 11.13 and $\tilde{\gamma}$ is defined by Eq. (11.108).

14.8 Equilibrium Conical Flow

Referring to our chapter road map in Fig. 14.1, we now return to the central column and treat the equilibrium high-temperature supersonic and hypersonic flow over a cone at zero degree angle of attack. This flow is sketched in Fig. 14.19, where θ_c and β are the cone half-angle and the shock-wave angle,

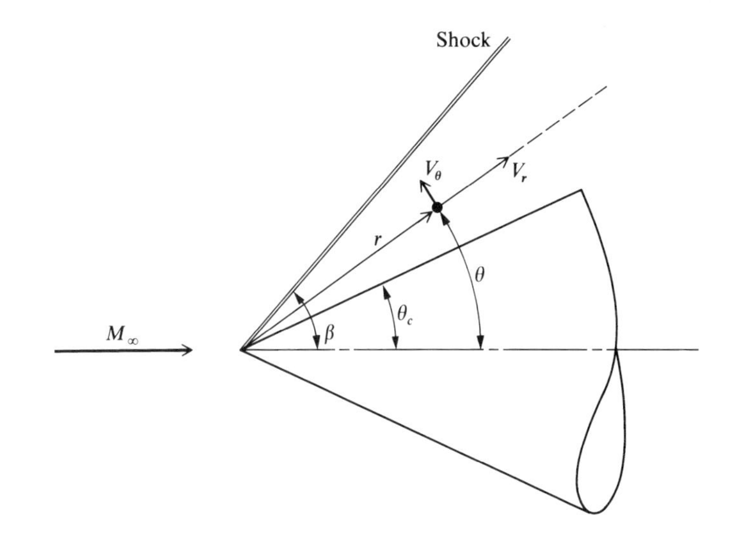

Fig. 14.19 Cone-flow geometry.

respectively. Any point in the flow between the shock and the body is located by the spherical coordinates r, θ, and Φ. The cone is a right-circular cone; hence, the flow is axisymmetric, where variations of properties in the azimuthal direction Φ are zero. (For this reason, Φ is not shown in Fig. 14.19.) Moreover, we take advantage of a property of conical flow, namely, that the flow properties along any conical ray (the r direction) are constant. This is the classical picture of conical flow (for example, see [4]); it is unchanged for the case of equilibrium, high-temperature flow.

The governing equations for axisymmetric conical flow can be obtained by writing Eqs. (14.2) and (14.3) in spherical coordinates, while setting $\partial/\partial\Phi = 0$ and $V_\Phi = 0$. From Eq. (14.2) for a steady flow

$$\nabla \cdot (\rho \mathbf{V}) = 0 \tag{14.55}$$

In spherical coordinates, with the axisymmetric assumption, Eq. (14.55) becomes

$$\frac{1}{r^2}\frac{\partial}{\partial r}(r^2 \rho V_r) + \frac{1}{r \sin\theta}\frac{\partial}{\partial \theta}(\rho V_\theta \sin\theta) = 0 \tag{14.56}$$

Expanding the derivatives in Eq. (14.56), and setting $\partial/\partial r = 0$ for conical flow, we obtain

$$2\rho V_r + \rho V_\theta \cot\theta + \frac{\mathrm{d}(\rho V_\theta)}{\mathrm{d}\theta} = 0 \tag{14.57}$$

Equation (14.57) is the continuity equation for conical flow. Note that, because θ is the only independent variable, Eq. (14.57) is an *ordinary* differential equation. For the momentum equation, we write Eq. (14.3) in spherical coordinates, taking components first in the r direction, and then in the θ direction.

r Direction:

$$V_r \frac{\partial V_r}{\partial r} + \frac{V_\theta}{r}\frac{\partial V_r}{\partial \theta} - \frac{V_\theta^2}{r} = -\frac{1}{\rho}\frac{\partial p}{\partial r} \tag{14.58}$$

Applying the conical flow assumption to Eq. (14.58), we obtain simply that

$$V_\theta = \frac{\mathrm{d}V_r}{\mathrm{d}\theta} \tag{14.59}$$

Equation (14.59) is the r-momentum equation for conical flow.

θ Direction:

$$V_r \frac{\partial V_\theta}{\partial r} + \frac{V_\theta}{r}\frac{\partial V_\theta}{\partial \theta} + \frac{V_r V_\theta}{r} = -\frac{1}{\rho r}\frac{\partial p}{\partial \theta} \tag{14.60}$$

Applying the conical flow assumption to Eq. (14.60), we have

$$V_\theta \frac{\mathrm{d}V_\theta}{\mathrm{d}\theta} + V_r V_\theta = -\frac{1}{\rho}\frac{\mathrm{d}p}{\mathrm{d}\theta} \tag{14.61}$$

Equation (14.61) is the θ-momentum equation for conical flow.

The energy equation will be invoked by stating that the flow is isentropic between the shock and the cone, and hence any change in pressure $\mathrm{d}p$ in any direction in the flowfield is related to a corresponding change in density $\mathrm{d}\rho$ via the speed of sound relation

$$a^2 = \left(\frac{\partial p}{\partial \rho}\right)_s = \frac{\mathrm{d}p}{\mathrm{d}\rho} \tag{14.62}$$

(Note that setting the ordinary differential expression $\mathrm{d}p/\mathrm{d}\rho$ equal to a^2 is valid *only* when the flowfield is isentropic, as in the present case.) With Eq. (14.62), we can write Eq. (14.57) as

$$2V_r + V_\theta \cot\theta + \frac{\mathrm{d}V_\theta}{\mathrm{d}\theta} + \frac{V_\theta}{\rho a^2}\frac{\mathrm{d}p}{\mathrm{d}\theta} = 0 \tag{14.63}$$

Examine Eqs. (14.61) and (14.63); they are two equations in terms of the derivatives $\mathrm{d}V_\theta/\mathrm{d}\theta$ and $\mathrm{d}p/\mathrm{d}\theta$. Solving Eqs. (14.61) and (14.63) for these derivatives, we obtain

$$\frac{\mathrm{d}V_\theta}{\mathrm{d}\theta} = \frac{a^2}{V_\theta^2 - a^2}\left(2V_r + V_\theta \cot\theta - \frac{V_r V_\theta^2}{a^2}\right) \tag{14.64}$$

and

$$\frac{\mathrm{d}p}{\mathrm{d}\theta} = -\frac{\rho V_\theta a^2}{V_\theta^2 - a^2}(V_r + V_\theta \cot\theta) \tag{14.65}$$

In summary, Eqs. (14.59), (14.64), and (14.65) constitute three coupled ordinary differential equations in terms of the five unknowns, V_θ, V_r, p, ρ, and a. This system is completed by adding the equilibrium high-temperature thermodynamic properties in the form of

$$\rho = \rho(p, s) \tag{14.66}$$

$$a = a(p, s) \tag{14.67}$$

Therefore, Eqs. (14.59) and (14.64–14.67) represent the governing equations for equilibrium flow over a cone. They stem from Eqs. (14.2) and (14.3), along with the basic speed of sound relation—all of which hold for chemically reacting flow. Thus, Eqs. (14.59) and (14.64–14.67) hold for equilibrium chemically reacting flow. For this case, they constitute the high-temperature analog of the classic Taylor–Maccoll equation for the conical flow of a calorically perfect gas, found in many compressible flow texts (for example, see Chapter 10 of [4]).

An iterative numerical solution of the preceding equations can be carried out as follows. Consider a given cone with specified θ_c in a specified freestream (p_∞, ρ_∞, V_∞ are known):

1) Assume a shock wave angle β.

2) Calculate the equilibrium flow properties immediately behind the oblique shock using the method described in Sec. 14.3.

3) Using these shock conditions as initial values, solve Eqs. (14.59) and (14.64–14.67) using any standard numerical solver for ordinary differential equations, for example, a Runge–Kutta method. This solution is a forward-marching solution in steps $\Delta\theta$, starting at the shock.

4) Integrate these equations until the cone surface is reached. Then check to see if $V_\theta = 0$ at θ_c. This is the proper flow-tangency boundary condition at the surface, namely, that the component of velocity normal to the surface is zero. If this boundary condition is not satisfied, return to step l, and assume a new value of β.

5) Continue this iterative process until convergence is obtained, that is, until $V_\theta = 0$ at the specified θ_c.

Such conical flow solutions have been obtained in equilibrium air by Romig [168] and Hudgins [169] and [170]. Indeed, [170] is a massive tabulation of equilibrium cone properties covering altitudes from sea level to 200,000 ft and Mach numbers up to 40. Some sample results obtained from [169] are shown in Figs. 14.20–14.22. In each of these figures, $M_\infty \sin\theta_c$ is the variable plotted along the abscissa. This is the familiar hypersonic similarity parameter studied in Part 1 of this book; $K_c = M_\infty \sin\theta_c$ is a useful parameter, even for chemically reacting flows. The ratio of cone surface pressure to freestream pressure p_c/p_∞ is given in Fig. 14.20 for three different altitudes and is compared with calorically perfect results for $\gamma = 1.4$. Note the rather dramatic result that p_c/p_∞ is virtually unaffected by chemical reactions. This is characteristic of chemically reacting flows involving compression processes; we noted the same behavior in Sec. 14.3 dealing with shock waves. The pressure is a "mechanical" variable, strongly dependent on the mechanical aspects of the flow and essentially uninfluenced by chemically reacting effects. In contrast, the ratio of cone surface density to freestream density ρ_c/ρ_∞ is greatly affected by chemical reactions, as shown in Fig. 14.21. Note that, consistent with our earlier discussions, the effect of chemical reactions is to increase the density ratio compared to the $\gamma = 1.4$ results; this implies that the shock-layer thickness will be smaller for the chemically reacting case. Finally, the ratio of cone surface temperature to freestream temperature is given in Fig. 14.22. As expected, the equilibrium temperatures are lower than the $\gamma = 1.4$ results and are progressively smaller as the altitude increases; this is because the higher altitudes (lower pressures) result in increased dissociation.

14.9 Equilibrium Blunt-Body Flows

The calculation of the inviscid flow over a supersonic or hypersonic blunt body was discussed in Sec. 5.3 in the context of a calorically perfect gas. In that section, emphasis was placed on a time-marching approach as the only viable technique for the solution of the problem. The same time-marching philosophy is used to calculate the high-temperature equilibrium inviscid flow over a blunt body. A recent example of such an approach is the work of Palmer [171].

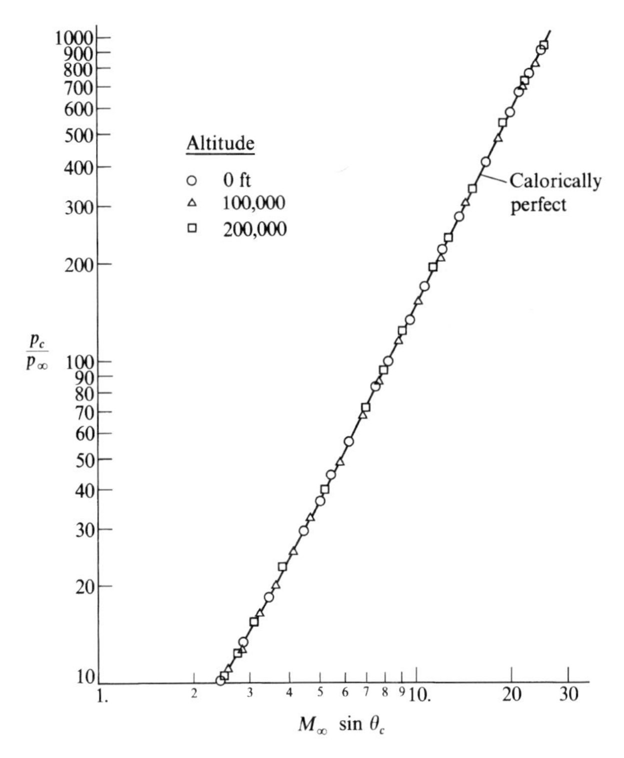

Fig. 14.20 Surface pressures on cones; comparison between equilibrium air and calorically perfect results (from Hudgins [169]).

Utilizing the equations for two-dimensional or axisymmetric flow in the strong conservation form given by

$$\frac{\partial Q}{\partial t} + \frac{\partial F}{\partial x} + \frac{\partial G}{\partial y} + rH = 0 \tag{14.68}$$

where

$$\boldsymbol{Q} = \begin{bmatrix} \rho \\ \rho u \\ \rho v \\ \rho\left(e + \dfrac{V^2}{2}\right) \end{bmatrix} \qquad \boldsymbol{F} = \begin{bmatrix} \rho u \\ \rho u^2 + p \\ \rho uv \\ \rho\left(e + \dfrac{V^2}{2} + \dfrac{p}{\rho}\right)u \end{bmatrix}$$

$$\boldsymbol{G} = \begin{bmatrix} \rho v \\ \rho uv \\ \rho v^2 + p \\ \rho\left(e + \dfrac{V^2}{2} + \dfrac{p}{\rho}\right)v \end{bmatrix} \qquad \boldsymbol{H} = \frac{1}{y}\begin{bmatrix} \rho v \\ \rho uv \\ \rho v^2 \\ \rho\left(e + \dfrac{V^2}{2} + \dfrac{p}{\rho}\right)v \end{bmatrix}$$

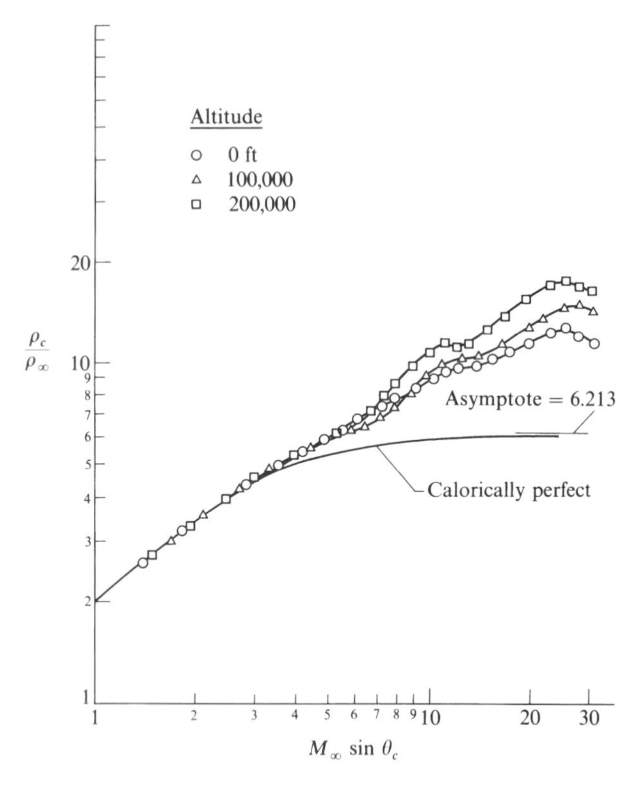

Fig. 14.21 Surface density on cones; comparison between equilibrium air and calorically perfect results (from [169]).

where $r = 0$ or 1 for two-dimensional or axisymmetric flow, respectively, and where $V^2 = u^2 + v^2$, Palmer carried out an implicit, finite difference, shock-capturing, time-marching solution of the equilibrium chemically reacting flow over cylinders and spheres, AOTV configurations, and hypersonic inlets. For details, see [171]. Equation (14.68) with the given forms of the vectors $\boldsymbol{Q}$, $\boldsymbol{F}$, $\boldsymbol{G}$, and $\boldsymbol{H}$ hold in general for an equilibrium chemically reacting gas; they can readily be obtained by a proper manipulation of Eqs. (14.2–14.5). Of course, the equilibrium high-temperature gas properties can be included in the calculation through relations such as $p = p(e, \rho)$ as correlated in Eqs. (11.107) and (11.108), and $T = T(e, \rho)$ as given by Eq. (11.109).

Some typical results are given in Figs. 14.23–14.25 obtained from [171]. In Fig. 14.23a, contours of ρ/ρ_∞ are shown for a calorically perfect gas with $\gamma = 1.4$ for flow over a sphere at Mach 20 and an altitude of 20 km; these are to be compared with similar results obtained from the equilibrium chemically reacting case shown in Fig. 14.23b. Clearly, the chemically reacting case exhibits higher densities and a thinner shock layer, as expected. In Figs. 14.24a and 14.24b, a similar comparison is made for the contours of T/T_∞. In Figs. 14.25a. and 14.25b, contours of N and NO mole fractions are given. Figure 14.25a

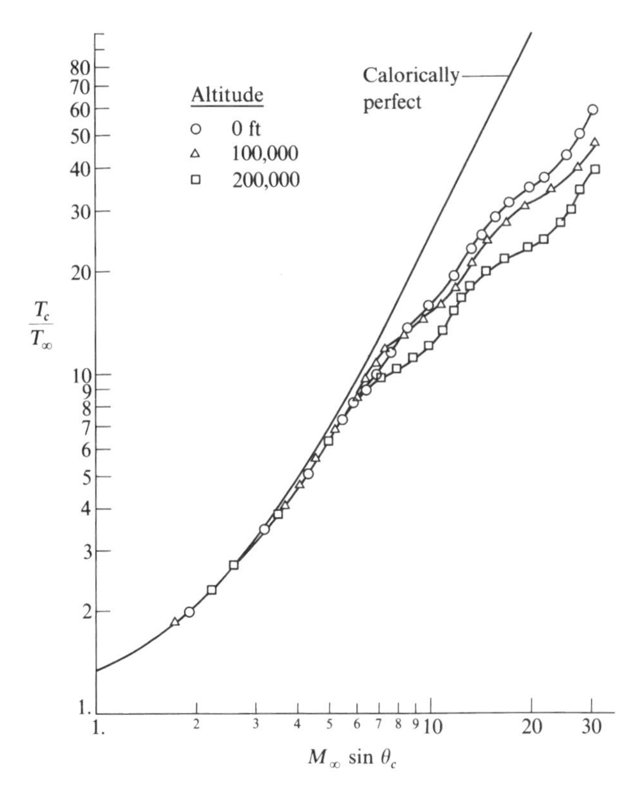

Fig. 14.22 Surface temperature on cones; comparison between equilibrium air and calorically perfect results (from [169]).

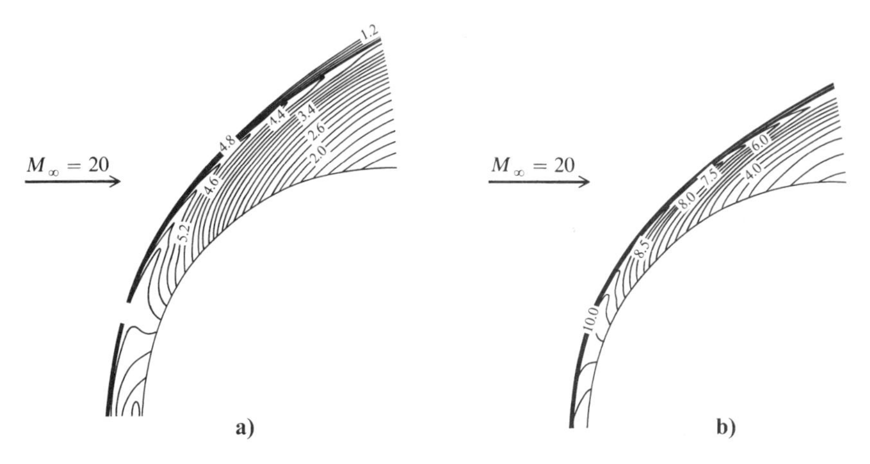

Fig. 14.23 Normalized density contours for blunt-body flow: a) calorically perfect-gas results and b) equilibrium chemically reacting air results. $M_\infty = 20$ and altitude = 20 km (from Palmer [171]).

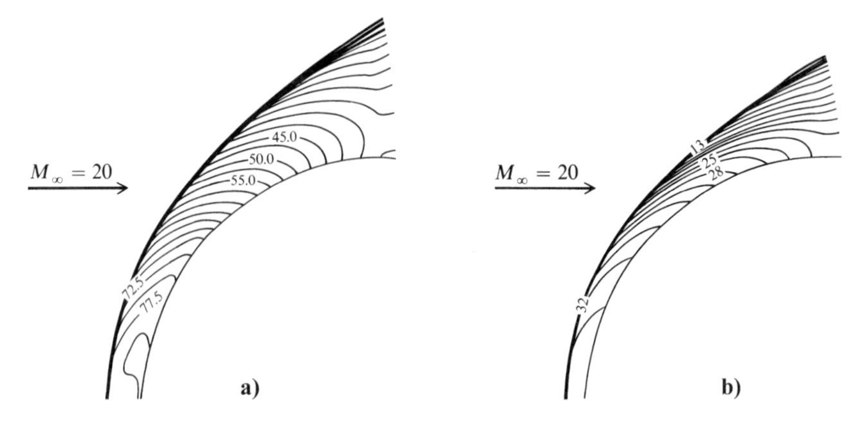

Fig. 14.24 Normalized temperature contours for blunt-body flow: a) calorically perfect gas and b) equilibrium chemically reacting air. $M_\infty = 20$ and altitude = 20 km (from [171]).

shows that the most intense dissociation occurs near the stagnation region, but Fig. 14.25b shows that the primary concentrations of NO occur further downstream, essentially beyond the sonic line.

We end this section with a rather interesting application of blunt-body equilibrium flowfield calculations. The three-dimensional equilibrium flow over the space shuttle was calculated by Maus et al. in [172]. A side and top view of the shuttle are shown in Fig. 14.26, obtained from [172]. The three-dimensional steady flowfield is calculated by using a time-marching finite difference solution in the blunt-nose region and then a spatial, downstream-marching finite-difference solution in the locally supersonic and hypersonic regions.

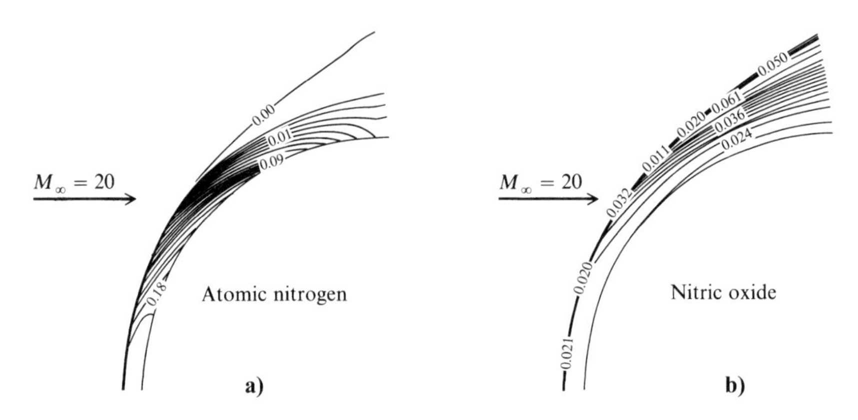

Fig. 14.25 Species mole fraction contours for blunt-body flow; equilibrium chemically reacting air. $M_\infty = 20$, and altitude = 20 km: a) X_N and b) X_{NO}.

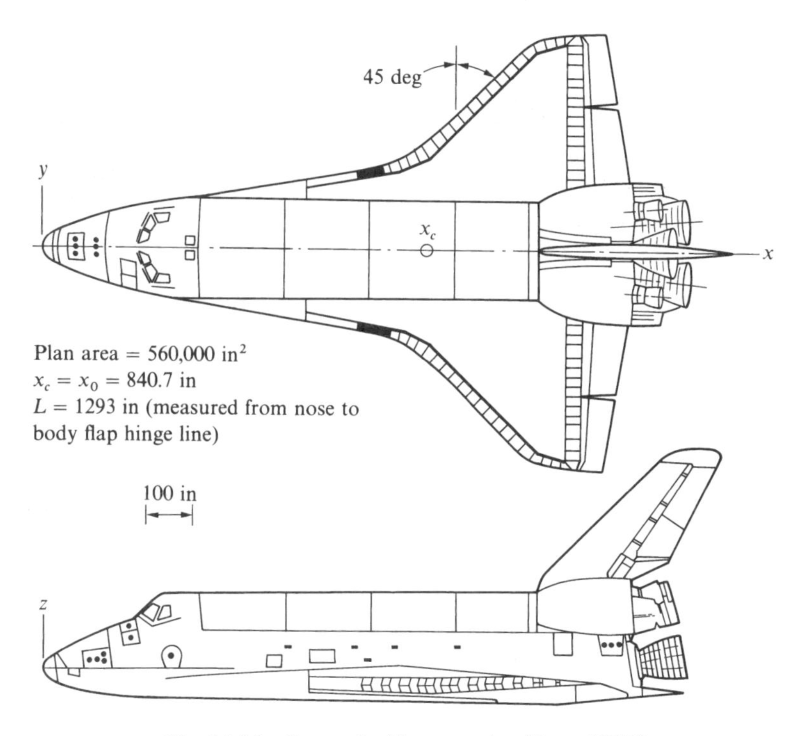

Fig. 14.26 Space shuttle geometry (from [172]).

High-temperature, equilibrium air thermodynamic properties were used. In [172], a rather interesting and unexpected effect of high-temperature flow on the space shuttle was pointed out. Flight experience with the shuttle has indicated a much higher pitching moment at hypersonic speeds than predicted; this has required the body flap deflection for trim to be more than twice that predicted. This discrepancy is considered to be one of the few major anomalies between design predictions and actual flight performance for the shuttle. Maus et al. argue that this discrepancy is caused by the effects of a chemically reacting shock layer, as follows. Figure 14.27, taken from [172], shows the calculated pressure distribution along the windward centerline of the shuttle; two sets of calculations are presented, one for a nonreacting shock layer with $\gamma = 1.4$, and the other for a reacting shock layer assuming local chemical equilibrium. At first glance, there appears to be little difference; indeed, pressure distributions are always somewhat insensitive to chemically reacting effects, as noted earlier. However, close examination of Fig. 14.27 shows that, for the chemically reacting flow, the pressures are slightly higher on the forward part of the shuttle and slightly lower on the rearward part. This results in a more positive pitching moment. Because the moment is the integral of the pressure through a moment arm, a slight change in pressure can substantially affect the moment. This is indeed the case here, indicated by Fig. 14.28, taken from [172]. Clearly, the pitching moment is substantially greater for the chemically reacting case.

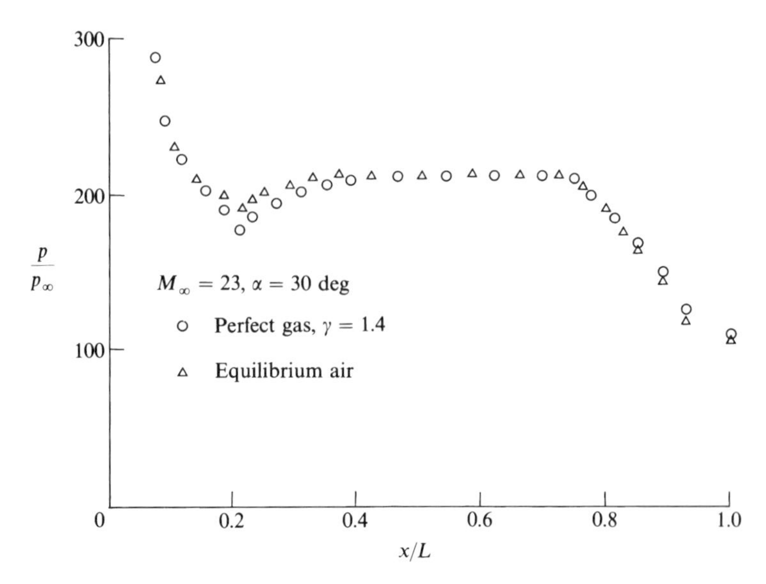

Fig. 14.27 Pressure distribution along the windward centerline of the space shuttle; comparison between a calorically perfect gas and equilibrium air calculations (from Maus et al. [172]).

This conclusion is a curious example of high-temperature effects having a rather unexpected but very important influence on the basic aerodynamics of the shuttle and serves to reinforce the importance of high-temperature flows in hypersonic aerodynamics.

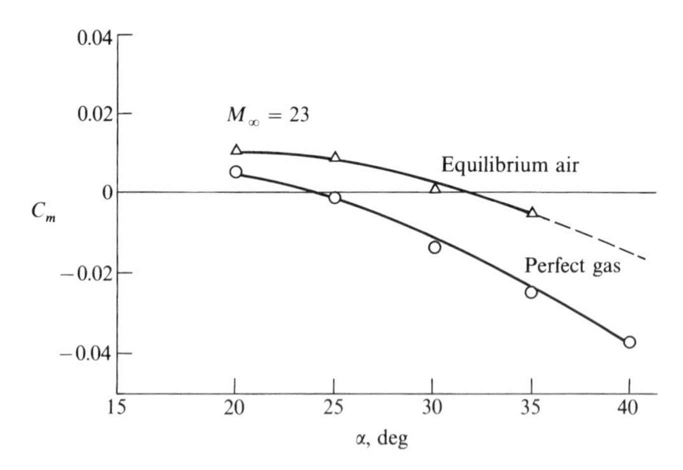

Fig. 14.28 Predicted pitching-moment coefficient for the space shuttle; comparison between a calorically perfect gas and equilibrium air calculations (from [172]).

14.10 Summary and Comments

In this chapter, we have examined various basic flows under the assumption of local thermodynamic and chemical equilibrium. In particular, we have studied the following:

1) Flow across normal and oblique shocks: Here we found that the ratios of properties across a normal shock depend on three upstream quantities (u_1, p_1, and T_1), and in addition, the oblique shock properties also depend on deflection angle θ or wave angle, β. A θ-β-V diagram for oblique shocks is given. This is in sharp contrast to shock properties for a calorically perfect gas, which depend only on M_1 for normal shocks and M_1 and (say) θ for oblique shocks. The increased density ratio across shock waves in the case of a chemically reacting flow results in smaller shock detachment distances and thinner shock layers on bodies, in comparison to the familiar constant $\gamma = 1.4$ case.

2) Quasi-one-dimensional flow: Here we found that flow properties at any given location depend on three quantities, such as p_0, T_0, and the local value of u. Indeed, the area ratio A/A^* depends on the same three quantities. Again, this is in contrast to the constant $\gamma = 1.4$ results, where flow properties can be related to local Mach number only.

3) Specific heat for a chemically reacting gas: Here the strong variations in c_p and c_v with T and p make the specific heats a rather awkward variable to deal with. For this reason, the energy equation is usually expressed in terms of the more fundamental variables, h or e.

4) Equilibrium speed of sound: This is also a strong function of T and p. In a chemically reacting gas, the speed of sound is an ambiguous quantity; the equilibrium speed of sound is always lower than the frozen speed of sound; this leads to the definition of two distinct Mach numbers at a given point in the flow—the equilibrium and frozen Mach numbers.

5) Flow over right-circular cones at zero angle of attack: Here, we found a system of equations for chemically reacting flow analogous to the classical Taylor–Maccoll equation for a calorically perfect gas. The equilibrium results show higher densities and lower temperatures than the constant $\gamma = 1.4$ results; however, p is virtually unaffected by chemical reactions.

6) Blunt-body flows: Here, the same time-marching philosophy as described in Chapter 5 for a calorically perfect gas is used, except using the proper governing equations for the chemically reacting flow, as well as the equilibrium thermodynamic properties.

Common to all of these flows is the fact that some type of numerical solution is required; analytical solutions for practical chemically reacting flows have not yet been obtained (and probably never will be). In many cases, these numerical solutions require some type of iterative approach. Also common to all of these flows is the fact that Mach number, which is so powerful in the analysis of the flow of a calorically perfect gas, is simply another flow variable when dealing with chemically reacting flows. Also, there are at least two Mach numbers that can be defined at any point—the equilibrium and frozen Mach numbers.

Design Example 14.1: Hypersonic Waveriders—Part 4

Question: How do high-temperature flows affect the design of the viscous optimized hypersonic waveriders discussed in three preceding Design Examples?

A major characteristic of the flowfields over hypersonic vehicles is the high enthalpy levels encountered in such flows; the attendant chemically reacting air is frequently important in such hypersonic flows. Is this true for the design of waveriders? This question was addressed by McLaughlin [265] and discussed in [255]. McLaughlin generated a new family of viscous optimized waveriders wherein the equilibrium chemically reacting flow over a cone (discussed in Sec. 14.8) is used as the generating flowfield. In this sense, chemically reacting effects are taken into account insofar as their impact on the *inviscid* flow aspects of waverider design are concerned.

As a precursor to the results, an examination of Fig. 14.22, which gives the ratio of cone surface temperature to freestream temperature, indicates rather moderate temperatures for slender cones. For example, for a 10-deg half-angle cone at Mach 25, the temperature ratio at an altitude of 200,000 ft is about 6, giving a cone surface temperature of about 1500 K—just barely on the verge of causing very mild oxygen dissociation. For a 15-deg half-angle cone, the surface temperature would be approximately 3000 K—in the range of oxygen dissociation but virtually no nitrogen dissociation. Therefore, at first glance we would not expect the inviscid flows used to generate slender waveriders to be dominated by chemically reacting effects. Indeed, such turns out to be the case.

For example, Table 14.2 gives the values of maximum lift-to-drag ratios for optimized waveriders from Mach 5 to Mach 50, all generated from the flow behind a conical shock wave of half-angle 15 deg. Each entry in Table 14.2 is for a waverider optimized within the flowfield of the given conical shock wave. They are not the "optimum of the optimum" obtained from using a variety of shock angles at the given Mach number as discussed in the earlier Design Examples. Rather, the relatively large cone half-angle of 15 deg was intentionally chosen so that the flow temperature would be high enough for chemical reactions to occur and therefore to underscore their effect. In

Table 14.2 Maximum L/D; comparison between the nonreacting and chemically reacting cases

Mach no.	Nonreacting	Reacting
M_{free}	L/D	L/D
5.0	4.62	4.65
10.0	3.67	3.64
15.0	3.48	3.42
20.0	3.46	3.35
30.0	3.45	3.21
40.0	3.48	3.17
50.0	3.47	3.18

Table 14.2, the nonreacting column is for a calorically perfect gas with $\gamma = 1.4$. The reacting column gives results for the case of a chemically reacting equilibrium flow. There is virtually no effect of chemically reacting flow on $(L/D)_{max}$ until freestream Mach numbers on the order of 40 and 50 are reached. Because lift and wave drag are caused by the pressure distribution exerted over the vehicle surface and because pressure is the flowfield property least affected by chemical reactions (see Fig. 14.20), such a result is not surprising. Also, the change in shape of the optimized waverider between the nonreacting and reacting cases is not great, as seen in Fig. 14.29 for the extreme case of Mach 50.

In conclusion, that part of the viscous-optimized waverider design process that depends on the inviscid flow is not greatly affected by chemical reactions. This is because of the slender-body geometry of the waverider, creating relatively weak shock waves, with temperature increases across the shocks that are too low

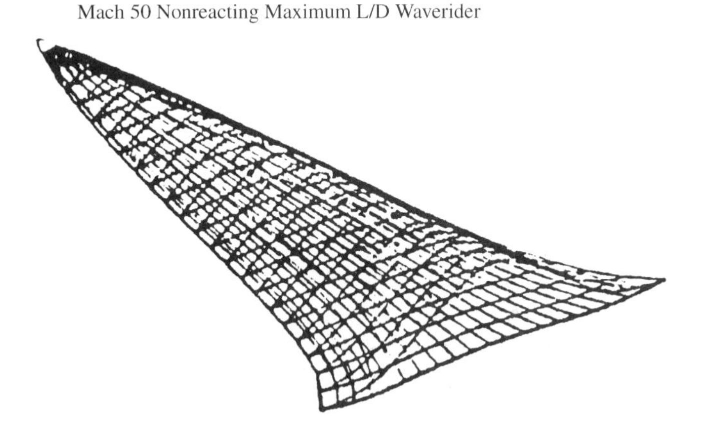

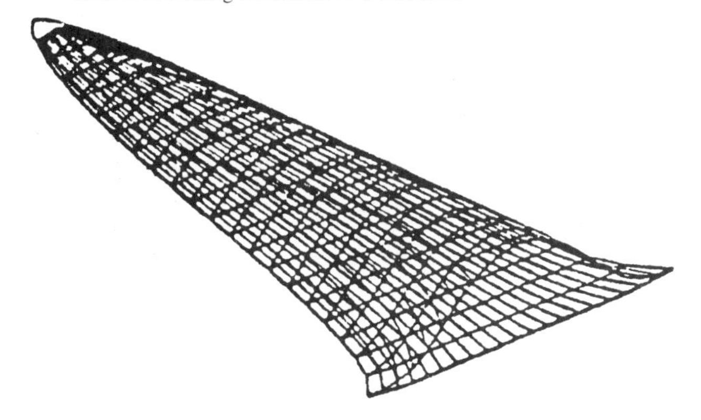

Fig. 14.29 Comparison of waverider shapes at Mach 50 for the nonreacting (top) and reacting (bottom) cases (Anderson et al. [255]).

to initiate strong chemically reacting effects. Hence, from this point of view, the waverider design process based on a calorically perfect gas appears to be reasonable for Mach numbers at least as high as 25.

As a caution, we note that higher temperatures will be generated by viscous dissipation in the boundary layer along the body surface, and this might be a region where chemically reacting effects are important in waverider design, in conjunction with the calculation of local skin friction and heat transfer. Viscous high-temperature flows is the subject of Chapter 17.

Problems

14.1 Consider the supersonic expansion of pure O_2 through a convergent-divergent nozzle. The reservoir temperature is 2000 K, and the velocity at the nozzle exit is 1500 m/s. Assuming local thermodynamic equilibrium, calculate the exit temperature and the exit-to-throat-area ratio A_e/A^*. Neglect the electronic energy.

14.2 Consider a stationary normal shock wave in pure diatomic nitrogen. The velocity and temperature upstream of the shock are 3000 m/s and 300 K, respectively. Calculate the temperature ratio across the shock, assuming local thermodynamic equilibrium. Neglect any chemical reactions and the electronic energy. For N_2, $h\nu/k = 3390$ K.

14.3 Consider a centered Prandtl–Meyer expansion wave in a chemically reacting gas. Assuming local chemical and thermodynamic equilibrium, describe how you would calculate the change in properties across this wave for given upstream conditions and a given expansion angle θ. Be as precise as you can. Describe the problem in general, and then give a step-by-step method for its solution.

15
Inviscid High-Temperature Nonequilibrium Flows

For low densities and high velocities, the molecular reaction rates within a fluid element do not keep pace with the rapid flow changes. In consequence, internal molecular energy modes, chemical dissociation, species ionization, and molecular radiation are all out of equilibrium. This essential nonequilibrium character complicates extensively the numerical computation of flowfields. It also has a significant impact on the relative roles of computation and experimentation in the vehicle design process.

From the National Academy of Sciences Report
Current Capabilities and Future Directions
in Computational Fluid Dynamics, 1986

Chapter Preview

The preceding chapter dealt with high-temperature flows through shock waves, nozzles, over cones, and over blunt-nosed bodies assuming local thermodynamic and chemical equilibrium. In the present chapter, we look at some of the same flows, but assuming local thermodynamic and chemical *nonequilibrium*. Does it make a difference? You bet your life it does. Nonequilibrium conditions change some of the basic physical characteristics of the flow. Oblique shock waves that you think should be straight become bent. The flow at the throat of a supersonic nozzle is no longer at Mach 1. The speed of sound now depends on the frequency of the sound waves. The surprises continue. Moreover, the calculation of nonequilibrium flows is mathematically *completely* different from that of an equilibrium flow. They say that variety is the spice of life. Compared to the preceding chapter, the present chapter provides plenty of variety. This is exciting stuff. So sit back and enjoy.

15.1 Introduction

For a given high-temperature flow, how can we judge whether it is close to local thermodynamic and/or chemical equilibrium (in which case the methods of Chapter 14 hold), or whether it substantially departs from such equilibrium conditions (in which case we have to do something different—the subject of the present chapter)? To answer this question, let us first be more precise about the definitions of equilibrium and frozen flows. In light of our preceding discussions, we can state the following!

Definition: A *frozen flow* is one where the reaction rate constants $k_f = k_b = 0$ and the vibrational relaxation time $\tau \to \infty$.

Definition: An *equilibrium* flow is one where $k_f = k_b \to \infty$ and $\tau = 0$.

These definitions make sense. For example, in a flowfield where the pressure, temperature, etc., change as a function of time and space, the only way that the internal energy modes and the chemical composition of a fluid element moving along a streamline can maintain their local equilibrium values at the local p and T is to be able to adjust *instantly* to the changing conditions, that is, to have infinitely fast reaction rates or, alternatively, a zero relaxation time. Similarly, for a frozen flow the only way that *no* changes in the internal energy modes and the chemical composition can occur is to have precisely zero reaction rates, or alternatively an infinitely long relaxation time. Of course, in practice, *neither* of the preceding flows actually occur *exactly*. However, let $\tau_f =$ characteristic time for a fluid element to traverse the flowfield of interest $\approx l/V_\infty$, where l is a characteristic length of the flowfield; and $\tau_c =$ characteristic time for the chemical reactions and/or vibrational energy to approach equilibrium.

Then, the following holds:

1) We can *assume* local *equilibrium* flow if

$$\tau_f \gg \tau_c$$

2) We can *assume frozen flow* if

$$\tau_f \ll \tau_c$$

3) For all other cases, the reacting and/or vibrationally excited flow is *nonequilibrium.*

To elaborate on the preceding criteria, visualize a fluid element moving through a flowfield (over a hypersonic vehicle, through a nozzle, etc.). Let l represent the characteristic length of the flowfield (the length of the body, or the length of the nozzle, etc.). Then τ_f is the approximate resident time of the fluid element in the flow, that is, the time it takes for the fluid element to flow past a body, flow through a nozzle, etc. Denote τ_f as the fluid-dynamic time. Similarly, τ_c is the time it takes for the internal energy modes and/or the chemical reactions to change. We will denote τ_c as the chemistry time. If in a given flowfield $\tau_f \gg \tau_c$, then the chemistry has plenty of time to adjust while the fluid element moves through the flowfield; in such a case, the flow can be *assumed* to be in local equilibrium. In contrast, if $\tau_f \ll \tau_c$, then the fluid element zips through the flowfield before any chemical changes can take place; in such a case, the

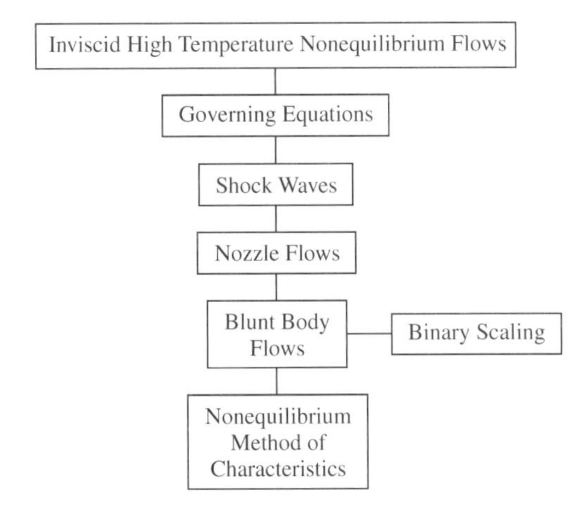

Fig. 15.1 Road map for Chapter 15.

flow can be assumed to be frozen. For all other situations, especially for $\tau_f \approx \tau_c$, a nonequilibrium flow exists. The analysis of such nonequilibrium flows is the subject of this chapter.

The road map for this chapter is shown in Fig. 15.1. It is a path similar to that taken in Fig. 14.1, except now we are dealing with *nonequilibrium* flows. We start with the governing equations for inviscid nonequilibrium flows and then apply these equations to study flows through shock waves, nozzle flows, and blunt-body flows. We close the chapter with a discussion on the method of characteristics as applied to a chemically reacting nonequilibrium flow. There is one side journey, namely, that to discuss binary scaling, an interesting feature that under special circumstances can be used to relate one nonequilibrium blunt-body flow to another.

15.2 Governing Equations for Inviscid, Nonequilibrium Flows

Equations (14.2–14.5) hold for nonequilibrium as well as equilibrium flows. However, for a nonequilibrium flow, in addition to the continuity equation given by Eq. (14.2), which we will now denote as the *global* continuity equation, we must also deal with a *species* continuity equation for each individual chemical species in the mixture. In the present book, we have not derived the basic governing equations of fluid dynamics, but rather have assumed such derivations to be prior knowledge on the part of the reader. (See, for example, [4] and [5] for such derivations.) However, here we make an exception in regard to the species continuity equation. Under the assumption that the reader might not be familiar with the species continuity equation, it is derived as follows.

Consider a fixed, finite-control volume in the nonequilibrium, inviscid, flow of a chemically reacting gas; such a control volume is sketched in Fig. 15.2. Let ρ_i

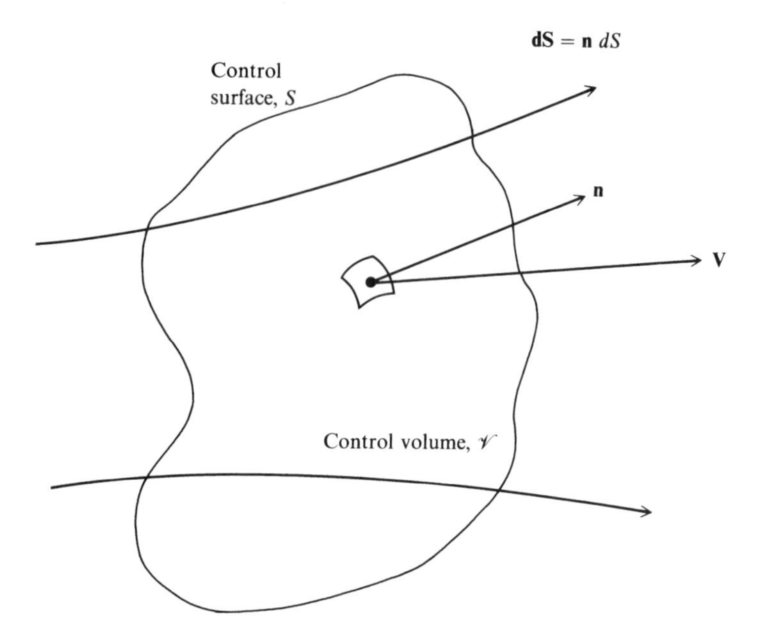

Fig. 15.2 Finite-control volume fixed in space, with the flow moving through it.

be the mass of species i per unit volume of mixture. Hence

$$\rho = \sum_i \rho_i$$

Examining Fig. 15.2, the mass flow of species i through the elemental surface area dS is $\rho_i\ \boldsymbol{V} \cdot \mathrm{d}\boldsymbol{S}$, where $\boldsymbol{V}$ is the local flow velocity and $\mathrm{d}\boldsymbol{S} = \boldsymbol{n}\mathrm{d}S$, where $\boldsymbol{n}$ is the unit normal vector. Hence, the net mass flow of species i out of the control volume is

$$\iint_S \rho_i \mathbf{V} \cdot \mathbf{dS}$$

The mass of species i inside the control volume is

$$\iiint_v \rho_i \, \mathrm{d}V$$

Let $\dot{w}_i$ be the local rate of change of ρ_i as a result of *chemical reactions inside the control volume*. Therefore, the net time rate of change of the mass of species i inside the control volume is caused by 1) the net flux of species i through the surface and 2) the creation or extinction of species i inside the control volume as a result of chemical reactions. Writing the preceding physical principle in terms of integrals over the control volume, we have

$$\frac{\partial}{\partial t}\iiint_{\mathscr{V}} \rho_i \, \mathrm{d}V = -\iint_S \rho_i \mathbf{V} \cdot \mathrm{d}\mathbf{S} + \iiint_{\mathscr{V}} \dot{w}_i \, \mathrm{d}V \tag{15.1}$$

Equation (15.1) is the integral form of the species continuity equation; you will note that its derivation is quite similar to the standard derivation of the global continuity equation given in most fluid-dynamics textbooks, such as [4] and [5]. In turn, using the divergence theorem, the differential form of the species continuity equation is obtained directly from Eq. (15.1) as

$$\frac{\partial \rho_i}{\partial t} + \nabla \cdot (\rho_i \mathbf{V}) = \dot{w}_i \tag{15.2}$$

[Recall that we are dealing with an inviscid flow. If the flow were viscous, Eqs. (15.1) and (15.2) would each have an additional term for the transport of species i by mass diffusion, and the velocity would be the mass motion of species i, which is not necessarily the same as the mass motion of the mixture $\boldsymbol{V}$. Such matters will be discussed in Chapter 17.]

In Eqs. (15.1) and (15.2) an expression for $\dot{w}_i$ comes from the chemical rate equation (13.31), couched in suitable dimensions. For example, assume that we are dealing with chemically reacting air, and we write Eqs. (15.1) and (15.2) for NO, that is, $\rho_i = \rho_{NO}$. The rate equation for NO is given by Eq. (13.45) in terms of

$$\frac{\mathrm{d}[NO]}{\mathrm{d}t} = -k_{f3a}[NO][O_2] + \cdots$$

The dimensions of this equation are moles per unit volume per unit time. However, the dimensions of $\dot{w}_{NO}$ in Eqs. (15.1) and (15.2) are the mass of NO per unit volume per unit time. Recalling that molecular weight is defined as the mass of species i per mole of i, we can write

$$\dot{w}_{NO} = \mathscr{M}_{NO} \frac{\mathrm{d}[NO]}{\mathrm{d}t}$$

where $\mathscr{M}_{NO}$ is the molecular weight of NO. Therefore, Eq. (15.2) written for NO is

$$\frac{\partial \rho_{NO}}{\partial t} + \nabla \cdot (\rho_{NO} \mathbf{V}) = \mathscr{M}_{NO} \frac{\mathrm{d}[NO]}{\mathrm{d}t}$$

where $\mathrm{d}[NO]/\mathrm{d}t$ is obtained from Eq. (13.45).

For a nonequilibrium chemically reacting mixture with n different species, we need $n - 1$ species continuity equations of the form of Eq. (15.2). These, along with the additional result that

$$\sum_i \rho_i = \rho$$

provide n equations for the solution of the instantaneous composition of a nonequilibrium mixture of n chemical species.

An alternative form of the species continuity equation can be obtained as follows. The mass fraction of species i, c_i is defined as $c_i = \rho_i/\rho$. Substituting this relation into Eq. (15.2)

$$\frac{\partial(\rho c_i)}{\partial t} + \nabla \cdot (\rho c_i \mathbf{V}) = \dot{w}_i \tag{15.3}$$

Expanding Eq. (15.3), we have

$$\rho\left[\frac{\partial c_i}{\partial t} + \mathbf{V} \cdot \nabla c_i\right] + c_i\left[\frac{\partial \rho}{\partial t} + \nabla \cdot (\rho \mathbf{V})\right] = \dot{w}_i \tag{15.4}$$

The first two terms of Eq. (15.4) constitute the substantial derivative of c_i. The second two terms (in brackets) result in zero from the global continuity equation (14.2). Hence, Eq. (15.4) can be written as

$$\boxed{\frac{\mathrm{D}c_i}{\mathrm{D}t} = \frac{\dot{w}_i}{\rho}} \tag{15.5}$$

In terms of the mole-mass ratio, $\eta_i = c_i/\mathscr{M}_i$, Eq. (15.5) becomes

$$\boxed{\frac{\mathrm{D}\eta_i}{\mathrm{D}t} = \frac{\dot{w}_i}{\mathscr{M}_i \rho}} \tag{15.6}$$

Equations (15.5) and (15.6) are alternative forms of the species continuity equation, couched in terms of the substantial derivative.

Recall that the substantial derivative of a quantity is physically the time rate of change of that quantity as we follow a fluid element moving with the flow. Therefore, from Eqs. (15.5) and (15.6), as we follow a fluid element of fixed mass moving through the flowfield, we see that changes of c_i or η_i of the fluid element are caused only by the finite-rate chemical kinetic changes taking place within the element. This makes common sense, and in hindsight, therefore, Eqs. (15.5) and (15.6) could have been written directly by inspection. We emphasize that in Eqs. (15.5) and (15.6) the flow variable inside the substantial derivative c_i or η_i is written per unit mass. As long as the nonequilibrium variable inside the substantial derivative is per unit mass of mixture, then the right-hand side of the conservation equation is simply caused by finite-rate kinetics, such as shown in Eqs. (15.5) and (15.6). In contrast, Eq. (15.2) can also be written as

$$\frac{D\rho_i}{Dt} = \dot{w}_i - \rho_i(\nabla \cdot \mathbf{V}) \tag{15.7}$$

The derivation of Eq. (15.7) is left to the reader. In it, the nonequilibrium variable inside the substantial derivative ρ_i is per unit volume. Because it is not per unit mass, an extra term in addition to the finite-rate kinetics appears on the right-hand side to take into account the dilation effect of the changing specific volume of the flow. (Recall from basic fluid mechanics that $\nabla \cdot \boldsymbol{V}$ is physically the time rate of change of volume of a fluid element per unit volume, as derived in [5].)

In addition to the species continuity equation, another equation must be added to the system given by Eqs. (14.2–14.5) if vibrational nonequilibrium is present. The finite-rate kinetics for vibrational energy exchange were discussed in Sec. 13.2, leading to Eq. (13.17) as the vibrational rate equation. Based on our earlier discussion, if we follow a moving fluid element of fixed mass, the rate of change of e_{vib} for this element is equal to the rate of molecular energy exchange inside the element. Therefore, we can write the vibrational rate equation for a moving fluid element as

$$\boxed{\frac{De_{\text{vib}}}{Dt} = \frac{1}{\tau}(e_{\text{vib}}^{\text{eq}} - e_{\text{vib}})} \tag{15.7a}$$

Note in Eq. (15.7a) that e_{vib} is the local nonequilibrium value of vibrational energy per unit mass of gas.

Let us now summarize the governing equations for an inviscid, nonequilibrium, high-temperature flow. In such a flow, we wish to solve for p, ρ, T, $\boldsymbol{V}$, h, e_{vib}, and c_i as functions of space and time. The governing equations that allow for the solution of these variables are as follows.

Global continuity:

$$\frac{\partial \rho}{\partial t} + \nabla \cdot (\rho \mathbf{V}) = 0 \tag{15.8}$$

Species continuity:

$$\frac{\partial \rho_i}{\partial t} + \nabla \cdot (\rho_i \mathbf{V}) = \dot{w}_i \tag{15.9}$$

or

$$\frac{Dc_i}{Dt} = \frac{\dot{w}_i}{\rho} \tag{15.10}$$

or

$$\frac{D\eta_i}{Dt} = \frac{\dot{w}_i}{\mathcal{M}_i \rho} \tag{15.11}$$

(Note that for a mixture of n species, we need $n-1$ species continuity equations; the nth equation is given by $\sum_i \rho_i = \rho$, or $\sum_i c_i = 1$, or $\sum_i \eta_i = \eta$.)

Momentum:

$$\rho \frac{\mathrm{D}\mathbf{V}}{\mathrm{D}t} = -\nabla p \tag{15.12}$$

Energy:

$$\rho \frac{\mathrm{D}h_0}{\mathrm{D}t} = \frac{\partial p}{\partial t} + \dot{q} \tag{15.13}$$

where

$$h_0 = h + \frac{V^2}{2} \tag{15.14}$$

In Eq. (15.13), $\dot{q}$ denotes a heat-addition term caused by volumetric heating (say, by radiation absorbed or lost from the gas). The term $\dot{q}$ does *not* have anything to do with chemical reactions. The energy exchanges as a result of chemical reactions are naturally accounted for by the heats of formation appearing in h in Eqs. (15.13) and (15.14), for example, Eqs. (11.105) or (11.106). In addition to the preceding equations, we also have the following.

Equation of state:

$$p = \rho R T \tag{15.15}$$

where

$$R = \frac{\mathscr{R}}{\mathscr{M}}$$

$$\mathscr{M} = \left(\sum_i \frac{c_i}{\mathscr{M}_i} \right)^{-1}$$

Enthalpy:

$$h = \sum_i c_i h_i \tag{15.16}$$

where

$$h_i = (e_{\text{trans}} + e_{\text{rot}} + e_{\text{vib}} + e_e)_i + R_i T + (\Delta h_f^{\circ})_i \tag{5.17}$$

In Eq. (15.16), for a nonequilibrium flow, c_i is obtained from the species continuity equation, say, Eq. (15.10). In regard to e_{vib_i} which appears in Eq. (15.17), there are some cases where the assumption of local thermodynamic equilibrium is appropriate even though chemical nonequilibrium prevails. (As noted in Sec. 13.1, far fewer molecular collisions are required for vibrational energy exchanges than for chemical reactions to occur. Hence, for some cases the molecular collision frequency might be high enough to allow near-equilibrium conditions for vibrational energy, but *not* high enough to provide near-equilibrium chemical conditions.) In such a case, e_{vib_i} is given by Eq. (11.61) for the species i. However, when both thermodynamic and chemical nonequilibrium prevail, e_{vib_i} is a nonequilibrium value that must be obtained from the vibrational rate equation (15.7a), written for species i as follows.

Vibrational energy:

$$\frac{\mathrm{D}(c_i e_{\text{vib}})}{\mathrm{D}t} = \frac{c_i}{\tau_i}(e^{\text{eq}}_{\text{vib}_i} - e_{\text{vib}_i}) \tag{15.18}$$

Before progressing further, look back over Eqs. (15.8–15.18), and make certain that you feel comfortable with them. *These equations are the governing equations for inviscid, nonequilibrium, high-temperature flow.* They will be used throughout the remainder of this chapter.

15.3 Nonequilibrium Normal and Oblique Shock-Wave Flows

Consider a strong normal shock wave in a gas. Moreover, assume the temperature within the shock wave is high enough to cause chemical reactions within the gas. In this situation, we need to reexamine the qualitative aspects of a shock wave, as sketched in Fig. 15.3. The thin region where large gradients in temperature, pressure, and velocity occur, and where the transport phenomena of viscosity and thermal conduction are important is called the *shock front.* For all of our preceding considerations of a calorically perfect gas, or equilibrium flow of a chemically reacting or vibrationally excited gas, this thin region is the shock wave. For these previous situations, the flow in front of and behind the shock front was uniform, and the only gradients in flow properties took place almost discontinuously within a thin region of no more than a few mean-free-paths thickness. However, in a nonequilibrium flow, all chemical reactions and/or vibrational excitations take place at a finite rate. Because the shock front is only a few mean free paths thick, the molecules in a fluid element can experience only a few collisions as the fluid element traverses the front. Consequently, the flow through the shock front itself is essentially *frozen.* In turn, the flow properties immediately behind the shock front are frozen flow properties, as discussed in Sec. 14.5 and as sketched in Fig. 15.3. Then, as the fluid element moves downstream, the finite-rate reactions take place, and the flow properties relax toward their equilibrium values, as also sketched in Fig. 15.3. With this picture in mind, the *shock wave* now encompasses both the shock front and the nonequilibrium region behind the front where the flow properties are changing as a result of the finite-rate reactions. For purposes of illustration, assume that the gas is pure diatomic nitrogen in front of the shock wave, that is, $(c_N)_1 = 0$ in Fig. 15.3. The properties immediately behind the shock front are obtained from frozen flow results, that is, the constant $\gamma = 1.4$ results from Chapter 2. Hence, the values of T_{frozen} and ρ_{frozen} shown in Fig. 15.3 can be obtained directly from standard compressible flow tables such as found in [4] and [5]

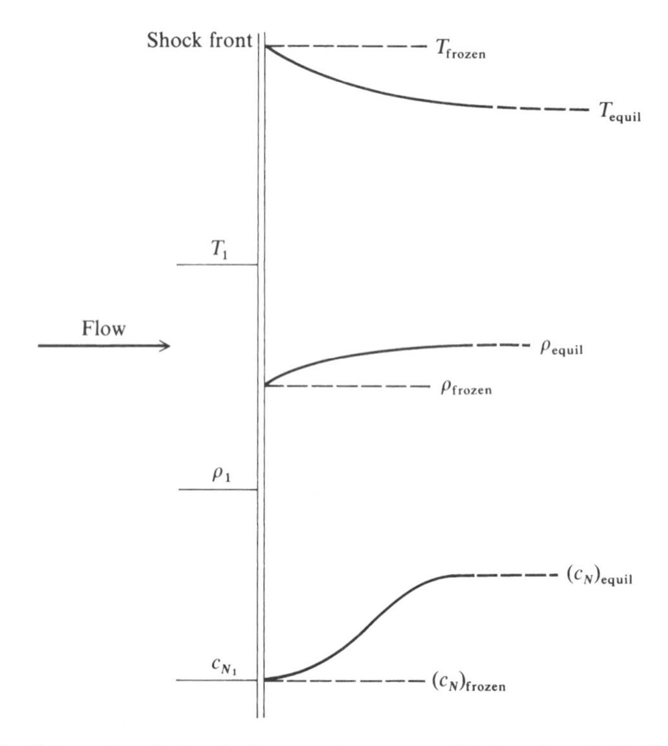

Fig. 15.3 Schematic of chemically reacting nonequilibrium flow behind a normal shock wave.

for air. In addition, c_N immediately behind the shock front is still zero because the flow is frozen. Downstream of the shock front, the nonequilibrium flow must be analyzed using the equations summarized in Sec. 15.2. In this region, the nitrogen becomes either partially or totally dissociated (depending on the strength of the shock wave), and c_N increases as sketched in Fig. 15.3. In turn, because this reaction is endothermic, the static temperature behind the shock front decreases, and the density increases. Finally, the downstream flow properties will approach their equilibrium values, as calculated from the technique described in Sec. 14.3.

A numerical calculation of the nonequilibrium region behind the shock front can be established as follows. Because the flow is one-dimensional and steady, the equations of Sec. 15.2 become the following.

Global continuity:

$$\rho \, \mathrm{d}u + u \, \mathrm{d}\rho = 0 \tag{15.19}$$

Momentum:

$$\mathrm{d}p = -\rho u \, \mathrm{d}u \tag{15.20}$$

Energy:

$$dh_0 = 0 \tag{15.21}$$

Species continuity:

$$u\,dc_i = \frac{\dot{w}_i}{\rho}dx \tag{15.21a}$$

In Eq. (15.21a) the x distance is measured from the shock front, extending downstream as shown in Fig. 15.4. Note that Eq. (15.21a) explicitly involves the finite-rate chemical reaction term $\dot{w}_i$, and that a distance dx multiplies this term. Hence, Eq. (15.21a) introduces a scale effect into the solution of the flowfield—a scale effect that is present solely because of the nonequilibrium phenomena. In turn, all flowfield properties become a function of distance behind the shock front, as sketched in Fig. 15.4. Equations (15.19–15.21a) can be solved by using any standard numerical technique for integrating ordinary differential equations, such as the well-known Runge–Kutta technique, starting right behind the shock front (point 1 in Fig. 15.4), and integrating downstream in steps Δx, as sketched in Fig. 15.4. The initial conditions at point 1 in Fig. 15.4 are obtained by assuming frozen flow across the shock front. If we are dealing with atmospheric flight, where the free stream conditions are those for cool, nonreacting air, then the chemical composition at point 1 is the same as the known composition ahead of the shock, and the local values of velocity, pressure, temperature, etc., at point 1 are the same as calculated for a normal shock wave in a calorically

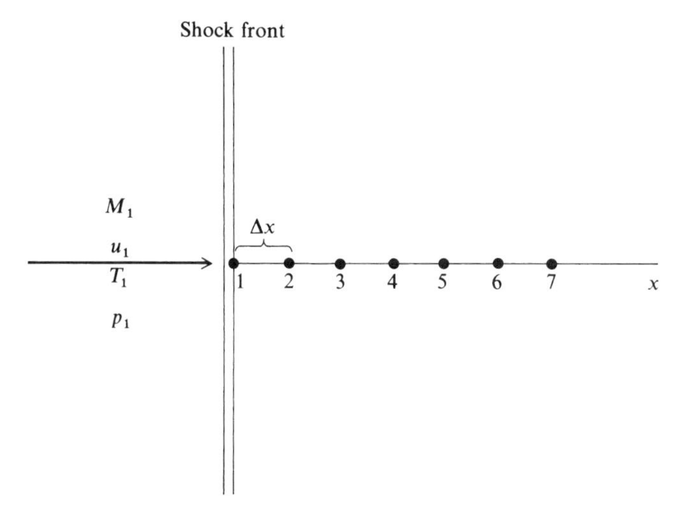

Fig. 15.4 Schematic of grid points for the numerical solution of nonequilibrium normal shock flows.

perfect gas with $\gamma = 1.4$, that is, the standard normal shock tables for air with $\gamma = 1.4$ (as found in [4] and [5]) yield the proper initial conditions at point 1.

Caution: In carrying out such a numerical solution of nonequilibrium flows, a major problem can be encountered. If one or more of the finite-rate chemical reactions are very fast [if $\dot{w}_i$ in Eq. (15.21a) is very large], then Δx must still be chosen very small even when a higher-order numerical method is used. The species continuity equations for such very fast reactions are called "stiff" equations and readily lead to instabilities in the solution. Special methods for treating the solution of stiff ordinary differential equations have been reviewed by Hall and Treanor (see [173]); such matters are still a state-of-the-art research problem today.

Typical results for the nonequilibrium flowfield behind a normal shock wave in air are given in Figs. 15.5 and 15.6, taken from the work of Marrone [174]. The Mach number ahead of the shock wave is 12.28, strong enough to produce major dissociation of O_2, but only slight dissociation of N_2. The variation of chemical composition with distance behind the shock front is given in Fig. 15.5. Note the expected increase in the concentration of O and N, rising from their frozen values (essentially zero) immediately behind the shock front, and monotonically approaching their equilibrium values about 10 cm downstream of the shock front. For the most part, the nonequilibrium flow variables will range between the two extremes of frozen and equilibrium values. However, in some cases, because of the complexities of the chemical kinetic

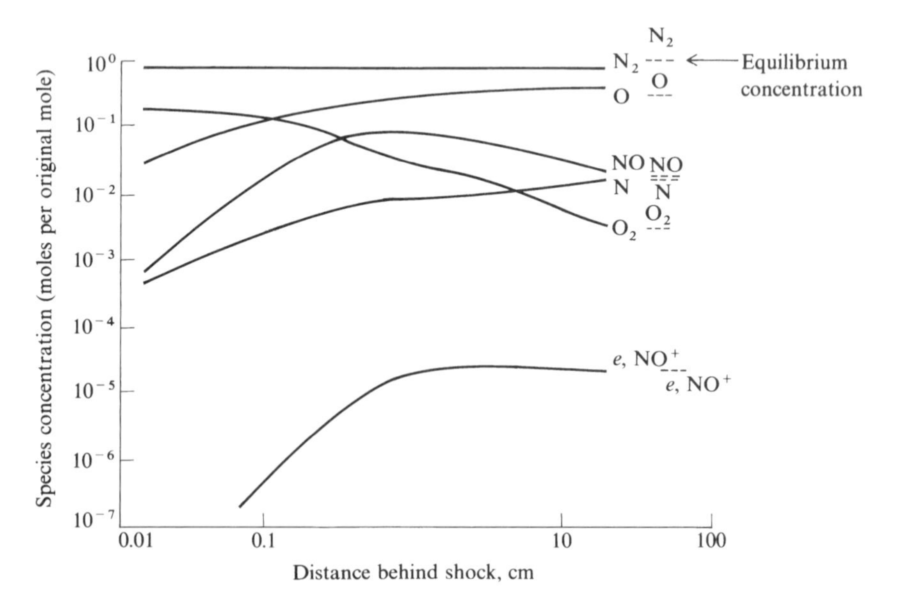

Fig. 15.5 Distributions of the chemical species for the nonequilibrium flow through a normal shock wave in air: $M_\infty = 12.28$, $p_\infty = 1.0$ mm Hg, and $T_\infty = 300$ K (from Marrone [174]).

mechanism, a species can exceed these two extremes. A case in point is the variation of NO concentration shown in Fig. 15.5. Note that it first increases from essentially zero behind the shock front and overshoots its equilibrium value at about 0.1 cm. Further downstream, the NO concentration approaches its equilibrium value from above. This is a common behavior of NO when it is formed behind a shock front in air; it is not just a peculiarity of the given upstream conditions in Fig. 15.5. The variations in temperature and density behind the shock front are shown in Fig. 15.6. As noted earlier, the chemical reactions in air behind a shock front are predominantly dissociation reactions, which are endothermic. Hence, T decreases and ρ increases with distance behind the front—both by almost a factor of 2.

Let us now consider the case of the nonequilibrium flow behind an oblique shock wave. First, consider the standard picture of a straight oblique shock front, as sketched in Fig. 15.7. Let x denote distance downstream of the shock front measured perpendicular to the front, as shown in Fig. 15.7. From the component of the momentum equation tangential to the shock front, we find that the tangential component of velocity V_t is preserved across the shock front, that is, $V_{t,2} - V_{t,1}$. This is a basic mechanical result, unaffected by chemical reactions. Moreover, for the same reason, V_t is constant everywhere behind the shock front; letting points 2 and 3 denote different x-wise locations in the flow behind the shock front, we have $V_{t,3} = V_{t,2} = V_{t,1}$. In contrast, the normal component of velocity V_n varies with x in the nonequilibrium flow

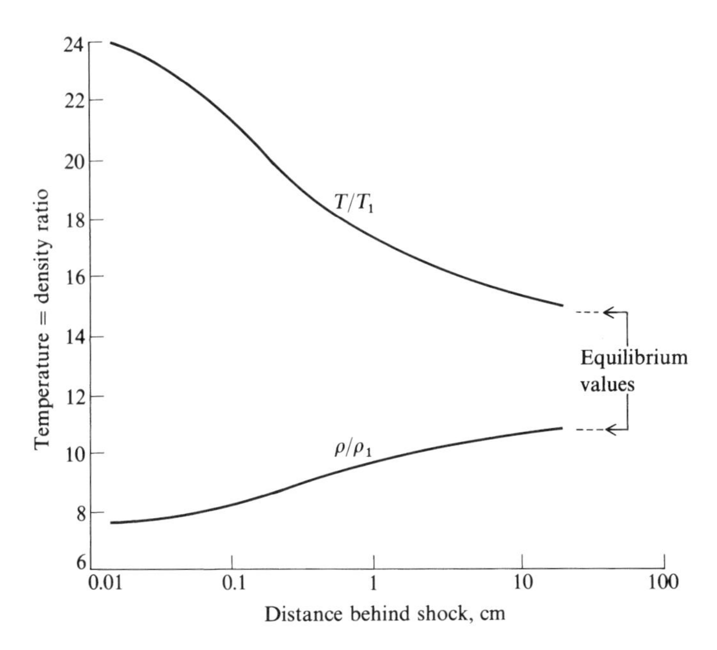

Fig. 15.6 Distributions of the temperature and density for the nonequilibrium flow through a normal shock wave in air: $M_\infty = 12.28$, $\rho_\infty = 1.0$ mm Hg, and $T_\infty = 300$ K (from [174]).

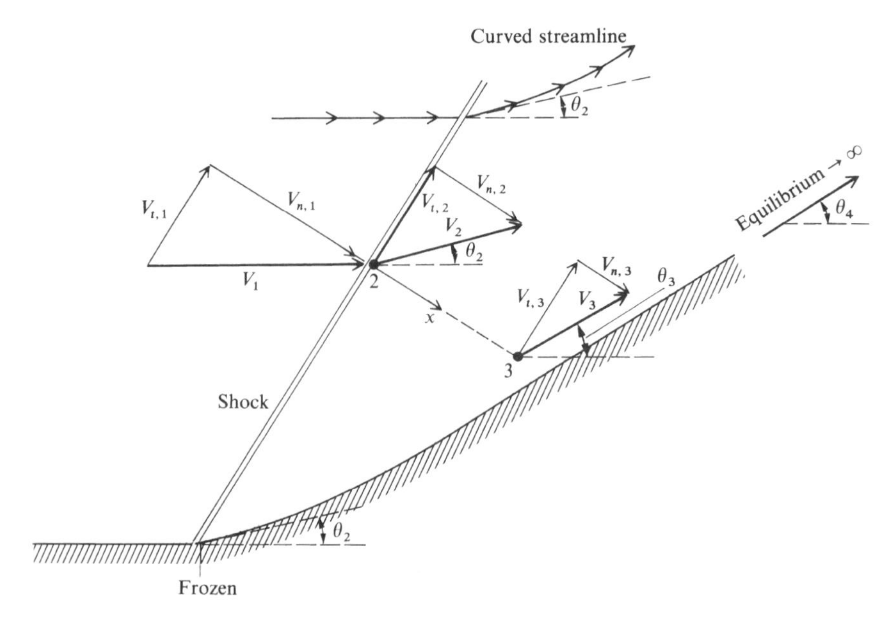

Fig. 15.7 Geometry for nonequilibrium flow behind a straight oblique shock wave.

behind the shock front. This can be explained as follows. The oblique shock properties are basically determined by normal shock properties based on an upstream velocity perpendicular to the shock front $V_{n,1}$ on which a constant tangential component V_t is superimposed throughout the flow. This is a familiar picture from fundamental oblique shock-wave theory, unaffected by chemical reactions. In the nonequilibrium flow behind a normal shock front, Fig. 15.6 shows that density increases with distance behind the front. Because ρV_n is a constant for flow across a normal shock, then V_n must *decrease* with distance behind the front. Hence, returning to Fig. 15.7, we see that V_n decreases with x, that is, $V_{n,3} < V_{n,2}$. Thus, because $V_{t,3} = V_{t,2}$, the flow deflection angle θ_3 is greater than θ_2. *Conclusion*: The streamlines in the nonequilibrium flow behind a straight oblique shock front are *curved* and continually increase their deflection angle until equilibrium conditions are reached far downstream. Therefore, in order to create a *straight* oblique shock front in a nonequilibrium flow, we have to have a compression corner that is shaped like the solid surface shown in Fig. 15.7. This compression surface, after its initial discontinuous deflection of θ_2 corresponding to frozen flow, must curve upward until equilibrium conditions are obtained far downstream, where θ_4 corresponds to the equilibrium deflection angle given by Fig. 14.8 (as calculated by the method discussed in Sec. 14.3). This curved, nonuniform flowfield in the nonequilibrium region behind a straight oblique shock front is an important difference from the familiar uniform flows obtained for calorically perfect and equilibrium oblique shock results.

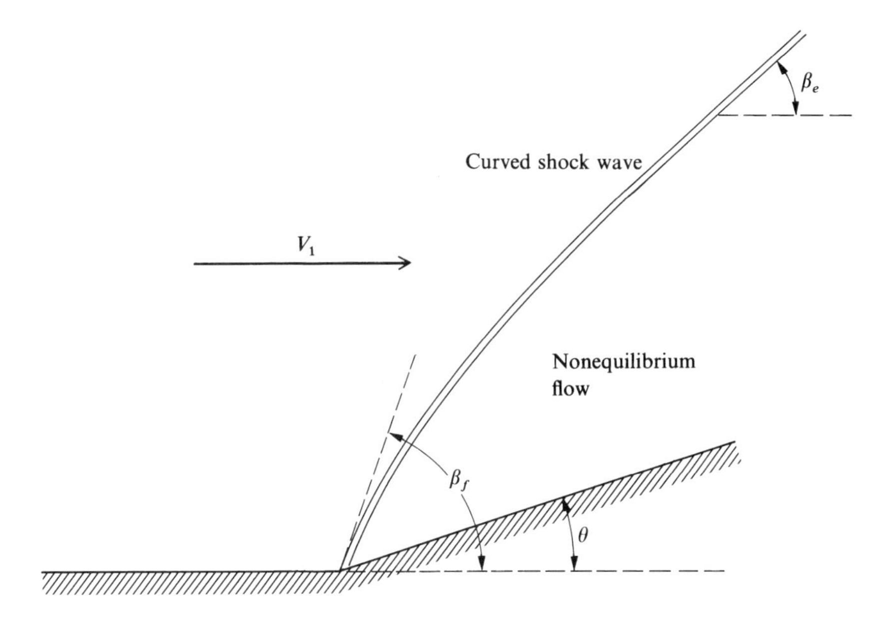

Fig. 15.8 Schematic of nonequilibrium flow over a compression corner.

Using the inverse of the preceding reasoning, for the supersonic or hypersonic nonequilibrium flow over a straight compression corner as sketched in Fig. 15.8, the shock wave will be curved. The wave angle right at the corner β_f corresponds to frozen flow. Far downstream, the wave angle approaches the equilibrium flow value β_e. Recall from Fig. 14.8 that, for a given deflection angle θ, the equilibrium shock-wave angle is always less than the frozen wave angle (for $\gamma = 1.4$).

In conclusion, the nonequilibrium flow behind a shock front, normal or oblique, varies with distance behind the front. This introduces a dimensional scale in such flows. For example, Bussing and Eberhardt [158] define a nonequilibrium length scale (or relaxation distance) as the distance downstream of the shock front required for the flow properties to reach 95% of their equilibrium values. For any given flow, the relaxation distances are different for different variables. Sample results from [158] are shown in Figs. 15.9 and 15.10. In Fig. 15.9, the relaxation distances behind a normal shock for T, X_{o}, and X_{N} are plotted vs freestream Mach number, where the upstream conditions at each Mach number correspond to the lower flight trajectory shown in Fig. 14.9 for a transatmospheric vehicle. In Fig. 15.10, the same quantities are given for the flow behind an oblique shock with $\beta = 30$ deg, also for the lower flight trajectory given in Fig. 14.9. In both Figs. 15.9 and 15.10, results are shown for two different sets of chemical rate data for high-temperature air, one set from Wray [159], and the other from Dunn and Kang [162]. The essential information to be derived from a comparison of Figs. 15.9 and 15.10 is that nonequilibrium distances behind a normal shock are much smaller (on the order of 1 cm) than behind a

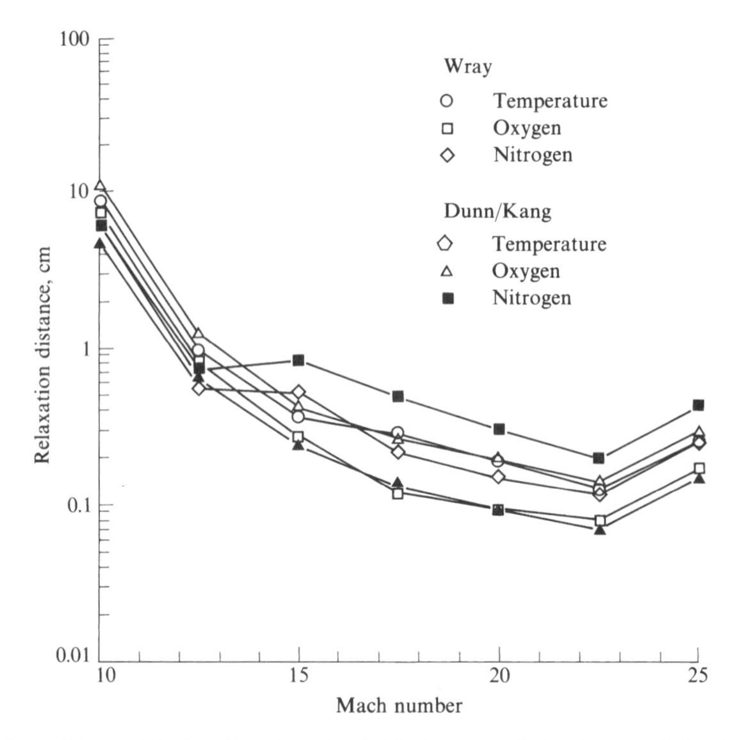

Fig. 15.9 Nonequilibrium length scales behind a normal shock wave, following the lower trajectory in Fig. 14.9 (from Bussing and Eberhardt [158]).

30-deg oblique shock (on the order of 700 cm) for the same flight conditions. This is because of the higher pressures and temperatures behind the normal shock wave, yielding much higher reaction rates and therefore producing lower relaxation distances. The results in Figs. 15.9 and 15.10 also demonstrate that nonequilibrium effects are important for hypersonic transatmospheric vehicles. For example, for a nose radius of 10 cm, the hypersonic shock detachment distance will be on the order of 1 cm or less [see Eqs. (14.19) and (14.4)]. Figure 15.9 indicates that a major portion of the blunt-body flow region will be nonequilibrium flow. Moreover, for the flow over slender bodies or wings, Fig. 15.10 predicts long regions of nonequilibrium flow downstream of the leading edges.

Finally, note that the analysis of nonequilibrium flows behind shock waves requires the numerical solution of *differential equations* [see Eqs. (15.19–15.21)]. This is in direct contrast to the solution of equilibrium flow behind shocks, which, although requiring a numerical solution, deals with a system of *algebraic* equations [see Eqs. (14.8–14.10)]. This is an example of the general nature of nonequilibrium flow solutions, namely, that the nonequilibrium behavior introduces a scale length into the flow, and the solution of such flows can only be treated by differential equations. This is true no matter how simple the fluid-dynamic aspects might be.

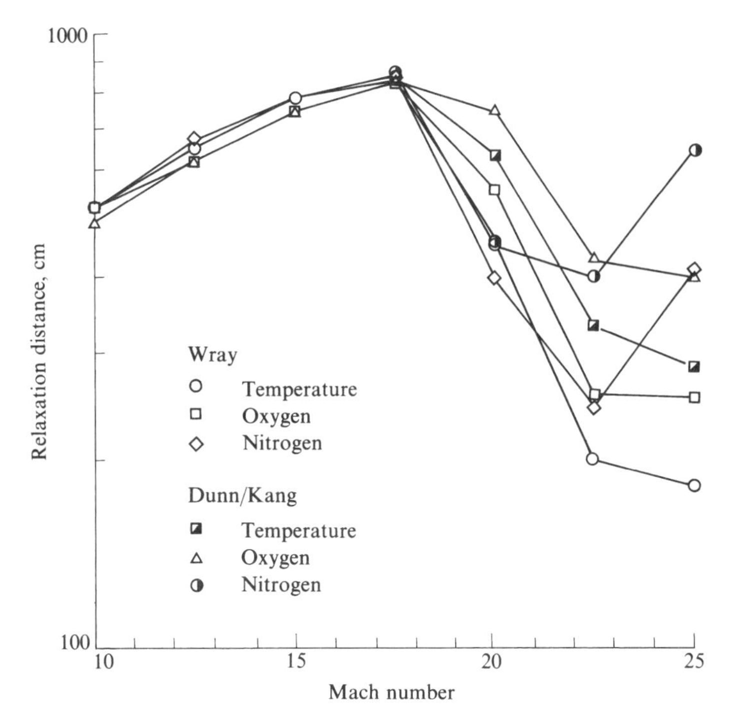

Fig. 15.10 Nonequilibrium length scales behind a 30-deg oblique shock wave, following the lower trajectory in Fig. 14.9 (from [158]).

15.4 Nonequilibrium Quasi-One-Dimensional Nozzle Flows

Because of the practical importance of high-temperature flows through rocket nozzles and high-enthalpy aerodynamic testing facilities, intensive efforts were made after 1950 to obtain relatively exact numerical solutions for the expansion of a high-temperature gas through a nozzle when vibrational and/or chemical nonequilibrium conditions prevail within the gas. In a rocket nozzle, nonequilibrium effects decrease the thrust and specific impulse. In a high-temperature wind tunnel, the nonequilibrium effects make the flow conditions in the test section somewhat uncertain. Both of the preceding are adverse effects, and hence rocket nozzles and wind tunnels are usually designed to minimize the nonequilibrium effects; indeed, engineers strive to obtain equilibrium conditions in such situations. In contrast, the gas dynamic laser (see [147]) creates a laser medium by intentionally fostering vibrational nonequilibrium in a supersonic expansion; here, engineers strive to obtain the highest degree of nonequilibrium possible. In any event, the study of nonequilibrium nozzle flows is clearly important.

Until 1969, all solutions of nonequilibrium nozzle flows involved steady-state analyses. Such techniques were developed to a high degree and are nicely reviewed by Hall and Treanor (see [173]). However, such steady-state analyses were not straightforward. Complicated by the presence of stiff chemical rate equations, such solutions encountered a saddle-point singularity in the vicinity of the nozzle throat, and this made it very difficult to integrate from the subsonic to the supersonic sections of the nozzle. Moreover, for nonequilibrium nozzle flows the throat conditions and hence the mass flow are not known a priori; the nozzle mass flow must be obtained as part of the solution of the problem. Therefore, in 1969 a new technique for solving nonequilibrium nozzle flows was advanced by Anderson (see [175] and [176]) using the time-marching finite difference method discussed in Sec. 5.3. This time-marching approach circumvents the preceding problems encountered with steady-state analyses and also has the virtue of being relatively easy and straightforward to program on the computer. Since its introduction in 1969, the time-marching solution of nonequilibrium nozzle flows has gained wide acceptance.

Consider the nozzle and grid-point distribution sketched in Fig. 15.11. The time-marching solution of nonequilibrium nozzle flows follows the general philosophy as described in Sec. 5.3, with the consideration of vibrational energy and chemical species concentrations as additional dependent variables. In this context, at the first grid point in Fig. 15.11, which represents the reservoir conditions, equilibrium conditions for e_{vib} and c_i at the given p_0 and T_0 are calculated and held fixed, invariant with time. Guessed values of e_{vib} and c_i are then arbitrarily specified at all other grid points (along with guessed values of all other flow variables); these guessed values represent initial conditions for the time-marching solution. For the initial values of e_{vib} and c_i, it is recommended that equilibrium values be assumed from the reservoir to the throat and then frozen values be prescribed downstream of the throat. Such an initial distribution of nonequilibrium variables is qualitatively similar to typical results obtained for nonequilibrium nozzle flows, as we will soon see.

The governing continuity, momentum, and energy equations for unsteady quasi-one-dimensional flow are given in Chapter 12 of [4] as follows.

Continuity:

$$\frac{\partial \rho}{\partial t} = -\frac{1}{A}\frac{\partial(\rho u A)}{\partial x} \tag{15.22}$$

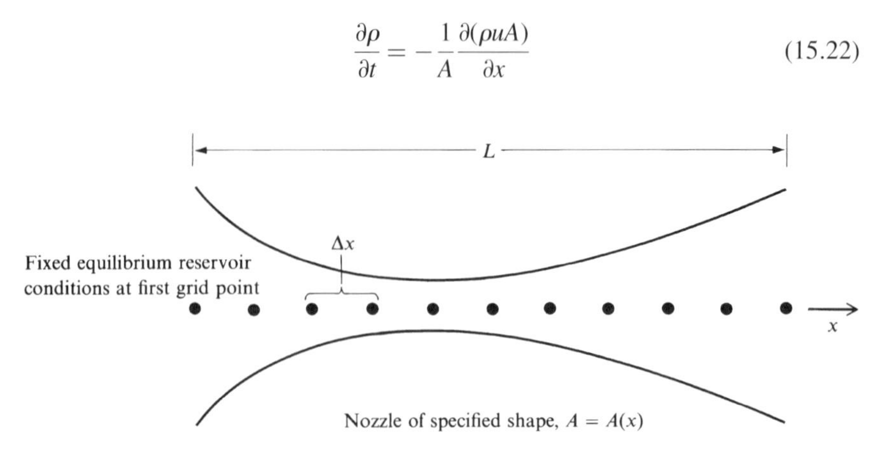

Fig. 15.11 Coordinate system and grid points for the time marching of quasi-one-dimensional flow through a nozzle.

Momentum:

$$\frac{\partial u}{\partial t} = -\frac{1}{\rho}\left(\frac{\partial p}{\partial x} + \rho u \frac{\partial u}{\partial x}\right) \tag{15.23}$$

Energy:

$$\frac{\partial e}{\partial t} = -\frac{1}{\rho}\left(p\frac{\partial u}{\partial x} + \rho u \frac{\partial e}{\partial x} + pu\frac{\partial \ell n A}{\partial x}\right) \tag{15.24}$$

where A is the local cross-sectional area of the nozzle. In addition to these equations, for a nonequilibrium flow the appropriate vibrational rate and species continuity equations are

$$\frac{\partial e_{\text{vib}}}{\partial t} = \frac{1}{\tau}(e_{\text{vib}}^{\text{eq}} - e_{\text{vib}}) - u\frac{\partial e_{\text{vib}}}{\partial x} \tag{15.25}$$

and

$$\frac{\partial c_i}{\partial t} = -u\frac{\partial c_i}{\partial x} + \frac{\dot{w}_i}{\rho} \tag{15.26}$$

Equations (15.22–15.26) are solved step by step in time using the finite difference predictor-corrector approach described in Sec. 5.3. Along with the other flow variables, e_{vib} and c_i at each grid point will vary with time; but after many time steps all flow variables will approach a steady state. It is this steady flowfield that we are interested in as our solution—the time-dependent technique is simply a means to achieve this end.

The nonequilibrium phenomena introduce an important new stability criterion for Δt in addition to the Courant–Friedrichs–Lewy (CFL) criterion discussed in Sec. 5.3. The value chosen for Δt must be geared to the speed of the nonequilibrium relaxation process and must not exceed the characteristic time for the fastest finite rate taking place in the system. That is,

$$\Delta t < B\Gamma \tag{15.27}$$

where $\Gamma = \tau$ for vibrational nonequilibrium, $\Gamma = \rho(\partial \dot{w}_i/\partial c_i)^{-1}$ for chemical nonequilibrium, and B is a dimensionless proportionality constant found by experience to be less than unity, and sometimes as low as 0.1. The value chosen for Δt in a nonequilibrium flow must satisfy both Eq. (15.27) and the usual CFL criterion, given here as

$$\Delta t < \frac{\Delta x}{u + a} \tag{15.28}$$

Which of the two stability criteria is the smaller, and hence governs the time step, depends on the nature of the case being calculated. If the local pressure and temperature are low enough everywhere in the flow, the rates will be slow, and Eq. (15.28) generally dictates the value of Δt. On the other hand, if some of the rates have particularly high transition probabilities and/or the local p and T are very high, then Eq. (15.27) generally dictates Δt. This is almost always encountered in rocket nozzle flows of hydrocarbon gases, where some of the chemical reactions involving hydrogen are very fast and combustion chamber pressures and temperatures are reasonably high.

The nature of the time-marching solution of a vibrational nonequilibrium expansion of pure N_2 is shown in Fig. 15.12. Here, the transient e_{vib} profiles at various time steps are shown; the dashed curve represents the guessed initial distribution. Note that during the first 250 time steps the proper steady-state distribution is rapidly approached and is reasonably attained after 800 time steps. Beyond this time, the time-dependent solution produces virtually no change in the results from one time step to the next. This steady-state distribution agrees with the results of a steady-flow analysis after Wilson et al. (see [177]), as shown in Fig. 15.13. Here, a local vibrational temperature is defined from the local nonequilibrium value of e_{vib} using the relation

$$e_{\text{vib}} = \left[\frac{h\nu/kT_{\text{vib}}}{e^{h\nu/kT_{\text{vib}}} - 1}\right] RT_{\text{vib}} \tag{15.29}$$

patterned after the equilibrium expression given by Eq. (11.61). (Recall that we first defined the vibrational temperature in Sec. 13.4.1.) Note that Eq. (15.29) is *not* a valid *physical* relationship for nonequilibrium flow; it is simply an equation that defines the vibrational temperature T_{vib} and that allows the calculation of a value of T_{vib} from the known value of e_{vib}. Hence, T_{vib} is simply an index for the local nonequilibrium value of e_{vib}. In Fig. 15.13, both the time-marching

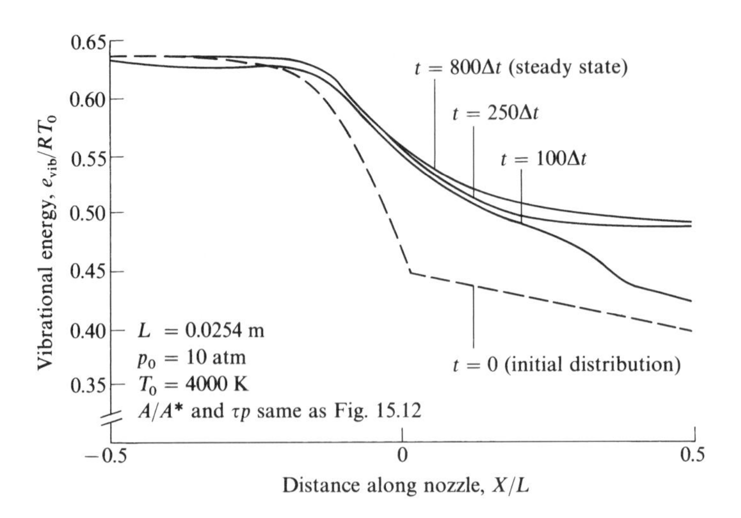

Fig. 15.12 Transient and final steady-state e_{vib} distributions for the nonequilibrium expansion of N_2 obtained from the time-marching analysis (from Anderson [175]).

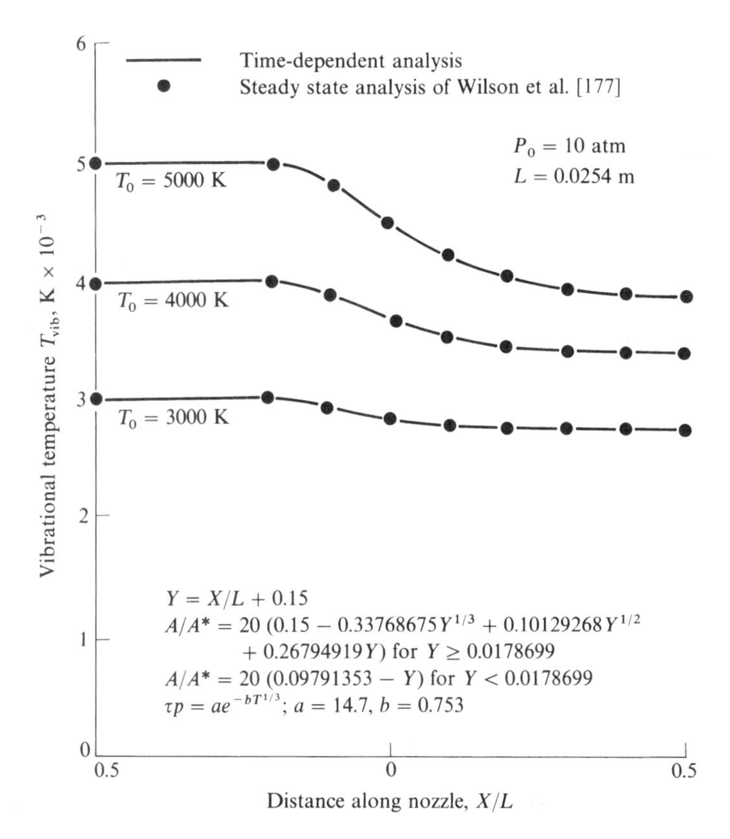

Fig. 15.13 Steady-state T_{vib} distributions for the nonequilibrium expansion of N_2; comparison of the time-marching analysis with the steady-flow analysts of Wilson et al. (from [175]).

calculations as well as the steady-flow analysis of Wilson et al. assume nonequilibrium flow at all points downstream of the reservoir, including the subsonic section. Very good agreement between the two techniques is obtained.

Many analyses of nonequilibrium nozzle flows in the literature assume local equilibrium to the throat and then start their nonequilibrium calculations downstream of the throat. In this fashion, the problems with the saddle-point singularity and the unknown mass flow, described earlier, are sidestepped. Examples of such analyses are given by Harris and Albacete [178] and by Erickson [179]. However, for many practical nozzle flows, nonequilibrium effects become important in the subsonic section of the nozzle, and hence a fully nonequilibrium solution throughout the complete nozzle is required.

Figures 15.12 and 15.13 illustrate an important qualitative aspect of nonequilibrium nozzle flows. Note that, as the expansion proceeds and the static temperature (T_{trans}) decreases through the nozzle, the vibrational temperature and energy also decrease to begin with. However, in the throat region, e_{vib} and T_{vib} tend to "freeze" and are reasonably constant downstream of the throat. This is a qualitative comment only; the actual distributions depend on pressure, temperature, and nozzle length. It is generally true that equilibrium flow is reasonably obtained throughout large nozzles at high pressures. Reducing both the size of the nozzle and the reservoir pressure tends to encourage nonequilibrium flows.

Results for a chemical nonequilibrium nozzle flow are given in Fig. 15.14, where the transient mechanism of the time-dependent technique is illustrated. Here, the nonequilibrium expansion of partially dissociated oxygen is calculated, where the only chemical reaction is

$$O_2 + M \rightleftharpoons 2O + M$$

In Fig. 15.14, the dashed line gives the initially assumed distribution for the atomic-oxygen mass friction, c_O. Note the rapid approach toward the steady-state distribution during the first 400 time steps. The final steady-state distribution is obtained after 2800 time steps. This steady-state distribution compares favorably with the results of Hall and Russo (solid circles), who performed a steady-flow analysis of the complete nonequilibrium nozzle flow (see [180]). Again, note the tendency of the oxygen mass fraction to freeze downstream of the throat.

A more complex chemically reacting nonequilibrium nozzle flow is illustrated by the expansion of a hydrocarbon mixture through a rocket engine, as calculated by Vamos and Anderson in [181]. The configuration of a rocket nozzle is given in Fig. 15.15. Here, for the time-marching numerical solution two grids are used along the nozzle axis: a fine grid of closely spaced points through the subsonic section and slightly downstream of the throat and a coarse grid of widely

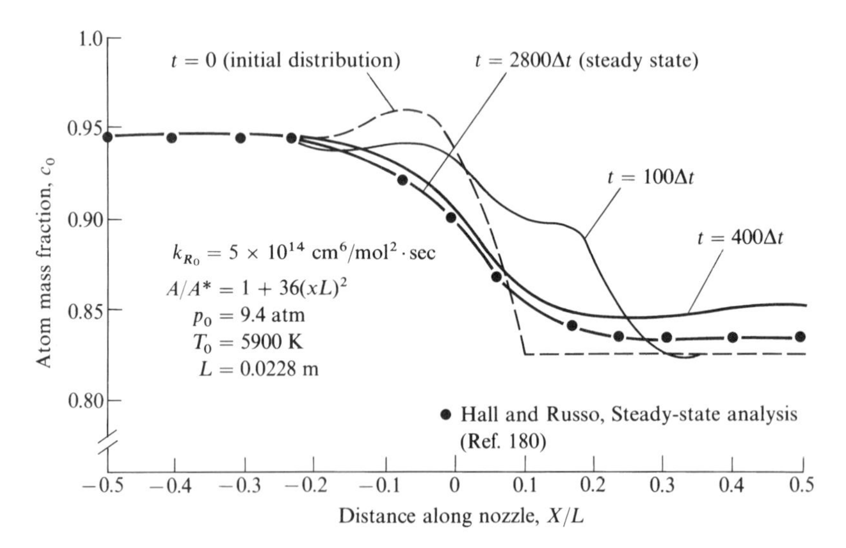

Fig. 15.14 Transient and final steady-state atom mass-fraction distributions for the nonequilibrium expansion of dissociated oxygen; comparison of the time-marching method with the steady-state approach of Hall and Russo (from [180]).

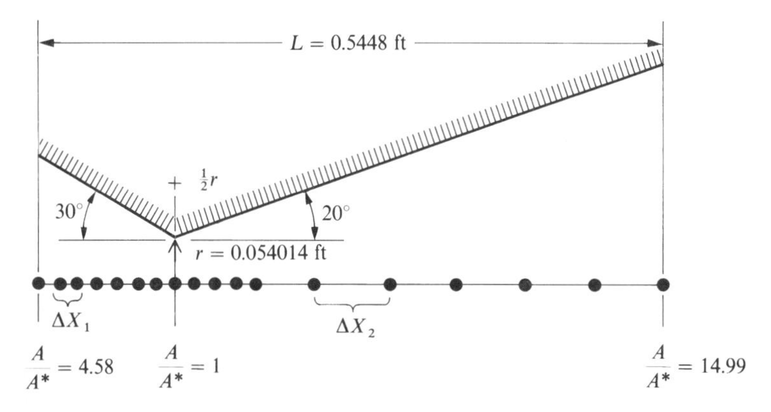

Fig. 15.15 Schematic representation of the rocket engine nozzle and grid-point system used by Vamos and Anderson (from [181]).

spaced points further downstream. Because most of the nonequilibrium behavior and the fastest reactions are occurring in the throat region, a fine grid is chosen here to maintain accuracy. In contrast, far downstream in the cooler supersonic region, the reactions are slower, the chemical composition is tending to freeze, and the grid spacing can be larger. (Parenthetically, we note that for any of the finite difference solutions discussed in this book, the grid spacings do not have to be constant. Indeed, the concept of adaptive grids, i.e., putting grid points only where you want them as dictated by the gradients in the flow, is a current state-of-the-art research problem of computational fluid dynamics.)

In Fig. 15.15, the reservoir conditions are formed by the equilibrium combustion of N_2O_4, N_2H_4, and unsymmetrical dimethyl hydrazine, with an oxydizer-to-fuel ratio of 2.25 and a chamber pressure of 4 atm. Results for the subsequent nonequilibrium expansion are shown in Fig. 15.16. Here, the transient variation of the hydrogen-atom mass fraction through the nozzle is shown. For convenience, the initial distribution is assumed to be completely frozen from the reservoir (the dashed horizontal line). Several intermediate distributions obtained during the time-marching calculations are shown, with the final steady state being achieved at a dimensionless time of 1.741. Note that, if the flow were in local chemical equilibrium, X_H would decrease continuously as T decreases, as shown in Fig. 15.16. In contrast, however, because of the complexities of the H-C-O-N chemical kinetic mechanism, X_H actually *increases* with distance along the nozzle. Here is another example (the first was given in Sec. 15.2) where a nonequilibrium variable falls outside the bounds of equilibrium and frozen flows. The variation of static temperature is given in Fig. 15.17; note that for nonequilibrium flow the temperature distribution is *lower* than the equilibrium value. This is because the nonequilibrium flow tends to freeze some of the dissociated products, hence locking up some of the chemical zero-point energy that would otherwise be converted to random molecular translational energy. The steady-state temperature distribution in Fig. 15.17 (at $t' = 1.741$) compares favorably with the steady-flow analysis of Sarli et al. (see [182]).

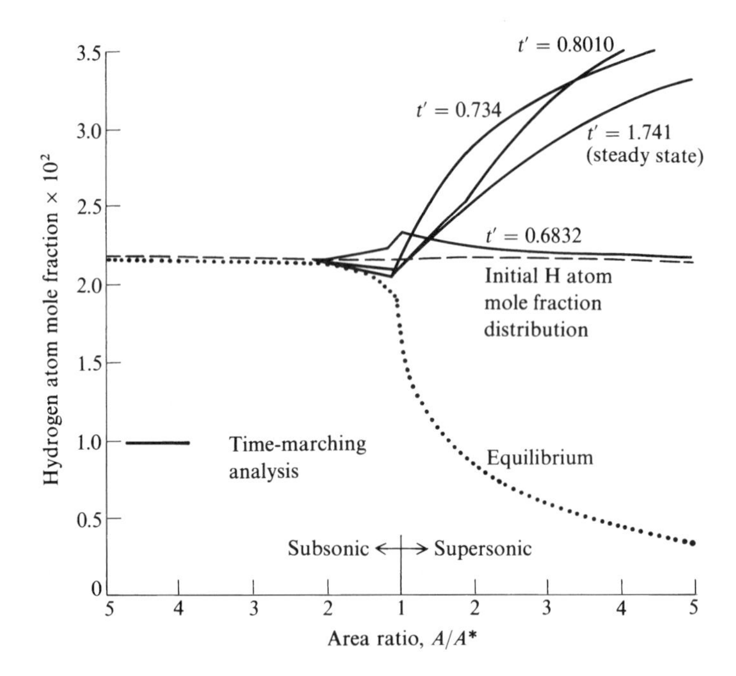

Fig. 15.16 Transient and final steady-state distributions of the hydrogen atom mole fraction through a rocket nozzle; nonequilibrium flows (from [181]).

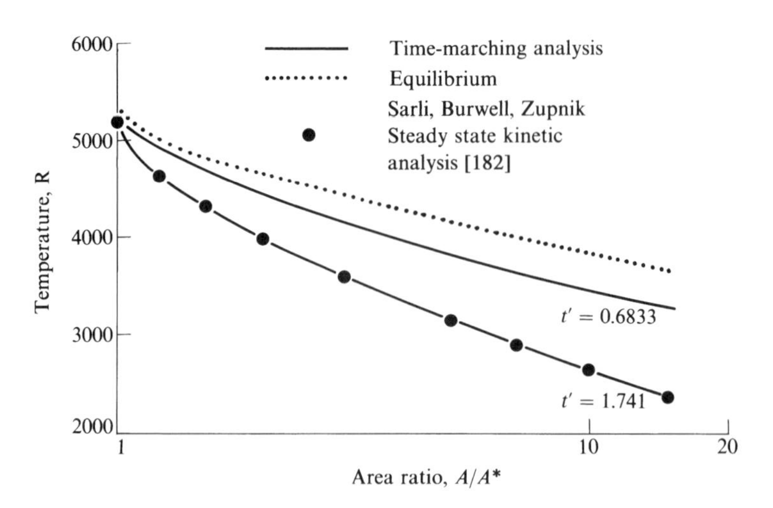

Fig. 15.17 Temperature distributions for the nonequilibrium flow through a rocket nozzle (from [181]).

As a final point concerning nonequilibrium, quasi-one-dimensional nozzle flows, note that any finite-rate phenomena are irreversible. Hence, an adiabatic, inviscid nonequilibrium nozzle flow is nonisentropic. Because the entropy of a fluid element increases as it moves through the nozzle, a simple analysis shows that the local velocity at the nozzle throat is *not* sonic. Indeed, in a nonequilibrium flow the speed of sound itself is not unique and depends on the frequency of the sound wave. However, if either the frozen or equilibrium speed of sound (see Sec. 14.7) is used to define the frozen or equilibrium Mach numbers at the nozzle throat, both Mach numbers will be less than unity. Sonic flow in a nonequilibrium nozzle expansion occurs slightly downstream of the throat.

Two-dimensional nonequilibrium nozzle flows can be calculated by finite difference methods or the method of characteristics. In regard to the latter, the characteristic lines through any point in the nonequilibrium supersonic flow are 1) the Mach lines based on the frozen speed of sound and 2) the streamlines because the entropy increases along a streamline in a nonequilibrium flow as a result of the irreversible aspects of the finite-rate processes. See [53] for details on the method of characteristics in a nonequilibrium flow.

15.5 Nonequilibrium Blunt-Body Flows

The general features of the inviscid flow over a supersonic or hypersonic blunt body were described in Sec. 5.3 for a calorically perfect gas and in Sec. 14.9 for an equilibrium chemically reacting gas; these sections should be reviewed before progressing further. In the case of the *nonequilibrium* flow over a blunt body, the flowfield resembles some of the nature of the previous cases, but also takes on some of the aspects of nonequilibrium flow behind shock waves, as discussed in Sec. 15.3. On a qualitative basis, the nonequilibrium flow over a blunt body behaves as sketched in Fig. 15.18. In the nose region, the chemical composition resembles that in the nonequilibrium region behind a normal shock wave, as discussed in Sec. 15.3. However, consider the streamline that goes through the stagnation point; this streamline is labeled *abc* in Fig. 15.18. Between *a* and *b*, the flow is compressed and slowed; it reaches zero velocity at the stagnation point *b*. In so doing, it can be shown that a fluid element takes an infinite time to traverse the distance *ab*. This means that local equilibrium conditions must exist at the stagnation point (point *b*) with its attendant highly dissociated and ionized state. The flow then expands rapidly downstream of the stagnation point; indeed, the surface streamline *bc* encounters very large pressure and temperature gradients in the region near the sonic point *c*, that is, $\mathrm{d}p/\mathrm{d}s$ and $\mathrm{d}T/\mathrm{d}s$ are large negative quantities. This is very similar to the nonequilibrium flow through a convergent-divergent nozzle discussed in Sec. 15.4, where it was indicated that sudden freezing of the flow can occur downstream of the throat. The same type of sudden freezing can be experienced near point *c* in Fig. 15.18. In turn, the surface of the body downstream of the sonic point can be bathed in a region of nearly frozen flow. Because the streamline started with a large amount of dissociation and ionization at point *b*, then this frozen flow is characterized by a thin region of high dissociated and ionized gas that flows downstream

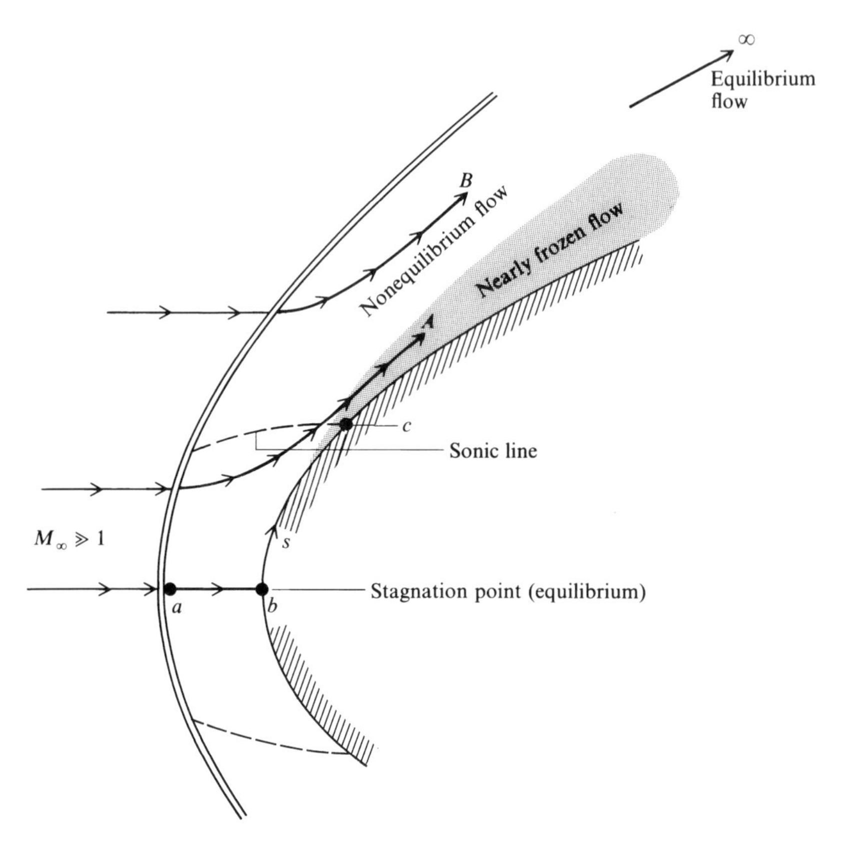

Fig. 15.18 Schematic of different regions in a high-temperature blunt-body flowfield.

over the body. Slightly away from the body, streamline A also passes through a strong portion of the bow shock and exhibits similar behavior to the stagnation-point streamline *abc*; that is, there is a region of highly dissociated and ionized nonequilibrium flow along A behind the shock as the chemistry is trying to move toward an equilibrium state and then fairly rapid freezing in the vicinity of the sonic line. Much further away from the body, streamline B passes through a weaker, more oblique portion of the bow shock. Consequently, the amount of dissociation and ionization is considerably smaller, but the nonequilibrium region extends much further downstream along B. (This effect is shown by comparing Figs. 15.9 and 15.10 for the relaxation distances downstream of a normal shock and oblique shock, respectively.) Finally, the entire flowfield will approach local equilibrium conditions a sufficiently far distance downstream of the nose.

Following the time-marching philosophy for the solution of blunt-body flowfields described in Secs. 5.3 and 14.9, a solution for the nonequilibrium blunt-body flow can be obtained by solving Eqs. (15.8–15.17) in steps of time. The first such time-marching solution for inviscid nonequilibrium blunt-body flows was carried out by Bohachevsky and Rubin [183] for a simplified model of a simple dissociating gas (the Lighthill model, described in [150]), using the Lax finite difference scheme (see [52]). Later the first time-marching inviscid nonequilibrium blunt-body solution for detailed air chemistry was obtained by Li [184] and [185]. In [185], the explicit MacCormack method using shock fitting—the same method described in detail in Sec. 5.3—was used to solve Eqs. (15.8–15.17) in steps of time, starting from a frozen flow solution used as the initial conditions at time $t = 0$. The governing equations are written with the time derivatives of ρ, u, v, w, h_0, and c_i on the left side and spatial derivatives on the right side. For example, Eq. (15.10) is written as

$$\frac{\partial c_i}{\partial t} = -u\frac{\partial c_i}{\partial x} - v\frac{\partial c_i}{\partial y} - w\frac{\partial c_i}{\partial z} + \frac{\dot{w}_i}{\rho}$$

The spatial derivatives are replaced by forward differences on the predictor step and rearward differences on the corrector step, and the new value of c_i at time $(t + \Delta t)$ is obtained (using the bar notation of Sec. 5.3) from

$$c_i(t + \Delta t) = c_i(t) + \frac{1}{2}\left(\frac{\partial c_i}{\partial t} + \overline{\frac{\partial c_i}{\partial t}}\right)\Delta t$$

See Sec. 5.3 for the details concerning MacCormack's predictor-corrector method. Its application to nonequilibrium blunt-body flows in this section is essentially the same; the details are left to the reader to construct.

As noted in Sec. 15.4, two different time steps are assessed, a fluid-dynamic time step based on the CFL criterion given by

$$\Delta t \leq \frac{\min(\Delta x, \Delta y, \Delta z)}{1.5[(u^2 + v^2 + w^2)^{1/2} + a]} \tag{15.30}$$

and a chemistry time step given by

$$\Delta t \leq 0.1 \min\left|\frac{\rho c_i}{\dot{w}_i}\right| \tag{15.31}$$

where Δx, Δy, and Δz are grid spacing in a Cartesian three-dimensional coordinate system. Note that Eq. (15.31) for the chemistry time step is slightly different from Eq. (15.27), which utilized $\rho(\partial\dot{w}_i/\partial c_i)^{-1}$ instead of $\rho c_i/\dot{w}_i$; the results for the chemistry Δt are essentially the same. In a given solution, the lower value of Δt obtained from Eqs. (15.30) and (15.31) is used to advance the flowfield in steps of time. In many cases, the chemistry Δt will be smaller than the fluid dynamic Δt, indeed sometimes orders of magnitude smaller. This is a ramification of the stiff nature of the rate equations, as described in Secs. 15.3 and 15.4. When this occurs, the time-marching solutions can require very large amounts of computer time to approach the steady state. To alleviate this situation somewhat, Li in [184] suggests advancing the fluid dynamics and the chemistry at their own respective timescales, that is, advance Eq. (15.10) using the Δt from Eq. (15.31), and simultaneously advance Eqs. (15.8), (15.12), and (15.13) using the Δt from Eq. (15.30). This can considerably reduce the number of time steps required to obtain convergence of the complete flowfield. Of course, with this method the transient variations obtained during the time-marching solution would not be time-wise accurate; however, if the steady state is the desired result, then the matter of time accuracy of the transients is not important. In Li's analysis, seven species were considered: N_2, O_2, NO, N, O, NO^+, and e^-. Although chemical nonequilibrium was treated, local thermodynamic equilibrium (involving the internal modes of vibrational and electronic energy) was assumed. Some sample results are shown in Fig. 15.19 for the variations of c_{O_2}, c_{NO}, and c_O as a function of distance along the stagnation streamline between the shock and the body (along streamline ab in Fig. 15.18) for flow

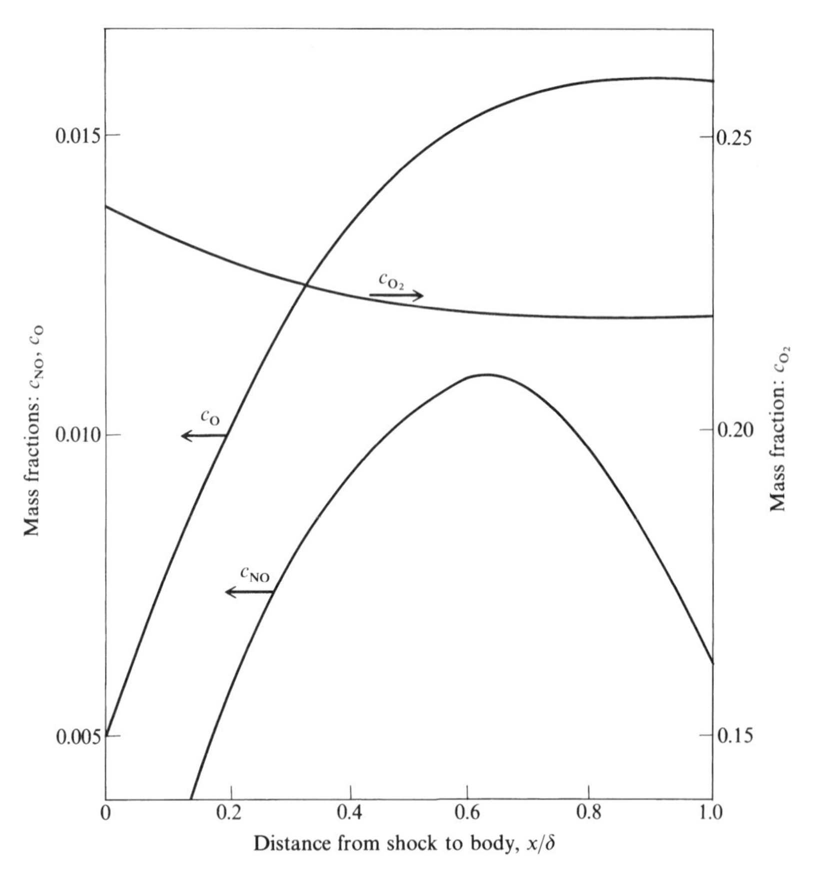

Fig. 15.19 Mass fraction distributions along the stagnation streamline of a sphere. Nonequilibrium flow. $V_\infty = 11{,}310$ ft/s, $p_\infty = 21.17$ lb/ft^2, $T_\infty = 540$°R, and sphere radius = 1 ft (from Li [185]).

over a sphere. For the freestream conditions given in Fig. 15.19, the shock layer is a mildly dissociated gas, with O_2 dissociation as the dominant mechanism. Note in Fig. 15.19 the overshoot of NO, similar to the normal shock wave results given in Fig. 15.5.

The first comprehensive nonequilibrium blunt-body analysis was carried out by Hall et al. [186] in 1962, well before the advent of time-marching solutions. In [186], the analytical technique was an inverse solution—assuming a given shock shape, integrating the nonequilibrium flowfield downstream of the given shock, and finding the body shape that supports the given shock. Setting up a coordinate system where x and y are tangential and normal to the shock respectively, Hall et al. replaced the partial derivatives in the x direction with a seven-point finite difference; hence, the x derivatives are known numbers, and the governing flow equations become ordinary differential equations in the y direction. In turn, these ordinary differential equations are integrated in the y direction by means of a standard Runge–Kutta method. (This smacks of the difference-differential approach used by Smith and Clutter in the same time period for the solution of the boundary-layer equations, as described in Sec. 6.6 and found in [97].) The details of the steady state numerical method used by Hall et al. in [186] are not important here; indeed, by present-day standards, the method is antiquated. However, [186] is a *classic* in its own right and is as important today as it was in 1962 because of the pioneering results obtained and the revealing manner in which they are discussed. For this reason, we will discuss these results at some length here. They provide an excellent picture of the physical nature of nonequilibrium blunt-body flows.

As in the case of Li [185], Hall et al. [186] use a seven-species, seven-reaction model for high-temperature air. The seven species are O_2, O, N_2, N, NO, NO^+, and e^-; the kinetic reaction mechanism is given by Eqs. (13.38–13.44), with rate constants from Wray [159]. Chemical nonequilibrium was the only finite-rate process treated; local thermodynamic equilibrium was assumed (see Sec. 14.1). Results are presented along two streamlines in the blunt-body flowfield, streamlines A and B shown in Fig. 15.20. This figure is drawn to scale, showing the assumed axisymmetric catenary shock in cylindrical coordinates, where z and r are coordinates parallel and perpendicular respectively to the freestream direction. The resulting body shape is nearly spherical, and is shown in Fig. 15.20 for the case of $V_\infty = 23{,}000$ ft/s, altitude equal to 200,000 ft, and a given shock radius of curvature at the point of symmetry $R_s = 0.0692$ ft. In Figs. 15.21–15.23 results are given for the variation of flow properties along streamlines A and B for the velocity-altitude point already given. Figure 15.21 shows the results for T, p, and ρ as a function of distance s along the streamlines, measured from the shock front. Streamline A crosses the shock near the stagnation region and is initially dominated by chemical nonequilibrium behavior similar to that behind a normal shock wave (see Sec. 15.3). The temperature along streamline A, T_A exhibits an initial rapid decrease behind the shock; this is because of the finite-rate dissociation of both O_2 and N_2. The more gradual decrease in T_A for $s/R_s > 0.2$ is caused primarily by the gas dynamic expansion around the body. Similarly, the initially slight increase in p_A and the substantial increase in ρ_A are caused by the nonequilibrium effects, and their subsequent decrease beyond $s/R_s = 0.2$ are indicative of the gas dynamic expansion

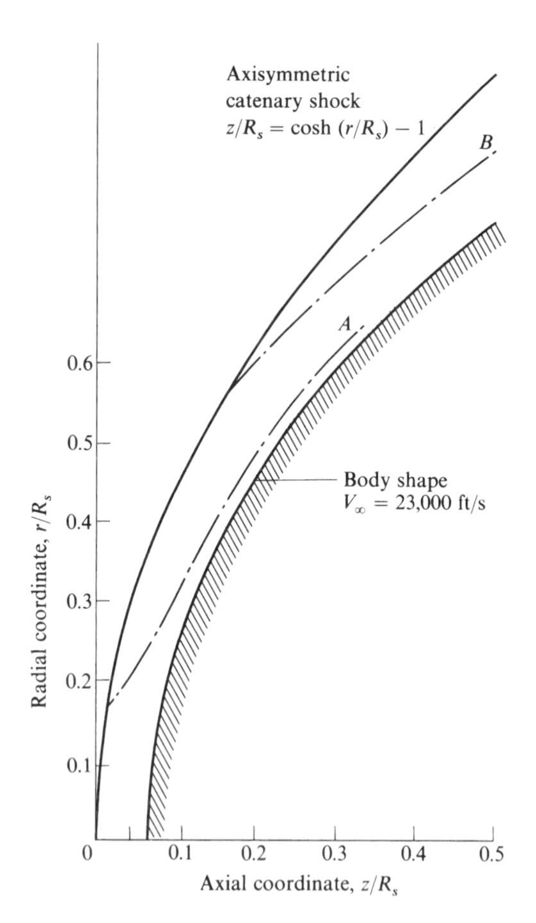

Fig. 15.20 Shock and body shapes and calculated streamlines for the nonequilibrium flow over a blunt body (from Hall et al. [186]).

around the body. In contrast, streamline B crosses a much weaker portion of the bow shock wave. As discussed in Sec. 15.3 comparing normal and oblique shock waves, the flow behind an oblique shock front experiences a much longer relaxation distance than a normal shock at the same upstream conditions, although at the same time the actual quantitative degree of dissociation behind the oblique shock is less because of the lower temperature. These comparisons from Sec. 15.3 carry over to the blunt-body flow. In Fig. 15.21, the behavior of T_B, p_B, and ρ_B along streamline B is a combination of the nonequilibrium effects and the gas dynamic expansion around the body—a combination that persists over the complete length of streamline B shown in Fig. 15.21. Also shown in Fig. 15.21 are the equilibrium (infinite-rate) values of p_A and p_B just behind the shock front, at $s = 0$. To be expected from our earlier discussions (for example, see Secs. 14.3, 14.8, and 14.9), the pressure is least affected by

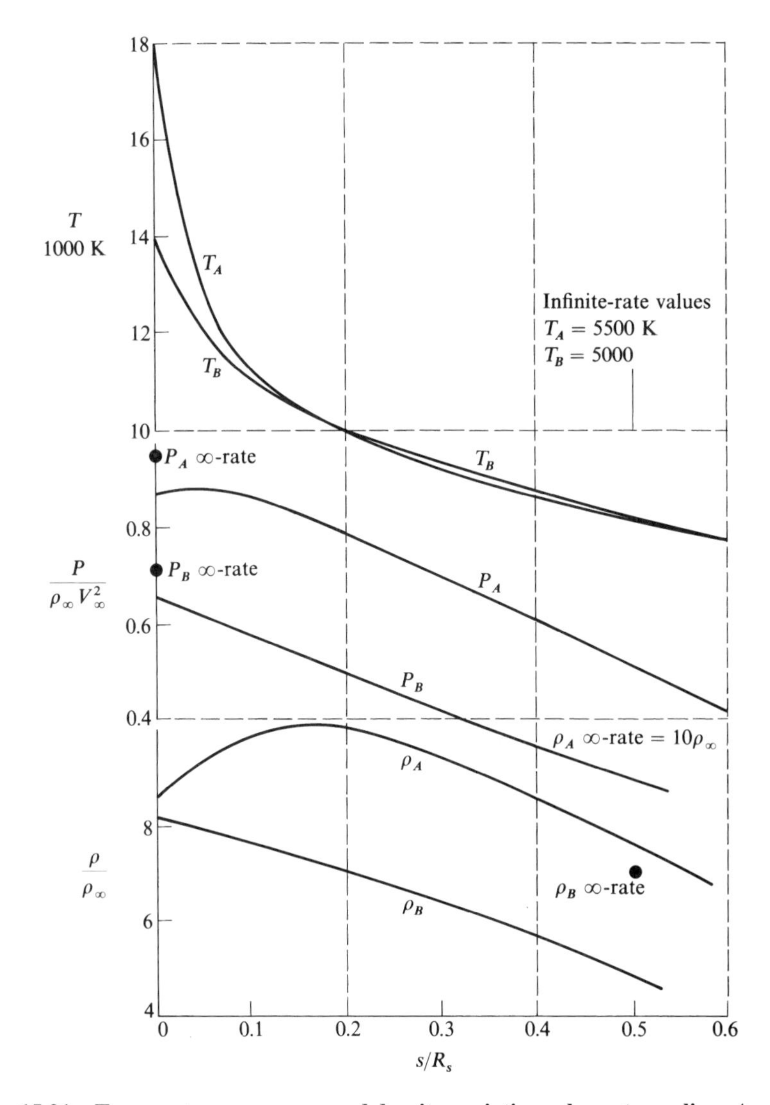

Fig. 15.21 Temperature, pressure, and density variations along streamlines *A* and *B* in the nonequilibrium blunt-body flowfield (from [186]).

chemically reacting effects in a compression region. Also shown in Fig. 15.21 are the values of ρ_A and ρ_B at $s/R_s = 0.5$ from a calculation of the blunt-body flowfield assuming local thermodynamic and chemical equilibrium (Sec. 14.9). Note that these infinite-rate values are considerably above the nonequilibrium results for density. As to be expected, the nonequilibrium effects are strongest on temperature. In Fig. 15.21, both T_A and T_B are far above the local equilibrium values shown.

The variation of atomic oxygen and nitrogen is given in Fig. 15.22 in terms of number of moles per original mole of air, which is related to our more familiar variables, mass fraction c_i, and mole-mass ratio η_i, through the relations $c_i \mathcal{M}_{\text{air}}/\mathcal{M}_i$ and $\eta_i \mathcal{M}_{\text{air}}$, where $\mathcal{M}_{\text{air}}$ is the molecular weight of the nonreacting air in the freestream. (Proof of these relations is left to the reader for a homework problem.) Because $\mathcal{M}_{\text{air}}$ and $\mathcal{M}_i$ are known constant values, we can visualize the ordinate in Fig. 15.22 as essentially the mass fraction or the mole-mass ratio. In Fig. 15.22, note that the amount of atomic oxygen denoted by $(O)_A$ increases

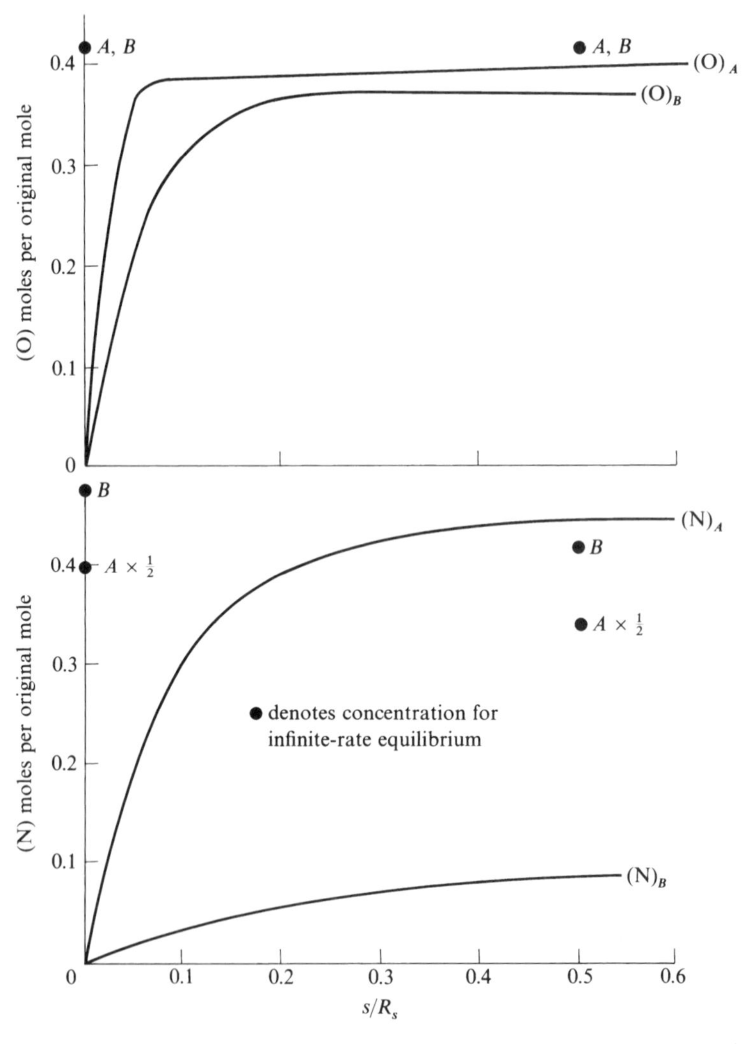

Fig. 15.22 Atomic oxygen and nitrogen concentrations along streamlines *A* and *B* in the nonequilibrium blunt-body flowfield (from [186]).

rapidly behind the shock front along streamline A; this is because of the nonequilibrium dissociation behind the strong shock front and is analogous to the normal shock results discussed in Sec. 15.3. However, for $s/R_s > 0.1$, the oxygen freezes because of the gas dynamic expansion and essentially plateaus at a value slightly less than the equilibrium value shown at $s/R_s = 0.5$. Along streamline B, the oxygen relaxation is slower, and $(O)_B$ freezes at a level even less than that for streamline A. [Note that the equilibrium values for both $(O)_A$ and $(O)_B$ shown at $s/R_s = 0$ and 0.5 are the same; this is because the temperatures along streamlines A and B are high enough such that, at local equilibrium

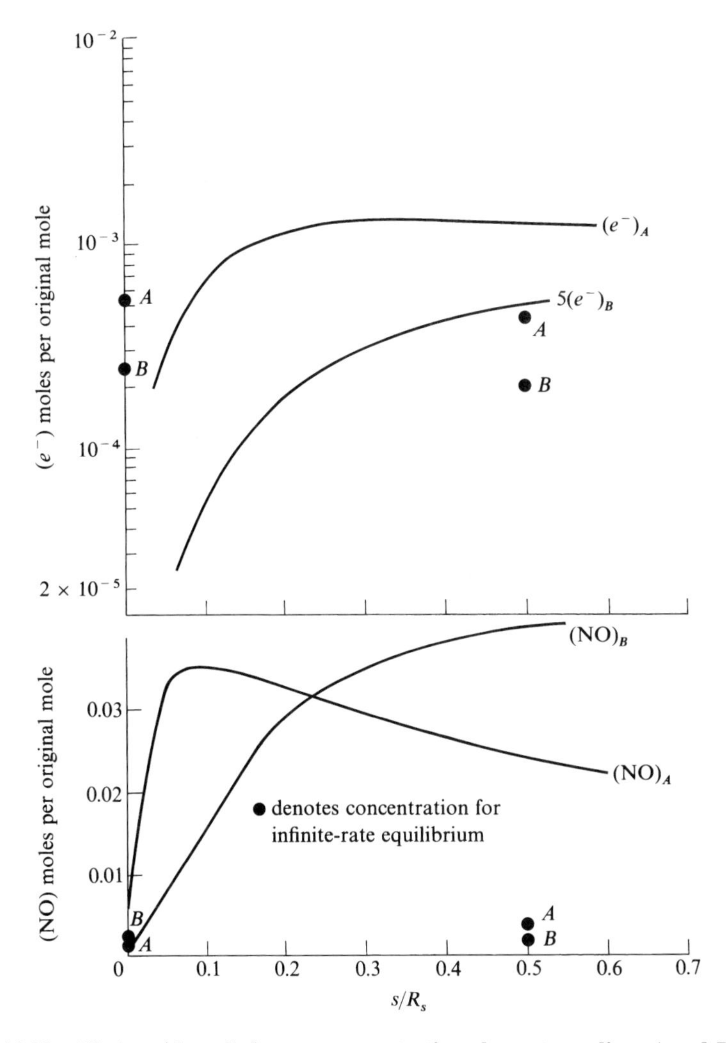

Fig. 15.23 Nitric oxide and electron concentration along streamlines *A* and *B* in the nonequilibrium blunt-body flowfield (from [186]).

conditions, the oxygen is fully dissociated.] Also shown in Fig. 15.22 is the variation of atomic nitrogen along streamlines A and B, denoted by $(N)_A$ and $(N)_B$, respectively. For N, the nonequilibrium relaxation distances are much longer than for O, and hence $(N)_A$ and $(N)_B$ exhibit strong nonequilibrium behavior. Note that $(N)_A$ is frozen at about one-half its local equilibrium value when compared at $s/R_s = 0.5$, and $(N)_B$ is about one-fourth its local equilibrium value at the same location. The variations of nitric oxide and electrons are shown in Fig. 15.23. Note that $(NO)_A$ exhibits the same type of overshoot observed behind a normal shock as discussed in Sec. 15.3, whereas $(NO)_B$ shows a monotonic increase. Also note that both $(NO)_A$ and $(NO)_B$ are considerably above their local equilibrium values. Examining the electron concentrations shown in Fig. 15.23, we see that $(e^-)_A$ freezes at a level *above* the local equilibrium value, but that $(e^-)_B$ is considerably below its equilibrium value.

The physical variations of the blunt-body flowfield variables just discussed are important and should be reread until you feel comfortable with the results. The conditions for the results just discussed were intentionally chosen by Hall et al. to accentuate the nonequilibrium effects, that is, a high altitude (hence low density with resulting large chemistry times) and a small body (hence small flow times). Thus, in reference to the discussion in Sec. 15.1. we have a situation where τ_f and τ_c are of the same relative magnitudes. If a lower altitude and/or a larger body is chosen, the relative nonequilibrium effects would diminish.

Once again, [186] is a classic presentation of the physical properties encountered in nonequilibrium blunt-body flows, with a much more extensive discussion than we have space to devote here. The reader is strongly encouraged to study [186] carefully.

15.6 Binary Scaling

For nonequilibrium processes involving two-body molecular collisions, an interesting and important scaling can be obtained for nonequilibrium flowfields. This scaling is called *binary scaling* and is derived and discussed in this section.

Assume that the dominant chemical reaction is caused by *dissociation*, such as

$$O_2 + M \longrightarrow 2O + M$$

where M is a collision partner. In the preceding chemical equation, dissociation is from left to right and is a two-body process. The reverse reaction from right to left is *recombination* and is a three-body process; hence, it has a lower probability of happening than the dissociation. Let us assume that, in a nonequilibrium flow, the dissociation reaction (from left to right) is the primary chemical reaction, and let us ignore the recombination (just for this discussion). Assume a steady, two-dimensional flow, for simplicity. For this flow, the species continuity equation for atomic oxygen is obtained from Eq. (15.10) as

$$u\frac{\partial c_O}{\partial x} + v\frac{\partial c_O}{\partial y} = \frac{\dot{w}_O}{\rho} = \frac{\mathscr{M}_O}{\rho}\frac{d[O]}{dt} \tag{15.32}$$

where, assuming just the dissociation reaction is taking place,

$$\frac{d[O]}{dt} = 2\,k_f\,[O_2][M] \tag{15.33}$$

The relation between concentration and mass fraction is

$$[i] = \frac{\rho c_i}{\mathcal{M}_i} \tag{15.34}$$

Combining Eqs. (15.32–15.34), we have

$$u\frac{\partial c_O}{\partial x} + v\frac{\partial c_O}{\partial y} = 2\frac{\mathcal{M}_O}{\rho}k_f\left(\frac{\rho c_{O_2}}{\mathcal{M}_{O_2}}\right)\left(\frac{\rho c_M}{\mathcal{M}_M}\right)$$

or

$$u\frac{\partial c_O}{\partial x} + v\frac{\partial c_O}{\partial y} = K_1 \rho c_{O_2} c_M \tag{15.35}$$

where

$$K_1 = 2\frac{\mathcal{M}_O}{\mathcal{M}_{O_2}\mathcal{M}_M}k_f = f(T)$$

Define the following nondimensional variables:

$$x' = \frac{x}{R} \quad y' = \frac{y}{R}$$

$$u' = \frac{u}{V_\infty} \quad v' = \frac{v}{V_\infty}$$

$$\rho' = \frac{\rho}{\rho_\infty}$$

where R is a characteristic length (for the blunt-body problem, R would be the nose radius) and V_∞ and ρ_∞ are the freestream velocity and density, respectively. Then Eq. (15.35) becomes

$$\frac{u}{V_\infty}\frac{\partial c_O}{\partial (x/R)}\left[\frac{V_\infty}{R}\right] + \frac{v}{V_\infty}\frac{\partial c_O}{\partial (y/R)}\left[\frac{V_\infty}{R}\right] = K_1\left(\frac{\rho}{\rho_\infty}\right)c_{O_2}c_M\rho_\infty$$

or

$$u'\frac{\partial c_O}{\partial x'} + v'\frac{\partial c_O}{\partial y'} = K_1\frac{(\rho_\infty R)}{V_\infty}\rho' c_{O_2} c_M \tag{15.36}$$

Equation (15.36) states the following: Consider two different flows with the same T_∞ and V_∞ (hence, with essentially the same value for K_1), but with different values of ρ_∞ and R. Plots of c_O (and all other mass fractions) vs x' or y' will be the *same* for the two flows *if* the product $\rho_\infty R$ is the same between the two flows.

This is a statement of *binary scaling*, where

$$\rho_\infty R = \text{binary scaling parameter}$$

This statement is illustrated qualitatively in Fig. 15.24. At the top of Fig. 15.24, the nonequilibrium variation of c_O vs s/R_1 is sketched for the flow over a body of radius R, with a freestream density of ρ_1. At the bottom of Fig. 15.24, the nonequilibrium variation of c_O vs s/R_2 is sketched for the flow over a body of radius R_2, where R_2 is three times larger than R_1, but the freestream density is one-third of ρ_1, such that the binary scaling parameter is the same between the two different flows, that is, $\rho_1 R_1 = \rho_2 R_2$. In this case, the curves of c_O vs s/R_1 and vs s/R_2 will be the *same*. This is the meaning of binary scaling.

Binary scaling for an actual application is dramatically illustrated in [186]. In Figs. 15.20–15.23 obtained from [186], the conditions were for $V_\infty = 23{,}000$ ft/s, altitude 200,000 ft where $\rho_\infty = 6 \times 10^{-7}$ slug/ft^3, and $R_s = 0.0692$ ft. For the case shown in these figures, $\rho_\infty\ R_s = 4.2 \times 10^{-8}$ slug/ft^2. Hall et al. in [186]

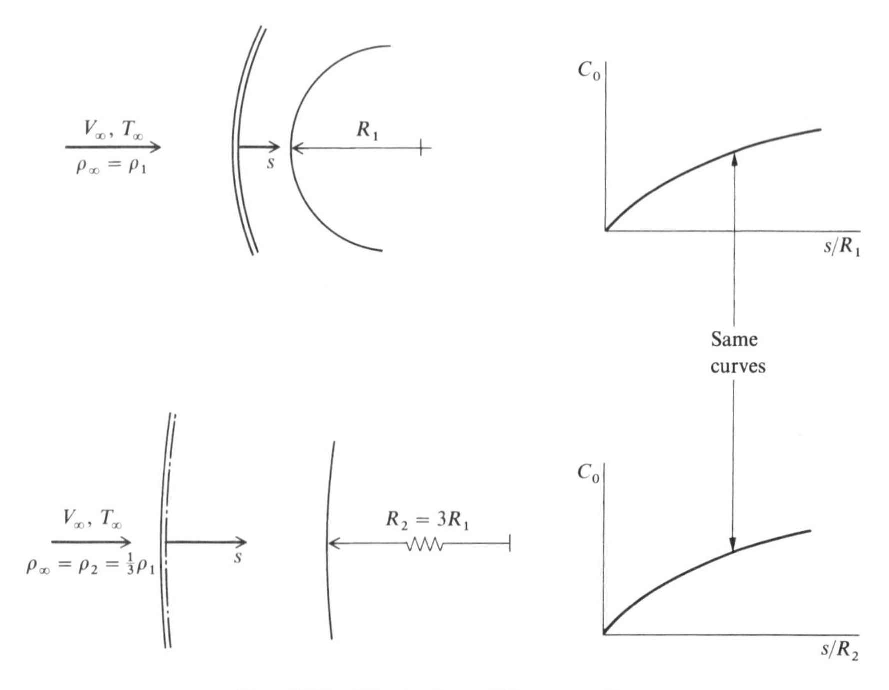

Fig. 15.24 Illustration of binary scaling.

also made calculations for the same V_∞ at an altitude of 250,000 ft where $\rho_\infty = 8 \times 10^{-8}$ slug/ft^3 and $R_s = 0.525$ ft. For this case, $\rho_\infty R_s = 4.2 \times 10^{-8}$, the same as just shown. According to binary scaling, as long as the flowfield associated with Figs. 15.20–15.23 is dominated by the two-body dissociation reactions rather than the three-body recombination reactions, then the results for this second case should fall right on top of the curves shown in Figs. 15.20–15.23. This is indeed the situation found by Hall et al.; hence, Figs. 15.20–15.23 hold also for the conditions $V_\infty = 23{,}000$ ft/s, altitude 250,000 ft, and $R_s = 0.525$ ft. This result clearly demonstrates the power and applicability of binary scaling for nonequilibrium flowfields. Indeed, Hall et al. go on to show that, for the most part in their calculations, the ratio of k_f/k_r is much larger than unity (on the order of 10^4 in certain sections of the flowfield); hence, it is no surprise that their results exhibit strong binary scaling.

15.7 Nonequilibrium Flow over Other Shapes: Nonequilibrium Method of Characteristics

The purpose of this section is to briefly examine the nonequilibrium flows over wedges, sharp cones, blunt cones, and a space-shuttle configuration. The computational details, which are many, are minimized here, and emphasis is placed on the physical results. In this manner, the reader will have some additional opportunities to obtain a physical understanding of nonequilibrium inviscid flows.

In Sec. 15.3, the nonequilibrium flow over a wedge was discussed qualitatively, and the reasons for the resulting curved shock wave were given in conjunction with the discussion of Fig. 15.8. This discussion should be reviewed before progressing further.

Because the nonequilibrium effects cause the flow over a wedge to have a variation in both the directions normal and tangential to the wedge surface, a two-dimensional flowfield solution must be used. These can take the form of finite difference calculations or a method-of-characteristics solution. For example, the method of characteristics has been used by Spurk et al. [187] and by Rakich et al. [188] to calculate nonequilibrium inviscid flows over wedges and cones. Some of these results will be discussed briefly in the present section. However, before examining the results, let us note the salient aspects of the method of characteristics for nonequilibrium flow.

In Sec. 5.2, the two-dimensional and axisymmetric irrotational and rotational methods of characteristics were discussed for a calorically perfect gas. Section 5.2 should be reviewed before progressing further. How does nonequilibrium flow modify the method of characteristics as discussed in Sec. 5.2? The major aspects are as follows:

1) In a nonequilibrium flow, the irreversible finite-rate mechanisms always increase the entropy [for example, see Eq. (10.72)]. Hence, the entropy of a fluid element increases as it moves along a streamline in a nonequilibrium flow. This causes all two- and three-dimensional nonequilibrium flows to be rotational, and therefore the *streamlines are characteristic lines in a nonequilibrium flow.*

2) The other characteristic lines are Mach lines based on the *frozen speed of sound* a_f, that is, lines that make an angle with the streamlines equal to

$\mu = \arcsin(1/M_f) = \arcsin(a_f/V)$. The use of the frozen speed of sound for the characteristic lines in a nonequilibrium flow (as opposed to, say, the equilibrium speed of sound, or the actual nonequilibrium speed of sound) comes from the theory itself and can be physically justified on the basis that the leading edge of a wave front is only a few mean free paths thick, and hence must propagate under locally frozen conditions.

The compatibility equations for the nonequilibrium method of characteristics are given in [187]. Letting s_1, s_2, and s denote distances along the left-running and right-running frozen Mach lines and the streamlines respectively, as sketched in Fig. 15.25, these compatibility equations are as follows:

Along s_1:

$$\frac{\partial\theta}{\partial s_1} + \frac{(M_f^2 - 1)^{1/2}}{\rho V^2}\frac{\partial p}{\partial s_1} + \sin\mu\left[\frac{j\sin\theta}{y} - \sum_i\left(\mathcal{M} - \frac{\rho\mathcal{R}}{p}\frac{H_i}{C_{pf}}\right)\frac{\partial\eta_i}{\partial s}\right] = 0 \quad (15.37)$$

Along s_2:

$$-\frac{\partial\theta}{\partial s_2} + \frac{(M_f^2 - 1)^{1/2}}{\rho V^2}\frac{\partial p}{\partial s_2} + \sin\mu\left[\frac{j\sin\theta}{y} - \sum_i\left(\mathcal{M} - \frac{\rho\mathcal{R}}{p}\frac{H_i}{C_{pf}}\right)\frac{\partial\eta_i}{\partial s}\right] = 0 \quad (15.38)$$

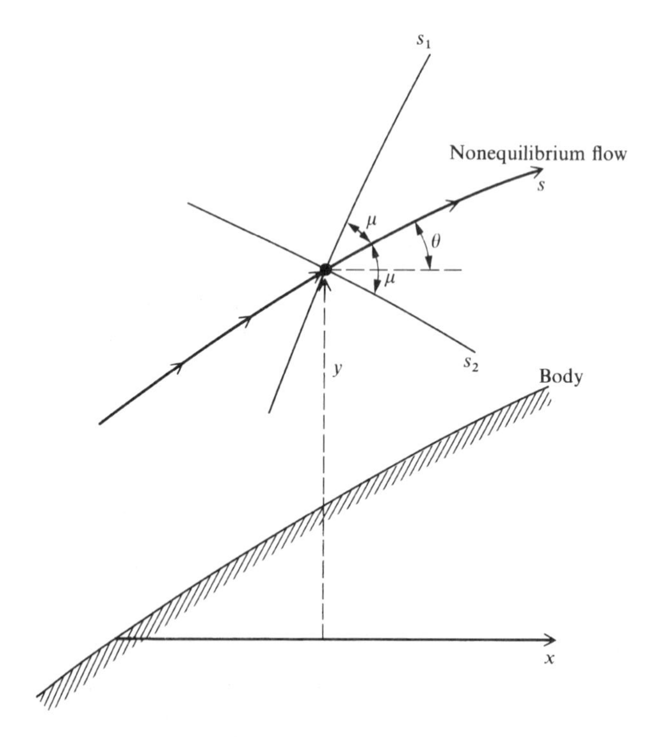

Fig. 15.25 Illustration of characteristic lines in a nonequilibrium flow.

Along *s:*

$$\rho V \frac{\partial \eta_i}{\partial s} = \frac{\dot{w}_i}{\mathscr{M}_i \rho} \tag{15.39}$$

Along *s*:

$$\rho V \frac{\partial V}{\partial s} = \frac{\partial p}{\partial s} \tag{15.40}$$

where θ is the angle between the streamline and the horizontal x axis shown in Fig. 15.25, M_f is the frozen Mach number defined as V/a_f, $j = 0$ or 1 for two-dimensional or axisymmetric flow, y is the vertical coordinate shown in Fig. 15.25, $\mathscr{M}$ is the local molecular weight of the mixture, H_i is the enthalpy of species i per mole of i, C_{pf} is a mean molar frozen specific heat defined as $C_{pf} = \sum_i \eta_i C_{pi} / \sum_i \eta_i$, which is similar to the frozen specific heat defined by Eq. (14.37), and C_{pi} is the specific heat of species i per mole of i. These compatibility equations are solved in a downstream-marching fashion at the grid points defined by intersections of the characteristics mesh, much in the same fashion as described in Sec. 5.2.

Results for the nonequilibrium flow of air over a wedge are shown in Fig. 15.26, taken from [187]. Here, the pressure and temperature distributions are shown as a function of distance along the wedge. The freestream conditions are $V_\infty = 6638$ m/s, $T_\infty = 273.16$ K, and $p_\infty = 0.01$ atm. Note that the pressure and temperature decrease with distance along the wedge surface. Also note that the *surface* conditions do *not* approach the equilibrium oblique shock results at large distances downstream. The reason for this can be seen by referring again to Fig. 15.8. The surface streamline comes through the frozen shock wave at the tip of the wedge, where the wave angle β is the greatest and hence the entropy increase is the largest. The entropy of the surface streamline is further increased by the irreversible finite-rate processes. Hence, the streamlines near the wedge surface experience a permanent increase in entropy that exceeds the predicted value for equilibrium flow over a wedge of the same angle (as calculated in Sec. 15.3). In this vein, the nonequilibrium flow over a wedge, with its curved shock wave, creates an entropy layer near the surface. This entropy layer results in the surface temperature approaching an asymptotic value far downstream, which is *higher* than the equilibrium shock value from Sec. 14.3; this is clearly seen in Fig. 15.26. Of course, the thickness of this entropy layer becomes a smaller percentage of the total shock-layer thickness for stations progressively further downstream. In the limit of an infinite distance downstream of the nose, the flow across the shock layer is in local equilibrium with uniform properties (as calculated in Sec. 14.3) with the exception of the surface properties, which have singular-like behavior at different values than the uniform, equilibrium flow.

Nonequilibrium flow over wedges and pointed and blunt-nosed cones are reported by Rakich et al. in [188]. Here, the method of characteristics for nonequilibrium flow is also employed for those regions of flow that are locally supersonic or hypersonic. Figure 15.27 illustrates the variation of shock-wave angle β as a function of distance along the surface of a 30-deg half-angle wedge for the

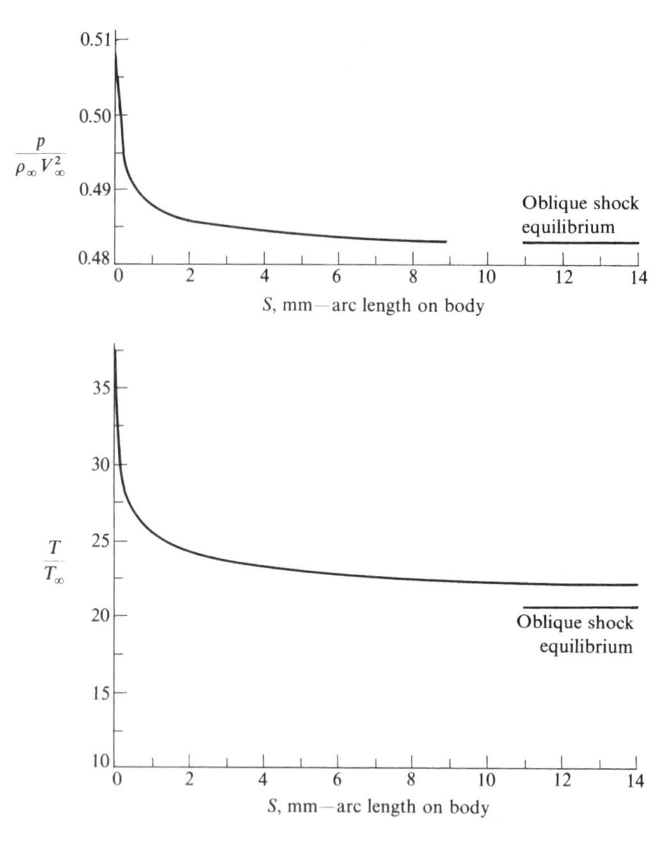

Fig. 15.26 Pressure and temperature along a wedge surface. Nonequilibrium flow. $V_\infty = 6638$ m/s, $T_\infty = 273.16$ K, $p_\infty = 0.01$ atm, and $\theta = 41.04$ deg (from Spurk et al. [187]).

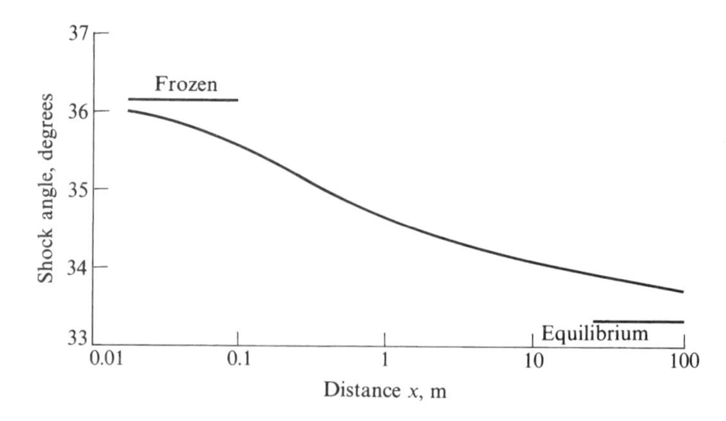

Fig. 15.27 Variation of shock-wave angle with distance in the nonequilibrium flow over a wedge (from Rakich et al. [188]).

case of $V_\infty = 6.7$ km/s at an altitude of 65.5 km (a trajectory point of interest because it corresponds to high laminar heat transfer to the space shuttle). Note that β changes from its frozen flow value and approaches its equilibrium flow value over a distance of 100 m, and that the change in shock angle is only a few degrees. Figure 15.28 shows the profiles of atomic oxygen as a function of a nondimensional normal distance $\bar{\eta}$ across the shock layer for a pointed and a blunt-nosed cone; $\bar{\eta} = (\eta - \eta_b)/(\eta_s - \eta_b)$, where η is the local coordinate of a point in the shock layer and η_b and η_s are the normal coordinates of the body and shock respectively, at the same streamwise station. The coordinate $\bar{\eta}$ is shown in Fig. 15.28. The two parts of Fig. 15.28 correspond to two different axial stations; Fig. 15.28a applies to $x/R_N = 3$, and Fig. 15.28b applies to $x/R_N = 6.8$, where R_N is the radius of the blunt nose. In Fig. 15.28, the dashed curves correspond to the pointed cone, and the solid curves correspond to the blunt-nosed cone. Note that, as $\bar{\eta}$ increases from zero, the sharp cone results show a rapid decrease in X_O in the region near the surface (near $\bar{\eta} = 0$), but then a more gradual decrease throughout the remainder of the shock layer toward the shock. This rapid change in X_O is related to the same type of nonequilibrium-induced entropy layer as discussed earlier in regard to a sharp wedge. The solid curves in Fig. 15.28 show that nose bluntness has a very strong effect on X_O. The strong entropy layer induced by the curved bow shock wave at the nose creates a marked dip in the value of X_O at some distance from the body surface. The location of this dip relative to the shock layer itself becomes closer to the body ($\bar{\eta}$ smaller) as distance downstream of the nose is increased. The comparison shown in Fig. 15.28 clearly demonstrates a strong coupling between the fluid-dynamic-induced entropy layer on a blunt-nosed hypersonic body (as discussed in various places in Part 1 of this book) and the finite-rate chemistry.

Finally, the nonequilibrium effect on the shock-wave shape on a space shuttle vehicle is shown in Fig. 15.29, also obtained from [188]. Note that in the nonequilibrium chemically reacting flow, the shock wave lies closer to the body than in a frozen flow case (essentially a calorically perfect gas with $\gamma = 1.4$).

15.8 Summary and Comments

Return again to the road map given in Fig. 1.24. With the end of this chapter, we complete our discussion of inviscid chemically reacting flows. Taken together, the material in Chapters 14 and 15 represents a study of basic flows—wedge flows, cone flows, nozzle flows, and blunt-body flows—in regard to how they are affected by high temperatures. The nonequilibrium flows discussed in the present chapter stand in stark contrast to the equilibrium flows considered in Chapter 14, primarily because of the importance of the scale effect introduced by the nonequilibrium phenomena. Unlike the equilibrium flows in Chapter 14, which are the same no matter how large the body might be, we have found that the nonequilibrium flow over a given shape depends critically on the size of the body. The size effect enters through the relative consideration of chemistry time τ_c and flow time τ_f, where $\tau_f = l/V_\infty$. Here, l represents the characteristic size of the flowfield. In contrast, τ_c does not depend on the size

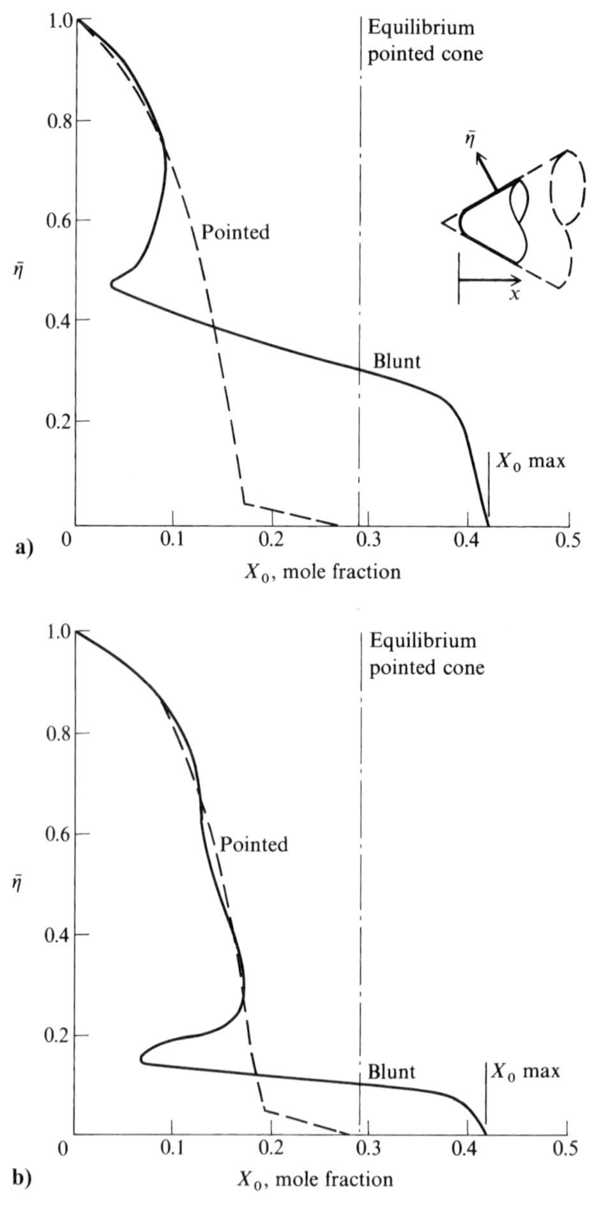

Fig. 15.28 Atomic-oxygen profiles for nonequilibrium flow over blunted and sharp cones: a) $x/R = 3$; b) $x/R_N = 6.8$. $V_\infty = 6.7$ m/s, altitude = 65.5 km, and $\theta_c = 30$ deg (from [188]).

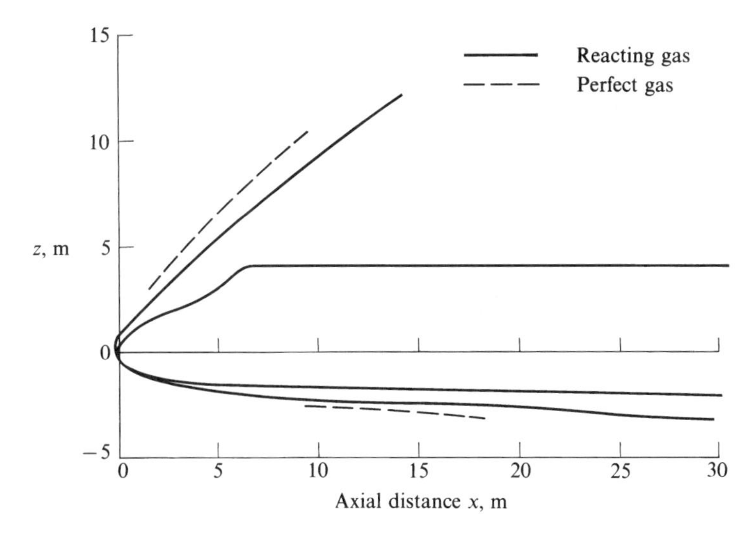

Fig. 15.29 Nonequilibrium effect on shock-wave shape for the space shuttle (from [188]).

of the body, but rather depends on the local density and temperature of the flow, for example, for the high-temperature air applications, on the flight altitude. Thus, the ratio τ_f/τ_c is strongly dependent on body size and the overall density and temperature levels of the flow. Hence, for flight applications, τ_f/τ_c depends on body size and altitude; nonequilibrium effects are accentuated by smaller bodies and higher altitudes, where τ_f/τ_c becomes small, and in contrast, equilibrium flows are approached for larger bodies and lower altitudes, where τ_f/τ_c becomes large.

Finally, we have seen that high-temperature inviscid flows, both equilibrium and nonequilibrium, stand in sharp contrast to the calorically perfect inviscid flows discussed in Part 1 of this book. High temperatures have a pronounced effect on the density and temperature profiles in a flow; to a lesser extent, the pressure is affected. Completely different methods must be used to properly calculate such high-temperature flows in comparison to the familiar calorically perfect-gas analyses. For example, virtually all high-temperature flows require some type of numerical solution; no closed-form analytic expressions are available for such flows. Also, the powerful role that M and γ play in the analysis of calorically perfect-gas flows is completely diluted at high temperatures; we can still define and identify M and γ for a high-temperature flow, but they are of no particular use in a calculation. Rather, high-temperature flows depend on more fundamental primitive variables such as velocity, density, pressure, and temperature (and for nonequilibrium flows, on the scale of the flow).

We can now appreciate the main thrust of Part 3 of this book. Our purpose is to present the basic physical chemistry background of high-temperature effects (Chapters 10–13) and to show the impact of these effects on some fundamental flow problems (Chapters 14 and 15). The nature and magnitude of the high-temperature effects are nicely brought out in the study of such basic flows, thus giving the reader important insight to applying such knowledge to more complex flows.

From here, we will turn our attention to high-temperature viscous flows in Chapters 16 and 17. Such flows introduce some additional physical considerations that are coupled with high-temperature effects.

Problems

15.1 Consider a normal shock wave in air. The flow upstream of the shock corresponds to a standard altitude of 200,000 ft at a Mach number of 25. Calculate the chemical species distributions in the nonequilibrium region behind the shock, as a function of distance behind the shock front. Plot your results graphically. Assume that the chemical species present are N_2, O_2, N, and O. (We will ignore NO and any ionization.)

15.2 In Figs. 15.22 and 15.23, the ordinate is given as number of moles of species i per original mole of air. Show how this ratio is related to our more familiar mass fraction c_i and mole-mass ratio η_i.

16
Kinetic Theory Revisited: Transport Properties in High-Temperature Gases

Victory is the beautiful, bright-colored flower. Transport is the stem without which it could never have blossomed.

Sir Winston Churchill, 1899

Chapter Preview

We are about to leave the world of inviscid flows behind and enter the world of *viscous* high-temperature flows. Viscous flows, being what they are, involve the physical processes of viscosity, thermal conduction, and mass diffusion. These are called transport properties in the gas. Add to that consideration the extra feature that the gas is at high temperature. How do we obtain the transport properties of a high-temperature, chemically reacting gas? This question has to be addressed before we can study viscous high-temperature flows, and that is the purpose of this chapter. For this, we retreat back to the world of basic physics and revisit the subject of kinetic theory. But do not worry. This is a fairly short chapter, and we will not be here very long. Our basic objective is to get to the study of viscous high-temperature *flows* as expeditiously as we can. Take this chapter seriously, and be prepared to be expedited.

16.1 Introduction

Taking a cue from the preceding quotation by Churchill referring to the role of transport in war, we can state in an analogous fashion that the physical processes of transport phenomena are the stems on which all viscous flows depends. By transport phenomena, we refer to the physical properties of viscosity, thermal conduction, and diffusion. These transport phenomena, particularly the first two, are important to the nonreacting viscous flows discussed in Part 2. All three are important in high-temperature, chemically reacting viscous flows and hence must be examined before we can discuss such flows. Therefore, the purpose of this chapter is to present the salient aspects of transport phenomena in high-temperature gases. In particular, our focus will be on the calculation of the viscosity coefficient, thermal conductivity, and diffusion coefficient for a chemically reacting mixture. A proper study of such matters constitutes an important part of classical kinetic theory. It is not our purpose to present the details here—such matters are far beyond the scope of the present book. Rather, we will briefly discuss the general philosophy and give results without detailed derivations; our purpose is to give the reader only enough understanding of the basic physical aspects to make him or her comfortable with our discussions of high-temperature viscous flows in Chapter 17. For authoritative presentations on transport phenomena in general, see [189–191].

Refer again to our road map in Fig. 1.24. With this chapter, we begin our leap into the subject of chemically reacting viscous flows, which plunges us deep into the heart of high-temperature flows in general.

16.2 Definition of Transport Phenomena

In this section, we will define the viscosity coefficient, thermal conductivity, and diffusion coefficient and show how simple equations for these transport phenomena can be obtained from the elementary kinetic theory introduced in Chapter 12. The essence of molecular transport phenomena in a gas is the *random* motion of atoms and molecules. When a particle (atom or molecule) moves from one location to another in space, it carries with it a certain momentum, energy, and mass associated with the particle itself. The transport of this particle momentum, energy, and mass through the gas as a result of the random particle motion gives rise to the transport phenomena of viscosity, thermal conduction, and diffusion, respectively.

To examine the transport of particle momentum, energy, and mass more closely, consider Fig. 16.1. At the left is sketched a gas in a two-dimensional (x, y) space, showing two particles crossing the horizontal line $y = y_1$ because of their random motion. Let ϕ denote some mean property carried by the particle, say, its momentum, energy, or mass-related property; moreover, assume that on the average ϕ has a variation in the y direction as shown at the right of Fig. 16.1. In Fig. 16.1, particle 1 crosses the line $y = y_1$ from above; let Δy be the distance above y_1 at which, on the average, particle 1 experienced its last collision before crossing y_1. Similarly, particle 2 crosses the line y_1 from below; let Δy be the average distance below y_1 at which particle 2 experienced its last collision before crossing y_1. In crossing y_1, particle 1 will carry with it a mean value of ϕ equal to $\phi(y + \Delta y)$, and particle 2 will carry a mean value equal to $\phi(y - \Delta y)$.

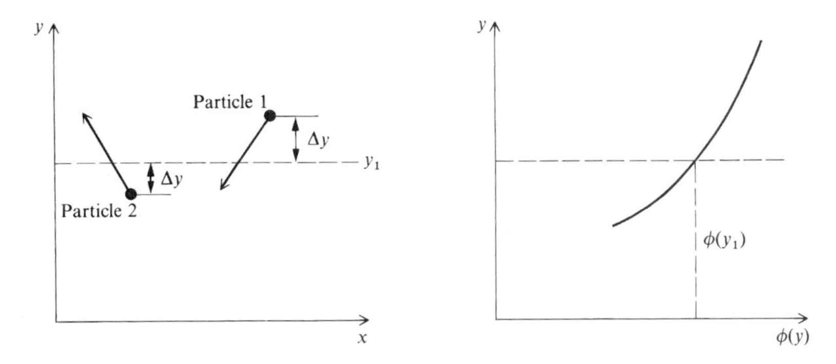

Fig. 16.1 Model for transport phenomena.

From our discussion in Chapter 12, a reasonable estimate for Δy would be the mean free path λ. Moreover, the flux of particles crossing y_1 from either above or below (particles per second per unit area) is proportional to $n\bar{C}$, where n is the number density and $\bar{C}$ is the mean particle speed given by Eq. (12.17). In turn, the flux of ϕ across y_1 caused by both directions is

$$\Lambda = an\bar{C}[\phi(y_1 - \lambda) - \phi(y_1 + \lambda)] \tag{16.1}$$

where the flux Λ is positive in the net upward direction and a is a proportionality constant. Expanding ϕ in a Taylor's series about $y = y_1$, we have

$$\phi(y_1 + \lambda) = \phi(y_1) + \frac{d\phi}{dy}\lambda + \frac{d^2\phi}{dy^2}\frac{\lambda^2}{2} + \cdots$$

and

$$\phi(y_1 - \lambda) = \phi(y_1) - \frac{d\phi}{dy}\lambda + \frac{d^2\phi}{dy^2}\frac{\lambda^2}{2} - \cdots$$

Substituting the preceding into Eq. (16.1) and neglecting terms of λ^2 and higher order, we have

$$\Lambda = -2an\bar{C}\lambda\frac{d\phi}{dy} \tag{16.2}$$

We can visualize Eq. (16.2) as a general transport equation for ϕ.

Let us now consider particular quantities for ϕ. First, let ϕ be the mean momentum of the particles. Because momentum is a vector quantity, let us examine only the x component of momentum, give by $m\bar{C}_x$, where m is the mass of the particle and $\bar{C}_x$ is the mean velocity in the x direction. Then, from Eq. (16.2) we have

$$\Lambda = -2an\bar{C}\lambda m\frac{d\bar{C}_x}{dy} \tag{16.3}$$

Referring to our knowledge of Newtonian mechanics, in a flow the flux in the y direction of the x component of momentum is simply the shear stress τ_{xy} (for example, see [5]). Thus, Eq. (16.3) becomes, with $\tau_{xy} = -\Lambda$,

$$\tau_{xy} = 2an\bar{C}\lambda m \frac{\mathrm{d}\bar{C}_x}{\mathrm{d}y} \tag{16.4}$$

Moreover, from our macroscopic fluid mechanics, we can write (for the picture given in Fig. 16.1 with gradients only in the y direction)

$$\tau_{xy} = \mu \frac{\partial u}{\partial y} = \mu \frac{\partial \bar{C}_x}{\partial y} \tag{16.5}$$

where u is the x component of the flow velocity and μ is the *viscosity coefficient*. In our kinetic theory picture, $\bar{C}_x = u$. Comparing Eqs. (16.4) and (16.5), we have

$$\boxed{\mu = 2anm\bar{C}\lambda} \tag{16.5a}$$

Now consider ϕ to be the mean energy of the particle, given by Eq. (12.9a) as $\frac{3}{2}k_1T$, where k_1 is the Boltzmann constant. (The reason for denoting the Boltzmann constant by k_1 here, as opposed to its normal symbol k, will become obvious next.) The flux of energy across y_1 is then obtained from Eq. (16.2) as

$$\Lambda = -3ak_1n\bar{C}\lambda \frac{\mathrm{d}T}{\mathrm{d}y}$$

or, denoting the constant $3ak_1$ by K,

$$\Lambda = -Kn\bar{C}\lambda \frac{\mathrm{d}T}{\mathrm{d}y} \tag{16.6}$$

From classical heat transfer, we know that the energy flux (energy per second per unit area) is given by

$$\dot{q} = -k \frac{\partial T}{\partial y} \tag{16.7}$$

where k is the *thermal conductivity*. Because, in our kinetic theory picture, Λ in Eq. (16.6) is $\dot{q}$, then, by comparing Eqs. (16.6) and (16.7), we have

$$\boxed{k = Kn\bar{C}\lambda} \tag{16.8}$$

Finally, let us consider the transport of molecular mass. Here, we will consider a binary gas mixture made up of A and B particles, with number densities n_A and n_B, respectively. In Eq. (16.2), let Λ be the flux of A particles across y_1, namely, the number of A particles crossing y_1 per second per unit area. In Eq. (16.2), n is the total number density, $n = n_A + n_B$. Therefore, the quantity ϕ being transported across y_1 must be a *probability* that the particle crossing y_1 is indeed an A particle.

This probability is the mole fraction X_A; thus, $\phi = X_A = n_A/n$. For this case, Eq. (16.2) is written as

$$\Lambda = -2an\bar{C}\lambda \frac{d(n_A/n)}{dy} = 2a\bar{C}\lambda \frac{dn_A}{dy} \tag{16.9}$$

On a macroscopic basis, we can define the flux of A particles per second per unit area as Γ_A and express it as

$$\Gamma_A = -D_{AB} \frac{dn_A}{dy} \tag{16.10}$$

where D_{AB} is the *binary diffusion coefficient* for species A into B. Comparing Eqs. (16.9) and (16.10), where $\Lambda = \Gamma_A$, we have

$$\boxed{D_{AB} = 2a\bar{C}\lambda} \tag{16.11}$$

The preceding results show that the transport coefficients μ, k, and D_{AB} depend on $\overline{C}$ and λ. Moreover, from Eqs. (12.21) and (12.29), we have

$$\lambda = -\frac{1}{\sqrt{2}\pi d^2 n} = \frac{1}{\sqrt{2}\sigma n}$$

and

$$\bar{C} = \sqrt{\frac{8RT}{\pi}}$$

where σ is the *collision cross section*. Thus, Eqs. (16.5), (16.8), and (16.11) can be written as

$$\mu = K_\mu \frac{\sqrt{T}}{\sigma} \tag{16.12}$$

$$k = K_k \frac{\sqrt{T}}{\sigma} \tag{16.13}$$

$$D_{AB} = K_D \frac{\sqrt{T}}{\sigma n} \tag{16.14}$$

where K_μ, K_k, and K_D are constants. Equations (16.12–16.14) demonstrate the important and familiar point that μ and k for a *pure gas* depend only on T, whereas D_{AB} depends on both T and n. The latter remark is worth emphasizing; the diffusion coefficient depends on *both* the temperature and the density of the gas. Using the equation of state in the form of $p = nk_1T$ (where again k_1 denotes the Boltzmann constant), Eq. (16.14) can be written as

$$\boxed{D_{AB} = K'_D \frac{\sqrt{T^3}}{p\sigma}} \tag{16.14a}$$

Equations (16.12–16.14) are simple results that come from a very elementary picture of kinetic theory. For a much more sophisticated analysis of these transport coefficients, see [189] and [190]. However, Eqs. (16.12–16.14) illustrate the important qualitative and quantitative aspects that we will need for our future discussions.

16.3 Transport Coefficients

In this section, some results for the calculation of transport coefficients will be given, without derivation. The following expressions are obtained from a more sophisticated kinetic theory treatment than was carried out in Sec. 16.2; see [189] and [190] for more details. These treatments take into account the relative motion of the molecules and replace our earlier "hard-sphere" model with a picture of particles moving under the influence of an intermolecular force field, which varies with distance r from the molecule. A common model for this force field is the Lennard–Jones (6–12) potential, which gives the intermolecular force as

$$F = -\frac{d\Phi_m}{dr}$$

where

$$\Phi_m(r) = 4\varepsilon\left[\left(\frac{d}{r}\right)^{12} - \left(\frac{d}{r}\right)^{6}\right] \tag{16.15}$$

and where d is a characteristic molecular diameter and ε is a characteristic energy of interaction between the molecules.

For a pure gas, μ and k can be obtained from (see [191])

$$\mu = 2.6693 \times 10^{-5} \frac{\sqrt{\mathcal{M}T}}{d^2\Omega_\mu} \tag{16.16}$$

and, for a monotomic gas,

$$k = 1.9891 \times 10^{-4} \frac{\sqrt{T/\mathcal{M}}}{d^2\Omega_k} \tag{16.17a}$$

whereas for a diatomic or polyatomic gas Eucken's relation that takes into account the additional energy modes of rotation, vibration, and electronic (see [150]) gives

$$k = \mu\left(\frac{5}{2}c_{v_{\text{trans}}} + c_{v_{\text{rot}}} + c_{v_{\text{vib}}} + c_{v_{\text{el}}}\right) \tag{16.17b}$$

where the units of μ are $\text{g cm}^{-1}\,\text{s}^{-1}$, T is in K, d is in Ångstrom units ($1\,\text{Å} = 10^{-8}$ cm), and k is in $\text{cal cm}^{-1}\,\text{s}^{-1}\,(\text{K})^{-1}$. In the preceding equations, $\mathcal{M}$ is the molecular weight. The quantities Ω_μ and Ω_k are collision integrals that give the variation of the effective collision diameter as a function of temperature (i.e., as a function of the relative energy between molecular collisions). Values for Ω_μ and Ω_k as a function of k_1T/ε are given in Table 16.1, obtained from [191]. Values for d and ε/k_1 (where k_1 is the Boltzmann constant) associated with the Lennard–Jones potential are tabulated in Table 16.2 for different gases. Note the similarity of Eqs. (16.16) and (16.17) to our simple results given by Eqs. (16.12) and (16.13).

For a binary mixture of species A and B, let us write an expression for the mass flux of species A (mass of A per second per unit area), denoted by $\boldsymbol{j}_A$, as

$$\boxed{\boldsymbol{j}_A = -\rho\mathcal{D}_{AB}\nabla c_A} \tag{16.18}$$

where c_A is the mass fraction of A and $\mathcal{D}_{AB}$ is the pertinent binary diffusion coefficient (sometimes called the diffusivity). Equation (16.18) is called *Fick's law*. In Eq. (16.18), $\mathcal{D}_{AB}$ can be obtained from

$$\mathcal{D}_{AB} = 0.0018583\frac{\sqrt{T^3((1/\mathcal{M}_A) + (1/\mathcal{M}_B))}}{pd_{AB}^2\Omega_{d,AB}} \tag{16.19}$$

where $\mathcal{D}_{AB}$ is in $\text{cm}^2\,\text{s}^{-1}$, T in K, p in atm, and d_{AB} in Å units. A fair approximation for d_{AB} is simply

$$d_{AB} = \tfrac{1}{2}(d_A + d_B)$$

Values for $\Omega_{d,AB}$ are given in Table 16.1 as a function of k_1T/ε_{AB}, where $\varepsilon_{AB} = \sqrt{\varepsilon_A\varepsilon_B}$. Note the similarity between Eq. (16.19) and our simple result given by Eq. (16.14a).

For a multicomponent gas, such as a chemically reacting mixture, the *mixture* values of μ and k must be found from the values of μ_i and k_i of each of the chemical species i by means *mixture rules*. A common rule for viscosity is *Wilke's rule*, which states that

$$\mu = \sum_i \frac{X_i\mu_i}{\sum_j X_j\phi_{ij}} \tag{16.20}$$

where

$$\phi_{ij} = \frac{1}{\sqrt{8}}\left(1 + \frac{\mathcal{M}_i}{\mathcal{M}_j}\right)^{-1/2}\left[1 + \left(\frac{\mu_i}{\mu_j}\right)^{1/2}\left(\frac{\mathcal{M}_j}{\mathcal{M}_i}\right)^{1/4}\right]^2$$

Table 16.1 Functions for prediction of transport properties of gases at low densities

k_1T/ε or k_1T/ε_{AB}	$\Omega_\mu = \Omega_k$ (for viscosity and thermal conductivity)	$\Omega_{\mathscr{D},AB}$ (for mass diffusivity)	$k_1\ T/\varepsilon$ or $k_1\ T/\varepsilon_{AB}$	$\Omega_\mu = \Omega_k$ (for viscosity and thermal conductivity)	$\Omega_{D,AB}$ (for mass diffusivity)
——	——	——	2.50	1.093	0.9996
0.30	2.785	2.662	2.60	1.081	0.9878
0.35	2.628	2.476	2.70	1.069	0.9770
0.40	2.492	2.318	2.80	1.058	0.9672
0.45	2.368	2.184	2.90	1.048	0.9576
0.50	2.257	2.066	3.00	1.039	0.9490
0.55	2.156	1.966	3.10	1.030	0.9406
0.60	2.065	1.877	3.20	1.022	0.9328
0.65	1.982	1.798	3.30	1.014	0.9256
0.70	1.908	1.729	3.40	1.007	0.9186
0.75	1.841	1.667	3.50	0.9999	0.9120
0.80	1.780	1.612	3.60	0.9932	0.9058
0.85	1.725	1.562	3.70	0.9870	0.8998
0.90	1.675	1.517	3.80	0.9811	0.8942
0.95	1.629	1.476	3.90	0.9755	0.8888
1.00	1.587	1.439	4.00	0.9700	0.8836
1.05	1.549	1.406	4.10	0.9649	0.8788
1.10	1.514	1.375	4.20	0.9600	0.8740
1.15	1.482	1.346	4.30	0.9553	0.8694
1.20	1.452	1.320	4.40	0.9507	0.8652
1.25	1.424	1.296	4.50	0.9464	0.8610
1.30	1.399	1.273	4.60	0.9422	0.8568
1.35	1.375	1.253	4.70	0.9382	0.8530
1.40	1.353	1.233	4.80	0.9343	0.8492
1.45	1.333	1.215	4.90	0.9305	0.8456
1.50	1.314	1.198	5.0	0.9269	0.8422
1.55	1.296	1.182	6.0	0.8963	0.8124
1.60	1.279	1.167	7.0	0.8727	0.7896
1.65	1.264	1.153	8.0	0.8538	0.7712
1.70	1.248	1.140	9.0	0.8379	0.7556
1.75	1.234	1.128	10.0	0.8242	0.7424
1.80	1.221	1.116	20.0	0.7432	0.6640
1.85	1.209	1.105	30.0	0.7005	0.6232
1.90	1.197	1.094	40.0	0.6718	0.5960
1.95	1.186	1.084	50.0	0.6504	0.5756
2.00	1.175	1.075	60.0	0.6335	0.5596
2.10	1.156	1.057	70.0	0.6194	0.5464
2.20	1.138	1.041	80.0	0.6076	0.5352
2.30	1.122	1.026	90.0	0.5973	0.5256
2.40	1.107	1.012	100.0	0.5882	0.5170

Table 16.2 Lennard–Jones parameters for various gases

Substance	Molecular weight $\mathscr{M}$	Lennard–Jones parameters σ, Å	ε/k_1, K
	Light elements		
H_2	2.016	2.915	38.0
He	4.003	2.576	10.2
	Noble gases		
Ne	20.183	2.789	35.7
Ar	39.944	3.418	124
Kr	83.80	3.498	225
Xe	131.3	4.055	229
	Simple polyatomic substances		
Air	28.97	3.617	97.0
N_2	28.02	3.681	91.5
O_2	32.00	3.433	113
O_3	48.00	——	——
CO	28.01	3.590	110
CO_2	44.01	3.996	190
NO	30.01	3.470	119
N_2O	44.02	3.879	220
SO_2	64.07	4.290	252
F_2	38.00	3.653	112
Cl_2	70.91	4.115	357
BR_2	159.83	4.268	520
I_2	252.82	4.982	550

In Eq. (16.20), μ is the viscosity coefficient for the mixture, μ_i is the viscosity coefficient for each species i from Eq. (16.16), $\mathscr{M}_i$ is the molecular weight of species i, X_i is the mole fraction of species i, and i and j are dummy subscripts denoting the various chemical species.

For the thermal conductivity of a mixture, Eq. (16.20) can be used again, replacing μ with k and μ_i with k_i, where k_i is obtained from Eq. (16.17).

For a gas with two species, the binary diffusion coefficient given by Eq. (16.19) and Fick's law given by Eq. (16.18) are sufficient to describe the diffusion processes. For a gas with more than two species, a *multicomponent diffusion coefficient* must be used, denoted by $\mathscr{D}_{\text{im}}$ for the diffusion species i through the mixture. The multicomponent diffusion coefficient $\mathscr{D}_{\text{im}}$ is related to the binary diffusion coefficients $\mathscr{D}_{ij}$ for the diffusion of species i into j by means of the approximate expression

$$\mathscr{D}_{\text{im}} = (1 - X_i) \Big/ \sum_j \frac{X_j}{\mathscr{D}_{ij}} \tag{16.21}$$

For the diffusion flux of species i, Fick's law is still reasonably applicable in the form of

$$\boldsymbol{j}_i = -\rho \mathscr{D}_{\text{im}} \nabla c_i \tag{16.22}$$

where $\boldsymbol{j}_i$ is the mass flux of species i diffusion through the mixture. For multi-component diffusion, Eq. (16.22) is an approximation that holds reasonably well for most high-temperature gas dynamic applications, at least for those applications discussed in this book.

16.4 Mechanism of Diffusion

In this section, we examine from a macroscopic point of view the mechanism of diffusion. It is common knowledge that if you are in a room and someone in the corner opens a bottle of ammonia, after a period of time you will smell the ammonia. This is because, over a period of time, some of the ammonia molecules will work their way over to you, just by virtue of their random motion in the gas. To be a little more precise, in the immediate vicinity of the ammonia bottle, after it is opened, there is a locally high concentration of ammonia, with a resulting concentration *gradient*. Under the influence of this gradient, the ammonia molecules will gradually diffuse away from the bottle. If you would imagine the ammonia molecules colored green, you would see a "green cloud" form in the vicinity of the bottle, and this green cloud would move toward you at some mean velocity; this velocity is defined as the diffusion velocity of the ammonia.

Let us now be more precise. Consider a stationary slab of gas mixture in which there exists a gradient in mass fraction of species i; the variation of c_i is sketched in Fig. 16.2a, and the resulting gradient ∇c_i is shown at a given point in the stationary slab in Fig. 16.2b. Because of this gradient, at the same point there is a mass motion of species i in the opposite direction; the velocity of this mass motion of species i is defined as the *diffusion velocity* of species i, denoted by $\boldsymbol{U}_i$. The corresponding mass flux of species i is $\rho_i \boldsymbol{U}_i$, which is shown in Fig. 16.2b and is given approximately by Fick's law [Eq. (16.22)] as

$$\boldsymbol{j}_i \equiv \rho_i \boldsymbol{U}_i = -\rho \mathscr{D}_{\text{im}} \nabla c_i \tag{16.23}$$

Let us now imagine that the slab in Fig. 16.2b is set into motion with the velocity $\boldsymbol{V}$, as sketched in Fig. 16.2c. The mass motion of species i, relative to us standing in the laboratory, is now $\boldsymbol{V}_i$, where

$$\underbrace{\boldsymbol{V}_i}_{\substack{\text{Mass motion of species} \\ i \text{ relative to the lab.} \\ \text{or just simply the } \textit{mass} \\ \textit{motion of species } i}} = \underbrace{\boldsymbol{V}}_{\substack{\text{Mass motion of} \\ \text{the } \textit{mixture} \\ \text{(relative to the lab)}}} + \underbrace{\boldsymbol{U}_i}_{\substack{\textit{Diffusion velocity} \\ \text{of species } i \\ \text{(relative to the} \\ \text{mass motion of} \\ \text{the mixture)}}} \tag{16.24}$$

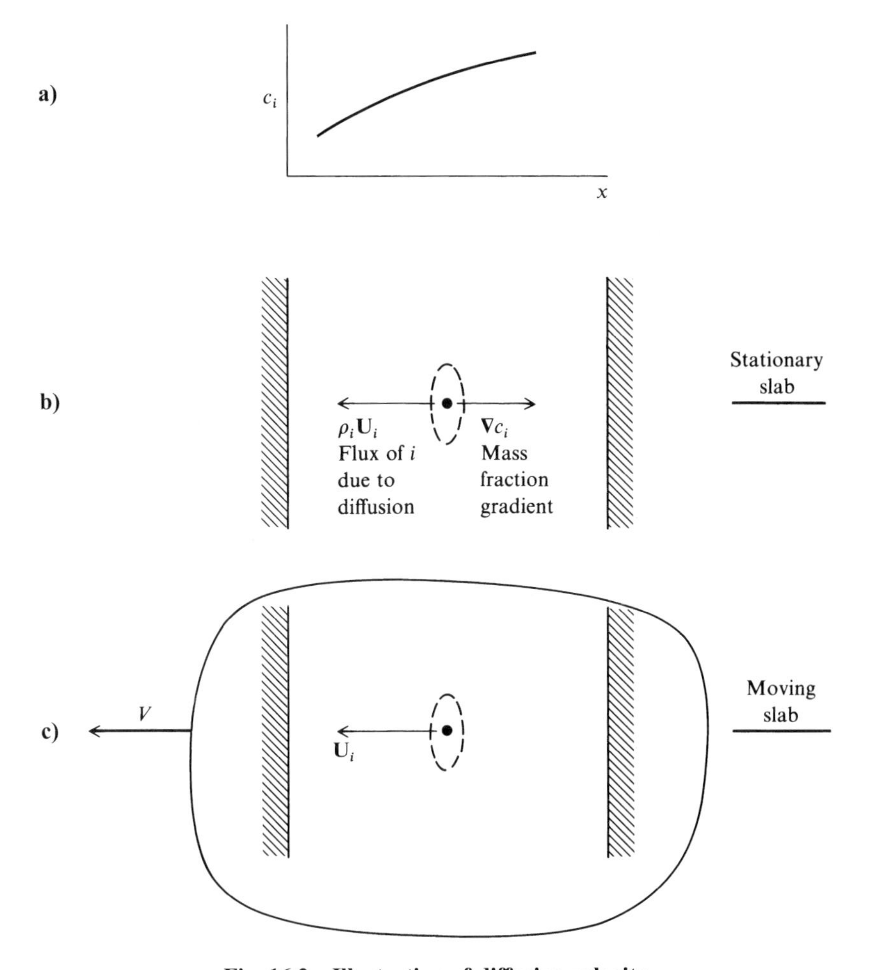

Fig. 16.2 Illustration of diffusion velocity.

In conjunction with a gas dynamic flow, $\boldsymbol{V}$ in Eq. (16.24) is the familiar flow velocity at a point in the flowfield; for a gas mixture, the flow velocity is in reality a mass average of all of the $\boldsymbol{V}_i$, that is,

$$\boldsymbol{V} = \sum_i c_i \boldsymbol{V}_i \tag{16.25}$$

Equation (16.24) is simply a statement that the mass motion of species i is equal to the flow velocity of the mixture plus the diffusion velocity $\boldsymbol{U}_i$, where $\boldsymbol{U}_i$ is *relative* to the mass motion of the mixture. If we multiply Eq. (16.24) by c_i and sum over all species, we have

$$\sum_i c_i \boldsymbol{V}_i = \boldsymbol{V} \sum_i c_i + \sum_i c_i \boldsymbol{U}_i \tag{16.26}$$

Recalling that $\sum_i c_i = 1$ and substituting Eq. (16.25) for $\boldsymbol{V}$ into Eq. (16.26), we obtain the important result that

$$\sum_i c_i \boldsymbol{U}_i = \frac{1}{\rho} \sum_i \rho_i \boldsymbol{U}_i = 0$$

or

$$\boxed{\sum_i \rho_i \boldsymbol{U}_i = 0} \tag{16.27}$$

As a final note in this section, we have to mention that the mass diffusion of species i through a mixture is also driven by pressure and temperature gradients, as well as by concentration gradients. Pressure diffusion caused by ∇p is extremely slight and is almost always neglected in gas dynamic problems. Thermal diffusion caused by ∇T is a more pronounced effect in regions of large temperature gradients. Thermal diffusion is a reflection that, at a given temperature, light particles have a higher mean molecular velocity than heavy particles, as discussed at the end of Sec. 12.2. Therefore, in a temperature gradient the light particles will tend to diffuse faster in the direction of decreasing temperature than will the heavy particles. (This is why dust in a heated room tends, over a period of time, to collect on the surface of radiators.) The mass flux of species i caused by thermal diffusion is given by (see [191])

$$\boldsymbol{j}_i^{(T)} = -D_i^T \nabla \ell n T \tag{16.28}$$

where D_i^T is the thermal diffusion coefficient. For most gas dynamic applications, thermal diffusion is small when compared with diffusion caused by concentration gradients and is usually neglected. (See [189–191] for more details on pressure and thermal diffusion.)

16.5 Energy Transport by Thermal Conduction and Diffusion: Total Thermal Conductivity

In the viscous flows discussed in Part 2, we used the fact that energy is transported by thermal conduction. Moreover, the flux of this energy (energy per second per unit area) is given by

$$\boldsymbol{q}_c = -k \nabla T \tag{16.29}$$

where k is the ordinary thermal conductivity as given by Eq. (16.17) for a pure species and Eq. (16.20) (with μ and μ_i replaced by k and k_i, respectively) for the mixture thermal conductivity.

For a chemically reacting mixture, there is also an energy transport caused by diffusion. This is easily seen by visualizing a chemical species i diffusing from location 1 to location 2, where at location 2 the species participates in a chemical reaction, thus exchanging some energy with the gas. That is, as species i diffuses through the gas, it carries with it the enthalpy of species i, h_i, which is a form of energy transport. (Keep in mind that h_i contains the heat of formation of species i.) Hence, at a point in the gas, we can write

$$\left\{\begin{array}{c}\text{Energy flux caused by} \\ \text{diffusion of species } i\end{array}\right\} = \rho_i \boldsymbol{U}_i h_i$$

In turn,

$$\left\{\begin{array}{c}\text{Energy flux caused by diffusion} \\ \text{of } \textit{all} \text{ species at the point}\end{array}\right\} = \boldsymbol{q}_D = \sum_i \rho_i \boldsymbol{U}_i h_i \tag{16.30}$$

Therefore, if we include the energy flux caused by radiation (which will be important for the applications discussed in Chapter 18), denote by $\boldsymbol{q}_R$, we can write for the total energy flux at a point in a high-temperature, chemically reacting gas

$$\boldsymbol{q} = \boldsymbol{q}_c + \boldsymbol{q}_D + \boldsymbol{q}_R$$

or

$$\boxed{\boldsymbol{q} = -k\nabla T + \sum_i \rho_i \boldsymbol{U}_i h_i + \boldsymbol{q}_R} \tag{16.31}$$

[Note that in Eq. (16.31) we are not including energy flux caused by convection in a flow; here, we are considering a stationary gas that has temperature and concentration gradients, and $\boldsymbol{q}$ is simply the energy transport at some point in the stationary gas as a result of transport phenomena and radiation.]

In some high-temperature flow applications, the concept of total thermal conductivity is used, as defined in [167] and [193], among others. This concept is developed as follows. Consider a flowfield with gradients of temperature and mass fractions in the y direction (such as in a boundary layer). The energy flux in the y direction (neglecting radiation) is obtained from Eq. (16.31) as

$$q_y = -k\frac{\partial T}{\partial y} + \sum_i \rho_i U_{i,y} h_i \tag{16.32}$$

where $U_{i,y}$ is the component of the diffusion velocity of species i in the y direction. From Eq. (16.22),

$$\rho_i U_{i,y} = -\rho \mathscr{D}_{\text{im}} \frac{\partial c_i}{\partial y} \tag{16.33}$$

Combining Eq. (16.32) and (16.33), we have

$$q_y = -k\frac{\partial T}{\partial y} - \rho \sum_i \mathscr{D}_{\text{im}} h_i \frac{\partial c_i}{\partial y} \tag{16.34}$$

Assume that the gas is in the local chemical equilibrium, such that $c_i = f(p, T)$, and hence

$$dc_i = \left(\frac{\partial c_i}{\partial T}\right)_p dT + \left(\frac{\partial c_i}{\partial p}\right)_T dp \tag{16.35}$$

Furthermore, assume that p is constant in the y direction (such as through a boundary layer). With this, Eq. (16.35) becomes

$$\frac{\partial c_i}{\partial y} = \frac{\partial c_i}{\partial T}\frac{\partial T}{\partial y} \tag{16.36}$$

Substituting Eq. (16.36) into (16.34), we have

$$q_y = -k\frac{\partial T}{\partial y} - \rho\left(\sum_i \mathscr{D}_{\text{im}}\, h_i \frac{\partial c_i}{\partial T}\right)\frac{\partial T}{\partial y}$$

or

$$q_y = -k\frac{\partial T}{\partial y} - k_r\frac{\partial T}{\partial y} = -k_T\frac{\partial T}{\partial y} \tag{16.37}$$

where k is the ordinary, familiar thermal conductivity, k_r is called the reaction conductivity (which is caused solely by diffusion) given by

$$k_r = \rho\sum_i \mathscr{D}_{\text{im}} h_i \frac{\partial c_i}{\partial T}$$

and k_T is the total conductivity, defined as

$$k_T = k + \rho\sum_i \mathscr{D}_{\text{im}} h_i \frac{\partial c_i}{\partial T} \tag{16.38}$$

Sometimes k_T is used to define an "equilibrium" Prandtl number as follows. For an equilibrium chemically reacting gas, $h = h(T, p)$, and we can write

$$dh = \left(\frac{\partial h}{\partial T}\right)_p dT + \left(\frac{\partial h}{\partial p}\right)_T dp \tag{16.39}$$

Assuming constant pressure in the y direction and noting that $(\partial h/\partial T)_p \equiv c_p$, Eq. (16.39) yields

$$\frac{\partial T}{\partial y} = \frac{1}{c_p}\frac{\partial h}{\partial y} \tag{16.40}$$

From Eqs. (16.37) and (16.40), we have

$$q_y = k_T\frac{\partial T}{\partial y} = \frac{k_T}{c_p}\frac{\partial h}{\partial y} = \frac{\mu}{Pr_{\text{eq}}}\frac{\partial h}{\partial y} \tag{16.41}$$

where Pr_{eq} is called the equilibrium Prandtl number, defined as

$$Pr_{eq} = \frac{\mu c_p}{k_T} \tag{16.42}$$

Note: The concepts of total conductivity and equilibrium Prandtl number have applications only in flows that 1) are in local chemical equilibrium and 2) the energy flux is being calculated in a direction in which the pressure is constant. Although this concept might sound very restrictive, there is a relatively large class of chemically reacting boundary-layer applications to which it has been applied. Indeed, for high-temperature air, values of k_T are Pr_{eq} have been calculated and tabulated by Hansen in [167].

16.6 Transport Properties for High-Temperature Air

High-temperature transport coefficients for air in chemical equilibrium have been calculated by Hansen [167] and Peng and Pindroh [193], among others. For example, Fig. 16.3 gives the variation of μ as a function of T with p as a parameter, obtained from [167]. In this figure, the reference value of μ_0 is given by

$$\mu_0 = 1.462 \times 10^{-5} \frac{T^{1/2}}{1 + 112/T} \frac{\text{gm}}{\text{cm s}}$$

where μ_0 is an approximate temperature variation of viscosity coefficient for nonreacting air, with a chemical composition frozen at standard conditions. Hence, the amount by which μ/μ_0 deviates away from unity in Fig. 16.3 is a reflection of the high-temperature, chemically reacting effect. The strong variation in μ/μ_0, which occurs above a temperature of 8000 K, is because of the important effects of ionization and hence free electrons on the transport properties. The total conductivity defined as k_T in Eq. (16.38) is given in Fig. 16.4, also from Hansen. Here, the reference value is

$$k_0 = 1.364\mu_0 \frac{\text{J}}{\text{(cm)(s)(K)}}$$

The large variations in k_T/k_0 are caused primarily by the reaction conductivity associated with diffusion; this shows the relatively large effect that diffusion can play in energy transport through chemically reacting gases. In Fig. 16.5, the equilibrium Prandtl number, defined by Eq. (16.42), is given as calculated by Hansen. Note that Pr_{eq} has a more benign variation than k_T because the temperature variations of k_T in Fig. 16.3 and c_p (similar to that shown for c_p in Fig. 14.16) tend to cancel each other. Note that, in the range of dissociation, Pr_{eq} varies between 0.6 and 0.8.

In passing, we note a fourth transport property of some importance in high-temperature gases, namely, the electrical conductivity. This property is important to the analysis of flow problems in the presence of electromagnetic fields. Such matters fall under the category of magnetohydrodynamics and are beyond the scope of this book.

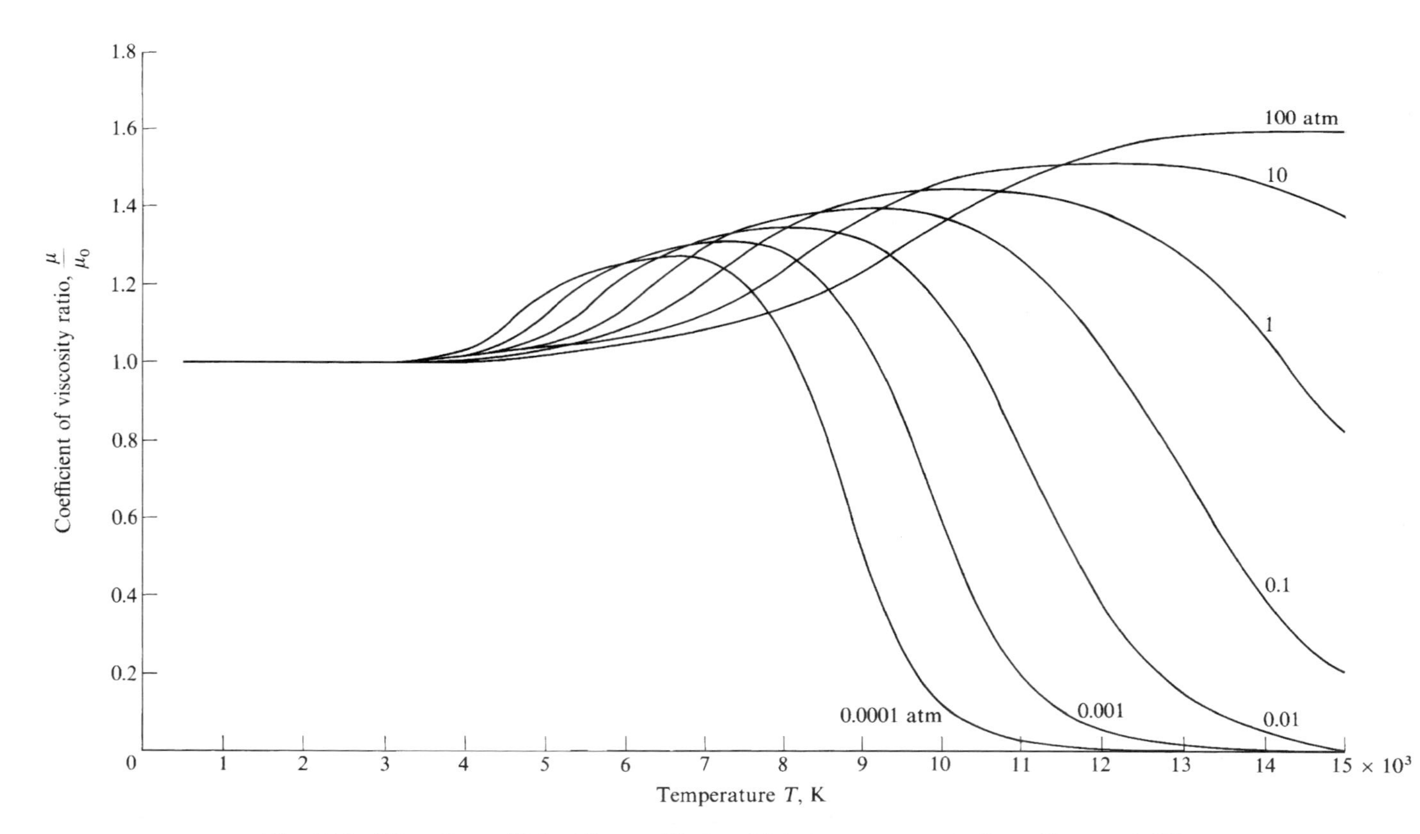

Fig. 16.3 Viscosity coefficient for equilibrium high-temperature air (from Hansen [167]).

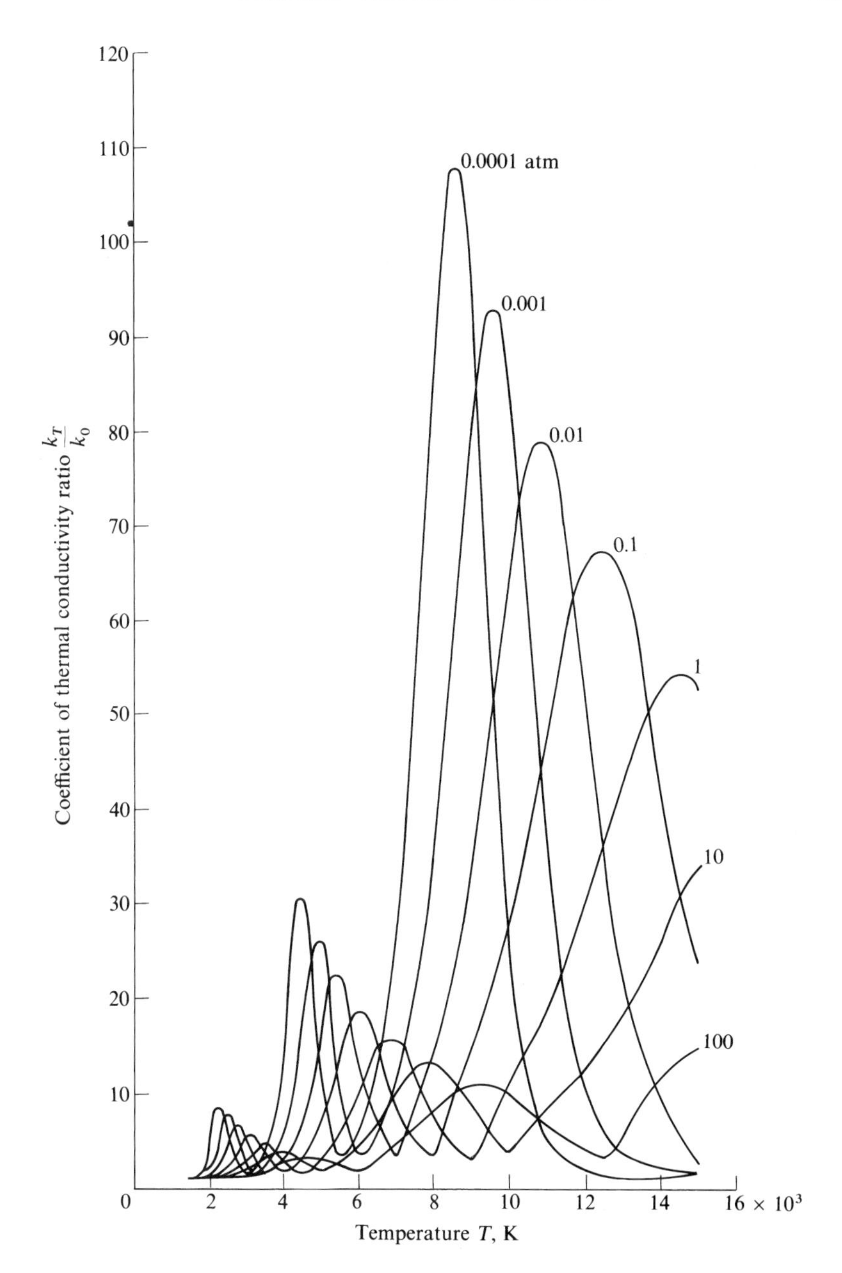

Fig. 16.4 Total thermal conductivity for equilibrium high-temperature air (from Hansen [167]).

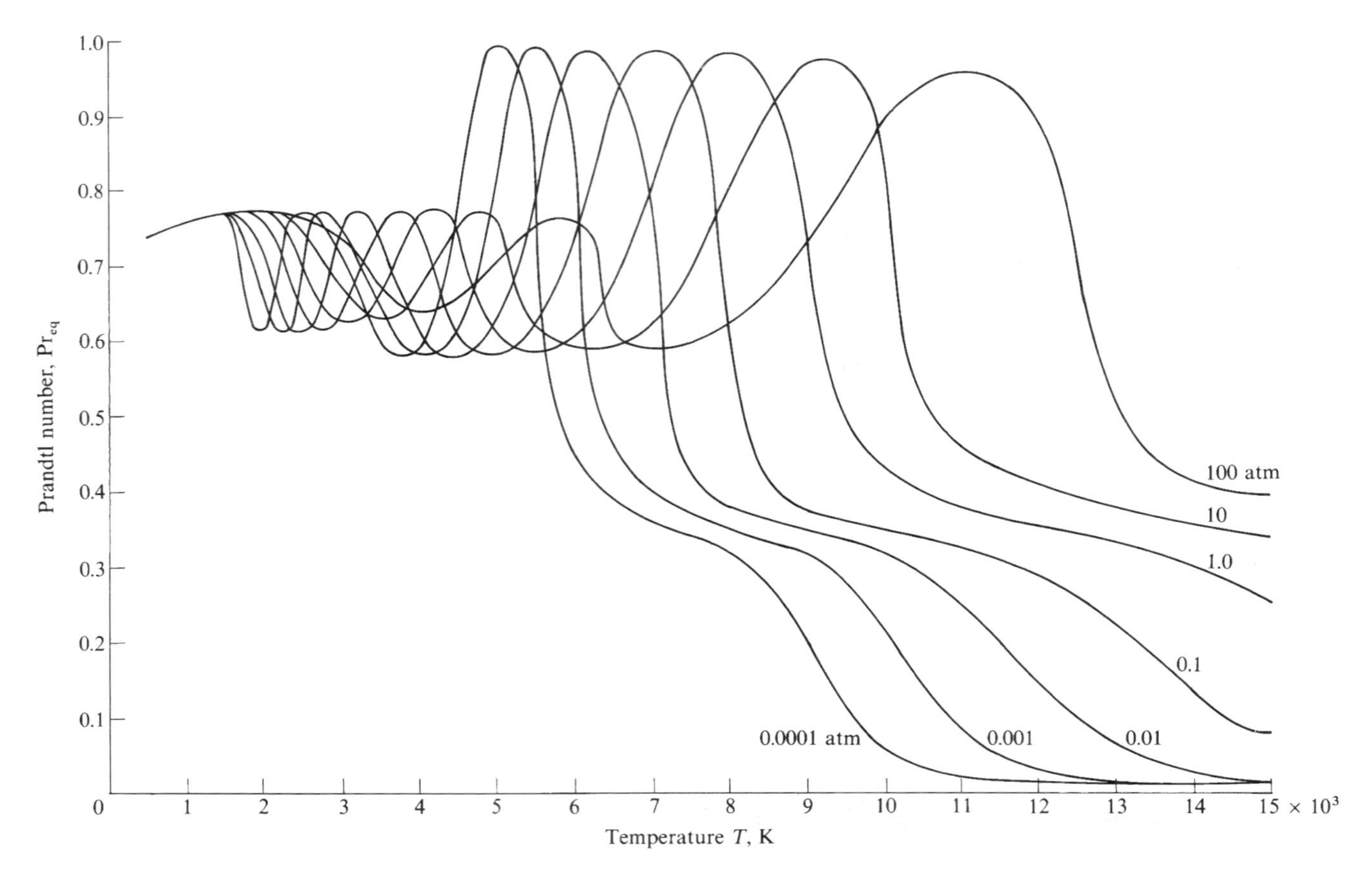

Fig. 16.5 Equilibrium Prandtl number for high-temperature air (from [167]).

Finally, simple correlations for the transport properties of high-temperature air, analogous to the Tannehill–Mugge correlations (for example) for the equilibrium thermodynamic properties, are hard to find. Perhaps one of the best examples are the correlations of Viegas and Howe [194], where polynomials for $\rho\mu$ and $\rho\mu/Pr_{eq}$ are given as a function of p and h.

16.7 Summary and Comments

The purpose of this chapter is to make the reader feel somewhat comfortable with the transport coefficients for a high-temperature, chemically reacting gas. In particular, from simple kinetic theory, we have seen that μ and k for a pure gas depend only on T, whereas the binary diffusion coefficient $\mathscr{D}_{AB}$ and, hence, the multicomponent diffusion coefficients D_{im} depend on both T and ρ (or T and p). Equations for the transport coefficients of a pure gas are given in terms of the collision diameter and the collision integrals, and various mixture rules are given to obtain the transport coefficients for a mixture from the individual values for each species i. Some results for high-temperature are discussed.

Caution: Published values for the transport coefficients for high-temperature gases are much more uncertain than for the equilibrium thermodynamic properties discussed in Chapter 12. Theoretical values of the transport coefficients depend critically on the assumption used for the intermolecular force potential, which is always uncertain. Experimental measurements at high temperatures are also uncertain and difficult to make. Therefore, when you are ready to perform a serious analysis of high-temperature viscous flows, it is good practice to scour the current literature for the most accurate data on transport coefficients.

17
Viscous High-Temperature Flows

Before engineers can reliably design devices to survive flight through or into an atmosphere at hypersonic speeds, they must somehow provide for, avoid, or otherwise accommodate the enormous heat-transfer rates to the vehicle engendered by such flight speeds.

William H. Dorrance, 1962

Chapter Preview

Reflecting on the old phrase "everything but the kitchen sink," in this chapter we have everything *including* the kitchen sink. Here, the kitchen sink is the inclusion of the viscous effects of viscosity, thermal diffusion, and mass diffusion in chemically reacting flows. *Question*: How do we calculate the heat transfer to a surface from a chemically reacting flowfield? This question is answered in the present chapter. Indeed, it is the major thrust of this chapter. Do you think the answer is different for nonequilibrium and nonequilibrium flows? Does the catalytic state of the surface make any difference? How do the viscous effects change the local chemical composition in a high-temperature flow? To find out, simply read on. This chapter is the pinnacle of our study of high-temperature gas dynamics. Climb to the top, and enjoy the view.

17.1 Introduction

The quotation by W. H. Dorrance on p. 711, an authoritative aerodynamicist and well-known author of a book on hypersonic viscous flow [195], clearly states one of the strongest forces that drives an interest in chemically reacting viscous flows, namely, the need to accurately predict and deal with the aerodynamic heating to hypersonic bodies. This motivation was a strong impetus behind our viscous-flow discussion in Part 2 of this book. However, by intent we did not include high-temperature effects in Part 2, because our emphasis there was the purely fluid-dynamic behavior of hypersonic viscous flows. In the present chapter, we now address the question: how are the viscous flows studied in Part 2 affected by high temperatures? We will find that many of the qualitative trends, analytical techniques, and numerical methods developed in Part 2 carry over directly to the high-temperature case. Therefore, to avoid unnecessary duplication, we will frequently refer to sections in Part 2; in such occasions, the reader is strongly encouraged to review the pertinent material in Part 2 as is necessary.

In the present chapter, our emphasis will be on the fundamental equations and the basic physical behavior of chemically reacting viscous flow. We will be "long" on basic fundamentals and "short" on details because the subject matter is so extensive that it is not possible to give a complete survey of the field within the length constraints of this book. Instead, our purpose will be to make the reader feel comfortable with the basic ideas and equations, and then we will discuss a few well-chosen examples to illustrate the physical nature of chemically reacting viscous flows. With the background provided in this chapter, the reader should be able knowledgeably to read the literature in the field and to embark on the solution of more complex problems. Also, in the present chapter we will consider simultaneously both equilibrium and nonequilibrium chemically reacting and vibrationally relaxing flows.

17.2 Governing Equations for Chemically Reacting Viscous Flow

The complete Navier–Stokes equations were given by Eqs. (6.1–6.6) for a nonreacting gas. The continuity equation (6.1) and the momentum equations (6.2–6.4) are purely mechanical in nature and are not affected by chemical reactions. Hence, for a chemically reacting flow, we have the following.

Global continuity:

$$\frac{\partial \rho}{\partial t} + \nabla \cdot (\rho \mathbf{V}) = 0 \tag{17.1}$$

x Momentum:

$$\rho \frac{\mathrm{D}u}{\mathrm{D}t} = -\frac{\partial p}{\partial x} + \frac{\partial \tau_{xx}}{\partial x} + \frac{\partial \tau_{yx}}{\partial y} + \frac{\partial \tau_{zx}}{\partial z} \tag{17.2}$$

y Momentum:

$$\rho \frac{\mathrm{D}v}{\mathrm{D}x} = -\frac{\partial p}{\partial y} + \frac{\partial \tau_{xy}}{\partial x} + \frac{\partial \tau_{yy}}{\partial y} + \frac{\partial \tau_{zy}}{\partial z} \tag{17.3}$$

z Momentum:

$$\rho \frac{Dw}{Dt} = -\frac{\partial p}{\partial z} + \frac{\partial \tau_{xz}}{\partial x} + \frac{\partial \tau_{yz}}{\partial y} + \frac{\partial \tau_{zz}}{\partial z} \tag{17.4}$$

The energy equation, given by Eq. (6.5), must be modified to include the effects of diffusion. Hence, we write the following:

Energy:

$$\rho \frac{D(e + V^2/2)}{Dt} = -\nabla \cdot \boldsymbol{q} - \nabla \cdot p\boldsymbol{V} + \frac{\partial(u\tau_{xx})}{\partial x} + \frac{\partial(u\tau_{yx})}{\partial y} + \frac{\partial(u\tau_{zx})}{\partial z} + \frac{\partial(v\tau_{xy})}{\partial x}$$
$$+ \frac{\partial(v\tau_{yy})}{\partial y} + \frac{\partial(v\tau_{zy})}{\partial z} + \frac{\partial(w\tau_{xz})}{\partial x} + \frac{\partial(w\tau_{yz})}{\partial y} + \frac{\partial(w\tau_{zz})}{\partial z} \tag{17.5}$$

where, from Eq. (16.31), the heat-flux vector is

$$\boldsymbol{q} = -k\nabla T + \sum_i \rho_i \boldsymbol{U}_i h_i + \boldsymbol{q}_R \tag{17.6}$$

In Eqs. (17.2–17.5), the expressions for the shear and normal stresses τ_{xy}, τ_{xx}, etc. are given by Eqs. (6.6a–6.6f), and μ and k are the *mixture* values for viscosity coefficient and thermal conductivity as discussed in Chapter 16. In terms of the substantial derivative, Eq. (17.5) says that the change in total energy, $e + V^2/2$, of a fluid element moving along a streamline is because of 1) thermal conduction across the surfaces of the fluid element; 2) transport of energy by diffusion into (or out of) the fluid element across its surfaces; 3) radiative energy emitted or absorbed by the element; 4) rate of work done by pressure forces exerted on the surfaces of the element; and 5) rate of work done by shear and normal stresses exerted on the surfaces.

For a viscous, chemically reacting flow in local chemical equilibrium, we add the equilibrium thermodynamic properties, given conceptually as

$$p = p(e, \rho) \tag{17.7}$$
$$T = T(e, \rho) \tag{17.8}$$

Hence, for such a flow, Eqs. (17.1–17.8) constitute the governing equations.

For chemical nonequilibrium flows, we found in Sec. 15.2 that the species continuity equations were also necessary. Such equations were derived in Sec. 15.2 for an *inviscid* flow using the model of the fixed, finite control volume shown in Fig. 15.2, leading to Eqs. (15.1), (15.2), (15.5), and (15.6).

> *Question:* How are this derivation and the resulting equations affected by *viscous* flow?

Physically, the answer is that mass transport of species i must be included in the species continuity equation. This comes about as follows. Return to Fig. 15.2, where the velocity $\boldsymbol{V}$ is shown at a point on the control surface. For an *inviscid* flow (no diffusion), *all* of the species i move with the *same* velocity, namely, the mixture velocity $\boldsymbol{V}$. In Fig. 15.2, $\boldsymbol{V}$ represents the *velocity of species i*, which is the same as the mixture velocity; this is why no subscript is placed on $\boldsymbol{V}$ in Fig. 15.2. In contrast, for a *viscous* flow, the mass motion velocities of species i, $\boldsymbol{V}_i = \boldsymbol{V} + \boldsymbol{U}_i$, all are *different*. Hence, in Fig. 15.2, which involves a control volume in which we are considering only the flow of species i, the velocity $\boldsymbol{V}$ must be replaced by $\boldsymbol{V}_i$. The derivation proceeds exactly the same as in Sec. 15.2, leading to the integral form

$$\frac{\partial}{\partial t}\iiint_{\mathscr{V}} \rho_i \,\mathrm{d}\boldsymbol{V} = -\iint_{S} \rho_i \boldsymbol{V}_i \,\mathrm{d}\mathbf{S} + \iiint_{\mathscr{V}} \dot{w}_i \,\mathrm{d}\mathscr{V} \tag{17.9}$$

Equation (17.9) is the viscous flow analog of Eq. (15.1). From this, we obtain directly

$$\frac{\partial \rho_i}{\partial t} + \nabla\cdot(\rho_i \boldsymbol{V}_i) = \dot{w}_i \tag{17.10}$$

in analogy to Eq. (15.2). Because $\boldsymbol{V}_i = \boldsymbol{V} + \boldsymbol{U}_i$, Eq. (17.10) becomes

$$\frac{\partial \rho_i}{\partial t} + \nabla\cdot[\rho_i(\boldsymbol{V} + \boldsymbol{U}_i)] = \dot{w}_i \tag{17.11}$$

Replacing ρ_i with $c_i\rho$ in Eq. (17.11), expanding the derivatives, and collecting terms, we have

$$c_i\left[\frac{\partial \rho}{\partial t} + \nabla\cdot(\rho\boldsymbol{V})\right] + \rho\frac{\partial c_i}{\partial t} + \rho\boldsymbol{V}\cdot\nabla c_i + c_i\nabla\cdot(\rho\boldsymbol{U}_i) + (\rho\boldsymbol{U}_i)\cdot\nabla c_i = \dot{w}_i \tag{17.12}$$

However, from the global continuity equation,

$$\frac{\partial \rho}{\partial t} + \nabla\cdot(\rho\boldsymbol{V}) = 0 \tag{17.13}$$

Also,

$$\rho\frac{\partial c_i}{\partial t} + \rho\boldsymbol{V}\cdot\nabla c_i \equiv \rho\frac{\mathrm{D}c_i}{\mathrm{D}t} \tag{17.14}$$

and

$$c_i\nabla\cdot(\rho\boldsymbol{U}_i) + (\rho\boldsymbol{U}_i)\cdot\nabla c_i \equiv \nabla\cdot(\rho c_i\boldsymbol{U}_i) \equiv \nabla\cdot(\rho_i\boldsymbol{U}_i) \tag{17.15}$$

Inserting Eqs. (17.13–17.15) into (17.12), we obtain

$$\boxed{\rho\frac{\mathrm{D}c_i}{\mathrm{D}t} + \nabla\cdot(\rho_i\boldsymbol{U}_i) = \dot{w}_i} \tag{17.16}$$

Equation (17.16) is a particularly useful form of the species continuity equation. In comparison to Eq. (15.5) for an inviscid flow, the viscous-flow version in Eq. (17.16) has an additional term involving the diffusion velocity. From Eq. (16.22), where $\boldsymbol{j}_i \equiv \rho_i\boldsymbol{U}_i$, Eq. (17.16) can be written as

$$\boxed{\rho\frac{\mathrm{D}c_i}{\mathrm{D}t} = \nabla\cdot(\rho\mathscr{D}_{\mathrm{im}}\nabla c_i) + \dot{w}_i} \tag{17.17}$$

In summary, for a *nonequilibrium* viscous chemically reacting flow, the governing equations are Eqs. (17.1–17.6), and either (17.16) or (17.17) written for each species in the mixture. [As noted in Sec. 15.2, if we have n chemical species, we really only need $(n-1)$ species continuity equations because we have the additional relation that $\sum_i c_i = 1$.] In addition, the usual expression for e holds, that is,

$$e = \sum_i c_i e_i \tag{17.18}$$

where

$$e_i = \frac{5}{2}RT + e_{\text{vib}_i} + e_{\text{el}} + (\Delta h_f)_i^{\text{o}} \tag{17.19}$$

If local thermodynamic equilibrium is assumed, then in Eq. (17.19)

$$e_{\text{vib}_i} = \frac{h\nu_i/kT}{e^{h\nu_i/kT} - 1}RT$$

for diatomic molecules. If vibrational nonequilibrium is present, then e_{vib_i} in Eq. (17.19) is given by

$$\frac{\mathrm{D}(c_i e_{\text{vib}_i})}{\mathrm{D}t} + \frac{1}{\rho}\nabla \cdot (\rho_i \boldsymbol{U}_i e_{\text{vib}_i}) = \frac{c_i}{\tau}(e_{\text{vib}_i}^{\text{eq}} - e_{\text{vib}_i}) \tag{17.20}$$

In comparison to the inviscid analog given in Eq. (15.18), the influence of diffusion is included in Eq. (17.20) for the vibrational rate equation in a viscous flow. This is because, in a fluid element moving along a streamline, the vibrational energy caused by species i per unit mass of mixture $c_i e_{\text{vib}_i}$, changes not only as a result of finite-rate vibrational relaxation, but also because species i diffuses into (or out of) the fluid element across the surface, carrying some vibrational energy with it.

In summary, the governing equations for a chemically reacting viscous flow are similar to those used in Part 2 for a nonreacting viscous flow. The only differences are 1) the use of the species continuity equation (which is needed for any chemical nonequilibrium flow, viscous or inviscid); 2) the inclusion of diffusion effects, which appear as terms in the species continuity equation and the energy equation; and 3) the inclusion of the heats of formation at absolute zero in the enthalpy (or internal energy).

17.3 Alternate Forms of the Energy Equation

In this section, we obtain several different forms of the energy equation for a chemically reacting viscous flow—all of which are frequently found in the literature and used in practice.

Equation (17.5) is written in terms of the total energy, $e + V^2/2$. Let us express it in terms of total enthalpy, $h_0 = h + V^2/2$, as follows. From the definition of enthalpy, $h = e + p/\rho$. Taking the substantial derivative, which follows the ordinary rules of differentiation, we have

$$\frac{\mathrm{D}e}{\mathrm{D}t} = \frac{\mathrm{D}h}{\mathrm{D}t} - \frac{1}{\rho}\frac{\mathrm{D}p}{\mathrm{D}t} + \frac{p}{\rho^2}\frac{\mathrm{D}\rho}{\mathrm{D}t} \tag{17.21}$$

From Eq. (17.1),

$$\frac{\partial \rho}{\partial t} + \nabla \cdot (\rho \mathbf{V}) = \frac{\partial \rho}{\partial t} + \mathbf{V} \cdot \nabla \rho + \rho \nabla \cdot \mathbf{V} = 0$$

or

$$\frac{\mathrm{D}\rho}{\mathrm{D}t} + \rho \nabla \cdot \boldsymbol{V} = 0 \tag{17.22}$$

Combining Eqs. (17.21) and (17.22), we have

$$\frac{\mathrm{D}e}{\mathrm{D}t} = \frac{\mathrm{D}h}{\mathrm{D}t} - \frac{1}{\rho}\frac{\mathrm{D}p}{\mathrm{D}t} - \frac{p}{\rho}\nabla \cdot \boldsymbol{V}$$

or

$$\rho\frac{\mathrm{D}e}{\mathrm{D}t} = \rho\frac{\mathrm{D}h}{\mathrm{D}t} - \frac{\partial p}{\partial t} - \boldsymbol{V} \cdot \nabla \rho - p\nabla \cdot \boldsymbol{V}$$

or

$$\rho\frac{\mathrm{D}e}{\mathrm{D}t} = \rho\frac{\mathrm{D}h}{\mathrm{D}t} - \frac{\partial p}{\partial t} - \nabla \cdot (p\boldsymbol{V}) \tag{17.23}$$

Now write the energy equation (17.5) in the form

$$\rho\frac{\mathrm{D}e}{\mathrm{D}t} = \rho\frac{\mathrm{D}(V^2/2)}{\mathrm{D}t} - \nabla \cdot \boldsymbol{q} - \nabla \cdot (p\boldsymbol{V}) + (\text{viscous terms}) \tag{17.24}$$

Substituting Eq. (17.23) into (17.24), we obtain, setting $h_0 = h + V^2/2$,

$$\boxed{\begin{aligned}\rho\frac{\mathrm{D}h_0}{\mathrm{D}t} = \frac{\partial p}{\partial t} - \nabla \cdot \boldsymbol{q} + \frac{\partial(u\tau_{xx})}{\partial x} + \frac{\partial(u\tau_{yx})}{\partial y} + \frac{\partial(u\tau_{zx})}{\partial z} + \frac{\partial(v\tau_{xy})}{\partial x} \\ + \frac{\partial(v\tau_{yy})}{\partial y} + \frac{\partial(v\tau_{zy})}{\partial z} + \frac{\partial(w\tau_{xz})}{\partial x} + \frac{\partial(w\tau_{yz})}{\partial y} + \frac{\partial(w\tau_{zz})}{\partial z}\end{aligned}} \tag{17.25}$$

Equation (17.25) is the form of the energy equation in terms of total enthalpy h_0.

Let us now obtain a form of the energy equation in terms of static enthalpy h. Multiply Eqs. (17.2), (17.3), and (17.4) by u, v, and w, respectively:

$$\rho\frac{\mathrm{D}(u^2/2)}{\mathrm{D}t} = -u\frac{\partial p}{\partial x} + u\frac{\partial \tau_{xx}}{\partial x} + u\frac{\partial \tau_{yx}}{\partial y} + \frac{\partial \tau_{zx}}{\partial z} \tag{17.26}$$

$$\rho\frac{\mathrm{D}(v^2/2)}{\mathrm{D}t} = -v\frac{\partial p}{\partial y} + v\frac{\partial \tau_{xy}}{\partial x} + v\frac{\partial \tau_{yy}}{\partial y} + v\frac{\partial \tau_{zy}}{\partial z} \tag{17.27}$$

$$\rho\frac{\mathrm{D}(w^2/2)}{\mathrm{D}t} = -w\frac{\partial p}{\partial z} + w\frac{\partial \tau_{xz}}{\partial x} + w\frac{\partial \tau_{yz}}{\partial y} + w\frac{\partial \tau_{zz}}{\partial z} \tag{17.28}$$

Adding Eqs. (17.26–17.28), we obtain

$$\rho\frac{\mathrm{D}(V^2/2)}{\mathrm{D}t} = -\boldsymbol{V}\cdot\nabla p + u\left(\frac{\partial \tau_{xx}}{\partial x} + \frac{\partial \tau_{yx}}{\partial y} + \frac{\partial \tau_{zx}}{\partial z}\right) + v\left(\frac{\partial \tau_{xy}}{\partial x} + \frac{\partial \tau_{yy}}{\partial y} + \frac{\partial \tau_{zy}}{\partial z}\right)$$
$$+ w\left(\frac{\partial \tau_{xz}}{\partial x} + \frac{\partial \tau_{yz}}{\partial y} + \frac{\partial \tau_{zz}}{\partial z}\right) \tag{17.29}$$

Subtract Eq. (17.29) from (17.25). We obtain

$$\boxed{\rho\frac{\mathrm{D}h}{\mathrm{D}t} = -\nabla\cdot\boldsymbol{q} + \frac{\mathrm{D}p}{\mathrm{D}t} + \Phi} \tag{17.30}$$

where Φ is the *dissipation function* given by

$$\Phi = \tau_{xx}\frac{\partial u}{\partial x} + \tau_{yy}\frac{\partial v}{\partial y} + \tau_{zz}\frac{\partial w}{\partial z} + \tau_{xy}\left(\frac{\partial u}{\partial y} + \frac{\partial v}{\partial x}\right)$$
$$+ \tau_{xz}\left(\frac{\partial u}{\partial z} + \frac{\partial w}{\partial x}\right) + \tau_{yz}\left(\frac{\partial v}{\partial z} + \frac{\partial w}{\partial y}\right)$$

Equation (17.30) is the form of the energy equation in terms of static enthalpy h. Recall that $\boldsymbol{q}$ is given by Eq. (16.31), and hence Eq. (17.30) can be written as

$$\rho\frac{\mathrm{D}h}{\mathrm{D}t} = \nabla\cdot(k\nabla T) - \nabla\cdot\sum_i \rho_i\boldsymbol{U}_i h_i - \nabla\cdot\boldsymbol{q}_R + \frac{\mathrm{D}p}{\mathrm{D}t} + \Phi \tag{17.31}$$

As has been stated before, the energy equation in the forms of Eqs. (17.5), (17.25), (17.30), and (17.31) does not contain an explicit term for the energy exchange because of chemical reactions. This is because e, h, and h_0 contain the effective zero-point energies in the form of the heats of formation of the species i, $(\Delta h_f)_i^{\mathrm{o}}$, that is, e, h, and h_0 are *absolute* values. In this fashion, the chemical energy exchanges are automatically taken into account. Recall that

$$h = \sum_i c_i h_i \tag{17.32}$$

where

$$h_i = \frac{7}{2}RT + e_{\text{vib}} + e_{\text{el}} + (\Delta h_f)_i^{\text{o}} \tag{17.33}$$

In Eq. (17.33), h_i is the *absolute* enthalpy of species i per unit mass of i. In turn, this can be written as

$$h_i = h_{i,\text{sens}} + (\Delta h_f)_i^{\text{o}} \tag{17.34}$$

where $h_{i,\text{sens}}$ is the *sensible* enthalpy of species i per unit mass of i, that is, the enthalpy measured *above the zero-point energy.* Note from Eqs. (17.34) and (17.35) that

$$h_{i,\text{sens}} = \frac{7}{2}RT + e_{\text{vib}} + e_{\text{el}}$$

If local thermodynamic equilibrium exists, then we can also write

$$h_{i,\text{sens}} = \int_0^T c_{pi}\,\mathrm{d}T \tag{17.35}$$

Returning to Eq. (17.32), the absolute enthalpy can be written as

$$h = \sum_i c_i h_{i,\text{sens}} + \sum_i c_i (\Delta h_f)_i^{\text{o}}$$

or

$$h = h_{\text{sens}} + \sum_i c_i (\Delta h_f)_i^{\text{o}} \tag{17.36}$$

Here, h_{sens} is the sensible enthalpy of the *mixture.*

There are occasions where the energy equation is expressed in terms of the sensible rather than the absolute enthalpy. Such an equation is developed as follows. From Eq. (17.36)

$$\frac{\mathrm{D}h}{\mathrm{D}t} = \frac{\mathrm{D}h_{\text{sens}}}{\mathrm{D}t} + \sum_i (\Delta h_f)_i^{\text{o}} \frac{\mathrm{D}c_i}{\mathrm{D}t} \tag{17.37}$$

Substitute Eq. (17.16) into (17.37). We obtain

$$\rho\frac{\mathrm{D}h}{\mathrm{D}t} = \rho\frac{\mathrm{D}h_{\text{sens}}}{\mathrm{D}t} + \sum_i \dot{w}_i (\Delta h_f)_i^{\text{o}} - \sum_i \nabla \cdot [\rho_i \boldsymbol{U}_i (\Delta h_f)_i^{\text{o}}] \tag{17.38}$$

Substituting Eq. (17.38) into (17.31), we have

$$\rho\frac{\mathrm{D}h_{\text{sens}}}{\mathrm{D}t}+\sum_i \dot{w}_i(\Delta h_f)_i^{\mathrm{o}}-\sum_i \nabla\cdot[\rho_i \boldsymbol{U}_i(\Delta h_f)_i^{\mathrm{o}}]$$

$$=\nabla\cdot(k\,\nabla T)-\nabla\cdot\sum_i \rho_i\boldsymbol{U}_i[h_{i,\text{sens}}+(\Delta h_f)_i^{\mathrm{o}}]-\nabla\cdot\boldsymbol{q}_R+\frac{\mathrm{D}p}{\mathrm{D}t}+\Phi \tag{17.39}$$

Cancelling terms, this yields

$$\boxed{\begin{aligned}\rho\frac{\mathrm{D}h_{\text{sens}}}{\mathrm{D}t}&=\nabla\cdot(k\nabla T)-\nabla\cdot\sum_i \rho_i\boldsymbol{U}_i h_{i,\text{sens}}\\ &\quad-\nabla\cdot\boldsymbol{q}_R+\frac{\mathrm{D}p}{\mathrm{D}t}+\Phi-\sum_l \dot{w}_i(\Delta h_f)_i^{\mathrm{o}}\end{aligned}} \tag{17.40}$$

Compare Eq. (17.40) with (17.31). Note that Eq. (17.40) is very similar to Eq. (17.31) with two very important exceptions. In Eq. (17.40), 1) the enthalpies appear as the *sensible* values, and 2) there is now an explicit term for the chemical energy exchange, namely, $-\sum_i \dot{w}_i(\Delta h_f)_i^{\mathrm{o}}$. From this, keep in mind the following fact. When you see an energy equation with an *explicit* term for chemical energy change, then the enthalpies (or internal energies) in that equation are *sensible* values. Similarly, if no such term appears explicitly, the enthalpies (or internal energies) are *absolute* values.

17.4 Boundary-Layer Equations for a Chemically Reacting Gas

By an order-of-magnitude analysis identical to that given in Sec. 6.4, the governing Navier–Stokes equations for a chemically reacting gas [Eqs. (17.1–17.6) and (17.17)] can be reduced to those for a chemically reacting boundary layer. The reader should review Sec. 6.4 before progressing further. In addition, for simplicity we will assume for the diffusion mechanism a binary gas, so that the diffusion coefficients are simply D_{12} and D_{21} for the diffusion of species 1 and 2 and vice versa. Because $\sum_i \rho_i\boldsymbol{U}_{\mathrm{i}}=0$, we have, for a binary gas,

$$\rho_1\boldsymbol{U}_1+\rho_2\boldsymbol{U}_2=0$$

From Fick's law, this becomes

$$\rho D_{12}\nabla c_1+\rho D_{21}\nabla c_2=0$$

Because $c_1=1-c_2$, this can be written as

$$\rho\nabla c_2(D_{21}-D_{12})=0$$

or

$$D_{21} = D_{12}$$

The binary-gas assumption for the diffusion mechanism has frequently been used for boundary-layer analyses in high-temperature air; in such a case, dissociated air is lumped into "heavy" molecules diffusing into "light" molecules and vice versa. With these assumptions, the resulting boundary-layer equations are as follows.

Global continuity:

$$\frac{\partial(\rho u r^j)}{\partial x} + \frac{\partial(\rho v r^j)}{\partial y} = 0 \tag{17.41}$$

Species continuity:

$$\rho u \frac{\partial c_i}{\partial x} + \rho v \frac{\partial c_i}{\partial y} = \frac{\partial}{\partial y}\left(\rho D_{12} \frac{\partial c_i}{\partial y}\right) + \dot{w}_i \tag{17.42}$$

x Momentum:

$$\rho u \frac{\partial u}{\partial x} + \rho v \frac{\partial u}{\partial y} = -\frac{\partial p}{\partial x} + \frac{\partial}{\partial y}\left(\mu \frac{\partial u}{\partial y}\right) \tag{17.43}$$

y Momentum:

$$\frac{\partial p}{\partial y} = 0 \tag{17.44}$$

Energy:

$$\rho u \frac{\partial h}{\partial x} + \rho v \frac{\partial h}{\partial y} = \frac{\partial}{\partial y}\left(k \frac{\partial T}{\partial y}\right) + \frac{\partial}{\partial y}\left(\rho D_{12} \sum_i h_i \frac{\partial c_i}{\partial y}\right) + \mu \left(\frac{\partial u}{\partial y}\right)^2 + u \frac{\partial p}{\partial x} \tag{17.45}$$

In Eq. (17.41), $j = 0$ or 1 depending on whether the boundary-layer flow two dimensional or axisymmetric, respectively. In the preceding form, the boundary-layer equations apply to either two-dimensional or axisymmetric cases.

It is sometimes useful to have the energy equation in terms of total enthalpy h_0. From an order-of-magnitude reduction of Eq. (17.25) we have, for the boundary-layer case,

$$\rho u \frac{\partial h_0}{\partial x} + \rho v \frac{\partial h_0}{\partial y} = \frac{\partial}{\partial y}\left(k \frac{\partial T}{\partial y}\right) + \frac{\partial}{\partial y}\left(\rho D_{12} \sum_i h_i \frac{\partial c_i}{\partial y}\right) + \frac{\partial}{\partial y}\left(\mu u \frac{\partial u}{\partial y}\right) \tag{17.46}$$

where $h_0 = h + u^2/2$; here, $v \ll u$ has been neglected.

For nonequilibrium flows, the energy equation in terms of T is particularly useful because the chemical rate constants depend on T. Therefore, let us couch the boundary-layer energy equation in terms of T as the dependent variable. We note that

$$h = \sum_i c_i h_i$$

Hence,

$$\frac{\partial h}{\partial x} = \sum_i c_i \frac{\partial h_i}{\partial x} + \sum_i h_i \frac{\partial c_i}{\partial x}$$

or

$$\frac{\partial h}{\partial x} = \sum_i c_i \left(\frac{\partial h_i}{\partial T}\right)\left(\frac{\partial T}{\partial x}\right) + \sum_i h_i \frac{\partial c_i}{\partial x} \tag{17.47}$$

However, by definition, because species i is a thermally perfect gas by itself, where $\mathrm{d}h_i = c_{pi}\,\mathrm{d}T$, then Eq. (17.47) can be written.

$$\frac{\partial h}{\partial x} = \left[\sum_i c_i c_{pi}\right]\left(\frac{\partial T}{\partial x}\right) + \sum_i h_i \frac{\partial c_i}{\partial x} \tag{17.48}$$

From Eq. (14.36), $\sum_i c_i c_{pi} \equiv c_{p_f}$, the frozen specific heat. Thus, Eq. (17.48) becomes

$$\frac{\partial h}{\partial x} = c_{p_f} \frac{\partial T}{\partial x} + \sum_i h_i \frac{\partial c_i}{\partial x} \tag{17.49}$$

Similarly,

$$\frac{\partial h}{\partial y} = c_{p_f} \frac{\partial T}{\partial y} + \sum_i h_i \frac{\partial c_i}{\partial y} \tag{17.50}$$

Substitute Eqs. (17.49) and (17.50) into (17.45). This yields

$$\begin{aligned} &\rho u c_{p_f} \frac{\partial T}{\partial x} + \rho v c_{p_f} \frac{\partial T}{\partial y} + \rho u \sum_i h_i \frac{\partial c_i}{\partial x} + \rho v \sum_i h_i \frac{\partial c_i}{\partial y} \\ &\quad = \frac{\partial}{\partial y}\left(k \frac{\partial T}{\partial y}\right) + \frac{\partial}{\partial y}\left(\rho D_{12} \sum_i h_i \frac{\partial c_i}{\partial y}\right) + \mu \left(\frac{\partial u}{\partial y}\right)^2 + u \frac{\partial p}{\partial x} \end{aligned} \tag{17.51}$$

or

$$\rho u c_{p_f}\frac{\partial T}{\partial x}+\rho v c_{p_f}\frac{\partial T}{\partial y}+\sum_i h_i\left(\rho u\frac{\partial c_i}{\partial x}+\rho v\frac{\partial c_i}{\partial y}\right)$$
$$=\frac{\partial}{\partial y}\left(k\frac{\partial T}{\partial y}\right)+\frac{\partial}{\partial y}\left(\rho D_{12}\sum_i h_i\frac{\partial c_i}{\partial y}\right)+\mu\left(\frac{\partial u}{\partial y}\right)^2+u\frac{\partial p}{\partial x} \tag{17.52}$$

From the species continuity equation (17.42), the third term in Eq. (17.52) can be expressed in terms of $\dot{w}_i$ and the diffusion coefficient yielding, from Eq. (17.52),

$$\rho u c_{pf}\frac{\partial T}{\partial x}+\rho v c_{p_f}\frac{\partial T}{\partial y}+\sum_i h_i\left[\dot{w}+\frac{\partial}{\partial y}\left(\rho D_{12}\frac{\partial c_i}{\partial y}\right)\right]$$
$$=\frac{\partial}{\partial y}\left(k\frac{\partial T}{\partial y}\right)+\frac{\partial}{\partial y}\left(\rho D_{12}\sum_i h_i\frac{\partial c_i}{\partial y}\right)+\mu\left(\frac{\partial u}{\partial y}\right)^2+u\frac{\partial p}{\partial x} \tag{17.53}$$

or

$$\rho u c_{p_f}\frac{\partial T}{\partial x}+\rho v c_{p_f}\frac{\partial T}{\partial y}=\frac{\partial}{\partial y}\left(k\frac{\partial T}{\partial y}\right)+\mu\left(\frac{\partial u}{\partial y}\right)^2+u\frac{\partial p}{\partial x}$$
$$+\left[\frac{\partial}{\partial y}\left(\rho D_{12}\sum_i h_i\frac{\partial c_i}{\partial y}\right)-\sum_i h_i\frac{\partial}{\partial y}\left(\rho D_{12}\frac{\partial c_i}{\partial y}\right)\right]-\sum_i h_i\dot{w}_i \tag{17.54}$$

Examine just the term in square brackets in Eq. (17.54)

$$\frac{\partial}{\partial y}\left(\rho D_{12}\sum_i h_i\frac{\partial c_i}{\partial y}\right)-\sum_i h_i\frac{\partial}{\partial y}\left(\rho D_{12}\frac{\partial c_i}{\partial y}\right)$$
$$=\sum_i h_i\frac{\partial}{\partial y}\left(\rho D_{12}\frac{\partial c_i}{\partial y}\right)+\sum_i\rho D_{12}\frac{\partial c_i}{\partial y}\frac{\partial h_i}{\partial y}-\sum_i h_i\frac{\partial}{\partial y}\left(\rho D_{12}\frac{\partial c_i}{\partial y}\right)$$
$$=\sum_i\rho D_{12}\frac{\partial c_i}{\partial y}\frac{\partial h_i}{\partial T}\frac{\partial T}{\partial y}$$
$$=\sum_i c_{p_i}\left(\rho D_{12}\frac{\partial c_i}{\partial y}\right)\left(\frac{\partial T}{\partial y}\right) \tag{17.55}$$

Substituting Eq. (17.55) into (17.54), we have

$$\boxed{\begin{aligned}\rho u c_{p_f}\frac{\partial T}{\partial x}+\rho v c_{p_f}\frac{\partial T}{\partial y}&=\frac{\partial}{\partial y}\left(k\frac{\partial T}{\partial y}\right)+\mu\left(\frac{\partial u}{\partial y}\right)^2\\&\quad+u\frac{\partial p}{\partial x}+\sum_i c_{p_i}\left(\rho D_{12}\frac{\partial c_i}{\partial y}\right)\left(\frac{\partial T}{\partial y}\right)-\sum_i h_i\dot{w}_i\end{aligned}} \tag{17.56}$$

Equation (17.56) is the boundary-layer energy equation for a chemically reacting gas in terms of temperature. Note that, in this form, an explicit term for the chemical energy exchange appears, namely, $\sum_i h_i \dot{w}_i$. However, also note that h_i is the *absolute* enthalpy of species i,

$$h_i = \int_0^T c_{pi}\,\mathrm{d}T + (\Delta h_f)_j^{\mathrm{o}}$$

and therefore the $\sum_i h_i \dot{w}_i$ term in Eq. (17.56) is distinctly different than the analogous term in Eq. (17.41), namely, $\sum_i \dot{w}_i (\Delta h_f)_i^{\mathrm{o}}$.

In analogy to Sec. 6.3, where the similarity parameters for a nonreacting viscous flow are obtained, let us obtain the appropriate similarity parameters for a viscous, chemically reacting boundary layer. The reader should review Sec. 6.3 before progressing further. We introduce the following nondimensional variables, denoted by a bar, where the subscript e denotes conditions at the edge of the boundary layer, and L is a characteristic length:

$$\bar{\rho} = \frac{\rho}{\rho_e} \qquad \bar{u} = \frac{u}{u_e} \qquad \bar{v} = \frac{v}{v_e}$$

$$\bar{h}_0 = \frac{h_0}{h_e} \qquad \bar{x} = \frac{x}{L} \qquad \bar{y} = \frac{y}{L}$$

$$\bar{T} = \frac{T}{h_e / c_{pfe}} \qquad \bar{k} = \frac{k}{k_e}$$

$$\bar{\mu} = \frac{\mu}{\mu_e} \qquad \bar{h}_i = \frac{h_i}{h_e} \qquad \bar{c}_i = c_i$$

$$\bar{D}_{12} = \frac{D_{12}}{(D_{12})_e}$$

Substituting the preceding into the energy equation in terms of h_0, Eq. (17.46) we obtain, after rearrangement,

$$\bar{\rho}\,\bar{u}\frac{\partial \bar{h}_0}{\partial \bar{x}} + \bar{\rho},\bar{v}\frac{\partial \bar{h}_0}{\partial \bar{y}} = \left(\frac{1}{Re\,Pr}\right)_e \frac{\partial}{\partial \bar{y}}\left(\bar{k}\frac{\partial \bar{T}}{\partial \bar{y}}\right) + \left(\frac{Le}{Re\,Pr}\right)_e \frac{\partial}{\partial \bar{y}}\left(\bar{\rho}\bar{D}_{12}\sum_i \bar{h}_i \frac{\partial \bar{c}_i}{\partial \bar{y}}\right)$$

$$+ \left(\frac{E}{Re}\right)_e \frac{\partial}{\partial \bar{y}}\left(\bar{\mu}\,\bar{u}\frac{\partial \bar{u}}{\partial \bar{y}}\right) \tag{17.57}$$

where the following holds:

Reynolds number:

$$(Re)_e = \frac{\rho_e u_e L}{\mu_e}$$

Prandtl number:

$$(Pr)_e = \frac{\mu_e c_{pfe}}{k_e}$$

Lewis number:

$$(Le)_e = \frac{\rho_e (D_{12})_e c_{pf\acute{e}}}{k_e}$$

Eckert number:

$$(E)_e = \frac{u_e^2}{h_e}$$

In the preceding, the subscript e emphasizes that the Reynolds, Prandtl, Lewis, and Eckert numbers in Eq. (17.57) are evaluated for conditions at the edge of the boundary layer. From the preceding analysis, and recalling the philosophy stated in Sec. 6.3, these numbers are *similarity parameters* for the flow. Both the Reynolds and Prandtl numbers are familiar parameters from Sec. 6.3 (and Part 2 in general); they carry over here for the case of a chemically reacting flow as well. But now, we have identified two new similarity parameters, namely, the Lewis and Eckert numbers. The Lewis number is an index of the energy transport caused by diffusion relative to thermal conduction, and the Eckert number is an index of flow kinetic energy relative to the thermal energy. In this sense, E is playing a similar role as M and γ did for a calorically perfect gas. [Indeed, for a calorically perfect gas, $E = (\gamma - 1)M^2$.] However, from the fact that M and γ do *not* appear in Eq. (17.57), we see once again that M and γ by themselves are *not* similarity parameters for a high-temperature, chemically reacting flow; this is proof again that M and γ lose their power and significance for such flows.

Let us use the preceding similarity parameters to obtain yet another (and for us, the final) form of the boundary-layer energy equation. This will be an important form where $(\partial T/\partial y)$ in the conduction term is replaced by $(\partial h_0/\partial y)$, thus explicitly eliminating T from the equation. Because $v \ll u$ for the boundary layer, $h_0 = h + u^2/2$. Thus,

$$\frac{\partial h_0}{\partial y} = \frac{\partial h}{\partial y} + u\frac{\partial u}{\partial y} \tag{17.58}$$

Substituting Eq. (17.50) into (17.58), we have

$$\frac{\partial h_0}{\partial y} - \frac{\partial u}{\partial y} = c_{pf}\frac{\partial T}{\partial y} + \sum_i h_i \frac{\partial c_i}{\partial y}$$

or

$$\frac{\partial T}{\partial y} = \frac{1}{c_{pf}}\left(\frac{\partial h_0}{\partial y} - u\frac{\partial u}{\partial y} - \sum_i h_i \frac{\partial c_i}{\partial y}\right) \tag{17.59}$$

Substituting Eq. (17.59) into (17.46), we obtain

$$\rho u\frac{\partial h_0}{\partial x} + \rho v\frac{\partial h_0}{\partial y} = \frac{\partial}{\partial y}\left[\frac{k}{c_{pf}}\left(\frac{\partial h_0}{\partial y} - u\frac{\partial u}{\partial y} - \sum_i h_i \frac{\partial c_i}{\partial y}\right)\right] + \frac{\partial}{\partial y}\left(\rho D_{12}\sum_i h_i \frac{\partial c_i}{\partial y}\right) + \frac{\partial}{\partial y}\left(\mu u\frac{\partial u}{\partial y}\right)$$

or

$$\rho u\frac{\partial h_0}{\partial x} + \rho v\frac{\partial h_0}{\partial y} = \frac{\partial}{\partial y}\left[\frac{\mu}{Pr}\frac{\partial h_0}{\partial y} + \left(1 - \frac{1}{Pr}\right)\mu u\frac{\partial u}{\partial y}\right] - \frac{\partial}{\partial y}\left(\frac{\mu}{Pr}\sum_i h_i \frac{\partial c_i}{\partial y} - \rho D_{12}\sum_i h_i \frac{\partial c_i}{\partial y}\right) \tag{17.60}$$

where Pr is the *local* Prandtl number, $Pr = \mu c_{pf}/k$. The last two terms (inside the Parentheses) in Eq. (17.60) can be combined as follows:

$$\begin{aligned}\frac{\mu}{Pr}\sum_i h_i \frac{\partial c_i}{\partial y} - \rho D_{12}\sum_i h_i \frac{\partial c_i}{\partial y} &= \left(\frac{\mu}{Pr} - \rho D_{12}\right)\sum_i h_i \frac{\partial c_i}{\partial y} \\ &= \rho D_{12}\left(\frac{k}{\rho D_{12} c_{pf}} - 1\right)\sum_i h_i \frac{\partial c_i}{\partial y} \\ &= \left(\frac{1}{Le} - 1\right)\rho D_{12}\sum_i h_i \frac{\partial c_i}{\partial y}\end{aligned}$$

Here, Le is the local Lewis number, $Le = \rho D_{12} c_{pf}/k$. Hence, Eq. (17.60) becomes

$$\boxed{\rho u\frac{\partial h_0}{\partial x} + \rho v\frac{\partial h_0}{\partial y} = \frac{\partial}{\partial y}\left[\frac{\mu}{Pr}\frac{\partial h_0}{\partial y} + \left(1 - \frac{1}{Pr}\right)\mu u\frac{\partial u}{\partial y} + \left(1 - \frac{1}{Le}\right)\rho D_{12}\sum_i h_i \frac{\partial c_i}{\partial y}\right]} \tag{17.61}$$

Equation (17.61) is a frequently employed form of the boundary-layer energy equation. In it, Pr and Le are local values that vary through the boundary layer. *Note*: For the special case of a constant $Le = 1$, the diffusion term in Eq. (17.62) drops out completely.

17.5 Boundary Conditions: Catalytic Walls

As in Part 2, the standard, no-slip boundary conditions on velocity at the wall hold for a chemically reacting viscous flow as well. If there is no mass transfer at the wall, that is, a solid wall with no ablation, then the following is true:

At $y = 0$,

$$u = v = 0 \qquad \text{(no mass transfer)}$$

If there is mass transfer, such as caused by transpiration of a gas through a porous wall, or caused by ablation, the wall boundary condition becomes the following:

At $y = 0$,

$$u = 0$$

$$v = v_w$$

where v_w is a known vertical velocity obtained from $\dot{m}_w = \rho v_w$, where $\dot{m}_w$ is the known mass flux of gas injected vertically into the boundary layer and ρ is the density of the gas mixture at the wall. (*Question*: Why is the gas *mixture* density at the wall, used in $\dot{m}_w = \rho v_w$, rather than the density ρ_i of the particular type of gas being injected into the flow? The answer is left to the reader as a homework problem.)

For a constant-temperature wall with known temperature T_w, we have the following at the wall.

At $y = 0$,

$$T = T_w \qquad \text{(specified)}$$

In contrast, for an adiabatic wall, the boundary condition must be obtained by setting q_y from Eq. (16.33) equal to zero at the wall.

At $y = 0$,

$$\left(k\frac{\partial T}{\partial y} + \rho \sum_i D_{\text{im}} h_i \frac{\partial c_i}{\partial y} \right)_w = 0$$

For binary diffusion, this becomes the following.

At $y = 0$,

$$\left(k\frac{\partial T}{\partial y} + \rho D_{12} \sum_i h_i \frac{\partial c_i}{\partial y} \right)_w = 0 \qquad \text{(adiabatic wall)} \quad (17.62)$$

Recall from Part 2 that, for a nonreacting gas, the adiabatic wall condition was simply $(\partial T/\partial y)_w = 0$. This is *not* the case for a chemically reacting flow, as seen from Eq. (17.63), because energy transport by diffusion must be included along with thermal conduction. Hence, in a chemically reacting flow for an adiabatic wall the normal temperature gradient is not necessarily zero.

In a chemically reacting flow, the mass fraction of species i is one of the dependent variables. Therefore, we need boundary conditions for c_i as well as for u, v, and T just discussed. At the wall, the boundary condition on c_i deserves some discussion because it involves, in general, a surface chemistry interaction with the gas at the wall. The wall can be made of a material that tends to catalyze (i.e., enhance) chemical reactions right at the surface. Such surfaces are called *catalytic walls*. This leads to the following definitions:

1) *Equilibrium catalytic wall* is wall at which chemical reactions are catalyzed at an *infinite rate*, that is, the mass fractions at the wall are their *local equilibrium values* at the local pressure and temperature at the wall.

2) *Partially catalytic wall* is wall at which chemical reactions are catalyzed at a *finite rate*.

3) *Fully catalytic wall* is wall where *all* atoms are recombined, irrespective of the mass fraction of atoms that would be allowed to exist at local chemical equilibrium conditions.

For an fully catalytic wall, the boundary condition is simply the following.

At $y = 0$,

$$c_A = 0 \qquad \text{(fully catalytic wall)}$$

where c_A is the mass fraction of an atomic species. *Note*: For many applications, the wall temperature is low enough that the equilibrium value of c_A is essentially zero. In this case, we have $c_A = 0$ as a boundary condition for both the fully catalytic and equilibrium wall cases.

For an equilibrium catalytic wall, the boundary condition is simply as follows.

At $y = 0$,

$$c_i = (c_i)_{\text{equil}} \qquad \text{(equilibrium catalytic wall)} \tag{17.63}$$

For a partially catalytic wall, the boundary condition can be developed as follows. For a wall with an arbitrary degree of catalyticity, the chemical reactions occur at a finite rate. Let $\dot{w}_c$ denote the catalytic rate at the surface. Then $(\dot{w}_c)_i =$ mass of species i lost at the surface per unit area per unit time caused by surface catalyzed chemical reaction. Right at the surface, the mechanism that feeds particles of species i from the gas to the surface is diffusion, as sketched in Fig. 17.1. The diffusion flux to the surface element of area dS is $-(\rho_i \boldsymbol{U}_i)_w$ dS. From Fick's law (assuming a binary gas)

$$-(\rho_i U_i)_w = \rho D_{12}\left(\frac{\partial c_i}{\partial y}\right)_w \tag{17.64}$$

For steady-state conditions, the amount of the amount species i "gobbled up" at the surface as a result of the catalytic rate $(\dot{w}_c)_i$ must be exactly balanced

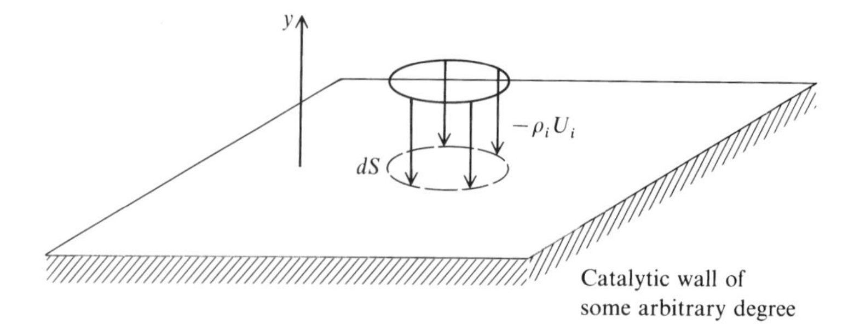

Fig. 17.1 Model for catalytic wall effects.

by the rate at which species i is diffused to the surface, given by Eq. (17.64). Hence

$$(\dot{w}_c)_i \, \mathrm{d}S = -(\rho_i U_i)_w \, \mathrm{d}S = \rho D_{12} \left(\frac{\partial c_i}{\partial y} \right)_w \mathrm{d}S$$

or

$$\boxed{(\dot{w}_c)_i = \rho D_{12} \left(\frac{\partial c_i}{\partial y} \right)_w} \tag{17.65}$$

Equation (17.65) is the *boundary condition* for a surface with finite catalyticity. It dictates the *gradient* of the mass fraction at the wall.

A *noncatalytic* wall is one where *no* recombination occurs at the wall, that is, $(\dot{w}_c)_i = 0$. For this case, from Eq. (17.65)

$$0 = \rho D_{12} \left(\frac{\partial c_i}{\partial y} \right)_w$$

or

At $y = 0$,

$$\left(\frac{\partial c_i}{\partial y} \right)_w = 0 \qquad \text{(noncatalytic wall)} \tag{17.66}$$

The subjects of surface catalyticity and the associated boundary conditions just discussed are serious matters for the analysis of chemically reacting viscous flows. We will see, for example, that a catalytic surface can experience a factor of two or more greater aerodynamic heating than a noncatalytic surface. Also, the matter of surface catalyticity, and its effect on chemically reacting viscous flow, is a current state-of-the art research area, especially in regard to obtaining knowledge about the *catalytic rates* $\dot{w}_c$ themselves. Knowledge of $\dot{w}_c$

for various gas-surface interactions is generally not accurate; indeed, because of the complex physical nature of such gas-surface interactions, values of $\dot{w}_c$ are usually more uncertain than the familiar gas reaction rate constants we have dealt with previously. Therefore, when you need to analyze a chemically reacting viscous-flow problem wherein surface catalytic effects are important, your first step should be to obtain the best possible values for $\dot{w}_c$ from the existing literature (or better yet, from your friendly local physical chemist).

Finally in regard to boundary-layer solutions, the boundary conditions at the outer edge of the boundary-layer are obtained from an independent knowledge of the inviscid flow over the given body flow (such as discussed in Chapters 14 and 15). That is,

At $y = \delta_u$,

$$u = u_e$$

At $y = \delta_T$,

$$T = T_e$$

At $y = \delta_c$,

$$c_i = (c_i)_e$$

In the preceding, we make a distinction between various different boundary-layer thicknesses: δ_u is the velocity boundary-layer thickness; δ_T the temperature boundary-layer thickness; and δ_c the species boundary-layer thickness. If the boundary-layer thickness is defined as that location above the surface where the flow property reaches 99% of its inviscid-flow value, then in a chemically reacting flow that location might be different for each of u, T, and c_i. Thus, in general, $\delta_u \neq \delta_T \neq \delta_c$.

17.6 Boundary-Layer Solutions: Stagnation-Point Heat Transfer for a Dissociating Gas

In Sec. 6.5, considerable attention was paid to self-similiar solutions of the compressible, nonreacting boundary-layer equations. The basic philosophy of self-similar solutions was described in the first part of Sec. 6.5; this should be reviewed before proceeding further. For chemically reacting boundary layers, the following cases lend themselves to such self-similar solutions: *local chemical and thermodynamic equilibrium* with flat plate, sharp right-circular cone, stagnation point; and *nonequilibrium* with stagnation point.

In the present section, we choose to study the stagnation-point flow for a dissociating gas, for two reasons: 1) it is an excellent example of a self-similar solution for chemically reacting flow—both equilibrium and nonequilibrium, and 2) the stagnation point heat-transfer results obtained from the solution are extremely important for hypersonic flow applications. We will follow the classic solution carried out by Fay and Riddell [196]; this work was a pioneering step forward in the analysis of chemically reacting viscous flow in 1958. The Fay and Riddell results for stagnation-point heat transfer in dissociated air are still in regular use today by industry and government for hypersonic vehicles analyses.

The physical model is sketched in Fig. 17.2. A blunt-body flow is sketched in Fig. 17.2a; we will concentrate on just the stagnation region near the nose, which is isolated and magnified in Fig. 17.2b. The following assumptions are made:

1) The flow conditions at the outer edge of the boundary layer are those for local thermodynamic and chemical equilibrium. The shock layer is partially dissociated.

2) Depending on the length of time spent by a fluid particle in the boundary layer and on the rate of chemical reaction (i.e., the comparison between τ_f

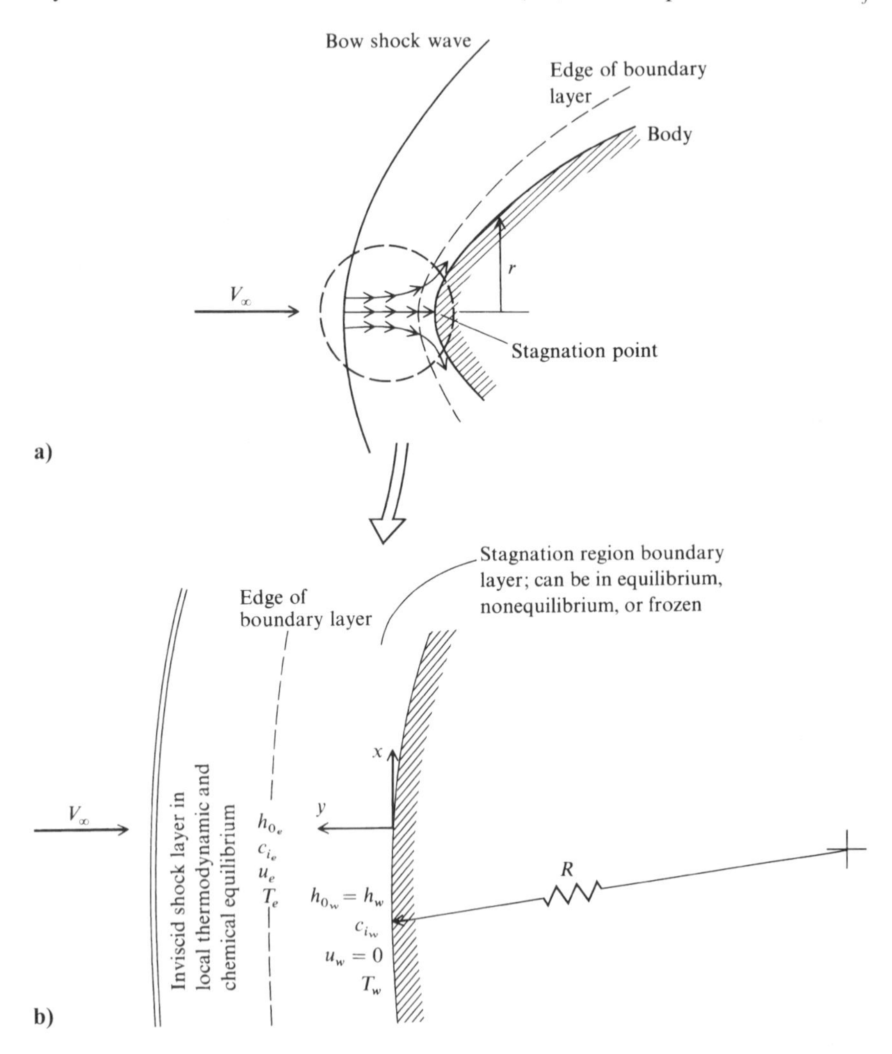

Fig. 17.2 Stagnation region flow model.

and τ_c, as discussed in Sec. 15.1), the boundary layer itself can have regions of equilibrium, nonequilibrium, or frozen flow. In this section, we will deal with all three cases.

3) The inviscid velocity distribution at the outer edge of the boundary layer in the stagnation region is given by the classical incompressible result

$$u_e = ax \tag{17.67}$$

where

$$a = \left(\frac{\mathrm{d}u_e}{\mathrm{d}x}\right)_s \tag{17.67a}$$

that is, $(\mathrm{d}u_e/\mathrm{d}x)_s$ is the velocity gradient at the stagnation point.

4) The wall can be equilibrium catalytic or noncatalytic; both cases will be discussed.

5) The gas is assumed to be a binary mixture, made up of "heavy particles" (molecules) and "light particles" (atoms). As explained in Sec. 17.4, this leads to a simplification in regard to the diffusion mechanism. Fortunately, for dissociated air, the principal molecules O_2 and N_2 are similar in terms of molecular weight and collision cross section (for the transport coefficients); the same can be said for the atoms O and N. Hence, average properties can be found for the molecules as one species, and for the atoms as the second species. See [196] for more details.

The boundary-layer equations for global continuity, species continuity, and momentum are given by Eqs. (17.41–17.44). For the analysis of the boundary layer in local thermodynamic and chemical equilibrium, the energy equation in the form of Eq. (17.61) is used. For the nonequilibrium cases, Eq. (17.56) in terms of T is used. For the wall boundary conditions, a constant-temperature wall is assumed with no transpiration or ablation at the surface. Cases are solved for both an equilibrium catalytic wall, where the boundary condition on c_i is given by Eq. (17.63), and a noncatalytic wall, where the boundary condition on the *gradient* of c_i is given by Eq. (17.66).

The independent variables in the boundary-layer equations are transformed by a version of the Lees–Dorodnitsyn transformation, that is, a slightly modified form of Eqs. (6.33) and (6.34) wherein the wall properties are used rather than the outer-edge properties. Specifically, we perform a transformation of x and y into ξ and η, where

$$\xi = \xi(x) = \int_0^x \rho_w \mu_w u_e r^2 \,\mathrm{d}x \tag{17.68}$$

$$\eta = \eta(x, y) = \frac{r u_e}{\sqrt{2\xi}} \int_0^y \rho \,\mathrm{d}y \tag{17.69}$$

The preceding equations are written for the axisymmetric case; for two-dimensional flow, the r in Eqs. (17.68) and (17.69) simply drops out. The dependent variables in the boundary-layer equations are transformed as follows:

$$\frac{u}{u_e} = f'(\xi, \eta) \equiv \frac{\partial f}{\partial n} \tag{17.70}$$

$$\frac{h_0}{h_{0_e}} = \left(h + \frac{u^2}{2}\right) h_{0_e} = g(\xi, \eta) \tag{17.71}$$

$$\frac{T}{T_e} = \theta(\xi, \eta) \tag{17.72}$$

$$\frac{c_i}{c_{i_e}} = s_i(\xi, \eta) \tag{17.73}$$

The details of the transformation are very similar to the process carried out in Sec. 6.5 and are left to the reader as a homework problem; therefore, only the results will be given here. At this point, the reader should review Sec. 6.5 before progressing further. The transformed boundary-layer equations are obtained from Eqs. (17.41–17.43) and (17.61) as follows.

Momentum:

$$(lf'')' + ff'' + 2\left[\frac{\rho_e}{\rho} - (f')^2\right]\frac{d(\ell n\, u_e)}{d\,\ell n\, \xi} = 2\xi\left(f'\frac{\partial^2 f}{\partial \eta\, \partial \xi} - f''\frac{\partial f}{\partial \xi}\right) \tag{17.74}$$

Species continuity:

$$\frac{\partial}{\partial \eta}\left[\frac{l}{Pr}(Le)s_i'\right] + fs_i' + \frac{2\xi \dot{w}_i}{\rho_w \mu_w u_e^2 r^2 \rho c_{i_e}} = 2\xi\left(f'\frac{\partial s_i}{\partial \xi} - \frac{\partial f}{\partial \xi}s_i'\right) + 2f's_i\frac{d(\ell n\, c_{i_e})}{d(\ell n\, \xi)} \tag{17.75}$$

Energy:

$$\frac{\partial}{\partial \eta}\left(\frac{l}{Pr}g'\right) + fg' + \frac{u_e^2}{h_{0_e}}\frac{\partial}{\partial \eta}\left[\left(1 - \frac{1}{Pr}\right)lf'f''\right] + \frac{\partial}{\partial \eta}\left[\frac{l}{Pr}\left(\sum_i \frac{c_{i_e}}{h_{0_e}}h_i\right)(Le - 1)s_i'\right] = 2\xi\left(f'\frac{\partial g}{\partial \xi} - \frac{\partial f}{\partial \xi}g'\right) \tag{17.76}$$

In Eqs. (17.74–17.76) the prime denotes *partial* differentiation with respect to η, and l denotes the ρ-μ ratio, $l = \rho\mu/\rho_w\mu_w$. These partial differential equations are analogous to Eqs. (6.55) and (6.58) for nonreacting flow. Equations (17.74–17.76), although transformed, are simply the full boundary-layer equations with the same significance as the original forms given by Eqs. (17.41–17.43) and (17.61). They are a system of coupled, partial differential equations where f, g, and s_i are functions of both ξ and η.

Let us now apply Eqs. (17.74–17.76) to the spherical stagnation region. Following the technique carried out in Sec. 6.5 for self-similar solutions, we first assume that f, g, and s_i are functions of η only, and we examine Eqs. (17.74–17.76) to see if all ξ dependency has dropped out. From Eq. (17.74), we obtain

$$(lf'')' + ff' + 2\left[\frac{\rho_e}{\rho} - (f')^2\right]\frac{d(\ell n\, u_e)}{d(\ell n\, \xi)} = 0 \tag{17.77}$$

The third term in Eq. (17.77) still appears to exhibit ξ dependency. Let us examine this term more closely. At the stagnation point, $r \approx x$, where x is very small. Inserting this and Eq. (17.67) into Eq. (17.68), we obtain, after integration

$$\xi = \rho_w \mu_w \left(\frac{du_e}{dx}\right)_s \frac{x^4}{4} \tag{17.78}$$

Also,

$$\left(\frac{du_e}{d\xi}\right)_s = \left(\frac{du_e}{dx}\right)_s \frac{dx}{d\xi} = \frac{(du_e/dx)_s}{d\xi/dx} \tag{17.79}$$

Obtaining $d\xi/dx$ by differentiating Eq. (17.78), Eq. (17.79) becomes

$$\left(\frac{du_e}{d\xi}\right) = \frac{1}{\rho_w \mu_w x^3} \tag{17.80}$$

In addition,

$$\frac{d(\ell n\, u_e)}{d(\ell n\, \xi)} = \frac{\xi}{u_e}\frac{du_e}{d\xi} \tag{17.81}$$

Substituting Eqs. (17.78) and (17.80) into (17.81), we have, for the spherical stagnation point,

$$\frac{d(\ell n u_e)}{d(\ell n \xi)} = \frac{1}{4} \tag{17.82}$$

Inserting Eq. (17.82) into (17.77), we obtain for the stagnation point,

$$\boxed{(lf'')' + ff'' + \frac{1}{2}\left[\frac{\rho_e}{\rho} - (f')^2\right] = 0} \tag{17.83}$$

We turn our attention now to the energy equation (17.16). Under the same assumption of $f = f(\eta)$ and $g = g(\eta)$, and noting that $u_e = 0$ at the stagnation point, we obtain directly

$$\left(\frac{l}{Pr} g'\right)' + fg' + \frac{d}{d\eta}\left[\frac{l}{Pr}\left(\sum_i \frac{c_{i_e}}{h_{0_e}} h_i\right)(Le - 1)s_i'\right] = 0 \tag{17.84}$$

Making the same assumption in Eq. (17.75), including $s_i = s(\eta)$, and also assuming c_{i_e} does not change with ξ, we obtain

$$\left(\frac{l}{Pr} Le\, s_i'\right)' + fs_i' + \frac{2\xi \dot{w}_i}{\rho_w \mu_w u_e^2 r^2 \rho c_{i_e}} = 0 \tag{17.85}$$

The third term in Eq. (17.85) still appears to exhibit ξ dependency. However, using Eq. (17.78), (17.76), and $r \approx x$, this term becomes (the details are left as a homework problem)

$$\frac{2\xi}{\rho_w \mu_w u_e^2 r^2 \rho c_{i_e}} = \frac{1}{2} \frac{\dot{w}}{(du_e/dx)_s \rho c_{i_e}} \tag{17.86}$$

Substituting Eq. (17.86) into (17.84), we have

$$\left(\frac{l}{Pr} Le\, s_i'\right)' + fs_i' + \frac{\dot{w}_i}{2(du_e/dx)_s \rho c_{i_e}} = 0 \tag{17.87}$$

Now examine Eqs. (17.83), (17.84), and (17.87) closely. They hold for the spherical stagnation-point boundary layer. Moreover, they are *ordinary* differential equations and hence prove that the stagnation-point flow is a self-similar flow in the spirit discussed in Sec. 6.5; the presence of chemical reactions, even finite-rate reactions, does not change this behavior. A corresponding form of the energy equation in terms of the transformed temperature, $\theta = T/T_e$, can be obtained by transforming Eq. (17.56) and applying the stagnation-point conditions; the resulting transformed ordinary differential equation for θ can be found in [196].

The transformed boundary conditions at the outer edge of the boundary layer are, as $\eta \to \infty$,

$$f' = 1 \quad g = 1 \quad \theta = 1 \quad \text{and} \quad s_i = 1$$

where $c_i = c_{i_e} = (c_{i_e})^{\text{equil}}$ is obtained from the locally equilibrium inviscid flow. The transformed boundary conditions at the wall are

$$f'(0) = 0 \qquad f(0) = 0 \qquad g(0) = g_w \qquad \text{and} \qquad \theta(0) = \theta_w$$

The wall boundary condition for s_i depends on whether the wall is catalytic or not and also on whether or not the boundary layer is in local chemical equilibrium. For example, for the boundary layer in local chemical and thermodynamic equilibrium, c_i always depends uniquely on the local pressure and temperature; hence, at the wall the value of $c_i(0)$ is the local equilibrium value for an equilibrium catalytic wall. The same condition holds for a nonequilibrium boundary layer with an equilibrium catalytic wall. As explained in Sec. 17.5, c_i takes on its local equilibrium value at the wall for such a case. *Thus, we have for 1) an equilibrium boundary layer and 2) a nonequilibrium or a frozen boundary layer with an equilibrium catalytic wall.*

At $\eta = 0$,

$$s_i(0) = \frac{c_i(0)}{c_{i_e}} = \frac{[c_i(0)]_{\text{equil}}}{c_{i_e}}$$

In contrast, for a noncatalytic wall, we have from Eq. (17.66) the fact that $(\partial c_i/\partial y)_w = 0$. Hence, for a *nonequilibrium boundary layer with a noncatalytic wall.*

At $\eta = 0$,

$$s_i'(0) = 0$$

Finally, the numerical solution of Eqs. (17.83), (17.84), and (17.87) can be carried out by the same type of "shooting technique" described in Sec. 6.5; see [196] for particular details. The numerical results give profiles of f' (hence u), g (hence h_0), θ (hence T), and s_i (hence c_i) as a function of η. Sample results for high-temperature air obtained by Fay and Riddell are shown in Figs. 17.3 and 17.4. These results assume constant values of $Le = 1.4$ and $Pr = 0.71$. (Note that such assumptions by Fay and Riddell are a convenience, not a necessity; because a numerical solution is being carried out, Pr and Le could easily be treated as variables.) In Fig. 17.3, the total enthalpy profiles (in the form of g vs η) are shown for both an equilibrium and a frozen boundary layer. Note that, at a given value of η, the equilibrium value of g is higher than the frozen value. This is consistent with the fact that, with a dissociated gas at the outer edge of the boundary layer, recombination will tend to occur locally within the boundary layer, along with the attendant chemical energy release; however, for a frozen flow no recombination will occur within the boundary layer. As a result, at a given location, g will be higher for the equilibrium case. For the same reason, θ (hence T) shown in Fig. 17.4 is higher for the equilibrium boundary layer. Also shown in Fig. 17.4 are profiles of the atom mass fraction c_A for both equilibrium and frozen flow. At first glance, the question can be asked: for the frozen flow, why is not c_A a constant value through the boundary layer? The answer lies in the fact that the wall is equilibrium catalytic, which means that the atom mass fraction must be in local equilibrium at the wall, no matter whether or not the boundary-layer flow is frozen. Moreover, the wall is cold, so that the equilibrium atom mass fraction at the wall is essentially zero. Thus, c_A for the frozen flow must be zero at the wall and c_{A_e} at the outer edge.

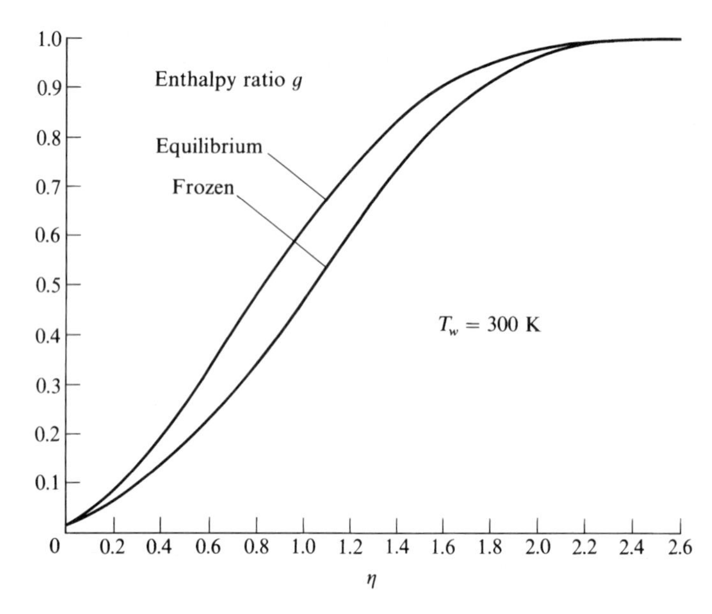

Fig. 17.3 Enthalpy profiles for an equilibrium and a frozen stagnation-point boundary layer. Equilibrium catalytic wall (from Fay and Riddell [196]).

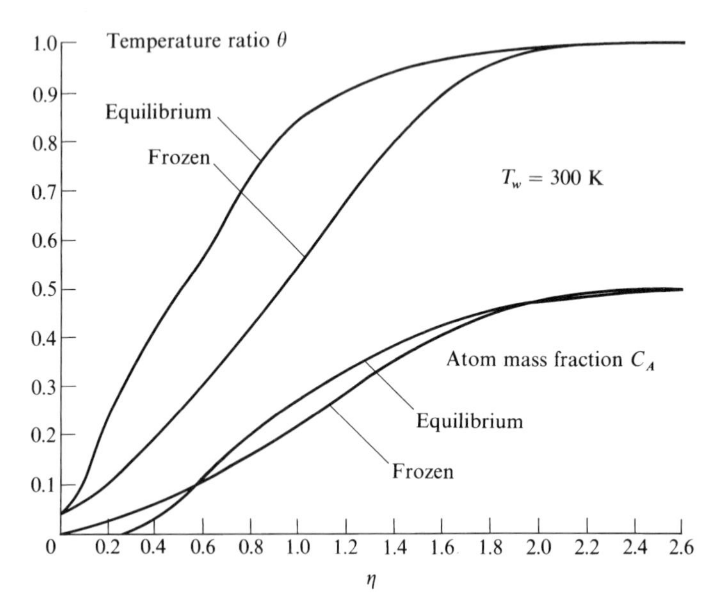

Fig. 17.4 Temperature and atom mass fraction profiles for an equilibrium and a frozen stagnation-point boundary layer. Equilibrium catalytic wall (from [196]).

The frozen flow profile of c_A between the wall and the outer edge is *not* a result of any chemical reactions within the boundary layer, but rather is completely a result of *diffusion* of the atoms from the outer edge to the wall.

The surface heat transfer is obtained from Eq. (16.33), written as

$$q_w = \underbrace{\left(k\frac{\partial T}{\partial y}\right)_w}_{\text{conduction}} + \underbrace{\left(\rho D_{12}\sum_i h_i \frac{\partial c_i}{\partial y}\right)_w}_{\text{diffusion}} \tag{17.88}$$

Note that, in contrast to the nonreacting viscous flows discussed in Part 2, the surface heat transfer in a chemically reacting viscous flow is caused not only by the familiar thermal conduction, but also by diffusion. In Eq. (17.88), the temperature and mass fraction gradients at the wall are obtained as part of the numerical solution of the boundary-layer equations described earlier. In [196], a large number of calculations were reported covering flight velocities from 5800 to 22,800 ft/s, altitudes from 25,000 to 120,000 ft, and wall temperatures from 300 to 3000 K. Fay and Riddell correlated these results in forms analogous to the nonreacting case given by Eq. (6.111), as follows.

1) Equilibrium boundary layer (spherical nose):

$$\boxed{\begin{aligned} q_w &= 0.76\,Pr^{-0.6}(\rho_e\mu_e)^{0.4}(\rho_w\mu_w)^{0.1}\sqrt{\left(\frac{\mathrm{d}u_e}{\mathrm{d}x}\right)_s}(h_{0_e} - h_w) \\ &\times \left[1 + (Le^{0.52} - 1)\left(\frac{h_D}{h_{0_e}}\right)\right] \end{aligned}} \tag{17.89}$$

where $h_D = \sum_i c_{i_e}(\Delta h_f)_i^{\circ}$.

2) Frozen boundary layer with an equilibrium catalytic wall (spherical nose):

$$\boxed{\begin{aligned} q_w &= 0.76\,Pr^{-0.6}(\rho_e\mu_e)^{0.4}(\rho_w\mu_w)^{0.1}\sqrt{\left(\frac{\mathrm{d}u_e}{\mathrm{d}x}\right)_s}(h_{0_e} - h_w) \\ &\times \left[1 + (Le^{0.63} - 1)\left(\frac{h_D}{h_{0_e}}\right)\right] \end{aligned}} \tag{17.90}$$

3) Frozen boundary layer with a noncatalytic wall (spherical nose):

$$\boxed{q_w = 0.76\,Pr^{-0.6}(\rho_e\mu_e)^{0.4}(\rho_w\mu_w)^{0.1}\sqrt{\left(\frac{\mathrm{d}u_e}{\mathrm{d}x}\right)_s}\left(1 - \frac{h_D}{h_{0_e}}\right)} \tag{17.91}$$

In Eqs. (17.89–17.91), the stagnation-point velocity gradient is given by Newtonian theory as Eq. (6.121), repeated here:

$$\left(\frac{\mathrm{d}u_e}{\mathrm{d}x}\right)_s = \frac{1}{R}\sqrt{\frac{2(p_e - p_\infty)}{\rho_e}}$$

Note the similarity between Eqs. (17.89–17.91) for a reacting gas and Eq. (6.111) for the nonreacting case. This similarity is somewhat interesting considering the major difference in the details of a reacting vs a nonreacting viscous flow. Note from Eqs. (17.89) and (17.90) that the driving potential for heat transfer is the enthalpy difference $(h_{0_e} - h_w)$ in the case of equilibrium flow or frozen flow with an equilibrium catalytic wall. This is similar to the driving potential $(h_{0_w} - h_w)$, in Eq. (6.111) for the nonreacting case. However, in Eqs. (17.89–17.91), the enthalpies are *absolute* values, that is, they contain the heats of formation, thus including the powerful chemical energy associated with the reacting gas. Also note that Eqs. (17.89) and (17.90) are essentially the same, varying only in the slightly different exponent on the Lewis number. This demonstrates that the surface heat transfer is essentially the same whether the flow is in local chemical equilibrium or is frozen with an equilibrium catalytic wall. In the former case (local chemical equilibrium), recombination occurs within the cooler regions of the boundary layer itself, releasing chemical energy throughout the interior of the boundary layer, most of which is transported by thermal conduction to the surface. In the latter case (frozen flow with an equilibrium catalytic wall), the chemical energy release as a result of recombination is right at the wall itself. Equations (17.89) and (17.90) indicate that the net heat transfer to the surface is essentially the same whether the chemical energy is released within the boundary layer or right at the surface. This trend is graphically illustrated in Fig. 17.5, which shows the heat-transfer coefficient $Nu/\sqrt{Re}$, [see

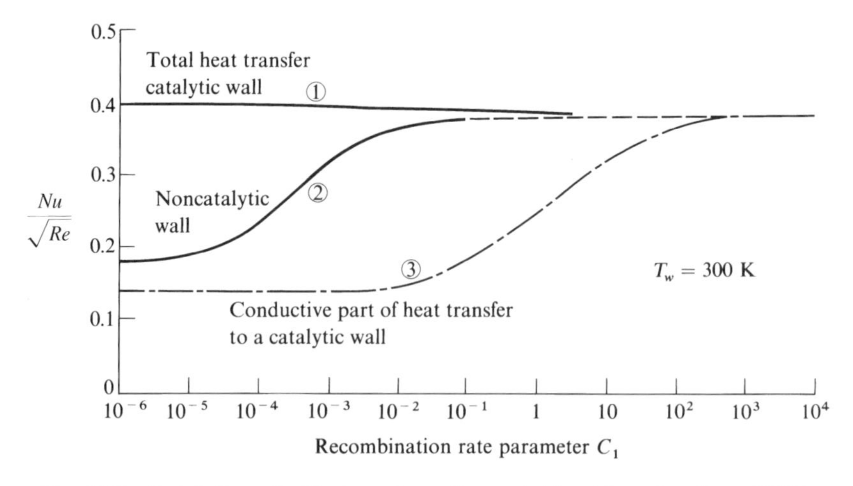

Fig. 17.5 Catalytic wall effect on stagnation-point heat transfer (from [196]).

Eq. (6.62) for the definition of Nusselt number Nu] as a function of a recombination rate parameter C_1, defined in [196]. On the abscissa, going from right to left in the direction of decreasing C_1, the chemical state of the boundary layer changes from equilibrium (large C_1) to frozen (small C_1). The results in Fig. 17.5 are from a large number of different nonequilibrium boundary-layer cases with different values of C_1 calculated by Fay and Riddell. Shown on this graph are two categories of solutions, one with an equilibrium catalytic wall (curve 1) and the other with a noncatalytic wall (curve 2). Curve 1 for the equilibrium catalytic wall gives the highest heat transfer, and it is essentially constant for all values of C_1. For large values of C_1, this curve corresponds to an equilibrium boundary layer, and for small values of C_1 it corresponds to a frozen boundary layer. Clearly, as long as the wall is equilibrium catalytic the heat transfer is essentially the same. This is precisely the observation we made earlier while comparing Eqs. (17.89) and (17.90). In contrast, curve 2 in Fig. 17.5 is for a noncatalytic wall. Here, as we move from right to left along this curve (and as the boundary layer becomes progressively more nonequilibrium, approaching a frozen flow), we see that the heat transfer drops by more than a factor of two. This is an important point: *For nonequilibrium and frozen flows, there is a substantial decrease in heat transfer if the wall is noncatalytic in comparison to a catalytic wall.* Finally, curve 3 in Fig. 17.5 goes along with curve 1 for a catalytic wall; curve 3 gives just the conductive part of the heat transfer to a catalytic wall [just the first term in Eq. (17.88)]. The difference between curves 1 and 3 represents the heat transfer caused by diffusion. Hence, for equilibrium flows, q is essentially all conductive; however, as we examine flows that progressively become more nonequilibrium with an equilibrium catalytic wall, diffusion progressively becomes a larger part of q.

As a final note in this section, the work of Fay and Riddell represents an excellent example of chemically reacting boundary-layer analysis, and their results convey virtually all of the important physical trends to be observed in chemically reacting viscous flows. Although this analysis is old—carried out more than 50 years ago—it is classical, and just as viable today as it was then. This is why we have chosen to highlight it here. The reader is strongly encouraged to study [196] closely, for more details and insight.

17.7 Boundary-Layer Solutions: Nonsimilar Flows

Various solutions of the boundary-layer equations for nonsimilar flows were discussed in Sec. 6.6. The reader should review Sec. 6.6 before progressing further. For example, the local-similarity method was illustrated by the work of Kemp et al. [95], with results for chemically reacting, dissociated air given in Fig. 6.19. Also, the implicit finite difference method was illustrated by the work of Blottner [94], with results for chemically reacting air given in Figs. 6.22 and 6.23.

As stated at the beginning of Sec. 17.6, self-similar solutions of the chemically reacting boundary-layer equations can be obtained for the equilibrium flow over flat plates, cones, and the stagnation point, and for nonequilibrium flow only the stagnation point. For all other cases, the nonsimilar boundary-layer equations must be used, that is, Eqs. (17.41–17.45), (17.46), and (17.74–17.76).

Perhaps the most classical example of the solution of the chemically reacting, laminar, nonsimilar boundary-layer equations is the early work of Blottner, which appeared in 1964 for nonequilibrium, dissociated air [197] and nonequilibrium ionized air [198]. Blottner used the same implicit finite difference numerical technique described in Sec. 6.6; hence, no further details will be given here. Some results are shown in Figs. 17.6–17.9 for the case of ionized, nonequilibrium flow of air over a 10-deg cone at $V_\infty = 21{,}590$ ft/s at an altitude of 100,000 ft, obtained from [198]. Here, Blottner considered only NO^+ as the ionized species; the temperature levels were below that for atomic ionization. The wall was equilibrium catalytic. The boundary-layer temperature profiles are given in Fig. 17.6 at the cone tip and at 15 ft downstream of the tip. The temperature gradually decreases in the downstream direction caused in part by the finite-rate dissociation and ionization reactions, which are endothermic and hence "absorb" some of the viscous dissipation energy into zero-point energy of the atoms and ions as opposed to it going into translational energy of the particles. As a result, the temperature profile (which is a frozen flow profile at $x = 0$) decreases with distance downstream of the tip. Profiles of the mass fraction of atomic nitrogen C_N are given in Fig. 17.7. Here, C_N increases with distance downstream of the tip, again because of the finite-rate dissociation. The corresponding electron densities are shown in Fig. 17.8. It is interesting to note that in each of Figs. 17.6–17.8, the difference in the profiles at different streamwise stations is an indication of the *nonsimilar* effect; if the flow were similar, then the profiles (vs η) would be exactly the same for all values of x. Also, for the case of a sharp cone illustrated here, *the nonsimilar effect is caused exclusively by the nonequilibrium flow*; if the flow were in local chemical equilibrium, or frozen, it would be a self-similar flow.

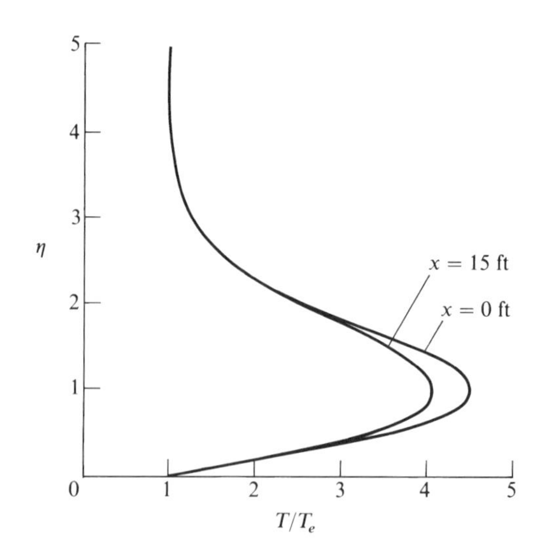

Fig. 17.6 Temperature profiles in the nonequilibrium boundary layer over a 10-deg cone (from Blottner [198]).

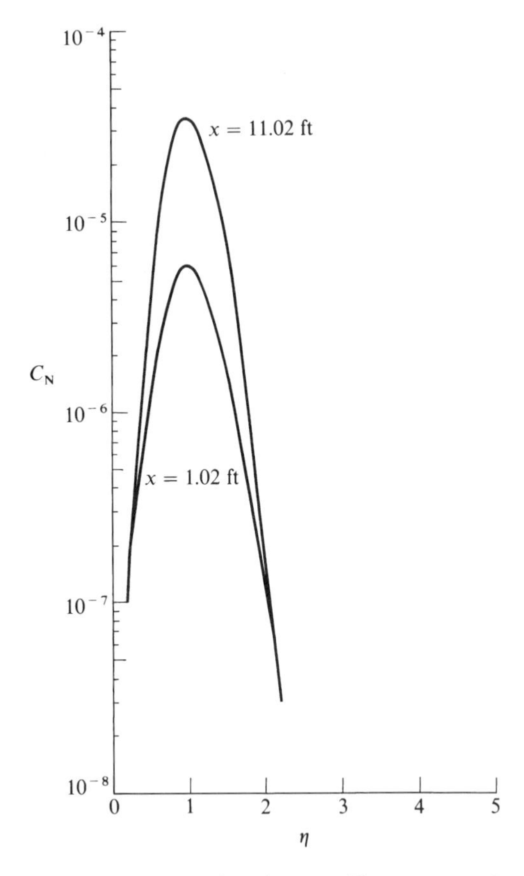

Fig. 17.7 Atomic-nitrogen mass fraction profiles across the nonequilibrium boundary layer on a 10-deg cone (from Blottner [198]).

17.8 Viscous-Shock-Layer Solutions to Chemically Reacting Flow

The viscous-shock-layer (VSL) technique—its philosophy and approach—was described in Sec. 8.2, which should be reviewed by the reader before progressing further. We will not be repetitive here. In the present section, we examine an application of the VSL technique to chemically reacting flow.

The governing VSL equations were given by Eqs. (8.1–8.4) for a non reacting gas. For the chemically reacting gas considered here, the VSL equations for global continuity [Eq. (8.1)], s momentum [Eq. (8.2)], and n momentum [Eq. (8.3)] carry over, unchanged. To these must be added a new energy equation and the species continuity equation. The VSL energy equation for a chemically reacting flow in terms of T is similar to the boundary-layer equation given by Eq. (17.57) and is given in [199] in terms of the shock-layer coordinates

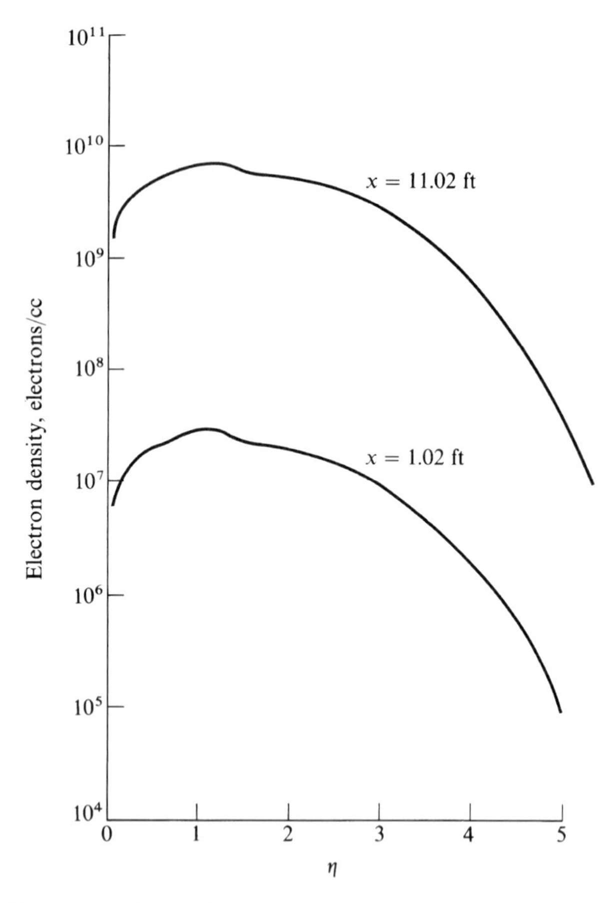

Fig. 17.8 Electron density profiles across the nonequilibrium boundary layer on a 10-deg cone (from [198]).

shown in Fig. 8.1 as

$$
\rho^* c^*_{pf}\left(\frac{u^*}{1+n^*\kappa^*}\frac{\partial T^*}{\partial s^*}+v^*\frac{\partial T^*}{\partial n^*}\right)-\left(\frac{u^*}{1+n^*\kappa^*}\frac{\partial p^*}{\partial s^*}+v^*\frac{\partial p^*}{\partial n^*}\right)
$$

$$
=\varepsilon^2\left[\frac{\partial^*}{\partial n^*}\left(k^*\frac{\partial T^*}{\partial n^*}\right)+\left(\frac{\kappa^*}{1+n^*\kappa^*}+\frac{m\cos\phi}{r^*+n^*\cos\phi}\right)k^*\frac{\partial T^*}{\partial n^*}\right.
$$

$$
\left.-\sum_i c^*_{pf}\left(\rho^* D^*_{12}\frac{\partial c_i}{\partial n^*}\right)\left(\frac{\partial T^*}{\partial n^*}\right)+\mu^*\left(\frac{\partial u^*}{\partial n^*}-\frac{\kappa^* u^*}{1+n^*\kappa^*}\right)^2\right]-\sum_i h^*_i\dot{w}^*_i \tag{17.92}
$$

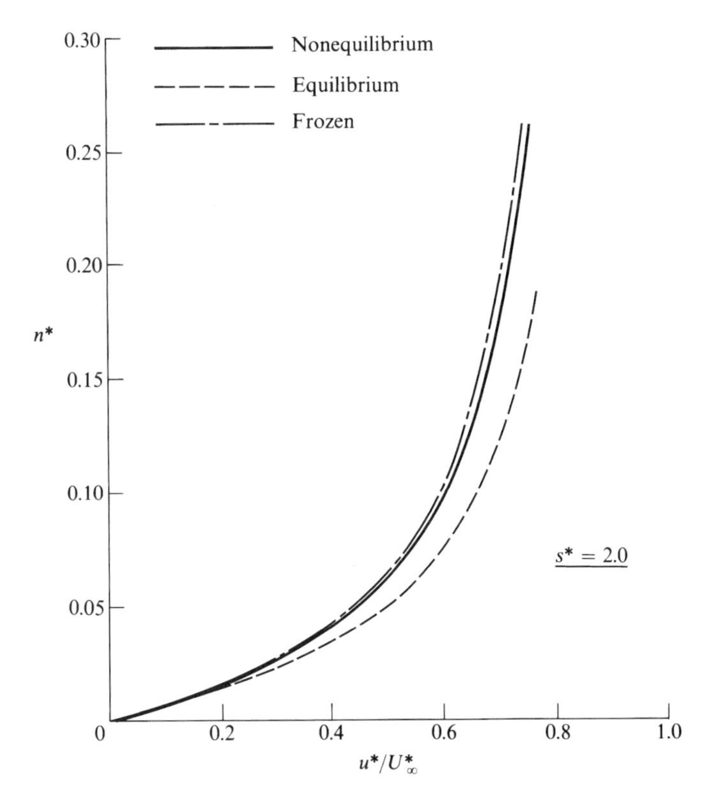

Fig. 17.9 Shock-layer velocity profiles on a hyperboloid. VSL calculations by Moss [199].

Similarly, the species continuity equation is [199]

$$\rho^*\left(\frac{u^*}{1+n^*\kappa^*}\frac{\partial c_i}{\partial s^*}+v^*\frac{\partial c_i}{\partial n^*}\right)=\dot{w}_i^*+\frac{\varepsilon^2}{(1+n^*\kappa^*)(r^*+n^*\cos\phi)^m}$$
$$\times\left\{\frac{\partial}{\partial n^*}\left[(1+n^*\kappa^*)(r^*+n^*\cos\phi)^m\rho^* D_{12}^*\frac{\partial c_i}{\partial n^*}\right]\right\} \tag{17.93}$$

In Eqs. (17.92) and (17.93), the asterisk denotes nondimensional variables defined in part in Sec. 8.2. In addition to those defined in Sec. 8.2, we have

$$c_{p_f}^* = \frac{c_{p_f}}{c_{p_{f\infty}}} \qquad k^* = \frac{k}{\mu_{\text{ref}}\, c_{p_{f\infty}}}$$
$$D_{12}^* = \frac{D_{12}}{\mu_{\text{ref}}/R\rho_\infty} \qquad h_i^* = \frac{h_i}{V_\infty^2}$$

Also $m = 0$ or 1 for two-dimensional or axisymmetric flow, respectively.

As described in Chapter 8, the VSL equations are approximate equations that describe the viscous flow *across the entire shock layer* and hence are a major improvement over the boundary-layer equations for applications where the flowfield is fully viscous between the body and the shock wave. This advantage, described in Sec. 8.2 for a nonreacting flow, carries over to chemically reacting flows. The numerical approach carries over as well. Equations (8.1–8.3), (17.92), and (17.93) are solved numerically by an implicit finite difference method similar to that described in Sec. 6.6 under the subsection entitled Finite Difference Method. The solution begins with an initial data line at the nose of the body and marches downstream in steps of s^*. In the stagnation region, the VSL equations reduce to ordinary differential equations, which are then solved for the stagnation line flow, thus producing the initial data for the downstream-marching solution. See [199] for details.

The work of Moss [199] is one of the first detailed investigations of a chemically reacting viscous shock layer using the VSL technique and in this sense is a classic contribution to the field. Moss considered the cases of frozen, equilibrium, and nonequilibrium laminar flow. Five chemical species were included: O_2, O, N_2, N, and NO. Surface catalysis and mass injection were also treated The viscous shock layer over hyperboloids with included angle of 20 and 45 deg were calculated. Sample results are shown in Figs. 17.9–17.15 for flow over a 45-deg hyperboloid with $R = 2.54$ cm, $V_\infty = 6.10$ km/s, $T_w = 1500$ K, and an altitude of 60.96 km. In Figs. 17.9 and 17.10, shock-layer profiles of velocity and temperature are shown respectively as a function of n^*, at a streamwise location of $s^* = 2$. Three cases are shown in each figure: frozen, equilibrium, and nonequilibrium flow. By comparing Figs. 17.9 and 17.10, note that the temperature is much more sensitive to chemically reacting flow than is the velocity—another indication that the thermodynamic properties rather than the more purely fluid-dynamic variables (such as velocity and pressure) are more affected by chemical reactions. The tops of the curves in Figs. 17.9 and 17.10 correspond to the location of the bow shock wave and hence give n^* at the shock. Once again, we see that the equilibrium shock layer is thinner than the frozen shock layer. Also, note that, for the flowfield conditions in these figures, the nonequilibrium flow is closer to frozen than to equilibrium. The surface-pressure distribution is given in Fig. 17.11 and graphically demonstrates the insensitivity of pressure to the chemically reacting effects. Indeed, the detailed, chemically reacting viscous-flow results are reasonably predicted by modified Newtonian theory (see Sec. 3.2). Catalytic wall effects on the chemical species profiles are shown in Fig. 17.12. The nonequilibrium flow is calculated for two cases, an equilibrium catalytic wall and a noncatalytic wall. The abscissa is the nondimensional distance across the shock layer n/n_s, where n_s is the coordinate of the shock wave. The effect of the catalytic wall reaches across more than 70% of the shock layer and, of course, is strongest near the wall. The catalytic wall effect on the heat-transfer distribution along the body surface is shown in Fig. 17.13. These results are consistent with our discussion surrounding Fig. 17.5 from Fay and Riddell. Note in Fig. 17.13 that the nonequilibrium heat transfer is reduced by a noncatalytic wall in comparison to an equilibrium catalytic wall. Also, note that the nonequilibrium, equilibrium catalytic wall

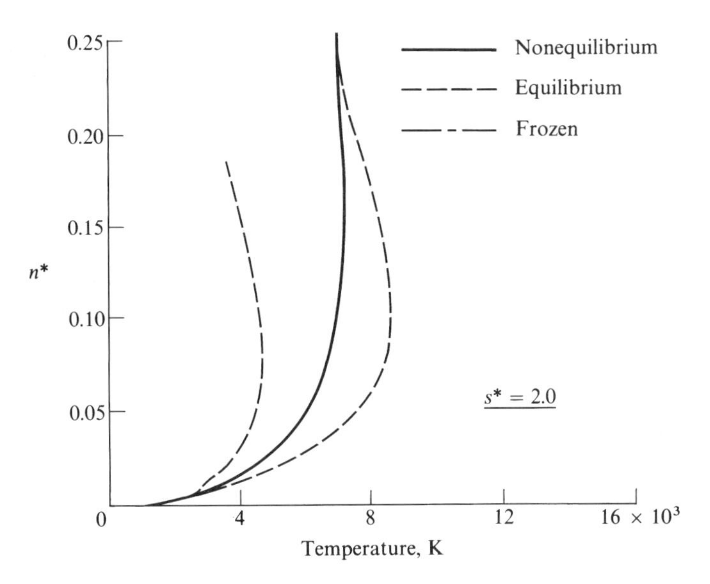

Fig. 17.10 Shock-layer temperature profiles on a hyperboloid. VSL calculation by Moss [199].

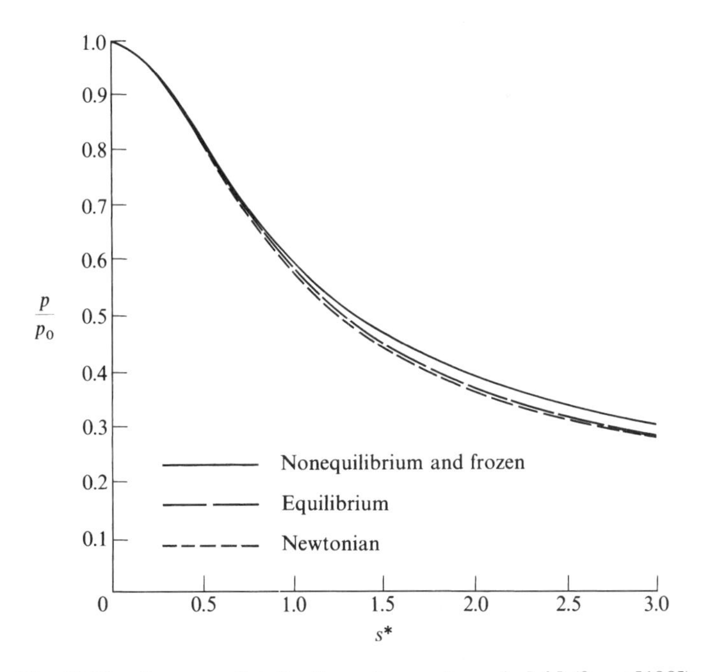

Fig. 17.11 Pressure distributions along a hyperboloid (from [199]).

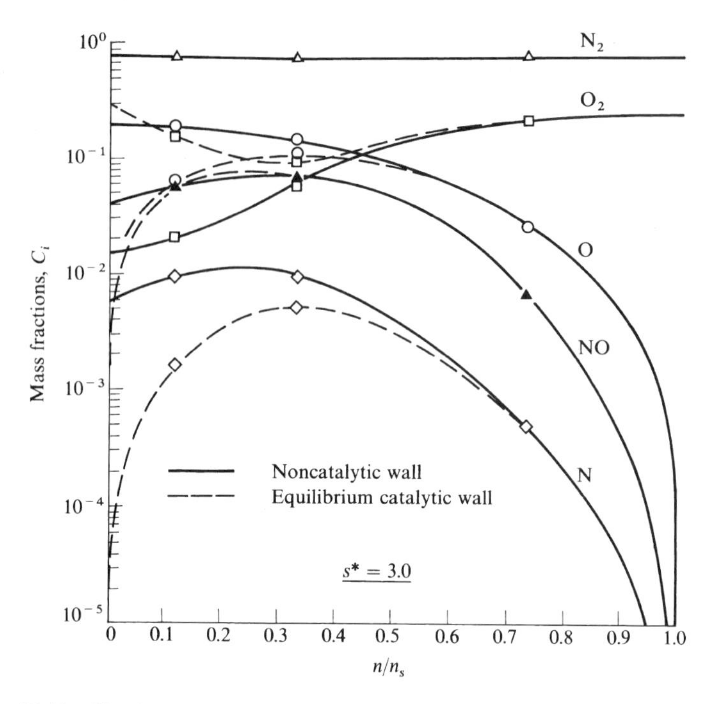

Fig. 17.12 Shock-layer mass fraction profiles on a hyperboloid (from [199]).

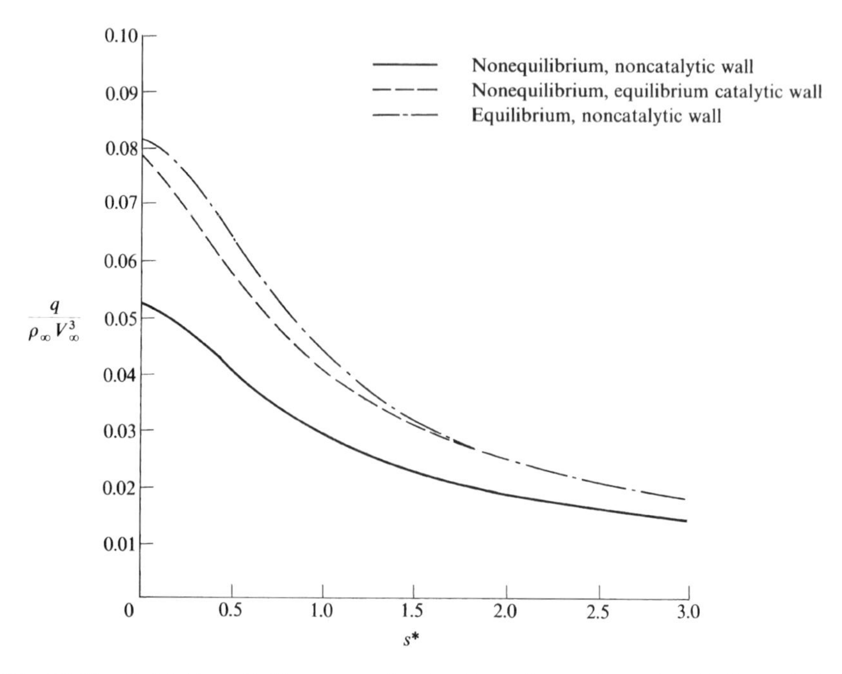

Fig. 17.13 Catalytic wall effects on surface heat transfer on a hyperboloid (from [199]).

case yields essentially the same heat transfer as the local chemical equilibrium flow. Finally, note that the relative influence of wall catalyticity diminishes as a function of downstream distance.

In [199], Moss carried out an interesting comparison of multicomponent vs binary diffusion. One set of nonequilibrium flow calculations was carried out using the detailed multicomponent diffusion coefficients (see Sec. 16.3), and another set assumed binary diffusion in the spirit set forth by Fay and Riddell described in Sec. 17.6. Figure 17.14 gives the results for chemical species profiles across the shock layer at a stream wise station of $s^* = 1$. The wall is assumed noncatalytic. Note that the two cases are in reasonable agreement with each other, thus indicating that the assumption of binary diffusion for high-temperature air (at least in the range of dissociation) is reasonable.

All of the preceding results were obtained with no mass injection into the shock layer through the wall. Moss examined the case of wall mass injection; Fig. 17.15 gives results for heat transfer at the stagnation point with mass injection q divided by its value with no mass injection, $(q)_{\dot{m}_0=0}$, vs the mass-injection rate $\dot{m}_0$. This figure demonstrates the important fact that mass injection dramatically reduces heat transfer to the surface. Indeed, when the mass-injection rate equals 0.4 of the freestream mass flux, the viscous layer

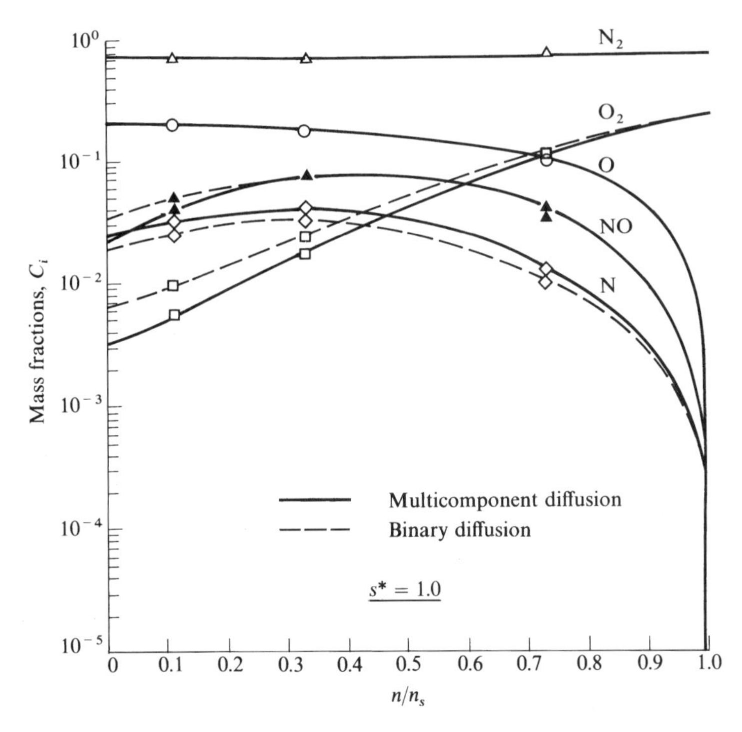

Fig. 17.14 Shock-layer profiles of species mass fractions; comparison between multicomponent and binary diffusion (from [199]).

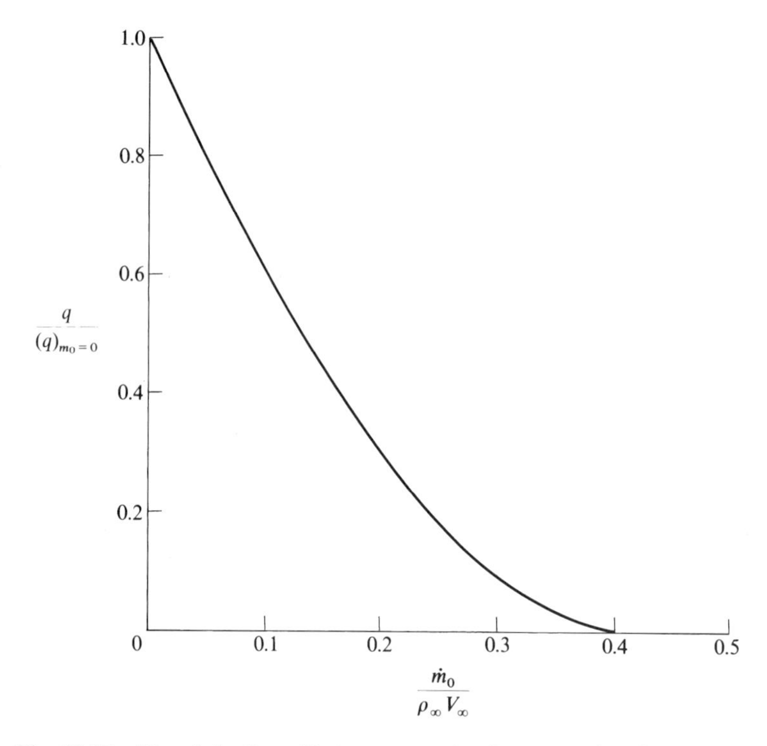

Fig. 17.15 Mass-injection effect on stagnation heat transfer (from [199]).

is blown completely off the surface, and the aerodynamic convective heat transfer is zero.

Reference [199] is a classic application of the VSL technique to chemically reacting viscous flows. The reader is strongly encouraged to study [199] for more details and insight. A large number of similar applications have been made by Lewis et al.; samples of this work are represented by [200] and [201], which deal with the VSL technique applied to the chemically reacting flow over the windward surface of the space shuttle. Another space shuttle application is discussed by Shinn et al. [202]. In this work, catalytic wall effects on convective heating along the centerline of the space shuttle were studied using the VSL technique. Figure 17.16 gives some typical results, obtained from [202]. Here, the convective heating rate is given as a function of distance along the windward centerline, for an altitude of 71 km and a velocity of 6.73 km/s. The solid line denotes equilibrium flow, the dash/dot line denotes nonequilibrium flow with a completely noncatalytic wall, and the dashed line is the result for a wall with finite catalytic rates. The open symbols are flight data from the shuttle. Clearly, the space shuttle experiences some finite wall catalytic effects, and the magnitude of this effect is not trivial.

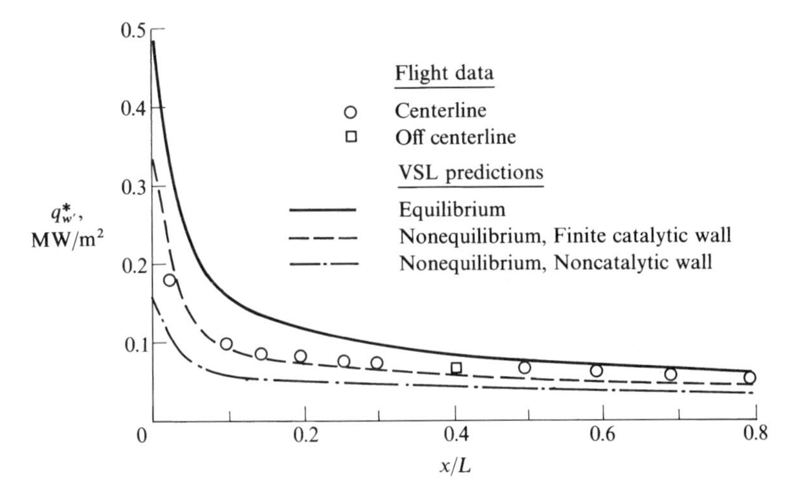

Fig. 17.16 Catalytic wall effects on convective heating along the windward centerline of the space shuttle: V_∞ = 6.73 km/s, and altitude = 71 km (from Shinn et al. [202]).

17.9 Parabolized Navier–Stokes Solutions to Chemically Reacting Flows

The calculation of a fully viscous, nonreacting shock layer by means of the parabolized Navier–Stokes equations was discussed in Sec. 8.3, which should be reviewed by the reader before progressing further. In this section, we will briefly treat a PNS application to equilibrium and nonequilibrium chemically reacting flow as carried out in the recent work of Prabhu and coworkers [203] and [204].

As discussed in Sec. 8.3, the parabolized Navier–Stokes equations are obtained from the full Navier–Stokes equations by dropping the viscous terms that involve derivatives in the streamwise direction. This approach carries over to chemically reacting flows, where the PNS global continuity and momentum equations carry over unchanged from Eqs. (8.14–8.17), and the PNS energy and species continuity equations are obtained from Eqs. (17.5), (17.6), and (17.17), wherein all x derivatives in the viscous terms (including diffusion) are neglected. In this manner, the resulting simplified equations are parabolic and can be numerically integrated by a steady-state finite difference technique marching in the streamwise direction.

In [203] the PNS equations were used to solve for the complete, viscous laminar flow over the space shuttle, assuming flow in local chemical equilibrium. Results are given for $V_\infty = 6.74$ km/s, altitude = 71.32 km, and an angle of attack of 40 deg. For comparison, results assuming a calorically perfect gas with $\gamma = 1.2$ (to simulate the high-temperature effects) were also obtained. Figure 17.17 gives the comparison between the calculated Mach-number contours for γ = constant = 1.2 (Fig. 17.17a) and the corresponding contours for

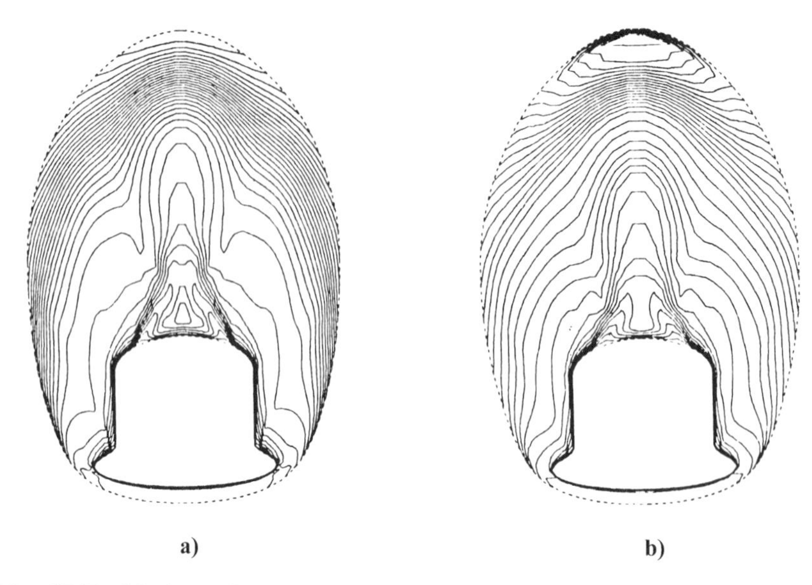

Fig. 17.17 Mach-number contours from PNS calculations of the viscous flow over the space shuttle where $x/L = 0.4$: a) calorically perfect gas with $\gamma = 1.2$ and b) equilibrium chemically reacting air (from Prabhu and Tennehill [203]).

equilibrium air (Fig. 17.17b), for a cross section of the shuttle flowfield located at a distance $x/L = 0.4$ downstream of the nose, where L is the body length. It is interesting to note that the $\gamma = 1.2$ results simulate the equilibrium air results fairly closely; this is an example where, if the right "effective value" of γ can be found, then a calorically perfect-gas calculation can simulate some aspects of a chemically reacting flowfield.

In [204], the PNS equations were used to solve for the nonequilibrium chemically reacting laminar flow over cones at angle of attack. Sample results are shown in Figs. 17.18 and 17.19 for the case of $V_\infty = 8100$ m/s, $T_w = 1200$ K, and altitude = 60.96 km. The wall is noncatalytic. Figure 17.18 shows the profiles of atomic oxygen mass fraction c_O as a function of distance across the shock layer at $x = 3.5$ m downstream of the tip. Three profiles are shown: 1) the solid line for zero degree angle of attack, 2) the solid circles for the windward meridian line $\alpha = 10$ deg, and 3) the open circles for the leeward meridian line at $\alpha = 10$ deg. To be expected, at angle of attack more dissociation occurs on the windward side than on the leeward side because of the higher temperature flow on the bottom of the cone. Also, we would expect the flow on the windward side (where p and T are higher) to be closer to equilibrium than the leeward-side flow (where p and T are lower). Heat-transfer results are given in Fig. 17.19 for the same case. Here, C_H is given as a function of x and shows the expected result of much higher heat transfer on the windward side compared to the leeward side.

The application of the PNS equations to chemically reacting flows is presently a state-of-the-art research problem. The discussion in the present section is

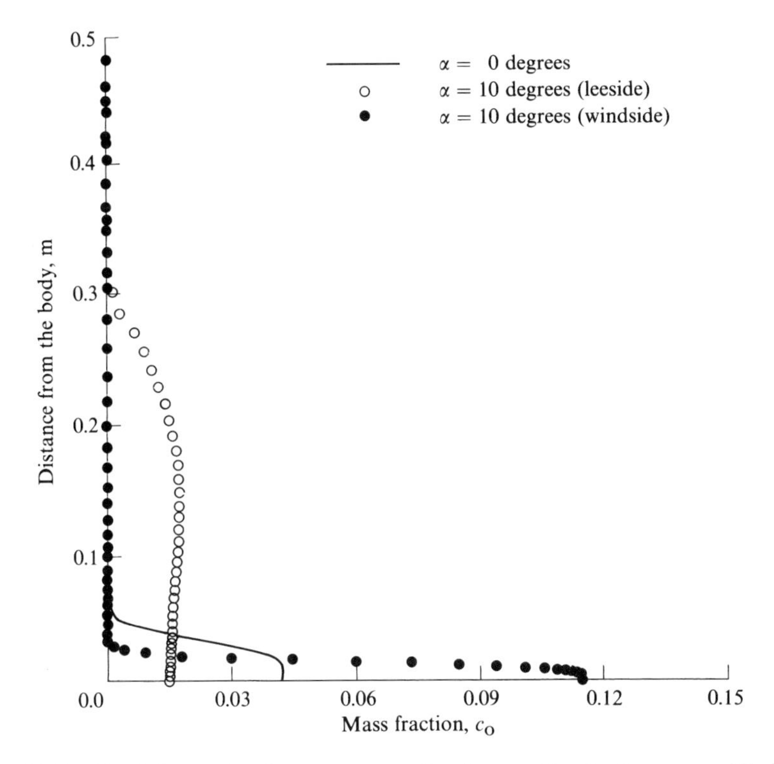

Fig. 17.18 Shock-layer atomic-oxygen mass fraction profiles for the nonequilibrium flow over a 10-deg cone at angle of attack. PNS calculations of Prabhu and Marvin [204].

intended to give only the flavor of such work. The reader is encouraged to keep track of the literature in this field. References [203] and [204] are just samples of the literature; the work of Bhutta et al. [205] is of particular note.

17.10 Full Navier–Stokes Solutions to Chemically Reacting Flows

The calculation of nonreacting viscous flows by means of the full Navier–Stokes equations is discussed in Sec. 8.4, which should be reviewed by the reader before progressing further. These calculations are time-marching solutions, and the basic approach and philosophy is carried over to the case of chemically reacting flow as well. The time-marching solution of the complete Navier–Stokes equations for chemically reacting flow involves the finite difference solution of the governing equations given by Eqs. (17.1), (17.6), and (17.17), and for alternate forms of the energy equation, Eqs. (17.25), (17.31), or (17.40). Examples of such time-marching solutions for complex reentry flowfields are given by the work of Gnoffo and coworkers represented by [206] and [207].

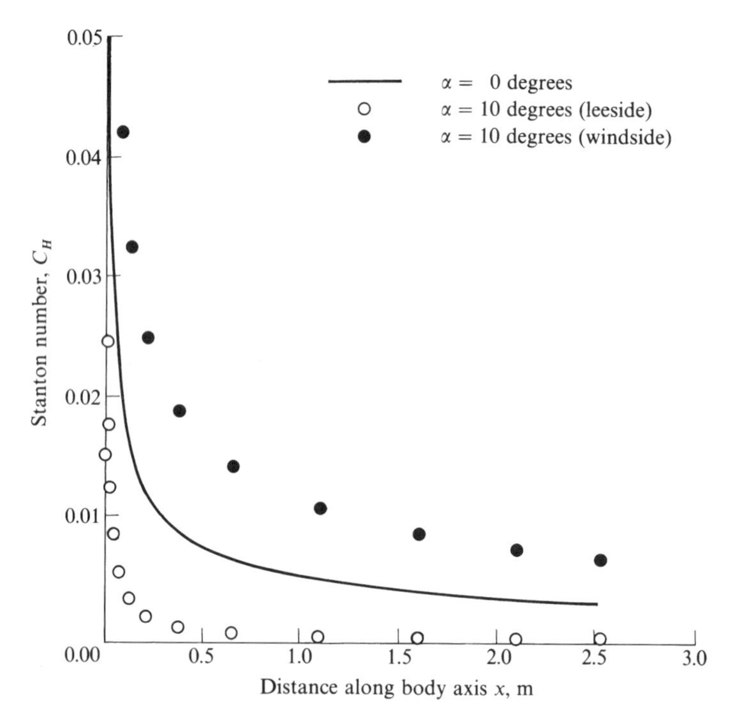

Fig. 17.19 Heat-transfer distributions on a 10-deg cone at angle of attack. Nonequilibrium flow (from [204]).

In [266], Gnoffo discusses recent Navier–Stokes calculations for the chemically reacting flowfield over the Mars Pathfinder vehicle. The front face of the entry vehicle is a spherically blunted, 70-deg half-angle cone, as shown in Fig. 17.20. In this figure, the changes in the sonic line shape and location are shown as the altitude and Mach-number change during entry in the Martian atmosphere. The light grey region is the subsonic flow region, and the darker regions are the zones of supersonic flow behind the bow shock. The Mach numbers range from 22.3 (left), 16 (center), and 9.4 (right), where the shock-layer flowfield changes from frozen flow, to nonequilibrium flow, and then to equilibrium flow. These calculations are at the cutting edge of the state of the art. For this reason, no details will be given here; rather, the reader is encouraged to keep up with the rapidly growing literature in the field.

To illustrate the application of the complete Navier–Stokes equations to a nonequilibrium chemically reacting flow, we choose to discuss here a simpler example, namely, the mixing flow in the cavity of an HF chemical laser. The principle of a high-energy chemical laser was discussed in Sec. 9.1. The first Navier–Stokes solutions to such flows were carried out by Kothari et al. as reported in [208] and [209] from which the following is taken.

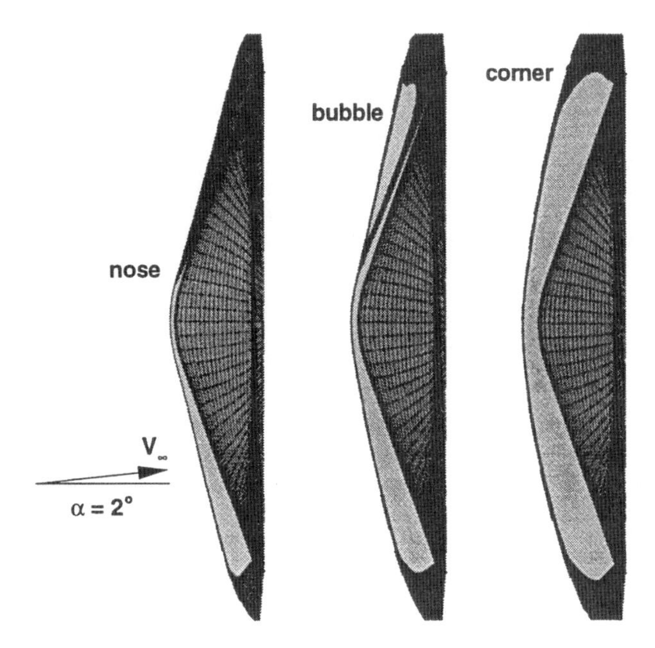

Fig. 17.20 Sonic line locations for the Mars Pathfinder at 2-deg angle of attack in the symmetry plane at Mach 22.3 (left), 16 (center), and 9.4 (right) showing the effect of gas chemisry (Gnoffo [266]).

The physical problem is sketched in Fig. 17.21, which shows two supersonic parallel streams mixing with each other, one stream of partially dissociated fluorine and the other of hydrogen. In the mixing region downstream of the nozzle exits, the following hypergolic chemical reactions take place:

$$F + H_2 \rightleftharpoons HF^*(v) + H$$
$$F_2 + H \rightleftharpoons HF^*(v) + F$$

In the preceding, $HF^*(v)$ denotes a vibrationally excited state of HF, where HF is formed directly in the excited vth vibrational energy level as a direct product of the chemical reaction. In turn, this vibrationally excited $HF^*(v)$ can contribute to a population inversion and hence laser action in the downstream mixing flow. (See [210] for a basic discussion of nonequilibrium effects associated with lasers and for the significance of a population inversion.) Therefore, in such a chemical laser flow, we are dealing with both thermal and chemical nonequilibrium.

In [209] and [210], the flowfield sketched in Fig. 17.21 was calculated by solving the two-dimensional form of Eqs. (17.1–17.6) and (17.17). In Eq. (17.17), each vibrational level of HF was treated as a different "species," so that the nonequilibrium system consisted of 14 species and 100

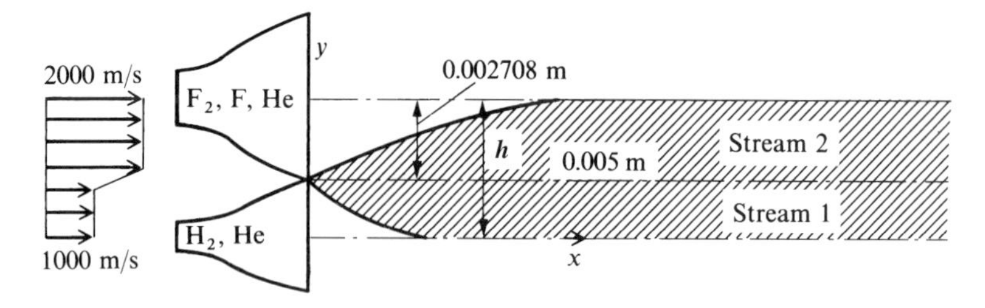

Nozzle configuration

	Stream 1	Stream 2
P, n/m^2	500	500
T, K	150	150
ρ, kg/m^3	1.2862×10^{-3}	2.4514×10^{-3}
ρ_{F_2}, kg/m^3		7.3128×10^{-4}
ρ_{H_2}, kg/m^3	3.2328×10^{-4}	
ρ_F, kg/m^3		2.4376×10^{-4}
ρ_H, kg/m^3		
ρ_{He}, kg/m^3	9.6288×10^{-4}	1.4764×10^{-3}

Fig. 17.21 Chemical laser model used by Kothari et al. [208].

elementary reactions. A time-marching solution of these equations was carried out using MacCormack's explicit, predictor-corrector, finite difference scheme, which is described in earlier sections of this book, starting with Sec. 5.3. An application of MacCormack's technique to nonequilibrium flows is discussed in Sec. 15.5. Hence, no computational details will be given here. However, we will mention one aspect of the calculational technique applied to nonequilibrium flows that is brought out in [209] and [210]. That is, in the time-marching calculation of a nonequilibrium flow it is suggested to advance the flowfield initially with the chemical reactions turned off, in order for the fluid-dynamic aspects of the flow to begin to establish themselves (for example, for shocks to form, for shear layers to occur, for mixing to take place, etc.). Then, at some intermediate time the nonequilibrium chemistry is switched on, and the coupled fluid-dynamic and chemical aspects of the flow subsequently evolve to the steady state at large times. This suggestion is illustrated in Fig. 17.22, which shows the variation of temperature at a particular grid point for the flow in Fig. 17.21. Cold flow (no chemical reactions) is calculated for about half the total elapsed nondimensional time [given by $t/(h/V)$, where V is a reference

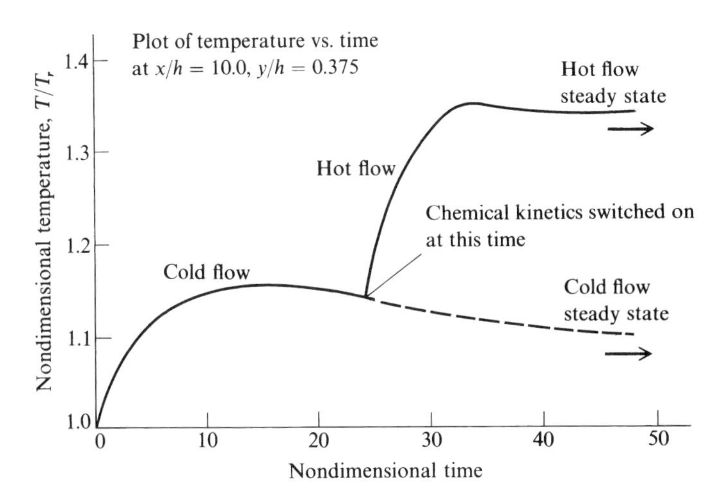

Fig. 17.22 Illustration of the cold-flow/hot-flow method used in the complete Navier–Stokes solution of chemical laser flows (from [208]).

velocity and h is the height of the mixing duct]; then, the chemical reactions are switched on, and the hot flow (with chemical reactions) is carried out to the steady state. This approach is recommended for the general solution of nonequilibrium viscous flows when time-marching methods are used. In this way, some extreme transients that might be induced by the finite-rate chemical reactions near time zero are avoided, thus avoiding possible numerical instabilities. For example, in the flow shown in Fig. 17.21, the initial conditions involve a slab of dissociated fluorine (stream 2) in direct contact with a slab of H_2 (stream 1). If this were to exist momentarily in real life inside the duct, an actual explosion would occur. Therefore, to avoid an analogous "numerical explosion," the cold flow is allowed to mix for the early time steps before the chemistry is turned on; in this fashion, when the chemistry is finally turned on, the chemical energy changes will be more evenly distributed over the whole flowfield rather than occurring in a local region as they would at $t = 0$.

A sample steady-flow result for this problem is shown in Fig. 17.23. Here, the density profiles of the first three vibrational energy levels ($v = 0$, 1, and 2) of HF are shown in the mixing region as a function of distance across the flow at a given streamwise location, $x/h = 5$. Note that near the middle of the flow (say, at $y/h = 0.4$). $\rho_{HF(2)} > \rho_{HF(1)} > \rho_{HF(0)}$, that is, there are more HF molecules in the higher-lying vibrational energy levels than in the lower levels. By definition, this is called a *population inversion* and is the basis of laser action in the gas. Recalling that the Boltzmann distribution for a gas in vibrational equilibrium called for $\rho_{HF(2)} < \rho_{HF(1)} < \rho_{HF(0)}$ we see that Fig. 17.23 illustrates a case of *high-vibrational nonequilibrium*. The reader is encouraged to read [208] and [209] for more details. Also, [211] is another source for such matters.

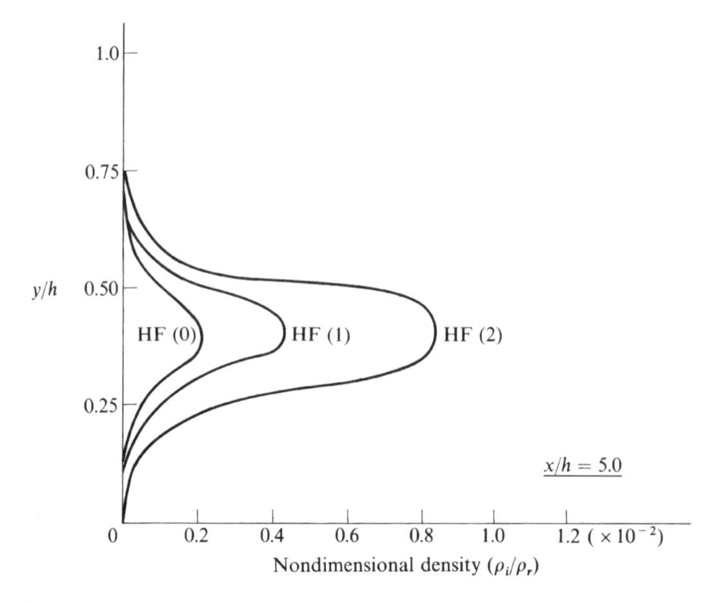

Fig. 17.23 Steady-state density profiles of three vibrational levels in HF in a chemical laser. Navier–Stokes calculations of [208].

17.11 Summary and Comments

This brings to an end our discussion of chemically reacting viscous flow. We have discussed the basic equations for such flows and some of the important physical mechanisms, such as diffusion and catalytic surfaces. We have applied these equations to a series of flows, progressing from boundary-layer solutions (self-similar and nonsimilar), through various viscous shock-layer techniques (VSL and PNS), and ending with Navier–Stokes solutions. The subject matter represented by this chapter is vast, and we have only scratched the surface here. Our purpose has been to discuss the fundamental principles and ideas and to illustrate these with selected examples of chemically reacting viscous flows. These examples have been carefully selected to illustrate the important *physical behavior* of high-temperature viscous flows, and the reader should interpret the thrust of this chapter accordingly.

With this chapter, we have completed a major milestone in our road map in Fig. 1.24. With this, we move on to our last item under high-temperature flow, namely, radiating flow.

Problems

17.1 Starting with the model of a finite control volume fixed in space (with the flow moving through the control volume), derive Eq. (17.20).

17.2 Consider a viscous flow over a porous surface. The flow can be a pure gas, or a mixture of different gases. A gas of a completely different species i is injected through the porous surface into the main viscous flow. The mass flux (mass per unit area per unit time) of this injected gas is $\dot{m}_w$. Prove that the proper boundary condition at the surface for the solution of the main viscous flow equations including mass injection is $\dot{m}_w = \rho v_w$, where ρ is the density of the *mixture* at the wall (*not* the density of the injected species ρ_i), and v_w is the vertical component of the gas mixture velocity at the wall. (This problem is an adjunct to the discussion of boundary conditions in Sec. 17.5.)

17.3 Derive Eqs. (17.74), (17.75), and (17.76), namely, the transformed equations for a chemically reacting viscous flow.

17.4 Derive Eq. (17.87), namely, the transformed species continuity equation for a viscous stagnation-point flow.

17.5 Consider the chemically frozen laminary boundary-layer flow of dissociated oxygen over a flat plate. The wall is fully catalytic to oxygen recombination, and the wall temperature is cool, say, less than 1000 K. If $Le = Pr = 1$, the profile of atomic mass fraction through the boundary layer is a function of the velocity profile only, that is, $c_O = f(u)$. Derive this function.

18
Introduction to Radiative Gas Dynamics

> Light seeking light doth light of light beguile.
>
> *William Shakespeare,*
> *Love's Labours Lost, 1594*

Chapter Preview

This chapter, for the most part, deals with extremes—extremes of temperature and the consequences. Depending on where you are as you are reading this page, presumably you are fairly comfortable temperature-wise. If you are in a room at some reasonable room temperature, the walls of the room are also at room temperature. Moreover, the walls of the room are emitting thermal radiation, some of which you are absorbing, even though you are not really aware of it. On the order hand, if the temperature of the walls of the room were to suddenly jump to 3000 K (hypothetically), you most certainly would be aware of it. The radiative energy from the walls would suddenly become unbearable. This is perhaps a ridiculous example, but it serves to get your attention about the large amount of thermal radiation that is associated with high temperatures.

As noted in Sec. 1.3.4, the shock layer in the nose region of the Apollo reentry capsule reached 11,000 K during its return through the Earth's atmosphere. Now this is an extreme temperature, and the gas in the shock layer emitted a lot of radiative energy. The energy radiated from the gas had two major physical consequences: 1) the shock layer lost energy to its surroundings, that is, the shock layer became *nonadiabatic*, and 2) radiative heat transfer to the body constituted over 30% of the total heating rate to the body. These are serious consequences.

In this chapter you will learn how to calculate these consequences. Radiative transport is a different physical mechanism than any we have studied so far in this book. So read on, and relish the difference.

18.1 Introduction

Consider a flowfield where the temperature is high enough that a fluid element *radiates* a substantial amount of energy. For air, this threshold temperature is about 10,000 K. In such a flow, the fluid element *loses* energy because of the *emission* of radiation, but it can also *gain* energy because of the *absorption* of radiation emitted from other fluid elements in the high-temperature flow. Two major consequences of this radiation are as follows:

1) The flowfield becomes *nonadiabatic*, that is, a flowfield wherein before we considered the total enthalpy as a constant (such as inviscid shock-wave flows, nozzle flows, etc.) now becomes one in which $h_0 = h + V^2/2$ is a *variable.*

2) In addition to the ordinary convective heat transfer q_c to a surface in the flow, we now have a new component, namely, *radiative heating of the surface* q_R. Thus, the total surface heat transfer becomes

$$q = q_c + q_R \tag{18.1}$$

where q_c includes both conduction and diffusion effects, as described in Chapters 16 and 17.

The radiation and the fluid flow are, in general, *coupled*, that is, the radiative intensity within the flowfield depends on ρ and T in the flow, and, in turn, the flowfield properties are influenced by the radiative intensity. This is the essence of the *radiative-gas dynamic interaction* effect, which is the main subject of this chapter.

In this conjunction, we introduce two definitions concerning the radiative nature of the gas:

1) *Transparent gas* is a gas that emits but does not absorb radiation. In such a gas, all radiation that is emitted from within the gas escapes to the surroundings.

2) *Self-absorbing gas* is a gas that emits and absorbs radiation. Some of the radiation that is emitted escapes to the surroundings, and some is self-absorbed by the gas, thus trapping some of the radiative energy within the flow.

The radiative nature of the gas can have a major impact on the proper analysis of a flow. For example, a self-absorbing gas is elliptic in nature, even though the flow might be inviscid and supersonic. This is because radiation that is emitted by downstream fluid elements can be absorbed by upstream fluid elements, thus feeding information *upstream* in the flow—a mathematically elliptic behavior. Consequently, such a flow would, in general, have to be calculated by a time-marching solution for the same mathematical reasons given in Sec. 5.3. On the other hand, this upstream effect does *not* occur for a transparent gas, which is affected only by local radiative emission.

In this chapter, we will give a brief picture of radiative gas dynamics. Our purpose is to make the reader feel comfortable with the basic concepts and appreciative of the physical trends.

18.2 Definitions of Radiative Transfer in Gases

There are two basic quantities that describe the transfer of radiation through a gas: radiative *intensity* and radiative *flux*. First, consider the definition of radiative intensity. Consider a given arbitrary direction r in a radiating gas, as sketched in

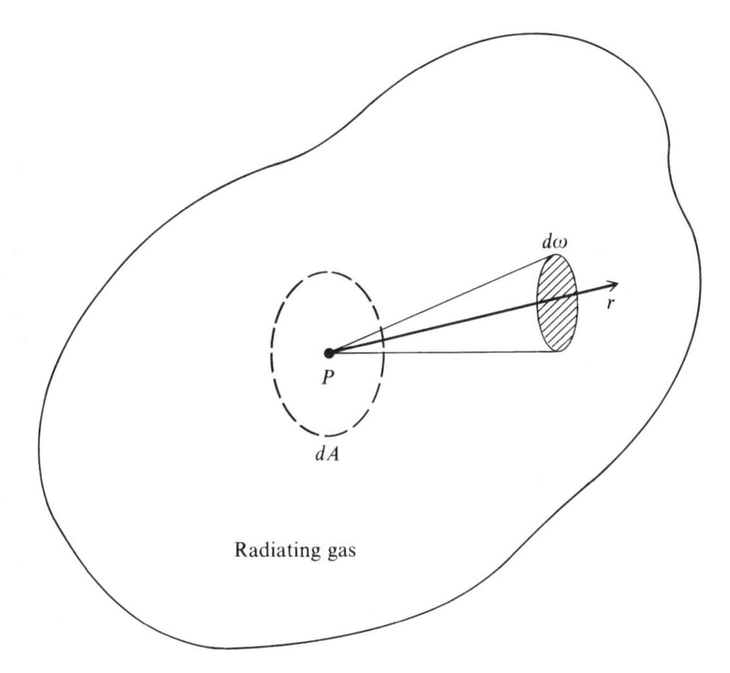

Fig. 18.1 Geometric model for radiative intensity.

Fig. 18.1. Consider also an area dA at point P *perpendicular* to the direction r and a solid angle dω about r Let dE' be the radiative energy in the frequency interval between ν and $\nu + \mathrm{d}\nu$ transmitted through dA during a time interval dt from all directions contained within the solid angle dω. Then, the *specific radiative intensity* I_ν is defined as

$$I_\nu \equiv \lim \left[\frac{\mathrm{d}E'}{\mathrm{d}A \ \mathrm{d}\omega \ \mathrm{d}\nu \ \mathrm{d}t} \right] \qquad (18.2)$$

$$\mathrm{d}A, \mathrm{d}\omega, \mathrm{d}\nu, \mathrm{d}t \to 0$$

that is, at a point P in the gas, I_ν is the radiative energy transferred in the r direction across a unit area perpendicular to r, per unit frequency, per unit time, per unit solid angle. Note that I_ν is directional in nature. When we talk about the intensity at point P, we also have to say the intensity in what *direction.*

The *radiative flux* is defined as the energy per second crossing a unit area caused by intensity coming from all directions. Let q_ν be the radiative flux per unit frequency. Then,

$$q_\nu = \int_\omega I_\nu \cos \theta \, \mathrm{d}\omega \qquad (18.3)$$

where θ is defined in Fig. 18.2 as the angle between an arbitrary direction L and a unit normal vector $\boldsymbol{n}$ perpendicular to the elemental area dA. We can write the directional variation of intensity as $I_\nu = I_\nu(\theta, \phi)$, where ϕ is also shown in Fig. 18.2. Also, using the geometry of Fig. 18.2, the solid angle dω is defined as the included area dσ divided by L^2:

$$\mathrm{d}\omega \equiv \frac{\mathrm{d}\sigma}{L^2} = \frac{(L\,\mathrm{d}\theta)(L\sin\theta\,\mathrm{d}\phi)}{L^2} = \sin\theta\,\mathrm{d}\theta\,\mathrm{d}\phi$$

Hence, Eq. (18.3) becomes

$$\boxed{q_\nu = \int_0^{2\pi}\int_0^{\pi} I_\nu(\theta, \phi)\cos\theta\sin\theta\,\mathrm{d}\theta\,\mathrm{d}\phi} \tag{18.4}$$

The *total* radiative flux, integrated over all frequencies, is

$$\boxed{q = \int_0^{\infty}\int_0^{2\pi}\int_0^{\pi} I_\nu(\theta, \phi)\cos\theta\sin\theta\,\mathrm{d}\theta\,\mathrm{d}\phi\,\mathrm{d}\nu} \tag{18.5}$$

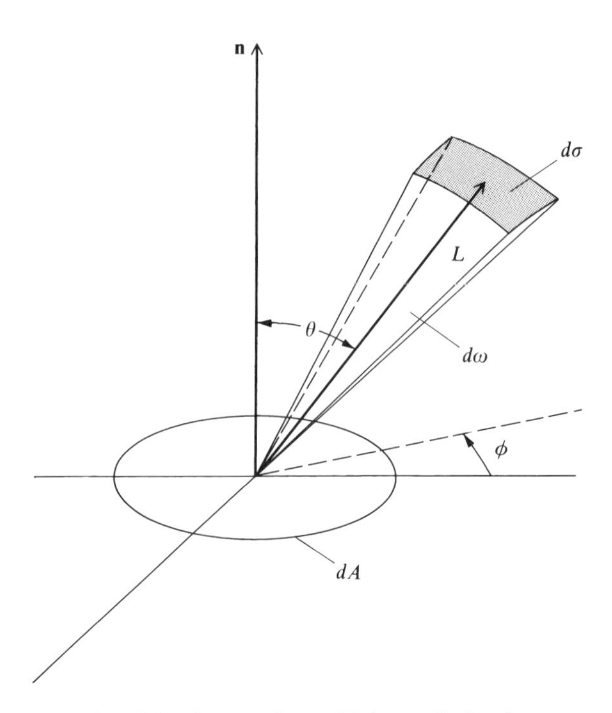

Fig. 18.2 Geometric model for radiative flux.

Finally, we note that a classical *blackbody* (see any good physics test) will emit radiative *intensity* of a value

$$B_\nu = \frac{2h\nu^3}{c^2(e^{h\nu/kT} - 1)} \tag{18.6}$$

where B_ν is the blackbody radiative intensity at frequency ν and temperature T, h is Planck's constant, k is the Boltzmann constant, and c is the speed of light.

18.3 Radiative-Transfer Equation

Consider an element in a radiating gas, as sketched in Fig. 18.3. Radiation of intensity I_ν in the s direction is incident on this element. Within the element, the local value of I_ν will be *increased* by *emission* and *decreased* by *absorption.* By definition, let

$$J_\nu \, \mathrm{d}s = \text{energy emitted}$$

and

$$\kappa_\nu I_\nu \, \mathrm{d}s = \text{energy absorbed}$$

where J_ν is called the *emission coefficient* and κ_ν is the *absorption coefficient.* Then the change in I_ν, namely, $\mathrm{d}I_\nu$, because of the element of gas shown in Fig. 18.2 is

$$\mathrm{d}I_\nu = J_\nu \, \mathrm{d}s - \kappa_\nu I_\nu \, \mathrm{d}s$$

or

$$\frac{\mathrm{d}I_\nu}{\mathrm{d}s} = J_\nu - k_\nu I_\nu \tag{18.7}$$

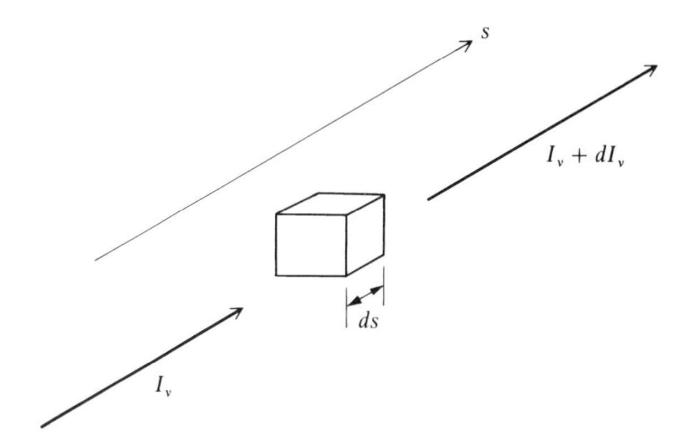

Fig. 18.3 Geometric model for the radiative-transfer equation.

Equation (18.7) is the fundamental form of the *radiative-transfer equation*. This equation allows us to calculate I_ν as a function of distance s through a gas. Moreover, the term $\mathrm{d}I_\nu/\mathrm{d}s$ has dimensions of energy per second per unit solid angle per unit frequency per *unit volume*. When integrated over all frequencies and all solid angles, it therefore represents the local change in energy of a fluid element per unit volume caused by radiation. For example, consider Eq. (17.31), which is the energy equation for a high-temperature, chemically reacting, radiating, viscous flow, repeated here:

$$\rho\frac{\mathrm{D}h}{\mathrm{D}t} = \nabla\cdot(k\nabla T) - \nabla\cdot\sum_i \rho_i \mathbf{U}_i h_i - \nabla\cdot\mathbf{q}_R + \frac{\mathrm{D}p}{\mathrm{D}t} + \Phi \tag{18.8}$$

In this equation, the radiation term is given by $\nabla\cdot q_R$; from our preceding discussion, this term is given by the radiative-transfer equation (18.7), as

$$\nabla\cdot\mathbf{q}_R = \int_0^\infty\int_{4\pi} J_\nu\,\mathrm{d}\omega\,\mathrm{d}\nu - \int_0^\infty\int_{4\pi}\kappa_\nu I_\nu\,\mathrm{d}\omega\,\mathrm{d}\nu \tag{18.9}$$

In Eq. (18.9), the two terms on the right-hand side represent local radiative emission and absorption, respectively. Because the fluid element emits energy equally in all directions, then

$$\int_0^\infty\int_{4\pi} J_\nu\,\mathrm{d}\omega\,\mathrm{d}\nu = 4\pi\int_0^\infty J_\nu\,\mathrm{d}\nu = 4\pi J$$

where J is the total radiative emission per second per unit volume. Hence, Eq. (18.9) can be written as

$$\boxed{\nabla\cdot\mathbf{q}_R = 4\pi J - \int_0^\infty\int_{4\pi}\kappa_\nu I_\nu\,\mathrm{d}\omega\,\mathrm{d}\nu} \tag{18.10}$$

In turn, Eq. (18.10) is the form of the radiation term that appears in the gas dynamic energy equation, such as Eq. (18.8).

Finally, we note that, if the gas were a blackbody with radiative intensity given by Eq. (18.6), the value of I_ν does not change, that is, I_ν is independent of distance through the blackbody, and is equal to B_ν. If we apply the radiative-transfer equation to a blackbody, Eq. (18.7) becomes

$$\frac{\mathrm{d}I_\nu}{\mathrm{d}s} = 0 = J_\nu - \kappa_\nu B_\nu$$

or

$$J_\nu = \kappa_\nu B_\nu \tag{18.11}$$

However, the radiation *emitted* by a gas (neglecting induced emission) is independent of the incident radiative intensity, blackbody or not. Therefore, Eq. (18.11) must be a *general* result for the emission coefficient, even though the surrounding radiation might not be blackbody. Thus, we can write, for the general form of the radiative-transfer equation

$$\boxed{\frac{dI_\nu}{ds} = \kappa_\nu B_\nu - \kappa_\nu I_\nu} \tag{18.12}$$

and for J in Eq. (18.10) we can write, in general,

$$\boxed{J = \int_0^\infty \kappa_\nu B_\nu \, d\nu} \tag{18.13}$$

18.4 Solutions of the Radiative-Transfer Equation: Transparent Gas

Consider an arbitrary volume of a radiating, transparent gas, as sketched in Fig. 18.4. We wish to calculate the radiative flux across the boundary of this volume at point P, because of all of the radiating gas inside the volume. Consider an infinitesimal volume element $d\mathcal{V}$, as sketched in Fig. 18.4. This elemental volume is at a distance r from an elemental surface area dA located at point P. The angle between r and the normal dA is β. Assuming the gas does not absorb (i.e., assuming a transparent gas), the energy emitted from $d\mathcal{V}$, which

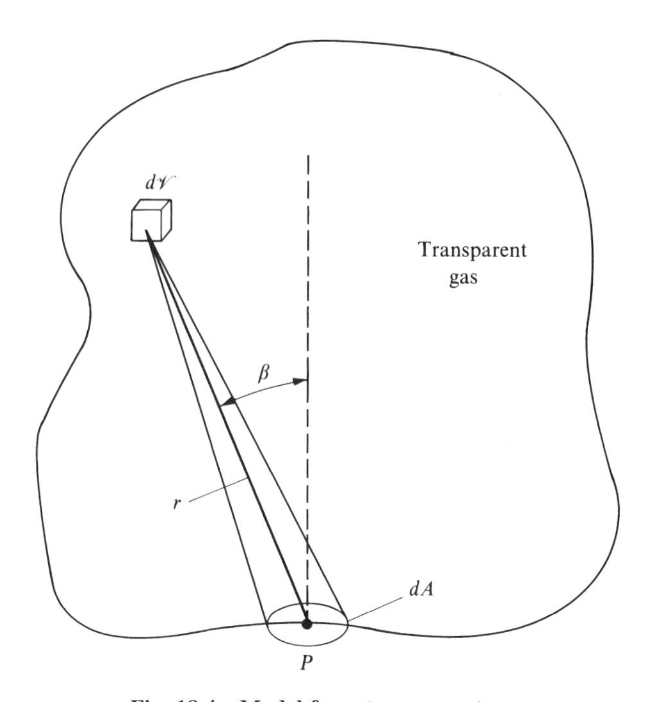

Fig. 18.4 Model for a transparent gas.

crosses $\mathrm{d}A$ per second, is

$$\underbrace{(J\,\mathrm{d}\mathscr{V})}_{\text{Energy per unit solid angle}} \times \underbrace{\frac{\mathrm{d}A\cos\beta}{r^2}}_{\text{Solid angle intercepted by }\mathrm{d}A}$$

Hence, integrating over all of the gas volume and dividing by $\mathrm{d}A$, we obtain the *radiative flux* (energy per second per unit area) crossing the boundary surface at point P:

$$q = \int_{\mathscr{V}} \frac{J\cos\beta}{r^2}\,\mathrm{d}\mathscr{V} = \frac{1}{4\pi}\int_{\mathscr{V}} \frac{E\cos\beta}{r^2}\,\mathrm{d}\mathscr{V} \tag{18.14}$$

where, by definition, E is the total energy emitted by the gas in all directions per second per unit volume, that is,

$$E = 4\pi J \tag{18.15}$$

Now consider a uniform slab of radiating gas at constant properties (e.g., constant T and ρ throughout), as sketched in Fig. 18.5. Thus, E is constant throughout

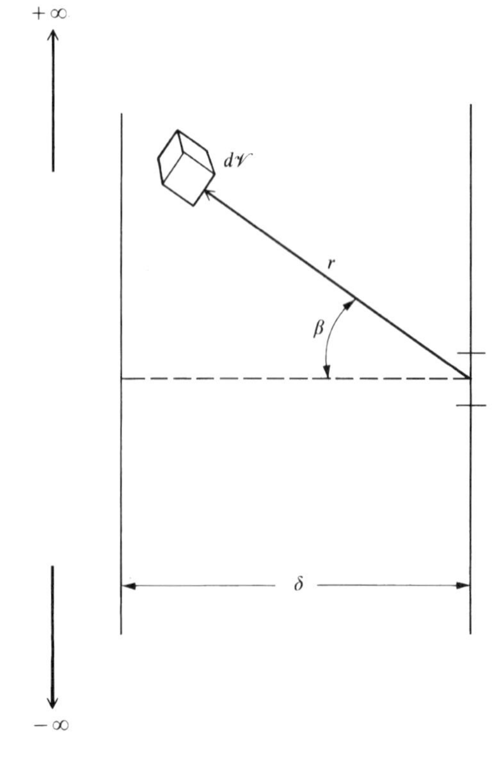

Fig. 18.5 Infinite slab geometry for a transparent gas.

the gas. The slab has thickness δ and stretches to plus and minus infinity. Consider an infinitesimal volume $d\mathscr{V}$ inside the slab. In spherical coordinates,

$$\mathrm{d}\mathscr{V} = r^2 \sin\beta \,\mathrm{d}\beta \,\mathrm{d}\phi \,\mathrm{d}r$$

The radiative flux across the right face of the slab is, from Eq. (18.14),

$$q = \frac{E}{4\pi}\int_0^{\pi/2}\int_0^{2\pi}\int_0^{\delta/(\cos\beta)} \frac{\cos\beta}{r^2} r^2 \sin\beta \;\mathrm{d}r\,\mathrm{d}\phi\,\mathrm{d}\beta$$

$$= \frac{E\delta}{4\pi}\int_0^{\pi/2}\int_0^{2\pi} \sin\beta\,\mathrm{d}\phi\,\mathrm{d}\beta = \frac{E\delta}{2}\int_0^{\pi/2} \sin\beta\,\mathrm{d}\beta$$

or

$$\boxed{q = \frac{E\delta}{2}} \tag{18.16}$$

Equation (18.16) gives the radiative flux across the surface of an infinite slab of radiating, transparent gas of thickness δ.

This result can be used to approximate the stagnation region radiative heating to a hypersonic blunt body, as shown in Fig. 18.6. Here, the stagnation region with a shock detachment distance δ appears to the stagnation point as essentially an infinite slab with a constant value of E, where $E = E(T_s, \rho_s)$, and where T_s and ρ_s are the temperature and density behind the normal shock wave. Hence, the stagnation-point radiative heat transfer $(q_R)_{\mathrm{stag}}$ is approximated by

$$\boxed{(q_R)_{\mathrm{stag}} = \frac{E\delta}{2}} \tag{18.17}$$

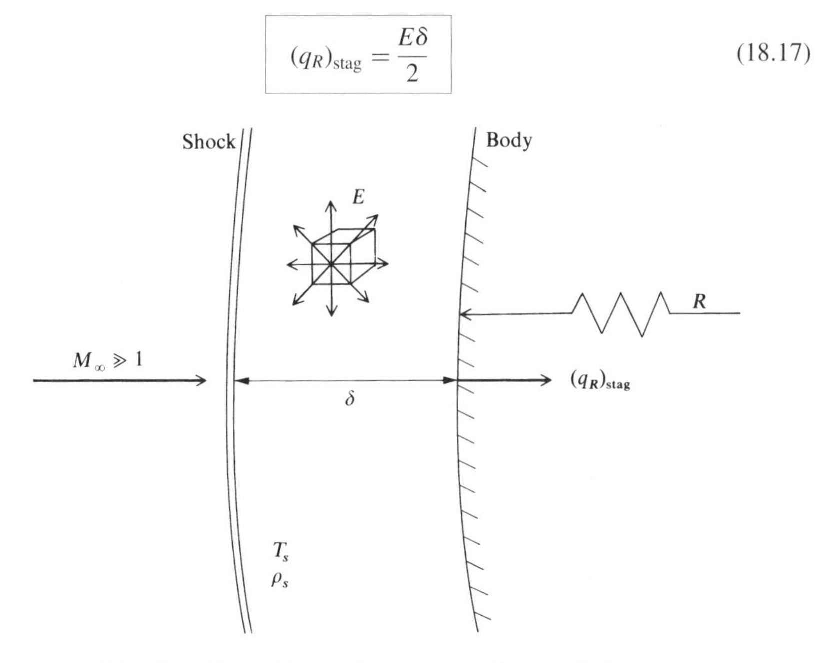

Fig. 18.6 Stagnation region geometry for a radiating gas.

If we further assume that δ is given by Eq. (14.19), namely,

$$\delta \approx \frac{R}{\rho_s/\rho_\infty}$$

then

$$(q_R)_{\text{stag}} = \frac{E}{2} R \left(\frac{\rho_\infty}{\rho_s}\right) \tag{18.18}$$

From Eq. (18.18), we see the important result that, for a transparent, radiating shock layer, the *radiative heat transfer is directly proportional to R.* This is in direct contrast to convective heating, where in Chapters 6 and 17 we saw that

$$(q_c)_{\text{stag}} \propto \frac{1}{\sqrt{R}} \tag{18.19}$$

For superorbital reentry vehicles, where radiative heating is important, Eqs. (18.18) and (18.19) form a classic compromise on the design of the nose radius; namely, to reduce convective heating, make R large, but to reduce radiative heating, make R small.

18.5 Solutions of the Radiative-Transfer Equation: Absorbing Gas

In Sec. 18.4, we considered an emitting but nonabsorbing gas (transparent gas). In the present section, we consider the direct opposite—an absorbing but non-emitting gas. For this case, from Eq. (18.12),

$$\frac{dI_\nu}{ds} = -\kappa_\nu I_\nu \tag{18.20}$$

Consider the volume of absorbing gas sketched in Fig. 18.7. Radiative intensity in the s direction is incident on the volume with value $I_{\nu_{\text{in}}}$. The intensity that emerges from the other side is $I_{\nu_{\text{out}}}$. The path length of the radiative intensity in the s direction through the volume is L. From Eq. (18.20),

$$\int_{I_{\nu_{\text{in}}}}^{I_{\nu_{\text{out}}}} \frac{dI_\nu}{I_\nu} = -\int_0^L \kappa_\nu \, ds \tag{18.21}$$

Assuming a constant property gas (hence κ_ν is constant throughout the volume), Eq. (18.21) yields

$$\boxed{I_{\nu_{\text{out}}} = I_{\nu_{\text{in}}} e^{-\kappa_\nu L}} \tag{18.22}$$

Equation (18.22) is called Lambert's law; it applies to a constant property gas.

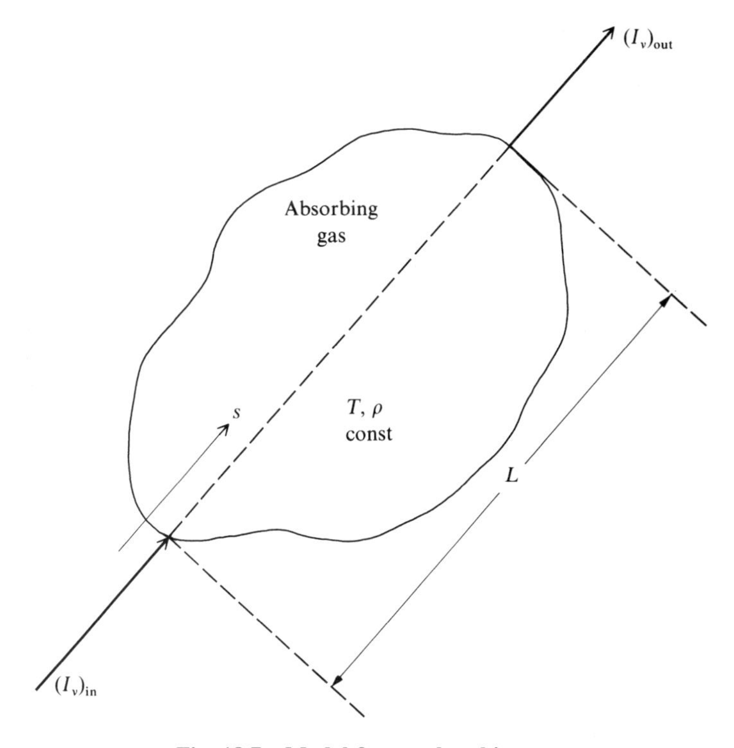

Fig. 18.7 Model for an absorbing gas.

If T and ρ are variable throughout the gas, then κ_ν is also a variable. For this case, Eq. (18.21) becomes

$$I_{\nu_{out}} = I_{\nu_{in}} \exp\left[-\int_0^L \kappa_\nu \, \mathrm{d}s\right] \tag{18.23}$$

Let us define the *optical thickness* τ_ν as

$$\tau_\nu \equiv \int_0^L \kappa_\nu \, \mathrm{d}s \tag{18.24}$$

Then, Eq. (18.23) can be written as

$$\boxed{I_{\nu_{out}} = I_{\nu_{in}} e^{-\tau_\nu}} \tag{18.25}$$

18.6 Solutions of the Radiative-Transfer Equation: Emitting and Absorbing Gas

Consider an emitting and absorbing gas, with variable properties (variable T and ρ, hence variable κ_ν), as sketched in Fig. 18.8. Let $I_\nu(0)$ be the incoming

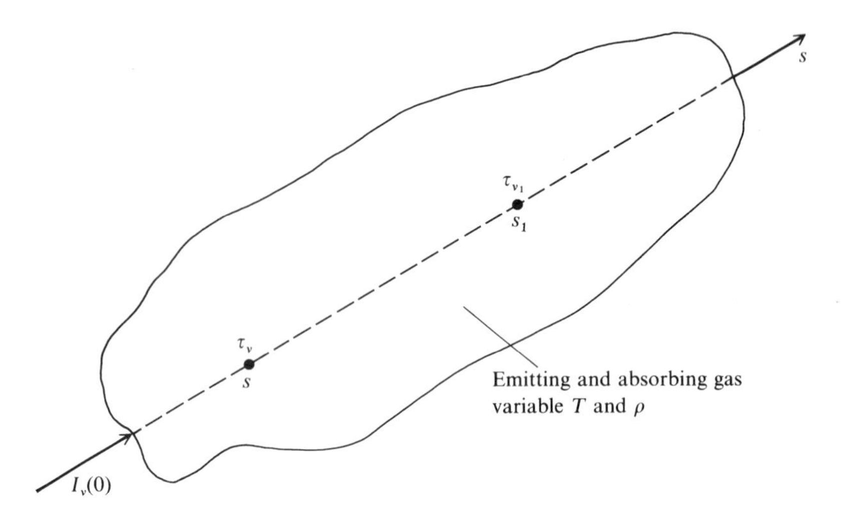

Fig. 18.8 Model for an emitting and absorbing gas.

intensity at the boundary of the system, $s = 0$. Let us calculate the local value of I_ν at $s = s_1$. Physically, the intensity at s_1 will consist of two parts: 1) the original incoming $I_\nu(0)$ after it has been attenuated by absorption from $s = 0$ to $s = s_1$, and 2) the emitted radiation from within the gas at any point s, after it has been attenuated by absorption between s and s_1. Let us quantify this picture.

From Eq. (18.12),

$$\frac{dI_\nu}{\kappa_\nu\, ds} = \beta_\nu - I_\nu \tag{18.26}$$

From the definition of optical thickness given by Eq. (18.24), we can write

$$d\tau_\nu = \kappa_\nu\, ds \tag{18.27}$$

Hence, Eq. (18.26) becomes

$$\frac{dI_\nu}{d\tau_\nu} = B_\nu - I_\nu \tag{18.28}$$

Now, *assume* a solution of Eq. (18.28) in the form of

$$I_\nu = c(\tau_\nu)e^{-\tau_\nu} \tag{18.29}$$

where c is a variable coefficient, a function of τ_ν. Differentiating Eq. (18.29), we have

$$\frac{dI_\nu}{d\tau_\nu} = -ce^{-\tau_\nu} + e^{-\tau_\nu}\frac{dc}{d\tau_\nu} \tag{18.30}$$

Substituting Eqs. (18.30) and (18.29) into the radiative-transfer equation (18.28), we have

$$-ce^{-\tau_\nu} + e^{-\tau_\nu}\frac{\mathrm{d}c}{\mathrm{d}\tau_\nu} = B_\nu - ce^{-\tau_\nu}$$

or

$$\frac{\mathrm{d}c}{\mathrm{d}\tau_\nu} = B_\nu e^{\tau_\nu} \tag{18.31}$$

Letting $\tau_\nu = 0$ at $s = 0$, and $\tau_\nu = \tau_{\nu_1}$ at $s = s_1$, Eq. (18.31) can be integrated as

$$\int_0^{\tau_{\nu_1}} \mathrm{d}c = \int_0^{\tau_{\nu_1}} B_\nu e^{\tau_\nu}\,\mathrm{d}\tau_\nu$$

or

$$c(\tau_{\nu_1}) - c(0) = \int_0^{\tau_{\nu_1}} B_\nu e^{\tau_\nu}\,\mathrm{d}\tau_\nu \tag{18.32}$$

However, from Eq. (18.29) evaluated at $s = s_1$ and $s = 0$, we have, respectively,

$$c(\tau_{\nu_1}) = I_\nu(\tau_{\nu_1})e^{\tau_{\nu 1}} \tag{18.33a}$$

and

$$c(0) = I_\nu(0) \tag{18.33b}$$

Substituting Eqs. (18.33a) and (18.33b) into Eq. (18.32), we have

$$\boxed{I_\nu(\tau_{\nu_1}) = I_\nu(0)e^{-\tau_{\nu 1}} + \int_0^{\tau_{\nu_1}} B_\nu e^{-(\tau_{\nu 1} - \tau_\nu)}\,\mathrm{d}\tau_\nu} \tag{18.34}$$

Equation (18.34) represents the general solution to the radiative-transfer equation for an emitting and absorbing gas with variable properties. The two terms on the right side of Eq. (18.34) physically represent the two parts of the intensity at point s_1 as described in the first paragraph of this section. With Eq. (18.34) in mind, reread the first paragraph before progressing further.

Now consider an infinite slab (sometimes called a plane layer) of gas, as sketched in Fig. 18.9. Let y denote the vertical distance through the slab. The upper boundary is located at $y = L$, and the lower boundary is at $y = 0$. The values of optical thickness at each of these locations are τ_{ν_2} and 0, respectively. The gas properties in the slab vary with y. Let I_ν^+ denote a *downward-directed* radiative intensity, as shown is Fig. 18.9. Hence, $I_\nu^+(L) = I_\nu^+(\tau_{\nu_2})$ is the downward intensity coming from the upper boundary (say, from radiation entering the slab from outside, or maybe the upper boundary might be a solid wall emitting some radiation of its own). We wish to calculate the downward-directed *radiative flux*

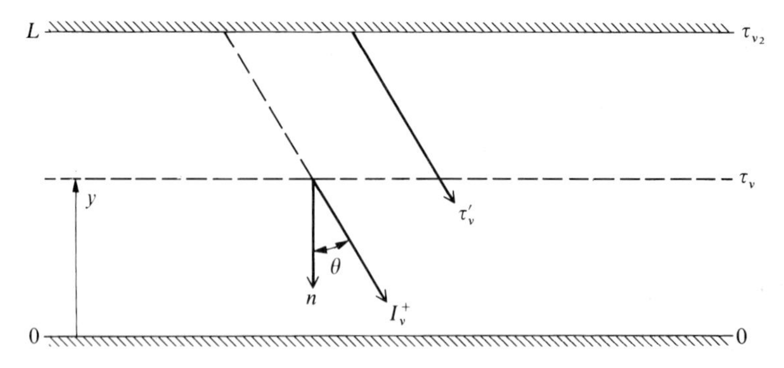

Fig. 18.9 Infinite slab geometry for an emitting and absorbing gas.

at the location y (hence τ_v) in the gas. Note that, in Eq. (18.34), the optical thickness is measured in the direction of the intensity, which, Fig. 18.9, is denoted by τ'_v. Hence, from Eq. (18.34),

$$I_v(\tau'_v) = I_v(0)e^{-\tau'_v} + \int_0^{\tau'_v} B_v(t)e^{-(\tau'_v - t)}\,\mathrm{d}t \tag{18.35}$$

where $I_v(0)$ is the intensity at $\tau'_v = 0$, which is the top of the slab, and t is a dummy variable of integration, ranging from 0 to τ'_v. However, by convention, τ_v is always directed upward, normal to the slab, as shown in Fig. 18.9. Note that τ'_v (measured from the top along the slant height shown in Fig. 18.9) is given by $(\tau_{v_2} - \tau_v)/\cos\theta$, where τ_v is measured in the normal direction from the bottom of the slab. Hence, Eq. (18.35) becomes

$$I_v^+(\tau_v) = I_v^+(\tau_{v_2})\exp\left[\frac{-(\tau_{v_2} - \tau_v)}{\cos\theta}\right] + \int_{\tau_v}^{\tau_{v2}} B_v(t)\exp[-(t - \tau_v)/\cos\theta]\frac{\mathrm{d}t}{\cos\theta} \tag{18.36}$$

Return to Eq. (18.5), and note that it gives the total radiative flux crossing a surface in both the upward and downward directions. For the radiation directed *downward only*, we have

$$q^+ = \int_0^\infty \int_0^{2\pi} \int_0^{\pi/2} I_v^+ \cos\theta \sin\theta\,\mathrm{d}\theta\,\mathrm{d}\phi\,\mathrm{d}v \tag{18.37}$$

Let $\mu = \cos\theta$; hence, $\mathrm{d}(\cos\theta) = -\sin\theta\,\mathrm{d}\theta = \mathrm{d}\mu$. Also, from the geometry of the slab, I_v is independent of ϕ. Hence, Eq. (18.37) becomes

$$q^+ = -2\pi \int_0^\infty \int_1^0 I_v^+ \mu\,\mathrm{d}\mu\,\mathrm{d}v \tag{18.38}$$

Note that I_ν^+ is a function of position y and direction μ, whereas q^+ is a function of position y only. Substituting Eq. (18.36) into (18.38), we obtain

$$q^+ = 2\pi \int_0^\infty \int_0^1 I_\nu^+(\tau_{\nu_2}) e^{-(\tau_{\nu_2}-\tau_\nu)/\mu} \mu \, \mathrm{d}\mu \, \mathrm{d}\nu$$

$$+ 2\pi \int_0^\infty \int_0^1 \int_{\tau_\nu}^{\tau_{\nu_2}} \frac{B_\nu(t)}{\mu} e^{-(t-\tau_\nu)/\mu} \mu \, \mathrm{d}\mu \, \mathrm{d}\nu \tag{18.39}$$

By definition, the *integro-exponential function of order n* is

$$\mathscr{E}_n(\omega) \equiv \int_0^1 \mu^{n-2} e^{-\omega/\mu} \, \mathrm{d}\mu$$

Hence, Eq. (18.39) can be written as:

$$\boxed{q^+ = 2\pi \int_0^\infty I_\nu^+(\tau_{\nu_2}) \mathscr{E}_3(\tau_{\nu_2} - \tau_\nu) \, \mathrm{d}\nu + 2\pi \int_0^\infty \int_{\tau_\nu}^{\tau_{\nu_2}} B_\nu(t) \mathscr{E}_2(t - \tau_\nu) \, \mathrm{d}t \, \mathrm{d}\nu} \tag{18.40}$$

Equation (18.40) gives the downward radiative flux at location y in Fig. 18.9. The first integral is the radiation from the upper surface, attenuated by absorption between L and y. The second integral is the radiation emitted locally by the gas between L and y and attenuated before it reaches y.

By a similar development, the total flux at y, $q = q^+ + q^-$, can be obtained. Then, the radiation term in the energy equation [Eq. (18.8)] becomes

$$-\nabla \cdot \mathbf{q}_R = -\frac{\mathrm{d}q}{\mathrm{d}y} = -4\pi J + 2\pi \int_0^\infty \kappa_\nu \int_0^{\tau_{\nu_S}} B_\nu(t) \mathscr{E}_1(|\tau_\nu - t|) \, \mathrm{d}t \, \mathrm{d}\nu \tag{18.41}$$

See [212] for more details on the derivation of Eq. (18.41).

18.7 Radiating Flowfields: Sample Results

A primary application of radiative gas dynamics has been in the area of planetary entry heating, especially reentry heating to space vehicles entering the Earth's atmosphere at speeds above 30,000 ft/s. Figure 18.10, taken from [213], illustrates the importance of radiative heating at such velocities; stagnation-point radiative heat transfer equals and exceeds the ordinary aerodynamic convective heating at high entry velocities. Reference [213] is a major survey of radiative shock-layer effects, and the reader is strongly encouraged to study it for more details.

In general, the values for the absorption coefficient κ_ν, as a function of temperature, density, and frequency, must be known in order to carry out a radiating flow calculation. For most gases, the frequency dependence of κ_ν is quite detailed and complex. For example, Fig. 18.11 (from [213]) shows the frequency variation of κ_ν for high-temperature air obtained from several different sources. This figure shows only the continuum radiation associated with free electron movement and

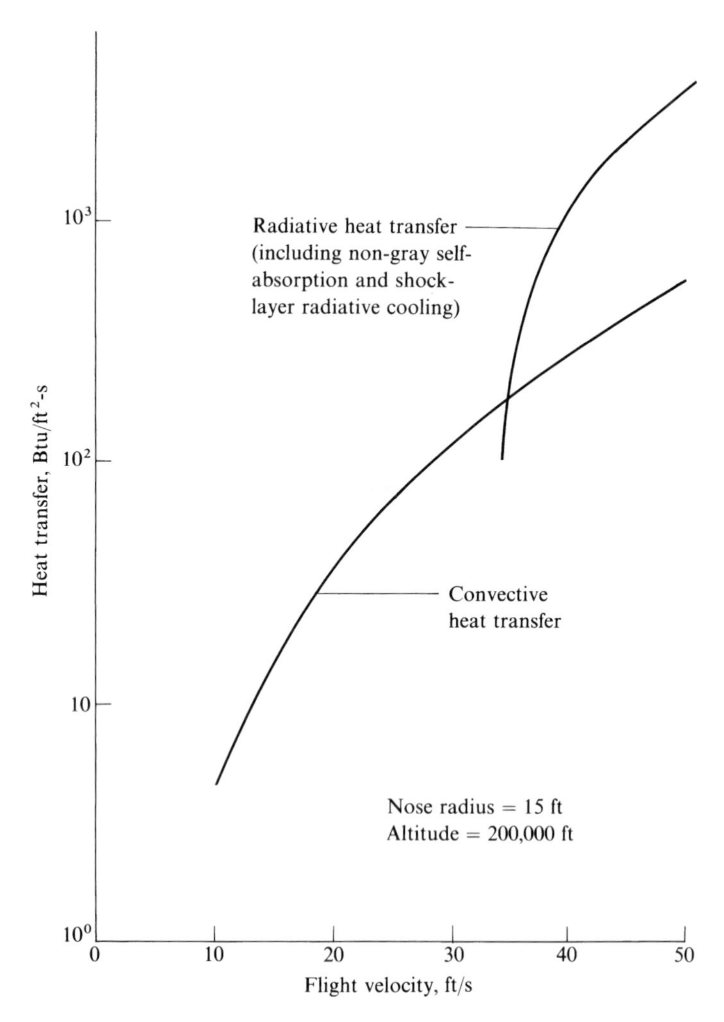

Fig. 18.10 Comparison of radiative and convective stagnation-point heat transfer (from Anderson [213]).

electron-ion recombination; to this must be added the detailed spectral line radiation from electronic transitions in the atoms and molecules. See [213] for more details and for a list of references from which absorption coefficient data can be obtained. Also, a recent survey of air radiation properties has been given by Sutton in [214].

As an example of a radiating shock-layer application, we take the results described in [212] for the viscous stagnation region of a blunt body. The flow problem is shown in Fig. 18.12, where the high-temperature viscous and radiating flow is calculated between the shock wave and the body. The flow is assumed to be in local thermodynamic and chemical equilibrium. For the radiative transport, the stagnation region is assumed to be an absorbing and emitting gas in an infinite slab, as described in Sec. 18.6, with properties that vary in the y direction across

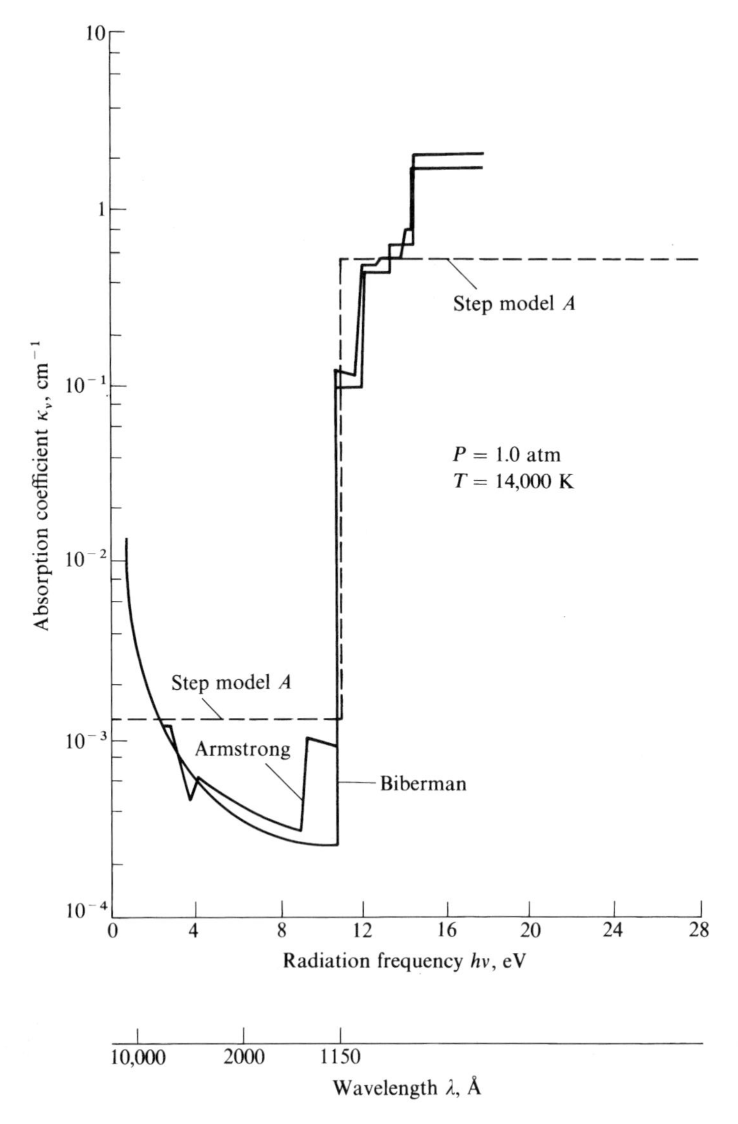

Fig. 18.11 Absorption coefficient variation with frequency (or wavelength) for high-temperature air (from [213]).

the shock layer. The governing equations that hold for a thin, viscous, shock layer near the stagnation region were first presented by Howe and Viegas [215] and are given as follows.

Continuity:

$$\frac{\partial}{\partial x}(\rho u r^m) + \frac{\partial}{\partial y}(K \rho v r^m) = 0 \tag{18.42}$$

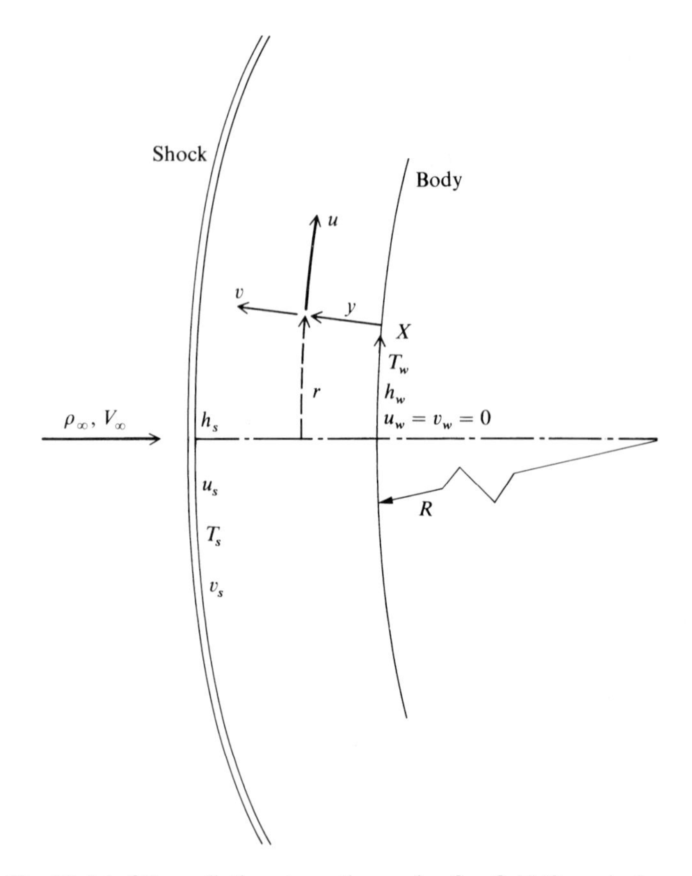

Fig. 18.12 Model of the radiating stagnation region flowfield (from Anderson [212]).

x Momentum:

$$\rho u \frac{\partial u}{\partial x} + K \rho v \frac{\partial u}{\partial y} = -\frac{\partial p}{\partial x} + K \frac{\partial}{\partial y}\left(\mu \frac{\partial u}{\partial y}\right) \tag{18.43}$$

y Momentum:

$$\frac{\partial p}{\partial y} = 0 \tag{18.44}$$

Energy:

$$\rho u \frac{\partial h_0}{\partial x} + K \rho v \frac{\partial h_0}{\partial y} = K \frac{\partial}{\partial y}\left(\frac{\mu}{Pr_{\text{eq}}} \frac{\partial h}{\partial y}\right) - K \frac{\partial q_R}{\partial y} \tag{18.45}$$

In the preceding,

$$h_0 = h + \frac{u^2 + v^2}{2}$$

$$K = 1 + \frac{y}{R}$$

$m = 0$ or 1 for the two-dimensional or axisymmetric flow.

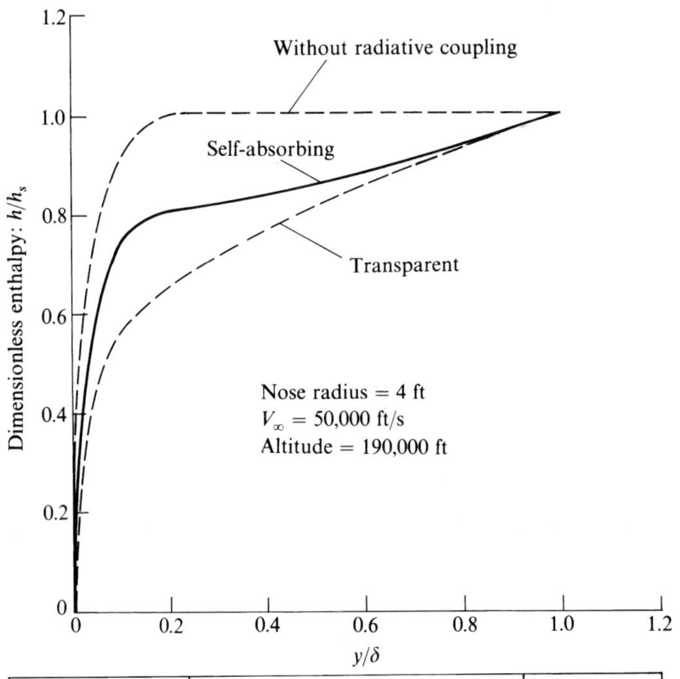

	Stagnation point heat transfer		Shock-detachment distance
	q_c Btu/ft^2-s	q_R Btu/ft^2-s	
Without radiative coupling	1262	12,970	0.172 ft
With radiative coupling—transparent gas	666	6,170	0.144 ft
With radiative coupling—nongray gas	990	5,220	0.154 ft

Fig. 18.13 Radiation coupling effects on stagnation region shock layer and heat transfer (from [212]).

Also, the Prandtl number Pr_{eq} is the equilibrium Prandtl number defined in Eqs. (16.41) and (16.42), which contains the influence of diffusion. The thermodynamic and transport properties are obtained from the correlations of Viegas and Howe [194]. In Eq. (18.45), the radiation term $\partial q_R/\partial y$ is given by Eq. (18.41).

Some typical results are shown in Figs. 18.13 and 18.14, taken from [212]. The static enthalpy profiles across the stagnation region shock layer are shown in Fig. 18.13. The upper curve (dashed line) is the result with no radiation, the lower dashed line corresponds to a transparent shock layer, and the middle solid curve is for a radiating, self-absorbing gas. The nonadiabatic flowfield effect caused by radiation is dramatically shown here; the radiation energy that is lost from the flow results in a cooling effect, thus lowering the enthalpy levels of the flow. The transparent gas case exhibits the strongest radiative cooling effect; the self-absorbing case retains some of the radiative energy caused by absorption within the shock layer and hence has a somewhat higher enthalpy level than the transparent gas case. The table at the bottom of Fig. 18.13 gives the convective and radiative heat transfer to the stagnation point, where q_R is calculated from Eq. (18.40) evaluated at the wall. Note that,

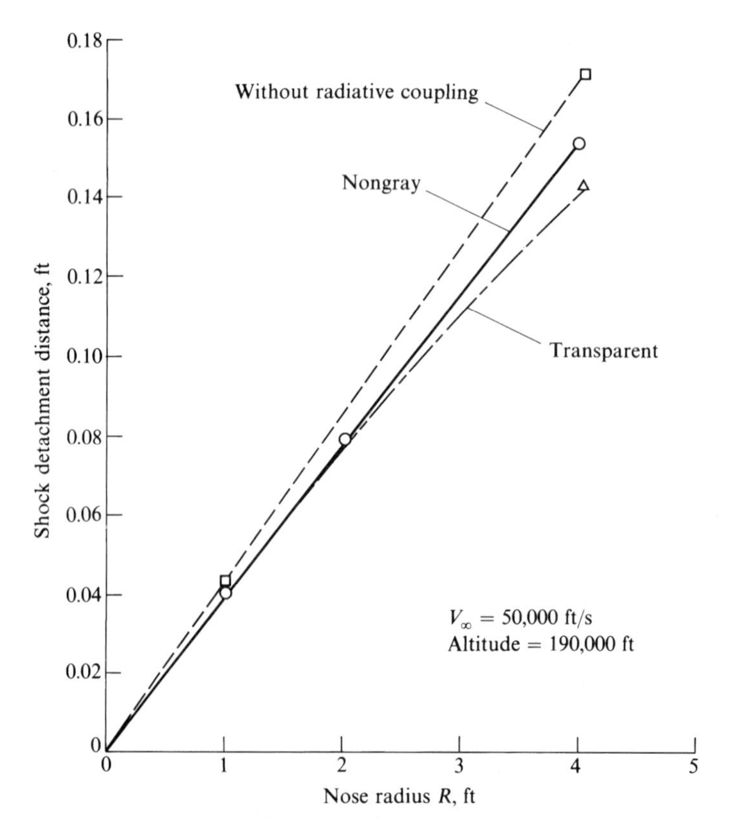

Fig. 18.14 Radiative coupling effects on shock detachment distance (from [212]).

for the very high velocity considered ($V_\infty = 50{,}000$ ft/s), q_R far exceeds q_c. The shock detachment distance is also affected by radiation, as seen in the table in Fig. 18.13 and in the plot of Fig. 18.14. Note that the cooling effect of shock-layer radiation (hence higher shock-layer density) results in a decrease in shock-detachment distance. Moreover, as seen in Fig. 18.14, this effect becomes stronger as R is increased. Here is a graphic illustration that larger nose radii increase the effect of radiation on the shock layer, as noted at the end of Sec. 18.4.

We end this section with some radiating shock-layer results obtained with the VSL technique by Moss. We have already noted in Sec. 17.8 that the VSL method was used to calculate chemically reacting shock layers by Moss (see [199]). The work of Moss has been progressively extended to include shock-layer radiation, ablating gases, turbulence, and foreign planetary atmospheres. Indeed, Moss's later work has led to a very exciting "first" in modern hypersonics—the design of the Galileo heat shield by means of detailed flowfield calculations. Previous reentry vehicles such as Apollo and the space shuttle were designed by means of a combination of wind-tunnel data and approximate calculations. However, the final design of the heat shield for the Galileo probe was performed on the basis of Moss's detailed viscous shock-layer calculations. These calculations,

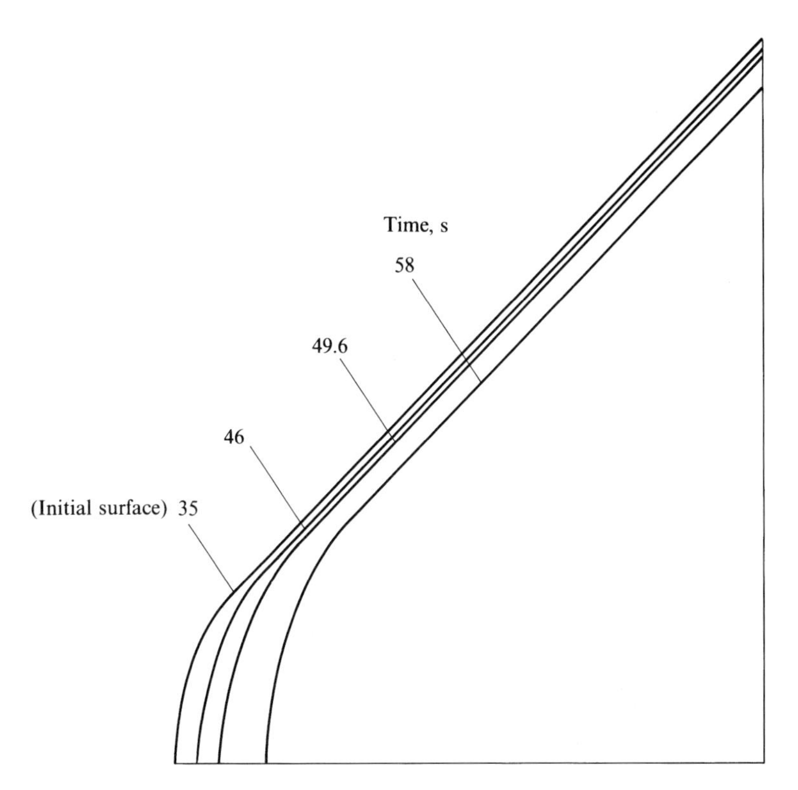

Fig. 18.15 Predicted time-varying contours for the shape of the Galileo probe during Jovian entry (from Moss and Simmonds [217]).

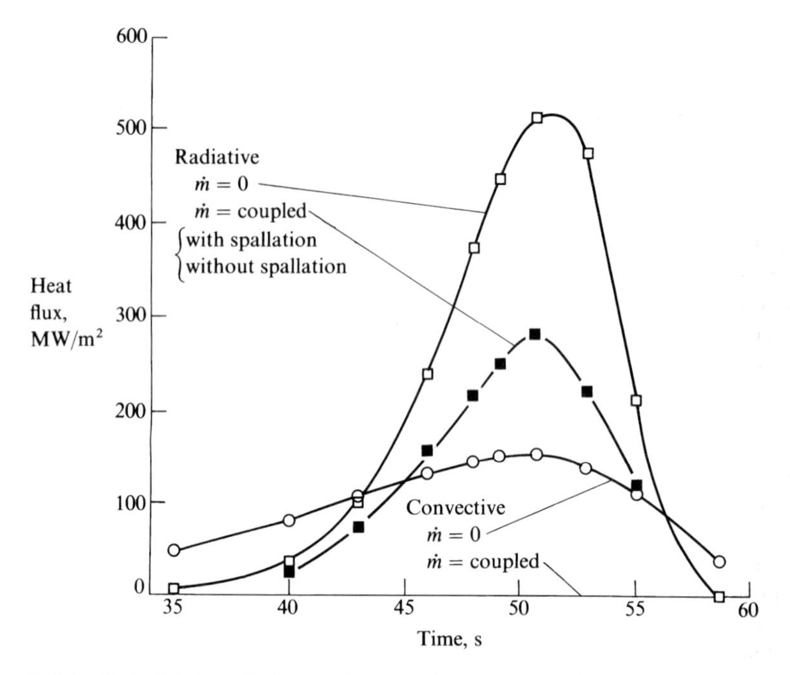

Fig. 18.16 Calculated radiative and convective heat transfer to the Galileo probe during Jovian entry (from [217]).

and their progressive development, have been extensively published; for the most recent summary, see [216] and [217] and the references contained therein. Some results in [216] and [217] were also obtained by means of a time-dependent viscous shock-layer analysis derived from the work of Kumar et al. [218]. Typical results from Moss's work are shown in Figs. 18.15 and 18.16, taken from [217]. In Fig. 8.15, the time-varying contours (caused by surface oblation) are given for the Galileo probe for various times during its Jovian entry trajectory. Figure 18.16 gives the calculated radiative and convective heat transfer to the stagnation point; note in particular that the heating to the Galileo probe is predicted to be virtually all radiative, because the convective heating is negligible as a result of massive ablation. Again, it is this type of data that has gone into the detailed design of the Galileo heat shield—truly a benchmark event in the development and use of detailed, modern, hypersonic flow calculations.

18.8 Surface Radiative Cooling

In this section we shift our focus to consider radiation from the surface rather than from the shock layer. The surface of a body radiates energy away from the surface at the rate

$$q_{RS} = \varepsilon \sigma T_w^4 \tag{18.46}$$

where q_{RS} is the heat flux radiated away from the surface, ε is the surface emissivity, σ is the Stefan–Boltzmann constant, and T_w is the wall surface temperature. The net heat transfer at the surface of the body q_w is equal to the sum of the convective and radiative heat transfer minus the energy radiated away from the surface q_{RS}, that is,

$$q_w = q_c + q_R - q_{\mathrm{RS}} \tag{18.47}$$

Radiation away from the surface can be a significant cooling mechanism for some hypersonic vehicles, especially hypersonic cruise vehicles flying for long times in the atmosphere. Hirschel [267] notes, for example, a case where the recovery temperature near the stagnation point of the HERMES reentry vehicle reaches about 6000 K at 70 km altitude, whereas with radiative cooling ($\varepsilon = 0.85$) the wall temperature drops to below 2000 K and remains below 2000 K for the whole entry flight trajectory. This is significant! Indeed, Hirschel discusses the subject of surface radiative cooling at great length, and you should consult [267] for more details.

In this section, we simply want to note the existence of radiative cooling of the surface, as calculated from Eq. (18.46). Also, we note that for a surface with significant radiative cooling the recovery temperature discussed in Chapter 6 loses its practical significance. Instead, when the surface reaches the equilibrium condition of an adiabatic wall, that is, no net heat transfer to or from the surface, the wall temperature reaches the radiation-adiabatic temperature T_{ra}, which is lower than the adiabatic wall temperature T_{aw}, defined in Chapter 6. When the surface is a radiative-adiabatic surface, from Eq. (18.47) we have

$$0 = q_c + q_R - q_{\mathrm{RS}} \tag{18.48}$$

The radiative-adiabatic temperature of the surface is obtained by combining Eqs. (18.46) and (18.48) as

$$0 = q_c + q_R - \varepsilon\sigma T_{\mathrm{ra}}^4 \tag{18.49}$$

or

$$T_{\mathrm{ra}}^4 = \frac{q_c + q_R}{\varepsilon\sigma} \tag{18.50}$$

The preceding calculation for T_{ra} assumes the simplest geometric case where the surface is completely convex, and surface radiation from other parts of the body is not incident on the point in questions. This is not the case for complex body shapes using nonconvex shapes where some parts of the body surface might face other parts. Taking this effect into account becomes a matter of geometry and is covered in detail by Hirschel. Such matters are beyond the scope of this section.

18.9 Summary and Comments

This chapter has covered the basic concepts necessary to analyze a radiating flowfield. Equations (18.2) and (18.3) define the radiative intensity

I_ν and the radiative flux q_ν; these two quantities are used to describe the propagation of radiative energy through a medium. The intensity is governed by the radiative-transfer equation

$$\frac{\mathrm{d}I_\nu}{\mathrm{d}s} = J_\nu - \kappa_\nu I_s = \kappa_\nu B_\nu - \kappa_\nu I_\nu$$

where B_ν is the blackbody radiative intensity

$$B_\nu = \frac{2h\nu^3}{c^2(e^{h\nu/kT} - 1)}$$

When the radiative intensity at a given point in the gas is known for all directions, then the radiative flux can be obtained as

$$q_\nu = \int_\omega I_\nu \cos\theta \,\mathrm{d}\omega$$

Various solutions for q_ν are given in this chapter, depending on whether the gas is transparent (Sec. 18.4), absorbing (Sec. 18.5), or both emitting and absorbing (Sec. 18.6). Finally, applications to the stagnation point of a hypersonic blunt body are made, contrasting the various possible cases.

With this chapter, we bring to an end our discussion of high-temperature gas dynamics: we are at the end of Part 3 of this book. At this stage, return to our road map in Fig. 1.4, and mentally walk down the items listed under high-temperature flows, thinking about the salient physical aspects as you encounter each subtopic. The intent of Part 3 has been to provide the necessary technical fundamentals along with a physical feeling for the reader to carry out and understand a high-temperature flowfield study. In Part 3, we have just scratched the surface, especially in the supporting areas of quantum mechanics, statistical mechanics, kinetic theory, spectroscopy, kinetic theory, and physical chemistry. The interested reader is encouraged to study deeper into these basic sciences because they provide the foundation for a true appreciation of high-temperature gas dynamics. However, with the material discussed in Part 3 you have a sufficient background to be labeled a "starter" in the field of high-temperature flows.

Design Example 18.1

The detailed calculation of shock-layer radiation and the resulting radiative heat transfer to the surface is an elegant and detailed process as described in this chapter. Indeed, it might be too elegant and too detailed when fast engineering calculations are desired for preliminary vehicle design studies. For such design studies, a simple algebraic equation for stagnation-point radiative heating is most desirable, based on correlations of detailed radiating flowfield calculations of the nature described earlier.

Tauber and Sutton [268] obtained a simple correlation for stagnation-point radiative heating in the form of

$$(q_R)_{\text{stag}} = C r_n^a \rho^b f(V) \tag{18.51}$$

where r_n is the hemispherical nose radius in meters and $f(V)$ is a velocity function tabulated vs flight velocity in Table 18.1 for both Earth and Mars entry. For Earth entry, in Eq. (18.51), where $(q_R)_{\text{stag}}$ is in W/cm^2, we have

$$C = 4.736 \times 10^4$$

$$A = 1.072 \times 10^6\, V^{-1.88} \rho^{-0.325}$$

if

$$1 \leq r_n \leq 2, \qquad a \leq 0.6$$

if

$$2 < r_n \leq 3, \qquad a \leq 0.5$$

$$b = 1.22$$

Table 18.1 Radiative heating functions for Earth and Mars entry

V, m/s	f_E (V)	V, m/s	f_M (V)
9,000	1.5	6,000	0.2
9,250	4.3	6,150	1.0
9,500	9.7	6,300	1.95
9,750	19.5	6,500	3.42
10,000	35	6,700	5.1
10,250	55	6,900	7.1
10,500	81	7,000	8.1
10,750	115	7,200	10.2
11,000	151	7,400	12.5
11,500	238	7,600	14.8
12,000	359	7,800	17.1
12,500	495	8,000	19.2
13,000	660	8,200	21.4
13,500	850	8,400	24.1
14,000	1,065	8,600	26.0
14,500	1,313	8,800	28.9
15,000	1,550	9,000	32.8
15,500	1,780	—	—
16,000	2,040	—	—

For Mars entry, where the atmosphere is assumed to be 97% CO_2 and 3% N_2,

$$C = 2.35 \times 10^4$$

$$a = 0.526$$

$$b = 1.19$$

Note in Table 18.1 that the velocity function $f_E(V)$ pertains to Earth entry and $f_M(V)$ pertains to Mars entry.

A comparison of results from Eq. (18.51) with the detailed radiating shock-layer calculations of Page et al. [269] is shown in Fig. 18.17, obtained from Tauber and Sutton [268]. Here, the radiative heat-transfer coefficient, defined as

$$C_{H_r} = \frac{(q_R)_{\text{stag}}}{1/2 \rho V^3}$$

is plotted vs flight velocity. The solid and dashed curves for a stagnation region pressure of 1.0 and 0.1 atm, respectively, are obtained from Eq. (18.51), and the squares and circles are from the shock-layer calculations reported in [269]. The average difference between the detailed shock-layer calculations and the results from Eq. (18.51) is 8.5%. Clearly Eq. (18.51) provides a useful engineering correlation from stagnation-point radiative heat transfer.

A recent example of an application of Eq. (18.51) can be found in [229], where the development and use of the configuration-based aerodynamics (CBAERO)

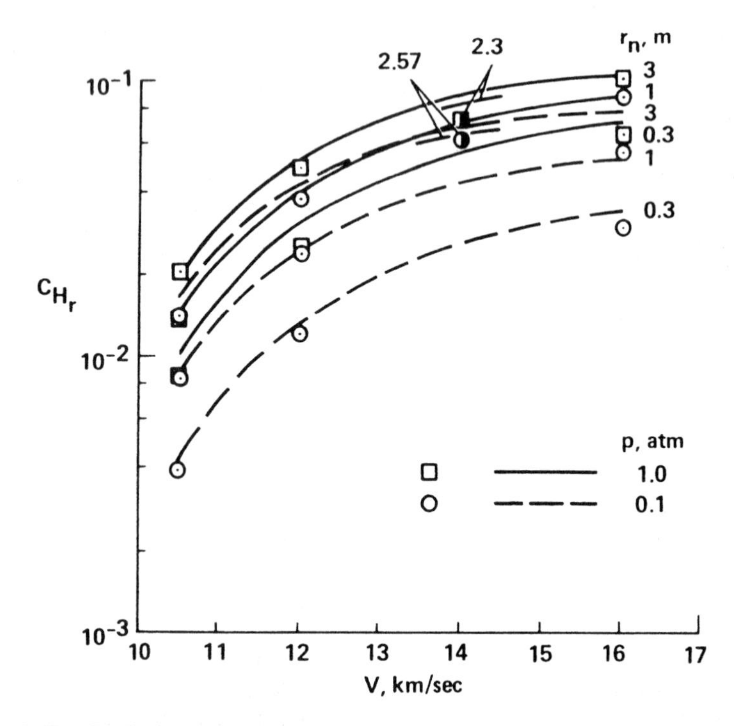

Fig. 18.17 Radiative heat-transfer comparison in air (Tauber and Sutton [268]).

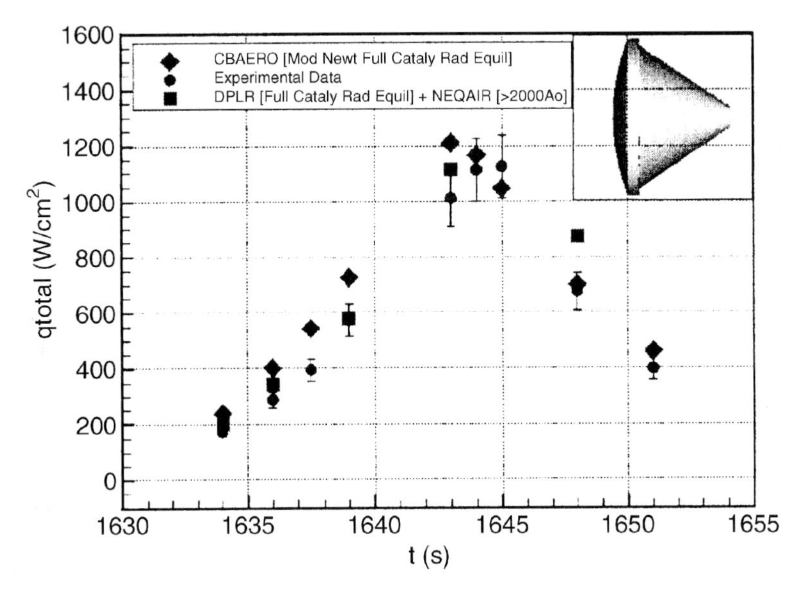

Fig. 18.18 Fire II total stagnation-point heat transfer. Comparison of calculations with flight-test results (Kinney and Garcia [229]).

software package at the NASA Ames Research Center is described. We have noted various aspects of CBAERO in Design Examples at the end of Chapters 3 and 6. Kinney and Garcia in [229] note that CBAERO uses Eq. (18.51) for an engineering prediction of stagnation-point radiative heat transfer. Results from [229] for the total stagnation-point heat transfer, $q_c + q_R$, are given in Fig. 18.18 for the Project Fire entry test vehicle. (The vehicle is shown in Fig. 3.34.) In Fig. 18.18 the CBAERO results (diamonds) are compared with flight-test data (circles) and with a detailed CFD calculation (squares). Clearly, the engineering calculations from CBAERO compare favorably with the flight-test data and detailed CFD calculations. This tends to further substantiate the value of Eq. (18.51) as an engineering method for predicting stagnation-point radiative heat transfer for preliminary vehicle design purposes.

Problems

18.1 Consider a radiating, transparent flowfield in the nose region of a hypersonic limit body. Define the radiation loss parameter Γ as

$$\Gamma = \frac{E_s \delta}{\frac{1}{2}\rho_\infty V_\infty^3}$$

where δ is the shock detachment distance and E_s is the value of E [given by Eq (18.15)] evaluated for properties immediately behind the normal portion of the shock wave. Prove that Γ is a similarity parameter for the flow.

18.2 In problem 18.1, let q_R be the stagnation-point radiative heating. Also, let $q_{R,0}$ denote the stagnation-point radiative heating if the shock layer were adiabatic, that is, if the shock-layer flowfield properties were calculated assuming no energy is lost as a result of radiation. Make a qualitative sketch of $q_R/q_{R,0}$ vs Γ, and explain your physical reasoning.

Postface

At this stage, return to the preface of this book, and review our stated intent in presenting the material to be found in Parts 1, 2, and 3. Recall that our purpose was to discuss the basic fundamentals of hypersonic and high-temperature gas dynamics. We could go on from here and discuss other aspects such as low-density aerodynamics, experimental techniques, and hypersonic vehicle considerations. However, to do so in the proper depth would far exceed the length constraints of the present book. Do not worry about this seemingly neglected material. Instead, rest assured that, with the fundamentals presented here, you are equipped to launch into any hypersonic or high-temperature problem with a minimum of confusion and questioning. This author hopes that you have been excited about what you have read here and that you will use this book as a launching pad to read much deeper into the subject matter. The world of hypersonic and high-temperature flows is still relatively new, with ever-expanding applications associated with hypersonic airplanes, manned space vehicles, etc. The purpose of the present book has been to help you chart a course through this new world, and to make your trip as meaningful and comfortable as possible. Bon voyage!

References

[1] Anderson, John D., Jr., *Introduction to Flight*, 5th ed., McGraw–Hill, New York, 2005.

[2] Hallion, Richard P., "The Path to the Space Shuttle: The Evolution of Lifting Reentry Technology," Air Force Flight Test Center, AFFTC Historical Monograph. Edwards Air Force Base, CA, Nov. 1983.

[3] Anderson, John D., Jr., "A Survey of Modern Research in Hypersonic Aerodynamics," AIAA Paper 84-1578, June 1984.

[4] Anderson, John D., Jr., *Modern Compressible Flow: with Historical Perspective*, 3rd ed. McGraw–Hill, New York, 2003.

[5] Anderson, John D., Jr., *Fundamentals of Aerodynamics*. McGraw–Hill, New York, 2007.

[6] Anderson, John D., Jr., "Hypersonic Viscous Flow over Cones at Nominal Mach 11 in Air," ARL Rept. 62-387, Wright-Patterson Air Force Base, OH, July 1962.

[7] Moss, J. N., and Bird, G. A., "Direct Simulation of Transitional Flow for Hypersonic Reentry Conditions," AIAA Paper 84-0223, Jan. 1984.

[8] Lees, Lester, "Hypersonic Flow," *Fifth International Aeronautical Conference, Los Angeles, 1955*, Inst. of Aeronautical Sciences, New York, pp. 241–276.

[9] Liepmann, H. W., and Roshko, A., *Elements of Gasdynamics*, Wiley, New York, 1957.

[10] Busemann, A., *Flussigkeits- und Gasbewegund.* Handworterbuch der Naturwissenschaften, Zweite Auflage (Gustav Fischer, Jena), 1933, pp. 275–277.

[11] Ivey, R. H., Klunker, E. R., and Bowen, E. N., "A Method for Determining the Aerodynamic Characteristics of Two- and Three-Dimensional Shapes at Hypersonics Speeds," NACA TN 1613, Aug. 1948.

[12] Eggers, A. J., Resknifoff, M. M., and Dennis, D. H., "Bodies of Revolution Having Minimum Drag at High Supersonic Airspeeds," NACA Rept. 1306, May 1957.

[13] Cox, R. N., and Crabtree, L. F., *Elements of Hypersonic Aerodynamics*, Academic Press, New York, 1965.

[14] Van Dyke, M. D., "The Supersonic Blunt Slender Body Problem—Review and Extension," *Journal of the Aerospace Sciences*, Vol. 25, 1958, pp. 485–495.

[15] Eggers, A. J., Syvertson, C. A., and Kraus, S., "A Study of Inviscid Flow About Airfoils at High Supersonic Speeds," NACA Rept. 1123, Feb. 1953.

[16] Chernyi, G. G., *Introduction to Hypersonic Flow* (translated from Russian by R. F. Probstein), Academic, New York, 1961.

[17] Kopal, Z., "Tables of Supersonic Flow Around Cones," M.I.T. Center of Analysis Tech., Report No. 1, U.S. Government Printing Office, Washington, DC, 1947.

[18] Sims, Joseph Z., "Tables for Supersonic Flow Around Right Circular Cones at Zero Angle of Attack," NASA SP-3004, 1964.

[19] Probstein, R. F., and Bray, K. N. C., "Hypersonic Similarity and the Tangent Cone Approximation for Unyawed Bodies of Revolution," *Journal of the Aeronautical Sciences*, Vol. 22, No. 1, Jan. 1955, pp. 66–68.

[20] Eggers, A. J., and Syvertson, C. A., "Inviscid Flow About Airfoils at High Supersonic Speeds," NACA TN 2646, Feb. 1952.

[21] Eggers, A. J., Savin, R. C., and Syvertson, C. A., "The Generalized Shock-Expansion Method and its Application to Bodies Travelling at High Supersonic Airspeeds," *Journal of the Aeronautical Sciences*, Vol. 22, March 1955, pp. 231–238.

[22] Gentry, A. E., "Hypersonic Arbitrary-Body Aerodynamic Computer Program (Mark III Version), Vol. 1—User's Manual," McDonnell-Douglas Corp., Rept. DAC 61552, Long Beach, CA, April 1968.

[23] Charters, A. C., and Thomas, R. N., "The Aerodynamic Performance of Small Spheres from Subsonic to High Supersonic Velocities," *Journal of the Aeronautical Sciences*, Vol. 12, June 1945, pp. 468–476.

[24] Hodges, A. J., "The Drag Coefficient of Very High Velocity Spheres," *Journal of the Aeronautical Sciences*, Vol. 24, July 1957, pp. 755–758.

[25] Stevens, V. I., "Hypersonic Research Facilities at the Ames Aeronautical Laboratory," *Journal of Applied Physics*, Vol. 21, pp. 1150–1155.

[26] Neice, Stanford, E., and Dorris, M. Ehret, "Similarity Laws for Slender Bodies of Revolution in Hypersonic Flows," *Journal of the Aeronautical Sciences*, Vol. 18, No. 8, Aug. 1951, pp. 527–530, 568.

[27] Van Dyke, M. D., "The Combined Supersonic-Hypersonic Similarity Rule," *Journal of the Aeronautical Sciences*, Vol. 18, No. 7, July 1951, pp. 499, 500.

[28] Tsien, H. S., "Similarity Laws of Hypersonic Flows," *Journal of Mathematics and Physics*, Vol. 25, 1946, pp. 247–251.

[29] Hayes, Wallace D., "On Hypersonic Similitude," *Quarterly of Applied Mathematics*, Vol. 5, No. 1, April 1947, pp. 105, 106.

[30] Van Dyke, M. D., "A Study of Hypersonic Small Disturbance Theory," NACA Rept. 1194, April 1954.

[31] Rasmussen, M. L., "On Hypersonic Flow past an Unyawed Cone," *AIAA Journal*, Vol. 5, No. 8, Aug. 1967, pp. 1495–1497.

[32] Doty, R. T., and Rasmussen, M. L., "Approximation for Hypersonic Flow Past an Inclined Cone," *AIAA Journal*, Vol. 11, No. 9, Sept. 1973, pp. 1310–1315.

[33] Sims, J. L., "Tables for Supersonic Flow Around Right Circular Cones at Small Angles of Attack," NASA SP-3007, Sept. 1964.

[34] Rasmussen, M. L., and Lee, H. M. "Approximation for Hypersonic Flow past a Slender Elliptic Cone," AIAA Paper 79-0364, Jan. 1979.

[35] Sedov, L. I., *Similarity and Dimensional Methods in Mechanics* (translated from Russian by M. Friedmann), edited by M. Holt, Academic Press, New York, 1959.

[36] Lukasiewicz, J., "Blast-Hypersonic Flow Analogy—Theory and Application," *American Rocket Society Journal*, Vol. 32, No. 9, Sept. 1962, pp. 1341–1346.

[37] Sakurai, A., "On the Propagation and Structure of the Blast Wave I," *Journal of the Physical Society of Japan*, Vol. 8, Sept.–Oct. 1953, p. 662.

[38] Sakurai, A., "On the Propagation and Structure of the Blast Wave II," *Journal of the Physical Society of Japan*, Vol. 9, March–April 1954, p. 256.

[39] Baradell, D. L., and Bertram, M. H., "The Blunt Plate in Hypersonic Flow," NASA TN-D-408, Oct. 1960.

[40] Bertram, M. H., and Baradell, D. L., "A Note on the Sonic-Wedge Leading Edge Approximation in Hypersonic Flow," *Journal of the Aeronautical Sciences*. Vol. 24, No. 8, Aug. 1957, pp. 627–629.

[41] Cheng, H. K., Hall, J. G., Golian, T. C., and Hertzberg, A., "Boundary-Layer Displacement and Leading-Edge Bluntness Effects in High-Temperature Hypersonic Flow." Inst. of the Aeronautical Sciences, Paper 60-38, Jan. 1960.

[42] Van Hise, V., "Analytic Study of Induced Pressure on Long Bodies of Revolution with Varying Nose Bluntness at Hypersonic Speeds," NASA TR-R-78, Jan. 1960.

[43] Lees, L., and Kubota, T., "Inviscid Hypersonic Flow over Blunt-Nosed Slender Bodies," *Journal of the Aeronautical Sciences*, Vol. 24, Feb. 1957, pp. 195–202.

[44] Weilmuenster, K. J., "High Angle of Attack Inviscid Flow Calculations over a Shuttle-Like Vehicle with Comparisons to Flight Data," AIAA Paper 83-1798, July 1983.

[45] Roe, P. L., "Thin Shock-Layer Theory," *Aerodynamic Problems of Hypersonic Vehicles*, AGARD Lecture Series No. 42, Vol. 1, 1972, Von Kármán Inst., Brussels, pp. 4-1–4-26.

[46] Hayes, W. D., and Probstein, R. F., *Hypersonic Flow Theory*, Academic Press, New York, 1959.

[47] Maslen, S. H., "Inviscid Hypersonic Flow past Smooth Symmetric Bodies," *AIAA Journal*, Vol. 2, No. 6, June 1964, pp. 1055–1061.

[48] Yakura, J. H., "A Theory of Entropy Layers and Nose Bluntness in Hypersonic Flow," *Hypersonic Flow Research*, Academic Press, New York, 1962, pp. 421–470.

[49] Sychev, V. V., "On the Theory of Hypersonic Gas Flow with a Power-Law Shock Wave," *Prikladnaga Matematika Mekhanika*, Vol. 24, 1960, pp. 756–764.

[50] Kubota, T., "Investigation of Flow Around Simple Bodies in Hypersonic Flow," Graduate Aeronautical Labs., California Inst. of Technology, Memo 40, Pasadena, CAO, 1957.

[51] Inouye, M., and Lomax, H., "Comparison of Experimental and Numerical Results for the Flow of a Perfect Gas About Blunt-Nosed Bodies," NASA TN D-1426, Nov. 1962.

[52] Anderson, John, D., Jr., *Computational Fluid Dynamics: The Basics with Applications*, McGraw-Hill, New York, 1995.

[53] Zucrow, M. J., and Hoffman, J. D., *Gas Dynamics, Vol. 2: Multidimensional Flows*, Wiley, New York, 1977.

[54] Ferri, Antonio, "Application of the Method of Characteristics to Supersonic Rotational Flow," NACA Technical Report 841, Aug. 1946.

[55] Sauerwein, H., "Numerical Calculation of Multidimensional and Unsteady Flows by the Method of Characteristics," *Journal of Computational Physics*, Vol. 1, No. 1, Feb. 1967, pp. 406–432.

[56] Chushkin, P. I., "Numerical Method of Characteristics for Three-Dimensional Supersonic Flows," *Progress in Aeronautical Sciences*, Vol. 9, edited by D. Kuchemann, Pergamon, Elmsford, New York, 1968.

[57] Butler, D. S., "The Numerical Solution of Hyperbolic Systems of Partial Differential Equations in Three Independent Variables," *Proceedings of the Royal Society*, Vol. A255, No. 1281, 1960, pp. 232–252.

[58] Rakich, John V., "A Method of Characteristics for Steady Three-Dimensional Supersonic Flow with Application to Inclined Bodies of Revolution," NASA TN D-5341, May 1969.

[59] Rakich, John V., and Cleary, Joseph W., "Theoretical and Experimental Study of Supersonic Steady Flow Around Inclined Bodies of Revolution," *AIAA Journal*, Vol. 8, No. 3, March 1970, pp. 511–518.

[60] Hayes, W. D., and Probstein, R. F., *Hypersonic Flow Theory*, Vol. 1: Inviscid Flows, Academic Press, New York, 1966.

[61] Moretti, G., and Abbett, M., "A Time-Dependent Computational Method for Blunt-Body Flows," *AIAA Journal*, Vol. 4, No. 12, 1966, pp. 2136–2141.

[62] Van Dyke, M. D., "The Supersonic Blunt-Body Problem Review and Extension," *Journal of the Aero Space Sciences*, Vol. 25, Aug. 1958, pp. 485–495.

[63] MacCormack, R. W., "The Effect of Viscosity in Hypervelocity Impact Cratering," AIAA Paper 69-354, Jan. 1969.

[64] Abbett, M. J., "Boundary Condition Calculation Procedures for Inviscid Supersonic Flow Fields," *Proceedings of the 1st AIAA Computational Fluid Dynamics Conference*, AIAA, New York, 1973, pp. 153–172.

[65] Courant, R., Friedrichs, K. O., and Lewy, H., "Uber die Differenzengleichungen der Mathematischen Physik," *Math. Ann.*, Vol. 100, 1928, p. 32.

[66] Lomax, H., and Inouye, M., "Numerical Analysis of Flow Properties About Blunt Bodies Moving at Supersonic Speeds in an Equilibrium Gas," NASA TR-R-204, May 1964.

[67] Moretti, G., and Bleich, G., "Three-Dimensional Flow Around Blunt Bodies," *AIAA Journal*, Vol. 5, No. 9, Sept. 1967, pp. 1557–1562.

[68] Weilmuenster, K. J., "High Angle of Attack Inviscid Flow Calculations over a Shuttle-Like Vehicle with Comparisons to Flight Data," AIAA Paper 83-1798, June 1983.

[69] Billig, F. S., "Shock-Wave Shapes Around Spherical- and Cylindrical-Nosed Bodies," *Journal of Spacecraft and Rockets*, Vol. 4, No. 6, June 1967, pp. 822, 823.

[70] Anderson, John D., Jr., Albacete, L. M., and Winkelmann, A. E., "Comment on Shock-Wave Shapes Around Spherical- and Cylindrical-Nosed Bodies," *Journal of Spacecraft and Rockets*, Vol. 5, No. 10, Oct. 1968, pp. 1247, 1248.

[71] Anderson, John D., Jr., and Albacete, L. M., "On Hypersonic Blunt-Body Flow Fields Obtained with a Time-Dependent Technique," Naval Ordnance Lab., NOLTR 68–129, White Oak, MD, Aug. 1968.

[72] Anderson, John D., Jr., "Introduction to Computational Fluid Dynamics," *Introduction to Computational Fluid Dynamics*, Short Course Lecture Notes, Von Kármán Inst. Belgium, 1986, pp. 1–267.

[73] Kutler, P., Warming, R. F., and Lomax, H., "Computation of Space Shuttle Flowfields Using Noncentered Finite-Difference Schemes," *AIAA Journal*, Vol. 11, No. 2, Feb. 1973, pp. 196–204.

[74] Cleary, J. W., "Hypersonic Shock-Wave Phenomena of a Delta-Wing Space-Shuttle Orbiter," NASA TN X-62,076, Oct. 1971.

[75] Maus, J. R., Griffith, B. J., and Szema, K. Y., "Hypersonic Mach Number and Real Gas Effects on Space Shuttle Orbiter Aerodynamics," *Journal of Spacecraft and Rockets*, Vol. 21, No. 2, March–April 1984, pp. 136–141.

[76] Mulder, W. A., and Van Leer, B., "Implicit Upwind Methods for the Euler Equations," *Proceedings of the AIAA 6th Computational Fluid Dynamics Conference*, AIAA, Washington, DC, 1988, pp. 303–310.

[77] Anderson, W. K., Thomas, J. L., and Van Leer, B., "A Comparison of Finite Volume Flux Vector Splittings for the Euler Equations," AIAA Paper 85-0122, Jan. 1985.

[78] Thomas, J. L., Van Leer, B., and Walters, R. W., "Implicit Flux-Split Schemes for the Euler Equations," AIAA Paper 85-1680, June 1985.

[79] Tauber, M. E., and Meneses, G. P., "Aerothermodynamics of Transatmospheric Vehicles," AIAA Paper 86-1257, June 1986.

[80] Bowcutt, K. G., Anderson John D., Jr., and Capriotti, D., "Viscous Optimized Hypersonic Wave riders," AIAA Paper 87-0272, Jan. 1987.

[81] Koppenwallner, G., "Fundamentals of Hypersonics: Aerodynamics and Heat Transfer," *Hypersonic Aerothermodynamics*, Short Course Notes, Von Kármán Inst. for Fluid Dynamics, Rhose Saint Genese, Belgium, Feb. 1984.

[82] Shapiro, A. H., *Shape and Flow*, Anchor Books, New York, 1961.

[83] White, Frank M., *Viscous Fluid Flow*, McGraw–Hill, 1974.

[84] Illingworth, C. R., "Steady Flow in the Laminar Boundary Layer of a Gas," *Proceedings of the Royal Society* (London), Ser. A, Vol. 199, 1949, pp. 533–558.

[85] Stewartson, K., "Correlated Incompressible and Compressible Boundary Layers," *Proceedings of the Royal Society* (London), Ser. A, Vol. 200, 1949, pp. 84–100.

[86] Howarth, L., "Concerning the Effect of Compressibility on Laminar Boundary Layers and Their Separation," *Proceedings of the Royal Society* (London), Ser. A, Vol. 194, 1948, pp. 16–42.

[87] Dorodnitsyn, A. A., "Laminar Boundary Layer in Compressible Fluid," *Doklady Akademii Nauk. SSSR* Vol. 34, 1942, pp. 213–219.

[88] Levy, S., "Effect of Large Temperature Changes (Including Viscous Heating) upon Laminar Boundary Layers with Variable Free-Stream Velocity," *Journal of Aeronautical Sciences*, Vol. 21, No. 7, July 1954, pp. 459–474.

[89] Lees, I., "Laminar Heat Transfer over Blunt-Nosed Bodies at Hypersonic Flight Speeds," *Jet Propulsion*, Vol. 26, 1956, pp. 259–269.

[90] Van Driest, F. R., "Investigation of Laminar Boundary Layer in Compressible Fluids Using the Crocco Method," NACA TN 2597, Jan. 1952.

[91] Cohen, C. B., and Reshotko, E., "Similar Solutions for the Compressible Laminar Boundary Layer with Heat Transfer and Pressure Gradient," NACA Report 1293, Sept. 1956.

[92] Van Driest, E. R., "The Problem of Aerodynamic Heating," *Aeronautical Engineering Review*, Oct. 1956, pp. 26–41.

[93] Beckwith, I. E., and Gallagher, J. J., "Local Heat Transfer and Recovery Temperatures on a Yawed Cylinder at a Mach Number of 4.15 and High Reynolds Numbers," NASA TR-R-104, Jan. 1962.

[94] Blottner, F. G., "Finite Difference Methods of Solution of the Boundary-Layer Equations," *AIAA Journal*, Vol. 8, No. 2, Feb. 1970, pp. 193–205.

[95] Kemp, N. H., Rose, R. H., and Detra, R. W., "Laminar Heat Transfer Around Blunt Bodies in Dissociated Air," *Journal of the Aerospace Sciences*, Vol. 26, No. 7, July 1959, pp. 421–430.

[96] Hartree, D. R., and Womersley, J. R., "A Method for the Numerical or Mechanical Solution of Certain Types of Partial Differential Equations," *Proceedings of the Royal Society*, Vol. 161A, Aug. 1937, p. 353.

[97] Smith, A. M. O., and Clutter, D. W., "Machine Calculation of Compressible Boundary Layers," *AIAA Journal*, Vol. 3, No. 4, April 1965, pp. 639–647.

[98] Schlichting, Hermann, *Boundary Layer Theory*, 7th ed., McGraw–Hill, New York, 1979.

[99] Reshotko, Eli, "Boundary-Layer Stability and Transition," *Annual Review of Fluid Mechanics*, Vol. 8, edited by M. Van Dyke, W. Vincenti, and J. Wehausen, Annual Reviews, Inc., Palo Alto, CA, 1976, pp. 311–349.

[100] Stetson, Kenneth, F., "On Predicting Hypersonic Boundary Layer Transition," Flight Dynamics Lab., Air Force Wright Aeronautical Lab., AFWAL-TM-84-160-FIMG, Wright-Patterson AFB, OH, March 1987.

[101] Beckwith, I. E., "Development of a High Reynolds Number Quiet Tunnel for Transition Research," *AIAA Journal*, Vol. 13, No. 3, March 1975, pp. 300–306.

[102] DiCristina, V., "Three-Dimensional Laminar Boundary-Layer Transition on a Sharp 8° Cone at Mach 10," *AIAA Journal*, Vol. 8, No. 5, May 1970, pp. 852–856.

[103] Stetson, K. F., "Mach 6 Experiments of Transition on a Cone at Angle of Attack," *Journal of Spacecraft and Rockets*, Vol. 19, No. 5, Sept.–Oct. 1982, pp. 397–403.

[104] Cebeci, T., and Smith, A. M. O., *Analysis of Turbulent Boundary Layers*, Academic Press, New York, 1972.

[105] Bradshaw, P., Cebeci, T., and Whitelaw, J., *Engineering Calculational Methods for Turbulent Flow*, Academic Press, New York, 1981.

[106] Dorrance, W. H., *Viscous Hypersonic Flow*, McGraw–Hill, New York, 1962.

[107] Marvin, Joseph G., "Turbulence Modeling for Computational Aerodynamics," *AIAA Journal*, Vol. 21, No. 7, July 1983, pp. 941–955.

[108] Baldwin, B. S., and Lomax, H., "Thin Layer Approximation and Algebraic Model for Separated Turbulent Flows," AIAA Paper 78-257, Jan. 1978.

[109] Wilcox, D. C., and Rubesin, M. W., "Progress in Turbulence Modeling for Complex Flow Fields Including the Effects of Compressibility," NASA TP-1517, July 1980.

[110] Donaldson, C. du P., and Sullivan, R. D., "An Invariant Second-Order Closure Model of the Compressible Turbulent Boundary Layer on a Flat Plate," Aeronautical Research Associates of Princeton, Inc., Rept. 178, Princeton, NA, June 1972.

[111] Van Driest, E., "Turbulent Boundary Layer in Compressible Fluids," *Journal of the Aerospace Sciences*, Vol. 18, No. 3, March 1951, pp. 145–160.

[112] Rubesin, M. W., and Johnson, H. A., "A Critical Review of Skin-Friction and Heat-Transfer Solutions of the Laminar Boundary Layer of a Flat Plate," *Transactions of the American Society of Mechanical Engineers*, Vol. 71, No. 4, May 1949, pp. 383–388.

[113] Eckert, E. R. G., "Engineering Relations for Heat Transfer and Friction in High-Velocity Laminar and Turbulent Boundary-Layer Flow over Surfaces with Constant Pressure and Temperature," *Transactions of the American Society of Mechanical Engineers*, Vol. 78, No. 6, Aug. 1956, p. 1273.

[114] Zoby, E. V., Moss, J. N., and Sutton, K., "Approximate Convective-Heating Equations for Hypersonic Flows," *Journal of Spacecraft and Rockets*, Vol. 18, No. 1, Jan.–Feb. 1981, pp. 64–70.

[115] Zoby, E. V., and Simmonds, A. L., "Engineering Flowfield Method with Angle-of-Attack Applications," AIAA Paper 84-0303, Jan. 1984.

[116] Bushnel, D. M., Jones, R. A., and Huffman, J. K., "Heat Transfer and Pressure Distributions on Spherically-Blunted 25° Half-Angle Cone at March 8 and Angles of Attack up to 90°," NASA TN D-4782, Oct. 1969.

[117] Hamilton, H. H., DeJarnette, F. R., and Weilmuenster, K. J., "Application of Axisymmetric Analogue for Calculating Heating in Three-Dimensional Flows," AIAA Paper 85-0245, Jan. 1985.

[118] DeJarnette, F. R., Hamilton, H. H., Weilmuenster, K. J., and Cheatwood, F. M., "A Review of Some Approximate Methods Used in Aerodynamic Heating Analyses," *Journal of Thermophysics and Heat Transfer*, Vol. 1, No. 1, Jan. 1987, pp. 5–12.

[119] Zoby, E. V., "Approximate Heating Analysis for the Windward Symmetry Plan of Shuttle-Like Bodies at Large Angles of Attack," *Thermodynamics of Atmospheric Entry, Progress in Astronautics and Aeronautics*, Vol. 82, edited by T. E. Horton, AIAA, New York, 1982, pp. 229–247.

[120] Goodrich, W. D., Li, C. P., Houston, D. K., Chiu, P. B., and Olmedo, L. "Numerical Computations of Orbiter Flow Fields and Laminar Heating Rates," *Journal of Spacecraft and Rockets*, Vol. 14, No. 5, May 1977, pp. 257–264.

[121] Rakich, J. V., and Lanfranco, M. J., "Numerical Computation of Space Shuttle Laminar Heating and Surface Streamlines," *Journal of Spacecraft and Rockets*, Vol. 14, No. 5, May 1977, pp. 265–272.

[122] Becker, J. V., "Results in Recent Hypersonic and Unsteady Flow Research at the Langley Aeronautical Laboratory," *Journal of Applied Physics*, Vol. 21, No. 7, July 1950, pp. 622–624.

[123] Stollery, J. L., "Viscous Interaction Effects and Re-entry Aerothermodynamics: Theory and Experimental Results," *Aerodynamic Problems of Hypersonic Vehicles*, Vol. 1, AGARD Lecture Series No. 42, July 1972, pp. 10-1–10-28.

[124] Probstein, R. F., "Interacting Hypersonic Laminar Boundary Layer Flow over a Cone," Div. of Engineering, Brown Univ., Technical Rep., AF 2798/1, Providence, RI, March 1955.

[125] Talbot, L., Koga, T., and Sherman, P. M., "Hypersonic Viscous Flow over Slender Cones," NACA TN 4327, Sept. 1958.

[126] Wilhite, A. W., Arrington, J. P., and McCandless, R. S., "Performance Aerodynamics of Aero-Assisted Orbital Transfer Vehicles," AIAA Paper 84-0406, Jan. 1984.

[127] Neumann, Richard D., "Special Topics in Hypersonic Flow," *Aerodynamic Problems of Hypersonic Vehicles*, AGARD Lecture Series No. 42, Von Kármán Inst., Brussels, 1972.

[128] Hallion, Richard P., *Test Pilots*, Doubleday and Co., 1981, pp. 258, 259.

[129] Reda, D. C., and Murphy, J. D., "Shock Wave Turbulent Boundary Layer Interactions in Rectangular Channels, Part II: The Influence of Sidewall Boundary Layers on Incipient Separation and Scale of Interaction," AIAA Paper 73-234, Jan. 1973.

[130] Marvin, J. G., Horstman, C. C., Rubesin, M. W., Coakley, T. J., and Kussoy, M. I., "An Experimental and Numerical Investigation of Shock-Wave Induced Turbulent Boundary-Layer Separation at Hypersonic Speeds," *Flow Separation*, AGARD Conference Proceedings No. 168, 1975.

[131] Oskam, B., Vas, I. E., and Bogdonoff, S. M., "Mach 3 Oblique Shock Wave/Turbulent Boundary Layer Interactions in Three Dimensions," AIAA Paper 76-336, Jan. 1976.

[132] Oskam, B., and Vas, J. E., "Experimental Study of Three-Dimensional Flow Fields in an Axial Corner at Mach 3," AIAA Paper 77-687, 1977.

[133] Knight, D. D., "A Hybrid Explicit-Implicit Numerical Algorithm for the Three-Dimensional Compressible Navier-Stokes Equations," *AIAA Journal*, Vol. 22, No. 8, Aug. 1984, pp. 1056–1063.

[134] Davis, R. T., "Numerical Solution of the Hypersonic Viscous Shock-Layer Equations," *AIAA Journal*, Vol. 8, No. 5, May 1970, pp. 843–851.

[135] Davis, R. T., and Flugge-Lotz, I., "Second-Order Boundary Layer Effects in Hypersonic Flow past Axisymmetric Blunt Bodies," *Journal of Fluid Mechanics*, Vol. 20, Pt. 4, 1964, pp. 593–623.

[136] McWherter, Mary, Noack, R. W., and Oberkampt, W. L., "Evaluation of Boundary-Layer and Parabolized Navier-Stokes Solutions for Re-entry Vehicles," *Journal of Spacecraft and Rockets*, Vol. 23, No. 1, Jan.–Feb. 1986, pp. 70–78.

[137] Baker, S. S., "Static Stability and Pressure Distribution Tests of a Slender Cone Model with Symmetric and Asymmetric Nosetips at Mach Number 6," Arnold Engineering Development Center, Rept. AEDC TR-75-56, Tullahoma, TN, March 1975.

[138] Boyland, D. E., Strike, W. T., and Shope, F. L., "A Direct Comparison of Analytical and Experimental Surface and Flow Field Data on a 4-deg Cone at Incidence in a Hypersonic Stream with Laminar Boundary Layers," Arnold Engineering Development Center, Rept. AEDC-TR-76-84, Tullahoma, TN, May 1976.

[139] Griffith, B. J., Strike, W. T., and Majors, B. M., "Ablation and Viscous Effects on the Force and Moment Characteristics of Slender Cone Models at Mach 10 Under Laminar Flow Conditions," Arnold Engineering Development Center, Rept. AEDC-TR-75-109, Tullahoma, TN, July 1975.

[140] Gnoffo, P. A., "Hypersonic Flows over Biconics Using a Variable-Effective-Gamma, Parabolized Navier-Stokes Code," AIAA Paper 83-1666, June 1983.

[141] Shang, J. S., and Scherr, S. J., "Navier-Stokes Solution for a Complete Re-Entry Configuration," *Journal of Aircraft*, Vol. 23, No. 12, Dec. 1986, pp. 881–888.

[142] Wannernwetsch, G. D., "Pressure Tests of the AFFDL X-24C-10D Model at Mach Numbers of 1.5, 3.0, 5.0 and 6.0," Arnold Engineering Development Center, AEDC TR-76-92, Tullahoma, TN, June 1976.

[143] Carver, D. B., "AFFDL X24C Flowfield Survey," Arnold Engineering Development Center, Project V41B-47, June 1979.

[144] Kuruvila, G., and Anderson, John D., Jr., "A Study of the Effects of Numerical Dissipation on the Calculation of Supersonic Separated Flow," AIAA Paper 85-0301, Jan. 1985.

[145] Berman, H. A., Anderson, John D., Jr., and Drummond, J. P., "Supersonic Flow over a Rearward Facing Step with Transverse Nonreacting Hydrogen Injection," *AIAA Journal*, Vol. 21, No. 12, Dec. 1983, pp. 1707–1713.

[146] Sullins, G. A., Anderson, John D., Jr., and Drummond, J. P., "Numerical Investigation of Supersonic Base Flow with Parallel Injection," AIAA Paper 82-1001. June 1982.

[147] Anderson, John D., Jr., *Gasdynamic Lasers: An Introduction*, Academic Press, New York, 1976.

[148] McBride, B. J., Heimel, S., Ehlers, J. G., and Gordon, S., "Thermodynamic Properties to 6000 K for 210 Substances Involving the First 18 Elements," NASA SP-3001, March 1963.

[149] Stull, D. R., *JANAF Thermodynamical Tables*, National Bureau of Standards, NSRDS-NBS 37, Washington, DC, 1971.

[150] Vincenti, W. G., and Kruger, C. H., *Introduction to Physical Gas Dynamics*, Wiley, New York, 1965.

[151] Herzberg, G., *Atomic Spectra and Atomic Structure*, Dover, New York, 1944.

[152] Herzberg, G., *Molecular Spectra and Molecular Structure*, D. Van Nostrand, New York, 1963.

[153] Davidson, N., *Statistical Mechanics*, McGraw–Hill, New York, 1962.

[154] Hilsenrath, J., and Klein, M., "Tables of Thermodynamic Properties of Air in Chemical Equilibrium Including Second Virial Corrections from 1500 to 15,000 K," Arnold Engineering Development Center, Rept. AEDC-TR-65-68, Tullahoma, TN, March 1965.

[155] Tannehill, J. C., and Mugge, P. H., "Improved Curve Fits for the Thermodynamic Properties of Equilibrium Air Suitable for Numerical Computation Using Time-Dependent or Shock-Capturing Methods," NASA CR-2470, Oct. 1974.

[156] Present, R. D., *Kinetic Theory of Gases*, McGraw–Hill, New York, 1958.

[157] Kennard, E. H., *Kinetic Theory of Gases*, McGraw–Hill, New York, 1938.

[158] Bussing, Thomas, and Eberhardt, Scott, "Chemistry Associated with Hypersonic Vehicles," AIAA Paper 87-1292, June 1987.

[159] Wray, Kurt, L., "Chemical Kinetics of High-Temperature Air," *Hypersonic Flow Research*, edited by F. Riddell, Academic Press, New York, 1962, pp. 181–204.

[160] Evans, J. S., Schexnayder, C. J., and Huber, P. W., "Computation of Ionization in Re-Entry Flow Fields," *AIAA Journal*, Vol. 8, No. 6, June 1970, pp. 1082–1089.

[161] Gnoffo, P. A., and McCandless, R. S., "Three-Dimensional AOTV Flow Fields in Chemical Nonequilibrium," AIAA Paper 86-0230, Jan. 1986.

[162] Dunn, M. G., and Kang, S. W., "Theroretical and Experimental Studies of Reentry Plasmas," NASA CR-2232, April 1973.

[163] Witthiff, C. E., and Curtiss, J. T., "Normal Shock Wave Parameters in Equilibrium Air," Cornell Aeronautical Lab. (now CALSPAN), Rept. CAL-111, Buffalo, NY, Nov. 1961.

[164] Marrone, P. V., "Normal Shock Waves in Air: Equilibrium Composition and Flow Parameters for Velocities from 26,000 to 50,000 ft/s," Cornell Aeronautical Lab. (now CALSPAN), Rept. AG-1729-A-2, Buffalo, NY, March 1962.

[165] Huber, Paul, W., "Hypersonic Shock-Heated Flow Parameters for Velocities to 46,000 Feet per Second and Altitudes to 323,000 Feet," NASA TR R-163, Dec. 1963.

[166] Moeckel, W. E., "Oblique-Shock Relations at Hypersonic Speeds for Air in Chemical Equilibrium," NACA TN 3895, Jan. 1957.

[167] Hansen, C. F., "Approximation for the Thermodynamic and Transport Properties of High-Temperature Air," NASA TR-R-50, Nov. 1959.

[168] Romig, Mary F., "Conical Flow Parameters for Air in Dissociating Equilibrium," Convair Scientific Research Lab., Research Rept. No. 7, San Diego, May 1960.

[169] Hudgins, H. E., "Supersonic Flow About Right Circular Cones at Zero Yaw in Air at Chemical Equilibrium. Part 1. Correlation of Flow Properties," Picatinny Arsenal, Technical Memorandum 1493, Dover, NJ, Aug. 1965.

[170] Hudgins, H. E., "Supersonic Flow About Right Circular Cones at Zero Yaw in Air at Thermodynamic Equilibrium. Parts II and III. Tables of Data," Picatinny Arsenal, Technical Memorandum 1493, Dover, NJ, Sept. 1964.

[171] Palmer, Grant, "An Implicit Flux Split Algorithm to Calculate Hypersonic Flowfields in Chemical Equilibrium," AIAA Paper 87-1580, June 1987.

[172] Maus, J. R., Griffith, B. J., Szema, K. Y., and Best, J. T., "Hypersonic Mach Number and Real Gas Effects on Space Shuttle Orbiter Aerodynamics," *Journal of Spacecraft and Rockets*, Vol. 21, No. 2, March–April 1984, pp. 136–141.

[173] Hall, J. G., and Treanor, C. E., "Nonequilibrium Effects in Supersonic Nozzle Flows," AGARDograph No. 124, 1968.

[174] Marrone, P. V., "Inviscid Nonequilibrium Flow Behind Bow and Normal Shock Waves, Part I. General Analysis and Numerical Examples," Cornell Aeronautical Lab. (now CALSPAN), Rept. QM-1626-A-12(1), Buffalo, NY, Oct. 1963.

[175] Anderson, John D., Jr., "A True-Dependent Analysis for Vibrational and Chemical Nonequilibrium Nozzle Flows," *AIAA Journal*, Vol. 8, No. 3, March 1970, pp. 545–550.

[176] Anderson, John D., Jr., "Time-Dependent Solutions of Nonequilibrium Nozzle Flows—A Sequel," *AIAA Journal*, Vol. 8, No. 12, Dec. 1970, pp. 2280–2282.

[177] Wilson, J. K., Schofield, D., and Lapworth, K. C., "A Computer Program for Nonequilibrium Convergent-Divergent Nozzle Flow," National Physical Lab., Rept. 1250, 1967.

[178] Harris, E. L., and Albacete, L. M., "Vibrational Relaxation of Nitrogen in the NOL Hypersonic Tunnel No. 4," Naval Ordnance Lab., TR 63-221, White Oak, MD, July 1964.

[179] Erickson, W. D., "Vibrational Nonequilibrium Flow of Nitrogen in Hypersonic Nozzles," NASA TN D-1810, June 1963.

[180] Hall, J. G., and Russo, A. L., "Studies of Chemical Nonequilibrium in Hypersonic Nozzle Flows," Cornell Aeronautical Lab. (now CALSPAN), Rept. AF-1118-A-6, Buffalo NY, June 1959.

[181] Vamos, J. S., and Anderson, John D., Jr., "Time-Dependent Analysis of Nonequilibrium Nozzle Flows with Complex Chemistry," *Journal of Spacecraft and Rockets*, Vol. 10, No. 4, April 1973, pp. 225, 226.

[182] Sarli, V. J., Burwell, W. G., and Zupnik, T. F., "Investigation of Nonequilibrium Flow Effects in High Expansion Ratio Nozzles," NASA CR-54221, Aug. 1964.

[183] Bohachevsky, I. O., and Rubin, E. L., "A Direct Method for Computation of Nonequilibrium Flows with Detached Shock Waves," *AIAA Journal*, Vol. 4, No. 4, April 1966, pp. 600–607.

[184] Li, C. P., "Time-Dependent Solutions of Nonequilibrium Dissociating Gases past a Blunt Body," *Journal of Spacecraft and Rockets*, Vol. 9, No. 8, Aug. 1972, pp. 571, 572.

[185] Li, C. P., "Time-Dependent Solutions of Nonequilibrium Airflow past a Blunt Body," AIAA Paper 71-595, Jan. 1971.

[186] Hall, J. G., Eschenroeder, A. A., and Marrone, P. V., "Blunt-Nose Inviscid Airflows with Coupled Nonequilibrium Processes," *Journal of the Aerospace Sciences*, Vol. 29, No. 9, Sept. 1962, pp. 1038–1051.

[187] Spurk, J. H., Gerber, N., and Sedney, R., "Characteristic Calculation of Flowfields with Chemical Reactions," *AIAA Journal*, Vol. 4, No. 1, Jan. 1966, pp. 30–37.

[188] Rakich, J. V., Bailey, H. E., and Park, C., "Computation of Nonequilibrium. Supersonic Three Dimensional Inviscid Flow over Blunt-Nosed Bodies," *AIAA Journal*, Vol. 21, No. 6, June 1983, pp. 834–841.

[189] Hirschfelder, J. O., Curtiss, C. F., and Bird, R. B., *The Molecular Theory of Gases and Liquids*, Wiley, New York, 1954.

[190] Chapman, S., and Cowling, T. G., *The Mathematical Theory of Nonuniform Gases*, 2nd ed., Cambridge Univ. Press, New York, 1958.

[191] Bird, R. B., Stewart, W. E., and Lightfoot, E. N., *Transport Phenomena*, Wiley, New York, 1960.

[192] Reid, R. C., and Sherwood, T. K., *The Properties of Gases and Liquids*, McGraw–Hill, New York, 1966.

[193] Peng, T. C., and Pindroh, A. L., "An Improved Calculation of Gas Properties at High Temperatures: Air," Boeing Airplane Co., Document No. D2-11722, Seattle, WA, Feb. 1962.

[194] Viegas, J. R., and Howe, J. T., "Thermodynamic and Transport Property Correlation Formulas for Equilibrium Air from 1000 K to 15,000 K," NASA TN D-1429, Oct. 1962.

[195] Dorrance, W. H., *Viscous Hypersonic Flow*. McGraw–Hill, New York, 1962.

[196] Fay, J. A., and Riddell, F. R., "Theory of Stagnation Point Heat Transfer in Dissociated Air," *Journal of the Aeronautical Sciences*, Vol. 25, No. 2, Feb. 1958, pp. 73–85.

[197] Blottner, F. G., "Chemical Nonequilibrium Boundary Layer," *AIAA Journal*, Vol. 2, No. 2, Feb. 1964, pp. 232–240.

[198] Blottner, F. G., "Nonequilibrium Laminar Boundary-Layer Flow of lonized Air," *AIAA Journal*, Vol. 2, No. 11, Nov. 1964, pp. 1921–1927.

[199] Moss, J. N., "Reacting Viscous-Shock-Layer Solutions with Multicomponent Diffusion and Mass Injection," NASA TR R-411, June 1974.

[200] Kim, M. D., Swaminathan, S., and Lewis, C. H., "Three-Dimensional Nonequilibrium Viscous Shock-Layer Flow over the Space Shuttle Orbiter," *Journal of Spacecraft and Rockets*, Vol. 21, No. 1, Jan.–Feb. 1984, pp. 29–35.

[201] Kim, M. D., Bhutta, B. A., and Lewis, C. H., "Three-Dimensional Effects upon Real Gas Flows Past the Space Shuttle," AIAA Paper 84-0225, Jan. 1984.

[202] Shinn, J. L., Moss, I. N., and Simmonds, A. I., "Viscous Shock Layer Heating Analysis for the Shuttle Windward Symmetry Plane with Surface Calalytic Recombination Rates," *Entry Vehicle Heating and Thermal Protection Systems: Space Shuttle. Solar Starprobe, Jupiter Galileo Probe,* edited by P. E. Bauer and H. E. Collicot, Vol. 85 Progress in Astronautics and Aeronautics Series, AIAA, New York, 1983.

[203] Prabhu, D. K., and Tannehill, J. C., "Numerical Solution of Space Shuttle Orbiter Flowfield Including Real-Gas Effects," *Journal of Spacecraft and Rockets*, Vol. 23, No. 3, May–June, 1986, pp. 264–272.

[204] Prabhu, D. K., Tannehill, J. C., and Marvin, J. G., "A New PNS Code for Three-Dimensional Chemically Reacting Flows," AIAA Paper 87-1472, June 1987.

[205] Bhutta, B. A., Lewis, C. H., and Kautz, F. A., "A Fast Fully-Interactive Parabolized Navier-Stokes Scheme for Chemically-Reacting Reentry Flows," AIAA Paper 85-0926, June 1985.

[206] Gnoffo, P. A., and McCandless, R. S., "Three-Dimensional AOTV Flowfield in Chemical Nonequilibrium," AIAA Paper 86-0230, Jan. 1986.

[207] Gnoffo, P. A., McCandless, R. S., and Yee, H. C., "Enchancement to Program LAURA for Computation of Three-Dimensional Hypersonic Flow," AIAA Paper 87-0280, Jan. 1987.

[208] Kothari, A. P., Anderson, John D., Jr., and Jones, E., "Navier-Stokes Solutions for Chemical Laser Flows," *AIAA Journal*, Vol. 15, No. 1, Jan. 1977, pp. 92–100.

[209] Kothari, A. P., Anderson, John D., Jr., and Jones, E., "Navier-Stokes Solutions for Chemical Laser Flows: Steady and Unsteady Flows," AIAA Paper 79-0009, Jan. 1979.

[210] Anderson, John D., Jr., *Gasdynamic Lasers: An Introduction*, Academic Press, New York, 1976.

[211] Anderson, John D., Jr., "Navier-Stokes Solutions of High Energy Laser Flows," *Gas-Flow and Chemical Lasers*, edited by J. F. Wendt, Hemisphere, New York, 1979, pp. 3–21.

[212] Anderson, John D., Jr., "Nongray Radiative Transfer Effects on the Radiating Stagnation Region Shock Layer and Stagnation Point Heat Transfer," U.S. Naval Ordnance Lab., NOLTR 67-104, White Oak, MD, July 1967.

[213] Anderson, John D., Jr., "An Engineering Survey of Radiating Shock Layers," *AIAA Journal*, Vol. 7, No. 9, Sept. 1969, pp. 1665–1675.

[214] Sutton, K., "Air Radiation Revisited," *Thermal Design of Aeroassisted Orbital Transfer Vehicles*, edited by H. F. Nelson, Progress in Astronautics and Aeronautics Series, Vol. 96, AIAA, New York, 1985, pp. 419–441.

[215] Howe, J. T., and Viegas, J. R., "Solution of the Ionized Radiating Shock Layer, Including Reabsorption and Foreign Species Effects and Stagnation Region Heat Transfer," NASA TR R-159, Nov. 1963.

[216] Moss, J. N., and Simmonds, A. L., "Galileo Probe Forebody Flowfield Predictions," *Entry Vehicle Heating and Thermal Prediction Systems: Space Shuttle, Solar Starprobe, Jupiter Galileo Probe*, edited by P. E. Bauer and H. E. Collicott, Progress in Astronautics and Aeronautics Series, Vol. 85, AIAA, New York, 1983.

[217] Moss, J. N., and Simmonds, A. L., "Galileo Probe Forebody Flowfield Predictions During Jupiter Entry," AIAA Paper 82-0874, June 1982.

[218] Kumar, A., Tiwari, S. N., Graves, R. A., and Weilmuenster, K. J., "Laminar and Turbulent Flow Solutions with Radiation and Ablation Injection for Jovian Entry," AIAA Paper 80-0288, Jan. 1980.

[219] Edney, B. E., "Anomalous Heat Transfer and Pressure Distributions on Blunt Bodies at Hypersonic Speeds in the Presence of an Impinging Shock," *Flygtekniska Forsoksanstalten*, The Aeronautical Research of Inst. of Sweden (FFA), Rept. 115, Stockholm, Sweden, 1968.

[220] Edney, B. E., "Effects of Shock Impingement on the Heat Transfer Around Blunt Bodies," *AIAA Journal*, Vol. 6, No. 1, Jan. 1968, pp. 15–21.

[221] Lind, C. A., "Effect of Geometry on the Unsteady Type-IV Shock Interaction," *Journal of Aircraft*, Vol. 34, No. 1, Jan.–Feb. 1997, pp. 64–71.

[222] Lind, C. A., and Lewis, M. J., "Computational Analysis of the Unsteady Type IV Shock Interaction of Blunt Body Flows," *Journal of Propulsion and Power*, Vol. 12, No. 1, Jan.–Feb. 1996, pp. 127–133.

[223] Hirsch, C., *Numerical Computation of Internal and External Flows*, Vols. 1 and 2, Wiley, New York, 1988.

[224] Tannehill, John C., Anderson, Dale A., and Pletcher, Richard H., *Computational Fluid Mechanics and Heat Transfer*, 2nd ed., Taylor and Francis, Bristol, PA, 1997.

[225] Fisher, Carren, M. E., "Experiences Using the Mark IV Supersonic Hypersonic Arbitrary Body Program," *Aerodynamics of Hypersonic Lifting Vehicles*, AGARD Conference Proceedings 428, 1987, pp. 31-1–31-18.

[226] Landrum, E. J., "Wind Tunnel Force and Flow Visualization at Mach numbers from 1.6 to 4.63 for a Series of Bodies of Revolution at Angles of Attack from -4° to 60°," NASA TM-78813, March 1979.

[227] Arrington, J. P., and Jones, J. J., "Shuttle Performance: Lessons Learned," NASA Conference Publication 2283, Part 1, March 1983.

[228] Kinney, David J., "Aero-Thermodynamics for Conceptual Design," AIAA Paper 2004-31, Jan. 2004.

[229] Kinney, David J., and Garcia, Joseph A., "Predicted Convective and Radiative Aerothermodynamic Environments for Various Reentry Vehicles Using CBAERO," AIAA Paper 2006-659, Jan. 2006.

[230] Kinney, David J., "Aerodynamic Shape Optimization of Hypersonic Vehicles," AIAA Paper 2006-239, Jan. 2006.

[231] Kothari, A. P., Tarpley, C., McLaughlin, T. A., Suresh Babu, B., and Livingston, J. W., "Hypersonic Vehicle Design using Inward Turning Flowfields," AIAA Paper 96-2552, July 1996.

[232] Billig, F. S., and Kothari, A. P., "Streamline Tracing, A Technique for Designing Hypersonic Vehicles," International Symposium on Air Breathing Engines, Paper 33.1, Sept. 1997.

[233] Dissel, Adam F., Kothari, Ajay P., and Lewis, Mark J., "Comparison of Horizontally and Vertically Launched Air-Breathing and Rocket Vehicles," AIAA Paper 2004-3988, July 2004.

[234] Rieger, H., "Solution of Some 3-D Viscous and Inviscid Supersonic Flow Problems by Finite-Volume Space-Marching Schemes," *Aerodynamics of Hypersonic Lifting Vehicles*, AGARD Conference Proceedings No. 428, 1987, pp. 17-1–17-17.

[235] Taylor, G. I., and Maccoll, J. W., "The Air Pressure on a Cone Moving at High Speed," *Proceedings of the Royal Society* (London), Series A, Vol. 139, 1933, pp. 278–311.

[236] Nonweiler, T. R., "Aerodynamic Problems of Manned Space Vehicles," *Journal of the Royal Aeronautical Society*, Vol. 63, 1959, pp. 521–528.

[237] Jones, J. G., "A Method for Designing Lifting Configurations for High Supersonic Speeds Using the Flow Fields of Nonlifting Cones," Royal Aeronautical Establishment Rept. Aero 2624, A.R.C. 24846, England, 1963.

[238] Jones, J. G., Moore, K. C., Pike, J., and Roe, P. L., "A Method for Designing Lifting Configurations for High Supersonic Speeds Using Axisymmetric Flow Fields," *Ingenieur-Archiv*, Vol. 37, Band, 1. Heft, 1968, pp. 556–572.

[239] Townend, L. H., "Research and Design for Lifting Reentry," *Progress in Aerospace Sciences*, Vol. 18, 1979, pp. 1–80.

[240] Rasmussen, M. L., "Waverider Configurations Derived from Inclined Circular and Elliptic Cones," *Journal of Spacecraft and Rockets*, Vol. 17, No. 6, Nov.–Dec. 1980, pp. 537–545.

[241] Kim, B. S., Rasmussen, M. L., and Jischke, M. D., "Optimization of Waverider Configurations Generated from Axisymmetric Conical Flows," AIAA Paper 82-1299, Jan. 1982.

[242] Broadway, R., and Rasmussen, M. L., "Aerodynamics of a Simple Cone Derived Waverider," AIAA Paper 84-0085, Jan. 1984.

[243] Anderson, John D., Jr., "Ludwig Prandtl's Boundary Layer," *Physics Today*, Vol. 58, Dec. 2005, pp. 42–48.

[244] Meader, William, E., and Smart, Michael K., "Reference Enthalpy Method Developed from Solutions of the Boundary-Layer Equations," *AIAA Journal*, Vol. 43, No. 1, Jan. 2005, pp. 135–139.

[245] Tauber, Michael E., "A Review of High-Speed Convective Heat Transfer Computation Methods," NASA TP 2914, June 1989.

[246] Bowcutt, Kevin G., Anderson, John D., Jr., and Capriotti, Diego, "Numerical Optimization of Conical Flow Waveriders Including Detailed Viscous Effects," *Aerodynamics of Hypersonic Lifting Vehicles*, AGARD Conference Proceedings No. 428, Nov. 1987, pp. 27-1–27-23.

[247] Nelder, J. A., and Meade, R., "A Simplex Method of Function Minimization," *Computer Journal*, Vol. 7, Jan. 1965, pp. 308–313.

[248] Bowcutt, Kevin G., "Optimization of Hypersonic Waveriders Derived from Cone Flows—Including Viscous Effects," Ph.D. Dissertation, Dept. of Aerospace Engineering, Univ. of Maryland, College Park, MD, May 1986.

[249] Kuchemann, D., *The Aerodynamic Design of Aircraft*, Pergamon Press, Oxford, England, UK, 1978, pp. 448–510.

[250] Corda, Stephan, and Anderson, John D., Jr., "Viscous Optimized Waveriders Designed from Axisymmetric Flowfields," AIAA Paper 88-0369, Jan. 1988.

[251] Vanmol, Denis O., and Anderson, John D., Jr., "Heat Transfer Characteristics of Hypersonic Waveriders with an Emphasis on Leading Edge Effects," AIAA Paper 92-2910, July 1992.

[252] Hamilton, Harris, "Approximate Method of Calculating Heating Rates at General Three-Dimensional Stagnation Points During Atmospheric Entry," NASA TM 84850.

[253] Vanmol, Denis O., "Heat Transfer Characteristics of Hypersonic Waveriders with an Emphasis on the Leading Edge Effects," M.S. Thesis, Dept. of Aerospace Engineering, Univ. of Maryland, College Park, MD, May 1991.

[254] Chang, J., "A Study of Viscous Interaction Effects on Hypersonic Waveriders," Ph.D. Dissertation, Dept. of Aerospace Engineering, Univ. of Maryland, College Park, MD, Dec. 1991.

[255] Anderson, John D., Jr., Chang, J., and McLaughlin, T. A., "Hypersonic Waveriders: Effects of Chemically Reacting Flow and Viscous Interaction," AIAA Paper 92-0302, Jan. 1992.

[256] Anderson, John D., Jr., Ferguson, F., and Lewis, M. J., "Hypersonic Waveriders for High Altitude Applications," AIAA Paper 91-0530, Jan. 1991.

[257] O'Neil, M. K., and Lewis, M. J., "Design Tradeoffs on Scramjet Engine Integrated Hypersonic Waverider Vehicles," *Journal of Aircraft*, Vol. 30, No. 6, 1993, pp. 943–952.

[258] Takashima, N., and Lewis, M. J., "Optimization of Waverider-Based Hypersonic Cruise Vehicles with Off-Design Considerations," *Journal of Aircraft*, Vol. 36, No. 1, 1999, pp. 235–245.

[259] O'Brien, T. F., and Lewis, M. J., "Rocket-Based Combined-Cycle Engine Integration on an Oscillating Cone Waverider Vehicle," *Journal of Aircraft*, Vol. 38, No. 6, Nov.–Dec. 2001, pp. 1117–1123.

[260] Meyer, Benjamin, Nelson, H. F., and Riggins, David W., "Hypersonic Drag and Heat-Transfer Reduction Using a Forward-Facing Jet," *Journal of Aircraft*, Vol. 38, No. 4, July–August 2001.

[261] Drummond, J. P., Rogers, R. C., and Hussaini, M. Y., "A Detailed Numerical Model of a Supersonic Mixing Layer," AIAA Paper 86-1427, June 1986.

[262] Shang, J. S., "Plasma Injection for Hypersonic Blunt-Body Drag Reduction," *AIAA Journal*, Vol. 40, No. 6, June 2002, pp. 1178–1196.

[263] Park, Chul, *Nonequilibrium Hypersonic Aerothermodynamics*, Wiley International, Wiley Interscience, New York, 1990.

[264] Eschenroeder et al., "Shock Tunnel Studies of High-Enthalpy Ionized Airflows," Cornell Aeronautical Lab., Rept. AF-1500 A1, Buffalo, NY, 1962.

[265] McLaughlin, T. A., "Viscous Optimized Hypersonic Waveriders for Chemical Equilibrium Flow," M.S. Thesis, Dept. of Aerospace Engineering, Univ. of Maryland, College Park, MD, May 1990.

[266] Gnoffo, Peter A., "Planetary-Entry Gas Dynamics," *Annual Reviews of Fluid Mechanics*, Vol. 31, 1999, pp. 459–494.

[267] Hirschel, Ernst H., *Basics of Aerothermodynamics*, Springer-Verlag, Heidelberg AIAA, Reston, VA, 2005.

[268] Tauber, M. E., and Sutton, K., "Stagnation-Point Radiative Heating Relations for Earth and Mars Entries," *Journal of Spacecraft and Rockets*, Vol. 28, No. 1, Jan.–Feb. 1991, pp. 40–42.

[269] Page, W. A., Compton, D. L., Borucki, W. J., Ciffone, D. L., and Cooper, D. M., "Radiative Transport in Inviscid Nonadiabatic Stagnation-Region Shock Layers," *Thermal Design Principles of Spacecraft and Entry Bodies*, edited by J. T. Bevans, Vol. 21, Progress in Astronautics and Aeronautics AIAA, New York, 1969, pp. 75–114.

[270] Dissel, A. F., Kothari, A. P., and Lewis, M. J., "Investigation of Two-Stage-to-Orbit Air Breathing Launch Vehicle Configurations," *Journal of Spacecraft and Rockets*, Vol. 43, No. 3, 2006, pp. 568–574.

Index

Supporting Materials

Many of the topics introduced in this book are discussed in more detail in other AIAA publications. For a complete listing of titles in the AIAA Education Series, as well as other AIAA publications please visit http://www.aiaa.org.